Remote Sensing for Geology and Mapping

Remote Sensing for Geology and Mapping

Guest Editors

Chao Chen
Tao Chen
Yanni Dong

MDPI

Basel • Beijing • Wuhan • Barcelona • Belgrade • Novi Sad • Cluj • Manchester

Guest Editors

Chao Chen
School of Geography Science and Geomatics Engineering
Suzhou University of Science and Technology
Suzhou
China

Tao Chen
School of Geophysics and Spatial Information
China University of Geosciences
Wuhan
China

Yanni Dong
School of Resource and Environment Sciences
Wuhan University
Wuhan
China

Editorial Office
MDPI AG
Grosspeteranlage 5
4052 Basel, Switzerland

This is a reprint of the Special Issue, published open access by the journal *Remote Sensing* (ISSN 2072-4292), freely accessible at: https://www.mdpi.com/journal/remotesensing/special_issues/NOFNZ699QZ.

For citation purposes, cite each article independently as indicated on the article page online and as indicated below:

Lastname, A.A.; Lastname, B.B. Article Title. *Journal Name* **Year**, *Volume Number*, Page Range.

ISBN 978-3-7258-3559-1 (Hbk)
ISBN 978-3-7258-3560-7 (PDF)
https://doi.org/10.3390/books978-3-7258-3560-7

Contents

About the Editors

Chao Chen

Chao Chen is currently a Full Professor at Suzhou University of Science and Technology, Suzhou, China. He received a B.E. degree in communication engineering and an M.S. degree in photogrammetry and remote sensing from Shandong University of Science and Technology, Qingdao, China, in 2005 and 2009, respectively. He received a Ph.D. degree in cartography and geography information systems from Peking University, Beijing, China, in 2013. His research interest focuses on remote sensing of coastal environments.

Tao Chen

Tao Chen , Associate Professor, PhD, and Doctoral Supervisor, is currently the Chair of the Department of Earth Information Science and Technology at China University of Geosciences. He serves as an Associate Editor (AE) and Editorial Board Member of *IEEE Journal of Selected Topics in Applied Earth Observations and Remote Sensing*, as an Editorial Board Member for *Remote Sensing* and *Journal of Mountain Science*, and as a Youth Editorial Board Member for *Natural Hazards Research*. He is a Senior Member of IEEE; the Chair of the IEEE GRSS Wuhan Chapter; a member of the Geological Society of China, the Chinese Society of Image and Graphics, the Remote Sensing Image Committee of the Chinese Society of Image and Graphics, the Youth Work Committee of the Chinese Association for Artificial Intelligence, the Chinese Society for Rock Mechanics and Engineering; and a committee member of the Academic Sub-Committee of the Henan Provincial Center for Science and Technology Innovation in Natural Resource (Information Sensing Technology Application Research). His research focuses on intelligent expression and computation of geoscience information and remote sensing information processing for geological disasters. He has received the Third Prize for Scientific and Technological Progress in Hubei Province. In the past five years, he has led and participated in more than ten major research projects, including four NSFC projects, one Hubei Provincial Natural Science Foundation general project, and one Hubei Provincial Key R&D Plan project. As the first or corresponding author, he has published over 50 academic papers in domestic and international journals, including one ESI hot paper and three ESI highly cited papers. He has also authored two monographs. From October 2015 to October 2016, he was funded by the China Scholarship Council for a one-year visit to the University of New South Wales in Australia.

Yanni Dong

Yanni Dong is a Professor and Ph.D. Supervisor at the School of Resources and Environmental Science, Wuhan University, and a recipient of funding from the National Excellent Young Scientist Fund. Her research primarily focuses on the intelligent interpretation of remote sensing imagery and geospatial intelligence applications. She has published more than 60 SCI-indexed papers in prestigious journals such as *IEEE Transactions on Image Processing* (*TIP*), *IEEE Transactions on Cybernetics* (*TCybern*), and *IEEE Transactions on Geoscience and Remote Sensing* (*TGRS*). She serves as an Associate Editor for the SCI journals *IEEE TGRS* and *IEEE GRSL*; a senior program committee member for international image analysis conferences such as ICCV, AAAI, and IJCAI; and a senior member of IEEE. She is also the Vice Chair of the IEEE GRSS Wuhan Chapter and the Vice President of the Excellent Doctoral Club of the China Society of Image and Graphics.

remote sensing

MDPI

Article

Advancing Skarn Iron Ore Detection through Multispectral Image Fusion and 3D Convolutional Neural Networks (3D-CNNs)

Jabir Abubakar [1], Zhaochong Zhang [1,2,*], Zhiguo Cheng [1], Fojun Yao [3] and Abdoul-Aziz Bio Sidi D. Bouko [4]

1 State Key Laboratory of Geological Processes and Mineral Resources, China University of Geosciences, Beijing 100083, China; jabir@mau.edu.ng (J.A.); zgcheng@cugb.edu.cn (Z.C.)
2 Frontiers Science Center for Deep-Time Digital Earth, China University of Geosciences, Beijing 100083, China
3 MNR Key Laboratory of Metallogeny and Mineral Assessment, Institute of Mineral Resources, Chinese Academy of Geological Sciences, Beijing 100037, China; yaofojun@mail.cgs.gov.cn
4 School of Water Resources and Environment, China University of Geosciences, Beijing 100083, China; adama9101200003@email.cugb.edu.cn
* Correspondence: zczhang@cugb.edu.cn

Abstract: This study explores novel techniques to improve the detection accuracy of skarn iron deposits using advanced image-processing methodologies. Leveraging the capabilities of ASTER image, band ratio (BR) images, and principal component analysis (PCA) alongside the power of 3D convolutional neural networks (3D-CNNs), the research aims to enhance the precision and efficiency of ore detection in complex geological environments. The proposed method employs a specific 3D-CNN architecture accepting input as a $7 \times 7 \times C$ image patch, where C represents the combined number of selected ASTER image bands, principal component (PC) bands, and computed BR images. To evaluate the accuracy of the proposed method, five distinct image band combinations, including the proposed band combination, were tested and evaluated based on the overall accuracy (OA), average accuracy (AA), and kappa coefficient. The results demonstrated that while the incorporation of BR images alongside ASTER bands initially seemed promising, it introduced significant confusion in certain classifications, leading to unexpected misclassification rates. Surprisingly, utilizing solely ASTER bands as input parameters yielded higher accuracy rates (OA = 93.13%, AA = 91.96%, kappa = 90.91%) compared with scenarios involving the integration with band ratios (OA = 87.02%, AA = 79.15, kappa = 82.60%) or the integration of BR images to PC bands (OA = 87.78%, AA = 82.39%, kappa = 83.81%). However, the amalgamation of ASTER bands with selected PC bands showed slight improvements in accuracy (OA = 94.65%, AA = 92.93%, kappa = 93.45%), although challenges in accurately classifying certain features persisted. Ultimately, the proposed combination of ASTER bands, PC bands, and BR images (proposed band combination) presented the most visually appealing and statistically accurate results (OA = 96.95%, AA = 94.87%, kappa = 95.93%), effectively addressing misclassifications observed in the other combinations. These findings underscore the synergistic contributions of each of the ASTER bands, PC bands, and BR images, with the ASTER bands proving pivotal for optimal skarn classification, the PC bands enhancing intrusions classification accuracy, and the BR images strengthening wall rock classification accuracy. In conclusion, the proposed combination of input image bands emerges as a robust and comprehensive methodology, demonstrating unparalleled accuracy in the remote sensing detection of skarn iron minerals.

Keywords: ASTER bands; band ratio (BR) images; principal component (PC) bands; 3D-CNN

Citation: Abubakar, J.; Zhang, Z.; Cheng, Z.; Yao, F.; Bio Sidi D. Bouko, A.-A. Advancing Skarn Iron Ore Detection through Multispectral Image Fusion and 3D Convolutional Neural Networks (3D-CNNs). *Remote Sens.* **2024**, *16*, 3250. https://doi.org/10.3390/rs16173250

Academic Editor: Amin Beiranvand Pour

Received: 14 April 2024
Revised: 4 August 2024
Accepted: 5 August 2024
Published: 2 September 2024

1. Introduction

The identification and classification of minerals constitute fundamental tasks within the many disciplines of geology, serving as the cornerstone for various scientific uses. The processes of mineral identification, detection, and mapping involve categorizing the mineral composition exposed in an area with diverse lithological composition and various geological settings. This serves the purpose of pinpointing specific geological areas for further survey and exploration. Over the years, diverse methods and techniques have been employed for mineral detection and mapping, encompassing physical and chemical methods as well as remote sensing techniques [1,2].

Various optical remote sensing methods have been employed for mapping and detection, specifically, iron oxides and hydroxyl minerals. These methods include image spectral band rationing (BR) and false-color composite (FCC) [3–8], principal component analysis (PCA) [9,10], spectral angle mapper (SAM) [11–13], linear spectral unmixing (LSU) [14], constrained energy minimization (CEM) [15], and convolutional neural network (CNN) methods [16–19], among others. BR and/or PCA in conjunction with FCC have been widely used in previous studies, largely due to their simplicity and their capacity for enhancing the spectral response of minerals as well as separating the altered mineralized rocks from all the other background materials. However, even with the series of refinements in these methods over the years, the interpretation of the final image largely requires significant expertise because the ratio image is not a classified image, even though the boundaries between some minerals can be seen when used as FCC. In the situation where thresholds are used to extract the pixels depicting the target minerals, there is no specific thresholding concept adopted across studies, making each study identify suitable values as the threshold. This introduces bias into the methods. In addition, there is lack of universality of the existing BRs, making it difficult to recycle them. For instance, refs. [20,21] identified different BRs (Landsat 8 BRs 4/2 and 6/2, respectively) for iron mineral enhancement. Similarly, ref. [22] identified Landsat 8 BRs 6/4 and 5/4 for ferrous and ferric iron minerals, which is different from those of [23] (6/5 and 7/5) for the same task.

For PCA, the resulting PCs may become very difficult to interpret if there are many variables in the datasets because it will be difficult to understand the underlying meaning of the PCs and how it relates to the variables [24]. In addition to the interpretability problem with the PCA method, the method works very well only with linear data, since it is based on the assumption that the relationship between variables is linear [25]. This makes the method very sensitive to outliers, distorting the capacity of the method to identify the pattern in the data [26]. Although some of the problems of PCA are partially solved by either of the CEM, SAM, and/or the CNN methods, the strength of this method in enhancing the spectral response of various minerals in conjunction with BR images is not fully utilized, sustained, or integrated in these methods. Applying the concept of BR may likely have an impact on the outcome of classification, if refined and integrated into any of the methods.

The integration of machine learning into feature detection and image classification has substantially overcome many challenges associated with traditional methods of mineral detection. Specifically, the use of 2D-CNN has displayed strength in generating abstract features during image classification [16,27]. However, the method convolves in only two dimensions, neglecting the adjacent slice or channels, which may have affected the efficiency of the feature extraction and classification due to the missing adjacent channel relationship information. This may also increase the number of convolutional operations, which may lead to an increase in the time of operation, since each channel has to be convolved separately. These problems were somewhat addressed by the introduction of 3D-CNN into image classification. The work of [17,28] and other similar studies has shown that 3D-CNN can effectively, efficiently, and rapidly detect and identify different types of minerals in an alteration zone using hyperspectral images. Similarly, the integration of both 3D and 2D-CNN by [29,30] has shown an improvement in accuracy against the use of either 2D or 3D in isolation. However, the application of this method and the other 3D methods seem to work well only with hyperspectral data that have many channels or spectra depths. In addition, these types of images contain a lot of noise and are very expensive to acquire and computationally complex to handle.

Consequently, subsequent to a meticulous evaluation of the merits and demerits inherent in the prevalent optical remote sensing methodologies employed for mineral mapping and detection, this study endeavors to capitalize on the efficacy of 3D convolutional neural networks (3D-CNNs) for swift, effective, and efficient mineral detection. This strategic approach involves the incorporation of a limited number of 3D convolutional layers, a methodology elucidated in prior works such as [17,28] in order to minimize the risk of misclassification by the 3D-CNN model, particularly in scenarios where minerals exhibit analogous textures. Noteworthy is the deliberate selection of multispectral imagery over hyperspectral counterparts, a decision rooted in the objective of mitigating operational costs and time constraints. To counterbalance the potential drawback of limited spectral depth inherent in multispectral imagery, this study proposes the augmentation of selected band ratios (BRs) images and principal component (PC) bands to the multispectral image bands. This augmentation strategy is underpinned by the demonstrated capacity of band ratios to enhance the spectral response of iron minerals, as evidenced in various investigations [4,22,31]. Additionally, the inclusion of principal component bands is rationalized by their efficacy in transforming highly correlated data into uncorrelated forms, thereby maximizing variance, a trait documented in studies such as [21].

The investigation was carried out on iron deposits situated within the Awulale Metallogenic Belt (AMB). The choice of this metallogenic belt emanated from its historical context, marked by notable volcanic activities within the belt [32].

2. Materials and Methods

2.1. Study Area and Geological Context

The study area is situated within the AMB, located in the western Tianshan Mountain, Xinjiang Uygur Autonomous Region, NW China. Specifically, the selected areas encompass three notable iron deposits positioned within the distinguished submarine volcanic-hosted iron oxide deposits belt (AMB) in China [32]. Geographically, the AMB spans a longitudinal range of approximately 82°E–86°E and a latitudinal range of 42°20′N–44°N, featuring Silurian and Devonian volcanoclastic successions in both its northern and southern portions (Figure 1a). In the western segment, the strata are dominated by Permian terrestrial clastic and volcanic rocks, while the eastern part prominently showcases Carboniferous volcanic and volcanoclastic rock formations. This geological context serves as the backdrop for the three iron deposits under investigation, namely, the Zhibo, Dunde, and Beizhan iron deposits, in conjunction with Chagangnuoer and Wuling iron deposits [33,34]. Notably, the iron mineralization in the AMB is closely associated both spatially and temporally with the Carboniferous volcanic rocks, which have been dated to around 310–330 Ma [35].

The Zhibo, Dunde, and Beizhan iron deposits, the focal points of this study, have attracted scientific attention due to their classification as submarine volcanic-hosted iron oxide deposits and their important economic value. The iron minerals in the ores within these deposits are majorly magnetite, and their wall rocks are Carboniferous calc-alkaline volcanic–volcanoclastic rocks, which are believed to have formed within a post-collisional setting subsequent to the break-off of a subducted slab [36,37]. It is noteworthy that certain research suggests that the formation may be attributed to subduction-related settings, owing to the "arc-like geochemical features" of the rocks [38–40], rather than the occurrence of slab "break-off".

The primary host rocks of the Zhibo iron deposit are andesite, basalt, basaltic andesite, and dacite compositions, with the ore district exhibiting the presence of intermediate-mafic volcanic rocks in its proximity, intruded predominantly by diorite and granodiorite formations (Figure 1c). The deposit is positioned within an oval-shaped collapsed caldera covering an approximate area of 15 km^2 [41]. However, Dunde's distinctiveness within the AMB lies in its remarkable mineralogical diversity. The deposit also contains large-scale amounts of Zn as well as medium-scale amounts of gold in addition to the iron content. A spectrum of minerals within the Dunde deposit includes magnetite, sphalerite, chalcopyrite, and other minerals. Intrusive rocks in the Dunde deposit consist of K-feldspar granite and diabase compositions (Figure 1b). The prevailing skarn predominantly comprises diopside, with garnet as a secondary constituent [42], and the zinc and gold mineralization are correlated with the products of superimposed mineralization on iron ore bodies during hydrothermal alteration related to granitic intrusions [43]. For the Beizhan deposit, the exposed volcanic rocks include rhyolite and dacite, along with andesite tuff and basaltic andesite formations in certain regions (Figure 1d). The intrusion of quartz-monzonite porphyry and diabase over a surface length of up to 4 km has induced alterations in the host rock, culminating in the mineralization of iron ore [33].

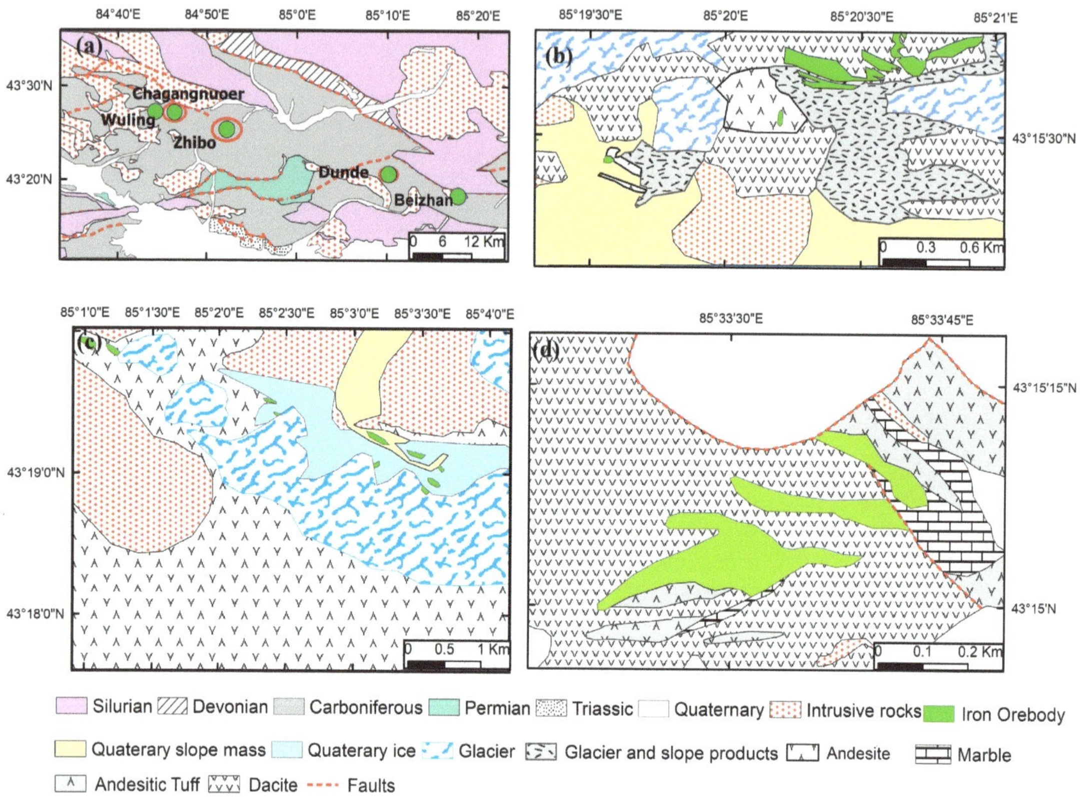

Figure 1. Map of the study area. (**a**) Regional geological map of Western Tianshan [33], (**b**) Geological map of Dunde [44], (**c**) Geological map of Zhibo [39], (**d**) Geological map of Beizhan [45].

2.2. Methodologies

The application of a spectral remote sensing approach for mineral mapping and detection necessitated the integration of the rock and mineral samples' spectra with satellite images covering the study area. Therefore, the requisite data for this study comprise multispectral images covering the study sites and rock samples representing the wall rocks (andesite, basalt, basaltic andesite, dacite, basaltic tuff, basaltic crystal tuff, and andesite tuff), iron ore (magnetite), intrusive rocks (diorite, granodiorite, granite, and diabase), and skarn minerals (epidote, diopside, chlorite, and calcite) within the study area, accompanied by their corresponding spectra and geographical positional information. Additional samples for vegetation and glacier were obtained to capture the other land covers within the study area.

2.2.1. Data Collection and Processing

ASTER data constitute multispectral data captured by the ASTER advanced multispectral imager, launched in 1999 onboard NASA's Terra spacecraft. With a spatial extent of 60 × 60 km, the sensor acquires data within 14 different spectral bands, including 3 VNIR bands (520–860 nm wavelength) at 15 m spatial resolution, 6 SWIR bands (1600–2430 nm wavelength) at 30 m spatial resolution, and 5 TIR bands (8125–11,650 nm wavelength) at 90 m spatial resolution. The ASTER Level 1 Precision Terrain Corrected Registered At-Sensor Radiance (AST_L1T_00309142002052127_20150425012155_14799) image, encompassing visible near infrared (VNIR) and shortwave infrared (SWIR) bands (bands 1 to 9), acquired on 14 September 2002, was employed as the primary dataset for this research. The image was obtained from the EROS Data Center's (EDC) Land Processes Distributed Active Archive Center (LP-DAAC) after preprocessing operations for terrain correction and calibration of at-sensor radiance were conducted on the image [46]. ENVI 5.5 facilitated image preprocessing tasks, encompassing radiometric and atmospheric corrections, image resampling, PCA, and band ratio computation. The atmospheric and radiometric corrections were performed to mitigate the atmospheric interference and transforming radiance images into reflectance [46], while the image resampling was implemented to align the SWIR bands with a resolution of 15 m (matching the VNIR bands) in order to preserve the essential resolution of the VNIR bands, crucial for enhancing the iron spectra.

Furthermore, a total of 103 rock samples comprising 34 iron ores, 34 skarns, 21 volcanic wall rocks, and 14 intrusive rocks were collected. Additionally, 45 other land cover samples were obtained, comprising 20 vegetation samples and 25 glacier samples. Disparities in the sample counts among these components were primarily attributed to the inherent variability in their occurrence within the study area (Figures A1–A4 in Appendix A). These samples were extracted from the three study iron deposits. It is important to highlight that the selection of samples deliberately accounted for diversity in both physical appearance and structural characteristics, enabling a comprehensive representation of the different components when determining the bands for band ratio computations. Furthermore, the selection of sampling locations was executed with precision based on the dominance of each mineral or ore type within a 30 m buffer, corresponding to the pixel size of the ASTER imagery. This approach aimed to minimize errors and enhance the sample size for image patch preparation. Additionally, the eight surrounding pixels within the buffer distance of sampling locations were incorporated as the center pixels for patch preparation. This further increased the number of training samples and was instrumental in averting or minimizing the occurrence of mixed pixels during the training or testing phases, thereby enhancing the spectral purity of the collected data.

The rock samples were subjected to bidirectional spectrum data collection, a task facilitated by OreXplorer field spectrometer SM-5400 209-9428 (Spectral Evolution, Haverhill, MA, USA). This high-resolution spectrometer operated within a spectral range of 350 to 2500 nm. The spectrometer has a spectral resolution of 2.7 nm to 5.8 nm across the various wavelengths. Simultaneously, spatial coordinates for the collected samples were recorded, ensuring precise geospatial information alignment. The spectral data underwent spectral resampling to facilitate a meaningful comparison with the spectral information derived from the ASTER images due to the few broader bands available in the ASTER image [47]. To harmonize the datasets, the average spectral profiles for each sample were resampled within the ENVI software (ENVI 5.5), aligning them with the corresponding wavelength ranges of the ASTER bands. This resampling process facilitated a detailed exploration of the behavior, including absorption and reflection characteristics, of the Fe minerals, rock intrusions, volcanic wall rock, and other background features, which guides the band ratio and PC band selections.

2.2.2. Band Ratio Selection

A total of 15 distinct band ratios were thoughtfully selected for comprehensive examination, with the overarching objective of identifying potential ratios for integration into the final imagery as inputs for the proposed model. For a comprehensive enumeration of the various band ratios designated for scrutiny, refer to Table 1. It is imperative to emphasize that each selected band ratio is specifically tailored to enhance the spectral response of a particular mineral. To elucidate, band ratios 5/3 + 1/2 and 4/8 were deployed to enhance ferrous iron and iron oxide, respectively, based on the findings of [13,48]. An array of additional band ratios, such as 2/3, (8/7)/((2 + 4)/8), (2/3)/((2 + 4)/8), and (2 + 4)/8, were meticulously devised in alignment with the absorption and reflection bands characteristic of iron minerals within the study area, with the explicit aim of intensifying the spectral response of iron ores during the classification process. Band ratios selected to enhance intrusions, skarn, and volcanic wall rock were also thoughtfully chosen for evaluation, as detailed in Table 1.

Table 1. Computed band ratios for examination.

S/No.	BR	Usage	Source
1	5/3 + 1/2	Ferrous iron enhancement	[48]
2	5/6	Skarn (chloride and calcite) enhancement	[3]
3	4/8	Iron oxides enhancement	[13]
4	(7 + 9)/8	Delineate skarn minerals (such as calcite and chloride)	[48]
5	(2 + 4)/8	Iron ore enhancement	From the observed iron ore spectra
6	(6 + 9)/(7 + 8)	Delineate skarn minerals	[49]
7	(2 + 4)/3	Iron ores and intrusions from skarn and wall rocks	[6]
8	((5 + 9)/7)/((6 + 9)/8)	Skarn minerals enhancement	[50]
9	2/3	Iron ore enhancement	From the observed iron ore spectra
10	(8/7)/((2 + 4)/8)	Iron ore enhancement	Modified from [50]
11	4/7	OH-bearing minerals	[51,52]
12	((6 + 8)/7)/((7 + 9)/8)	Distinguished skarn minerals	[3,50]
13	2/1	Ferric iron	[48]
14	4/2	Distinguished skarn from wall rocks and intrusions	[53]
15	(2/3)/((2 + 4)/8)	Ores from other components	From the observed iron ore spectra

2.2.3. Selection of Principal Components Bands

PC bands constitute an additional dimension of multispectral image derivatives employed in this study. The incorporation of PC bands as input data or as part of the input data in convolutional networks consistently demonstrates effectiveness in reducing data redundancy, enhancing classification accuracy, and minimizing processing time [27,54]. However, it is imperative to note that including only those PCs that emphasize specific features of interest can result in a reduction in processing time compared with using all PCs generated from the original input image. Consequently, this study leverages the PC selection concept employed by [18,27]. This approach revolves around selecting PCs that emphasize desired features, guided by the spectral information essential to those desired features.

2.2.4. Design of Proposed 3D-CNN Model Architecture

The proposed model is engineered to process a stacked image comprising ASTER image bands, BR images, and selected PC bands as its input. The architectural blueprint of the proposed model is founded on the framework introduced by [17], which refines the design put forth by [18]. The latter incorporates a few 3D convolutional layers, including a residual layer, to enhance the generalization capacity, mitigate overfitting, prevent gradient disappearance, and improve the learning rate. Notably, the proposed model adheres to a consistent use of a small receptive field (3 × 3 × 3) for all convolutional layers, as advocated by [55]; this is particularly effective for spatiotemporal feature extraction, especially in instances where the features possess limited spatial coverage. This uniformity in kernel size across layers is posited to yield more precise results compared with disparate kernel sizes for each layer [56]. The adoption of a small kernel size is necessitated by the modest size of the input patch required, set at 7 × 7 × C, where "C" denotes the number of bands (Figure 2). This configuration allows a 3 × 3 × 3 kernel to perform three convolutions (without maximum pooling, stride, or padding), thereby expediting operations. The choice of the 7 × 7 patch size aims to strike a balance between providing sufficient contextual information and maintaining classification precision, especially considering the 15 m × 15 m resolution of the resampled image and the nature of the deposits under investigation. Additionally, pixels within a small spatial vicinity generally reflect the same underlying material.

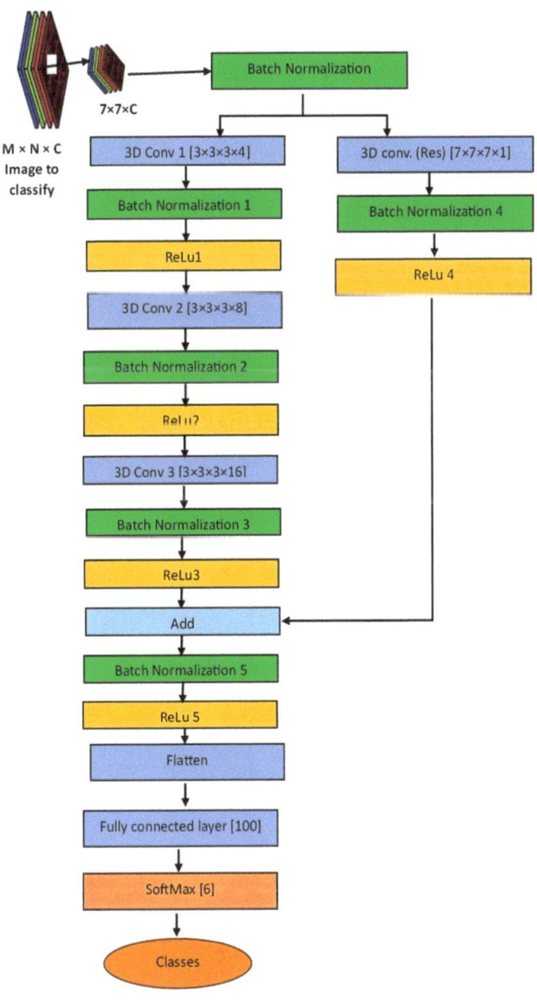

Figure 2. Proposed architecture.

The choice of the number of the hidden units used in the convolutional layer is pivotal, as it governs the network's capacity to learn complex representations by projecting input data into a higher-dimensional space. However, an excessive number of hidden layers may result in learning extraneous patterns, enhancing training accuracy while exerting minimal influence on the model's testing performance. In light of this, the current model embraces the paradigm of a shallow network with a small patch size, as advocated by [28,57]. The choice of a small patch size is further influenced by the coarse resolution of the ASTER images (30 m). Owing to the shallowness of the architecture, a residual layer is employed to convolve the input image, with its output merged with that of the third convolutional layer before progressing to the fully connected layer. This integration serves to smooth the data, enhance model stability, and augment the generalization capacity of the dense layer [17].

Given that the input image comprises stacked bands with disparate distributions, batch normalization (BN) assumes the role of the initial operation on the input to standardize all input channels to a uniform range. This normalization, computed for each channel (image band), precedes the application of the non-linearization function in the convolutional network. The ratio of filters between subsequent convolutions is set at 1:2, a configuration proven effective in various studies, including those by [28,57]. Consequently, the initial number of filters in the first convolutional layer is stipulated at 4. Finally, a SoftMax classifier is employed after the fully connected layer to categorize outputs into specified classes, with the probability of an input being classified as a particular class [27]. This configuration resulted in a model with only 26,490 parameters. It is noteworthy that out of the total parameters, a total of 26,398 are trainable, rendering the prediction of unclassified pixels within the entire image notably faster. The model predicts the category of the central pixel within a 7×7 patch, as the central pixel is considered the most representative of the local context, while the surrounding pixels provide additional context to enhance classification accuracy.

2.2.5. Model Training, Testing, and Evaluation

Following the preparation of training and testing data, 75% of the prepared image patches, centered on each sample location and its surrounding eight pixels, were allocated for training purposes, with the remaining 25% reserved for testing. Within the training dataset, 95% of the samples were further designated for actual training, while 5% were earmarked for validation. The classes were designated numerically from 1 to 6. Class 1 was assigned to iron ore, and classes 2 through 6 corresponded to skarn minerals, intrusions, volcanic wall rocks, vegetation, and glaciers, respectively. The number of training epochs was set at 100 to facilitate a comprehensive examination of model stability.

Upon completion of model training and testing, overall accuracy (OA) and average accuracy (AA) metrics were extracted to assess the model's efficacy. The OA of each class represents the classification accuracy for that specific class post-model testing. For each class, the overall accuracy of classification is calculated by dividing the number of samples correctly classified by the total number of samples in that class. The overall accuracy for all classes is determined by dividing the sum of samples correctly classified across all classes by the total number of samples tested. The average accuracy denotes the mean of overall accuracy across all classes and is calculated by dividing the overall accuracy for each class by the total number of classes. The values of these accuracy indicators signify how well the model is trained and is capable of distinguishing features it was trained to do.

2.2.6. Evaluation of the Impact of BR Images and PC Bands on the Accuracy

The model was specifically configured to operate on a concatenation of multispectral image bands, PC bands, and BR images. To comprehensively gauge the impact of these image derivatives on the accuracy of iron mineral detection and the classification of other minerals, the model underwent additional testing under various configurations: utilizing solely the ASTER image bands, combining ASTER image bands with PC bands, employing

ASTER image bands in conjunction with BR images, and utilizing only PC bands and BR images. This systematic evaluation allowed for a nuanced understanding of the contributions of BR images and PC bands to the model's accuracy. Accuracy indicators such as OA, kappa statistics, and AA were employed to quantify the model's performance (Figure A5 in Appendix A).

2.2.7. Comparison with Other Methods

To further validate the effectiveness of the proposed approach, we compared the classification results with those from existing methods: (1) a 3D-CNN method proposed by [28], which employs a few 3D convolutional layers (two convolutional layers and a fully connected layer), uses a small image cube size (5 × 5), and has very few parameters while taking full-image spectral bands as inputs; (2) a HybridSN model by [29], consisting of three 3D convolutions, a 2D convolution, and three fully connected layers, using a 25 × 25 image cube size as input, resulting in a large number of parameters; (3) a 3D inception model proposed by [58], which features four branches: the first branch has a 1 × 1 × 1 3D convolutional layer, the second branch has a 3 × 3 × 3 convolutional layer, the third branch includes a 1 × 1 × 1 and a 5 × 5 × 5 convolutional layer, and the fourth branch adds a 3 × 3 × 3 convolutional layer to the general inception architecture; and (4) a 2D-CNN method (AlexNet) proposed by [59], consisting of five convolution layers, three max-pooling layers, two normalization layers, two fully connected layers, and one SoftMax layer. For our data, the AlexNet architecture was modified by reducing the input size to 27 × 27 and omitting the first convolutional and pooling operations. The results were compared in terms of OA, AA, and the kappa coefficient.

2.2.8. Experimental Setting

This study introduces an innovative methodology aimed at enhancing the dimensionality of multispectral satellite image data. The approach integrates relevant band ratios and principal component bands into the existing image bands. Furthermore, it presents a 3D-CNN model characterized by a few convolutional layers, rapid convergence rate, and relatively high accuracy. The proposed model was implemented using Python incorporating the Keras and TensorFlow libraries along with other modules such as "sio", "os", and "matplotlib". This implementation was conducted within a Jupyter Notebook environment, a web-based open-source platform for code development and resource sharing. The "Keras" package provides the building blocks for model configuration and training, while the "sio" and "os" modules are used for handling external files. Data visualization and the generation of prediction maps were achieved using the plotting functionalities of "matplotlib".

All experiments were conducted on an ASUS Zenbook laptop (ASUS Computer Co., Ltd., Shanghai, China) equipped with an 11th Gen Intel(R) Core (TM) i7-1195G7 processor running at 2.90 GHz and 16 GB of RAM. During model configuration, the parameters were specified as follows: a batch size of 32 patches, a learning rate of 0.001, a dropout rate of 35%, a regularization parameter (L2 parameter) set at 0.001, and kernel and filter sizes in accordance with the specifications outlined in Figure 2. To ensure a thorough assessment of model stability and convergence, the number of training epochs was set at 100. A SoftMax classifier was introduced after the final fully connected layer to classify the outputs into specified classes.

The model's training, testing, and validation phases were conducted utilizing the prepared image data of the Zhibo deposit. This image comprises 19 bands, encompassing 9 ASTER multispectral bands and 10 bands derived from the original image transformations (4 PC bands and 6 BR images). This image comprises 19 bands, encompassing 9 ASTER multispectral bands and 10 bands derived from the original image transformations (4 principal component bands and 6 band ratio images). The image dimensions are 516 rows by 663 columns, resulting in an image shape of 516 × 663 × 19. The ground truth file was generated by associating the sample points' coordinates with six distinct classes, encapsulating both environmental and mineral/rock components of interest in the study area. During the ground truth image generation process, sample points were buffered to encompass the immediate adjacent pixels (within 15 m) in all directions. This data augmentation strategy aims to bolster the volume of training pixels available for generating input patches. Importantly, this strategy does not compromise the quality of the training data, as meticulous care was exercised during field sample collection to ensure each sample was acquired in locations where a conspicuous dominance of the target class was apparent within a 30 m proximity.

Subsequent to the data augmentation process, the ground truth image is systematically compiled, and each class is assigned a unique identifier ranging from 0 to 6. The "0" identifier denotes pixels unallocated to any specific class, signifying those to be classified after the training and validation phases. Table 2 presents the summary statistics of the total number of pixels for patch preparation for each category, which includes 949 training patches, 50 validation patches, and 333 testing patches. A visual representation of the selected prepared patches for model training, testing, and validation is displayed as a false-color composite in Figure A6 in Appendix A, using ASTER bands 3, 2, and 1 displayed through the red, green, and blue filters, respectively. The four rock components appear visually similar, making them difficult to distinguish with the naked eye, whereas vegetation and glaciers are distinct from the others, allowing for easier classification with minimal error. The evaluation of results was based on the assessment of OA, AA, and the kappa coefficient.

Table 2. Patches summary.

Class/Patches Category	Class ID	Number of Patches
Iron Ore	1	306
Skarn	2	306
Wall rock	3	126
Intrusion	4	189
Vegetation	5	180
Glacier	6	225
Total	6	1332
Training	-	949
Testing	-	333
Validation	-	50

3. Results

3.1. Processed Samples Spectra

The average ASTER-resampled spectra for some of the selected samples are presented in Figure 3. The resampled spectra of the iron ore exhibit pronounced absorptions within band 3 and band 8, as depicted in Figure 3a. Remarkably, these spectral characteristics bear a striking similarity to the spectra of skarn minerals, featured in Figure 3b. Notably, the spectral profile of iron ore showcases nearly uniform reflectance between bands 5 and 6, while skarn minerals manifest a distinct absorption between these two bands.

Conversely, the spectra of intrusive rock samples exhibit absorption in band 6, in addition to the absorptions observed in band 3, and reflections are present in bands 5 and 7, as noticeable in Figure 3c. This spectral behavior can be attributed to the presence of OH-bearing minerals within the composition of the intrusion rocks [60]. Given that some of the intrusions are identified as granites within the study area, it is notable that granites typically feature absorption features around 2200 nanometers in wavelength (corresponding to ASTER band 7) [61]. Nonetheless, it is noteworthy that specific intrusion samples (BZI-6, 7, and 11) do not display discernible absorption or reflection features across the measured wavelength range (ranging from 400 to 2500 nanometers) for this study.

These particular samples exhibit spectral behavior analogous to granodiorite, mirroring the observations in [62].

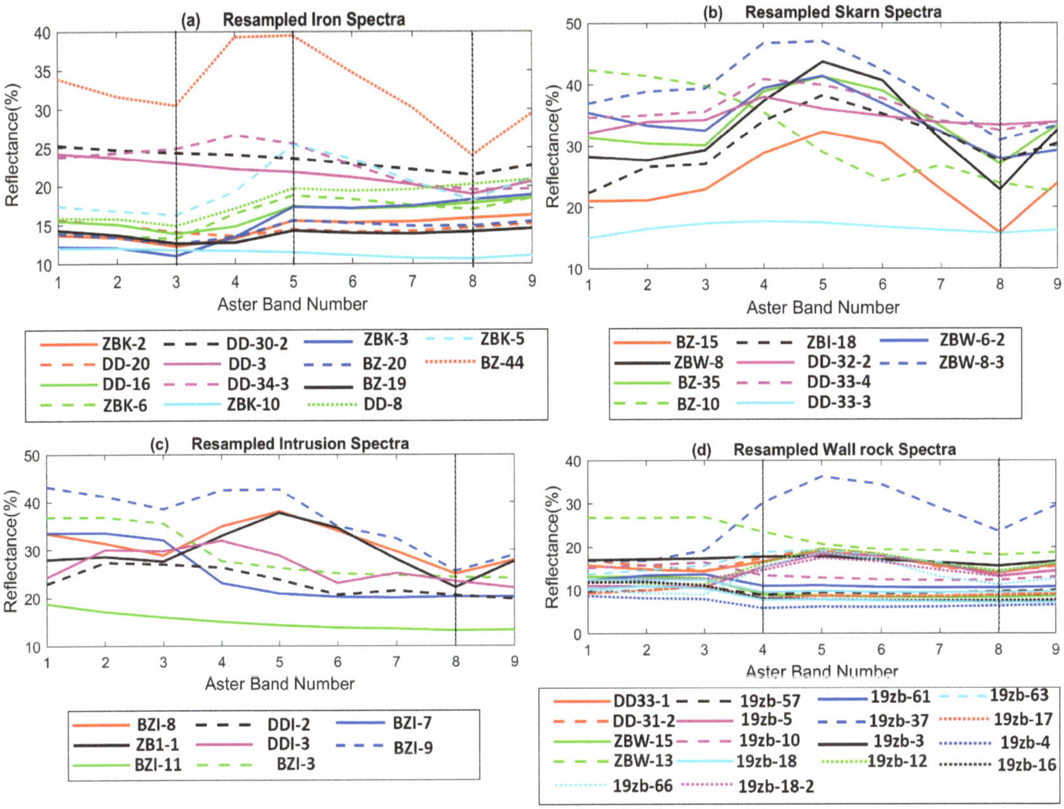

Figure 3. Resampled spectral profile of the collected samples. In the case of iron ore (**a**), notable absorptions are evident in bands 3 and 8, with prominent reflections observed in band 5. The majority of skarn samples (**b**) exhibit a pronounced absorption feature in band 8, with minimal variation observed between bands 2 and 3, yet uniform reflection is evident in bands 4 and 5. Certain wall rock samples (**d**), including those from Dunde deposits (basaltic tuff), display distinct reflectance patterns. The majority of the intrusive rock samples (**c**) exhibit absorptions in bands 3 and 6, accompanied by reflections in band 5.

The spectral plots of the wall rock samples (Figure 3d) exhibit distinguishable spectra between those sourced from Zhibo and Dunde (indicated by sample names beginning with "Z" and "D", respectively), reflecting the presence of subtly distinct wall rock types in these areas. Specifically, Zhibo deposits predominantly feature andesite, basaltic andesite, and basalt [39], while Dunde deposits comprise basaltic tuff, basaltic crystal tuff, and andesite [44] (Figure 1b,c). This discrepancy in mineral composition contributes to the observed dissimilarities in the spectra. Notably, wall rock spectra from Dunde (DD33-1 and DD-31-2, Figure 3d) closely resemble the spectra of skarn minerals. This resemblance can be attributed to the presence of calcite, epidote, or chlorite in most basaltic tuffs in Dunde (Figure A4 in Appendix A) [63], resulting in a spectral profile featuring a significant absorption around 2330 nm wavelength (ASTER band 8). Coincidentally, these minerals, integral to the basaltic tuff composition, also form part of the skarn minerals in the study deposits [33,63], explaining the observed spectral similarity. Despite this, discrepancies manifest in the VNIR range, where skarn exhibits absorption in band 3, contrasting with the reflection observed in the wall rock (Figure 3b,d).

The spectral profiles of ore samples demonstrate a pattern akin to that of intrusion samples, with a notable distinction in the region between band 6 and band 7, where intrusions exhibit reflection while ores maintain absorption. This divergence in bands 6 and 7 can be attributed to the presence of Al-OH in the mineral composition of both intrusions and ores. The similarities observed suggest common mineral composition as gangue components, owing to the formation of ore minerals between intrusions and wall rocks.

3.2. Selected Band Ratios

The analysis utilized scatter plots of two band ratios per plot (Figure 4) to evaluate their efficacy in differentiating between mineral components. The plot of BR 4/8 against 5/6 (Figure 4a) demonstrated the capability of these ratios to distinguish between ores and skarn minerals, with skarn minerals consistently displaying higher ratio values compared with ores. Additionally, iron ore samples exhibited uniform reflectance between bands 5 and 6, while skarn minerals displayed distinct absorption between these bands.

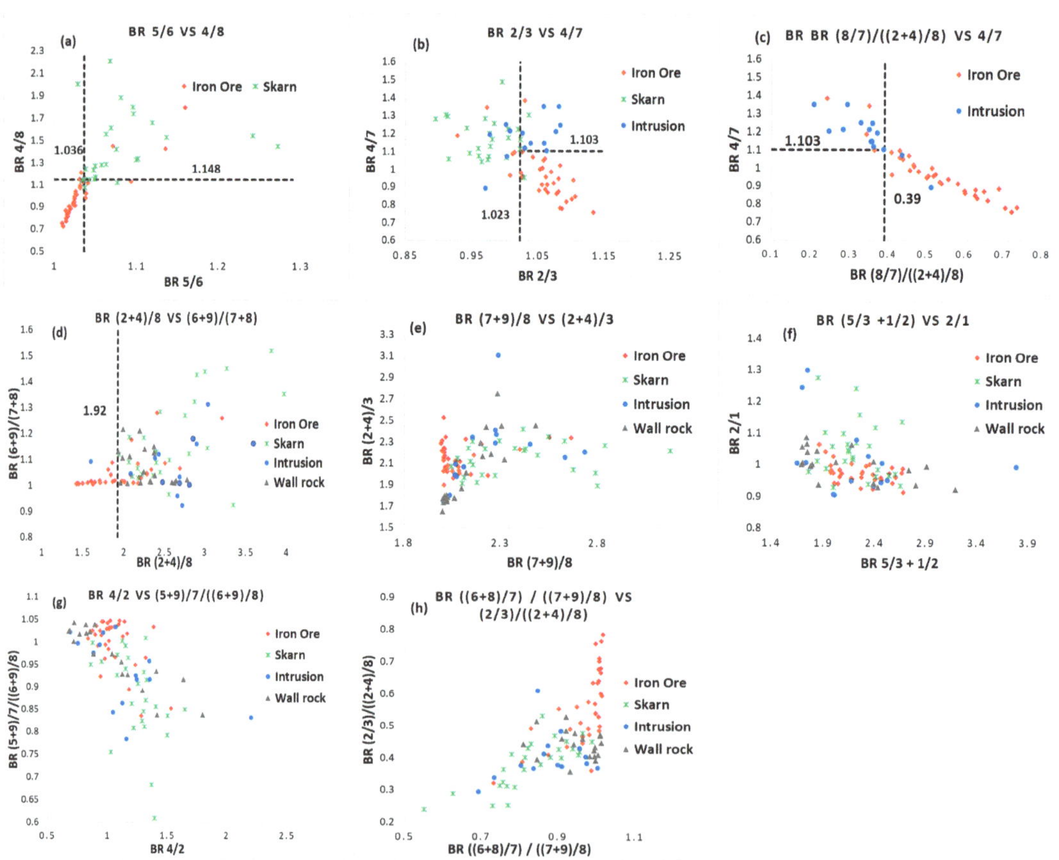

Figure 4. Band ratios assessment scatterplots. The initial four graphs (**a**–**d**) showcase selected band ratios distinguished by clear boundaries between at least two distinct components. Nevertheless, it is noteworthy that none of the individual band ratios exhibit a pronounced capability in effectively segregating each component from the others. Plots (**e**–**h**) indicate an absence of well-defined boundaries between two or more components.

The scatterplot of BR 2/3 against 4/7 (Figure 4b) indicates that these ratios collectively differentiate between three components, with BR 2/3 greater than 1.023 indicative of iron ore or intrusion rock and inverse values suggesting skarn minerals. Incorporating BR 4/7 provided further discrimination, with intrusions possessing higher BR 4/7 values compared with ore samples. Similarly, BR (8/7)/((2 + 4)/8) effectively separated ores from intrusions (Figure 4c), and BR (2 + 4)/8 proved valuable in resolving classification confusion between ores and wall rocks (Figure 4d), although it could not separate wall rocks from the other classes. However, BR (6 + 9)/(7 + 8) exhibited limited capability in distinguishing any one component from the others despite slight spectral pattern variations within the ratio across different classes. BR (5/3 + 1/2) and BR 2/1 displayed heightened sensitivity, making component differentiation challenging. Other tested band ratios encountered difficulties in isolating skarn samples, possibly due to the presence of calcite and chlorite in the skarn minerals across the study areas. Additionally, band ratios such as BR (2 + 4)/3 and (2/3)/((2 + 4)/8), aimed at enhancing iron content and differentiating iron ores from other components, showed no apparent clustering for the same samples. Table 3 provides the summary list of the selected BRs.

Table 3. Summary of the selected BRs.

BR	Ore	Skarn	Intrusions	Wall Rock	Remark
5/6	√	√			Differentiate iron ores from skarns
4/8	√	√			Separate iron ore from skarn
2/3	√	√	√		Separate skarn from iron ore and intrusion
(8/7)/ ((2 + 4)/8)	√		√		Separate iron ore from intrusions
4/7	√		√		Separate iron ore from intrusions
(2 + 4)/8	√	√	√	√	Separate iron ore from other components

3.3. Selected Principal Components Bands

The selection of the requisite PCs adheres to a concept widely adopted in various research works, including those by [18,27]. This concept involves identifying PC bands whose eigenvectors possess the highest opposing sign values corresponding to the major absorption and reflection bands of the targeted component, thereby characterizing that component. The eigenvectors matrix, detailed in Table 4, reveals that PC 3 exhibits moderate negative contributions in bands 2 and 1 (−0.32 and −0.31, respectively), coupled with a strong positive contribution in band 3 (0.67), which aligns with the reflection and major absorption bands within the spectra of the iron ore samples. Consequently, PC 3 is deemed a suitable selection for inclusion in the final set of image bands to be employed as input into the model, as it appears to contain vital information conducive to the simple detection of iron minerals.

Likewise, PC 4 demonstrates a moderate negative contribution in band 8 (−0.41) and a positive contribution in band 4 (0.59) (as outlined in Table 4), aligning with the absorption and reflection bands shared by the skarn and iron ore spectra. This characteristic makes PC 4 potentially useful for enhancing iron oxides, as previously proposed by [13]. PC 5, on the other hand, exhibits a positive contribution in band 4 (0.44) along with a negative contribution in band 5 (−0.60). This nuanced behavior corresponds to the spectral signature of volcanic wall rocks, which uniquely display absorption in band 4 followed by reflection in band 5, distinguishing them from other components, which generally exhibit reflection in both bands. Notably, PC band 7 displays its highest positive contribution from band 6 (0.63) and the most significant negative contribution from band 5 (−0.57), indicating the spectral characteristics of skarn samples within the study area, where band 5 serves as an absorption shoulder band, while band 6 functions as the primary absorption band.

Table 4. The eigenvector matrix from the PCA of the nine ASTER bands (VNIR and SWIR).

Eigenvectors	Band 1	Band 2	Band 3	Band 4	Band 5	Band 6	Band 7	Band 8	Band 9
PC 1	0.5547	0.5725	0.5841	0.0745	0.0604	0.0683	0.0695	0.0517	0.0453
PC 2	0.2119	−0.1415	0.0899	−0.5397	−0.3950	−0.3922	−0.3413	−0.3253	−0.3219
PC 3	**−0.3287**	**−0.3098**	**0.6716**	0.3413	−0.2152	−0.1120	−0.1800	−0.2223	−0.2963
PC 4	0.3354	−0.0532	0.3906	**0.5891**	0.1568	0.1203	−0.1589	**−0.4132**	−0.3873
PC 5	−0.0426	−0.2432	0.2046	**0.4391**	**−0.6036**	−0.3817	0.1846	**0.4002**	0.0052
PC 6	−0.6003	−0.6672	0.0657	0.0588	0.0917	0.1051	−0.3529	−0.1676	0.1207
PC 7	0.1629	−0.1326	0.0257	0.0261	**−0.5661**	**0.6326**	−0.3416	−0.0412	0.3400
PC 8	−0.1877	−0.1725	0.0155	−0.2053	−0.2049	0.4981	0.4797	0.0451	−0.6086
PC 9	−0.0540	0.0422	0.0015	0.0227	−0.1852	−0.0978	0.5619	−0.6924	0.3948

PC 3–5, and 7 are selected due to the opposite contributions observed from the bands corresponding to the reflection and absorption of the targeted minerals.

3.4. Mineral Detection Using the Proposed Model

In this study, we applied a novel model for the detection and classification of minerals within the Zhibo deposit. The methodology involved a comprehensive approach encompassing training, testing, and validation phases on stacked images of the deposit. The model was optimized using the "Adam" optimizer, with "categorical cross-entropy" serving as the loss function and "accuracy" as the evaluation metric. A best model was selected for subsequent classification and prediction tasks, culminating in the production of a classified image, as illustrated in Figure 5 (top center).

The classified image revealed the predominant presence of iron minerals in the northwestern region of the deposit, extending toward the southeast, particularly concentrated at the contact zone between the intrusion and wall rocks. This observation aligns with previous geological reports on the spatial distribution pattern of ore minerals within the deposit [39]. Individual class classification results demonstrated noteworthy accuracy rates, particularly with 100% accuracy achieved for iron ores, intrusions, vegetation, and glaciers. Skarn minerals and wall rocks exhibited accuracies of 84.62% and 95.24%, respectively (Table 5). The training phase lasted approximately 121.22 s, with subsequent prediction and testing phases completed efficiently within 0.828 s, indicating robust convergence of the model.

The classification result analysis reveals that the proposed methodology demonstrates commendable performance metrics. The OA stands at 96.95%, with an AA of 94.87% and a kappa coefficient of 95.93%. Notably, these results exhibit an improvement in the OA, AA, and kappa coefficient of 1.03, 1.27, and 1.49, respectively, in comparison with outcomes attained utilizing the 3D inception model introduced by [58] (Table 5). This disparity underscores a significant achievement of the proposed model despite its linear architectural design relative to the more complex parallel architecture of the 3D inception network. Relative to a 3D-CNN model proposed by [28], which shares a similarly shallow architectural design, the proposed model exhibits significant improvements of 9.17, 13.6, and 13.15 in OA, AA, and kappa coefficient accuracy, respectively. The disparity underscores the efficacy of the proposed model, raising inquiries into its underlying mechanisms. In contrast with the 2D-CNN model, our proposed model demonstrates improvements of 3.06%, 3.66%, and 4.06% in OA, AA, and kappa coefficient accuracy, respectively. Notably, the 2D-CNN model surpasses the 3D-CNN model proposed by [28], potentially due to the latter's comparatively shallow network architecture, as highlighted in [28]. Comparatively, against the HybridSN model [29], the proposed model exhibits a marginal shortfall of less than 0.06 in both OA and AA, despite possessing significantly fewer parameters and a less complex architectural design (Table 5).

In summary, the proposed methodology and model demonstrate exceptional performance in identifying and detecting skarn iron minerals within the Zhibo deposit, surpassing existing methodologies in accuracy while maintaining a simpler and more efficient architecture.

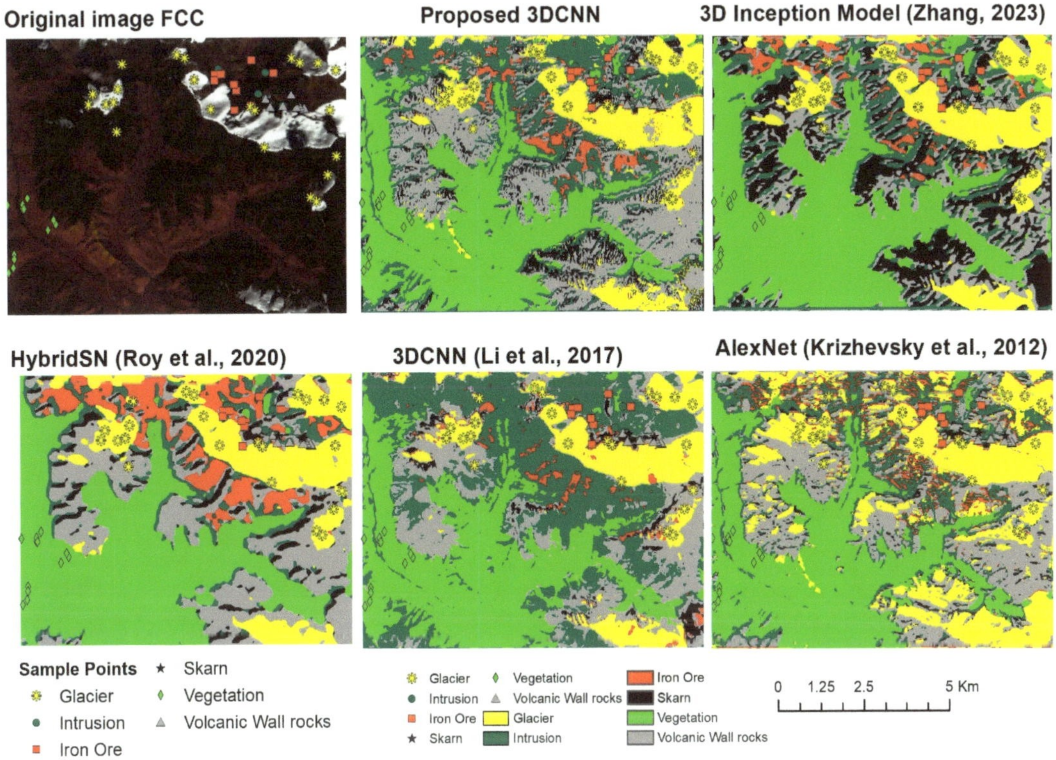

Figure 5. Classification maps using different methods, including the proposed method. Both the HybridSN and the 3D inception method produced smoother results, suppressing smaller classes, as can be observed in the vegetation and wall rocks [28,29,58,59].

Table 5. Comparison of the proposed method with the existing methods.

	Accuracy (%)								
Method	**Iron Ore**	**Skarn**	**Intrusion**	**Wall Rock**	**Vegetation**	**Glacier**	**OA**	**AA**	**Kappa**
Proposed model	100	84.62	100	95.24	100	100	96.95	94.87	95.93
HybridSN	100	100	92.86	85.71	100	100	97.01	94.93	96.21
3D-CNN	75.00	76.92	92.86	42.86	100	100	87.78	81.27	82.78
3D Inception Model	100	92.31	100	95.24	100	98.08	95.92	93.60	94.44
AlexNet	100	61.54	100	85.71	100	100	93.89	91.21	91.87

3.5. Evaluation of the Impact of the BR Images and PC Bands

The strategic integration of band ratios and principal component bands alongside the original ASTER image bands was undertaken not only to augment the number of bands/channels within the input image but also to capitalize on the unique strengths of these derivatives in enhancing discernible features. Therefore, the evaluation sought to scrutinize the effects of incorporating BR images and PC bands into the original ASTER image bands on mineral classification accuracy.

The results (Figure 6 and Table 6) indicated that the inclusion of selectively chosen band ratios alongside the original ASTER bands yielded an overall classification accuracy of 87.02%, with an average classification accuracy of 79.15% across all classes. While band ratios improved the classification of ores and volcanic wall rocks, they introduced significant confusion in differentiating skarn from intrusions and wall rocks, resulting in misclassification rates exceeding 80% for skarns and approximately 35% for intrusions.

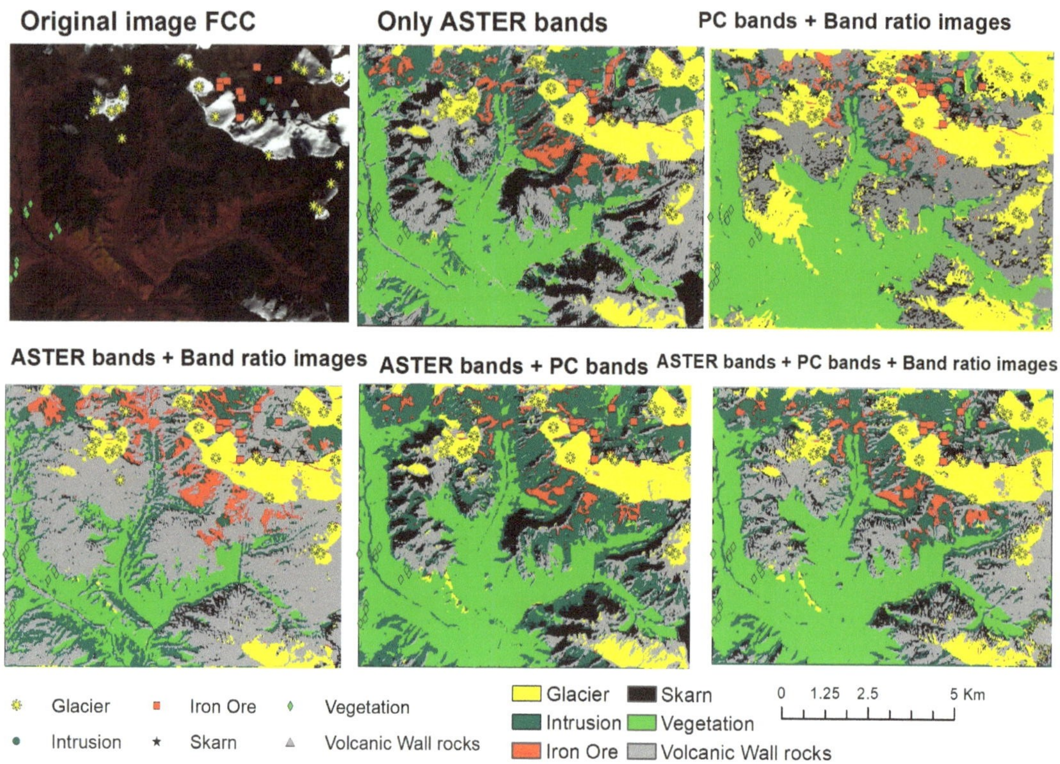

Figure 6. Classification results using the proposed model. The original image is presented in the top left section, while various classified images using different input band combinations are displayed in the remaining sections.

Table 6. Accuracy assessment results from five different input configurations.

	Accuracy (%)								
Image Used	**Iron Ore**	**Skarn**	**Intrusion**	**Wall Rock**	**Vegetation**	**Glacier**	**OA (%)**	**AA (%)**	**K (%)**
Only ASTER bands	100	**84.62**	92.86	76.19	100	98.08	93.13	91.96	90.91
PCs + BR	100	53.85	64.29	76.19	100	96.15	87.78	82.39	83.81
ASTER bands + BR	100	15.38	64.29	**95.24**	100	**100**	87.02	79.15	82.60
ASTER bands + PCs	100	76.92	**100**	85.71	100	98.08	94.65	92.93	93.45
ASTER bands + PCs + BR	100	**84.62**	**100**	**95.24**	100	**100**	**96.95**	**94.87**	**95.93**

The proposed band combination has retained the best classification results for each of the classes.

Remarkably, utilizing solely ASTER bands as the input parameters yielded significantly higher accuracy rates (OA = 93.13%, AA = 91.96%, and kappa = 90.91%) compared with scenarios involving the incorporation of ASTER bands with band ratios or PC bands with band ratios. The amalgamation of ASTER bands with selected PC bands yielded a slightly improved result, with an OA of 94.65%, AA of 93.45%, and kappa coefficient of 92.93%. However, this enhancement posed challenges in accurately classifying wall rocks, as some were misclassified as skarn or intrusions. Nevertheless, the inclusion of PC bands made a constructive contribution to both the overall and average classification accuracies.

The classified image derived from the combination of ASTER bands, PC bands, and band ratios (Figure 6) presented more visually appealing and statistically more accurate results when compared with the four alternative band combinations explored in this study, as summarized in Table 6. This particular combination notably amended the misclassification of intrusions evident in both the northern and central regions of the classified image when ASTER bands were used in conjunction with PC bands. Additionally, it addressed the misclassification of skarns observed in both the western and southeastern portions of the same image. This band combination further improved the classification accuracy beyond what was observed when exclusively employing ASTER bands in conjunction with PC bands. Specifically, the OA increased from 94.65% to 96.95%, reflecting a 2.3% improvement, while the average accuracy rose from 93.45% to 94.87%, indicating a 1.42% enhancement.

4. Discussion

4.1. Spectral Analysis and Data Selection

Spectral analysis unveiled distinct absorption and reflection patterns among the different mineral components. The majority of samples in each class exhibited absorption in band 8, likely due to the presence of calcite as a gangue mineral in the iron ore samples, part of the skarn mineral composition, or a constituent of the basaltic tuff in the wall rocks. This absorption, observed around 2330 nanometers, may be induced by the vibrational effect generated by the C-O bond in calcite [61] or the Fe-OH bonding [48]. Similarities between the spectra of the iron ore and skarn minerals (Figure 3a,b) suggest the presence of iron minerals in the majority of the skarn samples, which efficiently absorb radiation within the wavelength range of 500 to 1000 nanometers [50,64], particularly in the near-infrared region spanning 780 to 860 nanometers corresponding to ASTER band 3. These observed similarities and differences in electromagnetic radiation absorptions and reflections across the nine ASTER bands among the various minerals and rock components underscore the importance of band selection and derivation of ratios for accurate mineral identification and characterization.

The scatter plots of the band ratios (Figure 4) illustrate the efficacy of different ratios in distinguishing between mineral components with specific ratios, such as BR 4/8 and BR 2/3, showing consistent patterns in differentiating ores from skarn minerals. Similarly, the scatterplot of BR 5/6 against 4/8 convincingly highlights the ability of these two band ratios to distinguish ores from skarn minerals, with skarn minerals exhibiting higher ratio values in both ratios compared with ores (Figure 4a). BR (5/3 + 1/2) and BR 2/1, specifically identified by [48] for ferrous and ferric iron enhancement, could not perform this task efficiently. Notably, these ratios exhibited no distinct boundary between components. This may be attributed to the small and insignificant weight percentage of Fe in the skarn samples from all the sites.

The PCs, derived from PCA, constitute a significant aspect of this research's consideration. These bands are deemed valuable due to their potential to mitigate data redundancy, enhance classification accuracy, and simultaneously reduce processing time, as documented in studies such as [27,54]. Nevertheless, it is imperative to acknowledge that certain PCs, particularly those associated with higher PC bands, may contain minimal information and exhibit very low variance, appearing similar to noise, as noted by [65]. To optimize the model's performance, PCs 3, 4, 5, and 7 emerge as the only PCs among the nine computed PCs that contain pertinent information corresponding to the four components (iron ores, skarn, intrusion, and volcanic wall rocks) of interest in this research and are selected for inclusion in the image bands fusion.

4.2. Methods Comparison

During mineral detection, high accuracy was observed in iron ore classification, attributed to the complementary support provided by the multiple band ratio images integrated into the input data. However, relatively lower accuracy in distinguishing wall rocks from skarn minerals may be due to the presence of calcite and chlorite, significant constituents of both classes, particularly in the basaltic tuffs prevalent in the Zhibo deposit [39]. These mineral compositions posed challenges for the model, particularly when using broader wavelength range images like the ASTER image utilized in this study.

Relative to other models evaluated, the proposed model exhibited superiority, emphasizing that the depth of a network does not always correlate with superior performance, as deeper networks may capture extraneous patterns that may augment training accuracy but offer marginal benefits to testing performance [57]. The integration of residual layers within the proposed model architecture was identified as a key contributor to its superiority, enhancing generalization capacity, improving model stability, and mitigating overfitting, thereby preventing issues such as gradient disappearance and augmenting the learning rate [17,18,66]. Additionally, the utilization of larger patches in the other methods (HybridSN, AlexNet, and the 3D inception model) leads to overgeneralization in classification outcomes, resulting in smoother images that prevent subtle spatial variations typical of skarn iron formations, characterized by the sharp contact zonation between ores and surrounding rocks [38,67,68].

The close correspondence between the results from the proposed model and the HybridSN model underscores the pivotal role of PC bands in enhancing image classification accuracy. However, the proposed method demonstrated superior performance in identifying intrusive and wall rocks, with accuracies of 100% and 95.24%, respectively, compared with 92.86% and 85.71%, respectively, achieved by the HybridSN method (Table 5). This superiority may be attributed to the inclusion of band ratio images in the proposed model's input configuration, which aids in reducing confusion between wall rocks and iron ore or skarn minerals and enhances the spectral response of intrusions. Nonetheless, the HybridSN method exhibits slightly superior accuracy in skarn mineral classification.

4.3. Impact of Band Integration on Classification Accuracy

During the evaluation of the impact of the integration of PCs, band ratio images, and original ASTER bands, the incorporation of BRs into the ASTER bands notably influenced the classification outcomes, reducing misclassified pixels in the context of glaciers and wall rocks and thereby enhancing the accuracy of the classified image, particularly in glaciers and wall rock categorization. However, the remarkable results obtained while using only ASTER bands as input relative to the use of either ASTER bands with band ratios or PC bands with band ratios was unexpected, considering the anticipated favorable contribution of band ratios to accuracy, given their documented potential to enhance the spectral signatures of targeted features in prior studies [3,13,51–53]. Additionally, the classified image resulting from the integration of PC bands with ASTER bands exhibited notable visual and statistical improvements in mitigating misclassifications observed in the classified image derived from ASTER bands in conjunction with band ratios (Figure 6). However, this integration led to a reduction in the accuracy of wall rock classification, with some wall rocks erroneously classified as skarn or intrusions. Consequently, the pursuit of a solution through the simultaneous incorporation of all three sets of bands was prompted.

The results from the combination of all three sets of bands consistently retained the highest classification accuracy among the various combinations explored, underscoring the importance of considering the diversity of input channels in enhancing classification accuracy. Further investigation is warranted to understand the specific mechanisms driving this enhancement. However, the most plausible explanation for this enhancement could be attributed to the notion that the accuracy of the 3D-CNN model escalates in tandem with an increase in diversity afforded by a higher number of channels in the input layer. This expansion enables the capture of interdependencies and relationships between the various bands or channels, as posited by [69]. Notably, combinations featuring only PC bands showed reduced accuracy in classifying wall rocks and glaciers compared with combinations featuring only band ratios. This positive impact of different band combinations on classification outcomes underscores the importance of considering the diversity of input channels in enhancing classification accuracy.

5. Conclusions

This study introduces a novel 3D-CNN model architecture that incorporates the strategic integration of band ratios and principal component bands alongside the original ASTER image bands to enhance mineral detection accuracy. The proposed model architecture presents a streamlined, efficient, and effective framework distinguished by a notable reduction in parameters compared with prevailing models in the field. The incorporation of image derivatives notably improves classification accuracy and expedites model predictions. The fusion of image bands demonstrates synergistic advantages, resulting in the highest observed overall and average accuracy when compared with the alternative combinations of bands evaluated. Specifically, ASTER bands are found to play a pivotal role in optimizing accuracy for skarn classification, while principal component bands notably enhance intrusion classification accuracy. Additionally, band ratio images significantly contribute to the accuracy of wall rock classification. Each composite input image configuration achieves maximal accuracy for iron ore classification. Remarkably, the model exhibits versatility in handling both high- and low-spectral-resolution images without necessitating adjustments to kernel sizes. These findings underscore the importance of considering the diversity of input channels in neural network models to enhance mineral detection accuracy, warranting further investigation into the specific mechanisms underlying this enhancement.

Author Contributions: Conceptualization, J.A.; methodology, J.A.; software data curation, J.A.; writing—original draft, J.A.; writing—review and editing, Z.Z., Z.C. and F.Y.; visualization, J.A. and A.-A.B.S.D.B.; supervision, Z.Z.; funding acquisition, Z.Z. All authors have read and agreed to the published version of the manuscript.

Funding: This research endeavor was generously funded by the Chinese Scholarship Council (CSC) and the Federal Scholarship Board (FSB) of Nigeria, in accordance with the Bilateral Exchange Agreement (BEA) established between the People's Republic of China and the government of the Federal Republic of Nigeria and the "Deep-time Digital Earth" Science and Technology Leading Talents Team Funds for the Central Universities for the Frontiers Science Center for Deep-time Digital Earth, China University of Geosciences (Beijing) (Fundamental Research Funds for the Central Universities; grant number: 2652023001).

Data Availability Statement: The data presented in this study are available on request from the corresponding author. The data are not publicly available due to group policy.

Acknowledgments: We extend our sincere appreciation to Yang Wu, a representative of the Key Laboratory of Mineral Resources at the Institute of Geology and Geophysics, Chinese Academy of Sciences, Beijing, China, for generously supplying several rock samples employed in this research endeavor. We also express our appreciation to all those that contributed to the success of this research. Lastly, we extend our sincere gratitude to the anonymous reviewers for their insightful suggestions, which have significantly enhanced the quality of this work.

Conflicts of Interest: The authors declare no conflicts of interest.

Appendix A

Figure A1. A subset of iron ore samples employed for spectral analysis (sample names with letters beginning with "Z", "B", and "D" are for Zhibo, Beizhan, and Dunde deposits, respectively).

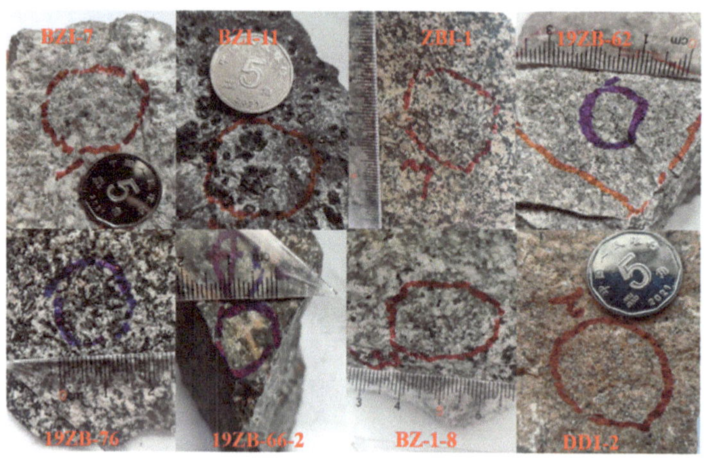

Figure A2. A subset of intrusion samples derived from the three deposits.

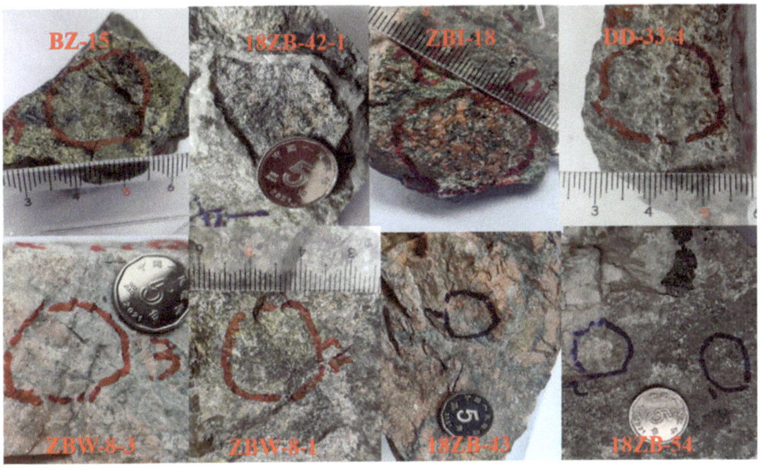

Figure A3. A subset of skarn deposit samples derived from the study sites.

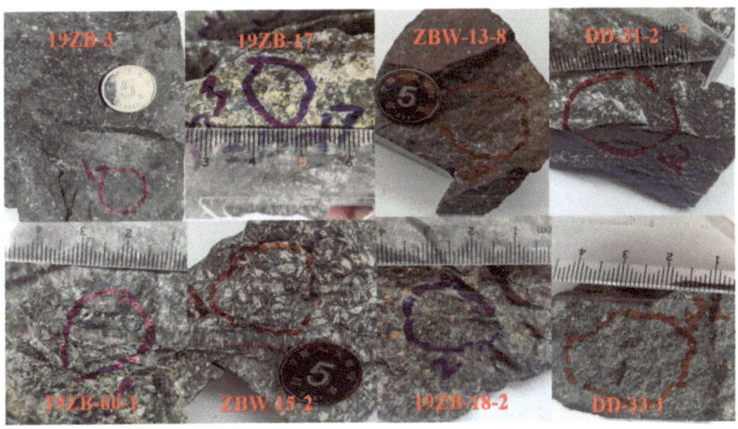

Figure A4. A subset of wall rock samples collected from the study deposit sites.

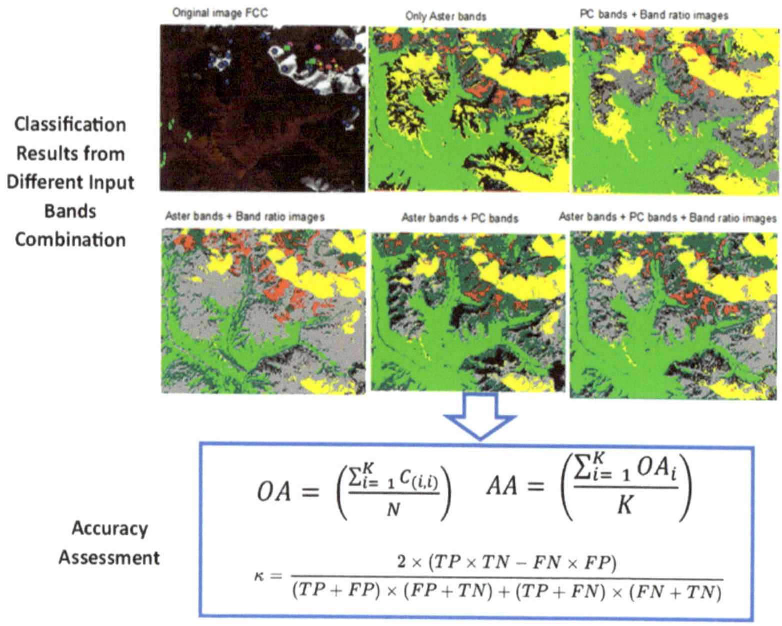

Figure A5. Evaluation of the impact of BR images and PCs bands. The meanings of the variables in the formula are as follows: the total number of classes (K); the total number of all samples tested (N); the total number of samples correctly classified ($C_{(i,i)}$); the overall accuracy of classification for the ith class (OA_i); true positive (TP); false positive (FP); false negative (FN); true negative (TN); and k, which represents the kappa coefficient.

Figure A6. *Cont.*

Figure A6. Selected input image patches for iron ore (**a**), skarn (**b**), wall rock (**c**), intrusive rock (**d**), glacier (**e**), and vegetation (**f**), displayed as a false-color composite with ASTER bands 3, 2, and 1 shown through the red, green, and blue filters, respectively.

References

1. Galdames, F.J.; Perez, C.A.; Estévez, P.A.; Adams, M. Rock lithological classification by hyperspectral, range 3D and color images. *Chemom. Intell. Lab. Syst.* **2019**, *189*, 138–148. [CrossRef]
2. Huang, S.; Chen, S.B.; Zhang, Y.Z. Comparison of altered mineral information extracted from ETM+, ASTER and Hyperion data in Águas Claras iron ore, Brazil. *IET Image Process.* **2019**, *13*, 355–364. [CrossRef]
3. Son, Y.S.; Lee, G.; Lee, B.H.; Kim, N.; Koh, S.M.; Kim, K.E.; Cho, S.J. Application of ASTER Data for Differentiating Carbonate Minerals and Evaluating MgO Content of Magnesite in the Jiao Liao-Ji Belt, North China Craton. *Remote Sens.* **2022**, *14*, 181. [CrossRef]
4. Rajendran, S.; Nasir, S. Mapping of Moho and Moho Transition Zone (MTZ) in Samail ophiolites of Sultanate of Oman using remote sensing technique. *Tectonophysics* **2015**, *657*, 63–80. [CrossRef]
5. Abdeen, M.; Thurmond, A.K.; Abdelsalam, M.; Stern, B. Use of TERRA ASTER band-ratio images for geological mapping in arid regions: The Neo-Proterozoic Allaqi suture, Egypt. *Egypt J. Remote Sens. Space Sci.* **2002**, *5*, 19–40.
6. Amer, R.; Kusky, T.; Ghulam, A. Lithological mapping in the Central Eastern Desert of Egypt using ASTER data. *J. Afr. Earth Sci.* **2010**, *56*, 75–82. [CrossRef]
7. Gad, S.; Kusky, T. ASTER spectral ratioing for lithological mapping in the Arabian-Nubian shield, the Neoproterozoic Wadi Kid area, Sinai, Egypt. *Gondwana Res.* **2007**, *11*, 326–335. [CrossRef]
8. Ninomiya, Y. A Stabilized Vegetation Index and Several Mineralogic Indices Defined for ASTER VNIR and SWIR Data. *IEEE Int. Geosci. Remote Sens. Symp. IGARSS'03* **2003**, *3*, 1552–1554.
9. Crósta, A.P.; Filho, C.R.D.S. Searching for gold with ASTER. *Earth Obs. Mag.* **2003**, *12*, 38–41.
10. Zhang, X.; Pazner, M.; Duke, N. Lithologic and mineral information extraction for gold exploration using ASTER data in the south Chocolate Mountains (California). *ISPRS J. Photogramm. Remote Sens.* **2007**, *62*, 271–282. [CrossRef]
11. Hecker, C.; Van Der Meijde, M.; Van Der Werff, H.; Van Der Meer, F.D. Assessing the influence of reference spectra on synthetic SAM classification results. *IEEE Trans. Geosci. Remote Sens.* **2008**, *46*, 4162–4172. [CrossRef]
12. Kruse, F.A.; Boardman, J.W.; Huntington, J.F. Comparison of airborne hyperspectral data and EO-1 Hyperion for mineral mapping. *IEEE Trans. Geosci. Remote Sens.* **2003**, *41*, 1388–1400. [CrossRef]
13. Gabr, S.; Ghulam, A.; Kusky, T. Detecting areas of high-potential gold mineralization using ASTER data. *Ore Geol. Rev.* **2010**, *38*, 59–69. [CrossRef]
14. Magendran, T.; Sanjeevi, S. Hyperion image analysis and linear spectral unmixing to evaluate the grades of iron ores in parts of Noamundi, Eastern India. *Int. J. Appl. Earth Obs. Geoinf.* **2014**, *26*, 413–426. [CrossRef]

15. Chang, C.I.; Wang, S. Constrained band selection for hyperspectral imagery. *IEEE Trans. Geosci. Remote Sens.* **2006**, *44*, 1575–1585. [CrossRef]
16. Xiao, D.; Le, B.T.; Ha, T.T.L. Iron ore identification method using reflectance spectrometer and a deep neural network framework. *Spectrochim. Acta A Mol. Biomol. Spectrosc.* **2021**, *248*, 119168. [CrossRef]
17. Xiao, D.; Vu, Q.H.; Le, B.T.; Ha, T.T.L. A method for mapping and monitoring of iron ore stopes based on hyperspectral remote sensing-ground data and a 3D deep neural network. *Neural. Comput. Appl.* **2023**, *35*, 12221–12232. [CrossRef]
18. Zhao, Y.; Wu, P.; Wang, J.; Li, H.; Navab, N.; Yakushev, I.; Weber, W.; Schwaiger, M.; Huang, S.C.; Cumming, P.; et al. A 3D Deep Residual Convolutional Neural Network for Differential Diagnosis of Parkinsonian Syndromes on 18F-FDG PET Images. In Proceedings of the 2019 41st Annual International Conference of the IEEE Engineering in Medicine and Biology Society (EMBC), Berlin, Germany, 23 July 2019; pp. 3531–3534.
19. Tomita, N.; Jiang, S.; Maeder, M.E.; Hassanpour, S. Automatic post-stroke lesion segmentation on MR images using 3D residual convolutional neural network. *Neuroimage Clin.* **2020**, *27*, 102276. [CrossRef]
20. Sabins, F.F., Jr. *Remote Sensing: Principles and Interpretation*, 2nd ed.; WH Freeman & Company: New York, NY, USA, 1987.
21. Ghoneim, S.M.; Salem, S.M.; El-Wahid, K.H.A.; Anwar, M.; Hegab, M.A.E.-R.; Soliman, N.M.; Ali, H.F. Application of remote sensing techniques to identify iron ore deposits in the Central Eastern Desert, Egypt: A case study at Wadi Karim and Gabal El-Hadid areas. *Arab. J. Geosci.* **2022**, *15*, 1596. [CrossRef]
22. Cardoso-Fernandes, J.; Teodoro, A.C.; Lima, A. Remote sensing data in lithium (Li) exploration: A new approach for the detection of Li-bearing pegmatites. *Int. J. Appl. Earth Obs. Geoinf.* **2019**, *76*, 10–25. [CrossRef]
23. Ahmadi, H.; Uygucgil, H. Targeting iron prospective within the Kabul Block (SE Afghanistan) via hydrothermal alteration mapping using remote sensing techniques. *Arab. J. Geosci.* **2021**, *14*, 183. [CrossRef]
24. Jolliffe, I.T. *Principal Component Analysis*; Springer: New York, NY, USA, 2002.
25. Abdi, H.; Williams, L.J. Principal component analysis. *Wiley Interdiscip. Rev. Comput. Stat.* **2010**, *2*, 433–459. [CrossRef]
26. Jackson, J.E. *A User's Guide to Principal Components*; John Wiley & Sons: Hoboken, NJ, USA, 2005.
27. Yue, J.; Zhao, W.; Mao, S.; Liu, H. Spectral-spatial classification of hyperspectral images using deep convolutional neural networks. *Remote Sens. Lett.* **2015**, *6*, 468–477. [CrossRef]
28. Li, Y.; Zhang, H.; Shen, Q. Spectral-spatial classification of hyperspectral imagery with 3D convolutional neural network. *Remote Sens.* **2017**, *9*, 67. [CrossRef]
29. Roy, S.K.; Krishna, G.; Dubey, S.R.; Chaudhuri, B.B. HybridSN: Exploring 3-D-2-D CNN Feature Hierarchy for Hyperspectral Image Classification. *IEEE Geosci. Remote Sens. Lett.* **2020**, *17*, 277–281. [CrossRef]
30. Hu, Y.; Tian, S.; Ge, J. Hybrid Convolutional Network Combining Multiscale 3D Depthwise Separable Convolution and CBAM Residual Dilated Convolution for Hyperspectral Image Classification. *Remote Sens.* **2023**, *15*, 4796. [CrossRef]
31. Gopinathan, P.; Priyadarsi, R.; Subramani, T.; Karunanidhi, D. Detection of iron-bearing mineral assemblages in Nainarmalai granulite region, south India, based on satellite image processing and geochemical anomalies. *Environ. Monit. Assess.* **2022**, *194*, 866.
32. Hou, T.; Zhang, Z.; Pirajno, F.; Santosh, M.; Encarnacion, J.; Liu, J.; Zhao, Z.; Zhang, L. Geology, tectonic settings and iron ore metallogenesis associated with submarine volcanism in China: An overview. *Ore Geol. Rev.* **2014**, *57*, 498–517. [CrossRef]
33. Zhang, Z.; Hong, W.; Jiang, Z.; Duan, S.; Xu, L.; Li, F.; Guo, X.; Zhao, Z. Geological Characteristics and Zircon U-Pb Dating of Volcanic Rocks from the Beizhan Iron Deposit in Western Tianshan Mountains, Xinjiang, NW China. *Acta Geol. Sin.-Engl. Ed.* **2012**, *86*, 737–747. [CrossRef]
34. Zhu, Y.; Zhou, J.; Song, B.; Zhang, L.; Guo, X. Age of the Dahalajunshan Formation in Xinjiang and its disintegration. *Geol. China* **2006**, *33*, 487–497.
35. Jiang, Z.; Zhang, Z.; Wang, Z.; Duan, S.; Li, F.; Tian, J. Geology, geochemistry, and geochronology of the Zhibo iron deposit in the Western Tianshan, NW China: Constraints on metallogenesis and tectonic setting. *Ore Geol. Rev.* **2014**, *57*, 406–424. [CrossRef]
36. Feng, W.; Zhu, Y. Petrogenesis and tectonic implications of the late Carboniferous calc-alkaline and shoshonitic magmatic rocks in the Awulale mountain, western Tianshan. *Gondwana Res.* **2019**, *76*, 44–61. [CrossRef]
37. Wang, X.S.; Zhang, X.; Gao, J.; Li, J.L.; Jiang, T.; Xue, S.C. A slab break-off model for the submarine volcanic-hosted iron mineralization in the Chinese Western Tianshan: Insights from Paleozoic subduction-related to post-collisional magmatism. *Ore Geol. Rev.* **2018**, *92*, 144–160. [CrossRef]
38. Li, H.-M.; Li, L.-X.; Yang, X.-Q.; Cheng, Y.-B. Types and geological characteristics of iron deposits in China. *J. Asian Earth Sci.* **2015**, *103*, 2–22. [CrossRef]
39. Luo, W.; Zhang, Z.; Duan, S.; Jiang, Z.; Wang, D.; Chen, J.; Sun, J. Geochemistry of the Zhibo submarine intermediate-mafic volcanic rocks and associated iron ores, Western Tianshan, Northwest China: Implications for ore genesis. *Geol. J.* **2018**, *53*, 3147–3172. [CrossRef]
40. Zhang, Z.; Li, H.; Li, J.; Song, X.Y.; Hu, H.; Li, L.; Chai, F.; Hou, T.; Xu, D. Geological settings and metallogenesis of high-grade iron deposits in China. *Sci. China Earth Sci.* **2021**, *64*, 691–715. [CrossRef]
41. Ping, S.; HongDi, P.; ChangHao, L.; HaoXuan, F.; Yang, W.; FuPin, S.; XinCheng, G.; WenGuang, L. Carboniferous ore-controlling volcanic apparatus and metallogenic models for the large-scale iron deposits in the Western Tianshan, Xinjiang. *Acta Petrol. Sin.* **2020**, *36*, 2845–2868. [CrossRef]

42. Li, C.; Shen, P.; Zhang, X.; Shi, F.; Feng, H.; Pan, H.; Wu, Y.; Li, W. Mineralogy and mineral chemistry related to the Au mineralization in the Dunde Fe-Zn deposit, western Tianshan. *Ore Geol. Rev.* **2020**, *124*, 103650. [CrossRef]
43. Li, H.; Zhang, Z.; Liu, B.; Jin, Y.; Santosh, M.; Pan, J. Superimposed zinc and gold mineralization in the Dunde iron deposit, western Tianshan, NW China: Constraints from LA-ICP-MS fluid inclusion microanalysis. *Ore Geol. Rev.* **2022**, *142*, 104713. [CrossRef]
44. Yan, S.; Niu, H.-C.; Zhao, J.; Bao, Z.-W.; Sun, W. Ore-fluid geochemistry and metallogeny of the Dunde iron–zinc deposit in western Tianshan, Xinjiang, China: Evidence from fluid inclusions, REE and C–O–Sr isotopes of calcite. *Ore Geol. Rev.* **2018**, *100*, 441–456. [CrossRef]
45. Duan, S.; Zhang, Z.; Wang, D.; Jiang, Z.; Luo, W.; Li, F. Pyrite Re–Os and muscovite 40Ar/39Ar dating of the Beizhan iron deposit in the Chinese Tianshan Orogen and its geological significance. *Int. Geol. Rev.* **2018**, *60*, 57–71. [CrossRef]
46. Duda, K.; Daucsavage, J.; Siemonsma, D.; Brooks, B.; Oleson, R.; Meyer, D.; Doescher, C. *Advanced Spaceborne Thermal Emission and Reflection Radiometer (ASTER) Level 1 Precision Terrain Corrected Registered At-Sensor Radiance Product (AST_L1T) AST_L1T Product User's Guide*; US Geological Survey: Reston, VA, USA, 2020.
47. Yamaguchi, Y.; Kahle, A.B.; Tsu, H.; Kawakami, T.; Pniel, M. Overview of advanced spaceborne thermal emission and reflection radiometer (ASTER). *IEEE Trans. Geosci. Remote Sens.* **1998**, *36*, 1062–1071. [CrossRef]
48. Rowan, L.C.; Mars, J.C. Lithologic mapping in the Mountain Pass, California area using Advanced Spaceborne Thermal Emission and Reflection Radiometer (ASTER) data. *Remote Sens. Environ.* **2003**, *84*, 350–366. [CrossRef]
49. Sadek, M.F.; Ali-Bik, M.W.; Hassan, S.M. Late Neoproterozoic basement rocks of Kadabora-Suwayqat area, Central Eastern Desert, Egypt: Geochemical and remote sensing characterization. *Arab. J. Geosci.* **2015**, *8*, 10459–10479. [CrossRef]
50. Rasouli Beirami, M.; Tangestani, M.H. A New Band Ratio Approach for Discriminating Calcite and Dolomite by ASTER Imagery in Arid and Semiarid Regions. *Nat. Resour. Res.* **2020**, *29*, 2949–2965. [CrossRef]
51. Rajendran, S.; Nasir, S. Hydrothermal altered serpentinized zone and a study of Ni-magnesioferrite–magnetite–awaruite occurrences in Wadi Hibi, Northern Oman Mountain: Discrimination through ASTER mapping. *Ore Geol. Rev.* **2014**, *62*, 211–226. [CrossRef]
52. Zoheir, B.; Emam, A. Integrating geologic and satellite imagery data for high-resolution mapping and gold exploration targets in the South Eastern Desert, Egypt. *J. Afr. Earth Sci.* **2012**, *66*, 22–34. [CrossRef]
53. Kalinowski, A.; Oliver, S. ASTER mineral index processing manual. *Remote Sens. Appl. Geosci. Aust.* **2004**, *37*, 36–37.
54. Wang, J.; Song, X.; Sun, L.; Huang, W.; Wang, J. A Novel Cubic Convolutional Neural Network for Hyperspectral Image Classification. *IEEE J. Sel. Top. Appl. Earth Obs. Remote Sens.* **2020**, *13*, 4133–4148. [CrossRef]
55. Simonyan, K.; Zisserman, A. Very deep convolutional networks for large-scale image recognition. *arXiv* **2014**, arXiv:1409.1556.
56. Tran, D.; Bourdev, L.; Fergus, R.; Torresani, L.; Paluri, M. Learning spatiotemporal features with 3D convolutional networks. In Proceedings of the IEEE International Conference on Computer Vision, Santiago, Chile, 7–13 December 2015; pp. 4489–4497.
57. Gui, Y.; Zeng, G. Joint learning of visual and spatial features for edit propagation from a single image. *Vis. Comput.* **2020**, *36*, 469–482. [CrossRef]
58. Zhang, X. Improved Three-Dimensional Inception Networks for Hyperspectral Remote Sensing Image Classification. *IEEE Access* **2023**, *11*, 32648–32658. [CrossRef]
59. Krizhevsky, A.; Sutskever, I.; Hinton, G.E. ImageNet classification with deep convolutional neural networks. *Adv. Neural Inf. Process. Syst.* **2012**, *25*, 1097–1105. [CrossRef]
60. Ge, W.; Cheng, Q.; Jing, L.; Armenakis, C.; Ding, H. Lithological discrimination using ASTER and Sentinel-2A in the Shibanjing ophiolite complex of Beishan orogenic in Inner Mongolia, China. *Adv. Space Res.* **2018**, *62*, 1702–1716. [CrossRef]
61. Guha, A.; Chatterjee, S.; Oommen, T.; Kumar, K.V.; Roy, S.K. Synergistic use of ASTER, L-band ALOS PALSAR, and hyperspectral AVIRIS-NG data for exploration of lode type gold deposit—A study in Hutti Maski Schist Belt, India. *Ore Geol. Rev.* **2021**, *128*, 103818. [CrossRef]
62. Guha, A.; Chakraborty, D.; Ekka, A.B.; Pramanik, K.; Vinod Kumar, K.; Chatterjee, S.; Subramanium, S.; Ananth Rao, D. Spectroscopic study of rocks of Hutti-Maski schist belt, Karnataka. *J. Geol. Soc. India.* **2012**, *79*, 335–344. [CrossRef]
63. Sun, M.D.; Xu, Y.G.; Chen, H.L. Subaqueous volcanism in the Paleo-Pacific Ocean based on Jurassic basaltic tuff and pillow basalt in the Raohe Complex, NE China. *Sci. China Earth Sci.* **2018**, *61*, 1042–1056. [CrossRef]
64. Abrams, M.J.; Rothery, D.A.; Pontual, A. Mapping in the Oman ophiolite using enhanced Landsat Thematic Mapper images. *Tectonophysics* **1988**, *151*, 387–401. [CrossRef]
65. Lever, J.; Krzywinski, M.; Altman, N. Points of significance: Principal component analysis. *Nat. Methods* **2017**, *14*, 641–642. [CrossRef]
66. Ioffe, S.; Szegedy, C. Batch normalization: Accelerating deep network training by reducing internal covariate shift. In Proceedings of the 32nd International Conference on Machine Learning, Lille, France, 6–11 July 2015; pp. 448–456.
67. Chang, Z.; Shu, Q.; Meinert, L.D. Chapter 6: Skarn deposits of China. In *Mineral Deposits of China*; Chang, Z., Goldfarb, R.J., Eds.; Society of Economic Geologists: Littleton, CO, USA, 2019; pp. 189–234.

68. Sun, W.; Yuan, F.; Jowitt, S.M.; Zhou, T.; Liu, G.; Li, X.; Wang, F.; Troll, V.R. In situ LA–ICP–MS trace element analyses of magnetite: Genetic implications for the Zhonggu ore field, Ningwu volcanic basin, Anhui Province, China. *Miner. Depos.* **2019**, *54*, 1243–1264. [CrossRef]
69. Xiang, C.; Wang, D.; Pan, Y.; Chen, A.; Zhou, X.; Zhang, Y. Accelerated topology optimization design of 3D structures based on deep learning. *Struct. Multidiscip. Optim.* **2022**, *65*, 99. [CrossRef]

remote sensing

MDPI

Article

Mapping of Clay Montmorillonite Abundance in Agricultural Fields Using Unmixing Methods at Centimeter Scale Hyperspectral Images

Etienne Ducasse [1,*], Karine Adeline [1], Audrey Hohmann [2], Véronique Achard [1], Anne Bourguignon [2], Gilles Grandjean [2] and Xavier Briottet [1]

1 ONERA/DOTA, Université de Toulouse, F-31055 Toulouse, France; karine.adeline@onera.fr (K.A.); veronique.achard@onera.fr (V.A.); xavier.briottet@onera.fr (X.B.)

2 BRGM, Risk Division, 3 Avenue Claude Guillemin, CEDEX 02, F-45060 Orléans, France; a.hohmann@brgm.fr (A.H.); a.bourguignon@brgm.fr (A.B.); g.grandjean@brgm.fr (G.G.)

* Correspondence: etienneducasse@yahoo.fr

Abstract: The composition of clay minerals in soils, and more particularly the presence of montmorillonite (as part of the smectite family), is a key factor in soil swell–shrinking as well as off–road vehicle mobility. Detecting these topsoil clay minerals and quantifying the montmorillonite abundance are a challenge since they are usually intimately mixed with other minerals, soil organic carbon and soil moisture content. Imaging spectroscopy coupled with unmixing methods can address these issues, but the quality of the estimation degrades the coarser the spatial resolution is due to pixel heterogeneity. With the advent of UAV-borne and proximal hyperspectral acquisitions, it is now possible to acquire images at a centimeter scale. Thus, the objective of this paper is to evaluate the accuracy and limitations of unmixing methods to retrieve montmorillonite abundance from very-high-resolution hyperspectral images (1.5 cm) acquired from a camera installed on top of a bucket truck over three different agricultural fields, in Loiret department, France. Two automatic endmember detection methods based on the assumption that materials are linearly mixed, namely the Simplex Identification via Split Augmented Lagrangian (SISAL) and the Minimum Volume Constrained Non-negative Matrix Factorization (MVC-NMF), were tested prior to unmixing. Then, two linear unmixing methods, the fully constrained least square method (FCLS) and the multiple endmember spectral mixture analysis (MESMA), and two nonlinear unmixing ones, the generalized bilinear method (GBM) and the multi-linear model (MLM), were performed on the images. In addition, several spectral preprocessings coupled with these unmixing methods were applied in order to improve the performances. Results showed that our selected automatic endmember detection methods were not suitable in this context. However, unmixing methods with endmembers taken from available spectral libraries performed successfully. The nonlinear method, MLM, without prior spectral preprocessing or with the application of the first Savitzky–Golay derivative, gave the best accuracies for montmorillonite abundance estimation using the USGS library (RMSE between 2.2–13.3% and 1.4–19.7%). Furthermore, a significant impact on the abundance estimations at this scale was in majority due to (i) the high variability of the soil composition, (ii) the soil roughness inducing large variations of the illumination conditions and multiple surface scatterings and (iii) multiple volume scatterings coming from the intimate mixture. Finally, these results offer a new opportunity for mapping expansive soils from imaging spectroscopy at very high spatial resolution.

Keywords: clay; montmorillonite; imaging spectroscopy; unmixing; spectral preprocessing; agricultural ploughed fields; very high spatial resolution

Citation: Ducasse, E.; Adeline, K.; Hohmann, A.; Achard, V.; Bourguignon, A.; Grandjean, G.; Briottet, X. Mapping of Clay Montmorillonite Abundance in Agricultural Fields Using Unmixing Methods at Centimeter Scale Hyperspectral Images. *Remote Sens.* **2024**, *16*, 3211. https://doi.org/10.3390/rs16173211

Academic Editors: Yanni Dong, Tao Chen and Chao Chen

Received: 15 July 2024
Revised: 11 August 2024
Accepted: 18 August 2024
Published: 30 August 2024

1. Introduction

The analysis of clay mineral composition in soils constitutes an important challenge because of the particular behavior of swelling clays (smectite, interstratified clay minerals, vermiculite and chlorite) that are sensitive to moisture variations (e.g., soil moisture content): they shrink in dry conditions and swell after heavy rains. These soil volume changes create cracks on buildings and damage infrastructures. The costs incurred by clay swelling–shrinking represent 38% of all natural hazards' costs in France, second after floods. Between 1990 and 2014, its total cost was more than EUR 370 million yearly [1]. The Association of British Insurers in the UK estimated the cost of swelling–shrinking disaster to more than GBP 400 million yearly [2], and in USA, it has been estimated to more than USD 15 billion per year [3]. Actually, the swelling behavior of soils also impacts off–road vehicle mobility, which depends on the ability of a terrain to sustain a large flow of vehicles in wet weather conditions. The knowledge of these vehicle mobility conditions allow (i) access to areas affected by natural or industrial disasters (for instance, the 2023 Turkey–Syria earthquake, 2021 European Floods, 2011 Fukushima nuclear disaster followed by a tsunami), (ii) access to agricultural machinery on cultivated fields and (iii) military deployment.

Clay mineral composition in soils is identified and quantified using well-known techniques such as ground geotechnical instrumentation measuring soil swelling potential [4] and X-ray diffraction (XRD) [5,6]. Refs. [7,8] defined four classes linking shrink–swell potential and montmorillonite content: low swelling potential (<10%), moderate swelling potential (between 10% and 50%), high swelling potential (between 50% and 70%) and very high swelling potential (>70%). Thus, a reasonable accuracy on montmorillonite abundance could be a target of 10%. However, these techniques can cover a very limited area with a small number of soil samplings. An alternative relies on the use of spectroscopic data which allow the mineral mapping of soils over large surfaces [9,10]. Actually, clays present spectral features in the Short-Wave InfraRed (SWIR) domain, at 1400 nm, 1900 nm and, more specifically, between 2100 and 2500 nm [11], corresponding to vibrational hydroxyl processes [12]. Hereafter, only the three most common clays in areas affected by swelling risk [13], namely kaolinite, illite and montmorillonite, are studied, and their main spectral absorption features are mentioned below:

- The Al–OH vibrational mode produces the 2200 nm absorption feature with a bandwidth around 100 nm whatever the clay type. Kaolinite also has a double absorption feature (2160 nm and 2206 nm), which is leftward asymmetric.
- OH-stretching bands combined with lattice vibrations produces absorption features at 2360 nm for both kaolinite and illite. This feature is shallow for illite and sharp for kaolinite. Kaolinite has also two absorption features at 2320 nm and 2380 nm.

Remote sensing data from multispectral sensors have already been used for mineralogy applications. It has been tested to separate argillic from non-argillic materials over a hydrothermal alteration area using a two-band ratio from the Landsat Thematic Mapper instrument at 30 m spatial resolution [10]. Sentinel 2 was also used to indirectly map clays minerals associated to the alteration of volcanic rocks [14]. Using the bands 5–9 (2.14–2.43 μm) from the Advanced Spaceborne Thermal Emission and Reflectance Radiometer sensor (ASTER), ref. [15] identified clay minerals in agricultural fields, but they were unable to quantify montmorillonite abundance below 16%. Ref. [16] classified clay minerals for gold mining application with ASTER, showing that a maximum of 70% pixels was correctly classified with the spectral angle mapper (SAM) and 22% for montmorillonite with the spectral information divergence. Nevertheless, the performances of all these previous studies have been limited by the weak spectral richness (e.g., number of bands, spectral resolution) of these multispectral sensors, essentially in the range from 2200 to 2500 nm, and globally by their low spatial resolutions that induce a large amount of mixed pixels.

Oppositely, hyperspectral sensors can give access to better mineral identification and quantification [10,17,18]. From one side, the Hyperion spaceborne sensor detected clay minerals related to hydrothermal activities in a volcanic and plutonic area [19]. Despite its low signal-to-noise ratio (SNR), this method showed a good correlation to existing geological maps. Recently, ref. [20] used PRISMA spaceborne data to accurately detect and map alteration minerals, such as illite, kaolinite and alunite, when compared to alteration maps realized from field mapping. However, a large number of mixed pixels were observed due to their coarse spatial resolution (30 m) [21]. From another side, hyperspectral airborne sensors have usually higher SNR (for instance around 500:1 for AVIRIS compared to 50:1 for Hyperion in the SWIR domain [21]) and a better spatial resolution [17,22]. This allows a better discrimination of the clay mineral spectral absorption feature characteristics (depth and width). For instance, ref. [23] built three clay maps (montmorillonite, illite, smectite) and their related abundance content from AVIRIS-NG images at 8 m spatial resolution over agriculture fields in India. Their method was based on a linear regression analysis between absorption peak depth and each clay type. Errors in abundance achieved, respectively, 6.20%, 8.05% and 12.08% for montmorillonite, illite and kaolinite. Previous projects (e.g., AGEOTHYP led by BRGM, the French Geological Survey) aimed at mapping and identifying swelling clays from airborne hyperspectral data [7,24].

Ref. [25] also used the same strategy to map these three clay minerals in Tuscan intermontane basins from the airborne Hyper SIM-GA sensor at a 1.2 m spatial resolution. However, no quantitative errors were provided. Actually, there are very few works that have quantified the comparison between clay mineral abundance estimations and ground truth laboratory measurements. This could be explained because such measurements can be very expensive. At last, the advent of new compact hyperspectral cameras onboard unmanned airborne vehicle (UAV) platforms and the easy access of proximal spectroscopic measurements in indoor (laboratory) and outdoor (in situ) conditions opens the way to map soil mineral composition at a centimeter scale [26]. Most studies have been carried out on mine mineralogy [26–28] or for soil texture [29], but to our knowledge, none of them have studied both clay mineral discrimination and quantification.

Actually, at this scale, mixed pixels still exist, and the results are very sensitive to the local soil roughness, particularly to the illumination conditions [30], but also the soil composition and the on-ground environmental context. Among these factors, one can quote: the fraction of vegetation (green or dry) or rocks present at the surface, the soil roughness linked to the topography inducing shadows, the presence of other soil minerals (carbonates, quartz, etc.) and soil organic carbon (SOC), as well as the soil moisture content (SMC). For instance, the 2200 nm absorption band of clay minerals is masked when the fraction of photosynthetic vegetation inside a pixel exceeds 40–50% with green vegetation and 60–70% for dry vegetation in a semi-desertic environment [22]. Thus, several corrections have been applied to this absorption band in order to suppress the vegetation effect for vegetal cover less than 60%, such as the use of spectral indices and multiple linear regressions from hyperspectral airborne data [31–33]. With the increase of soil roughness, light undergoes multiple reflections, leading to nonlinear mixtures [34–36]. On natural surfaces, multiple interactions may add up to 30% confusion for abundance estimation of minerals using linear spectral unmixing [37]. Moreover, the spectral reflectance decreases due to the presence of shadows, which can be reduced with a spectral derivative preprocessing [38]. Also, high SOC values decrease the global reflectance level and emphasize its broad absorption feature around 2300 nm due to C-H vibrational mode. The latter masks the clay 2200 nm absorption feature when SOC is more than 15% in laboratory mixtures containing montmorillonite, and 7% in natural soils [39]. Moreover, SMC affects spectra in two ways: (i) global reflectance level decreases nonlinearly [40] and (ii) 2200 nm absorption band depth is masked, whereas 1400 nm and 1900 nm absorption band depth and width increase [39,41]. For instance, the kaolinite absorption band at 2160 nm is totally hidden with a SMC superior to 15%, while the smectite absorption band at 2200 nm is masked for a SMC of 30% [39]. Ref. [42] used a multivariate approach to reduce this effect and improve the estimation of clays. Thus, the challenge is the choice of the most appropriate method able to overcome the contribution of these factors for the quantification of clay minerals.

Several physical models have been developed, such as the ones of Hapke [43] and Shkuratov [44], requiring a geometric description of the soil (optical index and relative distribution of minerals, granulometry, etc.), of which access is difficult to obtain. Linear and nonlinear unmixing methods, which are not site-dependent and do not require a training dataset, have proved their efficiency in estimating mineral abundance in case of laboratory mixtures [12,45–47]. Coupling a spectral preprocessing like the first derivative and the fully constrained least square (FCLS) unmixing method delivered the best method to estimate montmorillonite abundance in clay–quartz–carbonate mineral mixtures (Root Mean Square Error or RMSE of 15.5%) [12]. However, some limits can be identified: (i) the prior knowledge of the nature and number of endmembers (EM) inside a mixture [34], and (ii) the fact that EM spectra highly depend on the heterogeneity of the mineral chemistry and (iii) the data acquisition conditions [48]. Also, these methods have not been tested in field conditions at this scale.

The main objective of this study is to map clay montmorillonite abundances at a centimeter scale in ploughed field, using unmixing methods. This work is the continuation of this performed in laboratory conditions [12]. This new study is based on an outdoor experiment involving very-high spatial resolution hyperspectral acquisitions of 1.5 cm. Because montmorillonite usually cannot be detected independently of the smectite clay family, throughout the paper, we confound both terminologies, aware that this can be a source of error. The addressed scientific questions are the following:

(1) How can spectral libraries or automatic EM detection benefit from unmixing method performances?
(2) Which combined spectral preprocessing and unmixing algorithm is the most efficient strategy to estimate montmorillonite abundances in soil?
(3) What is the impact of soil composition and other factors influencing montmorillonite abundance accuracy of ploughed fields at the centimeter scale?

Fieldwork and methodology are presented in Section 2. Then, results are shown in Section 3 and discussed in Section 4, with the conclusions in Section 5.

2. Materials and Methods

2.1. Site Descriptions

The study area is located in France, in Loiret Department, northwest of Orléans. With its temperate climate, this region allows a broad variety of crops, mainly cereals (wheat, barley, corn) but also root vegetables (potatoes and sugar beets). A full description of Loiret's surface geology was given by Bouchut and Giot [49]. Previous projects (e.g., AGEOTHYP led by BRGM, the French Geological Survey) aimed at mapping and identifying swelling clays from airborne hyperspectral data [24,50]. Soil analysis from sampled data, including XRD and the methylene blue test (MBT), and maps of soil shrink–swelling are available on this area of 25×12 km^2.

In this region, sites have been selected, based on three criteria:

1. The knowledge of soil mineral composition:
    - Montmorillonite abundance from XRD measurements of the AGEOTHYP project [7,24,50] (Figure 1a). Selected samples had distinctive clays composition: each site has either montmorillonite, kaolinite or illite maximum content. Site areas were defined within a 600 m radius around these measurements.
    - Soil swelling hazard maps from the BRGM [8] (Figure 1b) were used to roughly locate the previous XRD measurements. Their three classes (low/medium/high swelling classes) were selected using geotechnical analysis (MBT) and dominant lithology inside stratigraphic formations. The sites were chosen within the swelling class entity where the XRD measurements were sampled.
2. An easy access to the selected sites given the field owner agreement.
3. A low vegetation cover (less than 20%) observed on agricultural fields (the experiment has been carried out after wheat harvest, on freshly ploughed fields) (Figure 1c–e).

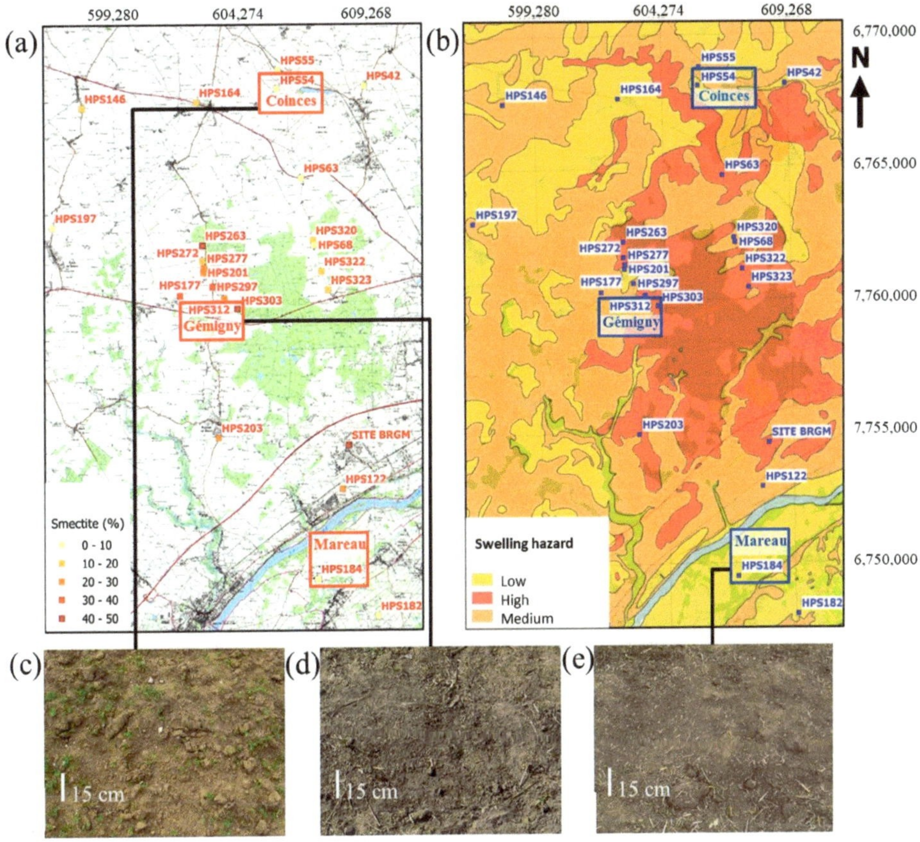

Figure 1. Site locations from AGLOTHYP project depicted with colored squares on: (**a**) topographic map by IGN (National Institute of Geographic and Forest Information) overlaid with smectite abundance from XRD analyses and (**b**) BRGM swelling hazard map. Soil digital photos of the three selected sites: (**c**) "Le Buisson" located in Coinces, (**d**) "Les Laps" located in Gémigny and (**e**) "La Malandière" located in Mareau.

Thus, three sites among the available ones have been selected:

- "Le Buisson" locality in Coinces municipality, hereafter named Coinces (WGS 84, 48.00901°N, 1.734826°E). This site lies on a stratigraphic formation of Quaternary loam and loess, clayey and carbonated. The nearest XRD measurements indicate a composition of 11% kaolinite, 7% illite and 2% smectite, presenting a low swelling risk (Figure 1c).
- "Les Laps" locality in Gémigny municipality, hereafter named Gémigny (WGS 84, 47.95422°N, 1.689848°E). This site lies on a stratigraphic formation lower Pliocene sand and clay with dominant sand and clayey sand with metric clay layers. The nearest XRD measurement indicates 2.9% kaolinite, 5% illite and 43.5% smectite, presenting a high swelling risk (Figure 1d).
- "La Malandière" locality in Mareau-aux-près municipality, hereafter named "Mareau" (WGS 84, 47.83964°N, 1.758915°E). This site lies on a stratigraphic formation of recent Holocene Loire alluvium, mainly siliceous with local imbrications of loam and clay deposits. The nearest XRD measurements indicate 14.6% kaolinite, 6.5% illite and 0% smectite, presenting a low swelling risk (Figure 1e).

2.2. Hyperspectral Data Acquisitions

The outdoor experiment was performed between 5 and 7 September 2017 with hyperspectral cameras installed on a bucket truck at a height of 12 m (Figure 2). Two hyperspectral cameras (HySpex-NorskElektroOptikk–Visible Near-Infrared VNIR-1600-e and HySpex-NorskElektroOptikk-SWIR-320m-e) (HySpex, Oslo, Norway) acquired the reflected signal, with, respectively, 160 and 256 spectral bands with a spectral resolution of 3.6 nm for the VNIR camera and 6 nm for the SWIR camera. The two cameras covered a spectral range from 0.4 to 2.5 µm. The VNIR camera had a field of view (FOV) of 17°, and the SWIR camera had a FOV of 13.5°, leading to a spatial resolution at Nadir of, respectively, 0.4 cm and 1.5 cm. A rotating system made it possible to cover an area of 12 × 3.7 m^2 approximatively. The experimental area was divided into 15 rectangular areas of 2.4 × 1.3 m^2, or "subzones", (called after as "SUB"). The acquisition setup was installed on each site for one day, in order to acquire several hyperspectral images at different periods along the day. However, due to high cloud cover and camera availability, only one image with the best illumination has been selected for each site. Gémigny (acquisition on September 6, local time: 11:20 a.m., solar elevation angle: 44°, sun azimuth angle: 147°) and Mareau (acquisition on September 7, local time: 3:13 p.m., solar elevation angle: 38.2°, sun azimuth angle: 227°) have been recorded with ideal weather condition. For Coinces, cloud cover was present all day, resulting in lower luminance measurements than the other sites (acquisition on September 5, local time: 3:48 p.m., solar elevation angle: 35.6°, sun azimuth angle: 235°). Signal-to-Noise Ratio (SNR) was obtained from the radiance images of the three test sites from a known reference panel (Spectralon® 20%, Labsphere, North Sutton, NH, USA) present in the scenes. Coinces had a mean SNR in the SWIR domain of 29, and Gemigny and Mareau had a higher mean SNR of 58 and 49, respectively.

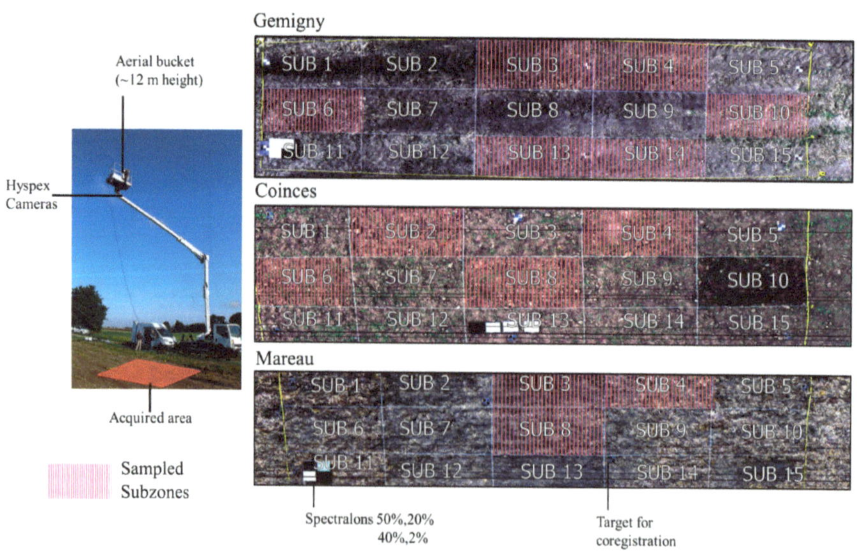

Figure 2. Acquisition setup with the HySpex cameras, RGB composite image from HySpex VNIR camera on Gémigny, Coinces and Mareau sites, with the sampling grid composed of 15 subzones (named after "SUB"), samples collected for laboratory soil characterization in subzones are delimited by red squares (**right**).

The acquired radiance images were then preprocessed. First, VNIR and SWIR images were coregistrated using Gefolki algorithm [51,52] with the SWIR image taken as the reference, leading to a coregistration RMSE less than the SWIR spatial resolution based on 30 random ground control points. Secondly, the conversion from radiance unit to reflectance one was performed using the empirical line method (ELM) with four calibrated known-reflectance targets (2%, 20%, 40%, 50% reflectance, Spectralon® Labsphere, US, Diffuse Reflectance Targets, see Figure 2) placed in the scene [53,54]. These target values have been chosen to match the reflectance levels expected in the field. The luminance data corresponding to each spectralon were retrieved using a square-shaped ROI (size: 144 pixels) manually centered on each target. An average spectrum corresponding to the pixel average was extracted for each target. Inversion was performed over the entire image for each wavelength. ELM retrieved reflectance with an average 5% RMSE and r^2 of 0.94 on all wavelengths. Thirdly, strong atmospheric gaseous absorption spectral intervals were removed from the spectra: 747 nm–766 nm for O_2, 905 nm–1019 nm, 1094 nm–1176 nm, 1339 nm–1465 nm and 1773 nm–2005 nm for H_2O. Finally, 306 bands were kept over the original 416 bands. Fourthly, the minimum noise fraction (MNF) transform [55] was applied to remove the noise [21,22,56,57], and the first 20 eigenvalues bands were selected for the spectral reconstruction of VNIR and SWIR data. At the end, 54 spectral bands from the SWIR (2100–2425 nm) were used to discriminate clay minerals, and the VNIR bands were used for masking the presence of shadow and dry/green vegetation.

For shadow masking, the Nagao index image (or intensity image) was computed based on [58]:

$$I = \frac{2 \times (\rho_{1000} + \rho_{650}) + \rho_{475} + \rho_{550}}{6} \tag{1}$$

With ρ_i the reflectance of image pixel at i nm (i = 1000, 650, 475 and 550). An arbitrary threshold has been manually fixed to 0.075 to discard shadowed pixels [59].

For vegetation masking, two vegetation spectral indices were combined:

- Normalized difference vegetation index (NDVI) to detect green vegetation [60]:

$$NDVI = \frac{\rho_{810} - \rho_{680}}{\rho_{810} + \rho_{680}} \tag{2}$$

- Chlorophyll absorption index (CAI) to detect dry vegetation [61]:

$$CAI = (0.5 \times (\rho_{2000} - \rho_{2100}) - \rho_{475}) \times 10 \tag{3}$$

In the CAI-NDVI space, pixels were classed as bare soils when NDVI < 0.25 and CAI < 0.03 (Figure 3). These thresholds have been empirically defined from the image. Hereafter, only bare soil pixels were processed.

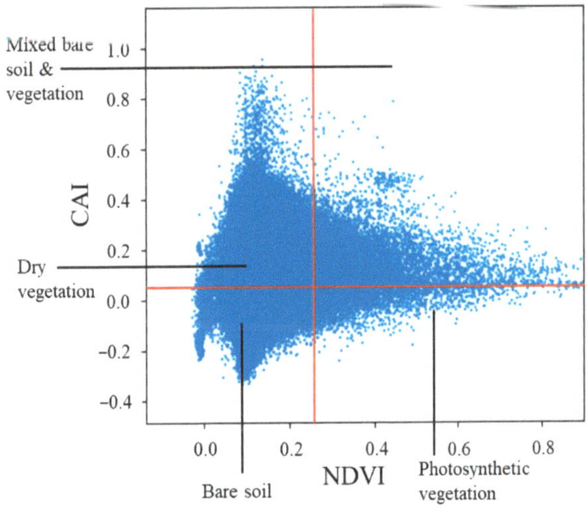

Figure 3. NDVI and CAI values for the Mareau hyperspectral image. In red: the thresholds chosen for each index in order to characterize four classes.

2.3. Field and Laboratory Measurements

Digital photos were acquired several minutes prior to the hyperspectral acquisition from a telescopic pole in order to build a digital elevation model (DEM) with MicMac software (version 1.0.beta13) [62] at a resolution of 1.3 mm, resampled to the hyperspectral images' resolution. Hilllshades maps were then obtained using Geographic Information System software (QGIS, version 2.18) from the DEM, using the illumination parameters (solar elevation and azimuth) for each scene. The resulting maps were then normalized in percent in order to model a level of illumination of the scene, designated as a roughness index.

Six samples have been collected post-acquisition of the hyperspectral images on the Gémigny field site (SUB 3, 4, 6, 10, 13 and 14), four on Coinces (SUB 2, 4, 6 and 8) and three on Mareau (SUB 3, 4 and 8) (Figure 2). The samples were taken on the surface of the overall subzone, allowing a single integrated measurement over the surface of the subzone.

Then, their mineral composition and abundance were quantitatively estimated in laboratory on the samples using XRD analysis [63] with a Bruker D8 °θ/°2θ diffractometer. The analysis was separated in two steps. First, the crystallized fraction of the samples was determined from the powder diffractogram (detection threshold inferior to 3%). Second, the oriented clay fraction (<2 µm) was analyzed after solvation with ethylene glycol and heating at 490 °C. Error could reach 5% for the clay minerals estimation. Mineralogy estimation was realized using the SiroQuant version 4 software and was based on the Rietvield method [64].

They had similar abundances of clay minerals, with smectite (containing montmorillonite) ranging between 13% and 23%, illite between 1% and 10% and kaolinite between 1% and 7%. Calcite was only present in the SUB 4 of the Coinces site, with an abundance of 7%. Quartz was present in high abundance for Gémigny and Coinces, accounting for more than 58%, while it was only around 30% for Mareau. The remaining part of soil mineral fraction for Mareau consisted mainly of potassium feldspars and plagioclases (Table 1). As they are spectrally inactive in the SWIR domain, they will be considered similar to the quartz in the current study.

Table 1. Mineral XRD analysis of samples collected on the 3 sites (statistics over the subzones per site).

Title 1	Gémigny		Coinces		Mareau	
Type of Mineral	**Min**	**Max**	**Min**	**Max**	**Min**	**Max**
Quartz	58	68	58	64	29	31
Smectite	13	20	18	23	17	20
Illite and/or micas	3	5	1	4	6	10
Kaolinite	1	5	2	3	5	7
Calcite	traces	traces	traces	7	traces	traces
Potassium Feldspars (Sanidine/Orthoclase)	9	10	5	7	12	16
Plagioclase Feldspars (Albite)	4	7	5	6	18	24
Quartz	58	68	58	64	29	31

Factors influencing the spectral behavior of soil such as SOC, soil granulometry (Figure 4) and SMC were analyzed on the same samples. Clay granulometry was similar for Mareau and Coinces (around 30%) and lower for Gémigny (around 23%). For Gémigny and Mareau, silt represented 25%, while silt represented more than 50% for Coinces. A high proportion of sand was found for Gémigny (around 64%) and Mareau (around 43%) compared to Coinces (around 15%). SOC varied between 2% and 3% whatever the site.

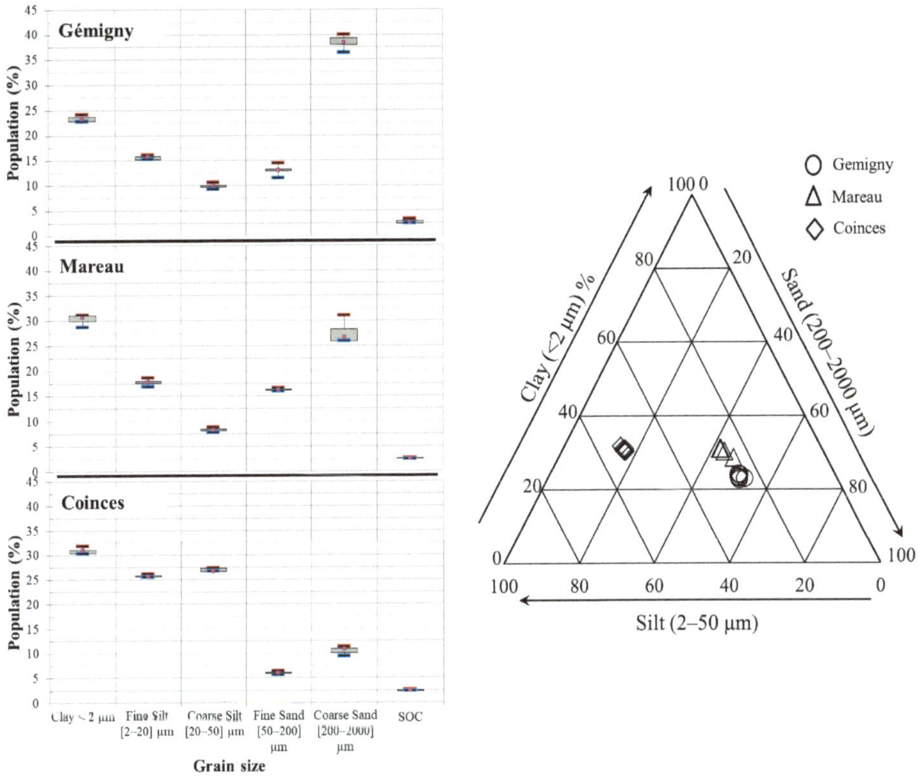

Figure 4. Grain size and SOC for each site (**left**), texture triangle for all samples (**right**).

SMC was measured a few minutes prior the acquisition on two areas: samples outside the hyperspectral acquisitions and samples in subzones that were manually wet (SUB 10 for Coinces and SUB 15 for Mareau). After the samples were collected, the mass was measured on site; then, samples were dried at 38 °C in an autoclave for 72 h in laboratory and measured for dry mass. Soils in the three sites were quite dry, with an SMC ranging between 8 and 10 wt% for Gémigny, between 8 and 13 wt% for Mareau, and between 2 and 8 wt% for Coinces. For the wet subzones, SMC reached 17 wt% for Mareau and 18 wt% for Coinces.

2.4. General Methodology

The processing flowchart is detailed in Figure 5. The first step defined the EM used as inputs for unmixing methods. Two options were tested: (i) detection from the hyperspectral image itself with automatic methods (Section 2.4.1) and (ii) selection from a spectral library. Then, after the application of spectral preprocessing (Section 2.4.2), linear and nonlinear unmixing methods were applied (Section 2.4.3) to give abundance maps that were compared with ground measured data in Section 2.4.4.

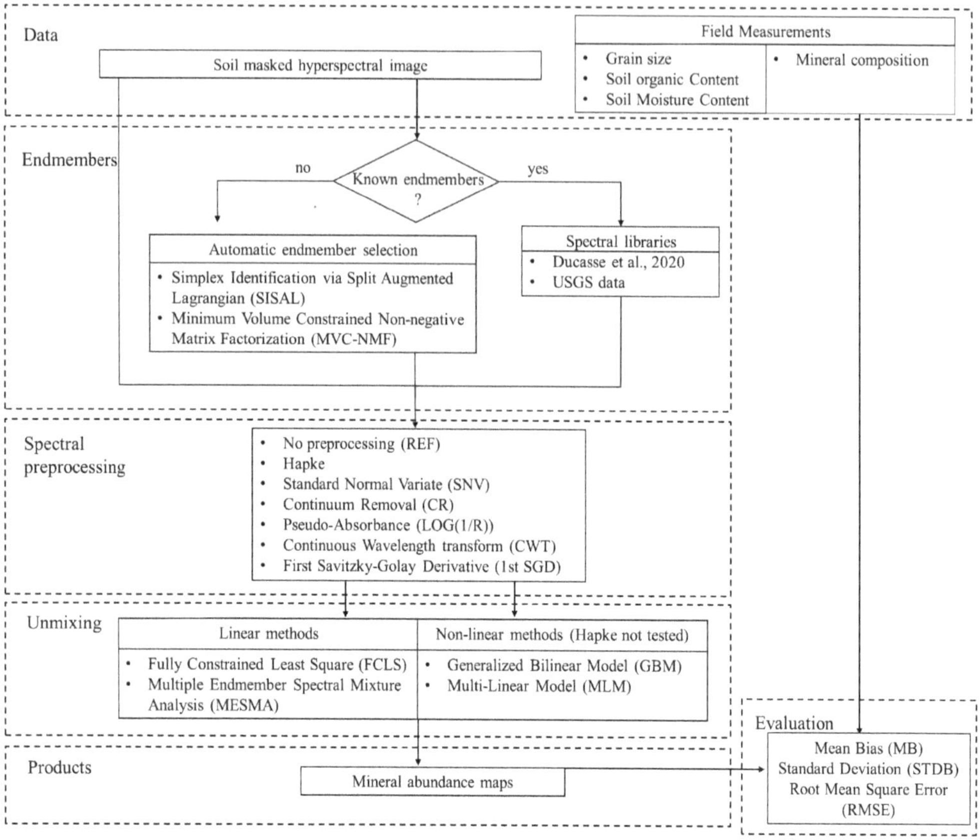

Figure 5. Processing scheme to estimate montmorillonite abundance.

2.4.1. Endmember Selection

On one side, if all minerals are known for a given soil sample, the EM will be chosen from a spectral library. In our study, two spectral libraries were tested and independently applied. The first one comes from the United States Geological Survey (USGS) with more than 1270 spectra [65] and is widely used by the soil community. We selected spectra from the purest sampled for each minerals, (i.e., 4 for montmorillonite, 4 for illite, 4 for kaolinite, 4 for quartz and 2 for calcite) and used the average spectra for each EM. The second one corresponds to pure samples of minerals acquired in a previous laboratory study [12], with the same sensor as the present study with XRD analysis. This dataset is composed of mixtures of montmorillonite, illite, kaolinite, quartz and calcite and is further named the Ducasse library. More than 25,000 spectra have been acquired per minerals, recording the spectral variability of these EM.

On the other side, if all minerals were unknown, an automatic EM detection was performed. Two geometric methods were selected considering no pure pixel existed in the image: minimum volume constrained non-negative matrix factorization (MVC-NMF; [66]) and simplex identification via split augmented Lagrangian (SISAL; [67]). They were both based on a simplex detection and supposed the number of EMs to be known. MVC-NMF differs from SISAL by its non-negativity constraint. Both methods were initialized with the vertex component analysis (VCA [68]), using 4 or 5 mineral spectra.

2.4.2. Spectral Preprocessings

We selected six spectral preprocessings commonly used by the soil scientific community and also in our previous laboratory study [12]: pseudo-absorbance (Log(1/R)), Hapke model, standard normal variate (SNV), continuum removal (CR), continuous wavelet transform (CWT) and first Savitzky–Golay derivative (1stSGD). The reflectance (no preprocessing) was also kept (REF) as input for the unmixing process. More details about their formulation are provided in [12] and were coded in Python.

2.4.3. Unmixing Methods

Four unmixing methods were applied since there were already used in [12]: fully constrained least square (FCLS) and multiple endmember spectral mixture analysis (MESMA) as linear methods, and generalized bilinear model (GBM) and multi-linear model (MLM) as nonlinear ones.

FCLS algorithm ([69]; Equation (4)) with f_i corresponds to the abundance of EMi, p the number of EM and ρ_i the reflectance of EMi.

$$\rho = \sum_{i=1}^{p}(f_i\rho_i)\ with\ 0 \le f_i \le 1\ and \sum_{i=1}^{p}(f_i) = 1 \tag{4}$$

MESMA ([70]; Equation (5)) accounts for the spectral variability of each EM by adding a new factor $\varepsilon_{\lambda k}$ that is the residual error, λ being the wavelength index. Therefore, for each EM, six spectra were considered: mean spectrum, mean spectrum $\pm$ standard deviation, mean spectrum $\pm$ 3 $\times$ standard deviation and median spectrum.

$$\begin{gathered}\rho = \sum_{i=1}^{p}(f_i\rho_i) + \varepsilon_\lambda\ with\ 0 \le f_i \le 1\ and \sum_{i=1}^{p}(f_i) = 1 \\ \text{and } RMS = \left(\frac{\sum_{\lambda=1}^{N}(\varepsilon_\lambda)^2}{N}\right)^{1/2}\end{gathered} \tag{5}$$

GBM ([71]; Equation (6)) takes into account first- and second-order interactions but does not consider self-interactions:

$$\rho = \sum_{i=1}^{p}(f_i\rho_i) + \sum_{i=1}^{p\ 1}\sum_{j=i+1}^{p}(\gamma_{ij}f_if_j\rho_i \odot \rho_j) \tag{6}$$

γ_{ij} is a free parameter corresponding to the nonlinear contribution where i (respectively, j) referred to the EMi (resp. EMj).

MLM ([72]; Equation (7)) considers the parameter P corresponding to the probability for a ray light to undergo multiple interactions.

$$\rho = \frac{(1-P)\sum_{i=1}^{p} f_i\rho_i}{1-P\sum_{i=1}^{p} f_i\rho_i} \tag{7}$$

The implementation of the FCLS algorithm comes from the pysptools Python's package. We coded our own version of MESMA based on [70]. GBM and MLM matlab implementation from [71] were used.

2.4.4. Evaluation Criteria

Three criteria evaluated the performances of these models and are expressed in the percentage of weight abundances (wt%). These criteria were made on a difference between the single measured montmorillonite value per subzone (measured, i.e., the XRD measurements, Section 2.3) and the abundance values estimated by unmixing. The measured value is therefore a single value for the entire zone; this is the difference with this validation value.

The mean bias (MB) will evaluate the mean error between estimated abundance $\hat{f}_n$ and the measured one f_n for a pixel n among the M pixels of each image region of interest (subzone) or the whole image:

$$MB = \frac{1}{M}\left(\sum_{n=1}^{M}\left(\hat{f}_n - f_n\right)\right) \tag{8}$$

The standard deviation is expressed as:

$$STDB = \sqrt{\frac{1}{M-1}\left(\sum_{n=1}^{M}\left(\left(\hat{f}_n - f_n\right) - MB\right)^2\right)} \tag{9}$$

The Root Mean Square Error (RMSE) is:

$$RMSE = \sqrt{MB^2 + STDB^2} \tag{10}$$

In addition, the RMSE and the spectral angle mapper (SAM, [73]) were used to compared spectral library spectra and hyperspectral image spectra. Let R be a reference spectrum, T be the test spectrum, with B the number of spectral bands; SAM is expressed as:

$$SAM = \cos^{-1}\left(\frac{\sum_{\lambda=1}^{B} T_\lambda R_\lambda}{\sqrt{\sum_{\lambda=1}^{B} T_\lambda}\sqrt{\sum_{\lambda=1}^{B} R_\lambda}}\right) \tag{11}$$

2.4.5. Validation Methodology

The validation was performed in three steps. First, spectra retrieved from automatic detection methods and from spectral libraries (USGS) were compared to the Ducasse library, visually and with the evaluation criteria (SAM and RMSE). Moreover, the position in the graph of mixtures is a good indicator of the accuracy of automatic detection methods.

Second, for the entire datasets, every preprocessing method coupled to unmixing methods, with different spectral libraries, was tested and presented. The evaluation criteria presented above (MB, STDB, RMSE) between XRD mineral composition of samples and estimated abundance values were used to choose the bests methods to retrieve montmorillonite abundances.

Then, the two best methods were compared at the subzone scale, then at the site scale, using visual interpretation.

3. Results

First, EM spectra derived from USGS and Ducasse libraries are compared with the results of the automatic EM detection (Section 3.1). Depending on this comparison, the best choice of EM spectra is further retained to be used with unmixing methods. Their performances are then described in three subsections: the montmorillonite estimates derived from all couples of spectral preprocessings and unmixing methods are compared to XRD smectite abundances as reference values (Section 3.2), the results for the best previous methodological couple are spatially interpreted over the three sites (Section 3.3), at subzone scale (Section 3.3.1) and image scale (Section 3.3.2), and the estimation of other soil minerals are compared with the best method (Section 3.4).

3.1. Endmember Detection and Comparison

USGS and Ducasse average spectra for every mineral were compared visually and then with SAM and RMSE (Figure 6). Generally, all mineral spectra presented the same spectral behavior (small SAM values) but differed in their absolute reflectance level (higher RMSE values). For montmorillonite, the SAM was less than 3°, and the spectral band at maximum absorption depth shifted by 10 nm (Figure 6a). For quartz, the RMSE was

17% with a SAM of 0.4°. The lowest RMSE (0.03%) was obtained for kaolinite. Calcite absorption features were deeper with the USGS library than with the Ducasse one.

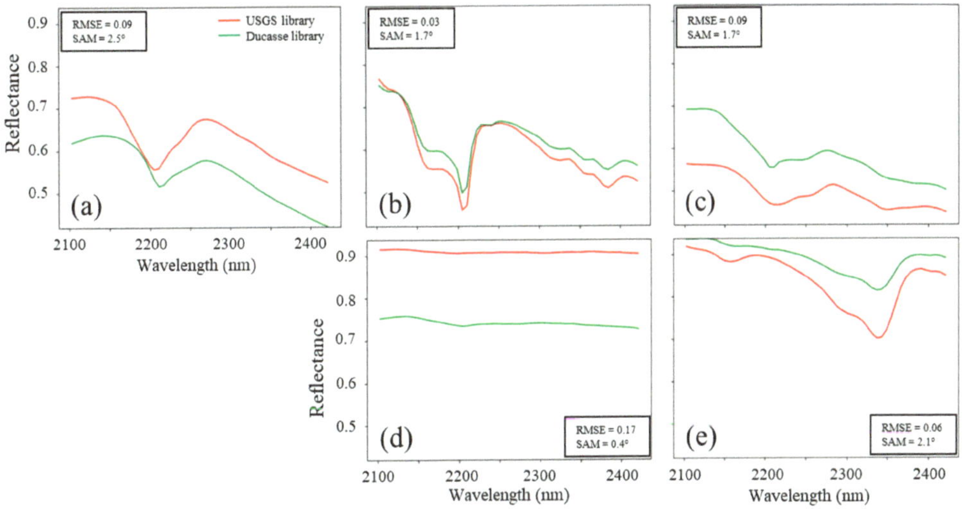

Figure 6. Endmembers from laboratory spectral libraries: (**a**) montmorillonite, (**b**) kaolinite, (**c**) illite, (**d**) quartz and (**e**) calcite.

The application of SISAL and MVC-NMF with VCA initialization showed similar results whatever the site; thus, only the results from Gémigny are shown (Figure 7). By applying SISAL with four or five EMs, the retrieved EMs did not exhibit specific clay or calcite absorption features (except slightly for EM2 and EM1 (Figure 7a,b, left). This was also confirmed by the fact that the detected spectra were outside the image simplex and far from the reference ones (Figure 7a,b, right) instead of being at the extremities of the simplex. MVC-NMF showed the worst results with all spectrally flat EMs whatever the initialization.

In conclusion, EM spectra were taken from the spectral libraries given the bad performances of the automatic EM detection methods.

3.2. Performance Analysis of Spectral Preprocessing Coupled with Unmixing Methods for Montmorillonite Abundance Estimation

The montmorillonite abundance estimated with FCLS, MESMA and GBM had most values close to either 0% or 100%, whatever the spectral library and the spectral preprocessing. Although GBM is a nonlinear model, its nonlinear fraction γ equals 0, meaning that GBM behaves like a linear unmixing method as FCLS or MESMA.

With MLM and USGS library as input, the best performances were obtained with REF whatever the site (the range of predictions encompasses this of measurements), with an RMSE between 2.2% and 13.3% (Table 2). It was also the case for Gémigny and 1stSGD on one side (RMSE between 1.4% and 11.8%) and for Coinces/Mareau and LOG(1/R) on another side (Figure 8 and Table 2). But the latter presented much higher prediction dispersions compared to Reflectance and 1stSGD. The results with theDucasse library were the same as the USGS library for REF (RMSE between 3% and 13%), better for 1stSGD (RMSE between 2.6% and 16.3%), particularly with Gémigny (2.6–5.4%), and a bit worse for LOG(1/R) (with higher prediction dispersions whatever the site) (Figure 9 and Table 3). Then, whatever the spectral library, SNV, CR and CWT led globally to poor estimations, all roughly close to either 0% or 100%.

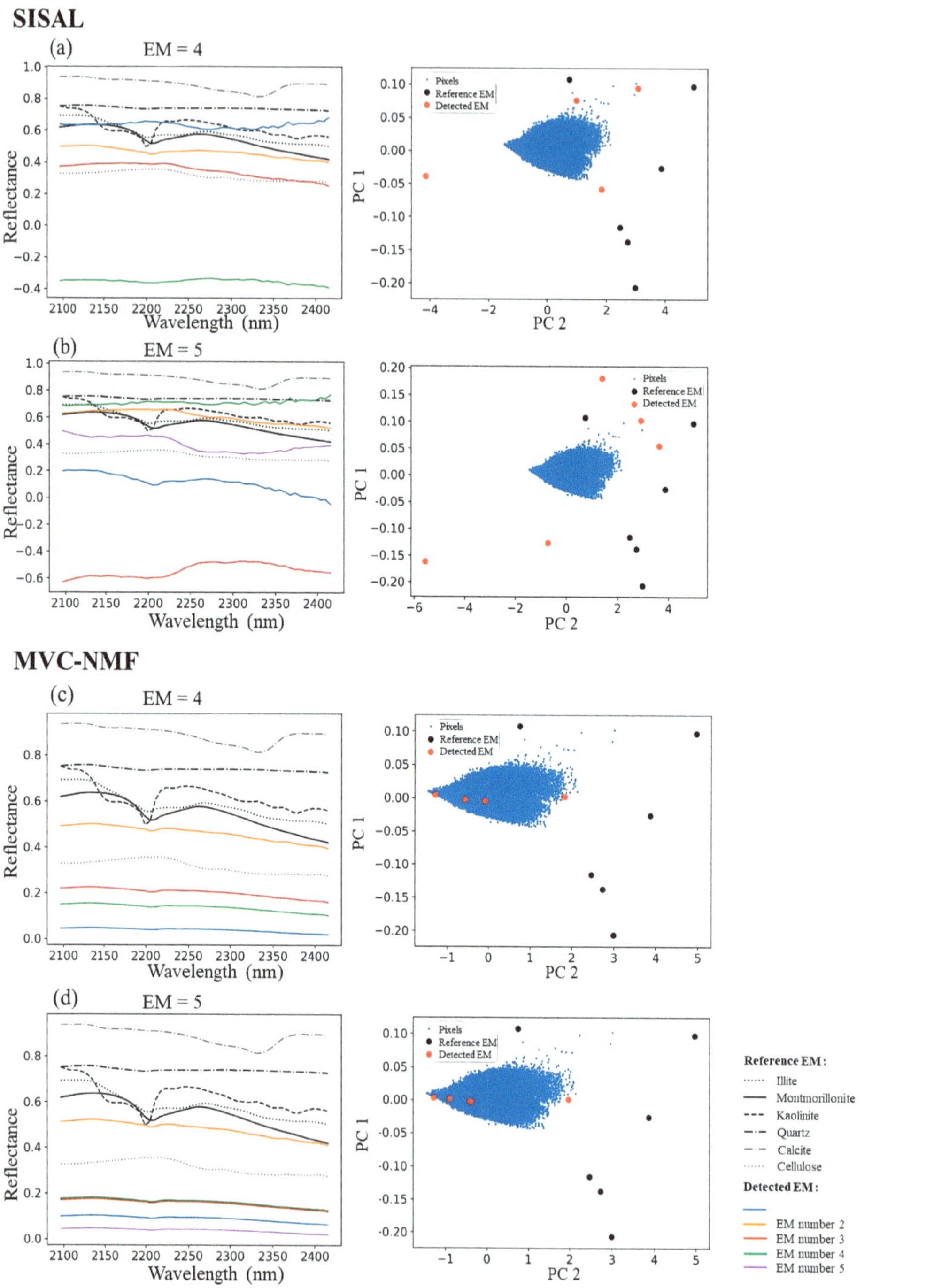

Figure 7. EM estimates over the Gémigny image. Comparison of the detected and Ducasse EM spectra and graphs of mixture simplex in the first two components space (PC 1 and PC 2) for (**a**) SISAL to detect 4 EM, (**b**) SISAL to detect 5 EM, (**c**) MVC-NMF to detect 4 EM and (**d**) MVC-NMF to detect 5 EM.

Table 2. Performances of Montmorillonite abundance estimations (wt%) obtained with REF-MLM and 1stSGD-MLM with the USGS library. In bold, the best RMSE, mean bias and standard deviation are presented for each sample.

	REF-MLM						1stSGD-MLM					
Sample	**Average Abundance**	**XRD Validation**	**MB**	**STDB**	**RMSE**	**R**	**Average Abundance**	**XRD Validation**	**MB**	**STDB**	**RMSE**	**R**
Coinces sub2	15.6	21	**−5.4**	4.3	**6.9**		7.6	21	−13.4	6.6	14.9	
Coinces sub4	7.4	20	**−12.6**	4.1	**13.3**		3.5	20	−16.5	5.1	17.3	
Coinces sub6	13.2	18	**−4.8**	4.3	**6.4**		3.7	18	−14.3	3.3	14.7	
Coinces sub8	10.5	23	**−12.5**	4.0	**13.1**		3.6	23	−19.4	3.5	19.7	
Coinces all subs	12.8	21	−7.8	6	9.8	−0.08	4.3	21	−16.3	5.4	17.1	0.05
Gémigny sub3	27.8	20	**7.8**	3.4	**8.5**		12.2	20	**−7.8**	4.3	9.0	
Gémigny sub4	25.0	18	7.0	2.9	**7.5**		12.3	18	**−5.7**	5.4	7.9	
Gémigny sub6	25.5	20	**5.5**	3.8	**6.6**		9.5	20	−10.5	5.5	11.8	
Gémigny sub10	23.9	13	10.9	2.5	11.1		9.3	13	**−3.7**	4.3	**5.7**	
Gémigny sub13	25.5	14	11.5	1.8	11.7		9.5	−14	**−4.5**	2.7	**5.3**	
Gémigny sub14	24.2	15	9.2	1.9	9.4		9.2	15	**−5.8**	3.4	**6.7**	
Gémigny all subs	25.6	17	8.6	3.4	9.2	0.27	10.7	17	−6.4	4.6	7.9	0.23
Mareau sub3	13.7	24	**−10.3**	1.6	**10.4**		6.6	24	−17.4	2.3	17.6	
Mareau sub4	13.6	18	**−4.4**	2.0	**4.9**		8.4	18	−9.6	8.4	12.7	
Mareau sub8	14.5	17	**−2.5**	1.8	**3.1**		4.7	17	−12.3	2.5	12.5	
Mareau all subs	14.0	20	−6.2	4.1	7.4	−0.17	6.0	20	−14.1	4.9	15.0	0.14
All subzones, all sites.	18.9	18	0.0	9.0	9.0	−0.44	18.9	−11.3	−11.3	6.7	13	−0.17

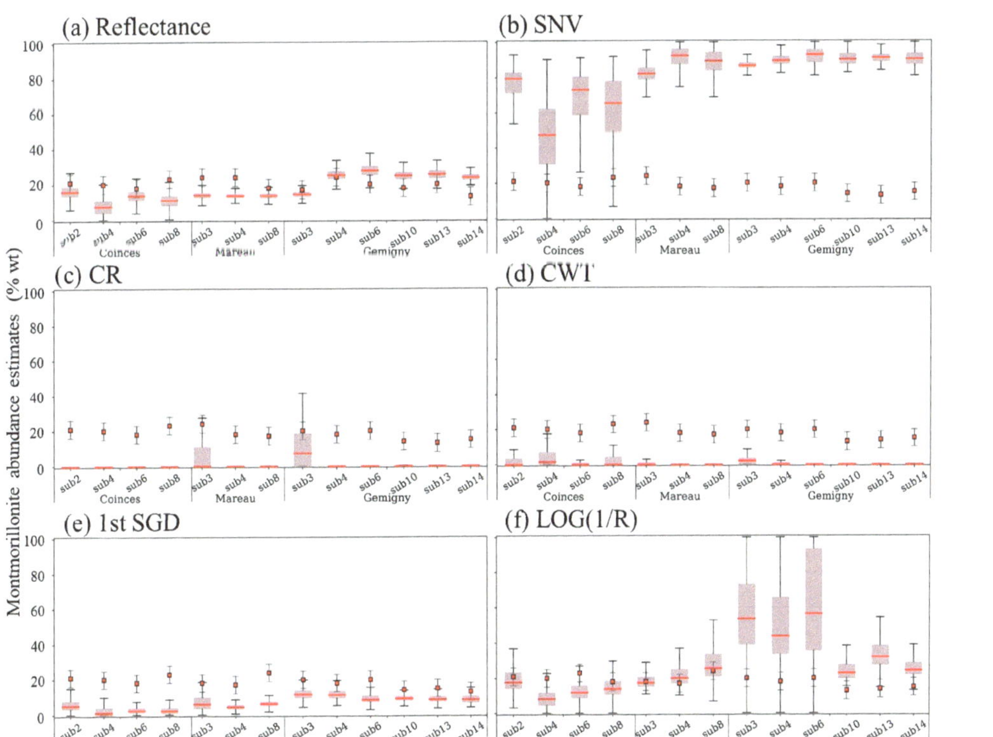

Figure 8. Montmorillonite abundance estimations over all the subzones per site (gray boxplots with the median highlighted by a red line) compared to the XRD dataset (boxplots with a red square depicting the median). The inputs are the USGS library, the six preprocessings and REF followed by MLM.

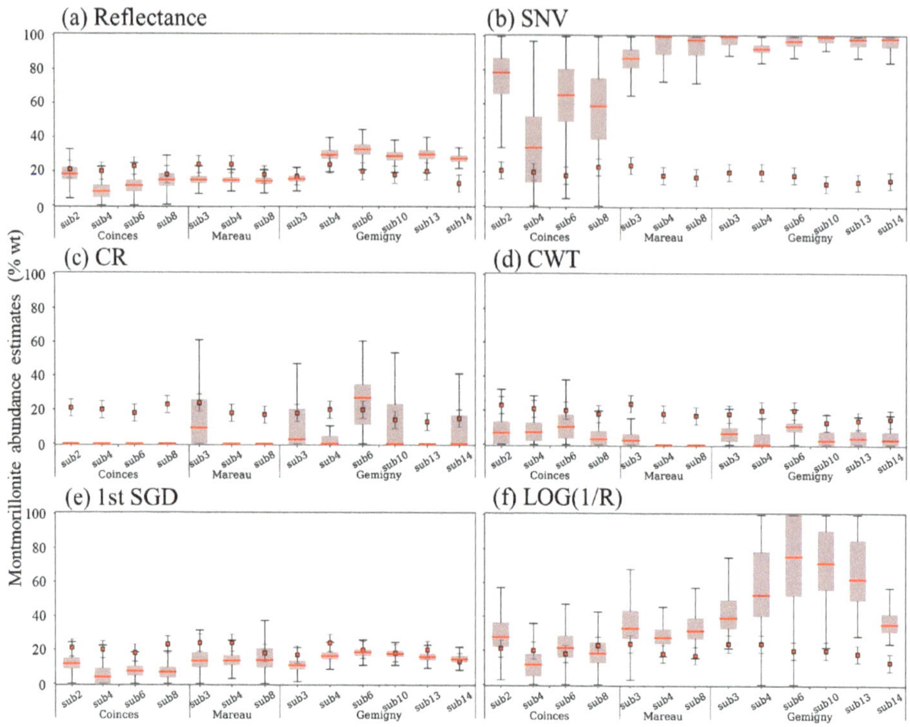

Figure 9. Montmorillonite abundance estimations over all the subzones per site (gray boxplots with the median highlighted by a red line) compared to the XRD dataset (boxplots with a red square depicting the median). The inputs are the Ducasse library, the six preprocessings and REF followed by MLM.

Table 3. Performances of Montmorillonite abundance estimations (wt%) obtained with REF-MLM and 1stSGD-MLM with the Ducasse library. In bold, the best RMSE, mean bias and standard deviation are presented for each sample.

	REF-MLM						1stSGD-MLM					
Sample	**Average Abundance**	**XRD Validation**	**MB**	**STDB**	**RMSE**	**R**	**Average Abundance**	**XRD Validation**	**MB**	**STDB**	**RMSE**	**R**
Coinces sub2	17.6	21	**−3.4**	5.9	**6.8**		12.3	21	−8.7	5.5	10.3	
Coinces sub4	6.9	20	**−13.1**	4.7	**14.0**		6.2	20	−13.8	6.5	15.3	
Coinces sub6	13.6	18	**−4.4**	6.3	**7.6**		7.9	18	−10.1	4.3	11.0	
Coinces sub8	10.2	23	**−12.8**	5.3	**13.8**		7.4	23	−15.6	4.6	16.3	
Coinces all subs	13.9	21	−6.6	7.4	9.9	−0.07	8.6	21	−12.0	6.4	13.6	0.05
Gémigny sub3	32.7	20	12.7	4.2	13.4		18.4	20	**−1.6**	2.9	**3.3**	
Gémigny sub4	28.9	18	10.9	3.6	11.5		17.9	18	**−0.1**	3.3	**3.3**	
Gémigny sub6	29.5	20	9.5	4.7	10.7		16.1	20	**−3.9**	3.7	**5.4**	
Gémigny sub10	27.7	13	14.7	3.2	15.0		15.0	13	**2.0**	3.1	**3.7**	
Gémigny sub13	29.8	14	15.8	2.2	16.0		16.1	14	**2.1**	2.2	**3.0**	
Gémigny sub14	28.3	**15**	13.3	**2.4**	13.5		15.5	15	0.5	2.5	**2.6**	
Gémigny all subs	29.9	17	12.8	4.00	13.4	0.26	16.9	17.1	−0.2	3.4	3.4	0.29
Mareau sub3	14.6	24	−9.4	2.4	**9.7**		14.3	24	−9.7	5.0	10.9	
Mareau sub4	14.1	18	−3.9	2.9	**4.9**		17.3	18	−0.7	13.2	13.2	
Mareau sub8	15.5	17	−1.5	2.6	**3.0**		11.5	17	−5.5	5.3	7.6	
Mareau all subs	14.9	20	−5.3	4.5	6.9	−0.11	14.9	20	−6.7	7.3	10.1	0.12
All subzones, all sites.	21.4	19	2.5	10.8	11.1	0.40	13.6	19	−5.3	7.6	9.3	−0.17

The best preprocessing–unmixing methods (i.e., 1stSGD-MLM and REF-MLM), compared to our validation dataset, give a negative r value for both methods, with a more pronounced negative trend (−0.44) for REF-MLM, due to montmorillonite overestimation for the Gemigny site (Figure 10a) and underestimation for Coinces and Mareau. Using different spectral libraries gave the same results (Figure 10). Noticeable trends were visible for Gemigny, with an r value between 0.23 and 0.29 for both methods and spectral libraries. This result suggests that unmixing could estimate montmorillonite abundances variations closely on this site.

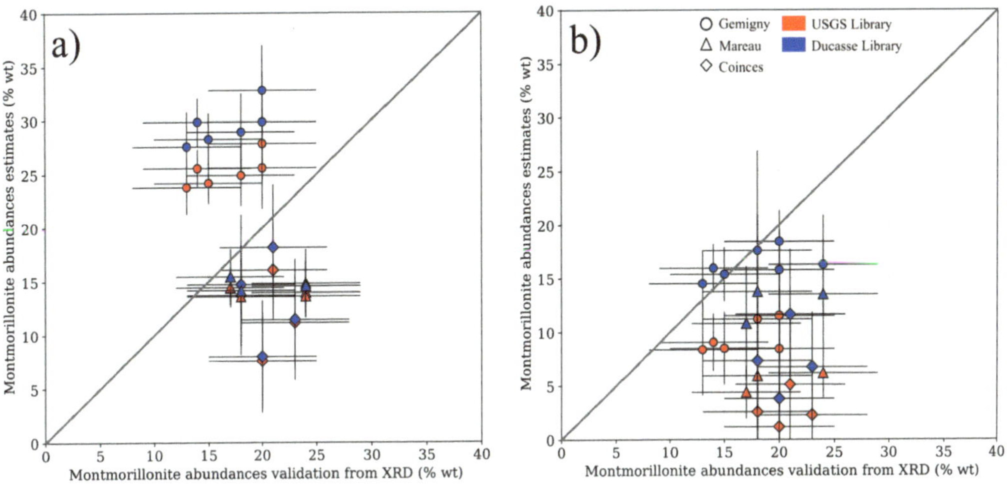

Figure 10. Performances of Montmorillonite abundance estimations (wt%) obtained with (**a**) REF-MLM and (**b**) 1stSGD-MLM with the USGS library (red) and Ducasse spectral library (blue). Bars in the x axis correspond to the accuracy of XRD analysis, and bars in the y axis correspond to the standard deviation of estimated montmorillonite abundances.

By considering the different assessment metrics in Tables 2 and 3 and Figure 10, it remains difficult to evaluate which spectral library is the best choice. Plus, as presented above (Figure 6), the libraries are very similar. Thus, USGS library was further chosen because it is a worldwide reference database for the soil community. In the following, only 1stSGD-MLM and REF-MLM were studied.

3.3. Analysis of Montmorillonite Abundance Maps

3.3.1. At the Subzone Scale

In order to understand the combined impact of topography (through the roughness index) and illumination conditions (through the solar elevation angle) on montmorillonite spatial estimation, two extrema subzone cases were selected among the three sites:

- Gémigny-SUB14, characterized by a weak spatial variability of the roughness index (median and standard deviation of 0.51 cm and 0.68 cm (Figure 11c) and by a solar elevation angle of 44°.
- Coinces-SUB2, characterized by a large spatial variability of the roughness index (median and standard deviation of 83% and 0.51 cm and 0.68 cm (Figure 12c), and a solar elevation angle of 35.6°.

In addition, the nonlinearity of MLM was analyzed by studying the variability of the parameter *P*.

For REF-MLM, the montmorillonite abundance estimations were rather spatially homogeneous for Gémigny-SUB14 with a spatial STDB of 1.9% (Figure 11d), but more heterogeneous for Coinces-SUB2 with 6.9% (Figure 12d). However, abundance values were overestimated for Gémigny-SUB14 (MB of 9.2%) and underestimated for Coinces-SUB2 (MB of −5.2%). Whatever the subzone, the *p*-value maps were very homogeneous (Figures 11e and 12e), leading to narrow distributions ranging between 0.75 and 1 (Figures 11f and 12f).

For 1stSGD-MLM, abundance estimations also had little variations on Gémigny-SUB14, with a spatial STDB of 3.4% (Figure 11g), while for Coinces-SUB2, they presented more variability with 6.6% (Figure 12g). Abundance values were underestimated for Gémigny-SUB14 (MB of −5.8%) and much more for Coinces-SUB2 (MB of −13.4%). The *p*-value range was larger with 1stSGD-MLM than with REF-MLM with negative values (between −0.4 and 0.6) (Figure 11f,i).

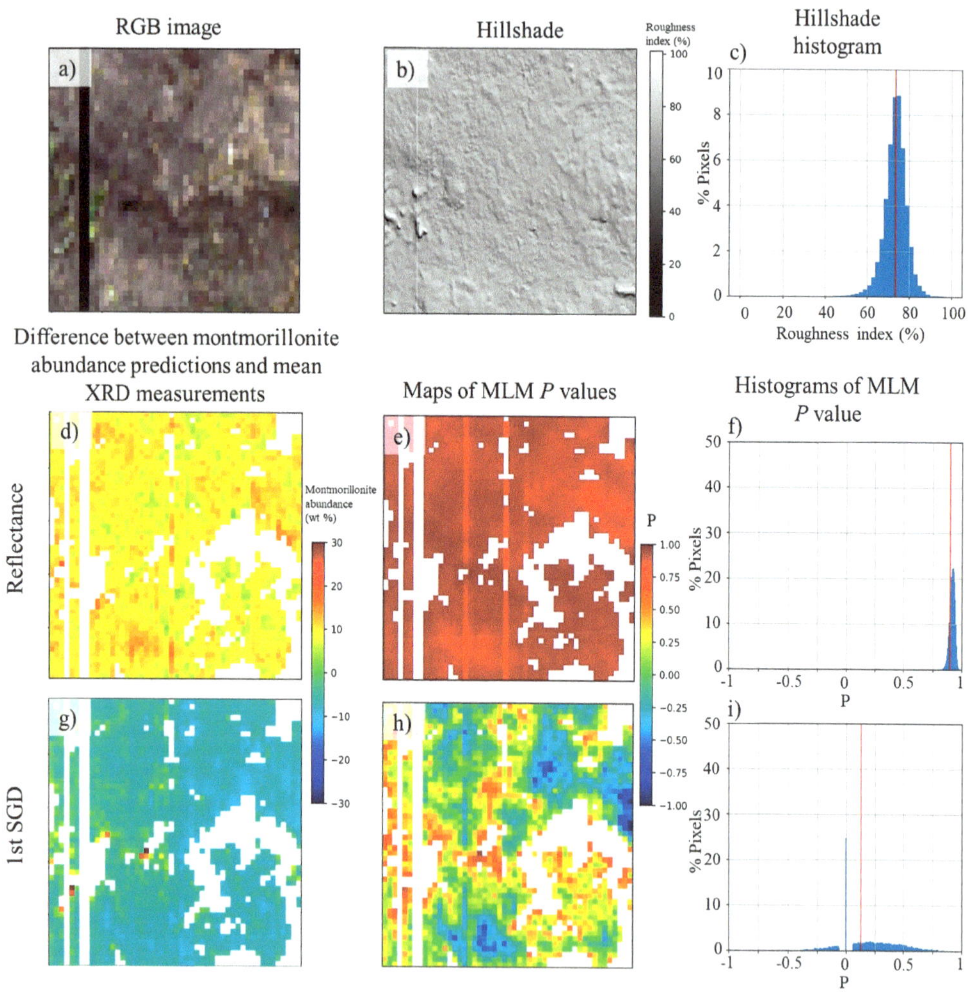

Figure 11. Results on Gémigny-SUB14: (**a**) RGB image (in black: masked areas), (**b**) hillshade map, (**c**) hillshade histogram (the red vertical line represents the median), (**d**) difference between the estimated montmorillonite abundance map obtained with REF-MLM and the XRD measured value (in white: masked areas), (**g**) the same for 1stSGD-MLM, (**e**) *p* value maps for REF-MLM (in white: masked areas), (**h**) the same for 1stSGD-MLM, (**f**) *p* value histogram for REF-MLM (the red vertical line represents the median) and (**i**) the same for 1stSGD-MLM.

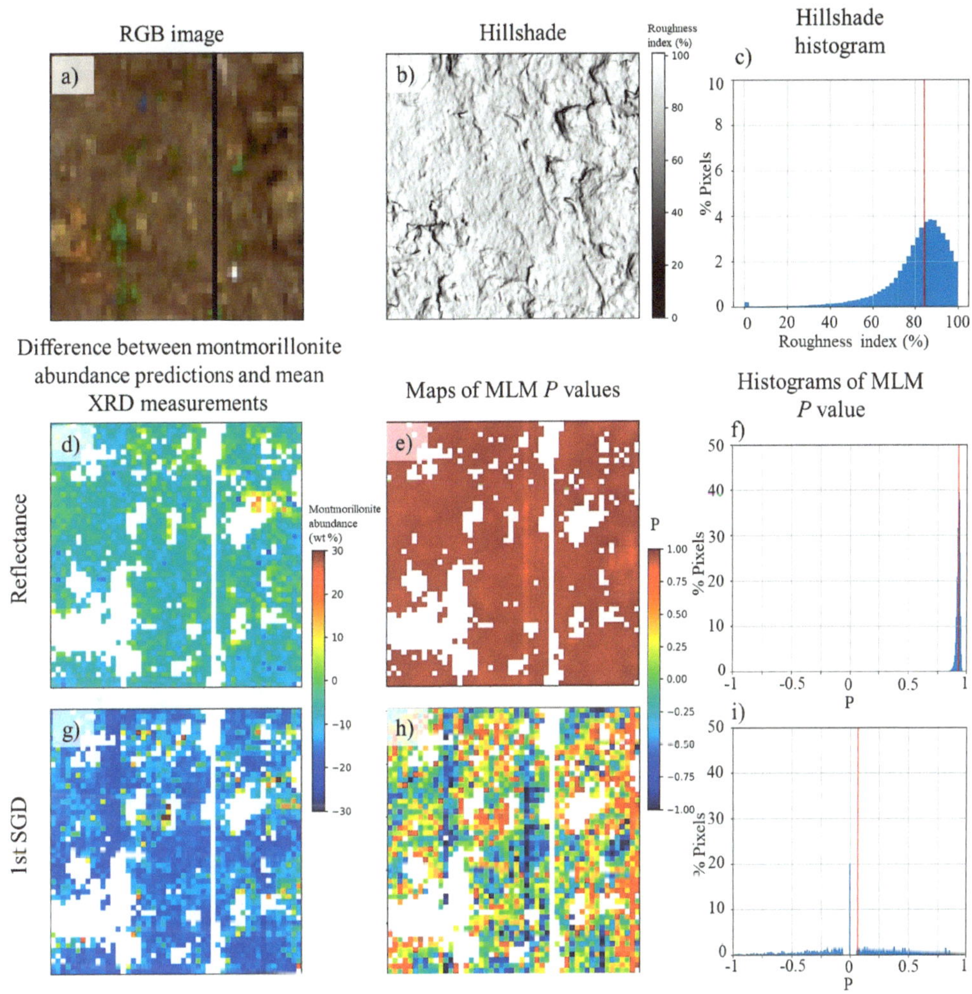

Figure 12. Results on Coinces-SUB2: (**a**) RGB image (in black: masked areas), (**b**) hillshade map, (**c**) hillshade histogram (the red vertical line represents the median), (**d**) difference between the estimated montmorillonite abundance map obtained with REF-MLM and the XRD measured value (in white: masked areas), (**g**) the same for 1stSGD-MLM, (**e**) *p* value maps for REF-MLM (in white: masked areas), (**h**) the same for 1stSGD-MLM, (**f**) *p* value histogram for REF-MLM (the red vertical line represents the median) and (**i**) the same for 1stSGD-MLM.

Globally, 1stSGD-MLM tended to estimate negative values of montmorillonite abundances, which was more marginal for REF-MLM. Therefore, for its simplicity, REF-MLM was selected to map the entire sites.

3.3.2. At the Image Scale

For Coinces, the average abundance of montmorillonite over the image was 15%, whereas XRD data were between 18% and 23% over the subzones (Figure 13a). The montmorillonite abundance was underestimated on all subzones, particularly on SUB4 and SUB8 (respectively, 7% and 10%). The subzone interquartile range (around 5%) was larger than that of the entire image (7%).

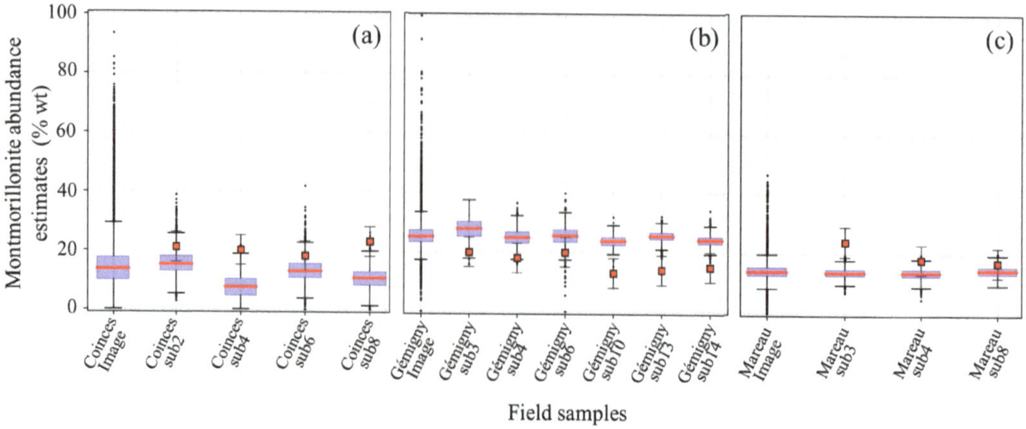

Figure 13. Performances for Montmorillonite abundance estimation with REF-MLM for all subsites (gray boxplots with the median highlighted by a red line) plotted with the XRD dataset (boxplots with a red square depicting the median).

For Gémigny, the average abundance of montmorillonite was between 24% and 28% for all subzones and 23% for the entire image (Figure 13b). In comparison, XRD data ranged between 13% and 20% for this site. Interquartile ranges were around 5% for all boxplots.

For Mareau, the average abundance of montmorillonite was between 14% and 15% for all subzones and 14% for the entire image (Figure 13c). In comparison, XRD data ranged between 17% and 24%. Interquartile ranges were around 3% for all boxplots. It seems that fewer outliers were observed for Mareau than for the other sites at the image scale, and also the montmorillonite abundance was globally underestimated with Coinces and Mareau, opposite to Gémigny.

On the three montmorillonite abundance maps, the estimates were spatially uniform (Figures 14–16). We can note:

- On the Gémigny map, the estimated abundance values of clays increased inside wheel tracks located in SUB1, SUB2, SUB3, SUB7, SUB8 and SUB9 (Figure 14c). They contained more estimated montmorillonite than the other subzones (from around 4% to 5%).
- On the Coinces map (Figure 15c), the montmorillonite estimation was higher in left subzones (SUB1, SUB2, SUB6, SUB7, SUB11, SUB12) than in the right ones (SUB5 and SUB15). This trend was not present in any validation data but may have been due to the presence of clouds during the acquisitions. In the wet subzone (SUB10), the majority of pixels was wrongly classified as "dry vegetation", and montmorillonite was estimated at around 40% in bare soils' pixels.
- On the Mareau map, the montmorillonite estimation was around 20% in the wet subzone (SUB15), whereas the montmorillonite estimation in other pixels was around 15% (Figure 16c).

The composite mask showed that only 53.6%, 59.2% and 42% of the image was composed of bare soil, respectively, for Gémigny, Coinces and Mareau. The main contribution in the masks came from dry vegetation, while the rate of mixed pixels was relatively low, except for Coinces (cf. Figures 14–16).

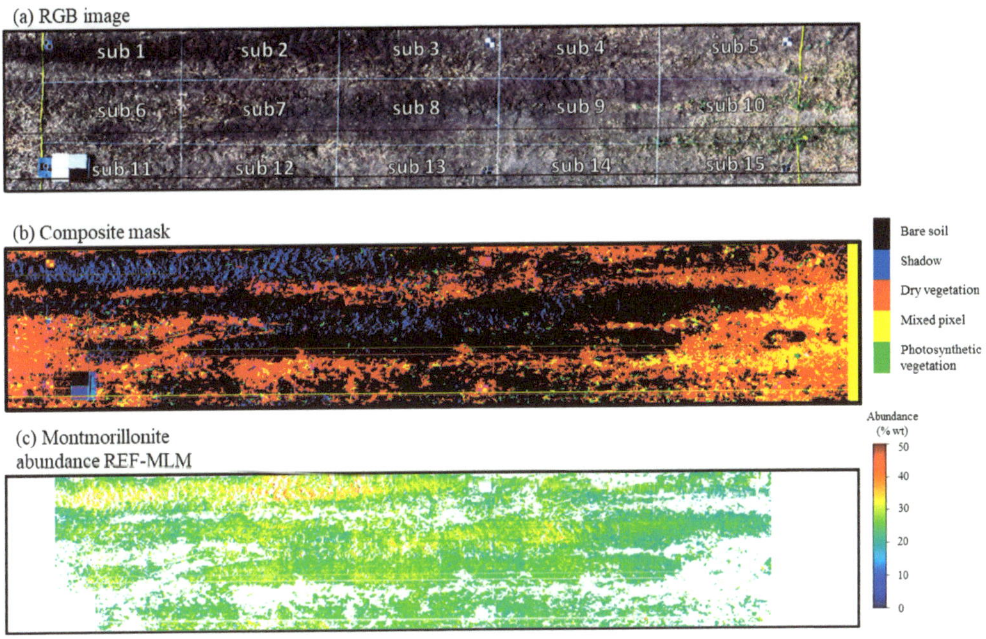

Figure 14. Maps for Gémigny site (**a**) RGB composite image, (**b**) composite mask and (**c**) abundance map of montmorillonite obtained with the REF-MLM and USGS library.

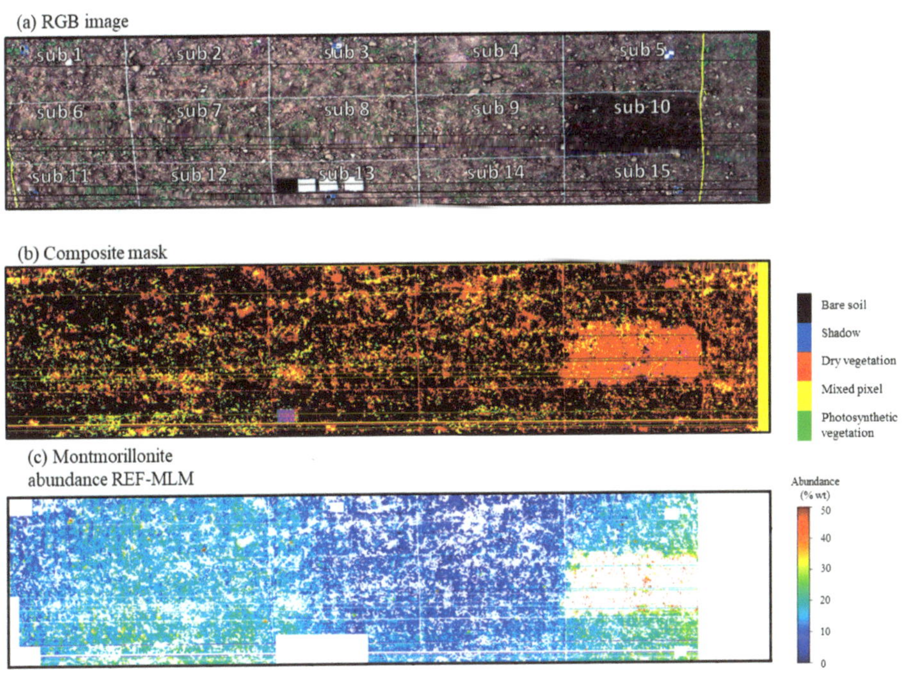

Figure 15. Maps for Coinces with wet area SUB10 site (**a**) RGB composite image, (**b**) composite mask and (**c**) abundance map of montmorillonite obtained with the REF-MLM and USGS library.

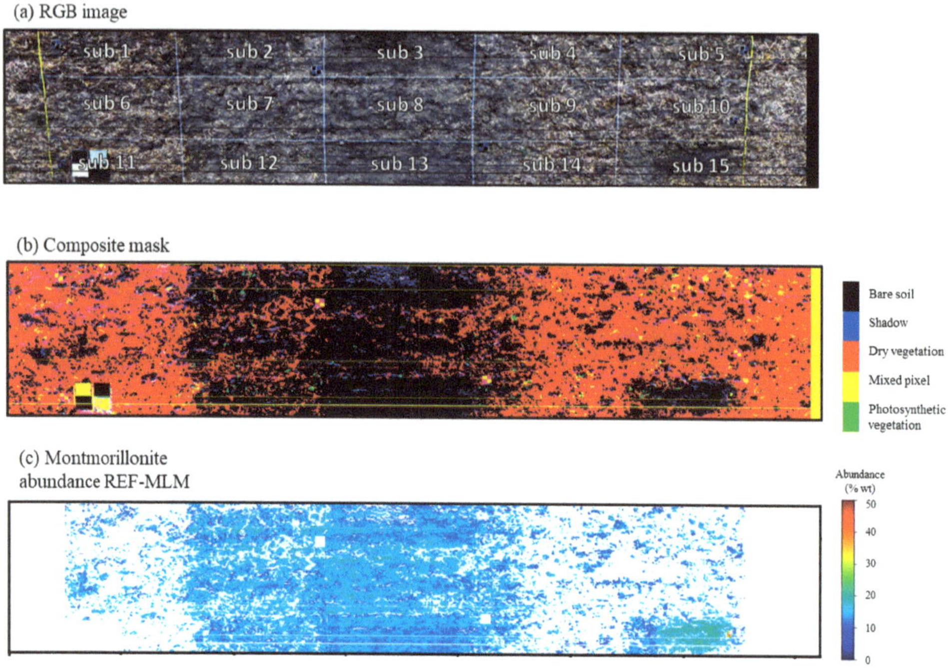

Figure 16. Maps for Mareau site with wet area SUB15 (**a**) RGB composite image, (**b**) composite mask and (**c**) abundance map of montmorillonite obtained with the REF-MLM and USGS library.

3.4. Estimation Performances of Other Mineral Abundance

As the estimation quality of montmorillonite abundance depends on that of the other minerals composing the soil, there is an interest to study it. Our results were close for all sites due to the soil composition uniformity (Figure 17a–c). Best performances for mineral estimation were obtained for kaolinite (RMSE between 1% and 7%). Quartz was estimated with the largest errors (RMSE range between 8% and 22%). Illite estimation for REF-MLM was equal to zero for Mareau and Gémigny, whereas there was between 6% and 10% illite in Mareau soil and between 2% and 5% illite in Gémigny soil (Figure 17b,c). Calcite mineral was estimated with low errors (RMSE range between 4% and 6%); however, this mineral has been detected in only one subzone (SUB4 for Coinces site). These previous RMSEs were computed for all sampled subplots of an image.

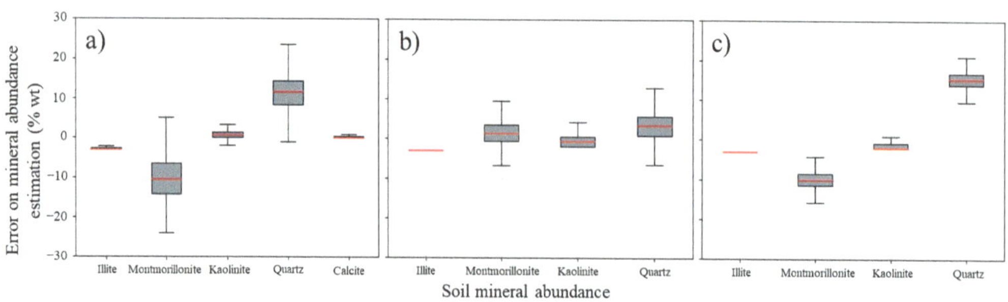

Figure 17. Comparison between mineral abundance estimations with REF-MLM and USGS library and the XRD dataset for each site: (**a**) Coinces, (**b**) Gémigny, (**c**) Mareau.

4. Discussion

4.1. Endmembers Selection for Unmixing

Considering the automated EM selection methods, the detected EMs with MVC or SISAL methods presented a low spectral behavior, explaining why they failed to detect the correct EMs (Figure 7) [34]. In our study, the detected EMs were far from the extremities of the simplex and also from the EMs of the libraries. Further, these methods were very sensitive to the initialization achieved here with VCA. Depending on the prior number of EMs, the likelihood of estimated EM highly differed, particularly with SISAL. In addition, one could expect better results if these methods were initialized with prior knowledge of the minerals present in the sampled soils. Unfortunately, this was not the case here; one reason could come from the high spatial intra-class variability of each mineral in the image. In this work, it was the first time that automatic endmember selection methods were used for clay mineral abundance estimation over a real rough soil at a centimeter scale. The derived results were mitigated. This work was a first attempt at using simple tools; thus, more sophisticated tools could be tested accounting for the intra-class variability, such as the inertia-constrained pixel-by-pixel nonnegative matrix factorization [74] or the genetic algorithm [75].

Considering the choice of EMs from a spectral library, the unmixing performances were sensitive to the upscaling from laboratory spectroscopic measurements to real case scenarios, since USGS and Ducasse libraries did not come directly from the study sites. Nevertheless, montmorillonite abundance estimations were rather coherent with the XRD measurements, and similar performances were obtained whatever the spectral library, suggesting that the leading factor was the spectral signature of minerals whatever the sampling site (cf. Figure 6, Tables 2 and 3). For the five studied minerals, the SAM was globally below 2.9°, while in the literature, standard threshold values were most often higher. For instance, ref. [76] discriminated two mineral spectra on average with an SAM equal to 5.7°, and ref [77] discriminated two clay spectra in the SWIR domain with 3.7°. Recently, ref. [78] developed an open synthetic dataset for the unmixing of 325 intimate mixtures from five clay powders. Overall, most available spectral libraries have focused on clay texture from the centimetric scale [79,80] to the global scale [81]. But, to our knowledge, except the two libraries we used, none of them have provided spectra associated with clay mineral discrimination, including montmorillonite.

4.2. Limitations of Preprocessings and Unmixing Methods for Montmorillonite Abundance Estimation

Considering the spectral preprocessings, they can be divided into three main classes [12]: those suppressing the spectral continuum and working at a local scale such as CWT and 1stSGD, those normalizing the global spectrum level such as SNV and CR and, at last those, that do not reduce the soil spatial variability such as Hapke and Log(1/R). Dealing with outdoor conditions in our case with the example of the MLM unmixing method (cf. Figures 8 and 9), SNV, CR and CWT globally had very poor montmorillonite estimations (roughly close to either 0% or 100%) whatever the spectral library. SNV was very sensitive to the high spectral and spatial variabilities of field soils, which might explain the difficulty of the unmixing methods to converge because SNV is the ratio between the spectral deviation compared to the mean over the standard deviation. For CR, an important reduction of clay absorption features was induced from the high proportion of quartz in soil, which modified their shape (depth and width), leading to difficulties for unmixing methods to find the proper EMs. For CWT, the main limitations came from the linear wavelet base decomposition of the spectra accounting only for the second to the fifth order (the first order corresponding to noise), which was probably not enough sufficient to model the nonlinearity due to multiple scatterings. Log(1/R) projects the spectral reflectance in a logarithmic space, allowing an increase of the linearity between the spectral data and their abundance [82,83]. But this transform remains very sensitive to the high spatial variability, leading to a strong overlap between the intra- and inter-classes. The W single scattering albedo of Hapke is the fraction of photons scattered by a particle over the total fraction of photons affected by this particle. This model considered the soil as a homogeneous mixture with the same granulometry for all the constituents composing this soil. Unfortunately, our ploughed soil was very heterogeneous with different granulometries (Figure 4) that explain the poor results obtained for linear unmixing methods. The best preprocessing was 1stSGD, with similar results as REF. This is in agreement with [12] noting that the best one was also 1stSGD for clay mixtures and reflectance when other minerals than clay were added. Ref. [84] found among several preprocessings, including SNV, that the best was first Log(1/R) combined with 1stSGD.

Considering the unmixing methods, FCLS, MESMA and GBM failed to estimate montmorillonite abundance. The main limitation for GBM is that it is not able to account for the complete nonlinear contribution of the signal with just second-order interactions between any two EMs, and it finally behaves as an FCLS (cf. Equation (6)). In the laboratory, similar results were obtained with clay and clay–calcite mixtures [12]. Also, GBM does not consider self-interactions and has no restriction on reflectance values that could be out of the [0–1] physical range [71]. On the contrary, MLM was the only unmixing method delivering correct results by accounting for all the multiple interactions found for ploughed fields. On contrary, in laboratory conditions, both linear and nonlinear unmixing methods have globally provided similar performances for montmorillonite abundance estimation [12].

Deeper explanations on MLM performances and p value analysis are further discussed. Initially, MLM takes as inputs albedos instead of reflectance spectra. The authors of [72] mentioned that they can be assumed to be equivalent for remote sensing imagery with a spatial resolution of 20 m or 2.4 m where nonlinear effects are relatively small (leading to small p values). But albedos should be preferred in cases of intimate mixtures due to a larger number of interactions (p values close to one), if EM spectra are known. However, for the latter case in laboratory conditions, [72] showed good results by only using reflectance for quartz abundance estimation, the same as [12] for montmorillonite. The trade-off most often relies on the P parameter of MLM having a physical meaning when its values are inside the range [0–1]. It indicates the probability that a light ray will undergo further reflections after each interaction and can thus be used to assess the size of the nonlinearity effect or the size of the deviation from a linear unmixing method. However, negative p values can occur if at least one material has a strong reflective behavior and/or a reflectance-albedo conversion has not been performed. Therefore, the observed spectra are larger in magnitude than any convex linear combination of EM spectra (given the same knowledge of EMs and abundances), and it is the opposite if the p values are positive [72]. Consequently, the analysis of p values is not straightforward, and very few studies have actually provided information about their distribution. In laboratory conditions, [12] found p values ranging [−0.2; 0.2] for montmorillonite–kaolinite mixtures and [−2; 0] for montmorillonite–calcite mixtures, while [72] found p values of 0.6, 0.71 and 0.88 for quartz–alunite mixtures. It seems that the presence of either quartz or calcite as reflective materials leads P to its lowest or highest values. From airborne hyperspectral data at varying spatial resolutions (2.4, 5.2 and 20 m) over desert, heathland and seashore areas, [72] found p value distribution with ranges of [−1.4; 0.6], [−1.2; 0.8] and [−2.5; 0.5], respectively. The areas with strongest nonlinear effects are highlighted by the pixels' highest P absolute values (ex: road, shadow, tree and coastline), and pixels' negative p values can come from additional illuminations due to adjacent topography outside the pixel field of view. Also, at coarse spatial resolutions, the unmixing problem is more due to areal mixtures than intimate ones. In our case, at the centimeter scale, both mixture types were encountered. Whatever the surface roughness, p values with reflectance exhibited a very narrow Gaussian distribution with a mean of 0.9 (Figures 11c and 12c) and were spatially homogeneous (Figures 11e and 12e). It might be attributed to the highest influence of the multiple scatterings from the predominance of intimate mixture effect and the quartz contribution in soils. It can also be attributed to the compensation between the high reflectance levels of the spectral library ranging from 50% up to 90% and those of the fields from 10 to 35%. A different behavior was observed when using 1stSGD instead of REF, which seems to be more sensitive to the topography and the surface composition, as if areal mixtures impacted more than intimate ones. This can be explained because reflectance bears the spectral continuum (and associated strong multiple scattering influence) oppositely to 1stSGD, which removes it (depends on local scale influences). For 1stSGD, p value distributions followed a very large Gaussian distribution over the range $[-1; 1]$, with a majority of positive values for the less rugged terrain in comparison with the most rugged one having more disperse negative and positive p values (Figures 11i and 12i). These distributions appear similar to those found at an airborne scale in [72]. Highest p values were mostly located around masked areas (Figures 11h and 12h). Indeed, artefacts can remain from the masking operation (based on empirical thresholds) because properly unmixing with spectral indices the presence of shade, vegetation and bare soil can result in a hard task. MLM can compensate the absence of EMs for shade and vegetation by varying the p value. This phenomenon has been observed at a larger scale, for instance, for improving unmixing in tree shadow pixels [72]. Although MLM was developed to be applied on reflectance, the performances obtained with 1stSGD preprocessing were similar that those with REF (RMSE of 11.6% with 1stSGD-MLM and 12.7% with REF-MLM).

In the recent years, several works have proposed some improvements of MLM, like [85], by exploiting the underlying manifold structure of the pixels' spectra; [86], by simulate the physics process of MLM with neural networks; [87], by proposing a coarse-to-fine scheme for unsupervised unmixing; and [88], by extending MLM to take into account multiple reflexions in the rough surface and different sources of illumination. Unfortunately, these extensions have only been validated either on synthetic datasets either on remote sensing data at least at a meter scale.

Other methods exist to retrieve the EMs and quantify their abundances. A large community has used linear based methods like PLSR, which are not adapted for complex soils (rough surface, intimate mixture, large variety of grain size) [72]. Recently, ref. [89] proposed another approach combining Fisher transformation and multiple endmember spectral mixture analysis to estimate lunar mineral abundance using the RELAB database. This method can address both the intra-class and inter-class variability of the spectral features of different lunar minerals and might be interesting to test on our dataset.

The best coupling preprocessing and unmixing methods are REF-MLM and 1stSGD-MLM. With laboratory clay mineral mixtures, best performances are obtained with 1stSGD-FCLS [12]. Indeed, intra-class variability has a higher amplitude than nonlinear effects in these mixtures, and derivative spectral preprocessing is the most efficient method to decrease intra-class variability. Using 1stSGD-FCLS, montmorillonite was estimated in mixtures of illite–montmorillonite–kaolinite with 10.8% RMSE, montmorillonite–calcite with 22.6% and montmorillonite–quartz with 55.2%. When upscaling to field conditions with more complex soil mineral composition, REF-MLM and 1stSGD-MLM methods provided RMSE performances within 2.2–13.3% and 1.4–19.7%, respectively, for REF-MLM and 1stSGD-MLM. The authors of [7] found an RMSE of 15.8% for fine fraction of soil samples coming from agricultural field using CR and a geometrical method trained with empirical relationships from synthetic laboratory mixtures. Ref. [90] estimated the abundance of smectite present in 77 samples collected in the Northern Morocco, with an RMSE of 18 wt%. After a CR, a multi-Gaussian method was applied on the resulting spectrum followed by a regression tree analysis to retrieve the abundance. These samples were dried, sieved and crushed to a powder (<20 μm) before the measurement of the spectral reflectance in the lab. Ref. [84] retrieved smectite abundance with a RMSE of 3.4% from kaolinite–illite–smectite–goethite–quartz samples in a laboratory. They coupled the use of Log(1/R) followed by the 1stSGD and then the bagging-PLSR method. Their samples accounted only for a size fraction smaller than 200 μm. In summary, most methods were applied on datasets coming either from synthetic samples mixed from mineral powders or from real samples collected in the field then measured in laboratory after being dried, sieved and flattened. For the first type of datasets, the RMSE of montmorillonite abundance estimations ranged between 3.4% and 55.2%, while for the second, it was between 1.4 and 19.7%. This highlights the high influence of the soil mineralogical composition.

4.3. Impact of Soil Mineralogical Composition

Performances of unmixing methods depend on soil mineralogy. Our study sites contained few clays (kaolinite < 7%, illite < 10% and montmorillonite < 24%) and calcite (<7%), but contained a large fraction of quartz (58–68% for Gémigny and Coinces, and 29–68% for Mareau). The estimated "quartz" class contains all spectrally inactive minerals between 2100–2500 nm: quartz, alkali feldspars and plagioclase feldspars. These minerals can be coated of alteration minerals (such as oxides) and change their spectral properties [90,91]. Some trends are interpreted depending on the mineral estimated.

On one hand, low errors obtained to estimate kaolinite fraction were probably due to the strong double absorption feature at 2200 nm. On the other hand, considering the low abundance of illite, this mineral was probably identified as montmorillonite because the feature corresponding to OH-stretching bands combined with lattice vibrations at approximately 2360 nm was shallow for illite and not visible in illite/montmorillonite mixtures with an illite fraction inferior to 20 wt% [12]. At last, poor results were obtained for the quartz class in all sites (RMSE between 8% and 22%), but the RMSE remained less than those obtained in the laboratory condition [12]. Furthermore, with the quartz class being the most abundant in our soil mineralogical composition, the estimation error of its abundances may be superior to those of the other minerals.

Ref. [84] obtained better results using the PLSR method to estimate abundances in mixtures simulating Australian soils, composed of smectite–illite–kaolinite–carbonate–quartz-organic matter (RMSE 3.4% for the three mineralogical clays). However, this approach is a learning-based method calibrated for soils of known composition, requiring extensive field sampling coupled with spectroscopic measurements. In our case, we are targeting an alternative with a partially unsupervised approach for automation purposes. This latter provides better results than [7] on real agricultural soils using index approaches, with an RMSE of 15.8% for montmorillonite, 25.2% for kaolinite and 29.8% for illite estimated over samples collected in the field, filtered (<2 mm) and then characterized in the lab.

4.4. Other Factors Influencing Montmorillonite Abundance Mapping

Compared to laboratory conditions where all the four unmixing methods (linear and nonlinear) deliver acceptable estimated montmorillonite abundances, the differences with outdoor conditions can come from the following explanations:

The obtained performances degraded compared to those obtained over in-lab mixtures as these samples were dried, sieved and crushed to a powder with a flattened surface,

The centimeter scale of spectroscopic acquisitions tended to exacerbate directional effects, both geometric (no local slope correction has been applied to retrieve surface reflectance) and optical (no anisotropic correction has been applied assuming only Lambertian materials), and so the spectral variability was increased, which could not be accounted by unmixing methods.

The better the spatial resolution is, the wider the spatial variability is; thus, at a centimeter scale, the variability was very high compared to the reflectance at a meter scale where the soil heterogeneities were more smoothed.

The presence of residual dry and wet vegetation and of shadows contributed to increasing the number of surface types seen by a pixel and then reducing the quality of the unmixing performances; an improved masking can be targeted in the future.

The roughness of the ploughed fields induced surface multiple scatterings and variations of illuminations (direct and diffuse downwelling irradiance), which were taken into account in our unmixing models; [36] and, more recently, [88] have proposed a physical approach to take into account these effects. In particular, [88] proposed an extension of MLM to this end. Unfortunately, these approaches have never been tested at a centimeter scale also including intimate mixtures.

The higher variability of mineral granulometry (10–35% for Gémigny, 10–31% for Mareau and 6–31% for Coinces accounting for textural clay and coarse sand, cf. Figure 4) impacted volume multiple scatterings and induced higher spectral variations, oppositely to the more homogeneous granulometry found for the laboratory mixtures, having either only clay minerals or clay minerals combined with calcite or quartz [12].
The important contribution of other minerals than clay ones, mainly quartz for Gémigny and Coinces (abundance more than 58%), and quartz, potassium feldspars and plagioclases for Mareau (global abundance more than 50%).
The presence of quartz in soils, such as noted by [11,21,92], that highlighted the difficulties to retrieve minerals and quantify their abundance when a mixture contains quartz, and also by [12] that confirmed this point in the laboratory where montmorillonite abundance estimation in the presence of quartz was very poor (RMSE more than 50%) whatever the unmixing method and the preprocessing.
The atmospheric conditions, such as for Coinces, for which the experiment was performed under partially cloudy conditions with varying illumination, while in the laboratory, these conditions were controlled and stable in time.
The soil water content led to an increase in montmorillonite abundance that could be explained by the decrease of global reflectance level due to the soil moisture content increase and the potential overestimation of the darkest EM (i.e., montmorillonite). However, in our case, the water content was low enough (<18%) to not mask the clay absorption band with soil moisture content, which happens for an SMC of 30% [39].

Concerning the Gémigny field site, changes in montmorillonite estimation along wheel tracks could be explain by several hypotheses. With these areas being darker on the RGB image, it may be possible that wheels stripped the topsoil horizon, leading eventually to a difference in physical properties of the mixture due to soil compaction (increase of soil roughness) and a change in mineral mixture.

Coinces montmorillonite abundance values were more variable than Gémigny and Mareau, because of cloud cover during data acquisition. To limit the induced changes of illumination by clouds, [30] recommend collecting hyperspectral images under direct illumination in order to avoid spurious absorption features in SWIR domain.

Mineral content was very similar for all sites, and grain size content was more similar between Mareau and Gémigny than Coinces (see Section 2.3). However, this montmorillonite underestimation could be explained by changes in solar illumination between sites (between 3:00 and 4:00 p.m. and between 36° and 38° solar illumination instead of 11:20 a.m. and 44° for Gémigny) may change the shape of an absorption feature (width and depth) [30], which can explain changes in montmorillonite estimation between our sites.

5. Conclusions

A comparative study was carried out to assess performances and limitations of several unmixing algorithms to estimate montmorillonite abundance in three ploughed fields at a very high spatial resolution (centimeter scale). To fulfill that goal, an experiment was performed using hyperspectral cameras on a bucket truck to record hyperspectral images in the VNIR–SWIR range (0.4–2.5 μm). Results showed that the two automatic EM detection methods (SISAL and MVC-NMF) were not efficient to identify pure minerals and that both used spectral libraries (USGS and Ducasse) could be used for the unmixing process. Performances of montmorillonite abundance estimations proved that linear unmixing models were not suitable (FCLS, MESMA and GBM with γ values equaling zero), but that REF-MLM and 1stSGD-MLM provided RMSEs within 2.2–13.3% and 1.4–19.7%. Compared to the target accuracy of 10% required for discriminating shrink–swell potential classes following [8,50], these performances hold great potential. Then, the major sources of error come from (i) the field roughness, (ii) local variations of mineralogical composition and soil moisture and (iii) illumination conditions. Despite that, the MLM model was able to mitigate them thanks to its P parameter, compensating their nonlinear effects.

Future work will then include further validation on a broader variation of soil composition. Quartz, which is featureless in the SWIR domain, has specific features in Long-Wave InfraRed (LWIR, 7.5–14 μm) [93–95]. We propose the combination of the SWIR and LWIR spectral domains to improve the estimation of both quartz and clays in soils. Automatic detection of EM is also an important step in order to automatize unmixing. Semi-supervised methods such as sparse regression [96,97] could be of interest in order to find the most significant EM inside a spectral library. Deep learning shows promising results to solve the endmember detection problem [78]. Also, future issues should address the moisture content in soils, which affects montmorillonite estimation. The use of moisture models to remove the moisture effect on soil spectra such as the one used by [98,99] or MARMIT [100,101] could improve mineral montmorillonite estimation in soils.

Author Contributions: Methodology and experiments, E.D., K.A. and X.B.; software, E.D. and V.A.; validation, K.A., X.B., G.G., A.H., V.A. and A.B.; writing—original draft preparation, E.D., K.A. and X.B; writing—review and editing, E.D., K.A. and X.B. All authors have read and agreed to the published version of the manuscript.

Funding: This work is funded by the French Government Defense procurement and technology agency (DGA) and the French geological survey (BRGM). It is also supported by the French Aerospace Lab (ONERA).

Data Availability Statement: The data presented in this study are available on request from the corresponding author. The data are not publicly available due to privacy.

Acknowledgments: We thank Philippe Déliot (ONERA) for performing the field acquisitions of hyperspectral images, and Stéphane Jacquemoud (IPGP) and Aurélien Bablet (IPGP) for their help on the field and photogrammetric acquisitions. We thank also Hubert Haas and Nicolas Maubec (BRGM) for their help in the sample analysis.

Conflicts of Interest: The authors declare no conflicts of interest.

References

1. Ministère de la Transition éCologique ET Solidaire, C. Général AU Développement Durable Retrait-Gonflement Des Sols Argileux: Plus de 4 Millions de Maisons Potentiellement Très Exposées. Available online: https://www.notre-environnement.gouv.fr/themes/risques/les-mouvements-de-terrain-et-les-erosions-cotieres-ressources/article/retrait-gonflement-des-sols-argileux-plus-de-4-millions-de-maisons (accessed on 17 August 2024).
2. Crilly, M.S.; Driscoll, R.M.C. The Behaviour of Lightly Loaded Piles in Swelling Ground and Implications for Their Design. *Proc. Inst. Civ. Eng.-Geotech. Eng.* **2000**, *143*, 3–16. [CrossRef]
3. Nelson, J.; Miller, D.J. *Expansive Soils: Problems and Practice in Foundation and Pavement Engineering*; John Wiley & Sons: Hoboken, NJ, USA, 1997; ISBN 978-0-471-18114-9.
4. Li, J.; Zou, J.; Bayetto, P.; Barker, N. Shrink-Swell Index Database for Melbourne. *Aust. Geomech. J.* **2016**, *51*, 17.
5. Kahle, M.; Kleber, M.; Jahn, R. Review of XRD-Based Quantitative Analyses of Clay Minerals in Soils: The Suitability of Mineral Intensity Factors. *Geoderma* **2002**, *109*, 191–205. [CrossRef]
6. Zhou, X.; Liu, D.; Bu, H.; Deng, L.; Liu, H.; Yuan, P.; Du, P.; Song, H. XRD-Based Quantitative Analysis of Clay Minerals Using Reference Intensity Ratios, Mineral Intensity Factors, Rietveld, and Full Pattern Summation Methods: A Critical Review. *Solid Earth Sci.* **2018**, *3*, 16–29. [CrossRef]
7. Dufréchou, G.; Grandjean, G.; Bourguignon, A. Geometrical Analysis of Laboratory Soil Spectra in the Short-Wave Infrared Domain: Clay Composition and Estimation of the Swelling Potential. *Geoderma* **2015**, *243–244*, 92–107. [CrossRef]
8. Chassagneux, D.; Stieltjes, L.; Mouroux, P. *Cartographie de l'aléa Retrait Gonflement Des Sols (Sécheresse/Pluie) Dans La Région de Manosque (Alpes de Haute-Provence). Echelle Communale et Départementale. Approche Méthodologique*; BRGM: Orléans, France, 1995.
9. Goetz, A.F.H. Three Decades of Hyperspectral Remote Sensing of the Earth: A Personal View. *Remote Sens. Environ.* **2009**, *113* (Suppl. S1), S5–S16. [CrossRef]
10. van der Meer, F.D.; van der Werff, H.M.A.; van Ruitenbeek, F.J.A.; Hecker, C.A.; Bakker, W.H.; Noomen, M.F.; van der Meijde, M.; Carranza, E.J.M.; Smeth, J.B.d.; Woldai, T. Multi- and Hyperspectral Geologic Remote Sensing: A Review. *Int. J. Appl. Earth Obs. Geoinf.* **2012**, *14*, 112–128. [CrossRef]
11. Yitagesu, F.A.; van der Meer, F.; van der Werff, H.; Hecker, C. Spectral Characteristics of Clay Minerals in the 2.5–14 Mm Wavelength Region. *Appl. Clay Sci.* **2011**, *53*, 581–591. [CrossRef]
12. Ducasse, E.; Adeline, K.; Briottet, X.; Hohmann, A.; Bourguignon, A.; Grandjean, G. Montmorillonite Estimation in Clay–Quartz–Calcite Samples from Laboratory SWIR Imaging Spectroscopy: A Comparative Study of Spectral Preprocessings and Unmixing Methods. *Remote Sens.* **2020**, *12*, 1723. [CrossRef]

13. Galán, E. Chapter 14 Genesis of Clay Minerals. In *Developments in Clay Science*; Elsevier: Amsterdam, The Netherlands, 2006; Volume 1, pp. 1129–1162. ISBN 978-0-08-044183-2.
14. van der Meer, F.D.; van der Werff, H.M.A.; van Ruitenbeek, F.J.A. Potential of ESA's Sentinel-2 for Geological Applications. *Remote Sens. Environ.* **2014**, *148*, 124–133. [CrossRef]
15. Bourguignon, A.; Delpont, G.; Chevrel, S.; Chabrillat, S. Detection and Mapping of Shrink–Swell Clays in SW France, Using ASTER Imagery. *Geol. Soc. Lond. Spec. Publ.* **2007**, *283*, 117–124. [CrossRef]
16. Amer, R.; Kusky, T.; El Mezayen, A. Remote Sensing Detection of Gold Related Alteration Zones in Um Rus Area, Central Eastern Desert of Egypt. *Adv. Space Res.* **2012**, *49*, 121–134. [CrossRef]
17. Ben-Dor, E.; Chabrillat, S.; Demattê, J.A.M.; Taylor, G.R.; Hill, J.; Whiting, M.L.; Sommer, S. Using Imaging Spectroscopy to Study Soil Properties. *Remote Sens. Environ.* **2009**, *113* (Suppl. S1), S38–S55. [CrossRef]
18. Grandjean, G.; Briottet, X.; Adeline, K.; Bourguignon, A.; Hohmann, A. Clay Minerals Mapping from Imaging Spectroscopy. In *Earth Observation and Geospatial Analyses [Working Title]*; IntechOpen: London, UK, 2019.
19. Bhattacharya, S.; Majumdar, T.J.; Rajawat, A.S.; Panigrahi, M.K.; Das, P.R. Utilization of Hyperion Data over Dongargarh, India, for Mapping Altered/Weathered and Clay Minerals along with Field Spectral Measurements. *Int. J. Remote Sens.* **2012**, *33*, 5438–5450. [CrossRef]
20. Bedini, E.; Chen, J. Application of PRISMA Satellite Hyperspectral Imagery to Mineral Alteration Mapping at Cuprite, Nevada, USA. *J. Hyperspectral Remote Sens.* **2020**, *10*, 87. [CrossRef]
21. Kruse, F.A.; Boardman, J.W.; Huntington, J.F. Comparison of Airborne Hyperspectral Data and Eo-1 Hyperion for Mineral Mapping. *IEEE Trans. Geosci. Remote Sens.* **2003**, *41*, 1388–1400. [CrossRef]
22. Chabrillat, S.; Goetz, A.F.H.; Krosley, L.; Olsen, H.W. Use of Hyperspectral Images in the Identification and Mapping of Expansive Clay Soils and the Role of Spatial Resolution. *Remote Sens. Environ.* **2002**, *82*, 431–445. [CrossRef]
23. Priya, S.; Ghosh, R. Soil Clay Minerals Abundance Mapping Using AVIRIS-NG Data. *Adv. Space Res.* **2024**, *73*, 1360–1367. [CrossRef]
24. Hohmann, A.; Bourguignon, A.; Grandjean, G. Cartographie Des Argiles Gonflantes En Milieux Tempérés à Partir de Données Hyperspectrales Aéroportées Couplées à Des Données in Situ et Laboratoire. In Proceedings of the 3Ème Colloque Scientifique du Groupe Hyperspectral de la Sfpt, Porquerolles, France, 14–16 May 2014.
25. Garfagnoli, F.; Ciampalini, A.; Moretti, S.; Chiarantini, L.; Vettori, S. Quantitative Mapping of Clay Minerals Using Airborne Imaging Spectroscopy: New Data on Mugello (Italy) from SIM-GA Prototypal Sensor. *Eur. J. Remote Sens.* **2013**, *46*, 1–17. [CrossRef]
26. Barton, I.F.; Gabriel, M.J.; Lyons-Baral, J.; Barton, M.D.; Duplessis, L.; Roberts, C. Extending Geometallurgy to the Mine Scale with Hyperspectral Imaging: A Pilot Study Using Drone- and Ground-Based Scanning. *Min. Metall. Explor.* **2021**, *38*, 799–818. [CrossRef]
27. Kurz, T.H.; Buckley, S.J. A Review of Hyperpsectral Imaging in Close Range Applications. *Int. Arch. Photogramm. Remote Sens. Spat. Inf. Sci.* **2016**, *XLI-B5*, 865–870. [CrossRef]
28. Kirsch, M.; Lorenz, S.; Zimmermann, R.; Tusa, L.; Möckel, R.; Hödl, P.; Booysen, R.; Khodadadzadeh, M.; Gloaguen, R. Integration of Terrestrial and Drone-Borne Hyperspectral and Photogrammetric Sensing Methods for Exploration Mapping and Mining Monitoring. *Remote Sens.* **2018**, *10*, 1366. [CrossRef]
29. Song, Q.; Gao, X.; Song, Y.; Li, Q.; Chen, Z.; Li, R.; Zhang, H.; Cai, S. Estimation and Mapping of Soil Texture Content Based on Unmanned Aerial Vehicle Hyperspectral Imaging. *Sci. Rep.* **2023**, *13*, 14097. [CrossRef] [PubMed]
30. Murphy, R.J.; Schneider, S.; Monteiro, S.T. Mapping Layers of Clay in a Vertical Geological Surface Using Hyperspectral Imagery: Variability in Parameters of SWIR Absorption Features under Different Conditions of Illumination. *Remote Sens.* **2014**, *6*, 9104–9129. [CrossRef]
31. Rodger, A.; Cudahy, T. Vegetation Corrected Continuum Depths at 2.20 Mm: An Approach for Hyperspectral Sensors. *Remote Sens. Environ.* **2009**, *113*, 2243–2257. [CrossRef]
32. Haest, M.; Cudahy, T.; Rodger, A.; Laukamp, C.; Martens, E.; Caccetta, M. Unmixing the Effects of Vegetation in Airborne Hyperspectral Mineral Maps over the Rocklea Dome Iron-Rich Palaeochannel System (Western Australia). *Remote Sens. Environ.* **2013**, *129*, 17–31. [CrossRef]
33. Ouerghemmi, W.; Gomez, C.; Naceur, S.; Lagacherie, P. Semi-Blind Source Separation for the Estimation of the Clay Content over Semi-Vegetated Areas Using VNIR/SWIR Hyperspectral Airborne Data. *Remote Sens. Environ.* **2016**, *181*, 251–263. [CrossRef]
34. Bioucas-Dias, J.M.; Plaza, A.; Dobigeon, N.; Parente, M.; Du, Q.; Gader, P.; Chanussot, J. Hyperspectral Unmixing Overview: Geometrical, Statistical, and Sparse Regression-Based Approaches. *IEEE J. Sel. Top. Appl. Earth Obs. Remote Sens.* **2012**, *5*, 354–379. [CrossRef]
35. Dobigeon, N.; Altmann, Y.; Brun, N.; Moussaoui, S. Linear and Nonlinear Unmixing in Hyperspectral Imaging. In *Resolving Spectral Mixtures-with Application from Ultrafast Spectroscopy to Super-Resolution Imaging*; Elsevier: Amsterdam, The Netherlands, 2016; Volume 45.
36. Meganem, I.; Deliot, P.; Briottet, X.; Deville, Y.; Hosseini, S. Linear–Quadratic Mixing Model for Reflectances in Urban Environments. *IEEE Trans. Geosci. Remote Sens.* **2014**, *52*, 544–558. [CrossRef]
37. Carmina, E. Impact Des Mélanges de Minéraux à Macro-éChelle Sur la Réflectance Spectrale de Surfaces Naturelles: ÉTude Empirique à Partir de Scénarios de Terrain. Ph.D. Thesis, University of Nantes, Nantes, France, 2012.

38. Wu, C.-Y.; Jacobson, A.R.; Laba, M.; Baveye, P.C. Accounting for Surface Roughness Effects in the Near-Infrared Reflectance Sensing of Soils. *Geoderma* **2009**, *152*, 171–180. [CrossRef]
39. Kariuki, P.C.; Woldai, T.; Meer, F.V.D. Effectiveness of Spectroscopy in Identification of Swelling Indicator Clay Minerals. *Int. J. Remote Sens.* **2004**, *25*, 455–469. [CrossRef]
40. Weidong, L.; Baret, F.; Xingfa, G.; Qingxi, T.; Lanfen, Z.; Bing, Z. Relating Soil Surface Moisture to Reflectance. *Remote Sens. Environ.* **2002**, *81*, 238–246. [CrossRef]
41. Lesaignoux, A.; Fabre, S.; Briottet, X.; Olioso, A.; Belin, E.; Cedex, T. Influence of Surface Soil Moisture on Spectral Reflectance of Bare Soil in the 0.4–15 μm Domain. In Proceedings of the 6th EARSeL SIG IS Workshop, Tel Aviv, Israel, 16–19 March 2009; p. 6.
42. Castaldi, F.; Palombo, A.; Pascucci, S.; Pignatti, S.; Santini, F.; Casa, R. Reducing the Influence of Soil Moisture on the Estimation of Clay from Hyperspectral Data: A Case Study Using Simulated PRISMA Data. *Remote Sens.* **2015**, *7*, 15561–15582. [CrossRef]
43. Hapke, B. Bidirectional Reflectance Spectroscopy: 1. Theory. *J. Geophys. Res.* **1981**, *86*, 3039–3054. [CrossRef]
44. Shkuratov, Y.; Starukhina, L.; Hoffmann, H.; Arnold, G. A Model of Spectral Albedo of Particulate Surfaces: Implications for Optical Properties of the Moon. *Icarus* **1999**, *137*, 235–246. [CrossRef]
45. Heylen, R.; Gader, P. Nonlinear Spectral Unmixing with a Linear Mixture of Intimate Mixtures Model. *IEEE Geosci. Remote Sens. Lett.* **2014**, *11*, 1195–1199. [CrossRef]
46. Robertson, K.M.; Milliken, R.E.; Li, S. Estimating Mineral Abundances of Clay and Gypsum Mixtures Using Radiative Transfer Models Applied to Visible-near Infrared Reflectance Spectra. *Icarus* **2016**, *277*, 171–186. [CrossRef]
47. Clark, R.N.; Swayze, G.A.; Livo, K.E.; Kokaly, R.F.; Sutley, S.J.; Dalton, J.B.; McDougal, R.R.; Gent, C.A. Imaging Spectroscopy: Earth and Planetary Remote Sensing with the USGS Tetracorder and Expert Systems. *J. Geophys. Res.* **2003**, *108*, 5131. [CrossRef]
48. Carvalho, A.O.J.; Guimarães, R.F. *Employment of the Multiple Endmember Spectral Mixture Analysis (MESMA) Method in Mineral Analysis*; JPL Publication: Pasadena, CA, USA, 2001; pp. 73–80.
49. Bouchut, J.; Giot, D. *Cartographie de L'aléa Retrait-Gonflement Des Argiles Dans Le Département du Loiret*; BRGM: Orléans, France, 2004.
50. Dufréchou, G.; Hohmann, A.; Bourguignon, A.; Grandjean, G. Targeting and Mapping Expansive Soils (Loiret, France): Geometrical Analysis of Laboratory Soil Spectra in the Short-Wave Infrared Domain (1100–2500 Nm). *Bull. De La Société Géologique De Fr.* **2016**, *187*, 169–181. [CrossRef]
51. Plyer, A.; Colin-Koeniguer, E.; Weissgerber, F. A New Coregistration Algorithm for Recent Applications on Urban SAR Images. *IEEE Geosci. Remote Sens. Lett.* **2015**, *12*, 2198–2202. [CrossRef]
52. Brigot, G.; Colin-Koeniguer, E.; Plyer, A.; Janez, F. Adaptation and Evaluation of an Optical Flow Method Applied to Coregistration of Forest Remote Sensing Images. *IEEE J. Sel. Top. Appl. Earth Obs. Remote Sens.* **2016**, *9*, 2923–2939. [CrossRef]
53. Smith, G.M.; Milton, E.J. The Use of the Empirical Line Method to Calibrate Remotely Sensed Data to Reflectance. *Int. J. Remote Sens.* **1999**, *20*, 2653–2662. [CrossRef]
54. Karpouzli, E.; Malthus, T. The Empirical Line Method for the Atmospheric Correction of IKONOS Imagery. *Int. J. Remote Sens.* **2003**, *24*, 1143–1150. [CrossRef]
55. Green, A.A.; Berman, M.; Switzer, P.; Maurice; Craig, D. A Transformation for Ordering Multispectral Data in Terms of Image Quality with Implications for Noise Removal. *IEEE Trans. Geosci. Remote Sens.* **1988**, *26*, 65–74. [CrossRef]
56. Rogge, D.M.; Rivard, B.; Zhang, J.; Sanchez, A.; Harris, J.; Feng, J. Integration of Spatial–Spectral Information for the Improved Extraction of Endmembers. *Remote Sens. Environ.* **2007**, *110*, 287–303. [CrossRef]
57. Bedini, E.; Meer, F.v.d.; Ruitenbeek, F.v. Use of HyMap Imaging Spectrometer Data to Map Mineralogy in the Rodalquilar Caldera, Southeast Spain. *Int. J. Remote Sens.* **2009**, *30*, 327–348. [CrossRef]
58. Nagao, M.; Matsuyama, T.; Ikeda, Y. Region Extraction and Shape Analysis in Aerial Photographs. *Comput. Graph. Image Process.* **1979**, *10*, 195–223. [CrossRef]
59. Adeline, K.R.M.; Chen, M.; Briottet, X.; Pang, S.K.; Paparoditis, N. Shadow Detection in Very High Spatial Resolution Aerial Images: A Comparative Study. *ISPRS J. Photogramm. Remote Sens.* **2013**, *80*, 21–38. [CrossRef]
60. Rouse, J., Jr.; Haas, R.H.; Schell, J.A.; Deering, D.W. Monitoring Vegetation Systems in the Great Plains with ERTS. *NASA Spec. Publ.* **1974**, *351*, 309.
61. Guerschman, J.P.; Hill, M.J.; Renzullo, L.J.; Barrett, D.J.; Marks, A.S.; Botha, E.J. Estimating Fractional Cover of Photosynthetic Vegetation, Non-Photosynthetic Vegetation and Bare Soil in the Australian Tropical Savanna Region Upscaling the EO-1 Hyperion and MODIS Sensors. *Remote Sens. Environ.* **2009**, *113*, 928–945. [CrossRef]
62. Rupnik, E.; Daakir, M.; Pierrot Deseilligny, M. MicMac—A Free, Open-Source Solution for Photogrammetry. *Open Geospat. Data Softw. Stand.* **2017**, *2*, 14. [CrossRef]
63. Brown, G. *Crystal Structures of Clay Minerals and Their X-ray Identification*; The Mineralogical Society of Great Britain and Ireland: Hampton, UK, 1982; ISBN 978-0-903056-08-3.
64. Bish, D.L.; Post, J.E. Quantitative Mineralogical Analysis Using the Rietveld Full-Pattern Fitting Method. *Am. Mineral.* **1993**, *78*, 932–940.
65. Kokaly, R.F.; Clark, R.N.; Swayze, G.A.; Livo, K.E.; Hoefen, T.M.; Pearson, N.C.; Wise, R.A.; Benzel, W.M.; Lowers, H.A.; Driscoll, R.L.; et al. *USGS Spectral Library Version 7*; Data Series; U.S. Geological Survey: Reston, VA, USA, 2017; p. 68.
66. Miao, L.; Qi, H. Endmember Extraction From Highly Mixed Data Using Minimum Volume Constrained Nonnegative Matrix Factorization. *IEEE Trans. Geosci. Remote Sens.* **2007**, *45*, 765–777. [CrossRef]

67. Bioucas-Dias, J. A Variable Splitting Augmented Lagrangian Approach to Linear Spectral Unmixing. In Proceedings of the 2009 First Workshop on Hyperspectral Image and Signal Processing: Evolution in Remote Sensing, Grenoble, France, 26–28 August 2009; pp. 1–4. [CrossRef]
68. Nascimento, J.M.P.; Dias, J.M.B. Vertex Component Analysis: A Fast Algorithm to Unmix Hyperspectral Data. *IEEE Trans. Geosci. Remote Sens.* **2005**, *43*, 898–910. [CrossRef]
69. Heinz, D.C.; Chang, C.-I. Fully Constrained Least Squares Linear Spectral Mixture Analysis Method for Material Quantification in Hyperspectral Imagery. *IEEE Trans. Geosci. Remote Sens.* **2001**, *39*, 529–545. [CrossRef]
70. Roberts, D.A.; Gardner, M.; Church, R.; Ustin, S.; Scheer, G.; Green, R.O. Mapping Chaparral in the Santa Monica Mountains Using Multiple Endmember Spectral Mixture Models. *Remote Sens. Environ.* **1998**, *65*, 267–279. [CrossRef]
71. Heylen, R.; Parente, M.; Gader, P. A Review of Nonlinear Hyperspectral Unmixing Methods. *IEEE J. Sel. Top. Appl. Earth Obs. Remote Sens.* **2014**, *7*, 1844–1868. [CrossRef]
72. Heylen, R.; Scheunders, P. A Multilinear Mixing Model for Nonlinear Spectral Unmixing. *IEEE Trans. Geosci. Remote Sens.* **2016**, *54*, 240–251. [CrossRef]
73. Kruse, F.A.; Lefkoff, A.B.; Boardman, J.W.; Heidebrecht, K.B.; Shapiro, A.T.; Barloon, P.J.; Goetz, A.F.H. The Spectral Image Processing System (SIPS)—Interactive Visualization and Analysis of Imaging Spectrometer Data. *Remote Sens. Environ.* **1993**, *44*, 145–163. [CrossRef]
74. Revel, C.; Deville, Y.; Achard, V.; Briottet, X. Inertia-Constrained Pixel-by-Pixel Nonnegative Matrix Factorisation: A Hyperspectral Unmixing Method Dealing with Intra-Class Variability. *arXiv* **2017**, arXiv:1702.07630.
75. Rommel, D.; Grumpe, A.; Felder, M.P.; Wöhler, C.; Mall, U.; Kronz, A. Automatic Endmember Selection and Nonlinear Spectral Unmixing of Lunar Analog Minerals. *Icarus* **2017**, *284*, 126–149. [CrossRef]
76. Crósta, A.P.; De Souza Filho, C.R.; Azevedo, F.; Brodie, C. Targeting Key Alteration Minerals in Epithermal Deposits in Patagonia, Argentina, Using ASTER Imagery and Principal Component Analysis. *Int. J. Remote Sens.* **2003**, *24*, 4233–4240. [CrossRef]
77. Ben-Dor, E. Characterization of Soil Properties Using Reflectance Spectroscopy. In *Hyperspectral Remote Sensing of Vegetation*; CRC Press: Boca Raton, FL, USA, 2011; pp. 513–558. ISBN 978-1-4398-4537-0.
78. Koirala, B.; Rasti, B.; Bnoulkacem, Z.; Ribeiro, A.d.L.; Madriz, Y.; Herrmann, E.; Gestels, A.; De Kerf, T.; Lorenz, S.; Fuchs, M.; et al. A Multisensor Hyperspectral Benchmark Dataset For Unmixing of Intimate Mixtures. *IEEE Sens. J.* **2023**, *24*, 4694–4710. [CrossRef]
79. Fasnacht, L.; Vogt, M.-L.; Renard, P.; Brunner, P. A 2D Hyperspectral Library of Mineral Reflectance, from 900 to 2500 Nm. *Sci. Data* **2019**, *6*, 268. [CrossRef]
80. Milliken, R.E.; Hiroi, T.; Scholes, D.; Slavney, S.; Arvidson, R. The NASA Reflectance Experiment LABoratory (RELAB) Facility: An Online Spectral Database for Planetary Exploration. In Proceedings of the Astromaterials Data Management in the Era of Sample-Return Missions Community Workshop, Virtual, 8–10 November 2021; Volume 2654.
81. Safanelli, J.L.; Hengl, T.; Parente, L.; Minarik, R.; Bloom, D.E.; Todd-Brown, K.; Gholizadeh, A.; Mendes, W.D.S.; Sanderman, J. Open Soil Spectral Library (OSSL): Building Reproducible Soil Calibration Models through Open Development and Community Engagement. *BioRxiv* **2023**, preprint. [CrossRef]
82. Rinnan, Å.; Berg, F.v.d.; Engelsen, S.B. Review of the Most Common Pre-Processing Techniques for near-Infrared Spectra. *TrAC Trends Anal. Chem.* **2009**, *28*, 1201–1222. [CrossRef]
83. Esquerre, C.; Gowen, A.A.; Burger, J.; Downey, G.; O'Donnell, C.P. Suppressing Sample Morphology Effects in near Infrared Spectral Imaging Using Chemometric Data Pre-Treatments. *Chemom. Intell. Lab. Syst.* **2012**, *117*, 129–137. [CrossRef]
84. Viscarra Rossel, R.A.; McGlynn, R.N.; McBratney, A.B. Determining the Composition of Mineral-Organic Mixes Using UV–Vis–NIR Diffuse Reflectance Spectroscopy. *Geoderma* **2006**, *137*, 70–82. [CrossRef]
85. Li, M.; Zhu, F.; Guo, A.J.; Chen, J. A Graph Regularized Multilinear Mixing Model for Nonlinear Hyperspectral Unmixing. *Remote Sens.* **2019**, *11*, 2188. [CrossRef]
86. Fang, T.; Zhu, F.; Chen, J. Nonlinear Hyperspectral Unmixing Based on Multilinear Mixing Model Using Convolutional Autoencoders. *arXiv* **2023**, arXiv:2303.08156.
87. Li, M.; Yang, B.; Wang, B. A Coarse-to-Fine Scheme for Unsupervised Nonlinear Hyperspectral Unmixing Based on an Extended Multilinear Mixing Model. *IEEE Trans. Geosci. Remote Sens.* **2023**, *61*, 5521415. [CrossRef]
88. Zhang, G.; Scheunders, P.; Cerra, D.; Muller, R. Shadow-Aware Nonlinear Spectral Unmixing for Hyperspectral Imagery. *IEEE J. Sel. Top. Appl. Earth Obs. Remote Sens.* **2022**, *15*, 5514–5533. [CrossRef]
89. Jin, M.; Ding, X.; Han, H.; Pang, J.; Wang, Y. An Improved Method Combining Fisher Transformation and Multiple Endmember Spectral Mixture Analysis for Lunar Mineral Abundance Quantification Using Spectral Data. *Icarus* **2022**, *380*, 115008. [CrossRef]
90. Mulder, V.L.; Plötze, M.; de Bruin, S.; Schaepman, M.E.; Mavris, C.; Kokaly, R.F.; Egli, M. Quantifying Mineral Abundances of Complex Mixtures by Coupling Spectral Deconvolution of SWIR Spectra (2.1–2.4 Mm) and Regression Tree Analysis. *Geoderma* **2013**, *207–208*, 279–290. [CrossRef]
91. Mathieu, M.; Roy, R.; Launeau, P.; Cathelineau, M.; Quirt, D. Alteration Mapping on Drill Cores Using a HySpex SWIR-320m Hyperspectral Camera: Application to the Exploration of an Unconformity-Related Uranium Deposit (Saskatchewan, Canada). *J. Geochem. Explor.* **2017**, *172*, 71–88. [CrossRef]

92. Jackisch, R.; Madriz, Y.; Zimmermann, R.; Pirttijärvi, M.; Saartenoja, A.; Heincke, B.H.; Salmirinne, H.; Kujasalo, J.-P.; Andreani, L.; Gloaguen, R. Drone-Borne Hyperspectral and Magnetic Data Integration: Otanmäki Fe-Ti-V Deposit in Finland. *Remote Sens.* **2019**, *11*, 2084. [CrossRef]
93. Feng, J.; Rogge, D.; Rivard, B. Comparison of Lithological Mapping Results from Airborne Hyperspectral VNIR-SWIR, LWIR and Combined Data. *Int. J. Appl. Earth Obs. Geoinf.* **2017**, *64*, 340–353. [CrossRef]
94. Notesco, G.; Ogen, Y.; Ben-Dor, E. Mineral Classification of Makhtesh Ramon in Israel Using Hyperspectral Longwave Infrared (LWIR) Remote-Sensing Data. *Remote Sens.* **2015**, *7*, 12282–12296. [CrossRef]
95. Sawut, M.; Ghulam, A.; Tiyip, T.; Zhang, Y.; Ding, J.; Zhang, F.; Maimaitiyiming, M. Estimating Soil Sand Content Using Thermal Infrared Spectra in Arid Lands. *Int. J. Appl. Earth Obs. Geoinf.* **2014**, *33*, 203–210. [CrossRef]
96. Bioucas-Dias, J.M.; Figueiredo, M.A.T. Alternating Direction Algorithms for Constrained Sparse Regression: Application to Hyperspectral Unmixing. *arXiv* **2010**, arXiv:1002.4527.
97. Iordache, M.-D.; Plaza, A.; Bioucas-Dias, J. *On the Use of Spectral Libraries to Perform Sparse Unmixing of Hyperspectral Data*; IEEE: New York, NY, USA, 2010; pp. 1–4.
98. Whiting, M.L.; Li, L.; Ustin, S.L. Predicting Water Content Using Gaussian Model on Soil Spectra. *Remote Sens. Environ.* **2004**, *89*, 535–552. [CrossRef]
99. Ou, D.; Tan, K.; Li, J.; Wu, Z.; Zhao, L.; Ding, J.; Wang, X.; Zou, B. Prediction of Soil Organic Matter by Kubelka-Munk Based Airborne Hyperspectral Moisture Removal Model. *Int. J. Appl. Earth Obs. Geoinf.* **2023**, *124*, 103493. [CrossRef]
100. Bablet, A.; Vu, P.V.H.; Jacquemoud, S.; Viallefont-Robinet, F.; Fabre, S.; Briottet, X.; Sadeghi, M.; Whiting, M.L.; Baret, F.; Tian, J. MARMIT: A Multilayer Radiative Transfer Model of Soil Reflectance to Estimate Surface Soil Moisture Content in the Solar Domain (400–2500 Nm). *Remote Sens. Environ.* **2018**, *217*, 1–17. [CrossRef]
101. Dupiau, A.; Jacquemoud, S.; Briottet, X.; Fabre, S.; Viallefont-Robinet, F.; Philpot, W.; Di Biagio, C.; Hébert, M.; Formenti, P. MARMIT-2: An Improved Version of the MARMIT Model to Predict Soil Reflectance as a Function of Surface Water Content in the Solar Domain. *Remote Sens. Environ.* **2022**, *272*, 112951. [CrossRef]

remote sensing MDPI

Article

Evidence of Dextral Strike-Slip Movement of the Alakol Lake Fault in the Western Junggar Based on Remote Sensing

Wenxing Yi [1], An Li [1,*], Liangxin Xu [2], Zongkai Hu [1] and Xiaolong Li [1]

1 State Key Laboratory of Earthquake Dynamics, Institute of Geology, China Earthquake Administration, Beijing 100029, China; yiwenxing@ies.ac.cn (W.Y.)
2 Academy of Seismic Engineering Surveys, Earthquake Administration of Shaanxi Province, Xi'an 710068, China
* Correspondence: lian@ies.ac.cn

Abstract: The NW-SE-trending dextral strike-slip faults on the north side of the Tian Shan, e.g., the Karatau fault, Talas–Fergana fault, Dzhalair–Naiman fault, Aktas fault, Dzhungarian fault, and Chingiz fault, play an important role in accommodating crustal shortening. The classic viewpoint is that these strike-slip faults are an adjustment product caused by the difference in the crustal shortening from west to east. Another viewpoint attributes the dextral strike-slip fault to large-scale sinistral shearing. The Alakol Lake fault is a typical dextral strike-slip fault in the north Tian Shan that has not been reported. It is situated along the northern margin of the Dzhungarian gate, stretching for roughly 150 km from Lake Ebinur to Lake Alakol. Our team utilized aerial photographs, satellite stereoimagery, and field observations to map the spatial distribution of the Alakol Lake fault. Our findings provided evidence supporting the assertion that the fault is a dextral strike-slip fault. In reference to its spatial distribution, the Lake Alakol is situated in a pull-apart basin that lies between two major dextral strike-slip fault faults: the Chingiz and Dzhungarian faults. The Alakol Lake fault serves as a connecting structure for these two faults, resulting in the formation of a mega NW-SE dextral strike-slip fault zone. According to our analysis of the dating samples taken from the alluvial fan, as well as our measurement of the displacement of the riser and gully, it appears that the Alakol Lake fault has a dextral strike-slip rate of 0.8–1.2 mm/a (closer to 1.2 mm/a). The strike-slip rate of the Alakol Lake fault is comparatively higher than that of the Chingiz fault in the northern region (~0.7 mm/a) but slower than that of the Dzhungarian fault in the southern region (3.2–5 mm/a). The Chingiz–Alakol–Dzhungarian fault zone shows a gradual decrease in deformation towards the interior of the Kazakhstan platform.

Keywords: Alakol Lake fault; dextral strike-slip; slip rate; UAV photogrammetry; satellite stereoimagery

Citation: Yi, W.; Li, A.; Xu, L.; Hu, Z.; Li, X. Evidence of Dextral Strike-Slip Movement of the Alakol Lake Fault in the Western Junggar Based on Remote Sensing. *Remote Sens.* **2024**, *16*, 2615. https://doi.org/10.3390/rs16142615

Academic Editor: Gianluca Groppelli

Received: 6 May 2024
Revised: 19 June 2024
Accepted: 26 June 2024
Published: 17 July 2024

1. Introduction

Central Asia has experienced extensive Cenozoic deformation and developed a large amount of active orogenic belts due to the impact of the India–Eurasia collision [1,2]. One of the most representative and significant orogenic belts is the Tian Shan, which underwent reactivation in the Cenozoic [3–6]. The Tian Shan is an active intracontinental orogenic belt that extends east–west for 2500 km and south–north for 250–350 km. In the Quaternary, the Tian Shan underwent strong tectonic deformation, such as compression and uplift, and rapid north–south crustal shortening [3,7–11]. Faults in the Tian Shan, including E-W-trending thrust faults [3,12–15], ENE-WSW-trending sinistral strike-slip faults [2,16,17], and NW-SE-trending dextral strike-slip faults [18–20], accommodated the crustal shortening. The GPS data indicate that the total shortening rate is approximately 20 mm/a across the western Tian Shan to the Kazakh platform, which is nearly half of the total convergence rate between India and Eurasia [21–24]. The NW-SE dextral strike-slip faults on the north side of the Tian Shan, e.g., the Karatau fault, Talas–Ferganas fault, Dzhalair–Naiman fault, Aktas fault, Dzhungarian fault (also named the Bolokenu–Aqikekuduk fault), and the Chingiz fault, play a crucial role in accommodating crustal shortening [2,19,25]. Some geodynamic mechanisms of the Tian Shan region suggest that the strike-slip faults are an adjustment product caused by the difference in east–west crustal shortening [2,3,26]. At the same time, based on GPS velocities, some researchers attribute the dextral strike-slip fault to large-scale sinistral shearing [27,28]. It is worth noting that other researchers have recognized the strike-slip characteristics of the fault related to block movement and rotation [17,21]. Therefore, it is significant to obtain and understand the activity characteristics of strike-slip faults for tectonic deformation in the Tian Shan.

The Dzhungarian fault (DZF) is a large NW-SE-trending dextral strike-slip fault, which extends for more than 700 km and constitutes the boundary fault between the north Tian Shan and the middle Tian Shan (Figure 1a) [19,20,25,29]. Previous studies revealed that the Dzhungarian fault had a slip rate of 2.2–3.2 mm/a in its middle segment [19,20] and a rate of ~1.7 mm/a at the end of the northwest segment in the late Quaternary [30]. The Chingiz fault (CF) is an NW-SE-trending dextral strike-slip fault that runs for over 600 km in the east–central Kazakhstan platform [31,32]. Distinct linear traces along the Chingiz fault can be observed from Landsat satellite images [32]. In the Cenozoic, the Chingiz fault has typical dextral strike-slip characteristics with a small thrust component [31]. The GPS data indicate ~0.7 mm/a as the dextral strike-slip rate in the Chingiz fault [33]. The Dzhungarian fault and the Chingiz fault are more than 100 km apart and arranged in right steps en echelon, with Lake Alakol in the stepover. A nearly 150 km long fault along the northeast of Lake Alakol to Dzhungarian gate is called the Alakol Lake fault (ALF).

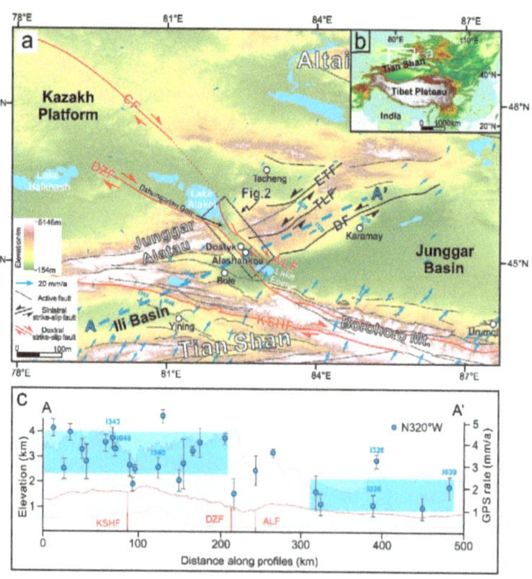

Figure 1. (**a**) The digital elevation model shows the distribution of the main Quaternary faults in the northern Tian Shan region (modified after Xu et al., 2016) [34]. Blue arrows show the GPS measurements from Wang and Shen (2020) [23]. The blue dashed lines (A–A′) show the locations of the GPS profiles. The white circles show the major cities. The black dashed boxes show the locations of Figure 2. DZF—Dzhungarian fault; ALF—Alakol Lake fault; CF—Chingiz fault; KSHF—Kashihe fault; ETF—East Tacheng fault; TLF—TuoLi fault; and DF—Daerbute fault. (**b**) The global digital elevation model shows the tectonic location of the research area (**a**). (**c**) Swath GPS profile A–A′ shows the velocity components parallel to (blue dots) the profile striking N320°W [20]. The brown line and gray shadow show the mean value and range of elevation with 50 km width along the profile A–A′. The blue-shaded rectangles are the visually fitted range of the GPS velocities. The blue letters and numbers represent the GPS observation stations' abbreviations.

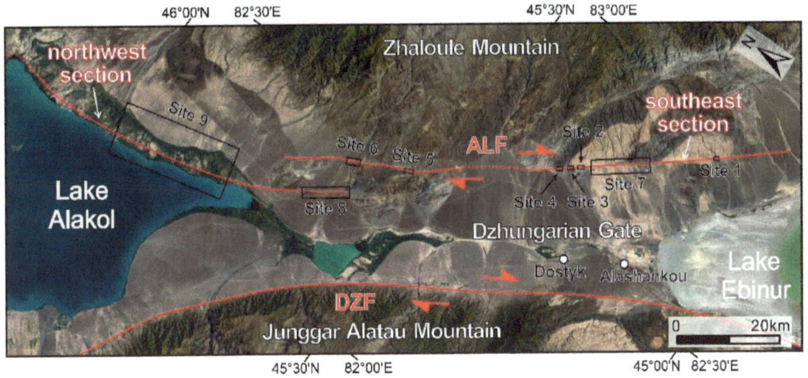

Figure 2. The extension of the Alakol Lake fault is shown on Google Earth. The red lines show the location of the fault trace. The red triangular arrows indicate dextral strike-slip movement. The black boxes are the study sites. The white circles show the major cities.

The GPS velocities with respect to Eurasia decrease from ~4 mm/a in the Ili Basin to less than 2 mm/a in the northwest margin of the Junggar Basin, indicating that the DZF and ALF accommodate around 2 mm/a of crustal shortening at N320°W (Figure 1c) [20]; however, it is still unclear what the geometric characteristics of the ALF are and how it absorbs crustal shortening. Furthermore, from the spatial distribution of the ALF, it connects the CF and the DZF (Figure 1a). All in all, it is significant to study the activity characteristics of these faults in the Quaternary to understand the tectonic deformation in the Tian Shan. Exploring the geometric structure and kinematic characteristics of the ALF is essential for understanding its structural relationship and block deformation with the two large strike-slip faults (DZF and CF). Otherwise, the Dzhungarian gate is a geographically and historically significant mountain pass between China and Central Asia. Furthermore, the Alakol Lake fault is located near the important trade ports of Alashankou in China and Dostyk in Kazakhstan. Therefore, investigating the kinematic properties of the fault can provide basic data for assessing seismic hazards in these cities and regions in the future.

In this study, we identified the dextral strike-slip characteristics of the ALF using high precision aerial photographs, satellite stereoimagery, and field observations. We then described the detailed fault traces and tectonic features, such as alluvial fan surfaces, gullies, offset stream channels, and fault scarps, based on the obtained data. Finally, we discussed the role of the ALF in the tectonic pattern formed in the north of the Tian Shan.

2. Geological Setting

The western Junggar is a significant part of the Central Asian orogenic belts, which lies between the Kazakh platform to the west, the Junggar Basin to the east, the Borohoro Mountain range to the south, and the Altai Shan to the north (Figure 1a). The western Junggar has undergone a multiple-stage orogenic evolution since the Early Paleozoic, including the multiple-phase subduction of the Paleo-Asian oceanic plate, crustal accretion, and continental growth [35–37]. This region primarily consists of a series of accretionary complex belts and Paleozoic magmatic arcs [38,39]. The main exposed strata in the western Junggar are volcanic sedimentary strata from the Ordovician to the Carboniferous, and the most widely distributed are Carboniferous thick pyroclastic sedimentary formations. The magmatic activity in the western Junggar area was particularly intense in the late Paleozoic, and granites with positive εNd(t) and low initial $^{87}Sr/^{86}Sr$ characteristics were widely distributed in the region [40–43]. In the western Junggar, large sinistral and dextral strike-slip faults are mainly developed. The sinistral strike-slip faults include the ENE-WSW-trending Daerbute fault (DF), TuoLi fault (TLF), and East Tacheng fault (ETF) (Figure 1a).

The dextral strike-slip faults include the NW-SE-trending Chingiz fault (CF), Dzhungarian fault (DZF), and Kashihe fault (KSHF) (Figure 1a).

The Junggar Alatau is located in the north of the western Tian Shan, and the overall trend is nearly east–west (Figure 1a). The Alatau Mountain is at the border between China and Kazakhstan, an important part of the Central Asian orogenic belts. The Junggar Alatau developed in the Upper Paleozoic and the Cenozoic and was absent in the Mesozoic [44,45]. The bedrock of the Junggar Alatau is well exposed, and the strata are mainly late Paleozoic clastic rocks and carbonate rocks. There were also frequent magmatic activities in the late Paleozoic, a large distribution area of igneous rocks, and diverse rock types in the Junggar Alatau [46].

The ALF lies in the western Junggar, which extends about 150 km from the northeast of Lake Ebinur (45°1′N, 82°59′E) to the eastern end of Lake Alakol (46°8′N, 82°1′E) (Figure 1a). The ALF can be divided into two sections by a compression ridge based on the distribution of the fault trace. The northwest section is located to the east of Lake Alakol, extending for ~70 km. The southeast section is distributed to the west of the Zhaloule Mountain, extending for ~80 km (Figure 2). The strike of the ALF is from 315° to 330°, and the overall strike of the fault is 320°. According to the 1:500,000 geological map compiled by China, previous research dates, and field observations, the bedrocks around the Alakol Lake fault are Paleozoic rocks, mainly including Ordovician marine sedimentary rocks and Carboniferous volcanic rocks [47]. Combined with the field investigation, the geomorphic surface of the area of the ALF is mainly Quaternary alluvial deposit and Gobi material.

3. Methods

3.1. Drone Digital Elevation Models and Other Image Data

In the research on the Alakol Lake fault, we require a set of quantitative parameters in order to elucidate its kinematic characteristics, activity intensity, and activity patterns. These parameters include the horizontal displacement of the alluvial fan surfaces, stream channels, gullies, and ridges; the vertical displacement of the fault scarps; and the age of these geomorphic surfaces [48–50]. High-precision and high-resolution topographic and geomorphic data that can reveal subtle geomorphic features are the basis of the quantitative study of active structures [49,51]. We achieved these conditions using the core concept of structure from motion (SfM), a photo-based technique for three-dimensional (3D) reconstruction to create a high-resolution digital elevation model (DEM) [52–54]. SfM is a technique with which to reconstruct a 3D scene from multiple flat images, and it can recover the relative position and attitude of the camera from a disordered image with a certain overlap degree via image feature extraction and matching [55]. Using the structure from motion (SfM) technique, we employ an unmanned aerial vehicle (UAV) (DJI Phantom 4 RTK) to create orthoimages and high-resolution DEMs with a pixel resolution of less than 0.1 m. In these areas where we use the UAV, the UAV flies at a height of 120–220 m, with 70% forward and side overlaps at each site. Through this process, we obtained hundreds of images with a resolution of 5472 pixels in width and 3648 pixels in height for each site. These images of each site were generated using high-resolution DEM software (Pix4Dmapper 4.4.10) integrated with SfM algorithms. DEMs with a resolution of less than 10 cm can clearly display various topographic indexes, such as hillshades, contour lines, and slopes.

When observing large-scale landforms, using a UAV photogrammetry technique based on SfM may not be sufficient to achieve the desired effect; therefore, utilizing Pleiades satellite images to generate high-resolution DEMs is another effective method. We use two stereo pairs of Pleiades images to construct a 0.5 m resolution DEM, and the detailed methods mainly followed those of Zhou et al. (2015) and Wang et al. (2019) [56,57]. The 0.5 m resolution of the DEM is high enough to identify hundreds of meters of horizontal offsets of alluvial fans and displacements of stream channels.

However, due to certain conditions, it is still difficult to obtain high-resolution DEMs in some regions, especially the ~70 km long fault of the northwest section (in Kazakhstan).

Therefore, using Google Earth to extract images to identify the kinematic evidence in the northwest section is an extremely efficient and convenient method to reveal the fault trace and displaced stream channels.

3.2. Offset Measurements of Landforms

The displacement of a fault can be measured by the distance between two correlation marker units projected onto the fault trace [49,58,59]. Additionally, confidence ratings express the reliability of the original morphology of the offset markers, and an offset range is assigned to every measurement to describe the physically plausible range of an offset measurement [49,60,61].

Alluvial fans are typically formed when flowing water interacts with mountains. The mountain on the north side of the ALF brought many material sources and formed multiple alluvial fans in front of the hill. The size and shape of these fans were identified using the orthoimage and DEM. Then, we checked each horizontal offset feature and traced the channel or the riser of the alluvial fans on both sides of the fault as a straight line. The next step was to measure the distances between the projections of upstream and downstream fitted straight lines onto the fault plane. Due to lateral erosion of the channel near the fault, it is difficult to trace an accurate offset landform [58]. We measured the maximum and minimum offset possibilities and expressed the form with the average value and error type to identify the offset range. As a result, the offset range does not represent a confidence interval.

The width of a gully and ridge is usually less than the width of a stream channel, so we can directly make linear fittings of the gully (bottom of the gully) and ridge (top of the ridge) on both sides of the fault trace. The projections of the fitted line onto the fault are measured as the horizontal offset. Their offset range is measured similarly to that of the stream channel.

Different fault scarps of varying heights were formed along the fault on various geomorphic surfaces due to the vertical displacement of the fault. To obtain the vertical displacement, well-preserved scarps were mapped and identified based on the high-resolution UAV DEM. Then, topographic profiles perpendicular to the fault scarp were extracted. The topographic profiles of the hanging wall and footwall were linearly regressed by two separate straight lines. The vertical distances of the two fitted lines projected onto the fault trace were measured to represent the vertical displacement.

In this paper, a method called "Back-slipping" was used to measure the lateral offsets of the gullies. After the fault trace and two separated geomorphic marker sections were identified, we back-slipped them along the fault trace to reconstruct the original morphology of the geomorphic markers and determine the offset [61–63].

3.3. Chronology Dating

Optically stimulated luminescence (OSL) dating is one of the methods of luminescence dating, which can determine the age of sedimentary strata by measuring the equivalent dose (De, which is the ionizing radiation dose corresponding to the natural OSL signal intensity of quartz and other minerals in the sediment) and the environmental dose rate received by the sample per year (D) [64–66]. The age of sedimentary strata is calculated as follows:

$$\text{Age} = \text{De}/\text{D}$$

The unit of age is ka. The unit of De is Gy. The unit of D is Gy/ka.

In this research, three OSL samples (AL-01, AL-04, and AL-05) were collected from the capping loess in the alluvial fans, and two OSL samples (AL-02 and AL-03) were collected from the trench wall in the alluvial fans. All the samples were carefully collected and sealed in opaque iron tubes. These samples were processed and measured at the National Institute of Natural Hazards, the Ministry of Emergency Management of China. The sample pretreatment followed the experimental procedure of Aitken (1998) and Lu et al. (2007), and then relatively pure quartz particles were obtained [67,68]. The equivalent dose (De) was measured using the methods of Murray et al. (2000, 2003) [65,69]. The effects of radionuclides and cosmic rays were measured to determine the dose rate received per year (D) [67,70].

4. Results

4.1. Distribution of the Study Site

Seven study sites were selected along the southeast section of the Alakol Lake fault in China, and two interpretation sites were selected near Lake Alakol in Kazakhstan (Figure 2). The high-resolution DEMs from aerial photography and some field observations were displayed at these seven sites to describe the southeast section of the ALF. At sites 7 and 8, we also utilized Pleiades satellite images to generate high-resolution DEMs over a large spatial range. In the northwest section, we observed geomorphological characters and displacements using the DEMs at site 8, while at site 9, we identified the fault trace and horizontal offsets using the Google Earth image and surveyed the vertical displacement in the DEMs.

4.2. Geomorphic Development and Dating

By identifying the height above the channel of the alluvial fan and assessing the subgully erosion on the alluvial fan surface, we divide these alluvial fans into four stages, which are named T1, T2, T3, and T4, from most recent to oldest. T1 is the lowest alluvial fan, which is 1–6 m high above the channel. T2 and T3 are 18–20 m and 24–50 m above the stream bed, respectively. T4 is the highest alluvial fan in the study area, preserved well and more than 50 m above the stream bed. The alluvial fan which is older than T4 has been eroded into pieces.

Five OSL samples were collected in order to date the geomorphic surface ages in this research. The sample AL-01 was collected from the coarse-grain sand layer at a ~0.8 m depth below the surface of the T2 alluvial fan at site 4 (Figure 3a). The OSL age of the sample AL-01 indicates a sedimentary age of 24.57 ± 1.72 ka. The sample AL-02 was collected from the coarse-grain sand layer at a depth of ~1 m below the surface of T2 at site 5, which dated the age of T2 at 22.23 ± 1.06 ka (Figure 3b). The sample AL-03 was also collected from the coarse-grain sand layer at a depth of ~5 m below the T3 surface at the riser, which estimated a minimum dating result of 76.01 ± 5.64 ka for T3 (Figure 3c). Because the burial depth of the sampling location is too deep below the surface, even the minimum dating result may overestimate the formation age of the T3 alluvial fan. Along the fault trace, sample AL-04 was collected from the T1 alluvial fan, ~1.5 km northwest of site 4. The sample AL-04 was collected from the fine-grain sand layer at ~1.0 m below the surface of the T1 alluvial fan, which OSL-dated it with an age of 2.68 ± 0.16 ka (Figure 3d). A fault scarp displaced this T1 alluvial fan, indicating the fault activity in the Holocene (Figure 3e). Sample AL-05 was collected from the T2 alluvial fan, ~1 km southeast of site 2. The sample AL-05 was collected from the coarse-grain sand layer at a ~1.8 m depth below the surface of the T2 alluvial fan, which is ~18 m high above the stream bed (Figure 3f). The dating of the sample AL-05 is 16.34 ± 1.06 ka.

Previous studies on the DZF, which is located on the southeast side of the Dzhungarian gate, have dated the alluvial fan. In cases where alluvial fans share a base level, the age of the geomorphic surfaces along the ALF can be referenced against the age of alluvial fans with a similar height above the channel in the area of the DZF. Campbell et al. (2013) determined the age of the alluvial fan by using optically stimulated luminescence (OSL) to date the fine-grained quartz-rich fluvial sediment. The sample site is only ~18 km west of our study area. The OSL age of the fan is 25.7 ± 5.8 ka, and the alluvial fan is ~12 m high above the modern channel [19]. In our study, T2 at site 5 is ~18 m high above the channel, which is close to the height of the alluvial fan studied by Campbell (~12 m). Furthermore, the age, 22.23 ± 1.06 ka, of the T2 alluvial fan at site 5 is close to the age, 25.7 ± 5.8 ka, of the alluvial fan that was dated by Campbell.

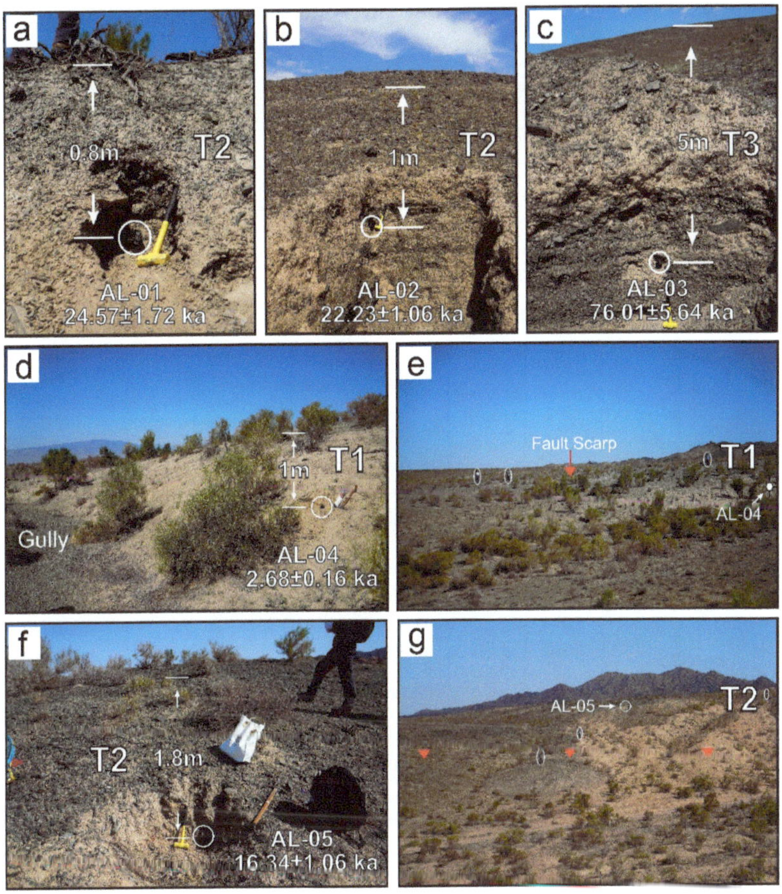

Figure 3. Field collection locations of the OSL samples: (**a**) sample AL-01; (**b**) sample AL-02; (**c**) sample AL-03; and (**d**) sample AL-04. (**e**) The field site shows a dextral alluvial fan and the fault scarp, which is also where the sample AL-04 was collected. (**f**) Sample AL-05. (**g**) The field site indicates the fault trace and fault scarp, which is also where the sample AL-05 was collected. The red triangular arrows indicate the fault trace.

Based on the ages of the sedimentary deposits and the heights of the alluvial fans above the channel, we believe that the alluvial fans, which have similar heights above the channel, have similar formation periods near the Dzhungarian gate. The formation ages of T1–T3 are 2–3 ka, ~20 ka, and less than 76 ka, respectively. Detailed OSL dating results for the five samples are shown in Table 1.

Table 1. OSL dating results.

Sample ID	Sample Coordinate	Depth (m)	U (ppm)	Th (ppm)	K (%)	Water Content (%)	D (Gy/ka)	De_{MAM} (Gy)	Age_{MAM} (ka)	De_{CAM} (Gy)	Age_{CAM} (ka)	De_{AVG} (Gy)	Age_{AVG} (ka)
AL-01	45°21′01″N 82°38′56″E	0.8	2.12 ± 0.01	6.57 ± 0.08	1.69 ± 0.01	5	2.73 ± 0.11	39.7 ± 3.15	14.57 ± 1.3	64.24 ± 3.91	23.57 ± 1.72	66.96 ± 3.81	24.57 ± 1.72
AL-02	45°32′24″N 82°26′29″E	1	1.55 ± 0.04	4.6 ± 0.11	1.59 ± 0.01	5	2.37 ± 0.1	49.53 ± 2.59	20.88 ± 1.39	52.62 ± 1.23	22.18 ± 1.05	52.74 ± 1.27	22.23 ± 1.06
AL-03	45°32′25″N 82°26′28″E	5	1.59 ± 0.03	4.5 ± 0.05	1.35 ± 0.01	5	2.05 ± 0.09	155.45 ± 9.57	76.01 ± 5.64	196.36 ± 7.61	96.02 ± 5.45	199.3 ± 7.09	97.46 ± 5.32
AL-04	45°21′38″N 82°38′21″E	1	1.69 ± 0.03	7.28 ± 0.05	1.89 ± 0.01	3	2.94 ± 0.12	5.65 ± 0.39	1.92 ± 0.16	7.7 ± 0.33	2.62 ± 0.16	7.87 ± 0.35	2.68 ± 0.16
AL-05	45°19′07″N 82°41′39″E	1.8	2.08 ± 0.04	9.45 ± 0.11	3.15 ± 0.01	5	4.26 ± 0.18	52.48 ± 4.6	12.31 ± 1.2	68.43 ± 3.43	16.05 ± 1.05	69.67 ± 3.43	16.34 ± 1.06

Note: OSL, optically stimulated luminescence; D is the environmental dose rate; De_{MAM} is the equivalent dose calculated by the minimum age model; De_{CAM} is the equivalent dose calculated by the median age model; De_{AVG} is the equivalent dose calculated by the average age model; Age_{MAM} is the optical luminescence age calculated by the minimum age model; Age_{CAM} is the optical luminescence age calculated by the median age model; and Age_{AVG} is the optical luminescence age calculated by the average age model.

4.3. The Southeast Section

The southeast section of the ALF extends for ~80 km from the northeast Lake Ebinur to study site 6 (Figure 2). Through remote sensing interpretation and field observations, it was revealed that there was a clear dextral strike-slip movement in this section. Study sites 1 to 7 were extensively investigated in order to describe their structures, as well as geomorphological characters, and to measure the vertical and horizontal displacements.

Site 1 is situated ~9 km northeast of Lake Ebinur, as shown in Figure 2. Fault strikes N30°W were found at site 1, with the main fault developing along the edge of the bedrock hill, while the branch cut through the alluvial fan (Figure 4a). There were four stages of alluvial fans (T1–T4) developed along the stream channel. T1 is the lowest geomorphic surface, only ~6 m above the stream bed, and developed on both sides of the fault. T2 is ~20 m above the stream bed, T3 is ~50 m above the stream bed, and T4 is ~60 m above the stream bed. T2 and T3 are only preserved on one side of the fault, but T4 is the highest and most widely geomorphic surface developed on both sides of the fault. The horizontal displacement of T4 was measured to be 10 ± 1 m along the main fault, and the obvious dextral offset of T4 was estimated to be 86 ± 4 m along the branch fault (Figure 4a,b); however, it was possible that the dextral displacement of 86 ± 4 m could have been exaggerated due to the collapse of the alluvial fan. Fault scarps clearly formed on T4, and the topographic profile A–A′ was extracted to measure the vertical displacement, which indicated that the vertical displacement of T4 across the fault scarp is 5 ± 0.5 m (Figure 4c). On the southeast side of the stream channel, red Carboniferous granite thrust on yellow Neogene sandstone and gray Quaternary deposits (Figure 4d) are shown. The Neogene sandstone also shows larger vertical dislocations on the riser (Figure 4e).

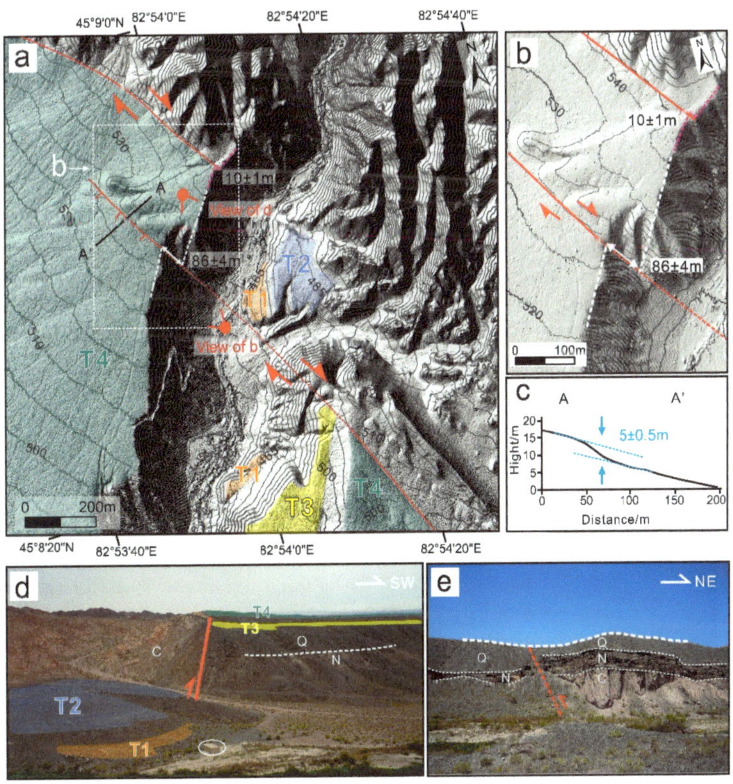

Figure 4. Site 1: (**a**) The hillshade map, which was built from a high-resolution UAV DEM using ArcGIS 10.8, shows structural and geomorphological characters of the Alakol Lake fault. The red

lines are the fault trace. The red dotted lines show that the fault traces are either unclear or covered. The red triangular arrows indicate dextral strike-slip movement. The black solid line (A–A′) marks the location of the profile in (**c**). Four stages of alluvial fans (T1–T4) are developed along the stream channel, and the shadows with different colors depict the corresponding alluvial fans. The white dashed lines and the pink dashed lines represent the fit lines of the T4 riser. The white dotted box represents the range of (**b**). (**b**) The image of the hillshade map displays a detailed view of the geomorphic surface in (**a**). The white and pink dashed lines in the image represent the fit lines. The white and pink arrows indicate the preferred offset. (**c**) Topographic profile across the fault extracted from the UAV DEM was used to measure the vertical offset. The blue dashed lines are fit lines. (**d**) The field photo shows the fault trace and alluvial fans. The white oval highlights the house for scale. (**e**) The field photo was shot in the northwest. The red dashed line represents the fault. Q represents the alluvial fan deposits (probably middle–upper Pleistocene), N represents the Neogene sandstone, and C represents the Carboniferous granite.

Site 2 is located ~25 km northwest of site 1 (Figure 2). At this site, the fault trace strikes N45°W. A series of gullies and ridges were dextrally displaced in the N40°W direction by the ALF (Figure 5a). The horizontal displacement of two large gullies were measured to be 20 ± 4 m and 22 ± 1 m, respectively, while the horizontal displacement of the ridge was measured to be 30 ± 6 m. The vertical displacement of the fault scarp is 1.0 ± 0.1 m in the profile A–A′ (Figure 5b). The black line B–B′ shows another dextral ridge, and a road crosses the ridge along the location of the fault scarp (Figure 5a). The vertical displacement of the dextral ridge is 2.1 ± 0.2 m in the profile B–B′ (Figure 5b).

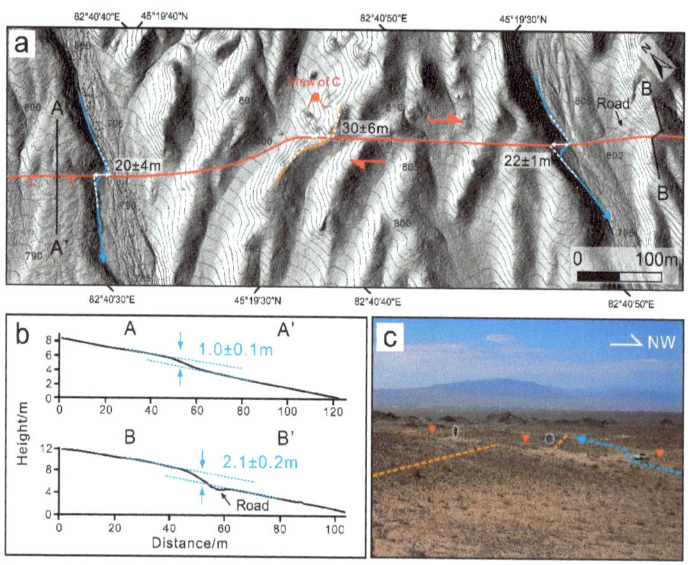

Figure 5. Site 2: (**a**) The hillshade map image, which was made using the high-resolution UAV DEM, shows the geomorphological expression of the Alakol Lake fault. The red lines are the fault trace. The red triangular arrows indicate dextral strike-slip movement. The two blue curves represent the horizontal offset of the gullies. The orange curve represents the dextrally displaced ridge. The black solid lines (A–A′ and B–B′) indicate the location of the extracted fault scarp. All of the white dashed lines represent the fit lines. The white arrows indicate the preferred offset. (**b**) Two topographic profiles across the fault extracted from the UAV DEM were used to measure the vertical offset. The blue dashed lines are fit lines. (**c**) The field photo shows the dextrally displaced ridge. The dotted orange lines show the location of the ridge. The red triangles indicate the fault trace. The blue dashed line indicates the dextral channel. The white ovals highlight the people for scale.

Site 3 is located ~1 km northwest of site 2 (Figure 2). Two stages of alluvial fans were interpreted at site 3. The fault was traced at T2 with an N40°W striking scarp and developed a micro-pull-apart basin at T1. The north edge of the T2 alluvial fan exhibits an apparent dextral offset of 34 ± 8 m at the erosional side, representing a minimum estimate of the displacement due to the possibility of the erosion of the channel (Figure 6a). Furthermore, a micro-pull-apart basin (~2 m deep, ~400 m long, and ~50 m wide) on T1 was controlled by a micro-stepover of the fault (Figure 6a,c). Two topographic profiles (A–A′ and B–B′) were extracted to measure the vertical displacement (Figure 6a,b). The topographic profile A–A′ of T1 indicates that the vertical displacement of the two fault scarps, in opposite directions, is 1.8 ± 0.1 m and 1.9 ± 0.4 m, respectively. The presence of two fault scarps in profile A–A′ primarily explains the existence of small pull-apart basins. Furthermore, the profile A–A′ across the pull-apart basin indicates almost no vertical displacement since T1 formed. The topographic profile B–B′ indicates that the vertical displacement across the fault scarps on T2 is 3.5 ± 0.9 m. Compared to the horizontal dislocations, the vertical displacement was not worth mentioning.

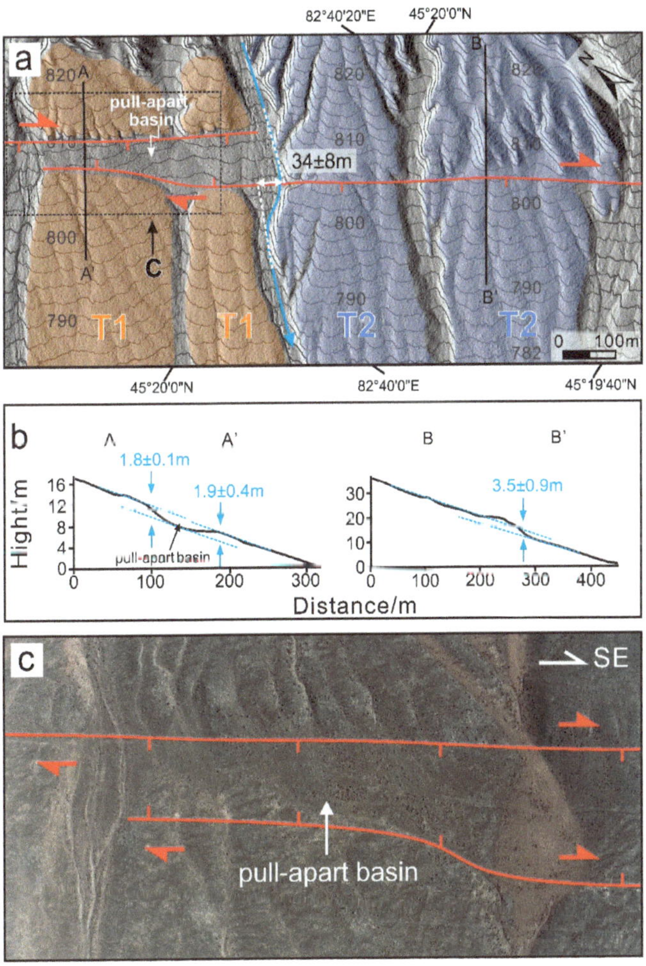

Figure 6. Site 3: (a) The hillshade map image, which was made using the high-resolution UAV DEM, shows the geomorphological expression of the Alakol Lake fault. The red lines are the fault trace,

and the scales indicate the slope direction of the fault scarp. The red triangular arrows indicate dextral strike-slip movement. The black dotted box represents the range of Figure c. The orange shadows represent T1, and purple shadows represent T2. The blue curves represent the dextrally displaced edge of the alluvial fans, and the white dashed lines represent the fit lines. The white arrows indicate the preferred offset. The black solid lines (A–A′ and B–B′) indicate the location of the extracted fault scarps. (**b**) Topographic profiles across the fault extracted from the UAV DEM were used to measure the vertical offset. The blue dashed lines are fit lines. (**c**) A photo taken by the drone shows the fault traces and a pull-apart basin.

Site 4 is located ~2 km northwest of site 3 (Figure 2). At site 4, two stages of alluvial fans were interpreted using the UAV DEM (Figure 7a). The horizontal displacement of two large subgullies was measured to be 20 ± 3 m and 18 ± 2 m, respectively, for the T2 alluvial fan (Figure 7a–c); however, 20 ± 3 m for the horizontal offset represented the minimum estimate of the displacement due to measurement on the erosion side of the channel. The topographic profiles (A–A′ and B–B′) indicate that the vertical displacements across the fault scarp are 0.7 ± 0.1 m and 2.6 ± 0.2 m, respectively (Figure 7d). The field photo shows the dextrally displaced gully and the fault scarp on the surface of the alluvial fan (Figure 7e). A clear fault scarp that has a distinct vertical displacement can be seen on the alluvial fan, as well as a series of subgullies (Figure 7f). In the field, a fault spring was found near the fault trace (Figure 7g). Many reeds and grasses grew around the spring and fixed a large amount of sand and soil.

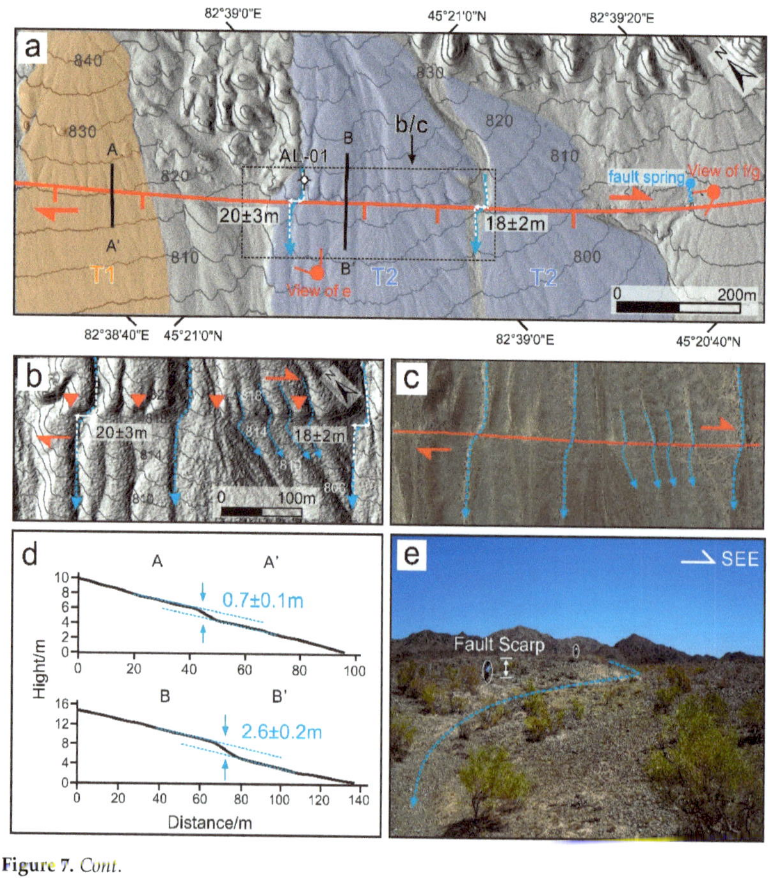

Figure 7. *Cont.*

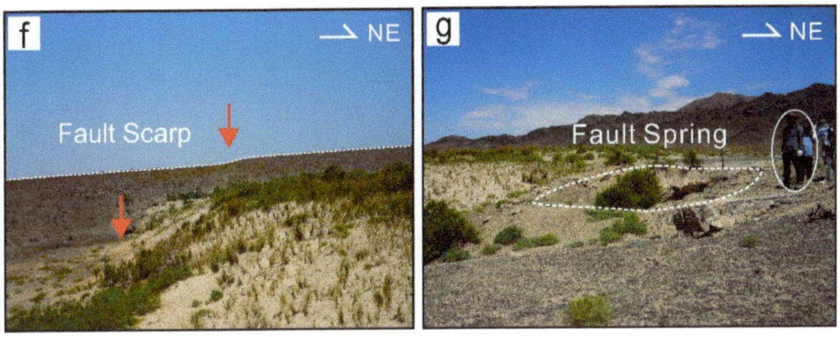

Figure 7. Site 4: (**a**) The hillshade map image, which was made using the high-resolution UAV DEM, shows the geomorphological expression of the Alakol Lake fault. The red line is the fault trace, and the scales indicate the slope direction of the fault scarp. Two red triangular arrows indicate dextral strike-slip movement. The orange shadow represents T1, and the purple shadows represent T2. The blue curves with arrows represent the dextrally displaced gullies, and the white dashed lines represent the fit lines. The white arrows indicate the preferred offset. The spring symbol composed of the blue circle and blue curve indicates the location of the fault spring. The black solid lines (A–A′, B–B′) indicate the location of the extracted fault scarps. The black dotted box represents the range of (**b**,**c**). (**b**) The enlarged hillshade map image shows the more detailed geomorphic surface in (**a**). Four identical red triangles indicate the fault trace. (**c**) A field photo of the range corresponding to (**b**). (**d**) Topographic profiles across the fault extracted from the UAV DEM were used to measure the vertical offset. The blue dashed lines are fit lines. (**e**) The field photo shows the dextrally displaced gully and fault scarp. (**f**) The field photo shows the fault scarp. The dotted white line indicates the geomorphic surface. Two red arrows show the fault scarp. (**g**) The picture shows the fault spring in the field.

Site 5 is located ~26 km northwest of site 4 (Figure 2). Three stages of alluvial fans (T1–T3) developed along the stream channel at site 5 (Figure 8a,c). T1 is the lowest alluvial fan and is severely eroded, which is ~14 m above the stream bed; T2 is ~18 m above the stream bed; and T3 is the highest alluvial fan, which is ~24 m above the stream bed. The ALF separated into two strands, forming the fault scarp on T2. The obvious horizontal displacement of the T2 riser was measured to be 27 ± 6 m on the west strand fault, and an OSL sample AL-02 was collected to calculate the slip rate of the ALF. The topographic profile A–A′ indicates that T2 has already experienced two vertical dislocations due to the active fault, and the vertical displacement across the two different scarps is 1.6 ± 0.1 m and 3.8 ± 0.8 m (Figure 8b). The topographic profile B–B′ also indicates that the vertical displacement of T2 is 1.5 ± 0.4 m. The field photograph shows the field view of the alluvial fans and the fault scarps across T2 (Figure 8c).

Site 6 is situated ~10 km northwest of site 5 (Figure 2). The Pleiades DEM shows the fault trace along strikes N30°W (Figure 9a). At the north side of site 6, there is a stream channel that crosses the ALF. Along the river, there are three stages of alluvial fans, named T1, T2, and T4 (inferred from residual ridges of a similar height). (Figure 9b). T1 is the lowest alluvial fan and is ~6 m above the stream bed. T2 is ~18 m above the stream bed, while T4 is the highest alluvial fan, ~58 m above the stream channel. The obvious dextral offset of T4 was estimated to be 238 ± 30 m, representing the minimum estimate of the displacement due to the possibility of the erosion of the channel. Additionally, two gullies developed on T4, which were estimated in their horizontal displacement to be 212 ± 20 m and 195 ± 15 m, respectively. The photos show the exposed fault breccia, which indicates the Carboniferous bedrock fault at this site (Figure 9c,d). The bedrock fault dips to the NW and shows that the Carboniferous bedrock has thrust over the Quaternary strata.

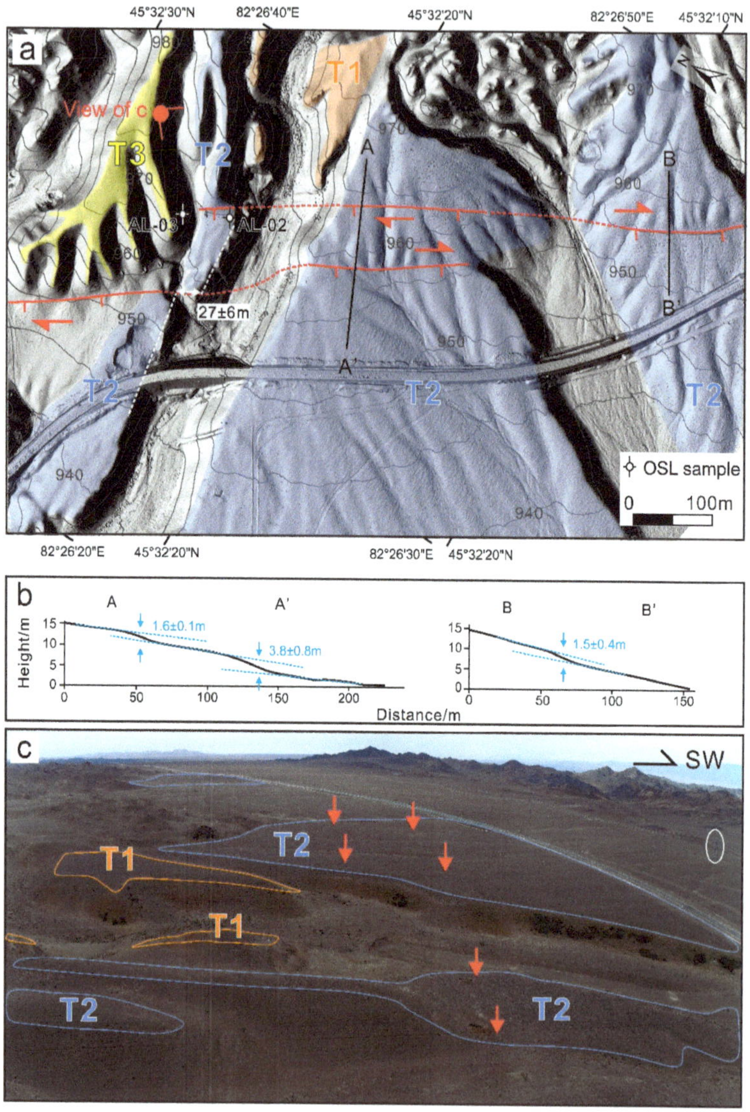

Figure 8. Site 5: (**a**) The hillshade map image, which was made using the high-resolution UAV DEM, shows the geomorphological expression of the Alakol Lake fault. Different colored shadows represent different alluvial fans. The red lines indicate clear fault traces, while the red dotted lines show fault traces that are either not visible or are covered. The white arrows indicate the preferred offset. The red triangular arrows indicate dextral strike-slip movement. The scales indicate the slope direction of the fault scarp. The white dashed lines represent the dextrally displaced T2 alluvial fan. The black solid lines (A–A′, B–B′) indicate the location of the extracted fault scarps. (**b**) Topographic profiles across the fault extracted from the UAV DEM were used to measure the vertical offset. The blue dashed lines are fit lines. (**c**) The field photo shows the fault trace and the different alluvial fans. The red arrows indicate the fault scarp and fault trace, and the white circle represents the iron tower as a reference.

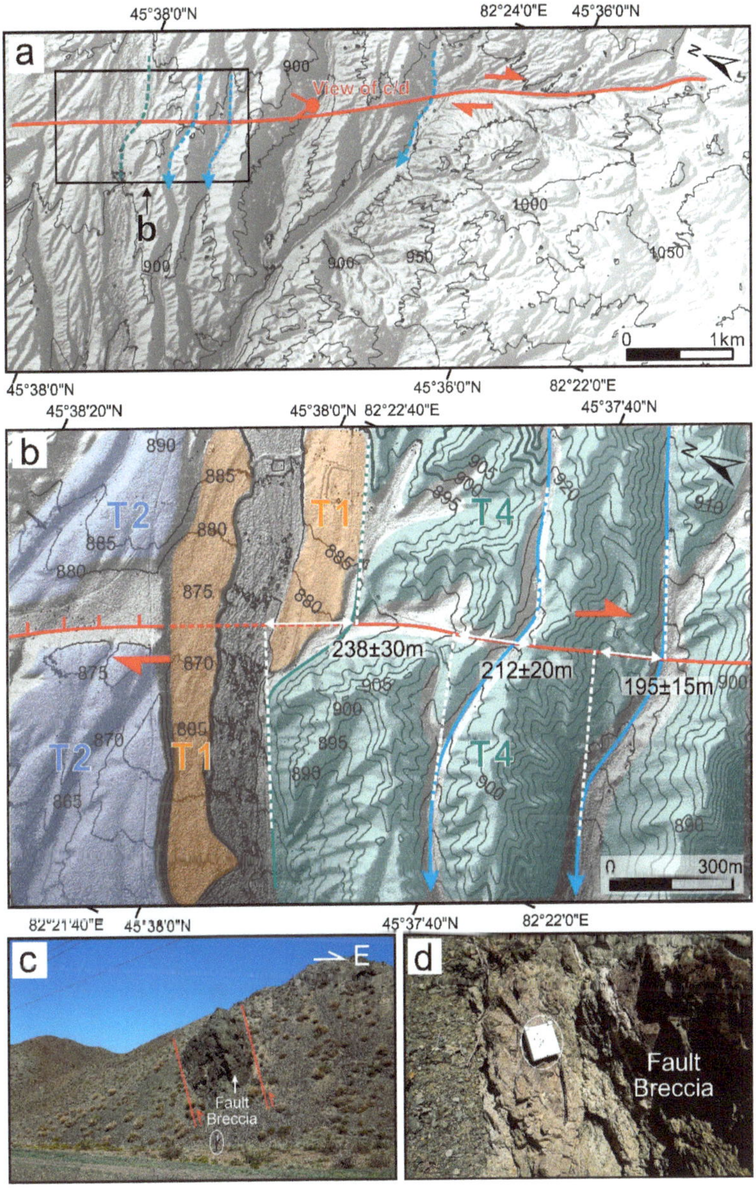

Figure 9. Site 6: (**a**) View of the Pleiades DEM shows the geomorphological expression of the Alakol Lake fault. (**b**) Hillshade map image, which was made using the high-resolution UAV DEM, shows the enlarged geomorphic surface. The red lines are the fault traces, and the red dotted lines show that the fault traces are not clear or are covered. The red triangular arrows indicate dextral strike-slip movement. The scales indicate the slope direction of the fault scarp. Different colored shadows represent different alluvial fans. The green curve represents the dextrally displaced alluvial fan. The blue curve represents the dextrally displaced gullies. All of the white dashed lines represent the fit lines. The white arrows indicate the preferred offset. (**c**,**d**) The fault breccia indicated by the white arrow revealed in the field.

Site 7 is between site 1 and site 2 (Figure 2), which shows four alluvial fans (at least equivalent to the T4 stage of the alluvial fan or older) at a large spatial scale from the view of the Pleiades DEM (Figure 10a). The fault strikes N42°W and develops along the edge of the bedrock hill. In the view, we mapped the range of the alluvial fans using the Pleiades satellite image and the DEM. The fan apex of the four alluvial fans was dextrally displaced relative to the stream channel. We roughly estimated the horizontal offset and uncertainty by measuring the distance from the stream bed to both sides of the alluvial fan edge. The horizontal offsets from fan 1 to fan 4 are 267 ± 60 m, 312 ± 131 m, 257 ± 73 m, and 233 ± 70 m, respectively (Figure 10).

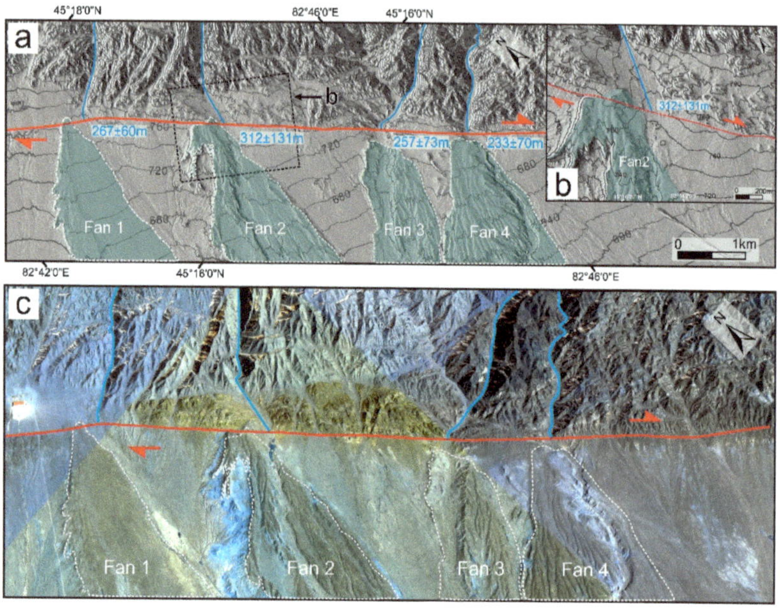

Figure 10. Site 7: (**a**) View of the Pleiades DEM shows some displaced alluvial fans. The translucent green shades are the alluvial fans. The red lines are the fault traces, and the red triangular arrows indicate dextral strike-slip movement. The blue curves represent the stream channels. (**b**) Hillshade map image, which was made by the high-resolution UAV DEM, shows the geomorphic surface of enlarged alluvial fan 2. (**c**) The outline of alluvial fans identified based on the texture and color characteristics in the Pleiades satellite image.

4.4. The Northwest Section

The northwest section is located on the eastern side of Lake Alakol and extends for ~70 km (Figure 2). To overcome the difficulty of obtaining information on the Kazakhstan section, Google Earth images and the Pleiades DEM are used to identify the fault trace and observe the horizontal displacement of the geomorphic surface.

Site 8 is located at the national boundary between China and Kazakhstan (Figure 2). The fault trace strikes N40°W, and some gullies that are dextrally displaced can be observed in the Pleiades DEM (Figure 11a,b). Because the alluvial fans are ~25 m above the channel, we estimated them to be T3 alluvial fans. The synchronous displacement of the three gullies and the offsets of T3 riser on the north side of the alluvial fans was estimated to be 56 ± 10 m according to the back-slipped method in the Pleiades DEM image (Figure 11c).

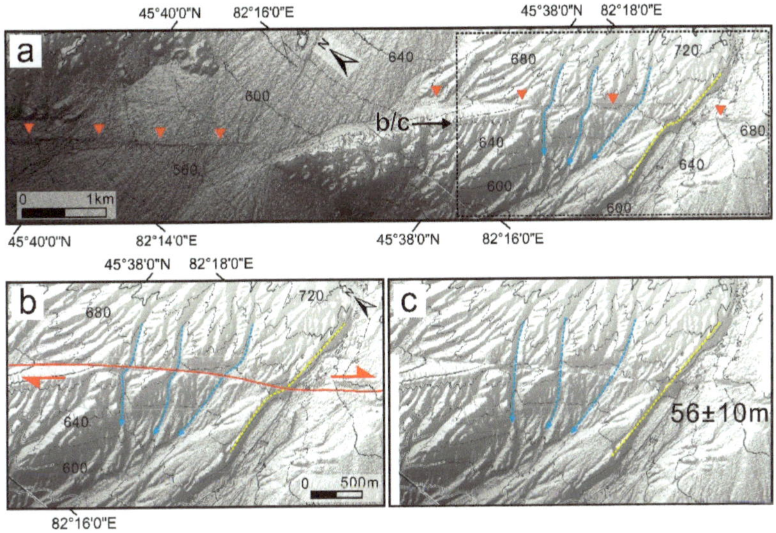

Figure 11. Site 8: (**a**) View of the Pleiades DEM shows the fault trace and the displaced geomorphic surface. The red triangles indicate the fault trace. The blue dashed lines indicate the displaced gullies, while the yellow dashed lines indicate the displaced T3 riser. (**b**) View of the enlarged geomorphic surface shows the displaced gullies. The red triangular arrows indicate dextral strike-slip movement. (**c**) The back-slipped view of the gullies.

Site 9 is located on the eastern shore of Lake Alakol (Figure 2). The northwest section of the ALF strikes N30°W and extends by ~25 km along the edge of Lake Alakol at this site (Figure 12a,b). Two topographic profiles (A–A′ and B–B′) across the fault extracted from the DEM indicate that the fault scarp is 5.6 ± 0.5 m and 2.6 ± 0.9 m, respectively (Figure 12a,c). A series of gullies were dextrally dislocated along the fault scarp, as interpreted from the Google Earth image. The horizontal displacement of these gullies was measured to be 76 ± 16 m, 68 ± 6 m, 51 ± 4 m, 27 ± 3 m, and 28 ± 1 m, respectively (Figure 12d,e); however, the lacustrine geomorphic surface near the lake shore does not correspond to the stages of the alluvial fans in our study.

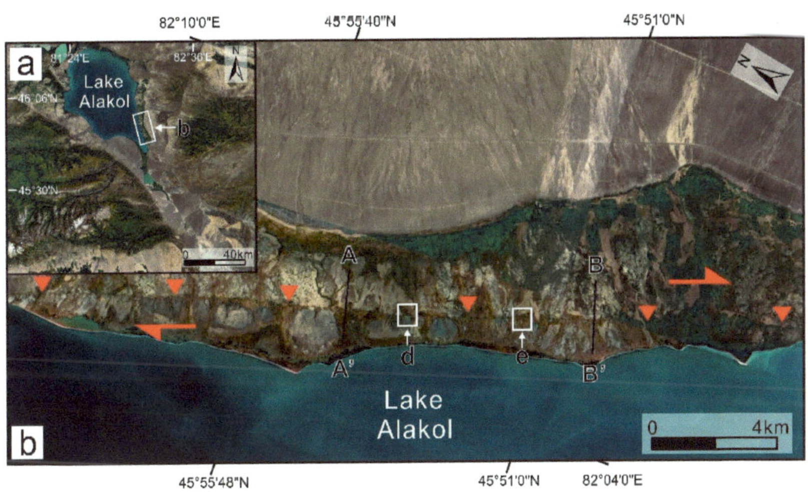

Figure 12. *Cont.*

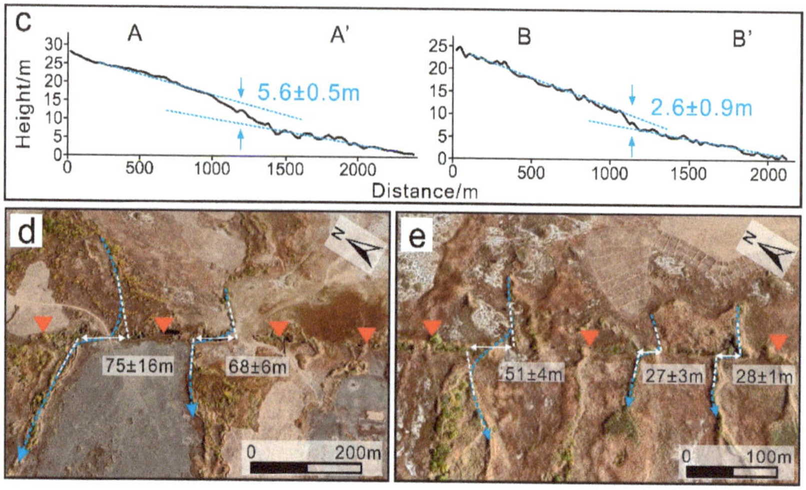

Figure 12. Site 9: (**a**) The location of site 9 is shown on Google Earth. (**b**) The fault trace and geomorphic surface are shown on Google Earth. Six identical red triangles indicate the fault trace. Two red triangular arrows indicate dextral strike-slip movement. The black solid lines (A–A′ and B–B′) show where the topographic profiles were extracted. The white boxes represent the viewing areas of (**d**,**e**). (**c**) Topographic profiles (A–A′ and B–B′) across the fault extracted from the DEM were used to measure the vertical offset. The blue dashed lines are fit lines. (**d**,**e**) Some images of the displaced gullies taken on Google Earth. The blue dashed lines indicate the displaced gullies. The white dashed lines represent the fit lines. The white arrows indicate the preferred offset.

5. Discussion

5.1. Geometric Distribution and Movement Characteristics

The ALF is a NW-SE-trending dextral strike-slip fault. The ALF is situated along the northern margin of the Dzhungarian gate, stretching roughly 150 km from Lake Ebinur to Lake Alakol. Lake Alakol is located within the pull-apart basin of the stepover between the DZF and CF. According to the geometric distribution, the ALF seems to be a link fault crossing the stepover and connecting the southeast end of the CF to the DZF (Figure 13). The straight northeast lakeshore of Lake Alakol indicates that the ALF affects the topography of the pull-apart basin.

Numerous landforms, which include alluvial fans, stream channels, gullies, and ridges along the fault, have been found to be dextrally displaced. From the measured displacement of the geomorphic surfaces, the horizontal displacement is mostly tens of meters, and the horizontal displacement of certain alluvial fans even reaches almost a hundred meters. In contrast, the vertical displacement is mostly less than 5 m; therefore, the movement characteristic of the ALF is mainly a dextral strike-slip. At site 1 and site 6, the Carboniferous bedrock thrust over the Quaternary young strata with a high dip angle (Figures 4b,d and 7c). This indicates the southeast section of the ALF is characterized by shearing and compression.

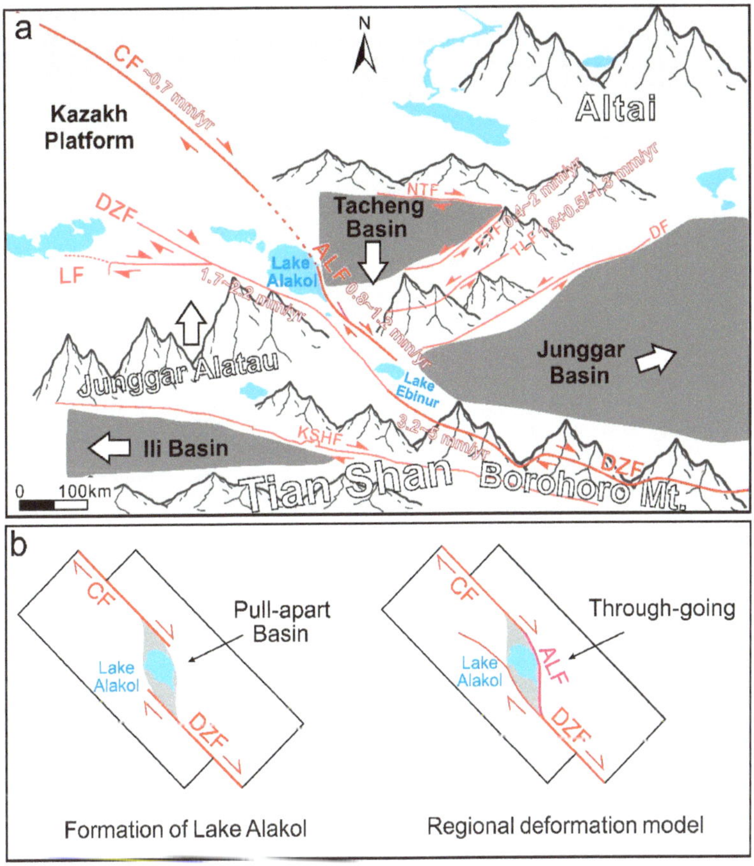

Figure 13. (**a**) Simplified geological model map of the new fault in the western Junggar. The white arrows indicate the relative movement directions of the Tacheng Basin, Junggar Alatau, and Junggar Basin. DZF—Dzhungarian fault; ALF—Alakol Lake fault; CF—Chingiz fault; LF—Lepsy fault; KSHF—Kashihe fault; DF—Daerbute fault, TLF—TuoLi fault; ETF—East Tacheng fault; and NTF—North Tacheng fault. (**b**) Evolutionary model map of Lake Alakol and the Alakol Lake fault.

5.2. Estimating the Slip Rate of the ALF

Many geomorphic surfaces were displaced dextrally along the ALF from site 1 to site 9. By measuring the displacement and dating these surfaces, the slip rate of the ALF can be estimated. At site 4, the dating age of the sample AL-01 represents the formation age of the T2 alluvial fan (Figures 3a and 7a,c), and the T2 riser displaced 20 ± 3 m of the horizontal offset (Figure 7a,c). We calculated a dextral strike-slip rate of 0.8 ± 0.2 mm/a. Considering the horizontal offset of 20 ± 3 m measured on the erosion side of the channel, the strike-slip rate of 0.8 ± 0.2 mm/a is lower than the actual value.

At site 5, the horizontal displacement of the T2 riser was measured to be 27 ± 6 m (Figure 8a), and an age for T2 of 22.23 ± 1.06 kyr was obtained by dating the sample AL-02 (Table 1); therefore, a dextral strike-slip rate of 1.2 ± 0.3 mm/a was calculated using the total horizontal offset and the T2 age.

Therefore, the Alakol Lake fault has a dextral strike-slip rate of 0.8–1.2 mm/a (closer to 1.2 mm/a). The strike-slip rate of the Alakol Lake fault is higher than that of the Chingiz fault in the northern region (~0.7 mm/a) (Figure 13) [33] but slower than that of the Dzhungarian fault in the southern region (3.2–5 mm/a) [19,20,30].

The slip rates of these dextral strike-slip faults indicate a gradual decrease from the Borohoro Mountain to the Kazakhstan platform. This result is consistent with the view that most of the shortening is absorbed within the Tian Shan based on the GPS data decreasing to the north and the view that the tectonic deformation of the Tian Shan decreases to the north [14,22–24].

5.3. *Differential Movement of the Block at the North Side of the Tian Shan*

The formation mechanism of the dextral strike-slip fault has great significance for the exploration of geodynamics. The GPS velocities display that the DZF and ALF accommodate ~2 mm/a of crustal shortening at the NE trending in the Dzhungarian gate (Figure 1c), and the horizontal-slip rates of the ALF (0.8–1.2 mm/a) and DZF (1.7–2.2 mm/a) at the NW trending are basically equivalent to the shortening of the NE trending. This indicates that the Junggar Alatau and the Tacheng Basin were oblique extruding. This movement characteristic is widespread on the north side of the Tian Shan and caused several NW-SE-trending dextral strike-slip faults to develop, including the Talas–Fergana fault (TFF) and the Dzhungarian fault, which have attracted much attention [2,19,20,27]. Campbell et al. (2013) and Hu et al. (2021) studied the slip rate of the Dzhungarian fault and concluded that the north–south crustal shortening in the Tian Shan is accommodated by the dextral strike-slip fault [19,20]. Rizza et al. (2019) studied the deformation mode of the Talas–Fergana fault based on the slip rate and concluded that the Talas–Fergana fault was subjected to the northward compression of the Pamir Plateau, resulting in counterclockwise rotation around the vertical axis of the fault [27]. We inferred that the ALF was also subjected to the northward compression and played an important role in accommodating the north–south crustal shortening. Affected by the Pamir Plateau wedging, the deformation of the Tian Shan decreases from west to east [2,3]; therefore, the Junggar Alatau moved northward with respect to the Tacheng Basin. By studying the East Tacheng fault (ETF), Yu et al. (2021) pointed out sinistral slip rates of 0.4–2.0 mm/a in the ETF, and Yuan et al. (2023) estimated sinistral slip rates of 1.8 + 0.5/-1.3mm/a in the TLF [17,71]. We can find that the slip rate of the ALF is close to these sinistral strike-slip faults. This implies that the Junggar Basin has moved northeastward with respect to the Tacheng Basin. Our study indicated that the Tacheng Basin moved southward and produced conjugate NE and NW strike-slip faults during the regional N-S shortening. The GPS velocities suggest that the Junggar Basin moves northeastward [17,28], which coordinates the southward movement of the Tacheng Basin and can also explain the formation of sinistral strike-slip faults in the western Junggar.

Based on preliminary estimates of the slip rate of the ALF, the ALF may replace the role of the northwest section of the DZF to transfer the crustal shortening to the further north Kazakhstan platform. Furthermore, on the south side of the northwest section of the DZF, the Lepsy fault also resolves part of the slip rate of the DZF [72]; therefore, the slip rate in the northwest section of the DZF shows a significant decrease compared with the east section. The northwest section of the DZF seems like a horsetail structure at the end of the strike-slip fault. Ultimately, we propose the model that the ALF serves as a connecting point for the Dzhungarian fault and the Chingiz fault, resulting in a mega NW-SE dextral strike-slip fault zone that extends more than 1400 km from the eastern side of the Borohoro Mountain to 500 km deep into the Kazakhstan platform (Figure 13).

6. Conclusions

The ALF extends at the northern margin of the Dzhungarian gate from Lake Alakol to Lake Ebinur for ~150 km. Numerous landforms, which include alluvial fans, stream channels, gullies, and ridges along the fault, are dextrally displaced, and the large horizontal/vertical offset ratio and straight fault traces indicated that the ALF is mainly a dextral strike-slip fault, which had significant activity in the late Quaternary. Based on the dislocation measurement and chronological dating of the geomorphic surface, the slip rates of the ALF are estimated at 0.8–1.2 mm/a (closer to 1.2 mm/a). The ALF serves as a connecting point for the Dzhungarian fault and the Chingiz fault, resulting in the formation of a mega NW-SE dextral strike-slip fault zone. According to our analysis of the dating samples taken from the alluvial fan, as well as our measurement of the displacement of the alluvial fan and gullies, it appears that the Alakol Lake fault has a dextral sinistral slip rate of 0.8–1.2 mm/a (closer to 1.2 mm/a). The strike-slip rate of the Alakol Lake fault is comparatively higher than that of the Chingiz fault in the northern region (~0.7 mm/a) but slower than that of the Dzhungarian fault in the southern region (3.2–5 mm/a). The Chingiz–Alakol–Dzhungarian fault zone shows a gradual decrease in deformation towards the interior of the Kazakhstan platform.

Author Contributions: Conceptualization, A.L.; methodology, W.Y. and A.L.; software, W.Y., A.L. and Z.H.; validation, A.L., L.X. and Z.H.; formal analysis, A.L. and L.X.; investigation, L.X. and X.L.; resources, A.L.; data curation, W.Y.; writing—original draft preparation, W.Y.; writing—review and editing, A.L.; visualization, W.Y. and A.L.; supervision, A.L.; project administration, A.L.; funding acquisition, A.L. All authors have read and agreed to the published version of the manuscript.

Funding: This research was funded by the State Key Laboratory of Earthquake Dynamics (grant no. LED2022A01), the Research Fund of the Institute of Geology, China Earthquake Administration ((grant no. IGCEA2115), and the Key Research and Development Project of China ((grant no. 2023YFC3012001).

Data Availability Statement: The data presented in this study is not available due to legal reason.

Acknowledgments: We are grateful to everyone who helped us successfully complete this research.

Conflicts of Interest: The authors declare no conflicts of interest.

References

1. Molnar, P.; Tapponnier, P. Cenozoic tectonics of Asia: Effects of a continental collision: Features of recent continental tectonics in Asia can be interpreted as results of the India-Eurasia collision. *Science* **1975**, *189*, 419–426. [CrossRef] [PubMed]
2. Tapponnier, P.; Molnar, P. Active faulting and Cenozoic tectonics of the Tien Shan, Mongolia, and Baykal regions. *J. Geophys. Res.* **1979**, *84*, 3245–3459. [CrossRef]
3. Avouac, J.P.; Tapponnier, P.; Bai, M.; You, H.; Wang, G. Active thrusting and folding along the northern Tien Shan and Late Cenozoic rotation of the Tarim relative to Dzungaria and Kazakhstan. *J. Geophys. Res. Solid Earth* **1993**, *98*, 6755–6804. [CrossRef]
4. Bande, A.; Sobel, E.R.; Mikolaichuk, A.; Torres Acosta, V. Talas–Fergana Fault Cenozoic timing of deformation and its relation to Pamir indentation. *Geol. Soc. Lond. Spec. Publ.* **2015**, *427*, 295–311. [CrossRef]
5. Bosboom, R.; Dupont-Nivet, G.; Huang, W.; Yang, W.; Guo, Z. Oligocene clockwise rotations along the eastern Pamir: Tectonic and paleogeographic implications. *Tectonics* **2014**, *33*, 53–66. [CrossRef]
6. Hendrix, M.S.; Dumitru, T.A.; Graham, S.A. Late Oligocene-early Miocene unroofing in the Chinese Tian Shan: An early effect of the India-Asia collision. *Geology* **1994**, *22*, 487–490. [CrossRef]
7. Burchfiel, B.C.; Brown, E.T.; Deng, Q.; Feng, X.; Li, J.; Molnar, P.; Shi, J.; Wu, Z.; Yu, H. Crustal Shortening on the Margins of the Tien Shan, Xinjiang, China. *Int. Geol. Rev.* **1999**, *41*, 665–700. [CrossRef]
8. Deng, Q.; Feng, X.; Zhang, P.; Xu, X.; Peng, S.; Li, J. *Active Tectonics of Tianshan*; Seismological Press: Beijing, China, 2000. (In Chinese)
9. Li, C.; Wang, S.; Li, Y.; Chen, Y.; Sinclair, H.; Wei, D.; Ma, D.; Lu, H.; Wang, X.; Wang, L. Growth of the Tian Shan Drives Migration of the Conglomerate-Sandstone Transition in the Southern Junggar Foreland Basin. *Geophys. Res. Lett.* **2022**, *49*, e2021GL097545. [CrossRef]
10. Yang, X.; Li, Z.; Wang, W.; Zhang, P.; Wu, C.; Chen, G.; Duan, L.; Wu, X.; Liu, K. Quaternary Crustal Shortening of the Houyanshan Structure in the Eastern Chinese Tian Shan: Constrained from Geological and Geomorphological Analyses. *Remote Sens.* **2023**, *15*, 1603. [CrossRef]
11. Li, J.; Yao, Y.; Li, R.; Yusan, S.; Li, G.; Freymueller, J.T.; Wang, Q. Present-Day Strike-Slip Faulting and Thrusting of the Kepingtage Fold-and-Thrust Belt in Southern Tianshan: Constraints From GPS Observations. *Geophys. Res. Lett.* **2022**, *49*, e2022GL099105. [CrossRef]
12. Hu, Z.; Yang, X.; Li, A.; Yang, H.; Miao, S.; Yuan, H. Late Pleistocene slip rate and subsurface fault-plane geometry of the Latgan Fold (Bole Basin, North Tian Shan, NW China) derived from a deformed cosmogenically dated alluvial fan. *Geomorphology* **2023**, *421*, 108526. [CrossRef]
13. Yin, A.; Nie, S.; Craig, P.; Harrison, T.; Ryerson, F.; Xianglin, Q.; Geng, Y. Late Cenozoic tectonic evolution of the southern Chinese Tian Shan. *Tectonics* **1998**, *17*, 1–27. [CrossRef]
14. Thompson, S.C.; Weldon, R.J.; Rubin, C.M.; Abdrakhmatov, K.; Molnar, P.; Berger, G.W. Late Quaternary slip rates across the central Tien Shan, Kyrgyzstan, central Asia. *J. Geophys. Res. Solid Earth* **2002**, *107*, ETG 7-1–ETG 7-32. [CrossRef]

15. Peng, Z.; Graveleau, F.; Vendeville, B.C.; Wang, X.; Averbuch, O. Interaction between basement inherited strike-slip structures and thrust wedge propagation in the northern Tianshan foreland basin: Insight from analogue modelling experiments. *J. Struct. Geol.* **2024**, *183*, 105143. [CrossRef]
16. Selander, J.; Oskin, M.; Ormukov, C.; Abdrakhmatov, K. Inherited strike-slip faults as an origin for basement-cored uplifts: Example of the Kungey and Zailiskey ranges, northern Tian Shan. *Tectonics* **2012**, *31*, 4. [CrossRef]
17. Yu, J.; Walker, R.T.; Rhodes, E.J.; Zhang, P.; Li, C.; Wang, S.; Wang, Y.; Liu, F.; Hao, Y. East Tacheng (Qoqek) Fault Zone: Late Quaternary Tectonics and Slip Rate of a Left-Lateral Strike-Slip Fault Zone North of the Tian Shan. *Tectonics* **2021**, *40*, e2020TC006377. [CrossRef]
18. Burtman, V.S.; Skobelev, S.F.; Molnar, P. Late Cenozoic slip on the Talas-Ferghana fault, the Tien Shan, central Asia. *Geol. Soc. Am. Bull.* **1996**, *108*, 1004–1021. [CrossRef]
19. Campbell, G.; Walker, R.; Abdrakhmatov, K.; Schwenninger, J.; Jackson, J.; Elliott, J.; Copley, A. The Dzhungarian fault: Late Quaternary tectonics and slip rate of a major right-lateral strike-slip fault in the northern Tien Shan region. *J. Geophys. Res. Solid Earth* **2013**, *118*, 5681–5698. [CrossRef]
20. Hu, Z.; Yang, X.; Yang, H.; Huang, W.; Wu, G.; Miao, S.; Zhang, L. Slip rate and paleoseismology of the Bolokenu-Aqikekuduk (Dzhungarian) right-lateral strike-slip fault in the northern Tian Shan, NW China. *Tectonics* **2021**, *40*, e2020TC006604. [CrossRef]
21. Abdrakhmatov, K.Y.; Aldazhanov, S.A.; Hager, B.H.; Hamburger, M.W.; Herring, T.A.; Kalabaev, K.B. Relatively recent construction of the Tien Shan inferred from GPS measurements of present-day crustal deformation rates. *Nature* **1996**, *384*, 450–453. [CrossRef]
22. Zubovich, A.V.; Wang, X.; Scherba, Y.G.; Schelochkov, G.G.; Reilinger, R.; Reigber, C.; Mosienko, O.I.; Molnar, P.; Michajljow, W.; Makarov, V.I.; et al. GPS velocity field for the Tien Shan and surrounding regions. *Tectonics* **2010**, *29*, e2010tc002772. [CrossRef]
23. Wang, M.; Shen, Z.K. Present-Day Crustal Deformation of Continental China Derived From GPS and Its Tectonic Implications. *J. Geophys. Res. Solid Earth* **2020**, *125*, e2019JB018774. [CrossRef]
24. Wu, C.; Zhang, P.; Zhang, Z.; Zheng, W.; Xu, B.; Wang, W.; Yu, Z.; Dai, X.; Zhang, B.; Zhang, K. Slip partitioning and crustal deformation patterns in the Tianshan orogenic belt derived from GPS measurements and their tectonic implications. *Earth-Sci. Rev.* **2023**, *238*, 104362. [CrossRef]
25. Tsai, C.H.; Abdrakhmatov, K.; Mukambayev, A.; Elliott, A.J.; Elliott, J.R.; Grützner, C.; Rhodes, E.J.; Ivester, A.H.; Walker, R.T.; Wilkinson, R. Probing the Upper End of Intracontinental Earthquake Magnitude: A Prehistoric Example From the Dzhungarian and Lepsy Faults of Kazakhstan. *Tectonics* **2022**, *41*, e2022TC007300. [CrossRef]
26. Li, A.; Ran, Y.; Gomez, F.; Thompson Jobe, J.A.; Liu, H.; Xu, L. Segmentation of the Kepingtage thrust fault based on paleoearthquake ruptures, southwestern Tianshan, China. *Nat. Hazards* **2020**, *103*, 1385–1406. [CrossRef]
27. Rizza, M.; Abdrakhmatov, K.; Walker, R.; Braucher, R.; Guillou, V.; Carr, A.S.; Campbell, G.; McKenzie, D.; Jackson, J.; Aumaître, G.; et al. Rate of Slip From Multiple Quaternary Dating Methods and Paleoseismic Investigations Along the Talas-Fergana Fault: Tectonic Implications for the Tien Shan Range. *Tectonics* **2019**, *38*, 2477–2505. [CrossRef]
28. Wu, C.; Wang, W.; Zheng, W.; Zhang, P.; Yu, Z. Opposite Sense of Strike-Slip Faulting and Crustal Rotation Accommodating Left-Lateral Shear Between the Tianshan Mountains and Kazakh Platform. *Geophys. Res. Lett.* **2021**, *48*, e2021GL096442. [CrossRef]
29. Shen, J.; Wang, Y.; Li, Y. Characteristics of the Late Quaternary right-lateral strike-slip movement of Bolokenu-Aqikekuduk fault in northern Tianshan Mountains, NW China. *Geosci. Front.* **2011**, *2*, 519–527. [CrossRef]
30. Tsai, C.H.; Walker, R.; Daout, S.; Abdrakhmatov, K.; Mukambayev, A.; Grützner, C.; Rhodes, E. Palaeo-earthquake magnitudes on the Dzhungarian fault, N. Tien shan, and implications for the rupture processes of intraplate strike-slip faults. In Proceedings of the EGU2020: Sharing Geoscience, Online, 4–8 May 2020. [CrossRef]
31. Samygin, S.G. *Chingiz Wrench Fault and Its Role in the Structure of Central Kazakhstan*; Nauka Publishing House: Moscow, Russia, 1974. (In Russian)
32. Mukambayev, A.S.; Mikhailova, N.N. Seismic hazard of the main Chingiz fault for the territory of Semipalatinsk test site. *NNC RK Bull.* **2015**, *3*, 82–86, (In Russian with English Abstract).
33. Li, Y.; Liu, M.; Hao, M.; Zhu, L.; Cui, D.; Wang, Q. Active crustal deformation in the Tian Shan region, central Asia. *Tectonophysics* **2021**, *811*, 228868. [CrossRef]
34. Xu, X.; Han, Z.; Yang, X.; Zhang, S.; Yu, G.; Zhou, B.; Li, F.; Ma, B.; Chen, G.; Ran, Y. *Seimotectonic Map of China and Its Adjacent Regions*; Seismological Publishing Press: Beijing, China, 2016. (In Chinese)
35. Xiao, W.; Han, C.; Yuan, C.; Sun, M.; Lin, S.; Chen, H.; Li, Z.; Li, J.; Sun, S. Middle Cambrian to Permian subduction-related accretionary orogenesis of Northern Xinjiang, NW China: Implications for the tectonic evolution of central Asia. *J. Asian Earth Sci.* **2008**, *32*, 102–117. [CrossRef]
36. Xiao, W.; Santosh, M. The western Central Asian Orogenic Belt: A window to accretionary orogenesis and continental growth. *Gondwana Res.* **2014**, *25*, 1429–1444. [CrossRef]
37. Wang, Y.; Wang, Y.; Yin, J.; Thomson, S.N.; Xiao, W.; He, Z.; Chen, W.; Cai, K.; Wu, M.; Meng, Y. Mesozoic exhumation of the northern West Junggar, NW China: Insights from low-temperature thermochronometers. *Tectonophysics* **2023**, *862*, 229939. [CrossRef]
38. Buckman, S.; Aitchison, J.C. Tectonic evolution of Palaeozoic terranes in west Junggar, Xinjiang, NW China. *Geol. Soc. Lond. Spec. Publ.* **2004**, *226*, 101–129. [CrossRef]

39. Choulet, F.; Faure, M.; Cluzel, D.; Chen, Y.; Lin, W.; Wang, B. From oblique accretion to transpression in the evolution of the Altaid collage: New insights from West Junggar, northwestern China. *Gondwana Res.* **2012**, *21*, 530–547. [CrossRef]
40. Wu, C.; Hong, T.; Xu, X.; Wang, C.; Dong, L. A-type granites induced by a breaking-off and delamination of the subducted Junggar oceanic plate, West Junggar, Northwest China. *China Geol.* **2022**, *4*, 457–474. [CrossRef]
41. Chen, B.; Jahn, B. Genesis of post-collisional granitoids and basement nature of the Junggar Terrane, NW China: Nd-Sr isotope and trace element evidence. *J. Asian Earth Sci.* **2004**, *23*, 691–703. [CrossRef]
42. Zhu, Y.; Chen, B.; Xu, X.; Qiu, T.; An, F. A new geological map of the West Junggar, north Xinjiang (NW China): Implications for Paleoenvironmental reconstruction. *Episodes* **2013**, *36*, 205–220. [CrossRef] [PubMed]
43. Zheng, B.; Han, B.F.; Wang, Z.Z.; Liu, B.; Feng, L.X. An example of Phanerozoic continental crustal growth: The West Junggar Orogenic Belt, Northwest China. *Lithos* **2020**, *376–377*, 105745. [CrossRef]
44. Hu, A.; Wang, Z.; Xu, G. *Geologic Evolution, Petrogenesis and Metallgenesis in Northern XinJiang*; Science Press: Beijing, China, 1997; pp. 9–104. (In Chinese)
45. Chen, Y.; Hu, A.; Zhang, G.; Zhang, Q. Zircon U-Pb age and Sr-Nd isotopic composition of granitic genesis and its geological implications from Percambrian window of western Tianshan, NW China. *Geochimica* **1999**, *28*, 515–519. (In Chinese with English Abstract)
46. Chen, J.; Zhou, T.; Xie, Z.; Zhang, X.; Guo, X. Formation of positive εNd(T) granitoids from the Alataw Mountains, Xinjiang, China, by mixing and fractional crystallization: Implication for Phanerozoic crustal growth. *Tectonophysics* **2000**, *328*, 53–67. [CrossRef]
47. Yin, J.; Wang, Y.; Hodges, K.V.; Xiao, W.; Thomson, S.N.; Chen, W.; Yuan, C.; Sun, M.; Cai, K.; Sun, J. Episodic Long-Term Exhumation of the Tianshan Orogenic Belt: New Insights From Multiple Low-Temperature Thermochronometers. *Tectonics* **2023**, *42*, e2022TC007469. [CrossRef]
48. Arrowsmith, J.R.; Zielke, O. Tectonic geomorphology of the San Andreas Fault zone from high resolution topography: An example from the Cholame segment. *Geomorphology* **2009**, *113*, 70–81. [CrossRef]
49. Zielke, O.; Arrowsmith, J.R.; Grant Ludwig, L.; Akciz, S.O. High-Resolution Topography-Derived Offsets along the 1857 Fort Tejon Earthquake Rupture Trace, San Andreas Fault. *Bull. Seismol. Soc. Am.* **2012**, *102*, 1135–1154. [CrossRef]
50. Bahrami, S.; Stokes, M. Analyzing drainage basin orientation and its relationship to active fold growth (Handun anticline, Zagros, Iran). *Geomorphology* **2023**, *426*, 108605. [CrossRef]
51. Bi, H.; Zheng, W.; Ren, Z.; Zeng, J.; Yu, J. Using an unmanned aerial vehicle for topography mapping of the fault zone based on structure from motion photogrammetry. *Int. J. Remote Sens.* **2017**, *38*, 2495–2510. [CrossRef]
52. James, M.R.; Robson, S. Straightforward reconstruction of 3D surfaces and topography with a camera: Accuracy and geoscience application. *J. Geophys. Res. Earth Surf.* **2012**, *117*, e2011jf002289. [CrossRef]
53. Johnson, K.; Nissen, E.; Saripalli, S.; Arrowsmith, J.R.; McGarey, P.; Scharer, K.; Williams, P.; Blisniuk, K. Rapid mapping of ultrafine fault zone topography with structure from motion. *Geosphere* **2014**, *10*, 969–986. [CrossRef]
54. Kovanič, L'.; Štroner, M.; Blistan, P.; Urban, R.; Boczek, R. Combined ground-based and UAS SfM-MVS approach for determination of geometric parameters of the large-scale industrial facility Case study. *Measurement* **2023**, *216*, 112994. [CrossRef]
55. Bemis, S.P.; Micklethwaite, S.; Turner, D.; James, M.R.; Akciz, S.; Thiele, S.T.; Bangash, H.A. Ground-based and UAV Based photogrammetry: A multi-scale, high-resolution mapping tool for structural geology and paleoseismology. *J. Struct. Geol.* **2014**, *69*, 163–178. [CrossRef]
56. Wang, S.; Ren, Z.; Wu, C.; Lei, Q.; Gong, W.; Ou, Q.; Zhang, H.; Ren, G.; Li, C. DEM generation from Worldview-2 stereo imagery and vertical accuracy assessment for its application in active tectonics. *Geomorphology* **2019**, *336*, 107–118. [CrossRef]
57. Zhou, Y.; Parsons, B.; Elliott, J.R.; Barisin, I.; Walker, R.T. Assessing the ability of Pleiades stereo imagery to determine height changes in earthquakes: A case study for the E Mayor-Cucapah epicentral area. *J. Geophys. Res. Solid Earth* **2015**, *120*, 8793–8808. [CrossRef]
58. McCalpin, J. *Paleoseismology*, 2nd ed.; Academic Press: San Diego, CA, USA, 1996.
59. Zielke, O.; Klinger, Y.; Arrowsmith, J.R. Fault slip and earthquake recurrence along strike-slip faults—Contributions of high-resolution geomorphic data. *Tectonophysics* **2015**, *638*, 43–62. [CrossRef]
60. Scharer, K.M.; Salisbury, J.B.; Arrowsmith, J.R.; Rockwell, T.K. Southern San Andreas Fault Evaluation Field Activity: Approaches to Measuring Small Geomorphic Offsets—Challenges and Recommendations for Active Fault Studies. *Seismol. Res. Lett.* **2014**, *85*, 68–76. [CrossRef]
61. Bi, H.; Zheng, W.; Lei, Q.; Zeng, J.; Zhang, P.; Chen, G. Surface Slip Distribution Along the West Helanshan Fault, Northern China, and Its Implications for Fault Behavior. *J. Geophys. Res. Solid Earth* **2020**, *125*, e2020jb019983. [CrossRef]
62. Sieh, K.E. Slip along the San Andreas fault associated with the great 1857 earthquake. *Bull. Seismol. Soc. Am.* **1978**, *68*, 1421–1448.
63. Zielke, O.; Arrowsmith, J.R.; Ludwig, L.G. Slip in the 1857 and earlier large earthquakes along the Carrizo Plain, San Andreas Fault. *Science* **2010**, *327*, 1119–1122. [CrossRef] [PubMed]
64. Wang, X.; Lu, Y.; Zhao, H. On the performances of the single-aliquot regenerative-dose (SAR) protocol for Chinese loess: Fine quartz and polymineral grains. *Radiat. Meas.* **2006**, *41*, 1–8. [CrossRef]
65. Murray, A.S.; Wintle, A.G. Luminescence dating of quartz using an improved single-aliquot regenerative-dose protocol. *Radiat. Meas.* **2000**, *32*, 57–73. [CrossRef]

66. Wintle, A.G.; Murray, A.S. A review of quartz optically stimulated luminescence characteristics and their relevance in single-aliquot regeneration dating protocols. *Radiat. Meas.* **2006**, *41*, 369–391. [CrossRef]
67. Aitken, M.J. *An Introduction to Optical Dating*; Oxford University Press: Oxford, Britain, 1998.
68. Lu, Y.; Wang, X.; Wintle, A.G. A new OSL chronology for dust accumulation in the last 130,000 yr for the Chinese Loess Plateau. *Quat. Res.* **2007**, *67*, 152–160. [CrossRef]
69. Murray, A.S.; Wintle, A.G. The single aliquot regenerative dose protocol: Potential for improvements in reliability. *Radiat. Meas.* **2003**, *37*, 377–381. [CrossRef]
70. Prescott, J.R.; Hutton, J.T. Cosmic ray contributions to dose rates for luminescence and ESR dating: Large depths and long-term time variations. *Radiat. Meas.* **1994**, *23*, 497–500. [CrossRef]
71. Yuan, H.; Li, A.; Huang, W.; Hu, Z.; Zuo, Y.; Yang, X. Geological deformation of the TuoLi fault in the west Junggar since the late Quaternary. *Seismol. Geol.* **2023**, *45*, 49–66, (In Chinese with English Abstract).
72. Campbell, G.E.; Walker, R.T.; Abdrakhmatov, K.; Jackson, J.; Elliott, J.R.; Mackenzie, D.; Middleton, T.; Schwenninger, J.L. Great earthquakes in low strain rate continental interiors: An example from SE Kazakhstan. *J. Geophys. Res. Solid Earth* **2015**, *120*, 5507–5534. [CrossRef]

remote sensing

MDPI

Article

Improving Mineral Classification Using Multimodal Hyperspectral Point Cloud Data and Multi-Stream Neural Network

Aldino Rizaldy *, Ahmed Jamal Afifi, Pedram Ghamisi and Richard Gloaguen

Helmholtz-Zentrum Dresden-Rossendorf (HZDR), Helmholtz Institute Freiberg for Resource Technology (HIF), 09599 Freiberg, Germany; a.afifi@hzdr.de (A.J.A.); p.ghamisi@hzdr.de (P.G.); r.gloaguen@hzdr.de (R.G.)
* Correspondence: a.rizaldy@hzdr.de

Abstract: In this paper, we leverage multimodal data to classify minerals using a multi-stream neural network. In a previous study on the Tinto dataset, which consisted of a 3D hyperspectral point cloud from the open-pit mine Corta Atalaya in Spain, we successfully identified mineral classes by employing various deep learning models. However, this prior work solely relied on hyperspectral data as input for the deep learning models. In this study, we aim to enhance accuracy by incorporating multimodal data, which includes hyperspectral images, RGB images, and a 3D point cloud. To achieve this, we have adopted a graph-based neural network, known for its efficiency in aggregating local information, based on our past observations where it consistently performed well across different hyperspectral sensors. Subsequently, we constructed a multi-stream neural network tailored to handle multimodality. Additionally, we employed a channel attention module on the hyperspectral stream to fully exploit the spectral information within the hyperspectral data. Through the integration of multimodal data and a multi-stream neural network, we achieved a notable improvement in mineral classification accuracy: 19.2%, 4.4%, and 5.6% on the LWIR, SWIR, and VNIR datasets, respectively.

Keywords: mineral classification; geology; multimodal; hyperspectral; point cloud; deep learning; graph-CNN; hyperclouds; data fusion; multi-stream

Citation: Rizaldy, A.; Afifi, A.J.; Ghamisi, P.; Gloaguen, R. Improving Mineral Classification Using Multimodal Hyperspectral Point Cloud Data and Multi-Stream Neural Network. *Remote Sens.* **2024**, *16*, 2336. https://doi.org/10.3390/rs16132336

Academic Editors: Yanni Dong, Tao Chen and Chao Chen

Received: 8 May 2024
Revised: 20 June 2024
Accepted: 24 June 2024
Published: 26 June 2024

1. Introduction

Several studies [1–5] have focused on identifying mineral classes using hyperspectral data. Traditionally, these studies treated hyperspectral data as 2D images, with the spectral information represented by the depth of the image. However, in tasks such as mineral classification, particularly in challenging environments like open-pit mines, representing the real-world situation accurately is difficult using 2D images alone. Instead, 3D data, such as point clouds, offer a more comprehensive representation of the complex environment. Recent research [6–8] has integrated hyperspectral data with point clouds, resulting in what is termed 'hyperclouds', for various applications in mineral exploration and mining monitoring. Similarly, our previous work [9] integrated the hyperspectral data of multiple sensors from different views with 3D point clouds, creating a multimodal dataset comprising three distinct modalities: XYZ coordinates, hyperspectral data, and RGB images.

While we achieved success in identifying mineral classes using various deep learning models for point clouds, our prior work [9] only utilized hyperspectral data as input, overlooking the potential benefits of exploiting the multiple modalities present in our dataset. In this study, we aim to investigate the prospect of enhancing mineral classification accuracy by leveraging the multimodal nature of the dataset.

Training a multimodal dataset is often approached as a data fusion problem in machine learning. Fusion models typically fall into categories such as early, late, and intermediate fusion [10–12]. In our investigation, we explore various fusion architecture designs to develop efficient and reliable models for our task. Our experiments indicate that intermediate fusion yields the best performance, highlighting the benefits of utilizing a multi-stream network for multimodal data.

Our network architecture consists of two main parts. The first part comprises multiple branches, which are each dedicated to a different modality within the dataset. In the second part, we incorporate a single-stream network that aggregates the learned features from the previous branches. This design facilitates the effective integration of information from diverse modalities, thereby enhancing the overall performance of the model.

With the advancements in deep learning models for point cloud classification, we have a wide range of architecture options to consider as the backbone of our framework. Many of these backbones rely on local feature aggregation techniques exemplified by models such as PointNet++ [13], DGCNN [14], and Point Cloud Transformer [15].

Our previous work [9] indicated that DGCNN consistently performs well across various hyperspectral datasets. The primary feature of DGCNN lies in its use of the EdgeConv operator, which is known for being lightweight and efficient. Therefore, we opted to employ the EdgeConv operator as the feature encoder in our multi-stream network.

Additionally, as evidenced in various studies where leveraging channel information proves beneficial [16–19], we integrate a channel attention module to exploit spectral information. Our implementation of the channel attention mechanism involves incorporating 3D CNNs on top of the EdgeConv operator within the hyperspectral stream of the network. This implementation draws inspiration from the Efficient Channel Attention Network (ECANet) [19], which employs 1D CNNs as gating mechanisms to modulate channel-wise features. This integration results in notable accuracy improvements.

Finally, we conduct a comparison between point-based classification and image-based classification for the same task. While our dataset primarily consists of 3D hyperclouds, it also includes 2D images as their direct counterpart. Notably, the hyperclouds were derived from backprojected hyperspectral images to 3D point clouds. To ensure equitable comparison with our 3D approach, we adapted the EdgeConv operator for image data. Our experimental results reveal a significantly superior accuracy in point-based classification compared to image-based classification.

The key contributions of our works are as follows.

- We demonstrate that leveraging multimodal data yields superior classification results compared to utilizing unimodal data for mineral classification tasks.
- Our study reveals the network's capability to learn various data modalities present in 3D hyperclouds through the expansion of the network into a multi-stream architecture.
- We introduce an approach of employing 3D CNNs on top of the EdgeConv operator to capture spectral patterns, leading to enhanced segmentation performance.
- We perform a direct comparison of segmentation results between point cloud and image data formats, highlighting the superior performance of point-based segmentation.

The rest of this paper follows the following structure: Section 2 discusses existing works on hyperspectral and point cloud data segmentation. Next, in Section 3, we introduce our hypercloud data used in this work along with the proposed framework. Section 4 presents the results of mineral classification on our dataset, including a comparison between a point-based segmentation and an image-based segmentation. In Section 5, we conduct ablation studies to further understand the significance of each component of our framework. Finally, Section 6 provides the conclusion of our work.

2. Related Works

2.1. Hyperspectral Data Segmentation

For a long time, hyperspectral features have been used as predictors for various image classification tasks, including remote sensing data for land cover classification [20,21]. Unlike standard three-channel RGB images, hyperspectral data often consist of hundreds of spectral channels. In the past, the exploitation of spectral features as the only predictors [22–26] has been the common way to classify pixels of hyperspectral data. Having identified the predictors, any machine learning models can be trained in supervised learning. Although it might result in good classification scores as was reported in some works [27,28], the spatial relationships are omitted. Given the fact that the semantic information of pixels is usually closely related to their neighbors, neglecting the spatial relationship of the pixels can be seen as wasting the full potential of the data. The spatial relationship should be exploited to achieve higher classification scores.

Some unsupervised learning segmentation techniques have been developed to benefit from the spatial relationships [29–31]. The segmentation is usually performed under the assumption that adjacent pixels in the metric domain have a higher probability of being in the same class. The classification can be followed by supervised learning or remains purely unsupervised.

The emergence of CNNs for natural images motivated the community to adopt 2D convolutional filters for the hyperspectral data [32,33]. Furthermore, instead of only using 2D CNNs, some studies [2,3,34,35] pair 2D CNNs with 1D CNNs. The former captures local spatial features, and the latter learns spectral features. However, this approach is not direct and therefore has been avoided lately.

Recent studies [1,4] suggest employing 3D CNN to learn spatial and spectral features simultaneously. It is inevitable since hyperspectral data are commonly presented as 3D cubes rather than 2D rasters. The 3D convolutional filters learn features naturally on the 3D cubes. Hence, the network is simpler and easier to train [4,36].

2.2. Point Cloud Segmentation

In recent years, there has been a significant increase in the popularity of deep learning networks designed for point cloud classification, as noted by researchers such as Guo et al. [37], Bello et al. [38], Zhang et al. [39], and Xie et al. [40]. The 3D point cloud format has become particularly attractive for deep learning applications, as it provides the possibility to extract more comprehensive features such as local and global 3D geometrical features.

Learning point features is quite different from learning features in images because point clouds are unordered, unstructured, and lack a grid format [37–39]. Nevertheless, various approaches have been developed for point cloud classification and segmentation. According to [37], some of these approaches can be categorized as follows:

- Multi view images [41,42], which project points onto images and use typical 2D convolution networks.
- Occupied voxels with 3D convolution networks, either using dense convolution [43,44] or sparse convolution [45,46].
- MLP-based architectures [13,47] that rely on the max-pooling layer to achieve permutation invariance.
- Graph-based architectures [14,48] that generate k-NN graphs and pass them to MLP-based networks.
- Convolutional-based architectures [49–52] that develop custom point-convolution operators.
- Transformer-based architectures [15,53] that solely use self-attention modules as the encoder.

2.3. Hyperspectral and Point Cloud Fusion

The fusion of hyperspectral data and point clouds has garnered increased attention in recent years. Some studies [54,55] have developed methods for integrating LiDAR and hyperspectral data. The MUUFL Gulfport dataset [56] not only provides an integrated product but also includes corresponding ground truth labels. Other research [57,58] utilizing this dataset has concluded that combining hyperspectral and LiDAR data yields the highest accuracy compared to the data alone. Spectral features enable the classifier to detect the materials of objects, while geometric features are suitable for distinguishing different object shapes. Additional studies [55,59–61] consistently showed that fusion leads to more promising results compared to working with each data modality separately.

Recent works [62,63] describe the application of deep learning models for the segmentation of hyperspectral point clouds using the urban scene GRSS18 dataset [64]. These models are based on various pixel and point convolution architectures. Surprisingly, an all-pixel convolution with the integration of hyperspectral and LiDAR data outperformed 2D and 3D combinations [62]. Another work [65] also demonstrated the performance of a deep learning model with a different architecture using random point sampling. These studies have motivated the exploration of deep learning and the development of specialized networks for multimodal data.

2.4. Fusion Models for Multimodal Data

In machine learning, fusion models for multimodal data are typically implemented as early fusion, late fusion, or intermediate fusion [10–12]. Early fusion, or data-level fusion, is the data merging from different sensors as input for the machine learning model. Late fusion, also known as decision-level fusion, involves using independent models for each modality and then combining their decisions to improve results. These fusion techniques are considered traditional and were commonly applied before the deep learning era, using classic machine learning models like Random Forest or Support Vector Machine.

Deep neural networks introduce intermediate fusion (feature-level fusion) [10,11]. The flexibility of neural network architecture allows for various fusion schemes, enabling models to fuse learned features at different levels of abstraction. This flexibility also allows different modalities to communicate with each other at different levels, as seen in recent works [66,67]. Additionally, research by [11] showed that intermediate fusion often resulted in higher classification scores than early or late fusion. In [68], different fusion strategies, including dense and sparse fusion approaches, were investigated.

Multimodal remote sensing data have also been widely explored within deep learning frameworks [69]. To improve classification results, some studies [70,71] have fused radiometric and geometric features of remote sensing images. Various early and late fusion networks (e.g., ResUNet, V-FuseNet, SegNet-RC) were briefly studied in [72,73]. The work in [12] goes beyond early, late, or intermediate fusion by extracting features in a cross-modal multi-scale manner. PerceiverIO [66], as a general model for multimodal data, has demonstrated its capabilities in remote sensing image segmentation when combined with 3D convolutions [74].

3. Dataset and Methods

3.1. Dataset Description

The Tinto dataset [9] is a point cloud dataset with additional hyperspectral information. The respective scene is located in the open-pit mine Corta Atalaya, Minas de Riotinto, Spain. There are three hyperspectral feature groups, i.e., Long-Wave Infrared (LWIR) with 126 bands, Short-Wave Infrared (SWIR) with 141 bands, and Visible and Near Infrared (VNIR) with 51 bands.

As mentioned before, we combined hyperspectral data with 3D point clouds, which is rare for common point cloud datasets. We utilized drone-borne and terrestrial sensors for the data acquisition. An aerial camera on the drone captured the RGB data. From there, dense 3D RGB point clouds were reconstructed using Structure from Motion (SfM) and a multi-view stereo technique. Meanwhile, hyperspectral data were captured using terrestrial sensors. Subsequently, the data were back-projected onto the point clouds to incorporate the additional hyperspectral features.

The dataset consists of 10 mineral classes: Saprolite, Chert, Sulfide, Shale, Purple Shale, Mafic A, Mafic B, Felsic A, Felsic B, and Felsic C. It has 297,968 points in the training set and 2,885,178 points in the testing set. The training samples were collected manually from the field by domain experts, ensuring the highest standards of accuracy. These samples were then meticulously evaluated in the laboratory to determine the precise classification of the minerals. Due to the challenging conditions at the mining site, 10% of the overall data available was collected. We can see the distribution of training and testing sets in Figure 1 where the training set is colored red and the testing set is colored blue. We can also see the class distribution of the training set. As we may notice, there is an imbalance class issue, which makes this dataset more challenging.

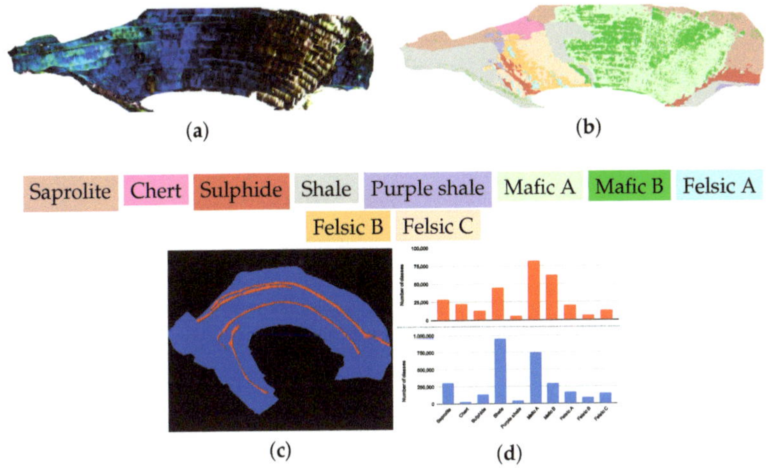

Figure 1. (**a**) False-color visualization of the LWIR hypercloud data at 10,114, 9181, and 8545 nm, (**b**) the corresponding semantic label, (**c**) training (red) and testing (blue) sets, and (**d**) the class distribution of training (red) and testing (blue) sets.

3.2. Data Preprocessing

Assuming both training and testing sets of point clouds denoted by P, we initially split these point clouds into blocks referred to as P_{block}. The size of each block is fixed. It is crucial to choose an appropriate block size that is large enough to capture the context of the points within the block while avoiding memory overloading.

To ensure that each block contains sufficient information, we set a threshold value as the minimum number of points required for each block. Before the training phase, we remove all blocks that contain a number of points below the threshold value. This step is essential to ensure that the network is trained on sufficient number of points.

After filtering the point clouds, we sample the points in each remaining P_{block} to obtain a fixed number of points, which is denoted by n. We implement the following conditional operation to check whether the number of points in each P_{block}, denoted as s, is greater or less than n. If s is less than or equal to n, we duplicate the points to obtain n points, while if s is greater than n, we downsample the points to obtain n points.

$$P_{block} = \begin{cases} \text{duplicate} & \text{if } s \leq n \\ \text{downsample} & \text{if } s > n \end{cases} \tag{1}$$

The choice of n depends on the point density of the point clouds itself. If it is set too low, then we lose the resolution of the data. But if it is set too high, then many points are duplicated, leaving the network to learn synthetic and meaningless features.

In summary, this preprocessing method ensures that the neural network trains on the same size of the input tensors for multi-batch multi-GPU training while reducing the computational complexity of the network.

The purpose of blocking and sampling points here is to have a consistent input size for the network during the training. It is similar to image-based networks where the input is always having a fixed size, e.g., 256 × 256, 512 × 512 pixels, or so on. Since the training phase is on batch processing, all batches need to have the same size if we use multiple batch sizes. During the testing, it is not mandatory to sample the number of points if we only use a batch size equal to one. Instead, we have to avoid sampling because we want to classify all points. However, it still needs to be split into blocks with the same size as in the training to ensure that the contextual information of the blocks is not different.

We observed that the 50 × 50 meters block size covers sufficient semantic information while keeping the block size reasonable for the computation cost. This size was also chosen to maximize the number of points per block. However, it should be noted that different datasets might need different sizes, depending on the object sizes and scene complexities. We then created blocks in 1000 random locations on the training set instead of generating non-overlapping systematic blocks as performed in previous works [47,49]. Next, the number of points for each block was set to 4096 points to have a similar order of magnitude between different blocks with different point density. Therefore, the downsample and upsample will only have a minor effect.

3.3. Proposed Network

Our network architecture utilizes several techniques to achieve its performance. The complete architecture can be seen in Figure 2. In this section, we will provide a brief overview of these techniques.

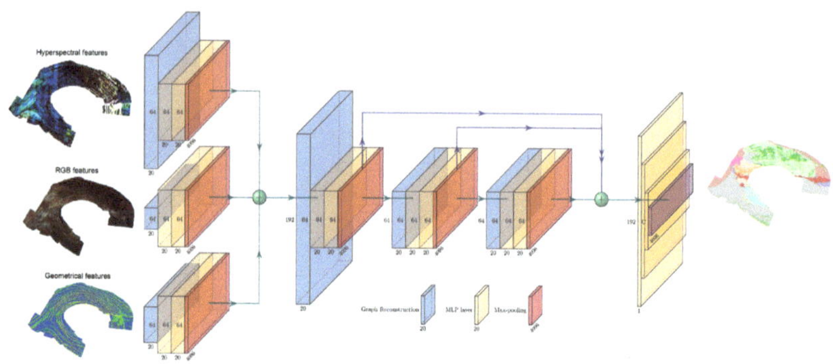

Figure 2. Illustration of the multi-stream network. The architecture consists of 3 individual streams in the earlier part of the network, which is followed by a typical single-stream network with skip connection links (illustrated as blue arrows) to bring the information from the earlier layers to the latter. The 3 individual streams are designed to work with 3 different modalities. Those are hyperspectral, RGB, and geometrical data. The channel-wise concatenation is utilized for merging the output of the 3 different streams. It is also used for concatenating the skip connection links. Graph reconstruction, MLP, and max-pooling layers are stacked together as the features encoder block. This block is utilized in each stream separately and then repeated three times for deeper layers. Finally, dense layers are applied in the last layers for the classification. Please note that height represents the number of features, width represents the number of k-NN points, and depth represents the number of points for each block.

3.3.1. Graph Reconstruction

In our previous work [9], we extensively analyzed various encoder architectures and concluded that EdgeConv (DGCNN) consistently delivers strong performance for mineral classification with hyperspectral data. Notably, EdgeConv's effectiveness is complemented by its lightweight operation, which is attributable to its reliance on graph MLP for feature aggregation. By employing EdgeConv, we ensure the equal treatment of different modalities, enabling the network to learn the amalgamation of features from all modalities. Further elaboration on this aspect is provided in the subsequent section.

EdgeConv layer is a graph-based operation. Suppose a point cloud P with d-dimensional features and p points. A point cloud P is then represented as $\{P_i \mid i = 1, \ldots, p\} \in \mathbb{R}^{p \times d}$.

The simplest point cloud always has $d = 3$ from the XYZ coordinates. The hyperspectral point clouds, however, have d equal to the number of hyperspectral channels plus the 3-dimensional of the coordinates itself. Therefore, P_i of each hyperspectral point is a vector $\mathbf{v} = (v_1, \ldots, v_{d-1}, v_d)$ where $v_1 = X\,coordinate, v_2 = Y\,coordinate, v_3 = Z\,coordinate$.

The EdgeConv layer creates a directed graph $G = (V, E)$ with V as vertices and E as edges. Since the graph allows edges to loop, the E is defined as $E \subseteq \{(x,y) \mid (x,y) \in V^2\}$ where (x,y) is the edge directed from x to y.

For each block, the graph G is constructed by finding the k-nearest neighbor (k-NN) of the point cloud P_{block} with the number of points n. It should be noted that n is the chosen number for the point size of the blocks. The distance to the k-NN could be defined as a Euclidean distance in the metric space (X, Y, Z) on simple point clouds plus the Euclidean distance in the feature space for the higher dimensionality data as in hyperspectral point clouds. In our implementation, the distance is calculated from both of them. However, only the normalized XYZ is used in the metric space to ensure that the distances are in the same range for all features.

Suppose a set of neighboring points where P_i is the center point and P_{ij} represents the k nearest points where $\{P_{ij} \mid j = 1, \ldots, k\} \in \mathbb{R}^{k \times d}$; then, the edges are simply defined as (P_i, P_{ij}). With Θ as a set of trainable weights and $h_\Theta : \mathbb{R}^d \times \mathbb{R}^d \to \mathbb{R}^{d'}$ as a non-linear function, EdgeConv [14] defines the edge features as $e_{ij} = h_\Theta(P_i, P_{ij})$. Explicitly, if $\Theta = (\theta_1, \ldots, \theta_m, \phi_1, \ldots, \phi_m)$, the output of the EdgeConv is expressed in Equation (2).

$$e'_{ijm} = (\theta_m \cdot (P_{ij} - P_i) + \phi_m \cdot P_i) \quad (2)$$

In a point cloud segmentation task, it is important to capture both local and global structures to learn extensive point features. To this end, EdgeConv extracts local neighborhood structures by $P_{ij} - P_i$ and combines them with the global features captured by P_i.

Suppose a tensor is taken as an input with the size of (b, d, n) where b is the batch size, d is the number of features, and n is the number of points in a block; then, the output of the graph features will be a tensor with the size of $(b, d \times 2, n, k)$ where k is the number of point neighbors.

3.3.2. Multi-Stream Architecture

Classification tasks with different modalities of the input data have frequently been investigated in image segmentation tasks by implementing multi-stream networks as in [75–77]. The idea of the multi-stream network is to let the network learn different features for different data modalities.

In contrast to learning multimodal data through multiple sub-networks, simply concatenating all different modalities to the same features vector is the naive solution. It treats all modalities equally, although some modalities are probably more informative than others. It leads to a dilution of useful information, as the simple concatenation may include irrelevant features [78,79].

In that sense, we argue that it would be beneficial to have a multi-stream network since our dataset has at least three different data types: XYZ, hyperspectral, and RGB. We present our proposed network in Figure 2. In this architecture, instead of stacking all features together and having a single-stream network, as in data-level fusion, the network is expanded into three independent streams. Despite its advantage, the construction of a large and heavy network is avoided, considering that the single stream network of hyperspectral features is already acknowledged as heavy. Dividing the network into three streams would ultimately triple the network's size. In that sense, the network is divided into three streams only for the first part. The output of each stream is then concatenated as the input for the second part of the network. While it would have been possible to have three fully independent streams and merge the features in the last layer, this approach would have led to a significant increase in the number of parameters. In our approach, the learning of different features from different data modalities is facilitated by the network without resulting in an excessive number of parameters.

In our model, an EdgeConv block is utilized for each stream. This block comprises a graph layer, two consecutive MLP layers, and max-pooling layers. In Figure 2, each layer is represented by a different color. The graph layer computes edge features, as explained in the previous section. The MLP layers act as feature encoders, each with 64 filters. These MLP layers are then followed by a max-pooling layer that operates on k-NN points. This means that the max-pooling layer selects the most significant points in the neighborhood. It is important to note that we define the distance for k-NN not only by the physical distance but also by the feature distance. The distance is computed using the Euclidean distance in multidimensional data. To enhance efficiency, the calculation is optimized in PyTorch version 1.13.0.

Each stream in the model produces a tensor of size (b, l, n), where b is the batch size, l is the number of filters of the MLP layer, and n is the number of points. To combine these streams, channel-wise concatenation is employed to create a tensor of size $(b, l \times 3, n)$, as shown in Figure 2. Here, the height represents the number of features, and the second graph layer in the network has a height three times the height of each previous max-pooling layer. After the concatenation, a single stream network is applied to process all the features. This network consists of three identical EdgeConv blocks stacked consecutively, each with a graph layer, two MLPs, and a max-pooling layer. Additionally, shortcut connections are used to obtain multi-scale features, again using channel-wise concatenation. The architecture does not use a downsampling mechanism, which is similar to PointNet [47]. As a result, the n number of points, or the block size in our case, is maintained for every layer, avoiding the downsampling and upsampling mechanisms seen in other semantic segmentation architectures [13,49], as illustrated in Figure 2.

Our architecture differs from the original Dynamic Graph CNN [14], as a global pooling layer is not used. The global pooling layer is commonly used to achieve permutation invariance given a set of points in object or shape classification tasks [47]. However, we argue that this pooling layer is not suitable for semantic segmentation tasks where points may have different labels.

The reason for this is that the pooling layer attaches the same pooled features to all points, which is useful for classification but not for segmentation. In segmentation tasks, the original encoded features of each point need to be maintained to preserve its semantic information. Therefore, the global pooling layer is avoided in our architecture for semantic segmentation tasks.

3.3.3. The 3D CNN as Channel-Attention Module

We enhance our model's ability to leverage spectral information by integrating a channel attention module. This module is designed to dynamically emphasize relevant features within the hyperspectral data. Our implementation strategy involves embedding 3D convolutional neural networks (CNNs) on top of the EdgeConv operator, which is a key component within the hyperspectral stream of the network. This innovative approach draws inspiration from the Efficient Channel Attention Network (ECANet) [19] that employs 1D CNNs as gating mechanisms to selectively modulate channel-wise features, effectively enhancing the model's discriminative power. By adapting this concept to our hyperspectral classification task, we aim to improve the model's ability to focus on salient spectral features, thereby boosting the overall classification performance.

Formally, the output matrix $\mathbf{O} \in \mathbb{R}^{b \times c_{out} \times d_{out} \times h_{out} \times w_{out}}$ of a 3D convolutional layer with input $\mathbf{I} \in \mathbb{R}^{b \times c_{in} \times d_{in} \times h_{in} \times w_{in}}$ can be described as

$$\mathbf{O}_{\mathbf{b_i},\mathbf{c_{out}}} = \mathbf{B}_{\mathbf{c_{out_j}}} + \sum_{\nabla=0}^{c_{in}-1} \mathbf{W}_{c_{out_j},\nabla} \bigstar \mathbf{I}_{b_i,\nabla} \tag{3}$$

where $\mathbf{W}$ is a weight matrix, $\mathbf{B}$ is a bias, and $\bigstar$ is a 3D cross-correlation operator with element-wise matrix multiplication.

While a 3D convolutional filter can extract spatial–spectral features, it is only suitable for grid-like data such as hyperspectral data in raster format and cannot be directly applied to hyperspectral point clouds. On the other hand, the EdgeConv layer works well for capturing local spatial structures, but it ignores the spectral patterns of the point clouds. To incorporate both spatial and spectral information, we need to integrate the 3D convolutional filter with the EdgeConv layer in the architecture. Here, we explain how we combine the 3D convolutional filter with the EdgeConv layer to achieve this integration.

First, graph reconstruction is performed on k-NN on points to gain the benefit of the local structures. Given n points and k nearest-neighbors for each point, the graph is a tensor of size $(b, d \times 2, n, k)$ with b as the batch size and d as the dimension of the features. We then add an empty dimension, so the tensor has a size of $(b, 1, d \times 2, n, k)$. Then, two identical 3D convolutional blocks are used; each has 4 filters with size $(32, 1, 1)$. As it suggests, the filters learn the spectral features by capturing the pattern for every 32 hyperspectral features. The latter $(1, 1)$ dimension works similarly to fully connected layers, as in MLP. The output tensors will be in the size of $(b, 4, d \times 2, n, k)$. In this manner, we can exploit the benefit of hyperspectral features on point cloud data. Figure 3 illustrates the operation of the hyperspectral stream using 3D CNNs.

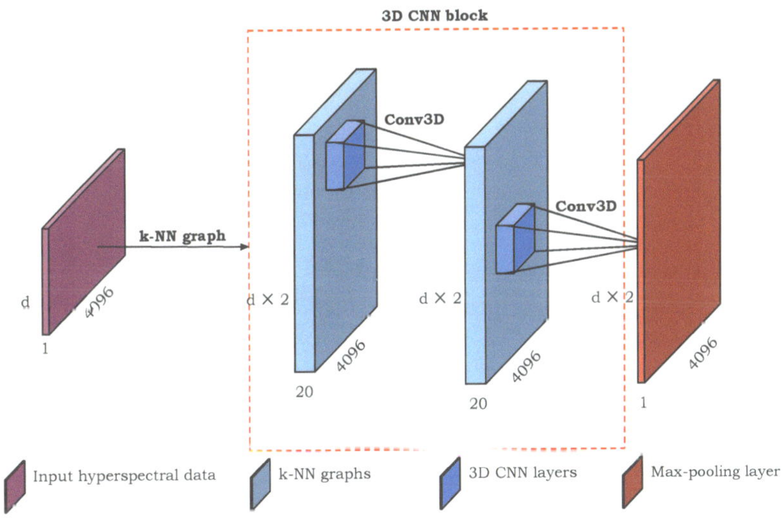

Figure 3. Schematic diagram of the 3D CNN block in the hyperspectral stream.

Subsequently, the dimension is reduced into $(b, d \times 2, n)$ by applying two max-pooling operators. The former is used to aggregate significant features among all filters, and the latter is used to aggregate the significant points in the neighborhood. Finally, the MLP-like network is used, as in other streams.

3.3.4. Geometric Features Transformation

In our experiments, we found that the graph network directly learning geometric features from the XYZ coordinates is not beneficial for detecting the mineral classes in the open-pit mine area. This is because minerals do not have distinct shapes that differ from one to another. Moreover, the training area spreads only in the horizontal direction, while the testing area has flat and sloped terrain, which is similar to a typical landscape in an open-pit mine. We provide more details about the dataset in Section 4.

However, we avoid completely ignoring the XYZ coordinates and still expect to gain some benefit from them. To further exploit the XYZ coordinates, we follow the Superpoint Graph (SPG) network [80] by transforming the XYZ coordinates into pre-computed geometrical features. These features capture the geometric characteristics of each point with respect to its neighbors, such as whether the point lies on a planar surface or not. Then, the geometric features are fed to the network as one of the streams. Unlike SPG where the geometric features are computed on the superpoint, the geometrical features for each point are computed before the graph reconstruction in our work. We followed the 14 geometric features as suggested by [81].

Let $\lambda_1 > \lambda_2 > \lambda_3 > 0$ be the eigenvalues and $\mathbf{e}_1, \mathbf{e}_2, \mathbf{e}_3$ be the eigenvectors of the covariance matrix of points to the neighbors; the 14 geometric features are formally described by [81] as in Table 1.

Since the computed features are highly dependent on the definition of the neighborhood, it is crucial to choose it carefully. In our architecture, a ball query is preferred to define the neighborhood rather than a k-NN. The reason behind this choice is that the ball query maintains the spatial extent of the neighborhood, while the spatial extent of the k-NN varies depending on the local density of the point cloud. Additionally, two different radii, 0.5 m and 1 m, are selected to capture multi-scale features. Overall, the point clouds are transformed into 28 geometrical features through the combination of EdgeConv blocks and multiscale features.

Table 1. The 14 transformed geometric features.

No	Feature	Description
1	Sum	$\lambda_1 + \lambda_2 + \lambda_3$
2	Omnivariance	$(\lambda_1 \cdot \lambda_2 \cdot \lambda_3)^{\frac{1}{3}}$
3	Eigenentropy	$-\sum_{i=1}^{3} \lambda_i \cdot ln(\lambda_i)$
4	Anisotropy	$(\lambda_1 - \lambda_3)/\lambda_1$
5	Planarity	$(\lambda_2 - \lambda_3)/\lambda_1$
6	PCA 1	$\lambda_1 - (\lambda_1 + \lambda_2 + \lambda_3)$
7	PCA 2	$\lambda_2 - (\lambda_1 + \lambda_2 + \lambda_3)$
8	Linearity	$(\lambda_1 - \lambda_2)/\lambda_1$
9	Surface Variation	$\lambda_3/(\lambda_1 + \lambda_2 + \lambda_3)$
10	Sphericity	λ_3/λ_1
11	Verticality	$1 - \lvert\langle[001], \mathbf{e}_3\rangle\rvert$
12, 13, 14	3 normal vectors	N_x, N_y, N_z

3.4. Comparison to Image Segmentation

The hyperspectral attributes were not originally part of the point clouds but were derived from hyperspectral data obtained by different sensors. These additional attributes were added by backprojecting the hyperspectral data onto the point clouds. It would be interesting to compare the performance of image-based and point-based deep learning networks by examining the segmentation results of both datasets.

To ensure a fair comparison between point cloud and image-based segmentation, we developed an image segmentation network based on the graph network architecture used for point cloud segmentation. The approach is similar with edge features computed for each pixel to its neighboring pixels using a fixed size kernel instead of k-NN to define the neighborhood and construct the graph. Specifically, the edge features are calculated as $e_{ij} = h_\theta(x_i, x_j - x_i)$, where x_i represents the central pixel and x_j represents the surrounding pixel. To obtain multi-scale graph features with different receptive field sizes, two different kernel sizes are used, 3×3 and 5×5 pixels. Once the graph is reconstructed, it passes through two shared MLP layers for each scale, and the results are concatenated before feeding them into the fully connected layers. Figure 4 illustrates the graph reconstruction and the architecture of the 2D-based network.

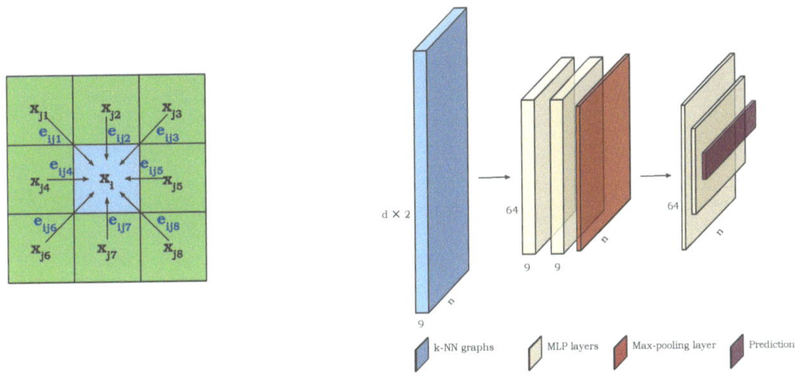

Figure 4. (**Left**): Graph reconstruction using an image (x_i represents the central pixel, x_j represents the surrounding pixel, and e_{ij} denotes the edge features). (**Right**): Illustration of the 2D-based network.

In addition to the graph-based network, we also train a simple MLP network that does not consider the relationship between pixels and their neighborhoods. This allows the network to learn solely from the spectral values of individual pixels.

4. Results

4.1. Training Setting

The network was trained from scratch using the SGD optimizer with momentum [82] and implementing it on Pytorch. The momentum was set to 0.9 with a learning rate of 0.001. A training batch size of 56 was used, and the network was trained for 100 epochs. The best models were selected based on the validation set.

A GPU cluster with 8 Nvidia A100s was utilized to speed up the training time; however, it is not necessary to use such a huge machine to train our network, as the size of the network and the dataset are reasonable to be trained with the regular GPUs. Depending on the input of the hyperspectral data, the numbers of parameters of the network are 311,424, 324,112, and 305,424 for the LWIR, SWIR, and VNIR sensors, respectively.

The class weights were also added during training to address the class imbalance. Given the number of instances for each class n_c and the number of all instances n_t, we computed the weight for each class w_c according to Equation (4) and passed the weights to the cross-entropy loss function.

$$w_c = 1 - \left(\frac{n_c}{n_t}\right) \tag{4}$$

The code for our work will be made available at https://github.com/aldinorizaldy/HyperspectralGraphNetwork, accessed on 7 May 2024.

4.2. Quantitative Results

The performance of our method is compared to the baseline methods as reported in our prior work [9]. Regarding the evaluation metric, in addition to the overall accuracy (OA), the F1-score per-class accuracy is used to show the detailed accuracy for each class. Furthermore, the class-averaged mean-F1 score (mF1) is also reported, with balanced (bal-mF1) and unbalanced (unbal-mF1) scores, to have a more comprehensive comparison of the performance of our network against the baselines.

In terms of the OA, as seen in Tables 2–4, the proposed method outperformed all comparative methods by a large margin, i.e., 19.2, 4.4, and 5.6 percentage points for each LWIR, SWIR, and VNIR dataset, respectively. In terms of the mean-F1 score, our method also shows similar results by surpassing the existing methods by 16.8, 4.0, and 4.8 percentage points for balanced scores and 16.6, 5.1, and 5.8 percentage points for unbalanced scores.

Table 2. Accuracy of LWIR dataset. The highest accuracy is indicated in bold.

Class	PointNet	PointNet++	PointCNN	ConvPoint	DGCNN	PT	PCT	Ours
Saprolite	38.5	39.1	46.0	41.9	39.4	40.5	39.3	**68.1**
Chert	16.3	17.4	17.5	18.4	17.8	25.9	17.8	**43.0**
Sulfide	50.1	50.0	45.9	46.8	45.9	41.8	49.3	**68.2**
Shale	45.5	42.4	58.7	47.9	46.8	44.5	48.1	**79.9**
Purple Shale	7.2	11.3	9.4	10.8	7.8	6.0	8.6	**27.0**
Mafic A	56.2	60.4	55.5	57.8	58.9	63.1	57.5	**68.9**
Mafic B	**40.8**	38.7	40.7	39.0	37.6	33.9	36.4	39.1
Felsic A	24.2	20.2	27.4	31.7	18.9	24.0	18.9	**52.7**
Felsic B	**20.6**	19.5	14.8	15.5	17.5	26.7	19.0	18.7
Felsic C	35.8	33.5	23.9	38.0	28.3	24.0	30.2	**48.2**
OA	43.2	43.0	47.3	45.1	42.7	45.3	43.3	**66.5**
bal-mF1	43.9	43.5	48.2	45.7	43.9	44.2	44.1	**65.0**
unbal-mF1	33.5	33.2	34.0	34.8	31.9	33.0	32.5	**51.4**

Table 3. Accuracy of SWIR dataset. The highest accuracy is indicated in bold.

Class	PointNet	PointNet++	PointCNN	ConvPoint	DGCNN	PT	PCT	Ours
Saprolite	55.0	63.9	49.6	58.9	53.7	51.0	66.4	**72.3**
Chert	32.7	**44.2**	15.3	21.0	37.2	26.8	43.7	43.9
Sulfide	51.5	54.4	36.6	54.3	47.9	29.0	49.6	**63.2**
Shale	72.3	76.4	73.7	74.3	76.3	71.8	78.1	**82.0**
Purple Shale	27.8	14.8	10.8	11.7	16.3	6.9	20.2	**36.8**
Mafic A	79.2	79.2	75.3	79.5	78.5	74.5	77.3	**81.6**
Mafic B	**77.8**	76.4	49.8	76.3	76.6	42.6	74.7	73.8
Felsic A	62.3	66.6	67.1	59.4	61.2	54.9	66.8	**72.3**
Felsic B	47.8	**59.0**	4.3	42.9	52.9	2.7	52.7	58.9
Felsic C	71.9	72.0	44.0	64.8	69.3	43.2	66.4	**72.9**
OA	67.7	71.5	59.7	67.7	68.9	60.2	71.8	**76.2**
bal-mF1	69.5	72.3	62.0	69.6	70.1	59.6	71.9	**76.2**
unbal-mF1	57.8	60.7	42.7	54.3	57.0	40.3	59.6	**65.8**

Table 4. Accuracy of VNIR dataset. The highest accuracy is indicated in bold.

Class	PointNet	PointNet++	PointCNN	ConvPoint	DGCNN	PT	PCT	Ours
Saprolite	60.3	64.0	52.7	55.1	62.8	43.2	68.5	**82.4**
Chert	23.1	29.3	14.3	13.6	28.9	25.4	28.7	**35.3**
Sulfide	51.9	47.3	39.5	49.0	51.0	25.4	49.4	**55.5**
Shale	69.4	62.0	62.1	57.3	62.6	59.5	65.5	**73.2**
Purple Shale	38.0	**43.4**	23.7	19.6	30.5	4.8	34.4	40.9
Mafic A	73.1	73.8	63.0	69.6	70.2	70.2	72.1	**76.6**
Mafic B	**54.4**	51.1	42.2	46.9	48.7	18.8	46.8	53.9
Felsic A	57.7	53.9	**57.9**	53.4	52.0	42.5	57.2	56.2
Felsic B	33.2	37.1	19.1	37.7	34.0	2.3	33.4	**40.9**
Felsic C	51.2	52.0	28.1	51.3	50.6	34.5	50.5	**56.8**
OA	62.1	60.7	51.8	55.4	59.1	51.2	61.2	**67.7**
bal-mF1	63.5	61.2	54.0	56.7	59.8	49.8	61.9	**68.3**
unbal-mF1	51.2	51.4	40.3	45.4	49.1	32.7	50.6	**57.2**

In the LWIR dataset, our method achieved an OA of 66.5%, while PointNet, PointNet++, PointCNN, ConvPoint, DGCNN, PointTransformer, and PCT achieved OAs of 43.2%, 43.0%, 47.3%, 45.1%, 42.7%, 45.3%, and 43.3%.

Similarly, in the SWIR dataset, our proposed method achieved an OA of 76.2%, while PointNet, PointNet++, PointCNN, ConvPoint, DGCNN, PointTransformer, and PCT achieved OAs of 67.7%, 71.5%, 59.7%, 67.7%, 68.9%, 60.2%, and 71.8%, respectively. Furthermore, our proposed network also shows its superiority on the third dataset, the VNIR dataset, where it obtained an OA of 67.7% while PointNet, PointNet++, PointCNN, ConvPoint, DGCNN, PointTransformer, and PCT achieved OAs of 62.1%, 60.7%, 51.8%, 55.4%, 59.1%, 51.2%, and 61.2%.

Our method also consistently achieved the highest class-averaged mean-F1 scores both for the balanced and unbalanced scores. The balanced mean-F1 shows the weighted accuracy of all classes with respect to the number of instances for each class. Meanwhile, the unbalanced mean-F1 simply averaged all per-class accuracy without taking into account the class imbalance. The proposed network obtained 65.0%, 76.2%, and 68.3% of the balanced mean-F1 and 51.4%, 65.8%, and 57.2% of the unbalanced mean-F1 for the LWIR, SWIR, and VNIR datasets, respectively. It is expected to have lower values for the unbalanced scores since some classes have a limited number of samples.

These results prove the ability of our method to infer the semantic information of mineral classes, which is measured by not only the overall point accuracy but also more importantly the class-averaged accuracy. Furthermore, if we look at the two classes with the smallest number of samples, Chert and Purple Shale, the proposed method achieved the highest per-class F1-score accuracy on all datasets. It only loses by PointNet++ on the VNIR dataset for the Purple Shale class.

4.3. Qualitative Results

We further compare the results of different methods by visually inspecting the prediction. Figures 5–7 plot the point clouds, color-coded by the predicted labels, on the LWIR, SWIR, and VNIR datasets.

As seen in Figure 5, our method generally resulted in less noisy labels than the comparison methods. Furthermore, all the comparison methods suffered from false positive errors of the Saprolite and Chert class on the left side of the area. Our method, on the other hand, predicted this particular area with the correct Shale class. Nevertheless, none of them has successfully predicted the Purple Shale class on the right side of the area.

In the SWIR dataset, as illustrated in Figure 6, our method also demonstrated superiority over the existing methods. We can clearly observe that our method avoids the false positive error of the Saprolite class, as can be seen in the results of the comparison methods. On the other hand, the existing methods suffered from the false negative error of the Sulfide class, while our method predicted this particular class more accurately.

The results in the VNIR dataset, however, show different errors where almost all the methods failed to predict the Shale class and wrongly labeled the area with the Felsic A class. Point Transformer, on the other hand, has the minimum false positive error of the Felsic A class, but it almost completely failed to detect the Mafic B class.

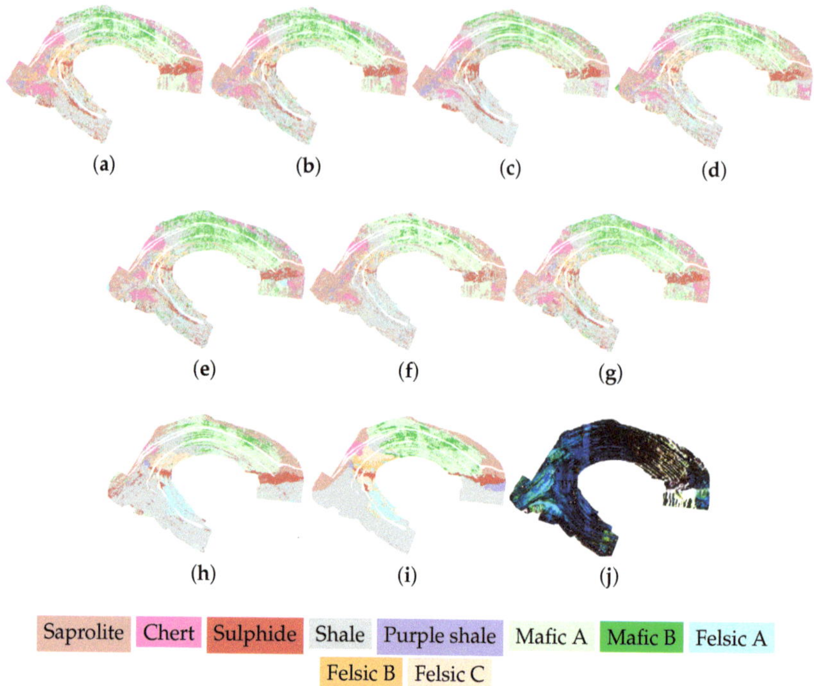

Figure 5. The predicted point clouds of the LWIR dataset and the corresponding ground truth. (**a**) PointNet; (**b**) PointNet++; (**c**) PointCNN; (**d**) ConvPoint; (**e**) DGCNN; (**f**) Point Transformer; (**g**) PCT; (**h**) ours; (**i**) ground truth; (**j**) LWIR hypercloud.

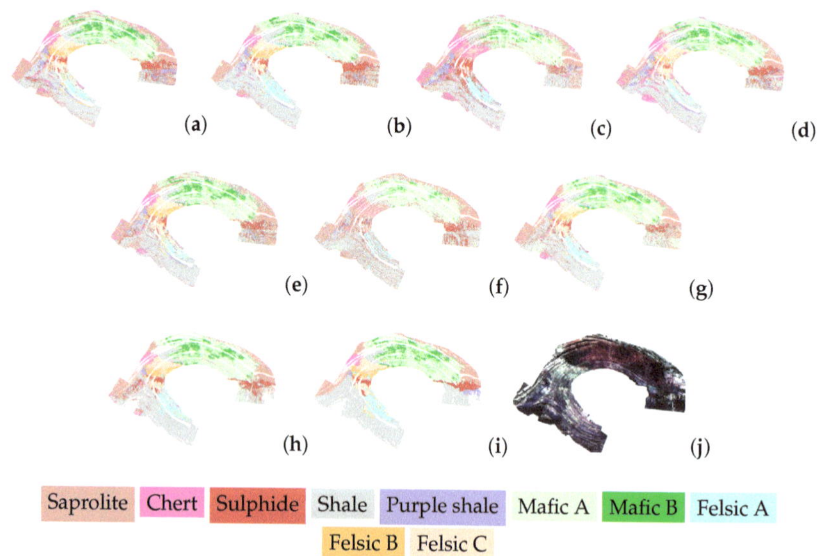

Figure 6. The predicted point clouds of SWIR dataset and the corresponding ground truth. (**a**) PointNet; (**b**) PointNet++; (**c**) PointCNN; (**d**) ConvPoint; (**e**) DGCNN; (**f**) Point Transformer; (**g**) PCT; (**h**) ours; (**i**) ground truth; (**j**) SWIR hypercloud.

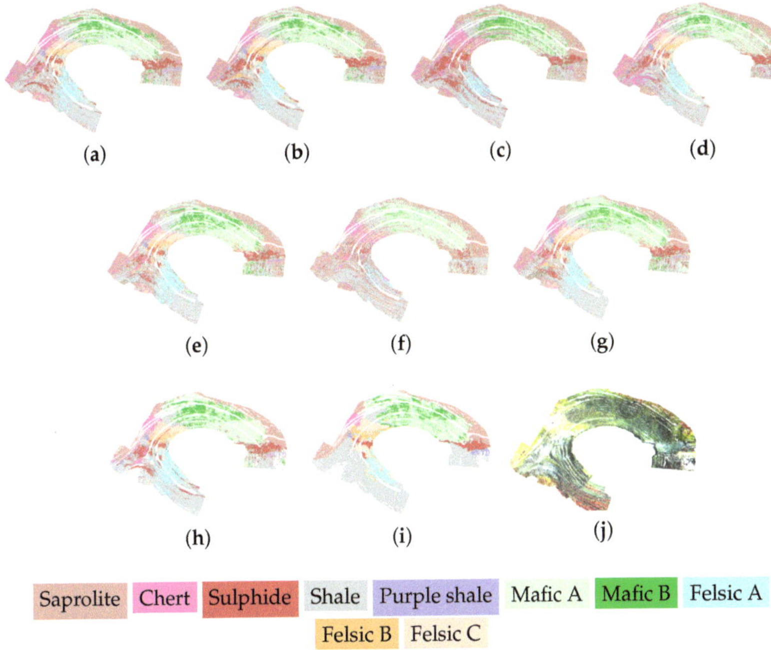

Figure 7. The predicted point clouds of VNIR dataset and the corresponding ground truth. (**a**) PointNet; (**b**) PointNet++; (**c**) PointCNN; (**d**) ConvPoint; (**e**) DGCNN; (**f**) Point Transformer; (**g**) PCT; (**h**) ours; (**i**) ground truth; (**j**) VNIR hypercloud.

4.4. Comparison of Point Cloud and Image Segmentation

After obtaining the hyperspectral attributes of the point clouds from the corresponding images, it is intriguing to observe the image-based segmentation's performance on the same dataset. We acknowledge that point clouds are in 3D format, providing a superior perception of the data to the network. However, deep learning has made significant strides in the field of image-based segmentation. Therefore, comparing the performance of the two methods can shed light on the ideal approach for achieving greater accuracy when both datasets are available. In Table 5, we show the F1-score per-class accuracy, overall accuracy, and F1-score of the image-based segmentation and point-based segmentation using the SWIR dataset.

Table 5. Comparison of image-based to point-based segmentation using the SWIR dataset.

Class	Image-Based Graph	Image-Based MLP	Point-Based Graph
Saprolite	31.6	–	72.3
Chert	72.8	56.3	43.9
Sulfide	42.1	–	63.2
Shale	67.1	31.2	82.0
Purple Shale	56.6	–	36.8
Mafic A	60.9	–	81.6
Mafic B	60.4	–	73.8
Felsic A	–	–	72.3
Felsic B	81.9	32.8	58.9
Felsic C	77.8	21.5	72.9
OA	67.3	37.1	76.2
bal-mF1	63.6	26.8	76.2
unbal-mF1	55.1	14.1	65.8

Table 5 highlights that not all classes are accurately predicted in the image segmentation result of the MLP network, whereas graph-based image segmentation and point-based segmentation predict all classes successfully. Although some classes of the image segmentation results display higher per-class accuracy, the whole point accuracy OA and the class-averaged accuracy F-1 scores of image segmentation are lower than those of point cloud segmentation.

Figure 8 shows the predicted images and the corresponding ground truth of the graph-based and MLP networks. As shown, the graph-based network clearly outperformed the MLP network.

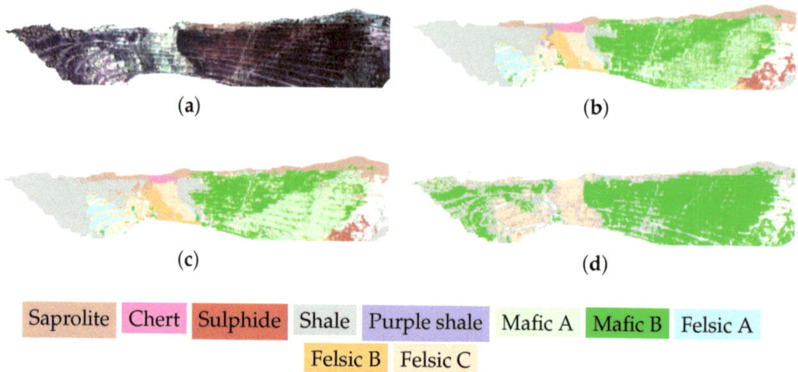

Figure 8. Image-based segmentation results using the SWIR dataset. (**a**) SWIR; (**b**) ground truth; (**c**) graph-network; (**d**) MLP.

5. Discussion

Our method relies on utilizing multimodal data on a multi-stream network. Here, we discuss the significance of each component of our method. We began by exploring the importance of integrating multiple modalities, which was followed by evaluating the significance of the multi-stream network design. Furthermore, we also reported our findings regarding the implementation of 3D CNNs as channel attention modules, demonstrating that integrating 3D CNNs significantly enhances the classification results.

5.1. Combining Different Data Modalities

We investigated the significance of combining various data modalities to the accuracy of the prediction. The DGCNN-like network was chosen as the basis of our experiments. We started the experiments with only hyperspectral features and added more features to examine which features vector results in the highest accuracy. The comparison results are shown in Table 6.

Our experiments are shown in Table 6 and suggest that combining more features resulted in higher segmentation accuracy. The benefit is more significant on the LWIR dataset, where the hyperspectral or RGB features alone are not sufficient to exceed 50% of the overall accuracy. Combining both features improves the result by 8.54 percentage points. Furthermore, adding the extra geometrical features boosts the OA by 3.40 percentage points. The results are consistent when fusing hyperspectral and RGB features in the SWIR dataset. The model's accuracy improves by a small margin (0.08 percentage points). Adding geometric features, however, does not help to increase the accuracy. Instead, it slightly reduces the accuracy. However, we argue that combining all data through feature-level fusion, such as using techniques like multi-stream networks, yields higher accuracy compared to combining data through data-level fusion as demonstrated in this section. In the subsequent section, we present evidence where all available data are leveraged, and fusion occurs after feature extraction in separate sub-networks. This configuration outperforms various configurations presented earlier.

Figure 9 shows the benefit of adding more modalities to the networks. The greater the number of modalities of data included in the networks, the less noise that is visible in the prediction.

Table 6. Different configuration of features as input to the network. The highest accuracy is indicated in bold.

Features	OA (%)
RGB	47.55
LWIR	45.84
LWIR, RGB	54.38
LWIR, RGB, geometric	**57.78**
SWIR	73.77
SWIR, RGB	**73.85**
SWIR, RGB, geometric	72.08

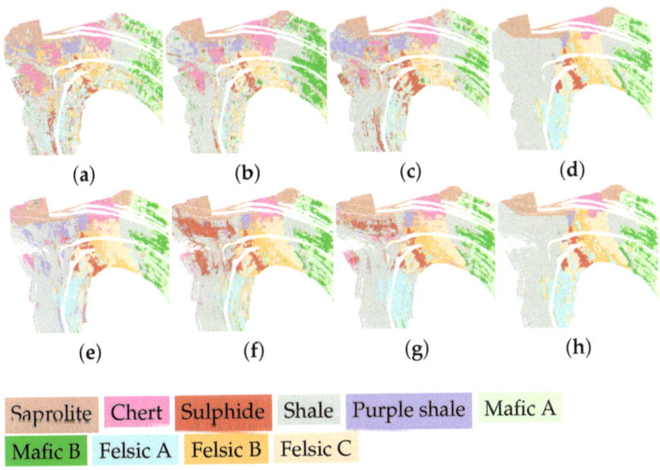

Figure 9. Different configuration of input features. (**a**) LWIR; (**b**) LWIR + RGB; (**c**) LWIR + RGB + geo; (**e**) SWIR; (**f**) SWIR + RGB; (**g**) SWIR + RGB + geo; (**d**,**h**) ground truth.

5.2. Multi-Stream Network

Although combining all different data modalities increases accuracy, we have investigated a more elegant fusing method to improve the results. While stacking all different features into one feature vector may seem like a simple solution, expanding the network into a multi-stream network is the most effective way to handle all different features. This approach yields better results, particularly on the LWIR dataset, where implementing the multi-stream network improves overall accuracy by 6.49 percentage points. A similar, but less significant, impact can also be observed on the SWIR dataset, where the overall accuracy increases by 3.37 percentage points. Table 7 compares the overall accuracy obtained by a one-stream network and a multi-stream network. In Figure 10, we notice the evidence where the multi-stream architecture reduces the mislabeled points, particularly in the Purple Shale class. Ultimately, separating different data modalities into distinct streams would be the optimal architecture for the dataset. We believe that enabling the network to learn various features independently enhances prediction results.

Table 7. The impact of implementing multi-stream network.

Features	Multi-Stream	OA (%)
LWIR, RGB, geometric	No	57.78
LWIR, RGB, geometric	Yes	64.27
SWIR, RGB, geometric	No	72.08
SWIR, RGB, geometric	Yes	75.45

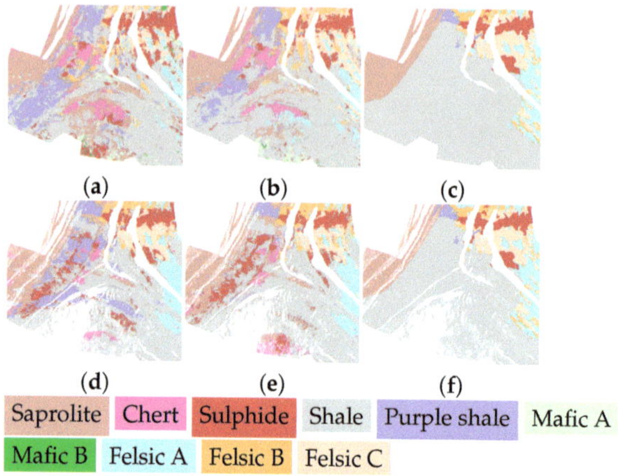

Figure 10. The impact of multi-stream (MS) network. (**a**) LWIR without MS; (**b**) LWIR with MS; (**d**) SWIR without MS; (**e**) SWIR with MS; (**c**,**f**) ground truth.

5.3. The 3D CNNs as Channel Attention Module

We added 3D convolutional filters following the graph reconstruction on the hyperspectral stream as the channel attention module to exploit the spectral information within the hyperspectral data. Table 8 shows our experiment results of the multi-stream network with the addition of 3D convolutional filters. The OA of the LWIR and SWIR datasets increased by 2.19 and 0.76 percentage points after employing the 3D convolutional filters. Again, we can witness the influence of adding 3D CNN by visually inspecting the predicted points in Figure 11. It is clear that after utilizing 3D CNN, the mislabeled classes Chert and Sulfide are then correctly predicted as Shale.

Table 8. The impact of adding 3D CNN in the hyperspectral stream.

Features	3D CNN	OA (%)
LWIR, RGB, geometric	No	64.27
LWIR, RGB, geometric	Yes	66.46
SWIR, RGB, geometric	No	75.45
SWIR, RGB, geometric	Yes	76.21

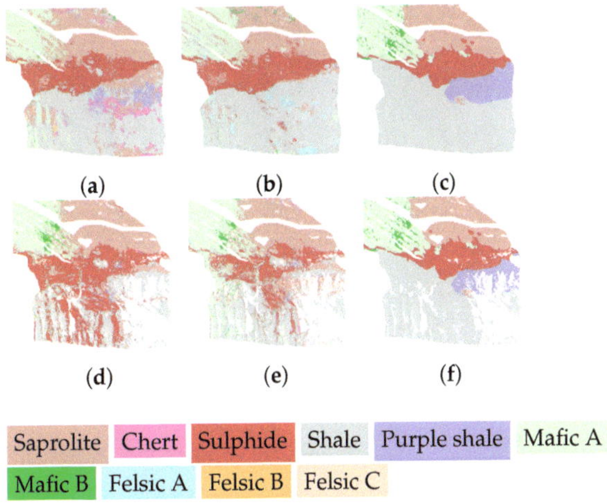

Figure 11. The impact of adding 3D CNN. (**a**) LWIR without 3DCNN; (**b**) LWIR with 3D CNN; (**d**) SWIR without 3DCNN; (**e**) SWIR with 3DCNN; (**c**,**f**) ground truth.

5.4. Geometric Features Transformation

We examined the significance of transforming the XYZ coordinates into the popular eigenvalues-based geometric features. The most primitive point cloud format only consists of XYZ coordinates, as that information is mandatory. Therefore, a point cloud is a data format to store 3D geometry information. In that sense, deep learning networks especially designed for point clouds would extract geometrical features from the XYZ coordinates. However, we faced a different situation here where we insisted that the spectral information of the dataset is a strong predictor for the segmentation, yet directly involving XYZ coordinates resulted in worse accuracy. Hence, omitting XYZ coordinates would be the straightforward solution. Rather than neglecting them, we propose to transform XYZ coordinates into eigenvalue-based features.

We compare the results from two different configurations. Firstly, we used XYZ coordinates and spectral features (hyperspectral and RGB) for the input data. Secondly, we transformed XYZ coordinates into geometric features and combined them with spectral features. We present the quantitative results in Table 9. Qualitatively, as Figure 12 suggests, we argue that our approach increases the true positive rate of the Sulfide class in the LWIR dataset (from 53.91% to 74.49%) and decreases the false positive rate of the same class in the SWIR dataset (from 7.72% to 5.47%).

Table 9. The impact of transforming XYZ coordinates into the eigenvalue-based geometric features.

Features	OA (%)
LWIR with XYZ	54.38
LWIR with geometric features	57.78
SWIR with XYZ	73.85
SWIR with geometric features	75.45

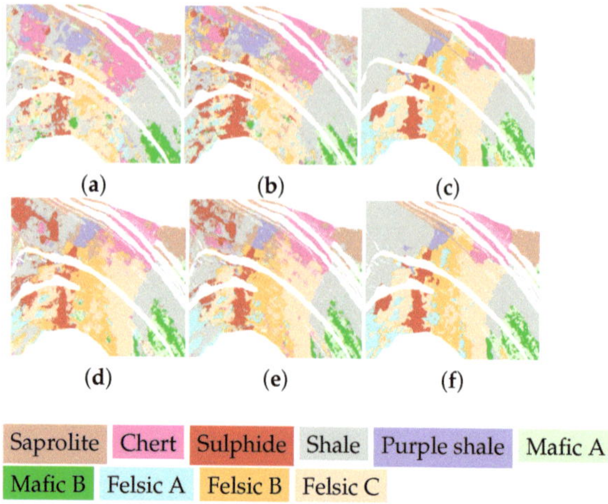

Figure 12. The impact of transforming XYZ into geometric features. (**a**) LWIR with XYZ; (**b**) LWIR with geometric features; (**d**) SWIR with XYZ; (**e**) SWIR with geometric features; (**c**,**f**) ground truth.

5.5. Choice of the Backbone

Finally, we examined the selection of the backbone for the multi-stream architecture. Specifically, we conducted a comparative analysis between multi-stream networks employing different encoders, namely EdgeConv and Self Attention (SA). The outcomes of this investigation are detailed in Table 10. The SA encoder is adopted from the global transformer layers as utilized in PCT [15]. Our experiments demonstrated that the EdgeConv-based multi-stream network outperformed the SA-based multi-stream network. Hence, we chose EdgeConv as the backbone of our proposed framework.

Table 10. The comparison of graph and transformer-based multi-stream network.

Backbone	OA (%)
LWIR with EdgeConv	64.27
LWIR with SA	54.65
SWIR with EdgeConv	75.45
SWIR with SA	65.80

6. Conclusions

This study presents a novel framework for mineral classification by employing multimodal point cloud data. Our approach involves a multi-stream architecture that combines data from different modalities, including XYZ, hyperspectral, and RGB. Our findings demonstrate that leveraging multimodal data yields superior performance compared to using unimodal data alone. Moreover, feeding distinct data modalities into separate streams enhances our model's performance. Additionally, we conducted an ablation study to gain insights into the significance of each component in our framework. Furthermore, we observed that point-based segmentation significantly outperforms image-based segmentation.

While the proposed framework has shown promising results, segmentation outcomes are subject to variations depending on the hyperspectral information. Specifically, the model faces challenges in effectively learning hyperspectral information for segmenting certain classes, such as Purple Shale in specific regions. In future research, we recommend exploring the fusion of multisensor hyperspectral data and developing a model capable of learning from diverse information sources. Additionally, we propose investigating various data fusion mechanisms and their optimal applications. Furthermore, we advocate for assigning higher weighting to sensors that capture more valuable information, aiming to enhance segmentation accuracy.

To summarize, the evidence presented in this study highlights the potential of a multi-stream architecture for processing multimodal hyperspectral point cloud data. By further investigating the fusion of hyperspectral and point cloud data, we can create more robust segmentation models that are suitable for geological applications.

Author Contributions: Conceptualization, A.R. and P.G.; methodology, A.R. and A.J.A.; validation, A.R.; formal analysis, A.R.; investigation, A.R., A.J.A. and P.G.; writing—original draft preparation, A.R.; writing—review and editing, A.J.A., P.G. and R.G.; visualization, A.R.; supervision, P.G. and R.G.; project administration, R.G. All authors have read and agreed to the published version of the manuscript.

Funding: This work was supported by the European Regional Development Fund and the Land of Saxony by providing the high specification Nvidia A100 GPU server that we used in our experiments.

Data Availability Statement: The dataset is available at https://rodare.hzdr.de/record/2256 (accessed on 23 June 2024).

Conflicts of Interest: The authors declare no conflicts of interest.

References

1. Chen, Y.; Jiang, H.; Li, C.; Jia, X.; Ghamisi, P. Deep Feature Extraction and Classification of Hyperspectral Images Based on Convolutional Neural Networks. *IEEE Trans. Geosci. Remote Sens.* **2016**, *54*, 6232–6251. [CrossRef]
2. Ben Hamida, A.; Benoit, A.; Lambert, P.; Ben-Amar, C. Deep Learning Approach for Remote Sensing Image Analysis. In Proceedings of the Big Data from Space (BiDS'16), Santa Cruz de Tenerife, Spain, 15–17 March 2016; Giorgio, S.P.M.P., Ed.; Publications Office of the European Union: Luxembourg, 2016; p. 133. [CrossRef]
3. Lee, H.; Kwon, H. Going Deeper With Contextual CNN for Hyperspectral Image Classification. *IEEE Trans. Image Process.* **2017**, *26*, 4843–4855. [CrossRef]
4. Li, Y.; Zhang, H.; Shen, Q. Spectral–Spatial Classification of Hyperspectral Imagery with 3D Convolutional Neural Network. *Remote Sens.* **2017**, *9*, 67. [CrossRef]
5. Li, S.; Song, W.; Fang, L.; Chen, Y.; Ghamisi, P.; Benediktsson, J.A. Deep Learning for Hyperspectral Image Classification: An Overview. *IEEE Trans. Geosci. Remote Sens.* **2019**, *57*, 6690–6709. [CrossRef]
6. Lorenz, S.; Salehi, S.; Kirsch, M.; Zimmermann, R.; Unger, G.; Vest Sørensen, E.; Gloaguen, R. Radiometric Correction and 3D Integration of Long Range Ground-Based Hyperspectral Imagery for Mineral Exploration of Vertical Outcrops. *Remote Sens.* **2018**, *10*, 176. [CrossRef]
7. Kirsch, M.; Lorenz, S.; Zimmermann, R.; Tusa, L.; Möckel, R.; Hödl, P.; Booysen, R.; Khodadadzadeh, M.; Gloaguen, R. Integration of Terrestrial and Drone-Borne Hyperspectral and Photogrammetric Sensing Methods for Exploration Mapping and Mining Monitoring. *Remote Sens.* **2018**, *10*, 1366. [CrossRef]
8. Thiele, S.T.; Lorenz, S.; Kirsch, M.; Cecilia Contreras Acosta, I.; Tusa, L.; Herrmann, E.; Möckel, R.; Gloaguen, R. Multi-scale, multi-sensor data integration for automated 3-D geological mapping. *Ore Geol. Rev.* **2021**, *136*, 104252. [CrossRef]
9. Afifi, A.J.; Thiele, S.T.; Rizaldy, A.; Lorenz, S.; Ghamisi, P.; Tolosana-Delgado, R.; Kirsch, M.; Gloaguen, R.; Heizmann, M. Tinto: Multisensor Benchmark for 3-D Hyperspectral Point Cloud Segmentation in the Geosciences. *IEEE Trans. Geosci. Remote Sens.* **2024**, *62*, 5501015. [CrossRef]
10. Ramachandram, D.; Taylor, G.W. Deep Multimodal Learning: A Survey on Recent Advances and Trends. *IEEE Signal Process. Mag.* **2017**, *34*, 96–108. [CrossRef]
11. Boulahia, S.; Amamra, A.; Madi, M. Early, intermediate and late fusion strategies for robust deep learning-based multimodal action recognition. *Mach. Vis. Appl.* **2021**, *32*, 121. [CrossRef]
12. Ma, X.; Zhang, X.; Pun, M.O. A Crossmodal Multiscale Fusion Network for Semantic Segmentation of Remote Sensing Data. *IEEE J. Sel. Top. Appl. Earth Obs. Remote Sens.* **2022**, *15*, 3463–3474. [CrossRef]
13. Qi, C.R.; Yi, L.; Su, H.; Guibas, L.J. Pointnet++: Deep hierarchical feature learning on point sets in a metric space. *Adv. Neural Inf. Process. Syst.* **2017**, *30*, 1–10.
14. Wang, Y.; Sun, Y.; Liu, Z.; Sarma, S.E.; Bronstein, M.M.; Solomon, J.M. Dynamic graph cnn for learning on point clouds. *ACM Trans. Graph. (tog)* **2019**, *38*, 146. [CrossRef]
15. Guo, M.H.; Cai, J.X.; Liu, Z.N.; Mu, T.J.; Martin, R.R.; Hu, S.M. Pct: Point cloud transformer. *Comput. Vis. Media* **2021**, *7*, 187–199. [CrossRef]
16. Woo, S.; Park, J.; Lee, J.Y.; Kweon, I.S. CBAM: Convolutional Block Attention Module. In *Computer Vision—ECCV 2018, Proceedings of the 15th European Conference, Munich, Germany, 8–14 September 2018*; Proceedings, Part VII; Springer: Cham, Switzerland, 2018; pp. 3–19. [CrossRef]

17. Hu, J.; Shen, L.; Albanie, S.; Sun, G.; Wu, E. Squeeze-and-Excitation Networks. *IEEE Trans. Pattern Anal. Mach. Intell.* **2020**, *42*, 2011–2023. [CrossRef] [PubMed]
18. Fu, J.; Liu, J.; Tian, H.; Li, Y.; Bao, Y.; Fang, Z.; Lu, H. Dual Attention Network for Scene Segmentation. In Proceedings of the 2019 IEEE/CVF Conference on Computer Vision and Pattern Recognition (CVPR), Long Beach, CA, USA, 15–20 June 2019; pp. 3141–3149. [CrossRef]
19. Wang, Q.; Wu, B.; Zhu, P.; Li, P.; Zuo, W.; Hu, Q. ECA-Net: Efficient Channel Attention for Deep Convolutional Neural Networks. In Proceedings of the 2020 IEEE/CVF Conference on Computer Vision and Pattern Recognition (CVPR), Seattle, WA, USA, 13–19 June 2020; pp. 11531–11539. [CrossRef]
20. Ghamisi, P.; Maggiori, E.; Li, S.; Souza, R.; Tarablaka, Y.; Moser, G.; De Giorgi, A.; Fang, L.; Chen, Y.; Chi, M.; et al. New frontiers in spectral-spatial hyperspectral image classification: The latest advances based on mathematical morphology, Markov random fields, segmentation, sparse representation, and deep learning. *IEEE Geosci. Remote Sens. Mag.* **2018**, *6*, 10–43. [CrossRef]
21. Ghamisi, P.; Plaza, J.; Chen, Y.; Li, J.; Plaza, A.J. Advanced spectral classifiers for hyperspectral images: A review. *IEEE Geosci. Remote Sens. Mag.* **2017**, *5*, 8–32. [CrossRef]
22. Goel, P.; Prasher, S.; Patel, R.; Landry, J.; Bonnell, R.; Viau, A. Classification of hyperspectral data by decision trees and artificial neural networks to identify weed stress and nitrogen status of corn. *Comput. Electron. Agric.* **2003**, *39*, 67–93. [CrossRef]
23. Ratle, F.; Camps-Valls, G.; Weston, J. Semisupervised Neural Networks for Efficient Hyperspectral Image Classification. *IEEE Trans. Geosci. Remote Sens.* **2010**, *48*, 2271–2282. [CrossRef]
24. Mou, L.; Ghamisi, P.; Zhu, X.X. Deep Recurrent Neural Networks for Hyperspectral Image Classification. *IEEE Trans. Geosci. Remote Sens.* **2017**, *55*, 3639–3655. [CrossRef]
25. Wu, H.; Prasad, S. Semi-Supervised Deep Learning Using Pseudo Labels for Hyperspectral Image Classification. *IEEE Trans. Image Process.* **2018**, *27*, 1259–1270. [CrossRef] [PubMed]
26. Wu, H.; Prasad, S. Convolutional Recurrent Neural Networks for Hyperspectral Data Classification. *Remote Sens.* **2017**, *9*, 298. [CrossRef]
27. Ham, J.; Chen, Y.; Crawford, M.; Ghosh, J. Investigation of the random forest framework for classification of hyperspectral data. *IEEE Trans. Geosci. Remote Sens.* **2005**, *43*, 492–501. [CrossRef]
28. Melgani, F.; Bruzzone, L. Classification of hyperspectral remote sensing images with support vector machines. *IEEE Trans. Geosci. Remote Sens.* **2004**, *42*, 1778–1790. [CrossRef]
29. Tarabalka, Y.; Chanussot, J.; Benediktsson, J. Segmentation and classification of hyperspectral images using watershed transformation. *Pattern Recognit.* **2010**, *43*, 2367–2379. [CrossRef]
30. Fauvel, M.; Tarabalka, Y.; Benediktsson, J.A.; Chanussot, J.; Tilton, J.C. Advances in Spectral-Spatial Classification of Hyperspectral Images. *Proc. IEEE* **2013**, *101*, 652–675. [CrossRef]
31. Aptoula, E.; Dalla Mura, M.; Lefèvre, S. Vector Attribute Profiles for Hyperspectral Image Classification. *IEEE Trans. Geosci. Remote Sens.* **2016**, *54*, 3208–3220. [CrossRef]
32. Makantasis, K.; Karantzalos, K.; Doulamis, A.; Doulamis, N. Deep supervised learning for hyperspectral data classification through convolutional neural networks. In Proceedings of the 2015 IEEE International Geoscience and Remote Sensing Symposium (IGARSS), Milan, Italy, 26–31 July 2015; pp. 4959–4962. [CrossRef]
33. Slavkovikj, V.; Verstockt, S.; De Neve, W.; Van Hoecke, S.; Van de Walle, R. Hyperspectral Image Classification with Convolutional Neural Networks. In Proceedings of the 23rd ACM International Conference on Multimedia, New York, NY, USA, 26–30 October 2015; pp. 1159–1162. [CrossRef]
34. Zhao, W.; Du, S. Spectral–Spatial Feature Extraction for Hyperspectral Image Classification: A Dimension Reduction and Deep Learning Approach. *IEEE Trans. Geosci. Remote Sens.* **2016**, *54*, 4544–4554. [CrossRef]
35. Yue, J.; Zhao, W.; Mao, S.; Liu, H. Spectral–spatial classification of hyperspectral images using deep convolutional neural networks. *Remote Sens. Lett.* **2015**, *6*, 468–477. [CrossRef]
36. Audebert, N.; Saux, B.L.; Lefèvre, S. Deep Learning for Classification of Hyperspectral Data: A Comparative Review. *IEEE Geosci. Remote Sens. Mag.* **2019**, *7*, 159–173. [CrossRef]
37. Guo, Y.; Wang, H.; Hu, Q.; Liu, H.; Liu, L.; Bennamoun, M. Deep learning for 3d point clouds: A survey. *IEEE Trans. Pattern Anal. Mach. Intell.* **2020**, *43*, 4338–4364. [CrossRef] [PubMed]
38. Bello, S.A.; Yu, S.; Wang, C.; Adam, J.M.; Li, J. Deep learning on 3D point clouds. *Remote Sens.* **2020**, *12*, 1729. [CrossRef]
39. Zhang, J.; Zhao, X.; Chen, Z.; Lu, Z. A Review of Deep Learning-Based Semantic Segmentation for Point Cloud. *IEEE Access* **2019**, *7*, 179118–179133. [CrossRef]
40. Xie, Y.; Tian, J.; Zhu, X.X. Linking Points With Labels in 3D: A Review of Point Cloud Semantic Segmentation. *IEEE Geosci. Remote Sens. Mag.* **2020**, *8*, 38–59. [CrossRef]
41. Su, H.; Maji, S.; Kalogerakis, E.; Learned-Miller, E. Multi-view Convolutional Neural Networks for 3D Shape Recognition. In Proceedings of the 2015 IEEE International Conference on Computer Vision (ICCV), Santiago, Chile, 7–13 December 2015; pp. 945–953. [CrossRef]
42. Guerry, J.; Boulch, A.; Le Saux, B.; Moras, J.; Plyer, A.; Filliat, D. SnapNet-R: Consistent 3D Multi-view Semantic Labeling for Robotics. In Proceedings of the 2017 IEEE International Conference on Computer Vision Workshops (ICCVW), Venice, Italy, 22–29 October 2017; pp. 669–678. [CrossRef]

43. Maturana, D.; Scherer, S. VoxNet: A 3D Convolutional Neural Network for real-time object recognition. In Proceedings of the 2015 IEEE/RSJ International Conference on Intelligent Robots and Systems (IROS), Hamburg, Germany, 28 September–2 October 2015; pp. 922–928. [CrossRef]
44. Le, T.; Duan, Y. PointGrid: A Deep Network for 3D Shape Understanding. In Proceedings of the 2018 IEEE/CVF Conference on Computer Vision and Pattern Recognition, Salt Lake City, UT, USA, 18–23 June 2018; pp. 9204–9214. [CrossRef]
45. Graham, B.; Engelcke, M.; Maaten, L. 3D Semantic Segmentation with Submanifold Sparse Convolutional Networks. In Proceedings of the 2018 IEEE/CVF Conference on Computer Vision and Pattern Recognition (CVPR), Los Alamitos, CA, USA, 18–23 June 2018; pp. 9224–9232. [CrossRef]
46. Choy, C.; Gwak, J.; Savarese, S. 4D Spatio-Temporal ConvNets: Minkowski Convolutional Neural Networks. In Proceedings of the 2019 IEEE/CVF Conference on Computer Vision and Pattern Recognition (CVPR), Los Alamitos, CA, USA, 15–20 June 2019; pp. 3070–3079. [CrossRef]
47. Qi, C.R.; Su, H.; Mo, K.; Guibas, L.J. Pointnet: Deep learning on point sets for 3d classification and segmentation. In Proceedings of the IEEE Conference on Computer Vision and Pattern Recognition, Honolulu, HI, USA, 21–26 July 2017; pp. 652–660.
48. Zhang, K.; Hao, M.; Wang, J.; Chen, X.; Leng, Y.; de Silva, C.W.; Fu, C. Linked Dynamic Graph CNN: Learning through Point Cloud by Linking Hierarchical Features. In Proceedings of the 2021 27th International Conference on Mechatronics and Machine Vision in Practice (M2VIP), Shanghai, China, 26–28 November 2021; pp. 7–12. [CrossRef]
49. Li, Y.; Bu, R.; Sun, M.; Wu, W.; Di, X.; Chen, B. Pointcnn: Convolution on x-transformed points. *Adv. Neural Inf. Process. Syst.* **2018**, *31*, 1–11.
50. Boulch, A. ConvPoint: Continuous convolutions for point cloud processing. *Comput. Graph.* **2020**, *88*, 24–34. [CrossRef]
51. Thomas, H.; Qi, C.R.; Deschaud, J.; Marcotegui, B.; Goulette, F.; Guibas, L. KPConv: Flexible and Deformable Convolution for Point Clouds. In Proceedings of the 2019 IEEE/CVF International Conference on Computer Vision (ICCV), Los Alamitos, CA, USA, 27 October–2 November 2019; pp. 6410–6419. [CrossRef]
52. Wu, W.; Qi, Z.; Fuxin, L. PointConv: Deep Convolutional Networks on 3D Point Clouds. In Proceedings of the 2019 IEEE/CVF Conference on Computer Vision and Pattern Recognition (CVPR), Long Beach, CA, USA, 15–20 June 2019; pp. 9613–9622. [CrossRef]
53. Zhao, H.; Jiang, L.; Jia, J.; Torr, P.H.; Koltun, V. Point transformer. In Proceedings of the IEEE/CVF International Conference on Computer Vision, Montreal, BC, Canada, 11–17 October 2021; pp. 16259–16268.
54. López, A.; Jurado, J.M.; Jiménez-Pérez, J.R.; Feito, F.R. Generation of hyperspectral point clouds: Mapping, compression and rendering. *Comput. Graph.* **2022**, *106*, 267–276. [CrossRef]
55. Brell, M.; Segl, K.; Guanter, L.; Bookhagen, B. 3D hyperspectral point cloud generation: Fusing airborne laser scanning and hyperspectral imaging sensors for improved object-based information extraction. *ISPRS J. Photogramm. Remote Sens.* **2019**, *149*, 200–214. [CrossRef]
56. Gader, P.; Zare, A.; Close, R.; Aitken, J.; Tuell, G. *Muufl Gulfport Hyperspectral and Lidar Airborne Data Set*; Technical Report REP-2013-570; University Florida: Gainesville, FL, USA, 2013.
57. Weinmann, M.; Weinmann, M. Fusion of hyperspectral, multispectral, color and 3d point cloud information for the semantic interpretation of urban environments. *Int. Arch. Photogramm. Remote Sens. Spat. Inf. Sci.* **2019**, *42*, 1899–1906. [CrossRef]
58. Weinmann, M.; Weinmann, M. Geospatial computer vision based on multi-modal data—How valuable is shape information for the extraction of semantic information? *Remote Sens.* **2017**, *10*, 2. [CrossRef]
59. Chen, B.; Shi, S.; Sun, J.; Gong, W.; Yang, J.; Du, L.; Guo, K.; Wang, B.; Chen, B. Hyperspectral lidar point cloud segmentation based on geometric and spectral information. *Opt. Express* **2019**, *27*, 24043–24059. [CrossRef] [PubMed]
60. Weidner, L.; Walton, G.; Krajnovich, A. Classifying rock slope materials in photogrammetric point clouds using robust color and geometric features. *ISPRS J. Photogramm. Remote Sens.* **2021**, *176*, 15–29. [CrossRef]
61. Nevalainen, O.; Honkavaara, E.; Tuominen, S.; Viljanen, N.; Hakala, T.; Yu, X.; Hyyppä, J.; Saari, H.; Pölönen, I.; Imai, N.N.; et al. Individual Tree Detection and Classification with UAV-Based Photogrammetric Point Clouds and Hyperspectral Imaging. *Remote Sens.* **2017**, *9*, 185. [CrossRef]
62. Decker, K.T.; Borghetti, B.J. Composite Style Pixel and Point Convolution-Based Deep Fusion Neural Network Architecture for the Semantic Segmentation of Hyperspectral and Lidar Data. *Remote Sens.* **2022**, *14*, 2113. [CrossRef]
63. Decker, K.T.; Borghetti, B.J. Hyperspectral Point Cloud Projection for the Semantic Segmentation of Multimodal Hyperspectral and Lidar Data with Point Convolution-Based Deep Fusion Neural Networks. *Appl. Sci.* **2023**, *13*, 8210. [CrossRef]
64. Xu, Y.; Du, B.; Zhang, L.; Cerra, D.; Pato, M.; Carmona, E.; Prasad, S.; Yokoya, N.; Hänsch, R.; Le Saux, B. Advanced Multi-Sensor Optical Remote Sensing for Urban Land Use and Land Cover Classification: Outcome of the 2018 IEEE GRSS Data Fusion Contest. *IEEE J. Sel. Top. Appl. Earth Obs. Remote Sens.* **2019**, *12*, 1709–1724. [CrossRef]
65. Mitschke, I.; Wiemann, T.; Igelbrink, F.; Hertzberg, J. Hyperspectral 3D Point Cloud Segmentation Using RandLA-Net. In Proceedings of the International Conference on Intelligent Autonomous Systems, Zagreb, Croatia, 13–16 June 2022; Springer: Berlin/Heidelberg, Germany, 2022; pp. 301–312.
66. Jaegle, A.; Borgeaud, S.; Alayrac, J.; Doersch, C.; Ionescu, C.; Ding, D.; Koppula, S.; Zoran, D.; Brock, A.; Shelhamer, E.; et al. Perceiver IO: A General Architecture for Structured Inputs & Outputs. In Proceedings of the Tenth International Conference on Learning Representations, ICLR 2022, Virtual Event, 25–29April 2022.

67. Li, P.; Gu, J.; Kuen, J.; Morariu, V.I.; Zhao, H.; Jain, R.; Manjunatha, V.; Liu, H. SelfDoc: Self-Supervised Document Representation Learning. In Proceedings of the IEEE/CVF Conference on Computer Vision and Pattern Recognition (CVPR), New Orleans, LA, USA, 18–24 June 2021; pp. 5652–5660.
68. Hazirbas, C.; Ma, L.; Domokos, C.; Cremers, D. FuseNet: Incorporating Depth into Semantic Segmentation via Fusion-Based CNN Architecture. In Proceedings of the Asian Conference on Computer Vision, Taipei, Taiwan, 20–24 November 2016.
69. Li, J.; Hong, D.; Gao, L.; Yao, J.; Zheng, K.; Zhang, B.; Chanussot, J. Deep learning in multimodal remote sensing data fusion: A comprehensive review. *Int. J. Appl. Earth Obs. Geoinf.* **2022**, *112*, 102926. [CrossRef]
70. Chen, K.; Weinmann, M.; Gao, X.; Yan, M.; Hinz, S.; Jutzi, B.; Weinmann, M. Residual Shuffling Convolutional Neural Networks for Deep Semantic Image Segmentation Using Multi-Modal Data. *ISPRS Ann. Photogramm. Remote. Sens. Spat. Inf. Sci.* **2018**, *IV-2*, 65–72. [CrossRef]
71. Chen, K.; Weinmann, M.; Sun, X.; Yan, M.; Hinz, S.; Jutzi, B.; Weinmann, M. Semantic Segmentation of Aerial Imagery via Multi-Scale Shuffling Convolutional Neural Networks with Deep Supervision. *ISPRS Ann. Photogramm. Remote. Sens. Spat. Inf. Sci.* **2018**, *IV-1*, 29–36. [CrossRef]
72. Diakogiannis, F.I.; Waldner, F.; Caccetta, P.; Wu, C. ResUNet-a: A deep learning framework for semantic segmentation of remotely sensed data. *ISPRS J. Photogramm. Remote Sens.* **2020**, *162*, 94–114. [CrossRef]
73. Audebert, N.; Le Saux, B.; Lefèvre, S. Beyond RGB: Very high resolution urban remote sensing with multimodal deep networks. *ISPRS J. Photogramm. Remote Sens.* **2018**, *140*, 20–32. [CrossRef]
74. Kieu, N.; Nguyen, K.; Sridharan, S.; Fookes, C. General-Purpose Multimodal Transformer meets Remote Sensing Semantic Segmentation. *arXiv* **2023**, arXiv:2307.03388.
75. Concha, D.T.; Maia, H.D.A.; Pedrini, H.; Tacon, H.; Brito, A.D.S.; Chaves, H.D.L.; Vieira, M.B. Multi-stream convolutional neural networks for action recognition in video sequences based on adaptive visual rhythms. In Proceedings of the 2018 17th IEEE International Conference on Machine Learning and Applications (ICMLA), Orlando, FL, USA, 17–20 December 2018; pp. 473–480.
76. Li, H.; Zech, J.; Hong, D.; Ghamisi, P.; Schultz, M.; Zipf, A. Leveraging openstreetmap and multimodal remote sensing data with joint deep learning for wastewater treatment plants detection. *Int. J. Appl. Earth Obs. Geoinf.* **2022**, *110*, 102804. [CrossRef] [PubMed]
77. Li, H.; Ghamisi, P.; Rasti, B.; Wu, Z.; Shapiro, A.; Schultz, M.; Zipf, A. A multi-sensor fusion framework based on coupled residual convolutional neural networks. *Remote Sens.* **2020**, *12*, 2067. [CrossRef]
78. Gao, J.; Li, P.; Chen, Z.; Zhang, J. A Survey on Deep Learning for Multimodal Data Fusion. *Neural Comput.* **2020**, *32*, 829–864. [CrossRef]
79. Baltrusaitis, T.; Ahuja, C.; Morency, L.P. Multimodal Machine Learning: A Survey and Taxonomy. *IEEE Trans. Pattern Anal. Mach. Intell.* **2019**, *41*, 423–443. [CrossRef]
80. Landrieu, L.; Simonovsky, M. Large-scale point cloud semantic segmentation with superpoint graphs. In Proceedings of the Proceedings of the IEEE Conference on Computer Vision and Pattern Recognition, Salt Lake City, UT, USA, 18–23 June 2018; pp. 4558–4567.
81. Hackel, T.; Wegner, J.D.; Schindler, K. Contour detection in unstructured 3D point clouds. In Proceedings of the IEEE Conference on Computer Vision and Pattern Recognition, Las Vegas, NV, USA, 27–30 June 2016; pp. 1610–1618.
82. Sutskever, I.; Martens, J.; Dahl, G.; Hinton, G. On the Importance of Initialization and Momentum in Deep Learning. In Proceedings of the 30th International Conference on International Conference on Machine Learning, Atlanta, GA, USA, 16–21 June 2013; Volume 28, pp. 1139–1147.

remote sensing

MDPI

Article

Uncovering a Seismogenic Fault in Southern Iran through Co-Seismic Deformation of the Mw 6.1 Doublet Earthquake of 14 November 2021

Peyman Namdarsehat [1,*], Wojciech Milczarek [1], Natalia Bugajska-Jędraszek [1], Seyed-Hani Motavalli-Anbaran [2] and Matin Khaledzadeh [3]

1 Department of Geodesy and Geoinformatics, Faculty of Geoengineering, Mining and Geology, Wroclaw University of Science and Technology, Wybrzeże Wyspiańskiego 27, 50-370 Wrocław, Poland
2 Institute of Geophysics, University of Tehran, Tehran P.O. Box 14155-6466, Iran
3 Institute for Advanced Studies in Basic Sciences (IASBS), Department of Earth Sciences, Zanjan P.O. Box 45137-66731, Iran
* Correspondence: peyman.namdarsehat@pwr.edu.pl; Tel.: +48-71-320-68-10

Abstract: On 14 November 2021, a doublet earthquake, each event of which had an Mw of 6.1, struck near Fin in the Simply Folded Belt (SFB) in southern Iran. The first quake occurred at 12:07:04 UTC, followed by a second one just a minute and a half later. The SFB is known for its blind thrust faults, typically not associated with surface ruptures. These earthquakes are usually linked to the middle and lower layers of the sedimentary cover. Identifying the faults that trigger earthquakes in the region remains a significant challenge and is subject to high uncertainty. This study aims to identify and determine the fault(s) that may have caused the doublet earthquake. To achieve this goal, we utilized the DInSAR method using Sentinel-1 to detect deformation, followed by finite-fault inversion and magnetic interpretation to determine the location, geometry, and slip distribution of the fault(s). Bayesian probabilistic joint inversion was used to model the earthquake sources and derive the geometric parameters of potential fault planes. The study presents two potential fault solutions—one dipping to the north and the other to the south. Both solutions showed no significant difference in strike and fault location, suggesting a single fault. Based on the results of the seismic inversion, it appears that a north-dipping fault with a strike, dip, and rake of 257°, 74°, and 77°, respectively, is more consistent with the geological setting of the area. The fault plane has a width of roughly 3.6 km, a length of 13.4 km, and a depth of 5.6 km. Our results revealed maximum displacements along the radar line of sight reaching values of up to −360 mm in the ascending orbit, indicating an unknown fault with horizontal displacements at the surface ranging from −144 to 170 mm and maximum vertical displacements between −204 and 415 mm. Aeromagnetic data for Iran were utilized with an average flight-line spacing of 7.5 km. The middle of the data observation period was considered to apply the RTP filter, and the DRTP method was used. We calculated the gradient of the residual anomaly in the N-S direction due to the direction of the existing faults and folds. The gradient map identified the fault and potential extension of the observed anomalies related to a fault with an ENE-WSW strike, which could extend to the ~ E-W. We suggest that earthquakes occur in the sedimentary cover of the SFB where subsurface faulting is involved, with Hormuz salt acting as an important barrier to rupture. The multidisciplinary approach used in this study, including InSAR and magnetic data, underscores the importance of accurate fault characterization. These findings provide valuable insights into the seismic hazard of the area.

Keywords: Fin earthquake; co-seismic deformation; simply folded belt; seismic inversion; magnetic anomaly; Hormuz salt; Handun anticline

Citation: Namdarsehat, P.; Milczarek, W.; Bugajska-Jędraszek, N.; Motavalli-Anbaran, S.-H.; Khaledzadeh, M. Uncovering a Seismogenic Fault in Southern Iran through Co-Seismic Deformation of the Mw 6.1 Doublet Earthquake of 14 November 2021. *Remote Sens.* **2024**, *16*, 2318. https://doi.org/10.3390/rs16132318

Academic Editors: Michele Saroli, Yanni Dong, Tao Chen and Chao Chen

Received: 3 May 2024
Revised: 4 June 2024
Accepted: 20 June 2024
Published: 25 June 2024

1. Introduction

The Zagros orogenic belt in southwestern Iran is a prime example of a fold-and-thrust belt at the leading edge of the Arabian–Eurasian continental collision zone. Seismic activity in the Zagros Fold-and-Thrust Belt (ZFTB) is interpreted from two distinct perspectives. One viewpoint posits that the majority of moderate to large mainshocks are due to basement faulting (e.g., [1–3]). Conversely, an alternative viewpoint primarily attributes these occurrences to faults within the middle and lower layers of the sedimentary cover (e.g., [4,5]). Most earthquakes occur along major thrust faults, such as the High Zagros Fault (HZF), the Mountain Front Fault (MFF), or transverse strike–slip faults. The recent perspective reinforces the notion that the basement is rarely affected by moderate to large earthquakes [4–6].

The ZFTB is generally divided into two parallel zones: the High Zagros (HZ) and the Simply Folded Belt (SFB) (Figure 1). This division is based on differences in topography, geomorphology, exposed stratigraphy, and seismicity [7]. The seismic activity in the SFB is primarily caused by blind thrust faults. These are not often associated with surface ruptures but are known for causing destructive earthquakes [3,7–11]. On 14 November 2021, at 12:07:04 UTC, a strong earthquake of Mw 6.1 struck southern Iran within the SFB. One and a half minutes later, another earthquake of similar magnitude occurred. The events were recorded by the Global Centroid Moment Tensor (GCMT), formerly known as the Harvard CMT project [12,13] (Figure 1).

The earthquakes caused 1 death and 99 injuries. The affected area comprises 1 city and 137 villages, with a population of approximately 9121 households (36,484 people). The events destroyed buildings in the affected area, known as Fin, and caused damage to various facilities and infrastructure. The earthquakes were also felt in neighboring countries, including the cities of Abu Dhabi, Dubai, and Ras al-Khaimah in the UAE.

Several earthquakes were reported in the nearby study area, which were attributed to the main faults of the ZFTB [9,14]. Roustaei et al. [14] used Interferometric Synthetic Aperture Radar (InSAR) to analyze the co-seismic displacements of the 25 March 2006 Fin earthquakes with Mws of 5.7, 5.5, 5.2, 5.0, and 4.9. Their findings revealed significant differences between the ~ENE–WSW strike of the fault identified by InSAR and the ~E–W strike indicated by body-wave solutions. The study also provided detailed insights into the well-resolved upper and lower segments of the fault, located at depths of 5–6 km and 9–10 km, respectively.

In reviewing the literature specifically concerned with the 14 November 2021 earthquakes, Fathian et al. [15] used radar data from the Sentinel-1 and ALOS-2 satellites and calculated co-seismic interferograms using the Differential Interferometric Synthetic Aperture Radar (DInSAR) method, which were then resampled and subjected to non-linear inversion to estimate the rupture mechanism and geometric parameters of the uniform slip source. On this basis, they identified two possible fault planes striking in a west–east direction but dipping oppositely towards the north and the south and with a maximum slip of 2 m.

Golshadi et al. [16] conducted an InSAR study combined with an analysis of seismicity data, including foreshocks and aftershocks. These data were recorded by sparse, locally established seismographs of the Iranian Seismological Center (IRSC), the International Institute of Earthquake Engineering and Seismology of Iran (IIEES), and a UAE national seismic network station.

Toker et al. [17] used teleseismically recorded broadband P-velocity waveforms in their analysis of finite-fault source inversion using data from the IRSC and identified two faults as the cause of the events.

Golshadi et al. [16] and Toker et al. [17] attributed the events to two different faults. However, these faults are approximately 50 km apart and have different orientations. Rezapour and Jamalreyhani [18] found that the distribution of relocated aftershocks from the IRSC data indicates that the fault(s) associated with the Fin doublet earthquake extend along a nearly west–east striking, SSE-dipping thrust fault.

Significant strides have been made in identifying earthquake precursors, with notable examples including advances in land surface temperature (LST) and ionospheric scintillation. In a recent investigation, Boudriki Semlali et al. [19] demonstrated that anomalies in LST occurring a few days before earthquakes could effectively be precursors for earthquake events. Molina et al. [20,21] established a positive correlation between ionospheric scintillation and the period preceding earthquakes.

Ommi and Smirnov [22] used the region–time–length (RTL) parameter and the time-to-failure method to analyze the earthquakes that occurred on 14 November 2021. They identified a seismic quiescence anomaly ten months before the main shock. This anomaly developed gradually and reached its minimum value only one month before the main quake.

The ZFTB is known for an abundance of diapirs that formed from the Ediacaran to early Cambrian Hormuz salt. These salt diapirs are particularly widespread in the eastern part of the ZFTB and the Persian Gulf. However, the central region of the ZFTB displays salt diapirs distributed along tear fault systems trending nearly N-S and NE-SW and crossing the mountain belt obliquely [6]. Many researchers have documented the diapirs in detail (e.g., [6,23–25]).

Nissen et al. [26] investigated a cluster of three large earthquakes (which occurred on 27 November 2005, 28 June 2006, and 10 September 2008) and numerous smaller shocks at Qeshm Island within the SFB, located 50 km south of our research area. Their findings suggest that the Hormuz salt acts as an important regional barrier to rupture and that subsurface faulting contributes to the occurrence of the largest earthquakes in the SFB.

Nissen et al. [7] reinterpreted the seismicity of the ZFTB based on InSAR, teleseismic data, and geomorphology. Their analysis revealed that the seismogenic zone extends to a depth of about 20 km, with most earthquakes of moderate magnitude (Mw 5–6) occurring in the lower sedimentary cover at a 5–10 km depth and not in the basement. This agrees with the estimations of the sedimentary cover thickness in the Persian Gulf and the eastern part of the ZFTB, ranging between about 12 and 17 km [6,24,25]. Importantly, these faults often have no expression at the surface because they are decoupled from the surface folding by incompetent strata at different levels of the sedimentary cover.

Nissen et al. [4] and Elliott et al. [5] suggested that the crystalline basement undergoes predominantly aseismic shortening, contributing to a more sophisticated understanding of the seismic behavior and structural dynamics of the ZFTB.

Most of our understanding of active faults in the ZFTB is based on seismology, involving solutions for focal mechanisms and earthquake source parameters [11]. Nevertheless, there are still considerable uncertainties in detecting the seismogenic faults responsible for triggering earthquakes in the region. The lack of a dense local seismological network suggests that the infrastructure for data collection in the area was suboptimal for capturing the full extent and details of the foreshock and aftershock sequences.

The inconsistencies in the reports regarding the causes of the doublet earthquake led us to reconsider the search for its origins. To fill in the gaps and verify the results of the identified causative faults, the study performed an integrated analysis that included both InSAR and magnetic interpretation, which have not been used before.

This study used the DInSAR method with radar data from Sentinel-1. The next step involved the decomposition of the signals into vertical and horizontal displacements in the east–west direction using data from ascending and descending paths. This approach provided a better insight into the observed effects of earthquakes on the ground surface. Subsequently, seismic inversion was performed, which plays an important role in our study, as the inversion allowed the orientation of the fault to be determined. Both the amount of the fault slip and the affected area are related to the magnitude of the occurrence. They can be modeled using finite-fault inversion based on the displacements obtained from InSAR observations, Global Navigation Satellite System (GNSS) data, and/or recorded shock waveforms to reconstruct the slip history. In this study, we used inversion theory to infer seismic source parameters based on the indirect measurements mentioned above (InSAR).

The overarching scientific investigations that guided this research raised the following key questions: Which major fault(s) contributed to the occurrence of the doublet earthquake on 14 November 2021? Can the seismic activity be attributed to a single fault, or was it a complex interaction of several faults?

We hypothesized that a multidisciplinary investigation combining InSAR and magnetic surveys could help us identify the fault(s) responsible for the earthquake, which originated in the sedimentary cover closely linked to the main fault trends in the SFB. It is assumed that the earthquake did not originate from a point source but from a fault of a finite zone. Based on the elastic half-space model [27], we performed a non-linear inversion and estimated the most probable geometric parameters of the fault plane where the rupture caused the Fin doublet earthquake.

The results were compared to subsurface data using magnetic surveying as an effective tool to map tectonic structures, including faults.

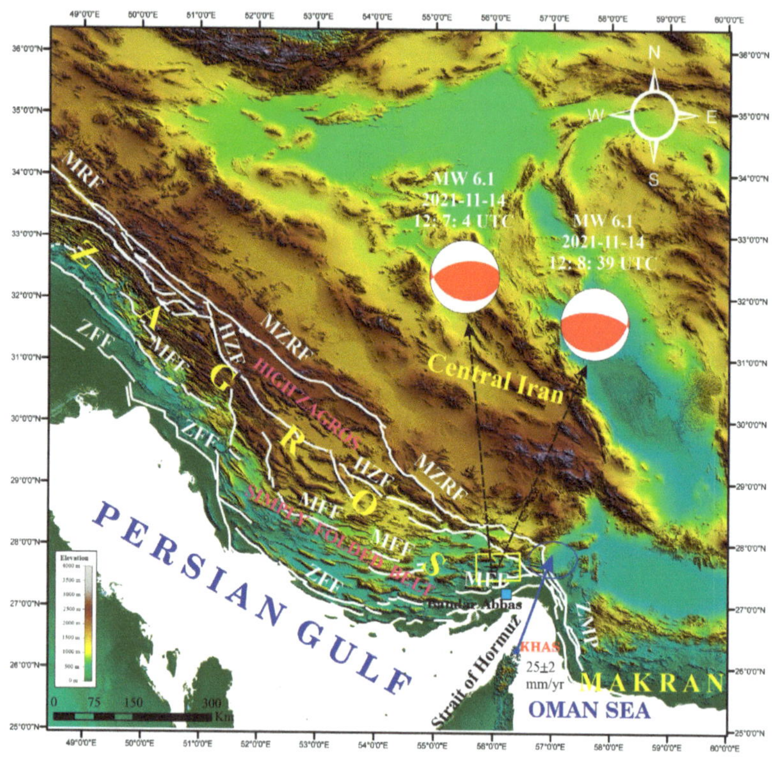

Figure 1. The base map shows the three geological zones of Iran (Zagros, Central Iran, and Makran). The yellow rectangle shows the study area. The active faults are marked as HZF, MFF, and ZFF in the ZFTB [2]. The ZMP marks the transition between the Zagros collision and the Makran subduction zone [28]. The beachballs show the mechanisms of the faults that caused the earthquakes (the data are available at https://www.globalcmt.org/CMTsearch.html, accessed on 19 October 2023). The blue arrow shows the GPS velocities relative to the Eurasian plate near the Strait of Hormuz (KHAS) [28]. The blue square is Bandar Abbas city. The suture between the Arabian margin and central Iran is labeled MZRF and MRF. The identification of faults on the map is based on data from the Iranian faults map [29].

2. Tectonic Setting

The study area lies within the ZFTB, which spans approximately 1800 km from northern Iraq to the Strait of Hormuz in Iran (Figure 1). The ZFTB is a significant part of the Alpine–Himalayan orogenic belt and the most seismically active fold-and-thrust belt in the world.

This region is an active continental collision zone that has promoted convergence between the Arabian and Eurasian plates for approximately 35 million years [30]. The ZFTB consists of sedimentary layers covering the entire Phanerozoic and has a thickness of 10–15 km [7,31–34], reaching up to 17 km in some parts of the Persian Gulf and the eastern part of the belt [6,24,25]. The stratigraphy, characterized by a combination of platform carbonates, evaporites, and marine shales, exerts a significant influence on the style of deformation [7]. The collision led to a shortening of the belt, and its prominent feature is the NW-SE double plunge folds at the surface, which are supported by steep reverse and thrust faults in the basement and NW-SE to N-S strike–slip faults [2,3]. This general trend deviates significantly in the easternmost part of the ZFTB, including the study area, where W-E and SW-NE folds with frequent fold-axis rotations are present.

The northeastern (and topographically highest) part of the ZFTB is referred to as the HZ, while the southwestern (and topographically lower) part is referred to as the SFB [14] (Figure 1). The HZ is bounded on its northeastern side by the Main Zagros Reverse Fault (MZRF) and on its southwestern side by the High Zagros Fault (HZF). The MZRF is commonly interpreted as a suture between the former rocks of the Arabian continental margin and the igneous and metamorphic rocks of central Iran [32]. While the MZRF is no longer active in reverse motion, the northern Zagros region of the same fault zone exhibits a right-lateral component of Arabian–Iranian convergence, known as the Main Recent Fault (MRF) [3,35], confirming a slip rate of 3 ± 2 mm per year [28]. The HZF is a large, NE-dipping thrust and marks the southern boundary of Paleozoic rocks in the Zagros at most locations [2]. The SFB extends from the HZF to the Persian Gulf. Most of the active deformation in the Zagros is concentrated in this zone and likely migrated southwestward from the HZF at an earlier stage [3,36–38]. A notable change in stratigraphy, often associated with elevation changes, occurs in certain folds within the SFB. These folds are closely related to prominent north-dipping basement faults, known as master blind thrusts, which play a crucial role in shaping structures in the region [2]. The most significant of these faults is the MFF (Figure 1), which marks the southwestern limit of surface exposures. The second major basement fault is the Zagros Foredeep Fault (ZFF) (Figure 1), which delineates the boundary of the Quaternary alluvium in the northern coastal plain of the Persian Gulf [7].

The study area, referred to as Bandar Abbas, is located in the SFB and serves as a transition zone between the Zagros and the Makran accretionary prism. The Zendan–Minab–Palami (ZMP) faults play a crucial role in adjusting the velocity gradient between the eastern ZFTB and western Makran areas (Figure 1) and delineate this transition zone [28].

3. Seismicity and Structural Geology

This section aims to enhance our understanding of seismic events and surface structures in the study area. The SFB is characterized by parallel double-plunge anticlines and synclines that dominate the short-wavelength topography and surface structure.

These features were initially categorized as detachment folds formed by buckling of the sedimentary cover above the Hormuz salt [39].

The study area displays parallel folds trending approximately east–west (Figure 2). The Hormuz salt has been exposed in the Handun, Namak, Guniz, and Anguru anticlines [40]. The original thickness of the Hormuz salt may have been several kilometers, although much of it was likely removed by diapirism and erosion [5].

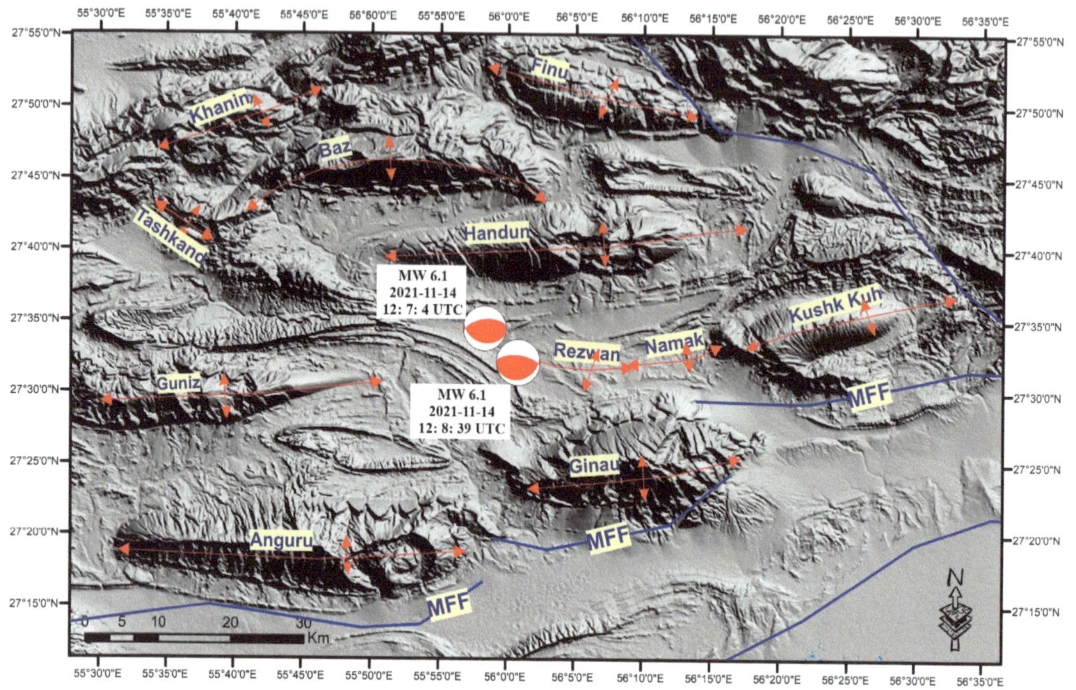

Figure 2. The parallel folds trend approximately east–west in the study area. The blue lines represent the distribution of faults based on the Iranian faults map [29]. Red arrows indicate the axial traces of anticlines.

The strongest earthquakes documented in the Bandar Abbas area of the SFB did not exceed a magnitude of Mw 6.7. These events are attributed to detachment levels comprising weak evaporitic and/or shale horizons [7]. The salt layers act as vertical separations within the overburden's seismogenic layer, hindering seismic rupture propagation [7,41]. The Hormuz salt is identifiable at the surface by numerous diapirs, primarily located in the southeastern part of the SFB [42–44]. Subsurface salt diapirs and pillows have been discovered through the analysis of seismic profiles in the Persian Gulf [24,25,45].

According to GCMT data [12,13], the earthquakes occurred at depths of 17.2 and 13.8 km at coordinates 27.57°N, 55.98°E and 27.54°N, 56.03°E. Both earthquakes exhibited a reverse fault mechanism with a minor strike–slip component.

The surface faults in the study area were digitized based on the geologic map of Bandar Abbas at a scale of 1:250,000 [40] and the Iranian faults map [29] (Figure 3). An examination of geological maps and relevant reports on the faults responsible for the earthquakes [46] revealed no identifiable faults in proximity to the earthquake epicenters.

The MFF is considered the main fault near the epicenter of these earthquakes and is associated with numerous folds along its segments [2]. The origin and tectonic significance of the MFF system and its sinuosity remain unclear. Most hypotheses suggest strong structural control from geological inheritances [47]. Several earthquakes were reported near our study area, attributed to the main faults of the ZFTB [9,14].

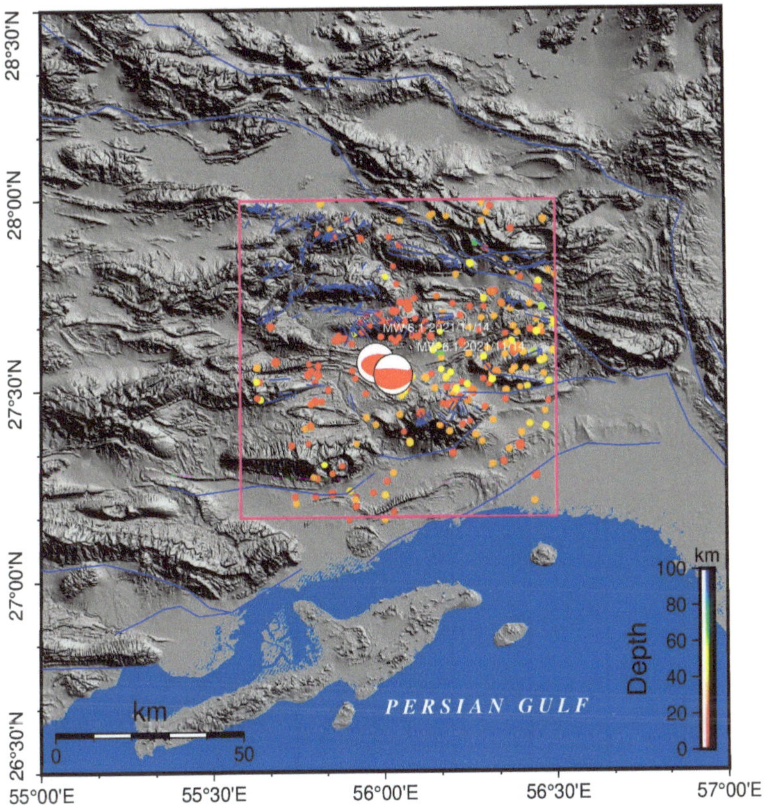

Figure 3. The pink rectangle shows our study area, while the blue lines depict the fault distribution based on the geological map of Bandar Abbas at a scale of 1:250,000 [40] and the Iranian faults map [29]. The colored circles represent the distribution of earthquake epicenters with magnitudes greater than four from 1900 to 13 November 2021, one day before the earthquakes (the data are available at https://earthquake.usgs.gov/earthquakes/search, accessed on 6 November 2023).

To analyze and compare historical earthquakes regardless of any adjustments or relocation to the epicenters, we utilized the seismic catalog for the study area obtained from the United States Geological Survey (USGS). The dataset contains all earthquakes with magnitudes greater than four that occurred from 1900 until 13 November 2021, one day before the earthquakes of 14 November 2021 (Figure 3).

4. Materials and Methods

This section details the materials and methods used to identify the causative fault(s) of the events. Our approach integrates InSAR, seismic inversion, and magnetic studies.

4.1. InSAR

The InSAR technique played a crucial role in this study, offering a valuable alternative for visualizing fault surface projections, and is essential for understanding earthquake behavior. The analysis of interferograms primarily reveals surface deformation over time. We utilized Sentinel-1 SAR images acquired before and after the earthquakes (Table 1) and calculated interferograms using the DInSAR method.

Table 1. Sentinel-1 pairs were used to detect the deformations caused by the doublet earthquake on 14 November 2021.

No.	Satellite	Flight Direction	Path	Master's Date	Slave's Date
1	Sentinel-1A	Ascending	57	13 November 2021	25 November 2021
2	Sentinel-1A	Descending	166	9 November 2021	21 November 2021

Interferogram calculations were performed using GMTSAR software (V6.2) [48], while phase unwrapping was executed with Snaphu software (v2.0.5) [48]. The Shuttle Radar Topography Mission (SRTM) version 1 [48] was employed to correct the wave phase relative to the ground surface (Figure 4).

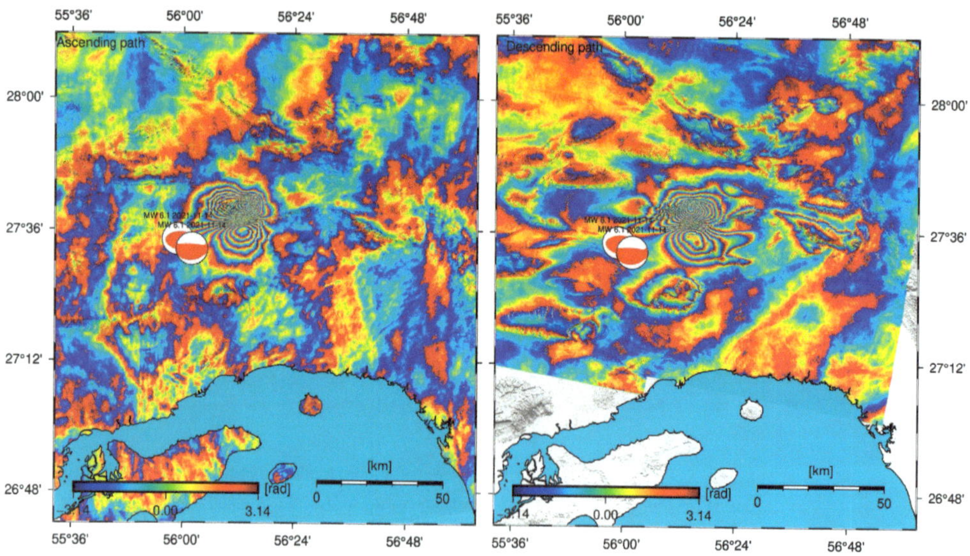

Figure 4. The left panel displays the interferogram of the ascending orbit, while the right panel depicts the interferogram of the descending orbit.

The ground surface displacement determined using the DInSAR method reflects the displacement along the satellite's line of sight (LoS).

Decomposition was used [49,50] to calculate two displacement components: the vertical component (*d_V*) and the horizontal component (*d_EW*) in the east–west direction.

$$\begin{bmatrix} d_{asc} \\ d_{dsc} \end{bmatrix} = \begin{bmatrix} -sin(\theta_{asc})cos(\theta_{asc}) & cos(\theta_{asc}) \\ -sin(\theta_{dsc})cos(\theta_{dsc}) & cos(\theta_{dsc}) \end{bmatrix} \begin{bmatrix} d_{EW} \\ d_V \end{bmatrix} \quad (1)$$

where d_{asc} and d_{dsc} are the LoS displacement vectors (d_{LOS}) from the descending and ascending paths, respectively, and θ is the incidence angle.

4.2. Seismic Inversion and Fault Modeling

We characterized the earthquake source by inferring the geometric parameters of possible fault planes through measurements and indirect observations made via InSAR, using the theory of seismic inversion. The probabilistic approach consisted of a combination of observations, theoretical data, and a priori information derived from the probability distributions of the earthquake. It was supported by the results of an a posteriori probability distribution. Building upon this foundation, we determined the most probable parameters of the earthquake source model, encompassing the offset, depth, length, and width of the fault plane, along with the slip, rake, dip, and direction of movement, which were consistent with the assumptions and observations.

Inversion of earthquake sources for a finite rectangular fault plane based on InSAR data was performed using Pyrocko scripts and Grond software (V1.6.1) employing the Bayesian bootstrap-based probabilistic joint inversion method. This technique estimates the distribution of estimation errors using multiple sampling with out-of-sample replacement [51]. The Bayesian method allowed us to determine the set of all likely source model parameters consistent with the observations and assumptions about the earthquake source and wave propagation.

The input data we used were unwrapped co-seismic interferograms obtained from the ascending and descending paths to determine the true course of the deformation field on the surface. Before the calculations, each of the interferograms was corrected for the influence of the troposphere with the use of the Generic Atmospheric Correction Model (GACOS) [52], which is based on the Iterative Tropospheric Decomposition Model (ITD). The data were divided into quadtrees to reduce the time required for calculations. In this process, four key parameters were determined: the variance threshold in each quadtree's tile, the percentage of allowed NaN pixels per tile, and the minimum and maximum size of the tiles. In the case of the ascending path interferogram, the following values were finally used for the mentioned parameters: 0.007, 0.90, 0.01°, and 0.35°, while for the data from the descending path, the values were equal to 0.010, 0.90, 0.01°, and 0.32°. Based on this analysis, a variance–covariance matrix, subsequently utilized in the optimization process, was generated using the Kite software (V1.3.0).

For the optimization process, in addition to the InSAR data obtained for both the ascending and descending paths, along with additional parameters such as timing (date, centroid time, and assumed duration), location (depth, longitude, and latitude), and seismic parameters (magnitude, scalar moment, moment tensor, and focal mechanism) derived from the GCMT catalog, Green's functions (precomputed for a regional CRUST2 Earth model) were used. Parameter estimation was conducted following data processing and the setting of search limits and the number of iterations.

4.3. Magnetic Study

To unveil the hidden fault(s) that caused the seismic events in the study area, we used aeromagnetic data for Iran within the longitude range of 55.60°–56.30°E and the latitude range of 27.20°–27.90°N. The data were surveyed by the Aeroservice Company of the USA under commission from the Geological Survey of Iran during 1974–1977, with an average flight-line spacing of 7.5 km [34,53].

To mitigate spurious frequencies that might arise during the grid generation process, we applied a low-pass filter with a cutoff wavelength of 15 km to the data. The survey was performed mostly at a constant barometric height of 2 km above mean sea level for the study area [53], which provides a suitable resolution for tectonic studies. The effects of the diurnal changes in the Earth's magnetic field and the effects of the Earth's core magnetic field had already been removed from the data [53]. Therefore, the data only needed to be corrected by reduction to the magnetic pole (RTP) filter.

RTP correction centers the asymmetrical magnetic anomalies over their sources, like the measured anomalies at the Earth's magnetic poles, thus making magnetic interpretation easier [54,55]. To apply the RTP filter, we considered the middle of the data observation period. Additionally, we used the Differential Reduction to Magnetic Pole (DRTP) method [56,57] to assess the variation in the inclination and declination of Earth's magnetic field. The reduced-to-pole magnetic anomaly of the study area is shown in Figure 5a.

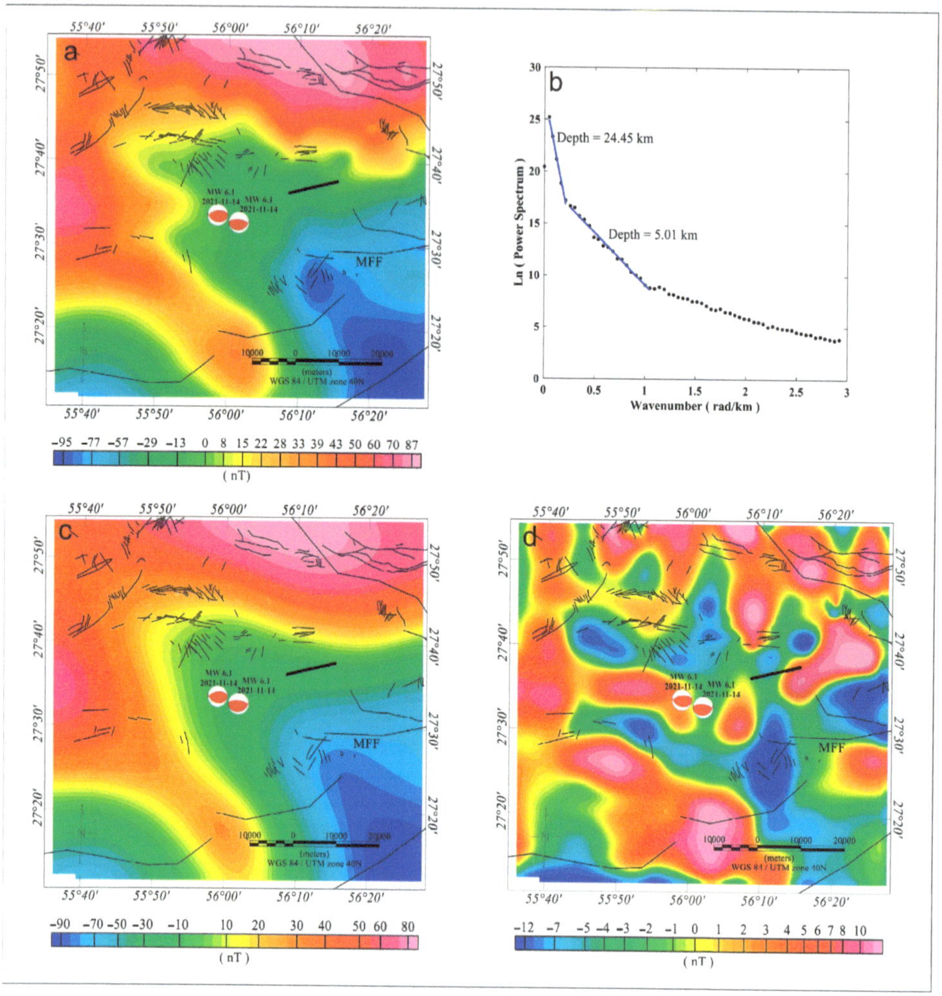

Figure 5. (**a**) Reduced-to-pole magnetic data. (**b**) Radially averaged power spectrum of the data in panel (**a**). (**c**) Regional magnetic data resulting from matched filtering. (**d**) Residual magnetic data resulting from matched filtering. The black solid lines represent the rupture obtained by the seismic inversion method, while the black thin lines represent the distributions of pre-identified faults in the area [29,40].

There are several methods that can be used to enhance potential field data to identify the edges of buried structural contacts and lineaments. One of the simplest and most efficient methods uses directional derivatives, such as horizontal derivatives, which are well suited for locating the lateral contrasts of density and magnetic susceptibility of structures placed next to each other [56]. In this method, the gradient of a field anomaly is calculated considering the orientation of the structures. The gradient of the potential field is highest in the direction perpendicular to the structures, facilitating better detection of the edges of anomalous bodies. Therefore, this perpendicular direction is preferred for calculating the gradient of the field anomaly. Before performing any process, such as calculating the derivative, it is better to separate the regional and residual anomalies originating from deeper and shallower depths, respectively (Figure 5c,d). This process facilitates the detection of structures located at different depths. There are many methods that can be used to separate the anomalies, such as removing the polynomial trends [58], upward continuation [59], and matched filtering [60–62].

We used the matched filtering method, which involves spectral analysis of field data. The power spectrum of reduced-to-pole magnetic anomalies shows the average depths of 5 and 25 km for structures that create the anomalies (Figure 5b). We designed a matched filter to extract the residual anomalies corresponding to shallower structures.

5. Results

This section presents the results obtained via the methods described in Section 4.

5.1. InSAR

The decomposition of the data into horizontal and vertical displacement components offered a significant advantage: it enabled precise determination of the movement along the primary earthquake fault zone [63].

We calculated the vertical and horizontal displacements in the east–west direction by analyzing data from two paths. The primary displacements along the LoS (Figure 6) revealed values of up to −360 mm for the ascending orbit. This indicates an unidentified fault with horizontal surface displacements ranging from −144 to 170 mm and maximum vertical displacements between −204 and 415 mm (Figure 7).

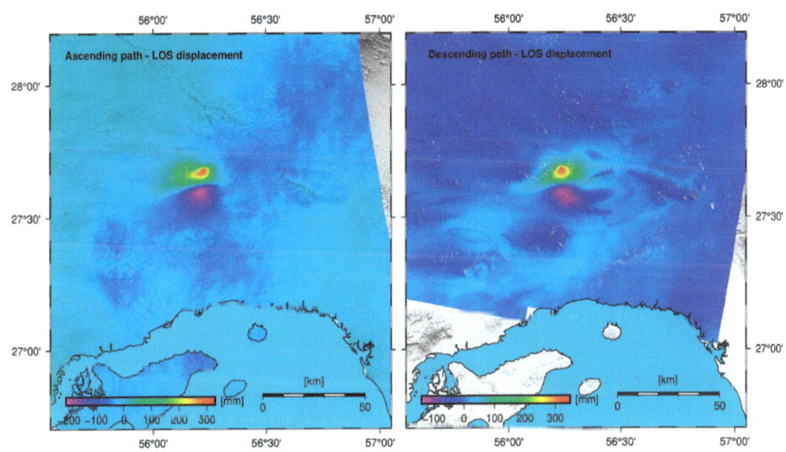

Figure 6. Displacement results for ascending and descending paths. The left panel shows the displacement results for the ascending path, while the right panel shows the displacement results for the descending path. Positive displacement indicates movement toward the satellite.

5.2. Seismic Inversion and Fault Modeling

We modeled the earthquake source by inferring the geometric parameters of a potential fault plane through measurements and indirect observations, specifically using InSAR and the theory of seismic inversion.

Based on the calculations, we derived two geodetic solutions for the fault plane that best fitted the displacements obtained from the InSAR data. The first solution features a north-dipping, ENE–WSW-oriented fault plane with the following parameters: strike: 257°, dip: 74°, and rake: 77°. The fault plane dimensions are approximately 3.6 km in width, 13.4 km in length, and 5.6 km in depth (Figure 8).

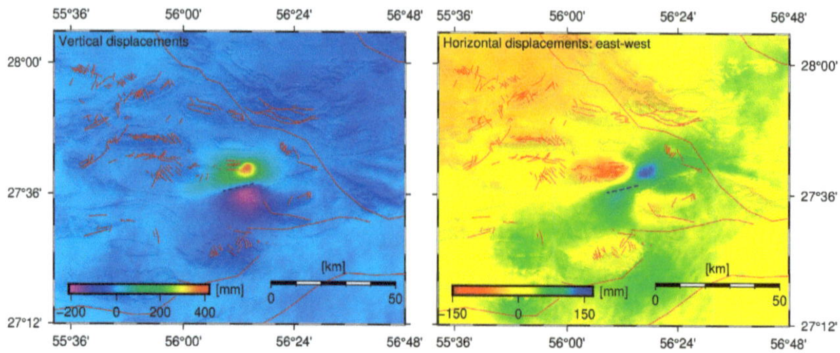

Figure 7. Decomposition of the displacements into vertical and horizontal components. The left panel displays the vertical displacement associated with the seismic event, while the right panel shows the horizontal displacement in the east–west direction. The red lines represent faults from geologic maps and serve as a reference for the known fault in the region. The black dashed lines indicate the fault resulting from seismic inversion.

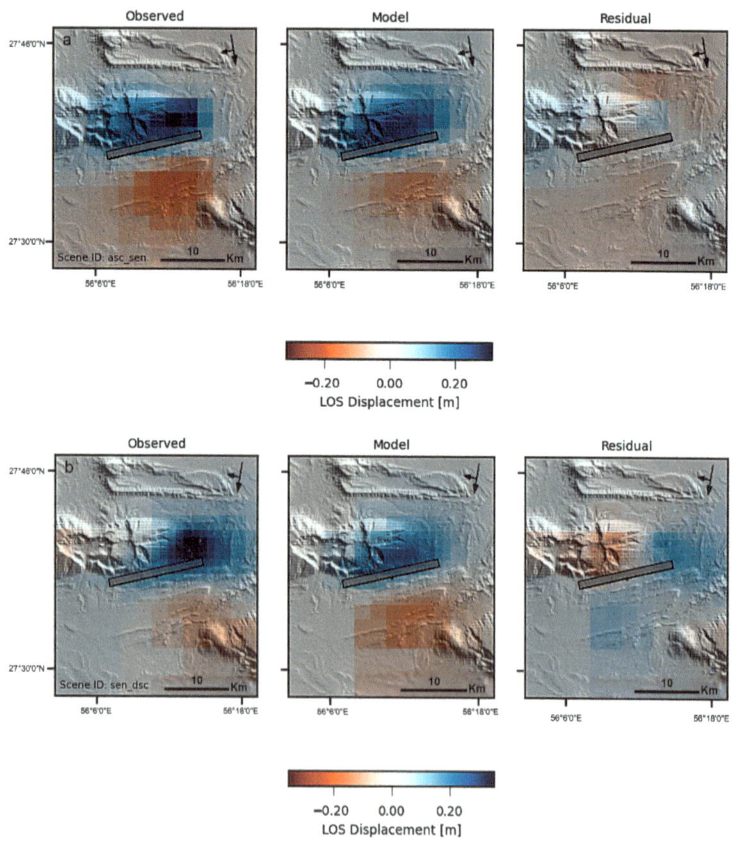

Figure 8. Projection of the north-dipping fault plane (gray rectangles) with the upper edge of the fault (bold black lines) overlaid on the background of LoS surface displacements. The displacements shown were derived from (**a**) the ascending orbit and (**b**) the descending orbit, with columns representing observed, modeled, and residual (observed minus modeled) displacements, respectively.

The second solution represents a south-dipping, ENE–WSW-oriented fault plane with the following parameters: strike: 80°, dip: 13°, and rake: 89°. The fault plane's dimensions are a depth of 4.0 km, a length of 13.4 km, and a width of 7.3 km (Figure 9).

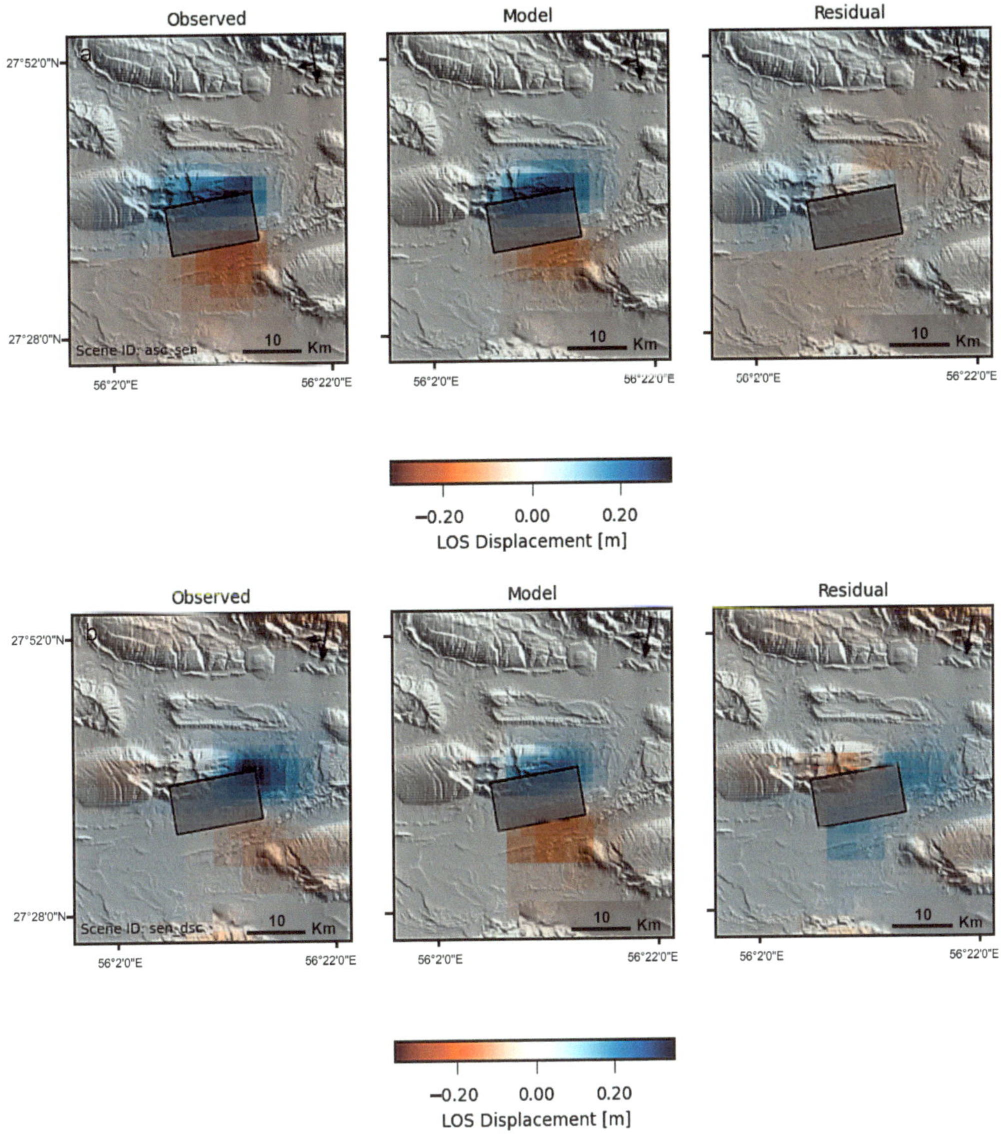

Figure 9. Projection of the south-dipping fault plane (gray rectangles) with the upper edge of the fault (bold black lines) overlaid on the background of LoS surface displacements. The displacements shown are derived from (**a**) the ascending orbit and (**b**) the descending orbit, with columns representing observed, modeled, and residual (observed minus modeled) displacements, respectively.

A graph with a probability density function (PDF) is presented for each parameter in both of the obtained solutions, north-dipping (Figure 10) and south-dipping (Figure 11).

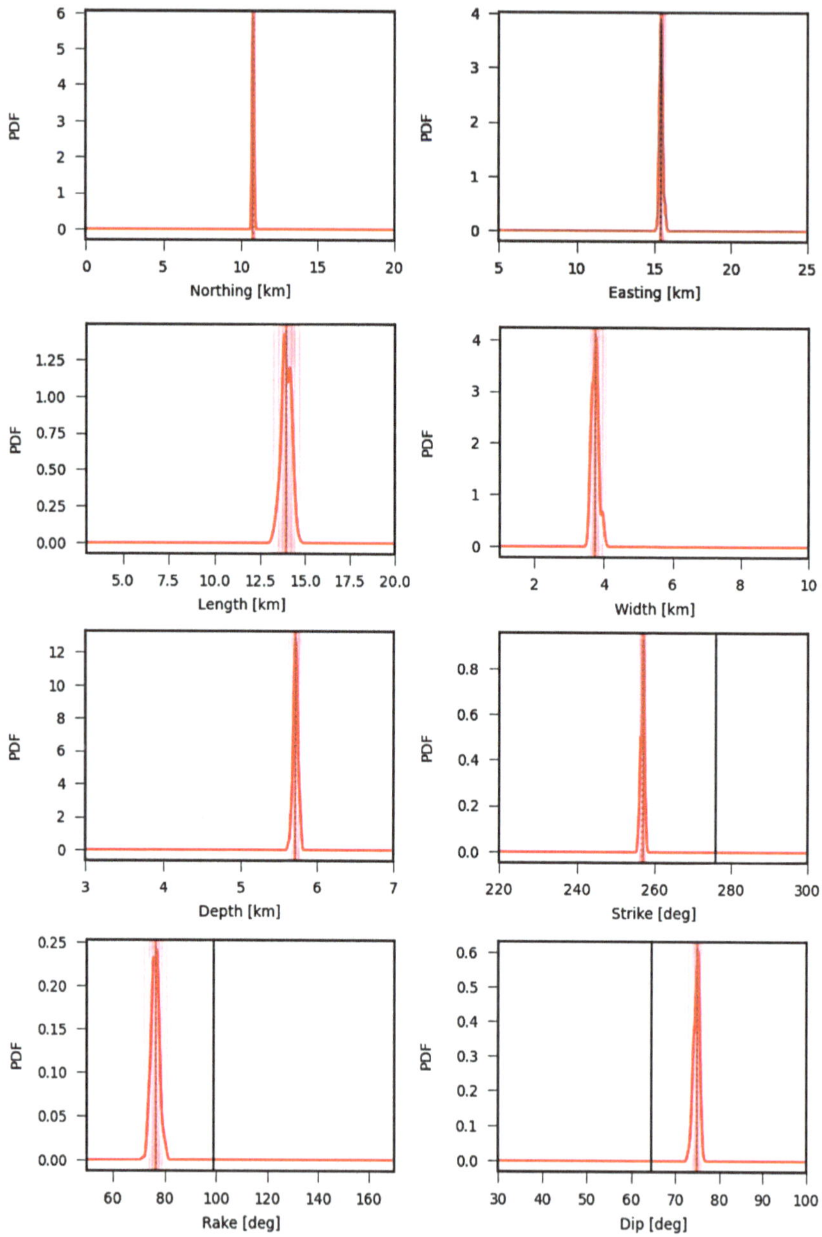

Figure 10. Distributions of the inferred parameters of the north-dipping fault plane solution presented as probability density functions (PDFs). The histograms are shown either as Gaussian kernel densities (red curved solid lines) or as bar plots (red shaded polygons). The medians of the probability distributions are marked with a solid red line, the mean values are marked with a red dashed line, and the reference solutions from the GCMT catalog are marked with a black line.

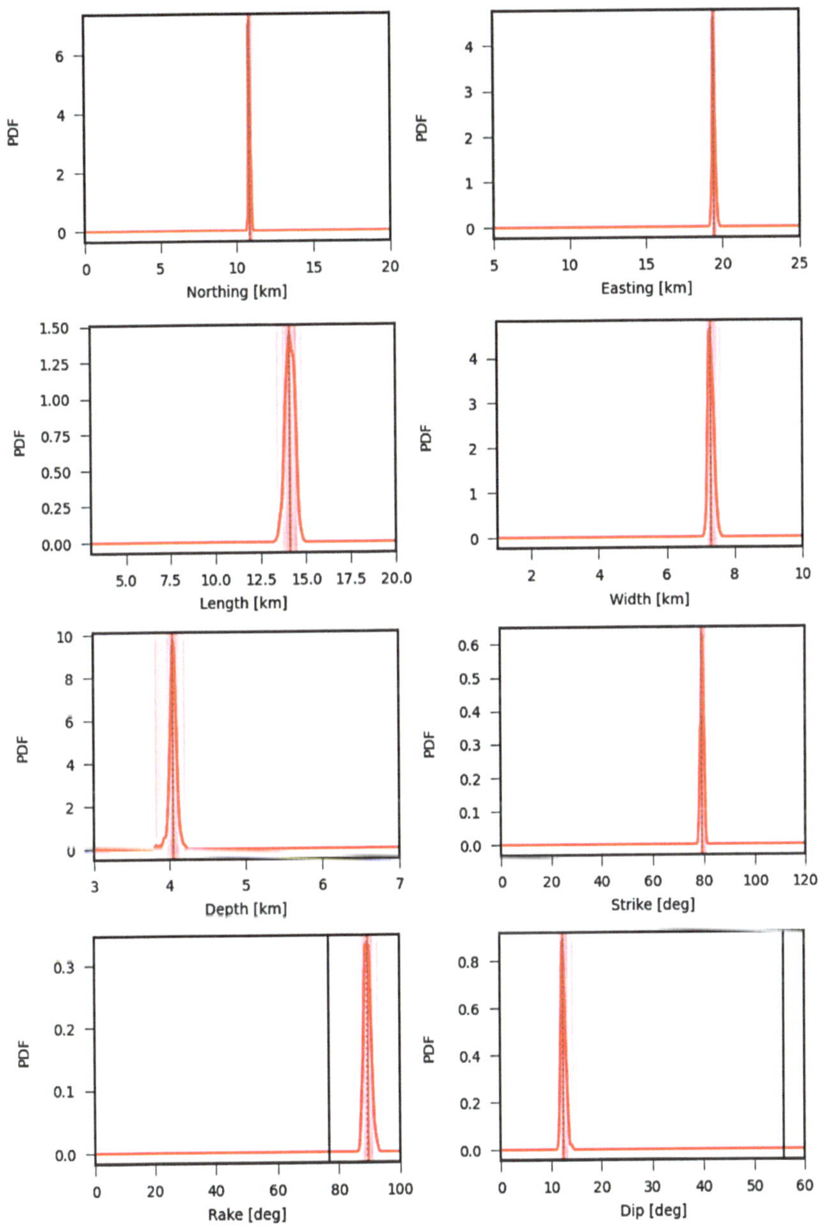

Figure 11. Distributions of the inferred parameters of the south-dipping fault plane solution presented as probability density functions (PDFs). The histograms are shown either as Gaussian kernel densities (red curved solid lines) or as bar plots (red shaded polygons). The medians of the probability distributions are marked with a solid red line, the mean values are marked with a red dashed line, and the reference solutions from the GCMT catalog are marked with a black line.

5.3. Magnetic Study

We calculated the gradient of the residual anomaly in the N-S direction, considering the orientation of the existing faults and folds. This analysis enabled us to identify the fault, as recognized by seismic inversion, through an examination of the horizontal gradient of the residual magnetic anomaly. The gradient map reveals a potential extension of the observed anomalies, attributed to a fault striking in an ENE-WSW direction, likely extending towards the ~ E-W(Figure 12). We determined the fault extension by simultaneously investigating the gradient map, the approximate locations of the main shocks, and the area's seismicity map. In this analysis, we focused on interpreting the major fault related to the seismic events and avoided evaluating all existing anomalies in the gradient map. The black dashed line depicts the proposed fault extension (Figure 12).

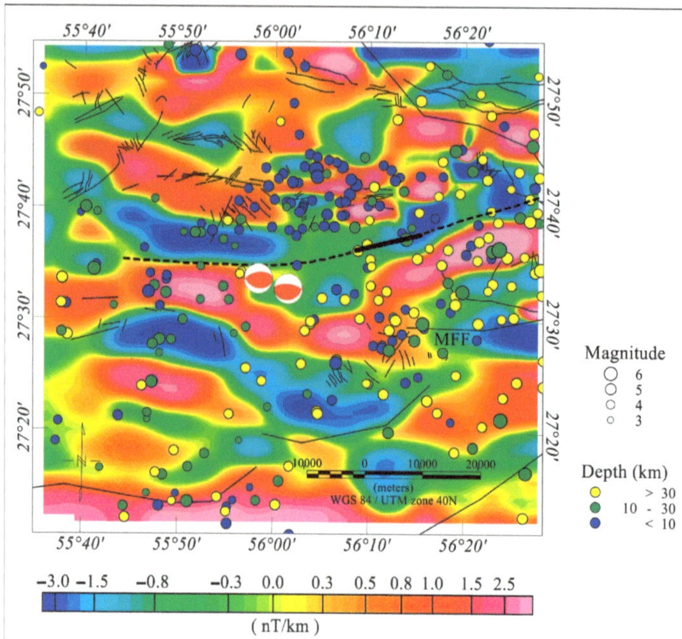

Figure 12. Horizontal gradient map of residual magnetic data in the N-S direction. Colored circles represent the distribution of seismic events (magnitude > 3) from 1900 to December 2023. The circles are colored and scaled to correspond to the depths and magnitudes of the events, respectively. Solid black lines indicate faults identified through seismic inversion, while the black dashed line depicts the proposed fault extension.

6. Discussion

The study aimed to investigate the underlying causes of the doublet earthquake that occurred in Fin on 14 November 2021. Seismic inversion results indicate the presence of two potential fault solutions: one dipping towards the north and the other towards the south. Both suggest a single fault and show no significant variation in strike and location.

The examination of the horizontal gradient of the residual magnetic anomaly confirms the fault identified in the seismic inversion results. The analysis suggests the potential extension of observed anomalies resulting from a fault striking in an ENE-WSW direction, likely extending towards the ~E-W. However, it remains unclear whether the fault with the north or south dip aligns with the geological setting of the area.

To address this, we thoroughly analyzed the history of structural sections in the region. Notably, our study area was limited to a few published regional structural cross-sections (Figure 13e,f) explicitly highlighted by Molinaro et al. [64] and Jahani et al. [45].

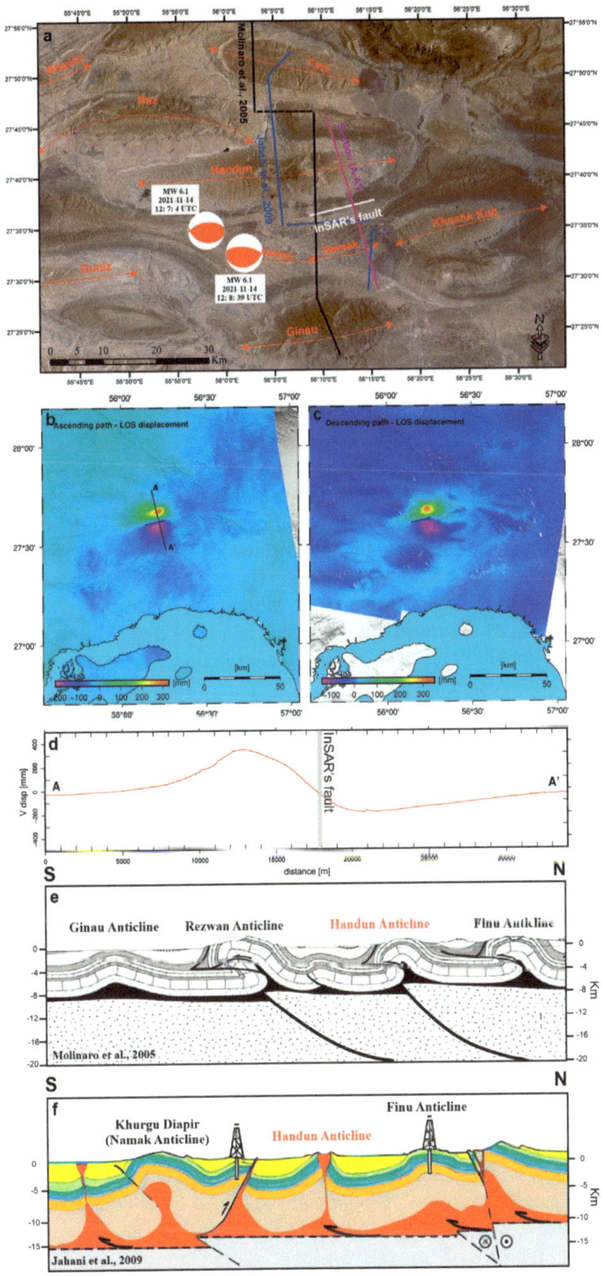

Figure 13. Comparison of two regional structural cross-sections presented by Molinaro et al. [64] (**e**) and Jahani et al. [23] (**f**). The InSAR profile is depicted in panel (**d**), with the ascending and descending tracks shown in panels (**b**,**c**), respectively. The InSAR fault is delineated by the white line in panel (**a**) and the gray line in panel (**d**).

Molinaro et al. [64] identified a low-dip thrust fault on the southern limb of the Handun anticline (Figure 13e). This fault poses challenges and uncertainties because the observation is based solely on surface data, which may not provide a comprehensive understanding.

The focal mechanisms of the events provide evidence for a thrust fault (Figure 13a). Furthermore, displacement distribution observed through the LoS indicates upward movement in the northern part of the proposed fault (Figure 13b,d), suggesting that the fault dips to the north rather than the south.

The strikes and dips for potential fault solutions in this study align closely with the results of Fathian et al. [15], who used radar data from the Sentinel-1 and ALOS-2 satellites to identify two fault planes. Both planes are oriented east–west and do not emphasize the source geometry. The first plane had a northward dip (71° dip, 263° strike), while the second plane had a southward dip (15° dip, 84° strike).

Previous studies, such as those conducted by Golshadi et al. [16], Toker et al. [17], and Rezapour and Jamalreyhani [41], predominantly relied on local seismic networks.

Golshadi et al. [16] reported that the first modeled plane corresponded to a NNE-dipping fault with parameter values of 284°, 53°, and 92° for the strike, dip, and rake, respectively. The second earthquake was associated with a SSE-dipping fault with parameter values of 85°, 6°, and 81°, with depths ranging between 7 and 14 km for the first event and between 1 and 7 km for the second. Golshadi et al. [16] endeavored to distinguish between a thrust and a back thrust in the Khurgu (Namak) and Handun anticlines. Toker et al. [17] presented the following results: the first fault plane was characterized by parameter values of 114°, 32°, and 91° for the strike, dip, and rake, with a SW dip, while the second fault plane had parameter values of 49°, 24°, and 27°, with a SE dip. They stated that the co-seismic displacement was confined to a depth range of 3 to 16 km. Rezapour and Jamalreyhani [41] suggested a more likely orientation of the fault plane with a west–east strike and a S-SE dip.

The investigations by Golshadi et al. [16] and Toker et al. [17] revealed significant discrepancies in fault locations attributable to differing interpretations and strike orientations of the causative faults. From our perspective, the identified faults, particularly the back thrust, are only approximations due to insufficiently reliable data. Further studies with robust references are needed to validate the authors' structural interpretations. Additionally, Toker et al. [17] and Rezapour and Jamalreyhani [41] relied on local seismic stations, resulting in limited data coverage and inconsistencies.

Discrepancies in the attribution of faults, differing interpretations by researchers, and challenges in defining seismic source characteristics underscore the need for further research. To verify and enhance our understanding of the causes of earthquakes, we conducted the present study using a novel method that combines InSAR and magnetic data.

We propose a single causative fault with a northward dip, suggesting that two earthquakes occurred along it. This aligns with typical seismic events observed in the SFB, as highlighted by Nissen et al. [65], who indicated that most regional earthquakes with magnitudes around 6 (Mw) occur along reverse faults with dips of approximately 50°. Furthermore, these faults do not reach the surface but are buried at depths of approximately 4 km. Our seismic inversion confirmed that the fault plane depth is 5.6 km, consistent with results from magnetic interpretation (Figure 5b). This is particularly relevant, as earthquakes in the region are triggered within the sedimentary cover [4–6,65].

We refute the possibility of a south-dipping fault or back-thrust mechanism due to the lack of concrete evidence supporting upward movement in the southern segment. Confirmation of upward movement in the southern part of the block is necessary to consider such mechanisms.

The research findings indicate that the earthquake resulted from a single fault dipping northward, which extends approximately E-W, through the interpretation of magnetic anomalies.

Our study emphasizes the importance of a multidisciplinary approach to understanding the complexities of seismic hazard assessment in tectonically active regions like the

ZFTB. Integrating different methodologies is crucial for developing effective risk mitigation strategies and fully understanding seismic hazards.

7. Conclusions

The current study employed a multidisciplinary approach to identify the causative fault and seismic sources of the doublet earthquake that occurred on 14 November 2021 in the Fin area. The study concluded that a single fault with a northward dip was responsible for both earthquakes.

The primary finding is that the fault has a strike, dip, and rake of 257°, 74°, and 77°, respectively. The fault plane measures approximately 3.6 km in width and 13.4 km in length. This previously unidentified fault showed horizontal surface displacements ranging from −144 to 170 mm and maximum vertical displacements between −204 and 415 mm during the earthquakes.

Complementing these findings, our magnetic study detected a concealed fault using aeromagnetic data and directional derivatives. The possible extension of observed anomalies was caused by a fault striking in an ENE-WSW direction, which likely extends towards the ~E-W, suggesting the presence of a major fault not previously identified.

The seismic inversion confirmed a fault plane depth of 5.6 km, consistent with the magnetic interpretation results. This is particularly relevant, as earthquakes in this region are triggered within the sedimentary cover.

While this study addressed some data gaps and uncertainties in previous research, future studies could incorporate structural fieldwork, structural cross-section analysis, non-seismic and microseismic investigations, and physical modeling.

Author Contributions: Conceptualization, P.N., and W.M.; Investigation, P.N.; Methodology, P.N., W.M, N.B.-J., S.-H.M.-A., and M.K.; Software, P.N., W.M, N.B.-J., and M.K.; Project administration, P.N.; Resources, P.N., W.M, N.B.-J., and M.K.; Supervision, P.N.; Validation, P.N.; Visualization, P.N.; Writing—original draft, P.N.; Writing—review and editing, P.N., W.M, N.B.-J., S.-H.M.-A., and M.K. All authors have read and agreed to the published version of the manuscript.

Funding: This research received no external funding.

Data Availability Statement: The original contributions presented in the study are included in the article, further inquiries can be directed to the corresponding author.

Acknowledgments: We appreciate the valuable feedback from the reviewers, which greatly helped improve this article. Peyman would like to express his gratitude to Behnam Hassanpour for his valuable insights and comments, which significantly contributed to improving this article. We are also grateful for the open-access software, including GMT, GMTSAR, Snaphu, Grond, and Kite, which played a central role in our analysis. This work was part of Peyman Namdarsehat's PhD dissertation, supported by Wroclaw University of Science and Technology.

Conflicts of Interest: The authors declare no conflicts of interest.

References

1. Ni, J.; Barazangi, M. Seismotectonics of the Zagros continental collision zone and a comparison with the Himalayas. *J. Geophys. Res. Solid Earth* **1986**, *91*, 8205–8218. [CrossRef]
2. Berberian, M. Master "blind" thrust faults hidden under the Zagros folds: Active basement tectonics and surface morphotectonics. *Tectonophysics* **1995**, *241*, 193–224. [CrossRef]
3. Talebian, M.; Jackson, J. A reappraisal of earthquake focal mechanisms and active shortening in the Zagros mountains of Iran. *Geophys. J. Int.* **2004**, *156*, 506–526. [CrossRef]
4. Nissen, E.; Jackson, J.; Jahani, S.; Tatar, M. Zagros "phantom earthquakes" reassessed—The interplay of seismicity and deep salt flow in the Simply Folded Belt? *J. Geophys. Res. Solid Earth* **2014**, *119*, 3561–3583. [CrossRef]
5. Elliott, J.; Bergman, E.; Copley, A.; Ghods, A.; Nissen, E.; Oveisi, B.; Tatar, M.; Walters, R.; Yamini-Fard, F. The 2013 Mw 6.2 Khaki-Shonbe (Iran) earthquake: Insights into seismic and aseismic shortening of the Zagros sedimentary cover. *Earth Space Sci.* **2015**, *2*, 435–471. [CrossRef]
6. Jahani, S.; Hassanpour, J.; Mohammadi-Firouz, S.; Letouzey, J.; de Lamotte, D.F.; Alavi, S.A.; Soleimany, B. Salt tectonics and tear faulting in the central part of the Zagros Fold-Thrust Belt, Iran. *Mar. Pet. Geol.* **2017**, *86*, 426–446. [CrossRef]

7. Nissen, E.; Tatar, M.; Jackson, J.A.; Allen, M.B. New views on earthquake faulting in the Zagros fold-and-thrust belt of Iran. *Geophys. J. Int.* **2011**, *186*, 928–944. [CrossRef]
8. Baker, C.; Jackson, J.; Priestley, K. Earthquakes on the Kazerun Line in the Zagros Mountains of Iran: Strike-slip faulting within a fold-and-thrust belt. *Geophys. J. Int.* **1993**, *115*, 41–61. [CrossRef]
9. Walker, R.T.; Andalibi, M.; Gheitanchi, M.; Jackson, J.; Karegar, S.; Priestley, K. Seismological and field observations from the 1990 November 6 Furg (Hormozgan) earthquake: A rare case of surface rupture in the Zagros mountains of Iran. *Geophys. J. Int.* **2005**, *163*, 567–579. [CrossRef]
10. Hatzfeld, D.; Molnar, P. Comparisons of the kinematics and deep structures of the Zagros and Himalaya and of the Iranian and Tibetan plateaus and geodynamic implications. *Rev. Geophys.* **2010**, *48*. [CrossRef]
11. Mirabedini, M.S.; Agh-Atabai, M.; Azimmohseni, M.; Zadeh, M.A. Temporal pattern of seismicity in the Zagros belt, SW Iran. *Arab. J. Geosci.* **2021**, *14*, 1707. [CrossRef]
12. Dziewonski, A.M.; Chou, T.A.; Woodhouse, J.H. Determination of earthquake source parameters from waveform data for studies of global and regional seismicity. *J. Geophys. Res. Solid Earth* **1981**, *86*, 2825–2852. [CrossRef]
13. Ekström, G.; Nettles, M.; Dziewoński, A. The global CMT project 2004–2010: Centroid-moment tensors for 13,017 earthquakes. *Phys. Earth Planet. Inter.* **2012**, *200*, 1–9. [CrossRef]
14. Roustaei, M.; Nissen, E.; Abbassi, M.; Gholamzadeh, A.; Ghorashi, M.; Tatar, M.; Yamini-Fard, F.; Bergman, E.; Jackson, J.; Parsons, B. The 2006 March 25 Fin earthquakes (Iran)—Insights into the vertical extents of faulting in the Zagros Simply Folded Belt. *Geophys. J. Int.* **2010**, *181*, 1275–1291. [CrossRef]
15. Fathian, A.; Atzori, S.; Svigkas, N.; Tolomei, C.; Shugar, D.H.; Reicherter, K. Source Characteristics of the Fin Doublet Earthquake of 14 November 2021 (Mw 6.2 and Mw 6.3) Utilizing InSAR Data. In Proceedings of IGARSS 2022-2022 IEEE International Geoscience and Remote Sensing Symposium, Kuala Lumpur, Malaysia, 17–22 July 2022; IEEE: Piscataway, NJ, USA; pp. 2119–2122.
16. Golshadi, Z.; Famiglietti, N.A.; Caputo, R.; SoltaniMoghadam, S.; Karimzadeh, S.; Memmolo, A.; Falco, L.; Vicari, A. Contemporaneous Thick-and Thin-Skinned Seismotectonics in the External Zagros: The Case of the 2021 Fin Doublet, Iran. *Remote Sens.* **2023**, *15*, 2981. [CrossRef]
17. Toker, M.; Durmuş, H.; Utkucu, M. Decoupled co-seismic deformation and stress changes during the 2021 (Mw 6.0, 6.4) North Bandar Abbas doublet earthquakes in Fin region, SE-syntaxis of Zagros, Iran: New insights into the tectonic deformation decoupling process. *Environ. Earth Sci.* **2023**, *82*, 533. [CrossRef]
18. Rezapour, M.; Jamalreyhani, M. Source fault analyses from InSAR data and aftershocks for the Fin doublet earthquakes on 14 November 2021 in Hormozgan province, South Iran. *J. Earth Space Phys.* **2023**, *48*, 87–97.
19. Boudriki Semlali, B.-E.; Molina, C.; Park, H.; Camps, A. First results on the systematic search of land surface temperature anomalies as earthquakes precursors. *Remote Sens.* **2023**, *15*, 1110. [CrossRef]
20. Molina, C.; Boudriki Semlali, B.-E.; Park, H.; Camps, A. A preliminary study on ionospheric scintillation anomalies detected using GNSS-R data from NASA CYGNSS mission as possible earthquake precursors. *Remote Sens.* **2022**, *14*, 2555. [CrossRef]
21. Molina, C.; Semlali, B.B.; González-Casado, G.; Park, H.; Camps, A. Ionospheric scintillation anomalies associated with the 2021 La Palma volcanic eruption detected with GNSS-R and GNSS-RO observations. In Proceedings of IGARSS 2022-2022 IEEE International Geoscience and Remote Sensing Symposium, Kuala Lumpur, Malaysia, 17–22 July 2022; IEEE: Piscataway, NJ, USA; pp. 7445–7448.
22. Ommi, S.; Smirnov, V.B. Seismicity patterns before the 2021 Fin (Iran) doublet earthquakes using the region-time-length and time-to-failure methods. *Earthq. Sci.* **2023**, *37*, 324–336.
23. Jahani, S.; Callot, J.-P.; de Lamotte, D.F.; Letouzey, J.; Leturmy, P. The salt diapirs of the eastern Fars Province (Zagros, Iran): A brief outline of their past and present. In *Proceedings of Thrust Belts and Foreland Basins: From Fold Kinematics to Hydrocarbon Systems*; Springer: Berlin/Heidelberg, Germany; pp. 289–308.
24. Hassanpour, J.; Muñoz, J.A.; Yassaghi, A.; Ferrer, O.; Jahani, S.; Santolaria, P.; SeyedAli, S.M. Impact of salt layers interaction on the salt flow kinematics and diapirism in the Eastern Persian Gulf, Iran: Constraints from seismic interpretation, sequential restoration, and physical modelling. *Tectonophysics* **2021**, *811*, 228887. [CrossRef]
25. Hassanpour, J.; Yassaghi, A.; Muñoz, J.A.; Jahani, S. Salt tectonics in a double salt-source layer setting (Eastern Persian Gulf, Iran): Insights from interpretation of seismic profiles and sequential cross-section restoration. *Basin Res.* **2021**, *33*, 159–185. [CrossRef]
26. Nissen, E.; Yamini-Fard, F.; Tatar, M.; Gholamzadeh, A.; Bergman, E.; Elliott, J.; Jackson, J.; Parsons, B. The vertical separation of mainshock rupture and microseismicity at Qeshm island in the Zagros fold-and-thrust belt, Iran. *Earth Planet. Sci. Lett.* **2010**, *296*, 181–194. [CrossRef]
27. Okada, Y. Surface deformation due to shear and tensile faults in a half-space. *Bull. Seismol. Soc. Am.* **1985**, *75*, 1135–1154. [CrossRef]
28. Vernant, P.; Nilforoushan, F.; Hatzfeld, D.; Abbassi, M.; Vigny, C.; Masson, F.; Nankali, H.; Martinod, J.; Ashtiani, A.; Bayer, R. Present-day crustal deformation and plate kinematics in the Middle East constrained by GPS measurements in Iran and northern Oman. *Geophys. J. Int.* **2004**, *157*, 381–398. [CrossRef]
29. Javadi, H.; Sheikholeslami, M.; Asadi Sarhar, M. *Iran Fault Map on Sedimentary-Structural Units*; Tectonic Group. Geological Survey of Iran: Tehran, Iran, 2013.
30. Allen, M.B.; Armstrong, H.A. Arabia–Eurasia collision and the forcing of mid-Cenozoic global cooling. *Palaeogeogr. Palaeoclimatol. Palaeoecol.* **2008**, *265*, 52–58. [CrossRef]

31. James, G.; Wynd, J. Stratigraphic nomenclature of Iranian oil consortium agreement area. *AAPG Bull.* **1965**, *49*, 2182–2245. [CrossRef]
32. Stocklin, J. Structural history and tectonics of Iran: A review. *AAPG Bull.* **1968**, *52*, 1229–1258. [CrossRef]
33. Falcon, N.L. Problems of the relationship between surface structure and deep displacements illustrated by the Zagros Range. *Geol. Soc. Lond. Spec. Publ.* **1969**, *3*, 9–21. [CrossRef]
34. Teknik, V.; Ghods, A. Depth of magnetic basement in Iran based on fractal spectral analysis of aeromagnetic data. *Geophys. J. Int.* **2017**, *209*, 1878–1891. [CrossRef]
35. Peyret, M.; Djamour, Y.; Rizza, M.; Ritz, J.-F.; Hurtrez, J.-E.; Goudarzi, M.; Nankali, H.; Chery, J.; Le Dortz, K.; Uri, F. Monitoring of the large slow Kahrod landslide in Alborz mountain range (Iran) by GPS and SAR interferometry. *Eng. Geol.* **2008**, *100*, 131–141. [CrossRef]
36. Hessami, K.; Koyi, H.A.; Talbot, C.J.; Tabasi, H.; Shabanian, E. Progressive unconformities within an evolving foreland fold–thrust belt, Zagros Mountains. *J. Geol. Soc.* **2001**, *158*, 969–981. [CrossRef]
37. Walpersdorf, A.; Hatzfeld, D.; Nankali, H.; Tavakoli, F.; Nilforoushan, F.; Tatar, M.; Vernant, P.; Chéry, J.; Masson, F. Difference in the GPS deformation pattern of North and Central Zagros (Iran). *Geophys. J. Int.* **2006**, *167*, 1077–1088. [CrossRef]
38. Oveisi, B.; Lavé, J.; Van Der Beek, P.; Carcaillet, J.; Benedetti, L.; Aubourg, C. Thick-and thin-skinned deformation rates in the central Zagros simple folded zone (Iran) indicated by displacement of geomorphic surfaces. *Geophys. J. Int.* **2009**, *176*, 627–654. [CrossRef]
39. Colman-Sadd, S. Fold development in Zagros simply folded belt, Southwest Iran. *AAPG Bull.* **1978**, *62*, 984–1003. [CrossRef]
40. Fakhari. *Geological Map of Bandar Abbas. 1: 250,000*; Exp-Directorate, NIOC: Tehran, Iran, 1996.
41. Jamalreyhani, M.; Pousse-Beltran, L.; Büyükakpınar, P.; Cesca, S.; Nissen, E.; Ghods, A.; López-Comino, J.Á.; Rezapour, M.; Najafi, M. The 2019–2020 Khalili (Iran) Earthquake Sequence—Anthropogenic Seismicity in the Zagros Simply Folded Belt? *J. Geophys. Res. Solid Earth* **2021**, *126*, e2021JB022797. [CrossRef] [PubMed]
42. Kent, P. The emergent Hormuz salt plugs of southern Iran. *J. Pet. Geol.* **1979**, *2*, 117–144. [CrossRef]
43. Edgell, H. Salt tectonism in the Persian Gulf basin. *Geol. Soc. Lond. Spec. Publ.* **1996**, *100*, 129–151. [CrossRef]
44. Letouzey, J.; Sherkati, S. *Salt Movement, Tectonic Events, and Structural Style in the Central Zagros Fold and Thrust Belt (Iran)*; SEPM Society for Sedimentary Geology: Broken Arrow, OK, USA, 2004. [CrossRef]
45. Jahani, S.; Callot, J.P.; Letouzey, J.; Frizon de Lamotte, D. The eastern termination of the Zagros Fold-and-Thrust Belt, Iran: Structures, evolution, and relationships between salt plugs, folding, and faulting. *Tectonics* **2009**, *28*. [CrossRef]
46. Eslami, A.P.S.; Mahshadnia, L. *Preliminary Report of Dual Earthquakes on 14 November 2021, Fin, Bandar Abbas, Hormozgan Province with Local Magnitudes of 6.3 and 6.4*; Technical report; International Institute of Earthquake Engineering and Seismology, Ministry of Science, Research and Technology: Tehran, Iran, 2021.
47. Tavani, S.; Camanni, G.; Nappo, M.; Snidero, M.; Ascione, A.; Valente, E.; Gharabeigli, G.; Morsalnejad, D.; Mazzoli, S. The Mountain Front Flexure in the Lurestan region of the Zagros belt: Crustal architecture and role of structural inheritances. *J. Struct. Geol.* **2020**, *135*, 104022. [CrossRef]
48. Sandwell, D.; Mellors, R.; Tong, X.; Wei, M.; Wessel, P. GMTSAR. An InSAR Processing System Based on Generic Mapping Tools. 2011. Available online: https://escholarship.org/uc/item/8zq2c02m (accessed on 2 May 2024).
49. Fuhrmann, T.; Garthwaite, M.C. Resolving three-dimensional surface motion with InSAR: Constraints from multi-geometry data fusion. *Remote Sens.* **2019**, *11*, 241. [CrossRef]
50. Milczarek, W. Application of a small baseline subset time series method with atmospheric correction in monitoring results of mining activity on ground surface and in detecting induced seismic events. *Remote Sens.* **2019**, *11*, 1008. [CrossRef]
51. Heimann, S.; Isken, M.; Kühn, D.; Sudhaus, H.; Steinberg, A.; Daout, S.; Cesca, S.; Bathke, H.; Dahm, T. *Grond: A Probabilistic Earthquake Source Inversion Framework*; GFZ Data Services: Potsdam, Germany, 2018. [CrossRef]
52. Yu, C.; Li, Z.; Penna, N.T.; Crippa, P. Generic atmospheric correction model for interferometric synthetic aperture radar observations. *J. Geophys. Res. Solid Earth* **2018**, *123*, 9202–9222. [CrossRef]
53. Saleh, R. *Reprocessing of Aeromagnetic Map of Iran*; Institute for Advanced Studies in Basic Sciences: Zanjan, Iran, 2006.
54. Blakely, R.J.; Simpson, R.W. Approximating edges of source bodies from magnetic or gravity anomalies. *Geophysics* **1986**, *51*, 1494–1498. [CrossRef]
55. Zhang, H.; Marangoni, Y.R.; Hu, X.; Zuo, R. NTRTP: A new reduction to the pole method at low latitudes via a nonlinear thresholding. *J. Appl. Geophys.* **2014**, *111*, 220–227. [CrossRef]
56. Cordell, L.; Grauch, V. Mapping basement magnetization zones from aeromagnetic data in the San Juan Basin, New Mexico. In *The Utility of Regional Gravity and Magnetic Anomaly Maps*; Society of Exploration Geophysicists: Houston, TX, USA, 1985; pp. 181–197. [CrossRef]
57. Arkani-Hamed, J. Differential reduction-to-the-pole of regional magnetic anomalies. *Geophysics* **1988**, *53*, 1592–1600. [CrossRef]
58. Telford, W.M.; Geldart, L.P.; Sheriff, R.E. *Applied Geophysics*; Cambridge university press: Cambridge, UK, 1990.
59. Jacobsen, B.H. A case for upward continuation as a standard separation filter for potential-field maps. *Geophysics* **1987**, *52*, 1138–1148. [CrossRef]
60. Phillips, J.D. Designing matched bandpass and azimuthal filters for the separation of potential-field anomalies by source region and source type. *ASEG Ext. Abstr.* **2001**, *2001*, 1–4. [CrossRef]

61. Sheriff, S.D. Matched filter separation of magnetic anomalies caused by scattered surface debris at archaeological sites. *Near Surf. Geophys.* **2010**, *8*, 145–150. [CrossRef]
62. Chen, G.; Cheng, Q.; Zhang, H. Matched filtering method for separating magnetic anomaly using fractal model. *Comput. Geosci.* **2016**, *90*, 179–188. [CrossRef]
63. Ma, C.; Shan, X.-J. Decomposing InSAR LOS displacement into co-seismic dislocation with a linear interpolation model: A case study of the Kunlun Mountain Ms = 8.1 earthquake. *Acta Seismol. Sin.* **2006**, *19*, 100–107. [CrossRef]
64. Molinaro, M.; Leturmy, P.; Guezou, J.C.; Frizon de Lamotte, D.; Eshraghi, S. The structure and kinematics of the southeastern Zagros fold-thrust belt, Iran: From thin-skinned to thick-skinned tectonics. *Tectonics* **2005**, *24*. [CrossRef]
65. Nissen, E.; Ghorashi, M.; Jackson, J.; Parsons, B.; Talebian, M. The 2005 Qeshm Island earthquake (Iran)—A link between buried reverse faulting and surface folding in the Zagros Simply Folded Belt? *Geophys. J. Int.* **2007**, *171*, 326–338. [CrossRef]

remote sensing

MDPI

Article

Aquaculture Ponds Identification Based on Multi-Feature Combination Strategy and Machine Learning from Landsat-5/8 in a Typical Inland Lake of China

Gang Xie [1,†], Xiaohui Bai [2,†], Yanbo Peng [1], Yi Li [1], Chuanxing Zhang [1], Yang Liu [1], Jinhui Liang [1], Lei Fang [2], Jinyue Chen [2], Jilin Men [2], Xinfeng Wang [2], Guoqiang Wang [2,3], Qiao Wang [2] and Shilong Ren [2,*]

1 Key Laboratory of Land and Sea Ecological Governance and Systematic Regulation, Shandong Academy for Environmental Planning, Jinan 250101, China; xiegang@sdaep.com (G.X.); pengyanbo@sdaep.com (Y.P.); liyi@sdaep.com (Y.L.); zhangchuanxing@sdaep.com (C.Z.); 201999900086@email.sdu.edu.cn (Y.L.); liangjinhui@sdaep.com (J.L.)

2 Academician Workstation for Big Data in Ecology and Environment, Environment Research Institute, Shandong University, Qingdao 266003, China; baixaiohui@mail.sdu.edu.cn (X.B.); leifang@sdu.edu.cn (L.F.); chenjinyue@sdu.edu.cn (J.C.); menj@sdu.edu.cn (J.M.); xinfengwang@sdu.edu.cn (X.W.); wanggq@bnu.edu.cn (G.W.); qiao.wang@sdu.edu.cn (Q.W.)

3 Center for Geodata and Analysis, Faculty of Geographical Science, Beijing Normal University, Beijing 100875, China

* Correspondence: slren@sdu.edu.cn

† These authors contributed equally to this work.

Abstract: Inland aquaculture ponds, as an important land use type, have brought great economic benefits to local people but at the same time have caused many environmental problems threatening regional ecology security. Therefore, understanding the spatiotemporal pattern of aquaculture ponds and its potential influence on water quality is vital for the sustainable development of inland lakes. In this study, based on Landsat5/8 images, three types of land features, namely spectral features, index features, and texture features, and five machine learning algorithms, namely random forest (RF), extreme gradient boosting (XGBoost), artificial neural network (ANN), k-nearest neighbor (KNN), and Gaussian naive Bayes (GNB), were combined to identify aquaculture ponds and some other primary land use types around a typical inland lake of China. The results demonstrated that the XGBoost algorithm that integrated the three features performed the best among all groups of the five machine learning algorithms and the three features, with an overall accuracy of up to 96.15%. In particular, the texture features provided additional useful information besides the spectral features to allow more accurately separation of aquaculture ponds from other land use types and thus improve the land use mapping ability in complex inland lakes. Next, this study examined the tendency of aquaculture ponds and found a segmented increase mode, namely sharp increase during 1984–2003 and then slow elevation since 2003. Further positive correlation detected between the area of aquaculture ponds and the phytoplankton population dynamics suggest a likely influence of aquaculture activity on the lake water quality. This study provides an important scientific basis for the sustainable management and ecological protection of inland lakes.

Keywords: aquaculture pond; land use classification; machine learning; spatiotemporal pattern; water quality

Citation: Xie, G.; Bai, X.; Peng, Y.; Li, Y.; Zhang, C.; Liu, Y.; Liang, J.; Fang, L.; Chen, J.; Men, J.; et al. Aquaculture Ponds Identification Based on Multi-Feature Combination Strategy and Machine Learning from Landsat-5/8 in a Typical Inland Lake of China. *Remote Sens.* **2024**, *16*, 2168. https://doi.org/10.3390/rs16122168

Academic Editor: Melanie Vanderhoof

Received: 29 April 2024
Revised: 12 June 2024
Accepted: 13 June 2024
Published: 15 June 2024

1. Introduction

Land use and cover change (LUCC) is currently a hot area of global environmental change, which closely links human society and natural ecological processes and has a profound impact on human survival and development [1,2]. Aquaculture ponds, one of the important land use types, serve as one of the main sources of animal protein and are increasingly contributing to food security in Asia's populous inland cities. Specifically,

China's aquaculture production accounted for approximately 60% of the world total until 2020 [3,4]. As the size of ponds has increased considerably, intensive aquaculture has caused serious destructive effects on local environments, such as the decrease in water quality, the decline in biodiversity, and the loss of services provided by aquatic ecosystems [5–7]. Therefore, understanding the expansion pattern of aquaculture ponds in inland lakes and its influences on local environments is of great importance to the healthy development of human–natural ecosystems.

In previous studies, scholars have devoted themselves to improving the accuracy of land use classification from two aspects, namely, the incorporation of multi-source features and the development of new classification algorithms. Multi-source features, mainly including band reflectance, remote sensing indices, and texture characteristics, and the rich fusion of these features can provide comprehensive surface information and thus enhance the ability of land use mapping. For example, Chen et al. (2017) [8] jointly employed Landsat-8 OLI, MODIS, HJ-1A, and ASTER DEM data to perform land cover classification in Beijing by integrating temporal, spectral, angular, and topographic information, which achieved a 4.53% higher overall accuracy (OA) than using only OLI data. Li et al. (2023) [9] deeply explored the scaling effect of image spatial resolution on land cover classification from the perspectives of hybrid image element decomposition and spatial heterogeneity based on GF-2, SPOT-6, Sentinel-2, and Landsat-8, and showed that GF-2 and SPOT-6 had the best classification performance with an OA of up to 92.81%. In addition, Feyisa et al. (2014) [10] proposed the innovative Automatic Water Extraction Index (AWEI), which improved the classification accuracy (kappa = 0.98) of shaded and dark surface areas that are usually difficult to classify correctly by normal methods in New Zealand. Huang et al. (2015) [11] successfully integrated texture features and DEM data using the BP artificial neural network and gained high accuracy in remote sensing image classification and land use change detection (OA = 95.08%).

In terms of algorithms applied to land use classification, traditional classification models such as the maximum likelihood method [12], the K-means method [13,14], and the k-nearest neighbor algorithm (KNN) [15] dominated early research. However, with the rapid development of pattern recognition and machine learning, some intelligent algorithms such as support vector machine (SVM) [16] and neural networks [17] have gradually come to the forefront, presenting higher accuracy and effectiveness than the traditional parametric methods in land use classification. Tree-based models, especially those equipped with learning methods such as random forest (RF) [18] and Extreme Gradient Boosting Tree (XGBoost) [19], have attracted widespread attention for their excellent performance and ease of use. In addition, feature selection plays a key role in improving classification accuracy [20]. By removing irrelevant or redundant features, model performance could be largely optimized without losing important information. It has been proven that proper feature selection has a significant impact on the final classification accuracy [21].

Inland lake aquaculture ponds are often overlooked or not included in the existing land use classification system. This is mainly due to the special nature and complexity of lake aquaculture ponds, which make them difficult to clearly delineate with traditional land use types. As a way of utilizing waters, lake aquaculture ponds have their own unique functions and characteristics, which are different from general land use types such as water bodies or agricultural land. Specifically, aquaculture ponds often appear as regularly shaped, isolated, and enclosed bodies of water [22,23]. The water quality of ponds is highly affected by aquaculture activities, such as feed delivery and fish excretion, which lead to a significant increase in the material circulation efficiency in the pond [24]. Therefore, a lot of aquaculture ponds are easily covered by some plant vegetation and phytoplankton in the growing season as an abundant nutrient supplement, which makes it hard to distinguish them from land vegetation.

Traditional land use classification mainly focuses on distinguishable land utilization modes, such as water bodies, agriculture, forests, and built-up areas, while it lacks a finer categorization of water use modes in the land–water transition zone. Normally, in the existing land classification system, lake aquaculture ponds are often categorized as an unspecified type or ignored. This fails to meet the needs of the scientific protection and sustainable development of lake resources. To accurately evaluate the intensity of aquaculture ponds, and to detect and quantify the distribution and change trends of aquaculture ponds, we selected a typical inland lake with a long history of fish pond culture in northern China to (1) implement different classifiers based on multilevel feature fusion for LUCC mapping and change detection for the selected 10 years of Landsat data from 1984 to 2022 and (2) explore the spatiotemporal pattern of aquaculture ponds and other associated land use types and their potential impacts on local water quality.

2. Materials and Methods

2.1. Study Area

Nansi Lake (116°34′–117°21′E, 34°27′–35°20′N) is one of the most important freshwater lakes in North China (Figure 1), which is not only the main fishery base of Shandong Province but also a critical intermediate lake on the east route of the South-to-North Water Transfer Project. It is approximately 126 km long from north to south and 5–25 km wide from east to west. The central part of the lake is slightly narrower, while the northern and southern parts are broader, forming a teardrop shape. The average water depth in the lake is 2 m. The study region belongs to a warm temperate semi-humid monsoon climate zone with an average annual temperature of about 13.7 °C and an average annual precipitation of 695.2 mm. More than 70% of the annual precipitation falls in the flood season from June to September. Pit-pond culture is the dominant fishery type in Nansi Lake and a vital part of the local economy. Aquaculture ponds constructed by setting up dike banks near the shore are mainly distributed in the water of Nansi Lake. However, with the continuous economic development, the land use structure of the lake area has undergone significant changes in the past several decades. The explosive growth in local population and economic development and the rapid expansion in aquaculture ponds have led to a continuous decrease in the arable land and waters and unavoidably resulted in a certain degree of ecological imbalance in this area.

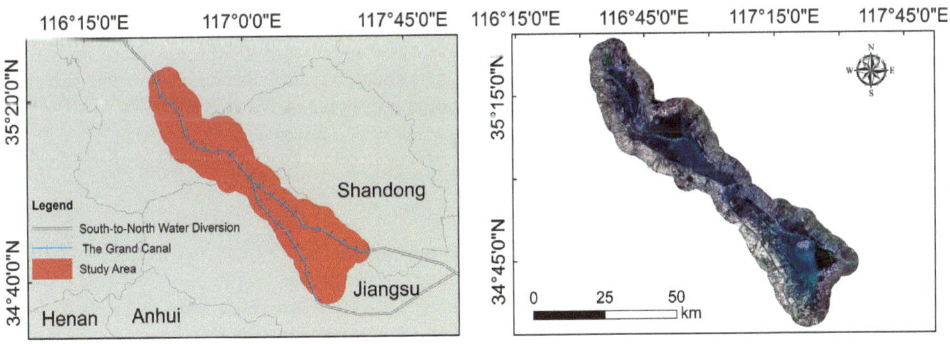

Figure 1. Geographical location and overview of Nansi Lake.

2.2. Data

Landsat TM/OLI (L2) images from 1984 to 2022 provided by the United States Geological Survey (USGS) were used in this study to carry out land cover classification. In order to accurately separate aquaculture ponds from nearshore vegetation, we elaborately selected a total of 10 winter cloud-free images to eliminate the effect of spectral convergence caused by plants growing in the ponds during growing season. The acquisition dates of these images were 7 February 1987, 27 January 1989, 24 December 1993, 5 February 1998, 31 January 2002, 20 December 2003, 26 January 2006, 29 November 2013, 10 December 2017, and 5 December 2021, respectively. In order to comprehensively analyze the long-term changes in LUCC around the lake, we set a 5 km buffer zone based on the vector extent of the lake. When determining the land cover types, we fully considered the actual land use status in the Nansi Lake area, and the potential relationship between different land use types and the dynamic changes in aquaculture ponds. Therefore, five land cover types were identified in the current study, namely farmland, water, aquaculture pond, built-up land, and others (primarily consists of forests and barens). In addition, the phytoplankton density data derived from Wang et al. (2024) [25] were adopted to study the possible influence of aquaculture activity on lake water quality, which allowed an RF-based model to be developed to quantify the ecological status in Nansi Lake by means of Landsat-8 OLI images and obtained a high prediction accuracy.

2.3. Samples Collection

In remote sensing image classification, sample quality is crucial to the final mapping accuracy. Following the principle of full frame selection, all Landsat TM/OLI images covering the study area were comprehensively visually interpreted. Special attention was paid to the selection of representative pixels for each type of land use to ensure that the samples could truly reflect the spectral characteristics and spatial distribution of each type of feature. After strict screening and calibration, 11,489 sample points evenly distributed throughout the study area were finally identified, including 1877 samples for farmland, 2649 samples for water, 3102 samples for aquaculture pond, 1122 samples for built-up land, and 2739 samples for others. This approach fully considered the balance of the scale of each LUCC type and thus could avoid classification bias due to excessive differences in the number of samples. Then, we randomly divided them into training and test datasets with an 8:2 ratio to ensure that the model could be adequately trained and its classification performance effectively evaluated.

2.4. Classification Features

First, this study selected the three visible bands, the near-infrared band and the two short-wave infrared bands, corresponding to the B1, B2, B3, B4, B5, and B7 bands of Landsat5 TM and the B2, B3, B4, B5, B6, and B7 bands of Landsat8 OLI, as the direct features to capture the difference among land use types in spectral characteristics. Second, we considered the Enhanced Vegetation Index (EVI) and the Modified Normalized Difference Water Index (MNDWI) as two other keys to enhance the discrepancy among the targeted objects. The EVI could reduce the atmospheric effects and address the saturation issue in the area of high vegetation coverage found in the traditional normalized difference vegetation index [26,27]. The expression is as follows:

$$\mathrm{EVI} = 2.5 \times \frac{\rho_{\mathrm{nir}} - \rho_{\mathrm{red}}}{\rho_{\mathrm{nir}} + 6 \times \rho_{\mathrm{red}} - 7.5 \times \rho_{\mathrm{blue}} + 1} \quad (1)$$

where ρ_{nir}, ρ_{red}, and ρ_{blue} are the atmospherically corrected reflectance values for the near-infrared and red and blue bands, respectively.

MNDWI can eliminate the effect of terrain difference and solve the problem of noise in water body identification [28,29]. The expression is as follows:

$$\mathrm{MNDWI} = (\mathrm{Green} - \mathrm{SWIR1})/(\mathrm{Green} + \mathrm{SWIR1}) \quad (2)$$

where Green and SWIR1 are the reflectance values in the green band and short-wave infrared band 1, respectively.

In addition, we also adopted the gray-level co-occurrence matrix (GLCM) method to extract texture information by calculating the gray-level spatial relationship between pixels for further improving the classification accuracy [30]. Texture features play a crucial role in the recognition of ground object types, especially in distinguishing ground objects with similar spectral features but different spatial features. For example, aquaculture pond and water body are similar in spectral reflection, but their texture features, such as regularity, roughness, and grain size, may differ significantly. By introducing texture features, we can more accurately depict the spatial structure of ground objects, thus improving the accuracy of classification. In total, eight texture metrics were calculated, including mean, variance, homogeneity, contrast, dissimilarity, entropy, angular second moment, and correlation. These variables provide an effective tool for quantifying surface irregularities and are essential for distinguishing different land cover categories (Table 1). To guarantee the accuracy and effectiveness of the texture analysis, we used a window size of 3 × 3 to traverse the entire image pixel by pixel and took the gray level of 64 to capture the detailed texture information in the image. With the above settings, we successfully extracted 48 texture features from the original image. However, too many features may lead to an increase in computational complexity and a decrease in classification performance. In order to reduce the feature dimension and extract the most important information, we performed principal component transform analysis on these texture features. Principal component analysis (PCA) is a statistical tool that transforms the original features into new, unrelated features through linear transformations. These new features are called principal components. The purpose of PCA is to identify the most important features from the data and aggregate them into a new, smaller set of features that explain the greatest degree of variance in the data. By calculating the covariance matrix of the texture feature matrix, PCA determines the direction that can maximize the variance in the data, that is, the main direction of the data change. Each principal component is a linear combination of the original features, with the first principal component explaining the largest variance in the data, the second principal component explaining the largest portion of the remaining variance, and so on. The first 5 principal components were selected for the subsequent land use classification.

Table 1. Characteristics and description of selected GLCM.

GLCM	Description
Mean	Reflects the degree of regularity of the texture.
Variance	Measures the dispersion of the gray-level distribution to emphasize the visual edges of land cover patches.
Homogeneity	Measures the local gray-level homogeneity of an image.
Contrast	Reflects the total amount of local gray level changes in an image.
Dissimilarity	Similar to contrast, if the local contrast is higher, the dissimilarity is also higher.
Entropy	Measures the amount of information contained in an image, representing the degree of non-uniformity or complexity of textures within the image.
Angular Second Moment	Measures the uniformity of the image gray-level distribution, reflecting the degree of uniformity of the image gray-level distribution and the coarseness of the texture.
Correlation	Measures the linear relationship of gray levels, describing the degree of similarity between elements in rows or columns.
Mean	Reflects the degree of regularity of the texture.

2.5. Classification Algorithms

In this study, we employed two tree-based machine learning models, RF and XGBoost; two classical models, KNN and Gaussian naive Bayes (GNB); and artificial neural network (ANN) for land cover classification. The RF model is the most widely used classification model in LUCC classification, with proven accuracy [31,32]. The XGBoost model stands out in the field of machine learning due to its efficient processing of large-scale data [33,34]. ANN is one of the most commonly used non-parametric classification techniques, renowned for its strong generalization capabilities [35]. KNN and GNB are both traditional machine learning algorithms that are computationally simple, run efficiently, and perform well in land object identification with high homogeneity [36,37].

2.5.1. RF

RF is an integrated classifier based on decision trees, each of which is independently generated with a user-defined number of features on which its node splits are based. The selection of these features is randomized to warrant model diversity. The training data and variables for each decision tree are generated through a bagging strategy, and the final classification results are derived through majority voting [38]. In this study, we set the number of trees (*n_estimators*) to 100 to ensure that the model had sufficient diversity; we set the maximum number of features per tree (*max_features*) to the square root of the total number of features to balance the complexity of the model and the risk of overfitting.

2.5.2. XGBoost

XGBoost, a remarkable machine learning algorithm based on the Gradient Boosting Decision Tree framework, stands out due to its superior flexibility, high efficiency, and outstanding performance in Kaggle machine learning competitions [19]. By introducing a regularization mechanism, XGBoost is able to smooth the weights of the final learning, effectively avoid the overfitting issue, and thus improve the learning accuracy. In addition, XGBoost is equipped with parallel and distributed computing capabilities, which significantly accelerates the learning speed. In this study, we set the number of iterations (*n_estimators*) to 100, the maximum depth of the decision tree (*max_depth*) to 10, and *learning_rate* to 1.

2.5.3. KNN

KNN is an instance-based classifier for classification and regression [39]. It does not rely on an explicit model training process, but instead finds the K closest training samples to an unknown sample by measuring the distance of that sample from all samples in the training set, and uses the category with the most votes as the prediction for the unknown sample based on the category labels of the K samples [40]. After experimental validation, K was set to 20 for ensuring that the model can make full use of the information from neighboring samples when classifying, while avoiding the influence of noisy data on the classification results.

2.5.4. GNB

GNB is a machine learning algorithm that uses probabilistic methods and relies on Gaussian distributions. Its principle is based on Bayes' theorem and the assumption of conditional independence between features, i.e., the feature variables of each category obey a normal distribution. By calculating the mean and variance of the feature variables of each category, the algorithm can estimate the probability that an unknown sample belongs to each category based on these statistics [41,42]. In our dataset, the distribution of most features is approximately normal, which provides a reasonable basis for the application of the GNB model (Figure S1). In this study, in GNB parameters, the prior was set to none and *var_smoothing* was set to 1×10^{-9}.

2.5.5. ANN

The ANN classification algorithm learns the relationship between input features and output categories through a training process. During training, the network calculates outputs through forward propagation and then computes output errors through the backpropagation algorithm, updating network weights based on these errors [43]. ANN classification algorithms typically consist of an input layer, one or more hidden layers, and an output layer, with each neuron using an activation function (Sigmoid, Tanh, ReLU, etc.) to determine whether to activate, introducing non-linear factors that enable the neural network to learn and model complex non-linear relationships [44]. In this study, we employed Multilayer Perceptron (MLP) as the neural network architecture, selected the *logistic* function as the activation function, and chose the *lbfgs* optimizer to refine the weights.

2.6. Analysis

Based on spectral features, index features, and texture features, we constructed three feature schemes (Table 2) and trained KNN, GNB, ANN, RF, and XGBoost classification models, respectively.

Table 2. Experimental feature schemes.

Schemes	Feature Variables
Scheme 1: spectral feature	Blue band, green band, red band, NIR band, SWIR1 band, SWIR2 band
Scheme 2: spectral feature +index feature	Blue band, green band, red band, NIR band, SWIR1 band, SWIR2 band, EVI, MNDWI
Scheme 3: spectral feature + texture feature	Blue band, green band, red band, NIR band, SWIR1 band, SWIR2 band, PC1, PC2, PC3, PC4, PC5
Scheme 4: index feature + texture feature	EVI, MNDWI, PC1, PC2, PC3, PC4, PC5
Scheme 5: spectral feature + index feature + texture feature	Blue band, green band, red band, NIR band, SWIR1 band, SWIR2 band, EVI, MNDWI, PC1, PC2, PC3, PC4, PC5

In order to objectively and systematically evaluate the performance of different classification algorithms and feature schemes, we used a variety of statistical metrics for quantitative analysis. Specifically, we calculated the confusion matrix of each model based on the training and testing datasets to visualize the model's classification effect on each category of samples. On this basis, we further quantified the OA, which can directly reflect the proportion of objects correctly classified by the model, providing us with an intuitive performance measurement.

In addition, in order to evaluate the model performance more comprehensively, we also introduced the Kappa consistency coefficient, which is an indicator describing the degree of consistency between the model's classification results and the actual situation. Meanwhile, we further calculated the Producer Accuracy (PA) and recall to assess model performance in terms of the prediction of positive examples of the classification results and the recall of real positive examples.

Finally, we adopted the F1 score as a comprehensive evaluation metric, which combines the information of precision and recall and can fully reflect the comprehensive performance of the model in the classification task. Through the comprehensive analysis of these metrics, we are able to more objectively and comprehensively assess the performance of different classification algorithms and feature schemes in LUCC classification.

3. Results

3.1. Feature Selection and Feature Importance

In this study, feature importance was assessed for the three feature schemes using RF (Figure 2). Among the 13 features assessed, the index feature EVI had the highest importance score of 0.16, indicating that EVI played a key role in classification prediction. The spectral features NIR and SWIR1 also exhibited high importance scores, reflecting their effectiveness in distinguishing different LUCC types. For texture features, the importance scores of the principal components PC1 and PC2 were relatively higher than PC3, PC4, and PC5. In comparison, the effects of visible bands are relatively weaker than those of index and texture features.

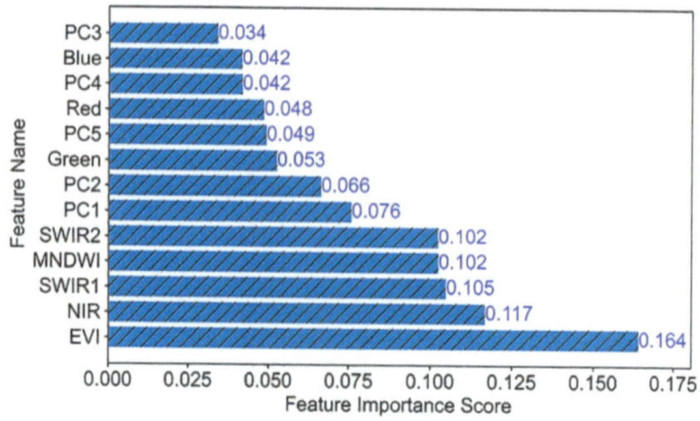

Figure 2. Importance ranking of 13 feature variables.

3.2. Feature Profile Comparison

In order to legibly understand the gaps among the five targeted objects, we systematically compared their differences in spectral features, index features, and principal components of texture features (Figure 3). In the visible light bands, the reflectance distributions of the five objects were relatively similar and generally in the range of 0~0.2. This indicated that the spectral characteristics of these classes did not differ too much in the visible light bands, making it difficult to effectively distinguish them by only relying on the spectral gaps in visible bands. In comparison, in the non-visible bands (NIR, SWIR1, and SWIR2), the spectral properties of the five classes showed significant differences. Specifically, farmland and built-up areas had relatively higher reflectance in the NIR and SWIR bands than aquaculture ponds, water, and others. Meanwhile, water in particular reflected less than aquaculture ponds, which would be helpful to distinguish them. Regarding index features, the highest EVI values and the smallest MNDWI values were detected in farmland. The EVI and MNDWI value of aquaculture ponds lies between water and built-up areas/others. As for texture profiles, only PC1 and PC2 exhibited obvious gaps among the five classes, providing an effective basis for LUCC classification. In contrast, the PC3-PC5 principal components largely overlapped with each other, suggesting that they were useless for improving classification accuracy.

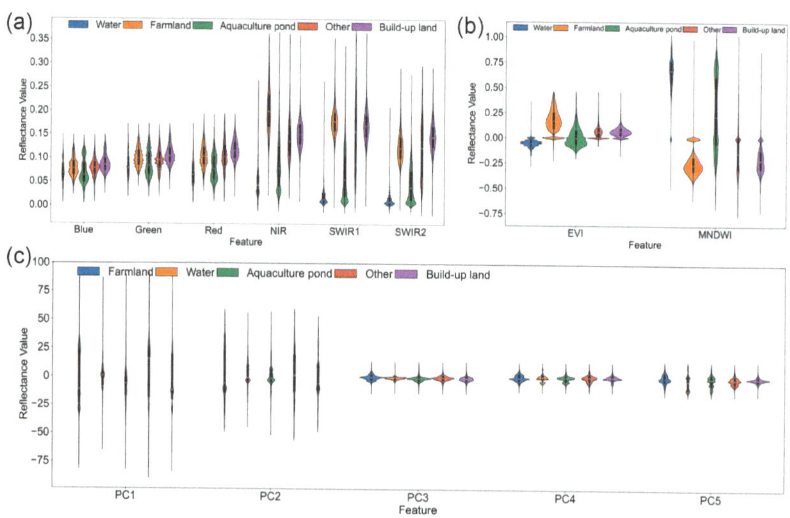

Figure 3. Quantity difference among five targeted objects in (**a**) spectral features, (**b**) index features, and (**c**) texture features.

3.3. Accuracy Comparation of Different Classification Models

The performance of classification models was explored in depth on the test data (Table 3). Among the classifiers, the XGBoost and RF models exhibited better performance in the classification task and reached higher accuracies of up to 96.15% and 95.92%, respectively. In contrast, the GNB model performed the worst with an accuracy below 65%. Among the three feature schemes examined, scheme 5 that incorporated texture features outperformed all other schemes overall, especially the XGBoost classifier. In comparison, scheme 3, which only included spectral and texture features, was slightly less effective than scheme 5, indicating the significance of spectral features in the classification process. scheme 4, which contained only index and texture features, had lower classification accuracy than scheme 5, further confirming the critical role of spectral features in classification tasks.

Table 3. Comparison of classification accuracy of different models.

Model	OA (%)	Kappa	PA (%)	Recall (%)	F1
KNN scheme 1	84.93	0.80	84.69	84.93	0.85
KNN scheme 2	85.26	0.80	84.99	85.26	0.85
KNN scheme 3	85.51	0.81	85.36	85.51	0.85
KNN scheme 4	85.74	0.81	85.59	85.74	0.86
KNN scheme 5	85.83	0.81	85.68	85.83	0.86
GNB scheme 1	59.68	0.48	60.68	59.68	0.58
GNB scheme 2	60.57	0.49	64.34	60.57	0.61
GNB scheme 3	58.01	0.45	59.79	58.01	0.55
GNB scheme 4	59.44	0.46	58.22	59.44	0.56
GNB scheme 5	60.90	0.50	65.83	60.90	0.61
ANN scheme 1	69.74	0.59	67.79	69.74	0.68
ANN scheme 2	70.85	0.60	69.72	70.85	0.68
ANN scheme 3	75.57	0.67	74.66	75.57	0.75
ANN scheme 4	76.50	0.68	75.76	76.50	0.76
ANN scheme 5	76.00	0.67	74.93	76.00	0.75
RF scheme 1	92.51	0.90	92.43	92.51	0.92
RF scheme 2	92.28	0.90	92.20	92.28	0.92
RF scheme3	**95.92**	**0.95**	**95.91**	**95.92**	**0.96**
RF scheme 4	92.03	0.89	91.98	92.03	0.92
RFscheme 5	95.66	0.94	95.65	95.66	0.96
XGBoost scheme 1	91.62	0.92	91.62	91.62	0.92
XGBoost scheme 2	92.01	0.89	91.93	92.01	0.92
XGBoost scheme 3	96.07	0.95	96.05	96.07	0.96
XGBoost scheme 4	91.32	0.88	91.27	91.32	0.91
XGBoost scheme5	**96.15**	**0.95**	**96.14**	**96.15**	**0.96**

This study employed the XGBoost scheme 5 classification scheme to evaluate the accuracy of land use types and presented the corresponding normalized confusion matrix (Figure 4). Among the different land use types, built-up land had the lowest classification accuracy, with PA and recall of 91.3% and 88.63%, respectively. In contrast, farmland had the highest recall, reaching 97.19%, while the 'other' type had the highest PA, at 97.24% (Figure 4b). Apart from built-up land, the correct classification ratio for other land use types was generally higher than 0.9. The correct classification ratio for aquaculture ponds was 0.93, which was relatively superior among all land use types. Further analysis of the misclassification of aquaculture ponds revealed that the highest proportion of errors was with water, amounting to 0.03 (Figure 4a). This indicates the high accuracy of XGBoost scheme 5 in classifying most types of land use.

In terms of visualization, compared to scheme 1, scheme 2, and scheme 4, the patch integrity of surface objects predicted under scheme 5 was significantly improved and the confusion of categories was significantly reduced (Figure 5). Especially in the categorization of aquaculture ponds, scheme 5-based prediction significantly refined the continuity and completeness of their distribution.

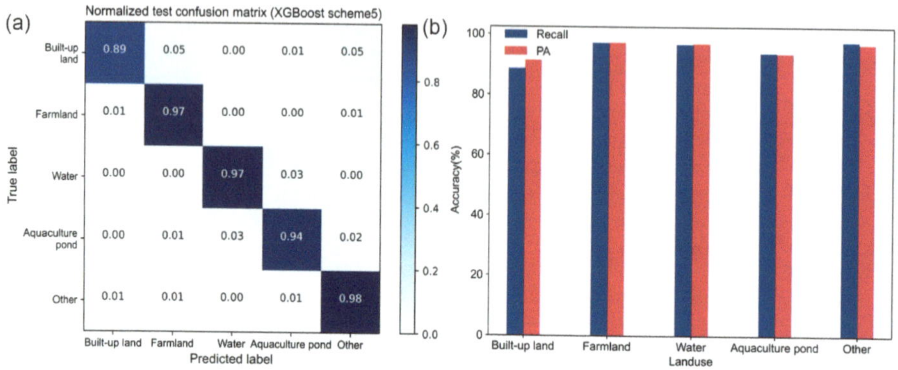

Figure 4. Comparison of classification accuracy for land use types identified with XGBoost scheme 5: (**a**) confusion matrix; (**b**) PA and recall for each land use type.

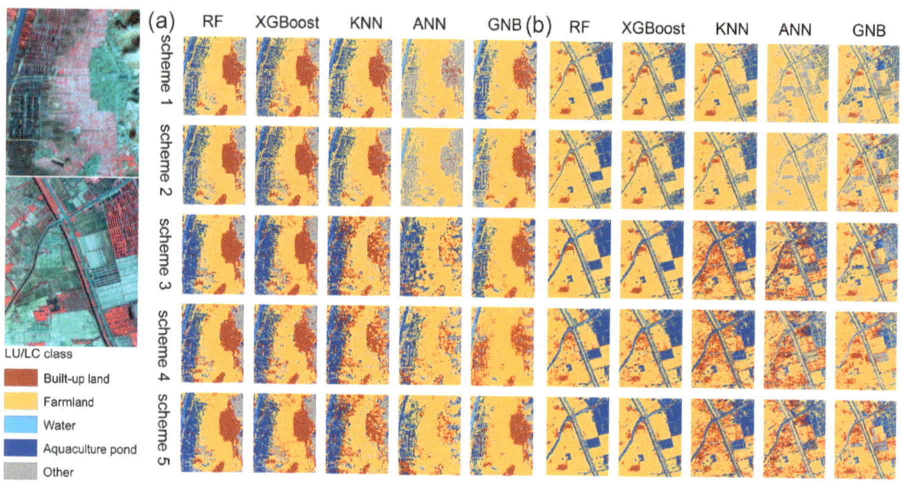

Figure 5. LUCC mapping and details comparison: (**a**,**b**) are two partial details presented to compare the land use classification performance among different models.

3.4. Land Cover Changes in Nansi Lake

Based on the results predicted by the XGBoost classifier under scheme 5, this study analyzed the development of the area of five land cover types around Nansi Lake between 1987 and 2021 (Figure 6b). Built-up land and aquaculture ponds have largely expanded since 1987. Specifically, the area of built-up land extended from 120 km^2 in 1987 to 296 km^2 in 2021, while the area of aquaculture ponds surged from 48 km^2 to 842 km^2, with a sharp increase during 1984–2003. On the contrary, the area of the 'other' type greatly shrank from the dominant cover type to about 215 km^2 by 2021. Farmland showed a slight expansion tendency overall with a sudden drop around 2003. The area of lake bodies fluctuated dramatically during the study period but no significant trends were found here. Spatially, the distribution of aquaculture ponds after 2002 has shown a pronounced characteristic of geographic clustering, primarily concentrated in the western and central regions of Nansi Lake (Figure 6a).

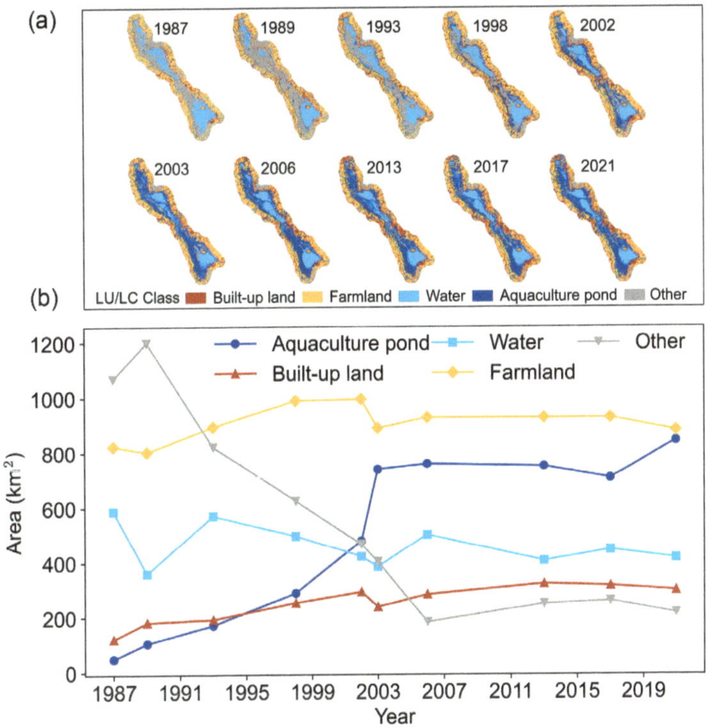

Figure 6. Land cover changes from 1984 to 2021 based on XGBoost scheme 5: (**a**) temporal variation; (**b**) spatial distribution.

3.5. Relationship between Water Quality and the Expansion of Aquaculture Ponds

This study analyzed the potential impact of aquaculture pond expansion on water quality, expressed through the amount of phytoplankton in Nansi Lake, by means of correlation analysis (Figure 7). We found the phytoplankton abundance decreased after 2003 with the increase in the area of aquaculture ponds. Nonetheless, it showed a positive correlation (R = 0.5) with annual fluctuation, suggesting that its synchronous relationship with the water quality of the lake may be partially affected by aquaculture activity.

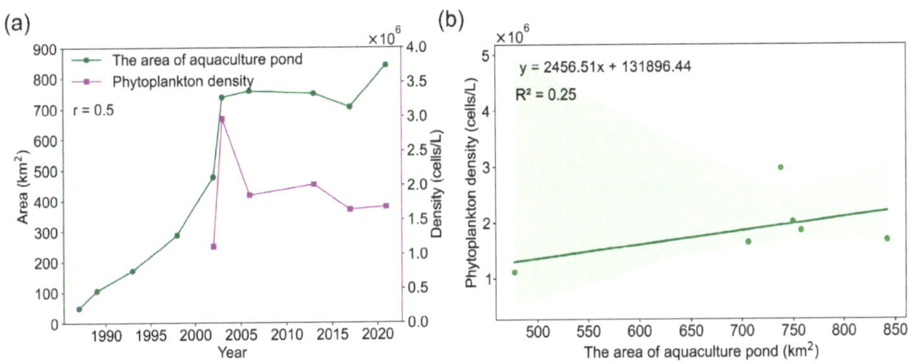

Figure 7. Correlation of aquaculture pond area and phytoplankton density. The correlation coefficient (**a**) and regression equation (**b**) between aquaculture pond area and phytoplankton density.

4. Discussion

In this study, we performed land use classification with Landsat 5/8 images around Nansi Lake by means of KNN, GNB, RF, and XGB algorithms under multilevel features. Our results showed that the XGB algorithm, especially when combined with texture features, achieved the highest classification accuracy of up to 96.15%. Compared with the existing literature, the classification accuracy in this study is significantly improved. For example, Talukdar et al. (2020) [35] used the RF algorithm to classify a riparian landscape in India and obtained a lower classification accuracy with a kappa coefficient of 0.89. Similarly, Abbas and Jaber (2020) [45] used WorldView-2 image and SVM algorithm to classify the land use in Hilla City in Babylon, Iraq, and obtained an overall classification accuracy of 94.48% and a kappa coefficient of 0.9, which are still less than in the current study. Xia et al. (2020) [46] extracted aquaculture ponds in Shanghai by integrating existing multi-source remote sensing data on the Google Earth Engine platform and combining multi-threshold connection component segmentation and random forest algorithm and reached an OA of 91.8%, which is still lower than the current study. This study not only confirms the key role of multiple feature integration in improving classification results, but also highlights the great potential of advanced machine learning algorithms in land use classification.

The key to achieving such a high classification accuracy in this study is the innovative introduction of texture features and effective dimensionality reduction. Texture features enable the model to better distinguish objects with similar spectra but large texture differences, especially for specific surface object types such as aquaculture ponds. The PCA method effectively reduced the information redundancy among features and improved the classification ability. The first PCA component typically accounts for the largest variance in the PCA analysis of texture data, while the second principal component explains the most variance among the remaining components, and so on. The first five principal components collectively represent approximately 99.99% of the shape information of all land cover types and contributed more to distinguishing aquaculture ponds and farmlands with a regular shape from others. The overall accuracy of RF scheme 4 is 92.03%, indicating that satisfactory classification can be achieved even when using only index and texture principal component features. This may be attributed to the EVI index, a vegetation index that integrates information from the near-infrared, red, and blue bands, outperforming single-band data in terms of classification performance. Similarly, the MNDWI, as a water body index, effectively distinguishes water from other land use types by utilizing information from the shortwave infrared and green bands. It is important to note that although index features (EVI and MNDWI) are included in this study, their contribution to the final classification accuracy when combined with spectral and texture features is not significant. However, in the feature importance evaluation, these indicators scored higher. The reason may be that the sufficiency of spectral information and its strong collinearity with index features factually caused no significant increment in useful information in the final classification. Nevertheless, the contribution of spectral features to the classification process remains substantial. The overall accuracy of the RF classification model based on spectral features reached 92.51%, highlighting the key role of spectral features in distinguishing different land cover types. Particularly in the feature importance evaluation, the NIR band, which scores highly, is greatly effective in differentiating vegetation types, while the SWIR band demonstrates its unique ability in identifying water types.

From the perspective of land use change, this study revealed dramatic changes in land feature type conversion in Nansi Lake. With the advancement of urbanization, natural land types such as bare land and forest land have been gradually converted to other uses. At the same time, the expansion of built land and aquaculture ponds reflects the increasing demand for land due to population growth and urban expansion. It is worth noting that since 2003, the expansion rate of aquaculture ponds in Nansi Lake has slowed down, and at the same time, the water quality of Nansi Lake has also shown a gradual improvement. This change may be closely related to the government's policy of keeping the lake natural, which forced a lot of farmers to reduce aquaculture activities to restore the integrity of the lake ecosystem. With the reduction in fertilizer application, a decrease in phytoplankton density was detected. The contrary yearly trends of aquaculture ponds and water quality are because the embankments of many aquaculture ponds are still there despite no fishery activity under the strong management of local government. However, we must also recognize the complexity of the relationship between aquaculture ponds and changes in water quality in the lake region. The change in water quality is the result of multiple factors, including climate, hydrology, land use, and human activities. Although the change in aquaculture pond area has a certain impact on water quality, it is only one of many factors affecting water quality changes. Therefore, to fully and deeply understand the causes of water quality evolution, more comprehensive investigation is needed in the future.

Despite the remarkable results of this study, there are still some limitations that need to be noted. First, high-precision classification relies heavily on the accurate selection of training and validation samples. In this study, we used visual interpretation to select samples, which is inevitably affected by subjective factors. Different interpreters may classify and categorize feature types in the same area differently according to their own experiences and judgments, leading to misclassification of sample types and thus adversely affecting the accuracy of the classification results. Meanwhile, due to the complexity of local land use types, especially in the transition areas, the unclear boundaries of these types often make it particularly difficult to accurately select high-purity pixels. Second, the current studies mainly rely on traditional feature selection methods, which may not be able to fully mine the potential information in the data. Future research could try to use deep learning techniques to automatically extract and select the most discriminative features.

5. Conclusions

Aiming at the difficult problem of identifying aquaculture ponds in the Nansi Lake region, this study integrated multi-level features into different machine learning algorithms to achieve high-precision land use classification with the highest accuracy of 96.15%, breaking through the limitations of traditional methods. This study shows that the land use pattern in the region has greatly transformed, and natural land such as bare land and forest land has largely been replaced by aquaculture ponds and built-up land. At the same time, we found that phytoplankton density was correlated with the changes in the area of aquaculture ponds, suggesting that the expansion of ponds and the reduction in local farming strength may have changed the hydrological environment of the lake.

Supplementary Materials: The following supporting information can be downloaded at: https://www.mdpi.com/article/10.3390/rs16122168/s1, Figure S1: The data distribution presents the texture features of five target objects prior to Principal Component Analysis (PCA). These texture features are derived from eight texture attributes extracted from the six selected spectral bands using the Gray Level Co-occurrence Matrix (GLCM) method. The attributes include Mean, Variance, Homogeneity, Contrast, Dissimilarity, Entropy, Angular Second Moment, and Correlation.

Author Contributions: Conceptualization, G.X. and S.R.; methodology, X.B.; formal analysis, X.B.; investigation, Y.P., Y.L. (Yi Li), C.Z., Y.L. (Yang Liu) and J.L.; writing—original draft preparation, G.X. and X.B.; writing—review and editing, S.R., L.F., J.C., J.M., X.W., G.W. and Q.W.; funding acquisition, S.R. All authors have read and agreed to the published version of the manuscript.

Funding: This study was specially funded by the National Key R&D Program of China (Grant No. 2022YFC3204400) and the Key Laboratory of Land and Sea Ecological Governance and Systematic Regulation.

Data Availability Statement: Data are contained within the article.

Conflicts of Interest: The authors declare no conflicts of interest.

References

1. Dehong, D.; Yinfang, S.; Lin, S.; Genxia, W. Remote Sensing Technology's Applied Research and Development Direction in Land-Use and Land-Cover Change (LUCC). In Proceedings of the 2012 2nd International Conference on Remote Sensing, Environment and Transportation Engineering, Nanjing, China, 1–3 June 2012; pp. 1–4.
2. Wang, X.; Dong, X.; Liu, H.; Wei, H.; Fan, W.; Lu, N.; Xu, Z.; Ren, J.; Xing, K. Linking land use change, ecosystem services and human well-being: A case study of the Manas River Basin of Xinjiang, China. *Ecosyst. Serv.* **2017**, *27*, 113–123. [CrossRef]
3. Cao, L.; Naylor, R.; Henriksson, P.; Leadbitter, D.; Metian, M.; Troell, M.; Zhang, W. China's aquaculture and the world's wild fisheries. *Science* **2015**, *347*, 133–135. [CrossRef] [PubMed]
4. Naylor, R.; Fang, S.; Fanzo, J. A global view of aquaculture policy. *Food Policy* **2023**, *116*, 102422. [CrossRef]
5. Jiang, Q.; Bhattarai, N.; Pahlow, M.; Xu, Z. Environmental sustainability and footprints of global aquaculture. *Resour. Conserv. Recycl.* **2022**, *180*, 106183. [CrossRef]
6. Luo, J.; Pu, R.; Ma, R.; Wang, X.; Lai, X.; Mao, Z.; Zhang, L.; Peng, Z.; Sun, Z. Mapping long-term spatiotemporal dynamics of pen aquaculture in a shallow lake: Less aquaculture coming along better water quality. *Remote Sens.* **2020**, *12*, 1866. [CrossRef]
7. Sun, Z.; Luo, J.; Gu, X.; Qi, T.; Xiao, Q.; Shen, M.; Ma, J.; Zeng, Q.; Duan, H. Policy-driven opposite changes of coastal aquaculture ponds between China and Vietnam: Evidence from Sentinel-1 images. *Aquaculture* **2023**, *571*, 739474. [CrossRef]
8. Chen, B.; Huang, B.; Xu, B. Multi-source remotely sensed data fusion for improving land cover classification. *ISPRS J. Photogramm. Remote Sens.* **2017**, *124*, 27–39. [CrossRef]
9. Li, R.; Gao, X.; Shi, F.; Zhang, H. Scale Effect of Land Cover Classification from Multi-Resolution Satellite Remote Sensing Data. *Sensors* **2023**, *23*, 6136. [CrossRef] [PubMed]
10. Feyisa, G.L.; Meilby, H.; Fensholt, R.; Proud, S.R. Automated Water Extraction Index: A new technique for surface water mapping using Landsat imagery. *Remote Sens. Environ.* **2014**, *140*, 23–35. [CrossRef]
11. Huang, D.-M.; Wei, C.-T.; Yu, J.-C.; Wang, J.-L. A method of detecting land use change of remote sensing images based on texture features and DEM. In Proceedings of the International Conference on Intelligent Earth Observing and Applications, Guilin, China, 23–24 October 2015; pp. 613–618.
12. Wang, H.; Zhao, H.; Li, W. Land-use Classification of Zhanghe River Basin Using the Maximum Likelihood and Decision Tree Method. In Proceedings of the 11th International Conference on Fuzzy Systems and Knowledge Discovery (FSKD), Xiamen, China, 19–21 August 2014; pp. 322–327.
13. Li, Y.; Wu, H. A Clustering Method Based on K-Means Algorithm. In Proceedings of the International Conference on Solid State Devices and Materials Science (SSDMS), Macao, China, 1–2 April 2012; pp. 1104–1109.
14. Papa, J.P.; Papa, L.P.; Pereira, D.R.; Pisani, R.J. A Hyperheuristic Approach for Unsupervised Land-Cover Classification. *IEEE J. Sel. Top. Appl. Earth Obs. Remote Sens.* **2016**, *9*, 2333–2342. [CrossRef]
15. Heydari, S.S.; Mountrakis, G. Effect of classifier selection, reference sample size, reference class distribution and scene heterogeneity in per-pixel classification accuracy using 26 Landsat sites. *Remote Sens. Environ.* **2018**, *204*, 648–658. [CrossRef]
16. He, T.; Sun, Y.-J.; Xu, J.-D.; Wang, X.-J.; Hu, C.-R. Enhanced land use/cover classification using support vector machines and fuzzy k-means clustering algorithms. *J. Appl. Remote Sens.* **2014**, *8*, 083636. [CrossRef]
17. Shakya, A.; Biswas, M.; Pal, M. Parametric study of convolutional neural network based remote sensing image classification. *Int. J. Remote Sens.* **2021**, *42*, 2663–2685. [CrossRef]
18. Breiman, L. Random forests. *Mach. Learn.* **2001**, *45*, 5–32. [CrossRef]
19. Chen, T.; Guestrin, C. Xgboost: A scalable tree boosting system. In Proceedings of the 22nd ACM SIGKDD International Conference on Knowledge Discovery and Data Mining, San Francisco, CA, USA, 13–17 August 2016; pp. 785–794.
20. Lu, D.; Weng, Q. A survey of image classification methods and techniques for improving classification performance. *Int. J. Remote Sens.* **2007**, *28*, 823–870. [CrossRef]
21. Pal, M.; Foody, G.M. Feature Selection for Classification of Hyperspectral Data by SVM. *IEEE Trans. Geosci. Remote Sens.* **2010**, *48*, 2297–2307. [CrossRef]
22. Duan, Y.; Li, X.; Zhang, L.; Chen, D.; Ji, H. Mapping national-scale aquaculture ponds based on the Google Earth Engine in the Chinese coastal zone. *Aquaculture* **2020**, *520*, 734666. [CrossRef]
23. Zeng, Z.; Wang, D.; Tan, W.; Yu, G.; You, J.; Lv, B.; Wu, Z. RCSANet: A full convolutional network for extracting inland aquaculture ponds from high-spatial-resolution images. *Remote Sens.* **2020**, *13*, 92. [CrossRef]
24. Zhang, M.; Dong, J.; Gao, Y.; Liu, Y.; Zhou, C.; Meng, X.; Li, X.; Li, M.; Wang, Y.; Dai, D. Patterns of phytoplankton community structure and diversity in aquaculture ponds, Henan, China. *Aquaculture* **2021**, *544*, 737078. [CrossRef]
25. Wang, W.; Chen, J.; Fang, L.; Yinglan, A.; Ren, S.; Men, J.; Wang, G. Remote sensing retrieval and driving analysis of phytoplankton density in the large storage freshwater lake: A study based on random forest and Landsat-8 OLI. *J. Contam. Hydrol.* **2024**, *261*, 104304. [CrossRef]
26. Gu, Z.; Zhang, Z.; Yang, J.; Wang, L. Quantifying the influences of driving factors on vegetation EVI changes using structural equation model: A case study in Anhui province, China. *Remote Sens.* **2022**, *14*, 4203. [CrossRef]
27. Pôças, I.; Calera, A.; Campos, I.; Cunha, M. Remote sensing for estimating and mapping single and basal crop coefficientes: A review on spectral vegetation indices approaches. *Agric. Water Manag.* **2020**, *233*, 106081. [CrossRef]
28. Bhangale, U.; More, S.; Shaikh, T.; Patil, S.; More, N. Analysis of surface water resources using Sentinel-2 imagery. *Procedia Comput. Sci.* **2020**, *171*, 2645–2654. [CrossRef]

29. Li, L.; Su, H.; Du, Q.; Wu, T. A novel surface water index using local background information for long term and large-scale Landsat images. *ISPRS J. Photogramm. Remote Sens.* **2021**, *172*, 59–78. [CrossRef]
30. Duan, M.; Song, X.; Liu, X.; Cui, D.; Zhang, X. Mapping the soil types combining multi-temporal remote sensing data with texture features. *Comput. Electron. Agric.* **2022**, *200*, 107230. [CrossRef]
31. Wu, H.; Lin, A.; Xing, X.; Song, D.; Li, Y. Identifying core driving factors of urban land use change from global land cover products and POI data using the random forest method. *Int. J. Appl. Earth Obs. Geoinf.* **2021**, *103*, 102475. [CrossRef]
32. Zhang, F.; Yang, X. Improving land cover classification in an urbanized coastal area by random forests: The role of variable selection. *Remote Sens. Environ.* **2020**, *251*, 112105. [CrossRef]
33. Abdi, A.M. Land cover and land use classification performance of machine learning algorithms in a boreal landscape using Sentinel-2 data. *GIScience Remote Sens.* **2020**, *57*, 1–20. [CrossRef]
34. Quan, Y.; Hutjes, R.W.; Biemans, H.; Zhang, F.; Chen, X.; Chen, X. Patterns and drivers of carbon stock change in ecological restoration regions: A case study of upper Yangtze River Basin, China. *J. Environ. Manag.* **2023**, *348*, 119376. [CrossRef]
35. Talukdar, S.; Singha, P.; Mahato, S.; Pal, S.; Liou, Y.-A.; Rahman, A. Land-use land-cover classification by machine learning classifiers for satellite observations—A review. *Remote Sens.* **2020**, *12*, 1135. [CrossRef]
36. Duran, Z.; Ozcan, K.; Atik, M.E. Classification of photogrammetric and airborne lidar point clouds using machine learning algorithms. *Drones* **2021**, *5*, 104. [CrossRef]
37. Jiang, W.; Zhang, M.; Long, J.; Pan, Y.; Ma, Y.; Lin, H. HLEL: A wetland classification algorithm with self-learning capability, taking the Sanjiang Nature Reserve I as an example. *J. Hydrol.* **2023**, *627*, 130446. [CrossRef]
38. Pal, M. Random forest classifier for remote sensing classification. *Int. J. Remote Sens.* **2005**, *26*, 217–222. [CrossRef]
39. Usman, M.; Ejaz, M.; Nichol, J.E.; Farid, M.S.; Abbas, S.; Khan, M.H. A Comparison of Machine Learning Models for Mapping Tree Species Using WorldView-2 Imagery in the Agroforestry Landscape of West Africa. *ISPRS Int. J. Geo-Inf.* **2023**, *12*, 142. [CrossRef]
40. Ge, G.; Shi, Z.; Zhu, Y.; Yang, X.; Hao, Y. Land use/cover classification in an arid desert-oasis mosaic landscape of China using remote sensed imagery: Performance assessment of four machine learning algorithms. *Glob. Ecol. Conserv.* **2020**, *22*, e00971. [CrossRef]
41. Islam, M.R.; Nahiduzzaman, M. Complex features extraction with deep learning model for the detection of COVID19 from CT scan images using ensemble based machine learning approach. *Expert Syst. Appl.* **2022**, *195*, 116554. [CrossRef] [PubMed]
42. Li, D.; Hu, S.; He, W.; Zhou, B.; Peng, J.; Wang, K. The area prediction of western North Pacific Subtropical High in summer based on Gaussian Naive Bayes. *Clim. Dyn.* **2022**, *59*, 3193–3210. [CrossRef]
43. Mao, W.; Lu, D.; Hou, L.; Liu, X.; Yue, W. Comparison of machine-learning methods for urban land-use mapping in Hangzhou city, China. *Remote Sens.* **2020**, *12*, 2817. [CrossRef]
44. Ghayour, L.; Neshat, A.; Paryani, S.; Shahabi, H.; Shirzadi, A.; Chen, W.; Al-Ansari, N.; Geertsema, M.; Pourmehdi Amiri, M.; Gholamnia, M. Performance evaluation of sentinel-2 and landsat 8 OLI data for land cover/use classification using a comparison between machine learning algorithms. *Remote Sens.* **2021**, *13*, 1349. [CrossRef]
45. Abbas, Z.; Jaber, H.S. Accuracy assessment of supervised classification methods for extraction land use maps using remote sensing and GIS techniques. *IOP Conf. Ser. Mater. Sci. Eng.* **2020**, *745*, 012166. [CrossRef]
46. Xia, Z.; Guo, X.; Chen, R. Automatic extraction of aquaculture ponds based on Google Earth Engine. *Ocean. Coast. Manag.* **2020**, *198*, 105348. [CrossRef]

remote sensing

MDPI

Article

InSAR-DEM Block Adjustment Model for Upcoming BIOMASS Mission: Considering Atmospheric Effects

Kefu Wu [1], Haiqiang Fu [1,*], Jianjun Zhu [1], Huacan Hu [1], Yi Li [1], Zhiwei Liu [1], Afang Wan [2] and Feng Wang [2]

[1] School of Geoscience and Info-Physics, Central South University, Changsha 410083, China; kefuwu@csu.edu.cn (K.W.); jjz@csu.edu.cn (J.Z.); csuhuacan@csu.edu.cn (H.H.); fysxjsxw@csu.edu.cn (Y.L.); liuzhiwei@csu.edu.cn (Z.L.)

[2] The First Institute of Surveying and Mapping of Hunan Province, Changsha 410002, China; wanafang@xcyy.net.cn (A.W.); wangfeng@xcyy.net.cn (F.W.)

* Correspondence: haiqiangfu@csu.edu.cn

Abstract: The unique P-band synthetic aperture radar (SAR) instrument, BIOMASS, is scheduled for launch in 2024. This satellite will enhance the estimation of subcanopy topography, owing to its strong penetration and fully polarimetric observation capability. In order to conduct global-scale mapping of the subcanopy topography, it is crucial to calibrate systematic errors of different strips through interferometric SAR (InSAR) DEM (digital elevation model) block adjustment. Furthermore, the BIOMASS mission will operate in repeat-pass interferometric mode, facing the atmospheric delay errors introduced by changes in atmospheric conditions. However, the existing block adjustment methods aim to calibrate systematic errors in bistatic mode, which can avoid possible errors from atmospheric effects through interferometry. Therefore, there is still a lack of systematic error calibration methods under the interference of atmospheric effects. To address this issue, we propose a block adjustment model considering atmospheric effects. Our model begins by employing the sub-aperture decomposition technique to form forward-looking and backward-looking interferograms, then multi-resolution weighted correlation analysis based on sub-aperture interferograms (SA-MRWCA) is utilized to detect atmospheric delay errors. Subsequently, the block adjustment model considering atmospheric effects can be established based on the SA-MRWCA. Finally, we use robust Helmert variance component estimation (RHVCE) to build the posterior stochastic model to improve parameter estimation accuracy. Due to the lack of spaceborne P-band data, this paper utilized L-band Advanced Land Observing Satellite (ALOS)-1 PALSAR data, which is also long-wavelength, to emulate systematic error calibration of the BIOMASS mission. We chose climatically diverse inland regions of Asia and the coastal regions of South America to assess the model's effectiveness. The results show that the proposed block adjustment model considering atmospheric effects improved accuracy by 72.2% in the inland test site, with root mean square error (RMSE) decreasing from 10.85 m to 3.02 m. Moreover, the accuracy in the coastal test site improved by 80.2%, with RMSE decreasing from 16.19 m to 3.22 m.

Keywords: interferometric synthetic aperture radar (InSAR); block adjustment; DEM; atmospheric effect; sub-aperture decomposition; multi-resolution analysis

Citation: Wu, K.; Fu, H.; Zhu, J.; Hu, H.; Li, Y.; Liu, Z.; Wan, A.; Wang, F. InSAR-DEM Block Adjustment Model for Upcoming BIOMASS Mission: Considering Atmospheric Effects. *Remote Sens.* **2024**, *16*, 1764. https://doi.org/10.3390/rs16101764

Academic Editor: Beatriz M. Funatsu

Received: 21 April 2024
Revised: 13 May 2024
Accepted: 13 May 2024
Published: 16 May 2024

1. Introduction

Interferometric synthetic aperture radar (InSAR) is a powerful tool for large-scale topographic mapping within remote sensing technologies due to its all-weather, all-day, and wide-area observation capability [1,2]. To further observe the three-dimensional structure of forests and estimate the subcanopy topography, the European Space Agency (ESA) proposed the BIOMASS mission [3], which has strong penetration capability due to operating in P-band. Compared with existing space-borne SAR satellites, BIOMASS holds excellent potential for global-scale inverting forest parameters [4] and estimating digital terrain models (DTMs) [5].

The BIOMASS mission will carry a P-band synthetic aperture radar (SAR) sensor with full polarimetric and tomographic capabilities, offering the first opportunity to measure global subcanopy topography [5]. The SAR will operate in repeat-pass interferometric mode and fly in a near-polar, Sun-synchronous orbit with a 637–666 km altitude. Due to instrument performance and high interferometric coherence requirements, the BIOMASS mission will achieve global coverage through interleave stripmap operations between three subswaths, which need a series of roll and repositioning maneuvers [5,6]. Therefore, the digital elevation models (DEMs) derived from BIOMASS will encounter systematic errors between different strips and atmospheric delay errors introduced by changes in atmospheric conditions [7–9]. To ensure the absolute height accuracy and the overall accuracy consistency of topography products, it is necessary to eliminate the above error sources.

For the systematic error calibration, one of the most effective methods is to establish the block adjustment model based on absolute elevation references from ground control points (GCPs) and elevation consistency constraints from tie-points (TPs) [10,11]. To achieve this, DLR (Microwaves and Radar Institute at the German Aerospace Center) utilized the spaceborne lidar, Ice, Cloud, and land Elevation Satellite-1 (ICESat-1) [12] as GCPs and proposed block adjustment using a two-dimensional polynomial as the function model [1,9–11], which generated global topography products TanDEM-X DEM. As the successor to ICESat-1, ICESat-2 offers GCPs with higher spatial resolution and higher repetition rates [13]. Due to the distribution and accuracy of GCPs and ensuring height matching being the keys for systematic error calibration [14,15], researchers began using ICESat-2 as GCPs and considered horizontal errors of DEMs, further improving the block adjustment methods [16,17]. However, the existing block adjustment methods aim to calibrate systematic errors in bistatic mode, whose atmospheric effects can be effectively eliminated through interferometry [18]. Therefore, for the BIOMASS mission, there is an urgent need for a systematic error calibration algorithm under the interference of atmospheric effects. In addition, for the P-band, the atmospheric delay errors significantly affect the elevation accuracy of DEM [7]. Moreover, it also interferes with the height matching between DEMs and GCPs, as well as between TPs, thus affecting the calibration of systematic errors. However, the existing atmospheric effects correction methods usually rely on external water vapor data [19] or require extensive SAR data to perform spatiotemporal analyses [20], limiting the real-time correction of atmospheric effects on a global scale.

Given the above issue, we utilize ICESat-2 ATL08 data as GCPs and propose the block adjustment model considering atmospheric effects for the BIOMASS mission. First, we use the block adjustment model to weaken atmospheric effects' interference in estimating systematic errors. Second, the forward-looking interferograms (F-Infs) and backward-looking interferograms (B-Infs) are composed by sub-aperture decomposition [21]. Then, the atmospheric delay errors can be detected through the multi-resolution weighted correlation analysis based on sub-aperture interferograms (SA-MRWCA), which can correct atmospheric effects without relying on external water vapor data. Finally, we establish the block adjustment model considering atmospheric effects and utilize robust Helmert variance component estimation (RHVCE) [22] to determine weight ratios for GCPs and TPs, overcoming the interference of gross errors and achieving fine systematic error calibration.

2. Methodology

InSAR-DEM block adjustment aims to calibrate systematic errors in different strips to fulfill height accuracy requirements and accomplish large-scale topographic mapping [9,23]. However, due to the error equation's lack of consideration for atmospheric delay errors' impact on height matching, current block adjustment methods prove inadequate for estimating systematic errors in repeat-pass interferometric mode. Hence, this section mainly describes the proposed block adjustment model considering atmospheric effects to address this issue.

2.1. InSAR-DEM Block Adjustment Based on Function Model

The interference factors such as baseline inaccuracies, random phase errors, and radar instruments residual offsets can cause residual systematic errors in different strips [10,11]. The above error sources primarily include offsets and tilts in range and azimuth, which can typically be modeled using a two-dimensional polynomial about image coordinates:

$$g_i(x,y) = a_i + b_i x + c_i y + d_i xy + e_i x^2 + f_i y^2 \tag{1}$$

where i is the index of the DEM acquisition, $g_i(x,y)$ is the height error function, and x, y are the range and azimuth coordinates. $\{a_i, b_i, c_i, d_i, e_i, f_i\}$ are the unknown error parameters of the block adjustment model, which can be estimated using height constraints from GCPs and height consistency constraints from TPs.

For the bistatic mode, since the atmospheric conditions of the two SAR images are the same, the atmospheric delay errors can be eliminated through interferometry [1]. Therefore, the error equations of block adjustment are as follows:

$$\begin{cases} V_{GCP} = \boldsymbol{H}_{GCP} - \left(\hat{\boldsymbol{H}}_i + g_i(x,y)\right) \\ V_{TP} = -\left(\hat{\boldsymbol{H}}_j + g_j(x,y)\right) + \left(\hat{\boldsymbol{H}}_k + g_k(x,y)\right) \end{cases} \tag{2}$$

where $\boldsymbol{V}_{GCP}$ and $\boldsymbol{V}_{TP}$ are residual vectors of GCPs and TPs, respectively. $\boldsymbol{H}_{GCP}$ is the height of GCPs, and $\hat{\boldsymbol{H}}$ is the initial height matched on InSAR-DEM. j and k are the indices of adjacent and overlapping DEMs. Then, we can use the least squares principle to calculate the unknown error parameters and correct the systematic errors in each DEM:

$$\begin{cases} \left|\boldsymbol{V}^T \boldsymbol{P} \boldsymbol{V}\right| = min \\ \boldsymbol{V} = \begin{bmatrix} \boldsymbol{V}_{GCP} & \boldsymbol{V}_{TP} \end{bmatrix}^T \quad \boldsymbol{P} = \begin{bmatrix} \boldsymbol{P}_{GCP} & 0 \\ 0 & \boldsymbol{P}_{TP} \end{bmatrix} \\ DEM^i_{post-adjust}(x,y) = DEM^i_{pre-adjust}(x,y) + g_i(x,y) \end{cases} \tag{3}$$

where T denotes the vector transposition, $\boldsymbol{V}$ is the residual matrix, and $\boldsymbol{P}$ is the stochastic model that characterizes the weight of observation equations in Equation (2).

2.2. Atmospheric Effects Correction Based on Correlation Analysis

Due to different atmospheric conditions in the repeat-pass interferometric mode, the propagation path and propagation speed of SAR ranging signals will change in the atmosphere, introducing atmospheric delay errors in InSAR-DEM [20,24]. Therefore, the core factors affecting DEM accuracy are systematic errors and atmospheric delay errors [24,25]. At this point, the height matching of the error equations shown in Equation (2) will be disturbed, which should transform into

$$\begin{cases} V_{GCP} = \boldsymbol{H}_{GCP} - \left(\hat{\boldsymbol{H}}_i + \Delta\boldsymbol{H}_i + g_i(x,y)\right) \\ V_{TP} = -\left(\hat{\boldsymbol{H}}_j + \Delta\boldsymbol{H}_j + g_j(x,y)\right) + \left(\hat{\boldsymbol{H}}_k + \Delta\boldsymbol{H}_k + g_k(x,y)\right) \end{cases} \tag{4}$$

where $\Delta\boldsymbol{H}$ represents atmospheric delay errors. A lightweight atmospheric effects correction method is urgently needed to estimate systematic errors accurately through block adjustment.

In the atmosphere, the troposphere and ionosphere severely impact SAR signals, resulting in troposphere delay errors and ionospheric delay errors in InSAR-DEM [26]. Among them, the troposphere mainly causes the delay of the SAR signal propagation path, which is manifested as the tropospheric delay phase φ_{trop} in the InSAR interferogram [24,26]:

$$\varphi_{trop} = \frac{4\pi}{\lambda} \frac{10^{-6}}{\cos\theta} \int_{z}^{z_{top}} N dl \tag{5}$$

where λ is the wavelength, θ is the SAR incidence angle, N is the zenith total delay, and z and z_{top} represent the surface elevation and tropospheric elevation, respectively. Under identical tropospheric conditions, shorter wavelengths reduce penetration into the troposphere and increase range delay. Therefore, long-wavelength SAR data can resist the interference of tropospheric delay for InSAR deformation measurements. However, for InSAR terrain mapping, the wavelength λ will be eliminated when the tropospheric delay phase φ_{trop} undergoes phase-to-height conversion, as shown in Equation (6). So, the tropospheric delay errors σ_{trop} are independent of wavelength λ.

$$\begin{cases} k_z = \frac{4\pi B_{\perp}}{\lambda R \sin\theta} \\ \sigma_{trop} = \frac{\varphi_{trop}}{k_z} \end{cases} \tag{6}$$

where k_z is the phase-to-height conversion factor determined by the baseline parameters, $B_{\perp}$ is the vertical baseline, R is the slant-range, and σ_{tropo} is the tropospheric delay errors after phase-to-height conversion. In particular, although BIOMASS will operate in a Sun-synchronous orbit to mitigate the effects of ionosphere disturbances, all measurements will still be influenced by the ionosphere [7]. The main impacts include Faraday rotation Ω and range delay φ_{iono}:

$$\begin{cases} \Omega = \frac{K}{f^2}\langle B\cdot\cos\psi\rangle\cdot\sec\theta\cdot TEC \\ \varphi_{iono} = 2\pi\frac{K}{cf}\cdot TEC \end{cases} \tag{7}$$

where constant $K = 40.28\text{m}^3/\text{s}^2$, f is the wave frequency, B is the magnetic field, ψ is the magnetic declination, and TEC is the Total Electron Content (TEC). Under identical ionospheric conditions, the longer the wavelength of the SAR signal, the more serious the ionospheric delay will be.

In summary, considering the needs of BIOMASS global terrain mapping, atmospheric correction is crucial. Liu et al. proposed the multi-resolution weighted correlation analysis (MRWCA) method, achieving atmospheric effects correction of single-baseline dual-polarization SAR data without external water vapor data [24]. To further alleviate the limitations imposed by the polarization mode, we propose the SA-MRWCA method for single-baseline single-polarization SAR data. Specifically, the SAR images are divided into F-Infs and B-Infs under different squint angles using sub-aperture decomposition. Based on the independence assumption between atmospheric effects and squint angles, the SA-MRWCA can detect similar atmospheric delay errors in sub-aperture interferograms. Although the sub-aperture decomposition changes the overall information characteristics of the interferograms, the atmospheric effects are hardly affected, manifesting the same attributes in the sub-aperture interferograms. This provides the possibility to detect the atmospheric delay phase (ADP) [21]. For stable ground scattering targets, the main components of the interferometric phase φ_{int} include

$$\varphi_{int} = \varphi_{flat} + \varphi_{topo} + \varphi_{orb} + \varphi_{atm} + \varphi_{noi} \tag{8}$$

where φ_{flat} is the flat-earth phase, φ_{topo} is the topographic phase, φ_{orb} is the orbit error phase caused by the inaccuracy of orbit parameters, φ_{atm} is the ADP, and φ_{noi} is the noise phase (NP). After the sub-aperture decomposition, the forward- and backward-looking interferometric phases $\varphi_{int}^{F/B}$ can be represented as

$$\varphi_{int}^{F/B} = \varphi_{flat}^{F/B} + \varphi_{topo}^{F/B} + \varphi_{orb}^{F/B} + \varphi_{atm}^{F/B} + \varphi_{noi}^{F/B} \tag{9}$$

where F/B represents forward- and backward-looking, respectively. Among the phase components shown in Equations (8) and (9), φ_{flat} is the systematic phase contributed by the reference ellipsoidal surface, and φ_{orb} is only related to the orbit parameters of SAR satellites, so they are the same in F-Infs and B-Infs. Additionally, the partial phase associated with the scattering characteristics of the ground target in φ_{topo} will be slightly different between F-Infs and B-Infs. At the same time, φ_{noi} will show randomness in sub-aperture interferograms. Most importantly, the changes in atmospheric conditions between F-Infs and B-Infs are insignificant compared with the distance from the satellite to the ground scattering target. Therefore, φ_{atm} is almost the same in F-Infs and B-Infs.

The SA-MRWCA is based on the MRWCA for multi-resolution analysis of F-Infs and B-Infs through wavelet transform [24], utilizing correlation analysis to estimate atmospheric effects, which requires that ADP is the only similar signal in the phase components. Therefore, the other identical or similar phase components must be eliminated. To achieve this purpose, we first employ SRTM DEM [2], generated based on InSAR, and ASTER GDEM [27], generated based on optical stereophotogrammetry, to perform differential interferometry of F-Infs and B-Infs:

$$\varphi_{int}^{F/B} = \varphi_{\varepsilon topo}^{F/B} + \varphi_{orb}^{F/B} + \varphi_{atm}^{F/B} + \varphi_{noi}^{F/B} \tag{10}$$

where $\varphi_{\varepsilon topo}^{F/B}$ represents the topographic error phase (TEP). Due to the different error sources of the above two DEMs, $\varphi_{\varepsilon topo}^{F}$ and $\varphi_{\varepsilon topo}^{B}$ are also different in forward-looking differential interferograms (F-DInfs) and backward-looking differential interferograms (B-DInfs). Subsequently, the same orbit error phase φ_{orb} can be removed through polynomial fitting [28] or block adjustment:

$$\varphi_{int}^{F/B} = \varphi_{\varepsilon topo}^{F/B} + \varphi_{atm}^{F/B} + \varphi_{noi}^{F/B} \tag{11}$$

At this point, the ADP is the only similar phase component between F-DInfs and B-DInfs. Moreover, the TEP and the NP primarily consist of high-frequency signals in the frequency domain, while the ADP predominantly includes low-frequency signals. Based on the differences in frequency characteristics, the SA-MRWCA decomposes F-DInfs and B-DInfs into building blocks through forward wavelet transform [24]. Then, it utilizes the weighted correlation analysis to estimate the wavelet coefficients associated with the ADP. Finally, the ADP is reconstructed by refining the wavelet coefficients in the inverse wavelet transform process to achieve the estimation of atmospheric delay errors. By substituting the atmospheric delay errors into the error equations shown in Equation (4), height matching between DEMs and GCPs, as well as between TPs, can be achieved for the repeat-pass interferometric mode of BIOMASS.

2.3. InSAR-DEM Block Adjustment Model Considering Atmospheric Effects

Based on the above analysis, we can establish the block adjustment model considering atmospheric effects to calibrate systematic errors under the interference of atmospheric delay errors. Figure 1 shows that the algorithm mainly includes the following three stages.

(1) Coarse Block Adjustment

The core task of this stage involves selecting GCPs and TPs and then utilizing the block adjustment model, as shown in Equations (1)–(3), to provide the orbit error phase φ_{orb} for atmospheric effects correction. Compared with fitting the orbit error phase φ_{orb} scene by scene, this method uses the overall observations of the block adjustment as constraints, weakening the interference of the ADP on the polynomial fitting orbit error phase φ_{orb}, thus separating the two as much as possible. High-quality GCPs and TPs are needed to achieve this goal [14,15], with the selection criteria shown in Table 1.

Table 1. Selection criteria for Ground control points (GCPs) and tie-points (TPs).

	Data Sources	Chip Grid	Selection Criteria
GCPs	ICESat-2 ATL08 Version 5	5 km × 5 km	1. Strong beam 2. Cloud cover less than 20% 3. Not located in the geometric distortion area of SAR images 4. InSAR coherence greater than 0.5
TPs	InSAR DEMs	3 km × 3 km	1. Slope less than 30° 2. Not located in the geometric distortion area of SAR images 3. InSAR coherence greater than 0.5

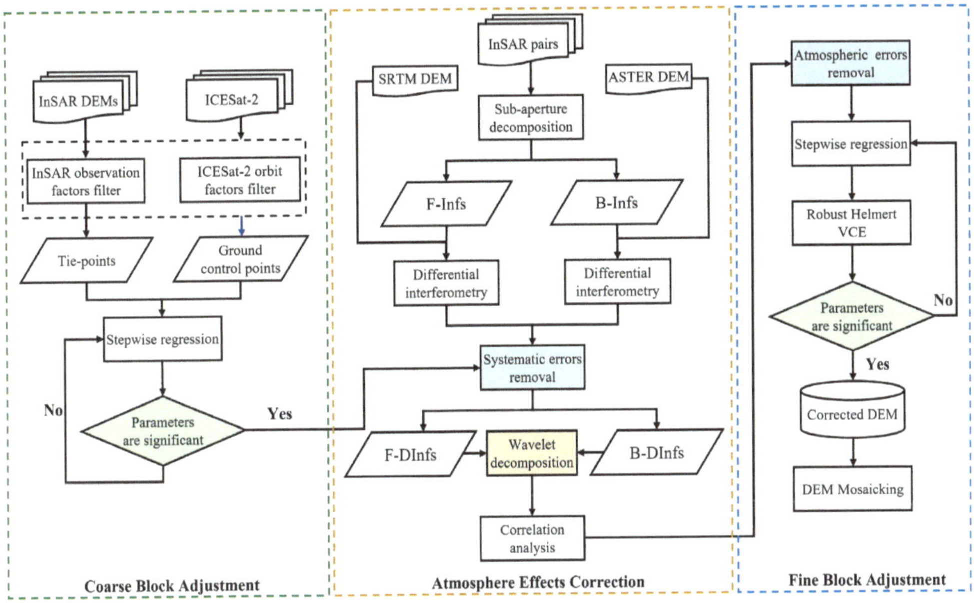

Figure 1. Flow chart of InSAR-DEM block adjustment considering atmospheric effects.

We use ICESat-2 ATL08 (Version 5) data [13] as GCPs and divide the overlapping DEMs uniformly into 3 km × 3 km grids, with the central points selected as TPs. Then, we integrate the ICESat-2 orbit, InSAR observations, and terrain factors as the selection criteria to refine the GCPs and TPs. The selection criteria determine the quality of the GCPs and TPs, playing a crucial role in DEM calibration. To further improve the quality of points and reduce the interference of atmospheric delay errors on height matching, a reselection based on the 3-sigma rule is carried out:

$$\begin{cases} abs(H_{GCP} - H_{TanDEM-X\ DEM}) < 3 * \sigma_{H_{GCP}-H_{TanDEM-X\ DEM}} \\ abs(H_{GCP} - H_{InSAR-DEM}) < 3 * \sigma_{H_{GCP}-H_{InSAR-DEM}} \\ abs(H_{TIE\_j} - H_{TIE\_k}) < 3 * \sigma_{H_{TIE\_j}-H_{TIE\_k}} \end{cases} \tag{12}$$

where H_{GCP}, $H_{TanDEM-XDEM}$, and $H_{InSAR-DEM}$ represent the height of GCPs, TanDEM-X DEM, and InSAR-DEM, respectively. H_{TIE_j} and H_{TIE_k} represent the height of adjacent and overlapping DEMs. Then, we downsample the GCPs using a 5 km × 5 km grid and select the GCPs with the smallest elevation uncertainty [13] within each grid as the GCP used for block adjustment. The remaining parts will serve as checkpoints after block adjustment.

By using the selected GCPs and TPs, we can establish the error equations shown in Equation (2). Then, the unknown parameters of the block adjustment model can be estimated by utilizing the least-squares principle iteratively to realize the adaptive correction of systematic errors by the function model [11]:

$$t = \frac{\hat{x}_{est}}{\sigma_{\hat{x}_{est}}} \tag{13}$$

where $\hat{x}_{est}$ and $\sigma_{\hat{x}_{est}}$ represent the unknown parameters and their standard deviation, respectively. If the significance of all parameters $t \geq 1.64$, then accept all parameters. Otherwise, iterate the estimation by reducing one parameter at a time until all remaining parameters are significant [11]. As shown in Figure 1, the systematic errors based on coarse block adjustment serve as the orbit error phase in atmospheric effect correction, making the ADP the only similar phase component in Equation (11).

(2) Atmosphere Effects Correction

Previous studies have reported that the ionosphere delay errors usually show a similar trend to the orbit error phase, causing the traditional polynomial fitting method to correct the orbit error to fail [29]. However, with the help of the systematic error estimated by the coarse block adjustment, the orbit error phase and the ADP in F-DInfs and B-DInfs can be more accurately separated. After removing the orbit error phase, the ADP is the only similar phase component between F-DInfs and B-DInfs, which can be detected through the SA-MRWCA, providing a basis for block adjustment considering atmospheric effects.

(3) Fine Block Adjustment

As shown in Figure 1, unlike the coarse block adjustment, the fine block adjustment utilizes the atmospheric delay errors detected by the SA-MRWCA to establish the error equations considering atmospheric effects, achieving height matching for the block adjustment of repeat-pass interferometric mode. Furthermore, since the GCPs and TPs come from two different observation systems, the priori stochastic model $\boldsymbol{P}$ in Equation (3) may not be appropriate, and the error equations shown in Equation (4) is also susceptible to interference from gross error. Therefore, after realizing systematic error adaptation through a t-test, we use RHVCE [22] in fine block adjustment to calculate the robust posterior stochastic model. In this way, the weight ratio of GCPs and TPs can be adjusted while overcoming the interference of gross errors. Combining the systematic errors estimated from block adjustment and the atmospheric delay errors detected by the SA-MRWCA, we can conduct precise error calibration:

$$\begin{cases} \left| \boldsymbol{V}^T \boldsymbol{P}_{RHVCE} \boldsymbol{V} \right| = min \\ DEM^i_{post-adjust}(x,y) = DEM^i_{pre-adjust}(x,y) + g^i_{RHVCE}(x,y) + \Delta H_{atm}(x,y) \end{cases} \quad (14)$$

where $\boldsymbol{P}_{RHVCE}$ represents the robust posterior stochastic model estimated by RHVCE, $g^i_{RHVCE}(x,y)$ represents the systematic errors estimated based on $\boldsymbol{P}_{RHVCE}$, and $\Delta H_{atm}(x,y)$ represents the atmospheric delay errors estimated by the SA-MRWCA at position (x,y).

3. Test Sites and Data Sets

To fully verify the algorithm's effectiveness, we conducted experiments in inland regions of Asia and coastal regions of South America with significant climate differences. Due to the lack of spaceborne P-band InSAR data, we emulated the BIOMASS data through L-band Advanced Land Observing Satellite (ALOS)-1 PALSAR data [30], which was also long-wavelength.

3.1. Test Sites

3.1.1. Inland Test Site

As shown in Figure 2, the inland test site lies at the junction of China and Mongolia. Situated in the midlatitude region of inland Asia. The test site exhibits a typical temperate continental climate: hot summers, cold winters, little precipitation, low air humidity, and a relatively stable atmosphere. In addition, hills and mountains dominate the topography of the area.

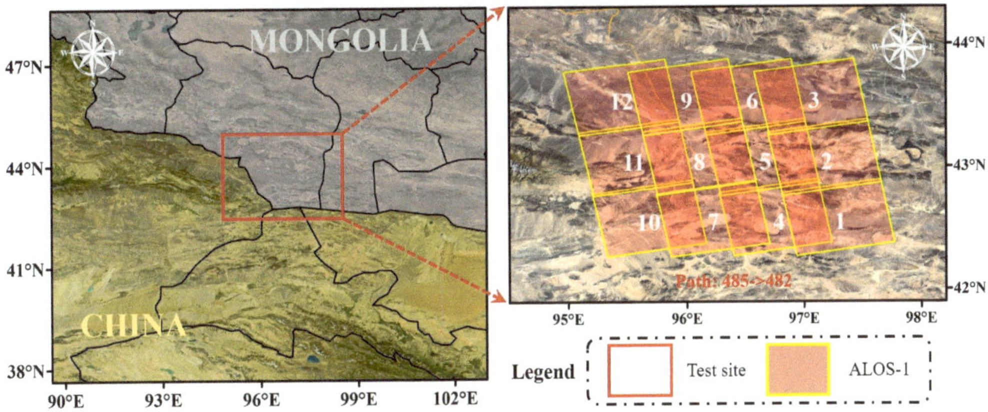

Figure 2. The location and the ALOS-1 footprint of the inland test site. The numbers are the index of the SAR data.

3.1.2. Coastal Test Site

Unlike the stable atmosphere in the inland test site, as shown in Figure 3, the coastal test site lies in South America's midlatitude eastern coastal area, facing the Atlantic Ocean to the east. The ocean brings a humid climate and an active atmosphere. Additionally, hills and plains dominate the test site.

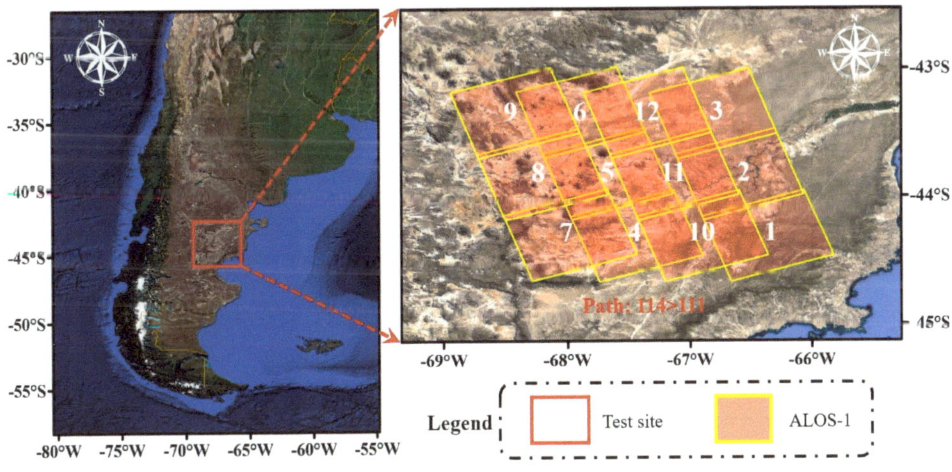

Figure 3. The location and the ALOS-1 footprint of the coastal test site. The numbers are the index of the SAR data.

3.2. Data Sets

3.2.1. ALOS-1 PALSAR

In 2006, the Japan Aerospace Exploration Agency (JAXA) launched the ALOS-1 satellite carrying the L-band PALSAR for research in Earth resource survey, environmental monitoring, and natural disaster management [30]. ALOS-1 has a 46-day repeat cycle and features multiple polarization and multimode capabilities. We used HH-polarization data from fine-beam dual-polarization (FBD) mode to emulate block adjustment considering atmospheric effects for the BIOMASS mission. Table 2 shows the data information in the inland and coastal test sites.

Table 2. The information of ALOS-1 PALSAR data used in this study.

Test Sites	Time	Number *	Path *	$\bar{B}$perp(m)	Btemp	Pixel Spacing (m)	Orbit
Inland	25 July 2010	1–3	482	557.5	92	3.16 × 9.37	ASC
	11 August 2010	4–6	483	435.6	46	3.16 × 9.37	ASC
	28 August 2010	7–9	484	317.3	46	3.16 × 9.37	ASC
	14 September 2010	10–12	485	465.3	46	3.16 × 9.37	ASC
Coastal	4 June 2010	1–3	111	131.4	46	3.26 × 9.37	ASC
	8 July 2010	4–6	113	160.9	46	3.26 × 9.37	ASC
	25 July 2010	7–9	114	537.5	46	3.26 × 9.37	ASC
	6 August 2010	10–12	112	489.3	46	3.26 × 9.37	ASC

"*" indicates correspondence to Figures 1 and 2.

3.2.2. ICESat-2 ATL08 Product (Version 5)

The National Aeronautics and Space Administration (NASA) has released the fifth version of the land elevation product ATL08 through photon cloud filtering and classification of ICESat-2 ATL03 data [13]. The ATL08 data provide terrain and canopy heights at a 20 m along-track spacing, which has been widely used for DEM systematic error correction or forest height deduction [16,31]. In actual quality assessments, the accuracy of ATL08 terrain elevation can reach 0.73 m [32]. Previous studies have shown that the quality of GCPs is crucial for block adjustment [14,15]. To ensure the reliability of GCPs, we established extraction rules and refined ATL08 data within the two test sites from 14 October 2018 to 15 July 2021, as shown in Table 1.

3.2.3. Global Digital Elevation Products

We introduced the SRTM DEM and ASTER GDEM to remove the topographic phases of F-Infs and B-Infs, respectively. In addition, the higher-precision TanDEM-X DEM was utilized to verify the correction of atmospheric delay errors and systematic errors in continuous space. Table 3 shows the information of the above DEMs. Before using the SRTM DEM and ASTER GDEM, it is necessary to convert their vertical datum to WGS84.

Table 3. The information of the DEMs used in this study.

Dataset	Horizontal Datum	Vertical Datum	Time (Year)	Resolution (m)	Absolute Vertical Accuracy (m)	Technique
TanDEM-X DEM	WGS84	WGS84	2016	30	10	InSAR
SRTM DEM V003	WGS84	EGM96	2013	30	16	InSAR
ASTER GDEM V2	WGS84	EGM96	2011	30	17	Optical stereophotogrammetry

4. Results and Analysis

4.1. Inland Test Site

Aiming to construct an observation equation for block adjustment that considers both systematic errors and atmospheric delay errors, the reliability of GCPs and TPs is essential. Figure 4 shows the distribution of GCPs and TPs in the inland test site, where Figure 4a represents the control point database established using the selection rules in Table 1, and Figure 4b displays the GCPs and TPs after refinement and thinning. The uniform distribution provides support for systematic error calibration.

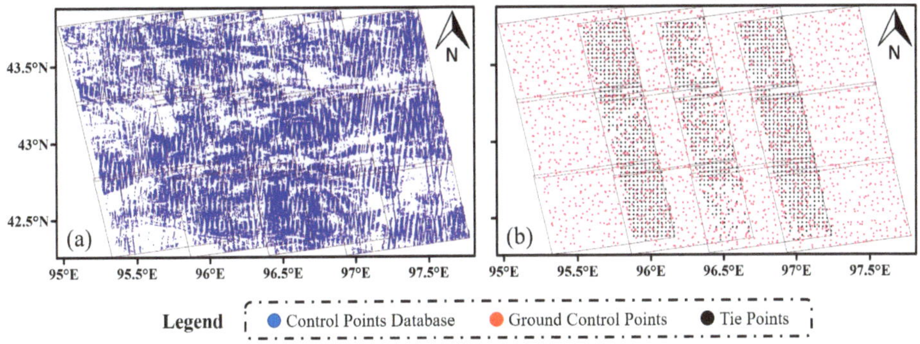

Figure 4. (**a**) The control point database of the inland test site; (**b**) the distribution of GCPs and TPs in the inland test site.

The existing research indicated that atmospheric delay errors mainly include tropospheric delay errors and ionospheric delay errors [26]. Since the ionospheric delay error is a long-wavelength signal usually manifested as the orbit error, it will cause the scene-by-scene polynomial fitting method to correct the orbit error phase to fail [29]. To overcome the influence of trend errors of ionospheric delay errors on polynomial fitting orbit error, we used the coarse block adjustment to provide the SA-MRWCA with more accurate orbit error phases for correcting atmospheric effects. Sub-aperture decomposition was employed to form F-Infs and B-Infs. Then, we utilized SRTM DEM and ASTER GDEM to perform differential interferometry, respectively. Subsequently, the orbit error phase of F-DInf and B-DInf was removed to guarantee the ADP's unique similarity. Figure 5 compares the above two methods for removing the orbit error phase, where Figure 5a–d shows scene-by-scene polynomial fitting, and Figure 5e–h displays coarse block adjustment estimates.

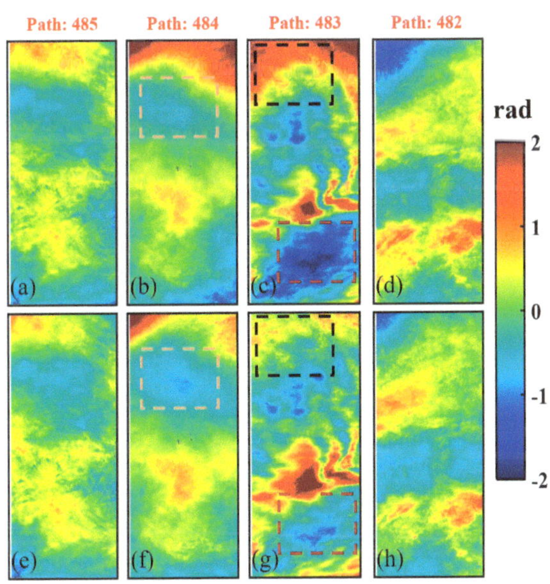

Figure 5. Comparison of different methods for removing orbit error phase in inland test site: (**a**–**d**) scene-by-scene polynomial fitting; (**e**–**h**) block adjustment estimates. The boxes show the difference.

Apparent trend errors are observed at the top of Figure 5b–d but are weakened in Figure 5f–h. In the orange frame of Figure 5, Figure 5f shows an ionospheric delay phase similar to the orbit error phase. We can find that the scene-by-scene polynomial fitting orbit error phase was interfered with by the ionospheric delay phase and absorbed part of it, resulting in a residual orbit error signal at the top of Figure 5b. In addition, comparing the black frame and red frame in Figure 5, it can be seen that the trend errors in Figure 5c are weakened in Figure 5g. The above differences prove that benefiting from the elevation control of GCPs and the elevation consistency constraint of TPs, the systematic errors estimated by coarse block adjustment can overcome the interference of atmospheric delay errors to a certain extent and better separate the two. As a result, this provides the basis for the SA-MRWCA to correct atmospheric effects. The remaining phase components after removing the orbit error phase are shown in Equation (11).

Since the TEP comes from different DEMs and the NP is random, the ADP is the only similar phase component between F-DInfs and B-DInfs. In the frequency domain, the TEP and the NP are mainly composed of high-frequency signals, while the ADP mostly appears as low-frequency signals. The above phase components are mixed in the differential interferograms and are difficult to separate effectively through spatial domain filtering. However, the different signal characteristics provide the basis for detecting the ATP. Therefore, we proposed the SA-MRWCA method to decompose F-DInfs and B-DInfs into multiple phase signals of different wavelengths through forward wavelet transform. Then, the SA-MRWCA used weighted correlation analysis to reduce the TEP and NP signals, reconstructing the ADP through inverse wavelet transform. Figure 6a–h displays the detected ADP and the phase after atmospheric correction of the inland test site, respectively. Combining with Figure 5e–h, we find that the SA-MRWCA correctly detected atmospheric delay errors.

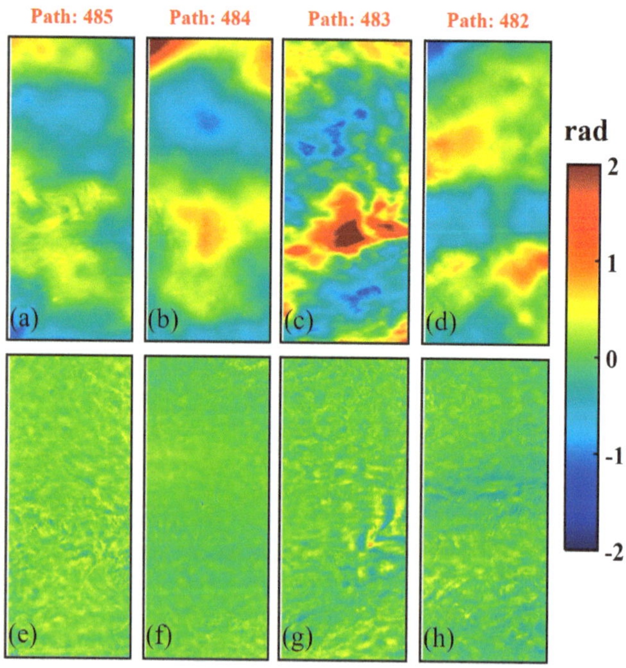

Figure 6. The atmospheric correction results of the inland test site: (**a**–**d**) ATP estimated by the SA-MRWCA; (**e**–**h**) residual phase after atmospheric correction.

After obtaining the atmospheric delay errors, the error equations considering the atmospheric delay errors can be established, achieving height matching between DEMs and GCPs, as well as between TPs. In addition, to construct a more stable block adjustment solution method in fine block adjustment, we utilized the *t*-test to test the significance of model parameters and realized the adaptation of systematic errors. On this basis, since GCPs and TPs came from two types of observation systems, and the error equations were affected by gross errors, the RHVCE [22] was to establish an accurate posterior stochastic model for calculating the parameters. Finally, we used the error correction equation shown in Equation (14) to correct the InSAR-DEM in the inland test site.

The results and accuracy statistics are shown in Figure 7, where Figure 7a displays the corrected DEM, and Figure 7b shows the error statistics histogram of the DEMs corrected by block adjustment considering and ignoring atmospheric effects compared with ICESat-2 elevation validation points. The block adjustment considering atmospheric effects markedly enhanced error concentration, reducing the root mean square error (RMSE) from 10.85 m to 3.02 m and resulting in a 72.2% increase in height accuracy. In order to spatially evaluate whether systematic errors and atmospheric delay errors have been removed, we introduced TanDEM-X DEM as elevation reference data and conducted a local comparison at the red frame in Figure 7a. Figure 7c partially enlarges the view of the corrected DEM. The difference map between the InSAR-DEM of the area and TanDEM-X DEM under block adjustment considering and ignoring atmospheric effects is shown in Figure 7d,e, respectively. The red frame contains three strips. Figure 7d displays calibration considering atmospheric effects, which removes the atmospheric delay errors and systematic errors completely. However, Figure 7e exhibits significant residual error trends, indicating that block adjustment ignoring atmospheric effects cannot accurately estimate the systematic errors of each strip.

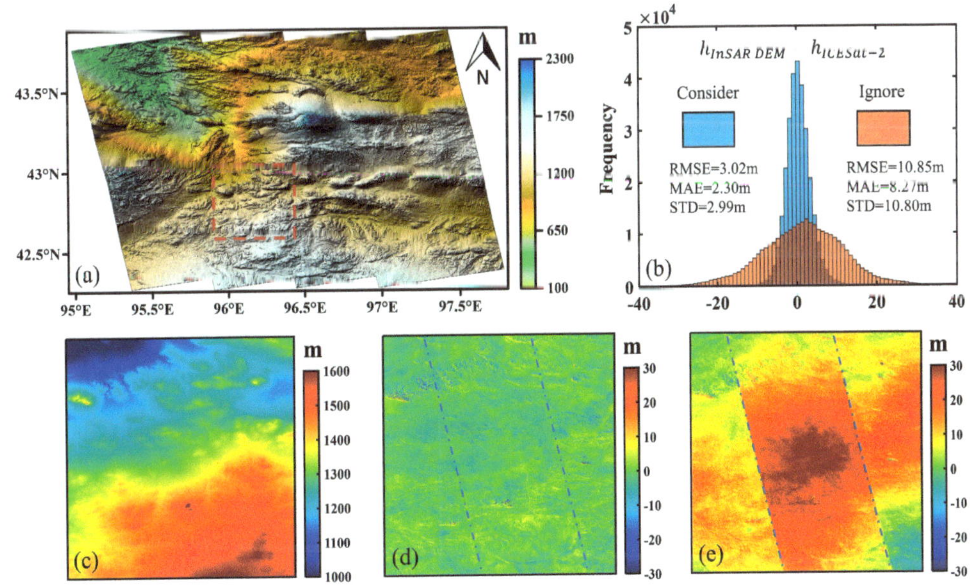

Figure 7. DEM calibration results and accuracy verification in the inland test site: (**a**) corrected DEM; (**b**) error statistical histogram of the corrected DEMs by block adjustment considering and neglecting atmospheric effects. (**c**–**e**) are partially enlarged views of the red frame in (**a**): (**c**) partial map of the corrected DEM; (**d**) difference map between (**c**) and TanDEM-X DEM; (**e**) difference map between the partial DEM ignoring atmospheric effects and TanDEM-X DEM.

4.2. Coastal Test Site

The coastal test site lies east of the Atlantic Ocean, characterized by scattered hills with elevations ranging from approximately 100 to 750 m. Compared with the inland test site, the coastal test site has a more humid atmospheric layer and a more undulating terrain. Additionally, as shown in Table 2, the vertical baselines in the inland test site are concentrated between 300 and 550 m. In contrast, the coastal test site exhibits significant differences in vertical baselines between adjacent strips, with variations exceeding 300 m, implying that each strip has different sensitivities to the ADP and the TEP. Therefore, the coastal test site has more difficulty detecting atmospheric delay errors and estimating systematic errors using the proposed algorithm. The distribution of GCPs and TPs in the coastal test site is shown in Figure 8, where Figure 8a represents the control point database, and Figure 8b represents the GCPs and TPs after refinement and thinning.

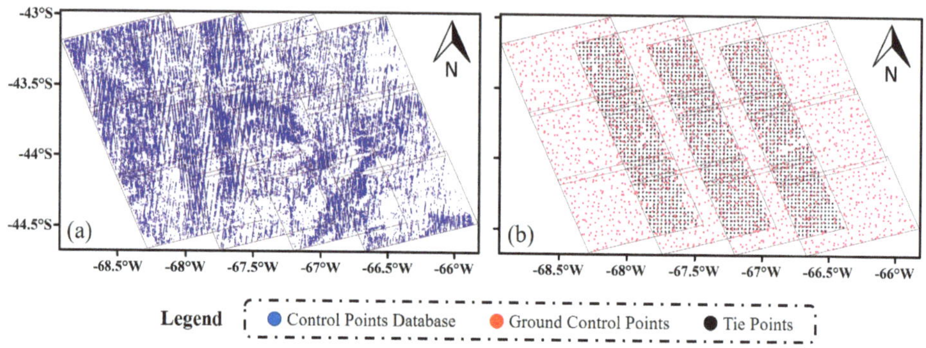

Figure 8. (**a**) The control point database of the coastal test site; (**b**) the distribution of GCPs and TPs in the coastal test site.

After making the ADP the only similar phase component between F-DInfs and B-DInfs, we can detect the ADP based on the SA-MRWCA and correct the atmospheric delay errors, shown in Figure 9a–h, respectively. Previous studies have indicated that the tropospheric delay errors, as one of the main components of atmospheric effects, can be divided into topography-dependent vertical stratified delay and topography-independent turbulent delay [26]. When comparing Figure 6a–d with Figure 9a–d, it is evident that the ADP extracted in the coastal test site appears relatively discrete. This discreteness can be attributed to the dispersed hills in the coastal test site, where the topography undergoes rapid changes, resulting in the incorporation of topography-dependent signals in the atmospheric delay. In addition, considering that the ground surface will not deform in a short period, the positive phase at the bottom of Figure 9c should be topography-independent turbulence signals. After atmospheric correction, the residual phases shown in Figure 9e–h have no apparent atmospheric delay signals. Due to the relatively shorter vertical baselines of Path 111 and Path 113, the height of ambiguity (height change corresponding to one phase cycle) is 500 m and 410 m, respectively, leading to smaller TEP values, as depicted in Figure 9f,h. Height matching was achieved between DEMs and GCPs, as well as between TPs, through the atmospheric delay error estimated by the SA-MRWCA. On this basis, we utilized RHVCE in fine block adjustment to establish an accurate posterior stochastic model for resisting the influence of gross errors on the observation equations. Finally, we calculated the model parameters to correct the InSAR-DEM in the coastal test site.

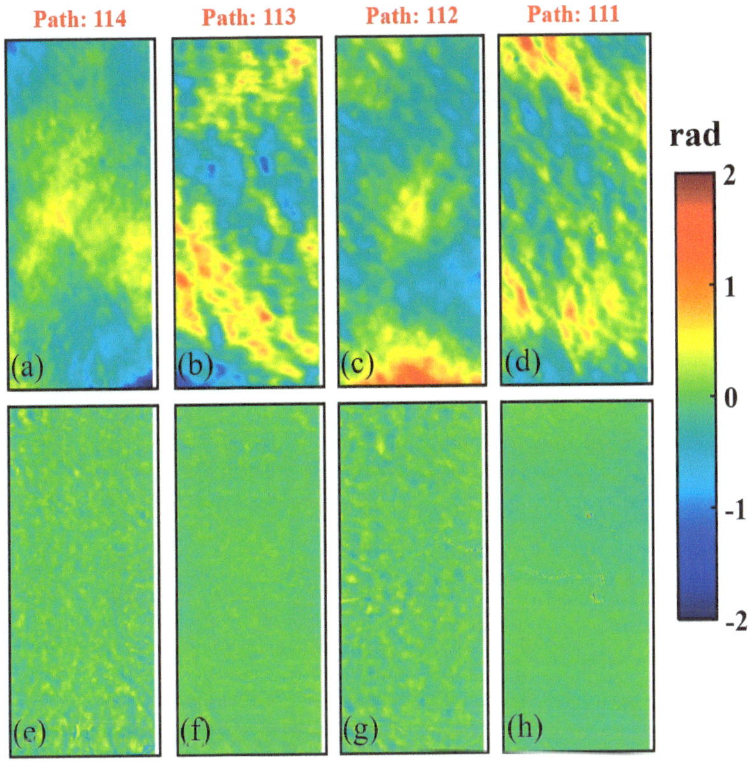

Figure 9. The atmospheric correction results of the coastal test site: (**a**–**d**) ATP estimated by the SA-MRWCA; (**e**–**h**) residual phase after atmospheric correction.

Figure 10 shows the results and the height accuracy statistics, where Figure 10a displays the corrected DEM, and Figure 10b displays the error statistics histogram of the DEMs corrected by block adjustment considering and ignoring atmospheric effects compared with ICESat-2 elevation validation points. It can be seen that the errors after correction of the block adjustment considering atmospheric effects are more concentrated, the RMSE reduced from 16.19 m to 3.22 m, and the height accuracy increased by 80.2%. In addition, we also introduced TanDEM-X DEM to test the calibration effect and conducted a local comparison in the red frame in Figure 10a. Figure 10c is a partially enlarged view of the corrected DEM. The difference map between the InSAR-DEM of the area and TanDEM-X DEM under the block adjustment considering and ignoring the atmospheric effects is shown in Figure 10d,e. The red frame contains three strips. When comparing Figure 10d with Figure 10e, it can be seen that the block adjustment considering atmospheric effects completely corrected systematic errors and atmospheric delay errors. However, the block adjustment ignoring atmospheric effects shows apparent elevation discontinuities in the overlapping area, indicating a misestimation of systematic errors. The correction results in the coastal test area illustrate that our algorithm is still effective when the water vapor content is rich and the baselines between adjacent strips differ.

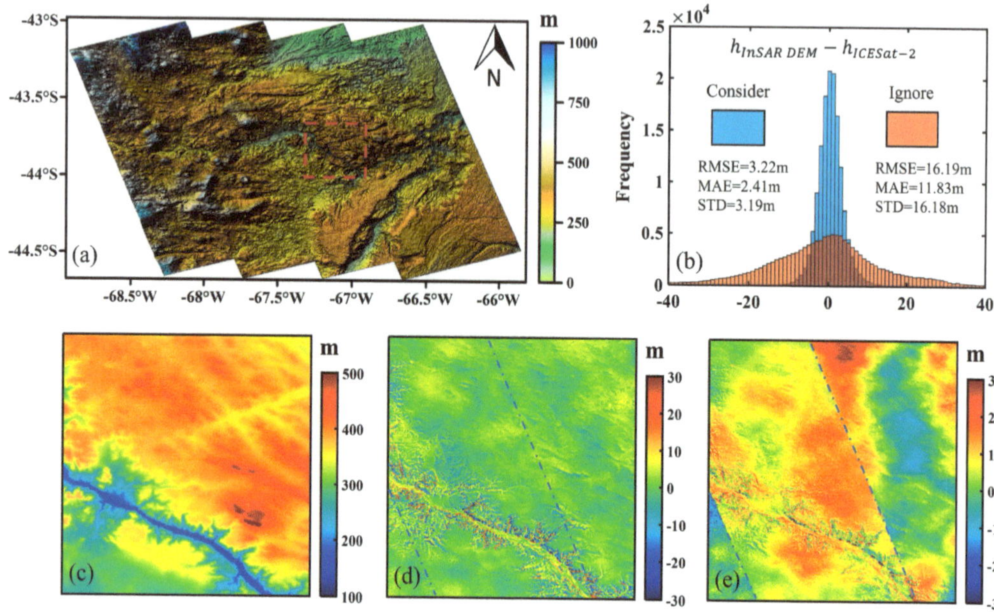

Figure 10. DEM calibration results and accuracy verification in the coastal test site: (**a**) corrected DEM; (**b**) error statistical histogram of the corrected DEMs by block adjustment considering and neglecting atmospheric effects. (**c**–**e**) are partially enlarged views of the red frame in (**a**): (**c**) partial map of the corrected DEM; (**d**) difference map between (**c**) and TanDEM-X DEM; (**e**) difference map between the partial DEM ignoring atmospheric effects and TanDEM-X DEM.

5. Discussion

5.1. The Impact of Atmospheric Effects on the Estimation of Block Adjustment

Through the above analysis, we can find that the impact of atmospheric effects on block adjustment mainly includes two aspects.

(1) Height matching between InSAR-DEM and GCPs, as well as between TPs

DLR proposed the block adjustment based on the function model, aiming to calibrate the systematic errors of the bistatic mode [9,10]. Therefore, the error equations of this model only observe systematic errors. In the repeat-pass interferometric mode, changes in the state of the atmosphere lead to the introduction of atmospheric delay errors in the DEM. When the relative humidity of the atmosphere changes by 20%, the resulting height measurement deviation can reach 80~290 m [26]. In addition to systematic errors, the height differences between DEMs and GCPs, as well as between TPs, also include atmospheric delay errors. In order to achieve height matching, it is crucial to establish an error equation that considers systematic errors and atmospheric delay errors. As shown in Figures 7e and 10e, without considering the atmospheric effects, the estimated model parameters through block adjustment are inaccurate, resulting in residual trend signals in DEM. In contrast, Figures 7d and 10d depict the difference maps between the DEMs corrected by our algorithm and TanDEM-X DEM, which observe no residual trend errors. It demonstrates the effectiveness of our algorithm in overcoming the atmospheric effects on height matching and accurately rectifying systematic errors.

(2) The long-wavelength signals of atmospheric effects are mixed with systematic errors

Since the trajectory of SAR satellites in space is relatively smooth, systematic errors caused by inaccurate orbit positions usually exist in the form of long-wavelength signals [28]. However, the ionospheric delay errors of the atmospheric effects are also long-wavelength signals, which have a similar distribution to the systematic errors [29]. Therefore, it will interfere with the estimating of systematic errors. However, combined with the height control of GCPs and the height consistency constraint of TPs, block adjustment is more resistant to the interference of atmospheric delay errors than scene-by-scene polynomial fitting.

Systematic errors exist as the orbit error phase at the interferometric phase. As shown in Figure 5b,f, the ionospheric delay phase in the red frame is absorbed by the polynomial of the fitted orbit error phase when correcting the systematic error scene by scene, causing an underestimation of the top system error. When the orbit error phase is removed using the systematic error estimated by the block adjustment, the ionospheric delay phase appears in the remaining phase component, indicating that the two are well separated. Taking the inland test site as an example, the systematic errors estimated by scene-by-scene fitting and block adjustment are shown in Figure 11a–h, respectively. Due to the limited orbit position accuracy, the systematic error trends shown in Figure 11 are relatively changeable. Benefiting from the significance test based on the *t*-test, the systematic error estimation of block adjustment can achieve adaptive estimation for each strip. The systematic errors estimated by the above methods are pretty different. Figure 11i–l displays the difference maps between the two. Due to the interference of atmospheric delay errors, the systematic error estimated scene by scene in Figure 11c shows local overfitting. Combining Figures 5c,g and 11g, it can be seen that the block adjustment overcomes the problem of local overfitting. The above analysis demonstrates that the block adjustment considering atmospheric effects can still estimate the complex systematic errors under the interference of atmospheric delay errors.

5.2. The Impact of Slope on DEM Calibration

In the block adjustment model considering atmospheric effects, the slope is another influencing factor besides atmospheric effects. Since the SAR uses side-looking imaging, geometric distortion can occur in steep slopes, resulting in limited DEM accuracy [26]. In addition, areas with large slopes often include mountains and hills, which are likely to contain topography-dependent tropospheric delay errors [26]. Based on the above factors, we divided the established control point database into five groups according to slope and then conducted height accuracy verification. The grouping situation is shown in Table 4.

Table 4. The count of checkpoints in each category after being grouped by the slope.

Test Sites	0~5°	5~10°	10~15°	15~20°	>20°
Inland	151,680	102,129	27,047	6448	2403
Coastal	71,292	50,441	15,840	3641	1064

We used the grouped checkpoints to assess InSAR-DEMs' accuracy in the inland and coastal test sites and then compared them with TanDEM-X DEM, SRTM DEM, and ASTER GDEM, respectively. The results are shown in Figure 12, where Figure 12a,c,e and Figure 12b,d,f are the RMSE, mean absolute error (MAE), and standard deviation (STD) of the inland and coastal test sites, respectively. With increasing slopes, the RMSE, MAE, and STD of the above DEMs decrease. However, compared with the original DEM, the accuracy indicators of the corrected DEM are significantly improved at each slope. In addition, the accuracy indicators of the corrected DEM are closer to TanDEM-X DEM and SRTM DEM produced by the zero-time baseline, which can prove the reliability of our algorithm in calibrating atmospheric delay errors and systematic errors.

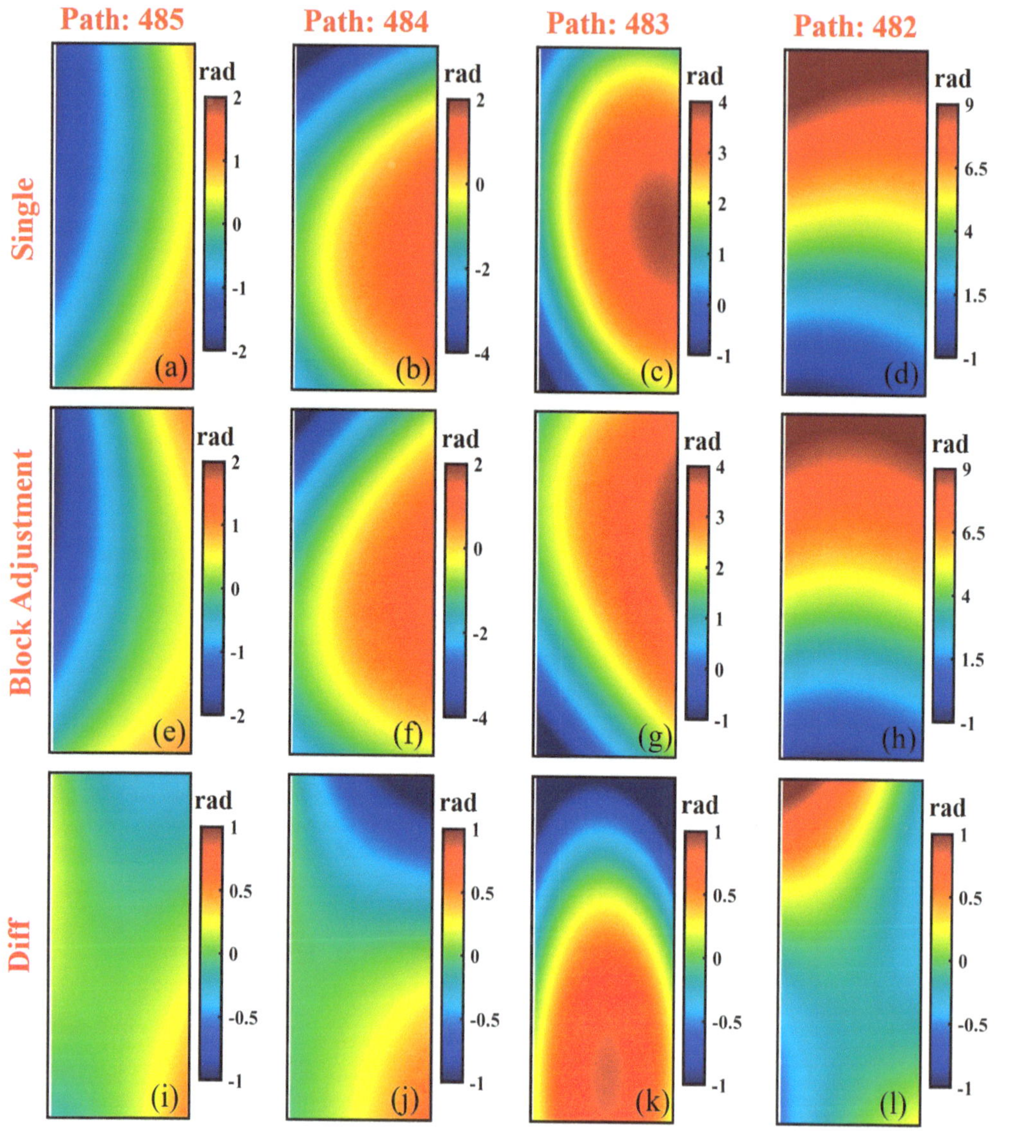

Figure 11. Comparison of systematic errors between scene-by-scene fitting and block adjustment fitting in the inland test site: (**a**–**d**) scene-by-scene fitting; (**e**–**h**) block adjustment fitting; (**i**–**l**) difference maps in estimated systematic errors between scene-by-scene fitting and block adjustment fitting.

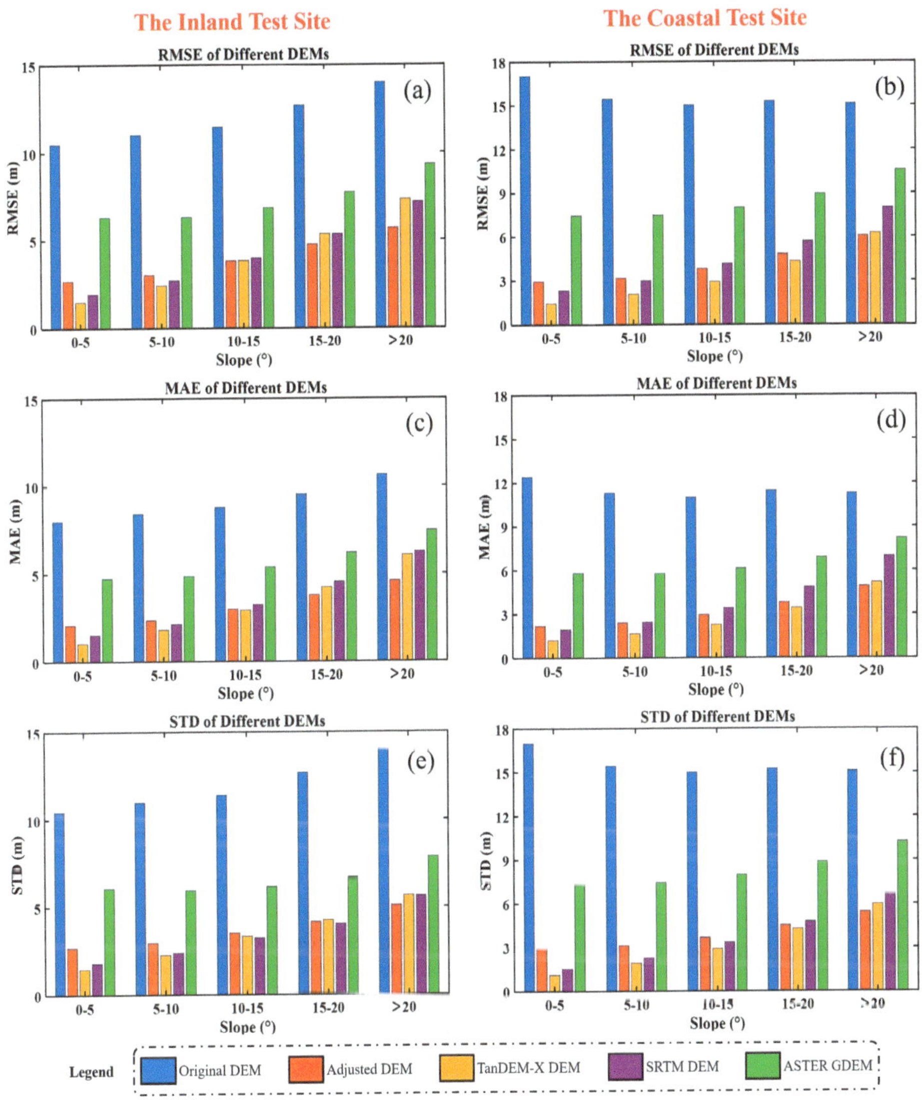

Figure 12. Comparison between the original DEM, corrected DEM, TanDEM-X DEM, SRTM DEM, and ASTER GDEM at varying slopes: (**a**,**c**,**e**) the RMSE, MAE, and STD of different DEMs in inland test site; (**b**,**d**,**f**) the RMSE, MAE, and STD of different DEMs in coastal test site.

6. Conclusions

Aiming at the block adjustment being interfered with by atmospheric effects for the BIOMASS mission, we proposed a block adjustment model considering atmospheric effects. This method separated systematic errors and atmospheric delay errors and then utilized the SA-MRWCA to detect atmospheric effects without relying on external water vapor data. Subsequently, the error equations containing the systematic errors and atmospheric delay errors can be established, achieving height matching between GCPs and DEMs, as well as between TPs. Finally, we used a significance test based on the t-test to achieve adaptive systematic error correction. Building on this foundation, we assessed the weights of GCPs and TPs by using the stochastic model estimated by RHVCE and then estimated the model parameters of block adjustment. Therefore, the atmospheric delay errors and systematic errors can be calibrated simultaneously, achieving block adjustment that overcomes the influence of atmospheric effects.

Through experiments with different climate conditions, we substantiated the algorithm's effectiveness. Furthermore, a precision comparison was conducted between the original DEM, corrected DEM, and global DEM products at varying slopes. Compared with the original DEM, the height accuracy of the corrected DEM is closer to TanDEM-X DEM and SRTM DEM at each slope, indicating the algorithm can simultaneously correct atmospheric delay errors and systematic errors. In addition, the computational complexity and processing time depend on the quality of the satellite orbit, the reliability of the GCPs and TPs, and the distribution of systematic errors and atmospheric delay errors.

Since there are no available spaceborne P-band data, L-band data are the best means of simulation. The P-band has a longer wavelength than the L-band, which will cause it to experience more severe ionospheric delay errors. Furthermore, the P-band and L-band have different penetration capabilities into the forests, resulting in different forest responses to SAR signals. Therefore, future research should validate the proposed model using the actual BIOMASS data. What is particularly important is that to avoid forest signal interference with systematic error observations, existing studies mainly select the GCPs and TPs on bare earth. However, the core monitoring object of BIOMASS is the forest. An even distribution of the GCPs and TPs is necessary to ensure accurate observation of systematic errors. Therefore, it is urgent to establish a P-band scattering model that considers differences in penetration depths to forests and then overcomes their impact on DEM calibration.

Author Contributions: Conceptualization, K.W., H.F. and J.Z.; methodology, K.W. and H.F.; software, K.W. and H.H.; validation, K.W., Z.L. and Y.L.; formal analysis, K.W. and A.W.; investigation, K.W.; resources, K.W.; data curation, K.W. and F.W.; writing—original draft preparation, K.W.; writing—review and editing, K.W., H.F. and Y.L.; visualization, K.W., H.H. and Y.L.; supervision, H.F. and J.Z.; project administration, H.F. and J.Z.; funding acquisition, H.F., J.Z. and A.W. All authors have read and agreed to the published version of the manuscript.

Funding: This research was partly funded by the National Natural Science Foundation of China (No. 42227801). Additionally, this research was partly funded by the Natural Science Foundation of Hunan Province (No. 2023JJ20061) and the Research Foundation of the Department of Natural Resources of Hunan Province (No. 湘自资科 20240109CH).

Data Availability Statement: The ICESat-2 ATL08 datasets (Version 5) are available from https://search.earthdata.nasa.gov/search (accessed on 7 March 2023). The TanDEM-X DEM is available from https://tandemx-science.dlr.de/ (accessed on 10 March 2023). The SRTM DEM is available from https://earthexplorer.usgs.gov/ (accessed on 10 March 2023). The ALOS-1 data are available from https://search.asf.alaska.edu/#/ (accessed on 26 February 2023). The ASTER GDEM is available from https://asterweb.jpl.nasa.gov/gdem.asp (accessed on 26 February 2023).

Acknowledgments: Many thanks to NASA, DLR, ESA, USGS, and JAXA for providing free datasets.

Conflicts of Interest: The authors declare no conflicts of interest.

References

1. Rizzoli, P.; Martone, M.; Gonzalez, C.; Wecklich, C.; Borla Tridon, D.; Bräutigam, B.; Bachmann, M.; Schulze, D.; Fritz, T.; Huber, M.; et al. Generation and Performance Assessment of the Global TanDEM-X Digital Elevation Model. *ISPRS J. Photogramm. Remote Sens.* **2017**, *132*, 119–139. [CrossRef]
2. Rodriguez, E.; Morris, C.; Belz, J.; Chapin, E.; Martin, J.; Daffer, W.; Hensley, S. An Assessment of the SRTM Topographic Products. 2005. Available online: https://www.researchgate.net/publication/235704654_An_assessment_of_the_SRTM_topographic_products_Technical_Report_JPL_D-31639 (accessed on 12 May 2024).
3. Le Toan, T.; Quegan, S.; Davidson, M.W.J.; Balzter, H.; Paillou, P.; Papathanassiou, K.; Plummer, S.; Rocca, F.; Saatchi, S.; Shugart, H.; et al. The BIOMASS Mission: Mapping Global Forest Biomass to Better Understand the Terrestrial Carbon Cycle. *Remote Sens. Environ.* **2011**, *115*, 2850–2860. [CrossRef]

4. Fatoyinbo, T.; Armston, J.; Simard, M.; Saatchi, S.; Denbina, M.; Lavalle, M.; Hofton, M.; Tang, H.; Marselis, S.; Pinto, N.; et al. The NASA AfriSAR Campaign: Airborne SAR and Lidar Measurements of Tropical Forest Structure and Biomass in Support of Current and Future Space Missions. *Remote Sens. Environ.* **2021**, *264*, 112533. [CrossRef]
5. Quegan, S.; Chave, J.; Dall, J.; Le Toan, T.; Papathanassiou, K.; Rocca, F.; Saatchi, S.; Scipal, K.; Shugart, H.; Ulander, L.; et al. The Science and Measurement Concepts Underlying the BIOMASS Mission. In Proceedings of the 2012 IEEE International Geoscience and Remote Sensing Symposium, Munich, Germany, 22–27 July 2012; pp. 5542–5545.
6. Banda, F.; Giudici, D.; Le Toan, T.; d'Alessandro, M.M.; Papathanassiou, K.; Quegan, S.; Riembauer, G.; Scipal, K.; Soja, M.; Tebaldini, S.; et al. The BIOMASS Level 2 Prototype Processor: Design and Experimental Results of Above-Ground Biomass Estimation. *Remote Sens.* **2020**, *12*, 985. [CrossRef]
7. Quegan, S.; Lomas, M.; Papathanassiou, K.P.; Kim, J.-S.; Tebaldini, S.; Giudici, D.; Scagliola, M.; Guccione, P.; Dall, J.; Dubois-Fenandez, P.; et al. Calibration Challenges for the Biomass P-Band SAR Instrument. In Proceedings of the IGARSS 2018—2018 IEEE International Geoscience and Remote Sensing Symposium, Valencia, Spain, 22–27 July 2018; pp. 8575–8578.
8. Ferretti, A.; Prati, C.; Rocca, F. Permanent Scatterers in SAR Interferometry. *IEEE Trans. Geosci. Remote Sens.* **2001**, *39*, 8–20. [CrossRef]
9. Wessel, B.; Gruber, A.; Huber, M.; Roth, A. TanDEM-X: Block Adjustment of Interferometric Height Models. In Proceedings of the ISPRS Hannover Workshop 2009 "High-Resolution Earth Imaging for Geospatioal Information", International Archives of the Photogrammetry, Remote Sensing and Spatial Information Sciences, Hannover, Germany, 2–5 June 2009; pp. 1–6. Available online: https://elib.dlr.de/62358/ (accessed on 12 May 2024).
10. Wessel, B.; Gruber, A.; Gonzalez, J.H.; Bachmann, M.; Wendleder, A. TanDEM-X: DEM Calibration Concept. In Proceedings of the IGARSS 2008—2008 IEEE International Geoscience and Remote Sensing Symposium, Boston, MA, USA, 7–11 July 2008; Volume 3, pp. III-111–III-114.
11. Gruber, A.; Wessel, B.; Huber, M.; Roth, A. Operational TanDEM-X DEM Calibration and First Validation Results. *ISPRS J. Photogramm. Remote Sens.* **2012**, *73*, 39–49. [CrossRef]
12. Schutz, B.E.; Zwally, H.J.; Shuman, C.A.; Hancock, D.; DiMarzio, J.P. Overview of the ICESat Mission. *Geophys. Res. Lett.* **2005**, *32*, 2005GL024009. [CrossRef]
13. Markus, T.; Neumann, T.; Martino, A.; Abdalati, W.; Brunt, K.; Csatho, B.; Farrell, S.; Fricker, H.; Gardner, A.; Harding, D.; et al. The Ice, Cloud, and Land Elevation Satellite-2 (ICESat-2): Science Requirements, Concept, and Implementation. *Remote Sens. Environ.* **2017**, *190*, 260–273. [CrossRef]
14. Huber, M.; Wessel, B.; Kosmann, D.; Felbier, A.; Schwieger, V.; Habermeyer, M.; Wendleder, A.; Roth, A. Ensuring Globally the TanDEM-X Height Accuracy: Analysis of the Reference Data Sets ICESat, SRTM and KGPS-Tracks. In Proceedings of the 2009 IEEE International Geoscience and Remote Sensing Symposium, Cape Town, South Africa, 12–17 July 2009; IEEE: Piscataway, NJ, USA, 2009; pp. II-769–II-772.
15. Hueso Gonzalez, J.; Bachmann, M.; Scheiber, R.; Krieger, G. Definition of ICESat Selection Criteria for Their Use as Height References for TanDEM-X. *IEEE Trans. Geosci. Remote Sens.* **2010**, *48*, 2750–2757. [CrossRef]
16. Wang, R.; Lv, X.; Zhang, L. A Novel Three-Dimensional Block Adjustment Method for Spaceborne InSAR-DEM Based on General Models. *IEEE J. Sel. Top. Appl. Earth Obs. Remote Sens.* **2023**, *16*, 3973–3987. [CrossRef]
17. Wang, R.; Lv, X.; Chai, H.; Zhang, L. A Three-Dimensional Block Adjustment Method for Spaceborne InSAR Based on the Range-Doppler-Phase Model. *Remote Sens.* **2023**, *15*, 1046. [CrossRef]
18. Huber, S.; de Almeida, F.Q.; Villano, M.; Younis, M.; Krieger, G.; Moreira, A. Tandem-L: A Technical Perspective on Future Spaceborne SAR Sensors for Earth Observation. *IEEE Trans. Geosci. Remote Sens.* **2018**, *56*, 4792–4807. [CrossRef]
19. Xiao, R.; Yu, C.; Li, Z.; He, X. Statistical Assessment Metrics for InSAR Atmospheric Correction: Applications to Generic Atmospheric Correction Online Service for InSAR (GACOS) in Eastern China. *Int. J. Appl. Earth Obs. Geoinf.* **2021**, *96*, 102289. [CrossRef]
20. Dou, F.; Lv, X.; Chai, H. Mitigating Atmospheric Effects in InSAR Stacking Based on Ensemble Forecasting with a Numerical Weather Prediction Model. *Remote Sens.* **2021**, *13*, 4670. [CrossRef]
21. Jung, H.-S.; Won, J.-S.; Kim, S.-W. An Improvement of the Performance of Multiple-Aperture SAR Interferometry (MAI). *IEEE Trans. Geosci. Remote Sens.* **2009**, *47*, 2859–2869. [CrossRef]
22. Yang, Y.; Xu, T.; Song, L. Robust Estimation of Variance Components with Application in Global Positioning System Network Adjustment. *J. Surv. Eng.* **2005**, *131*, 107–112. [CrossRef]
23. Gonzalez, J.H.; Bachmann, M.; Krieger, G.; Fiedler, H. Development of the TanDEM-X Calibration Concept: Analysis of Systematic Errors. *IEEE Trans. Geosci. Remote Sens.* **2010**, *48*, 716–726. [CrossRef]
24. Liu, Z.; Fu, H.; Zhu, J.; Zhou, C.; Zuo, T. Using Dual-Polarization Interferograms to Correct Atmospheric Effects for InSAR Topographic Mapping. *Remote Sens.* **2018**, *10*, 1310. [CrossRef]
25. Fu, H.Q.; Zhu, J.J.; Wang, C.C.; Zhao, R.; Xie, Q.H. Atmospheric Effect Correction for InSAR With Wavelet Decomposition-Based Correlation Analysis Between Multipolarization Interferograms. *IEEE Trans. Geosci. Remote Sens.* **2018**, *56*, 5614–5625. [CrossRef]
26. Hanssen, R.F. *Radar Interferometry: Data Interpretation and Error Analysis*; Kluwer: Norwell, MA, USA, 2001.
27. Abrams, M.; Crippen, R.; Fujisada, H. ASTER Global Digital Elevation Model (GDEM) and ASTER Global Water Body Dataset (ASTWBD). *Remote Sens.* **2020**, *12*, 1156. [CrossRef]

28. Shirzaei, M.; Walter, T.R. Estimating the Effect of Satellite Orbital Error Using Wavelet-Based Robust Regression Applied to InSAR Deformation Data. *IEEE Trans. Geosci. Remote Sens.* **2011**, *49*, 4600–4605. [CrossRef]
29. Liao, H.; Meyer, F.J.; Scheuchl, B.; Mouginot, J.; Joughin, I.; Rignot, E. Ionospheric Correction of InSAR Data for Accurate Ice Velocity Measurement at Polar Regions. *Remote Sens. Environ.* **2018**, *209*, 166–180. [CrossRef]
30. Rosenqvist, A.; Shimada, M.; Ito, N.; Watanabe, M. ALOS PALSAR: A Pathfinder Mission for Global-Scale Monitoring of the Environment. *IEEE Trans. Geosci. Remote Sens.* **2007**, *45*, 3307–3316. [CrossRef]
31. Li, Y.; Fu, H.; Zhu, J.; Wu, K.; Yang, P.; Wang, L.; Gao, S. A Method for SRTM DEM Elevation Error Correction in Forested Areas Using ICESat-2 Data and Vegetation Classification Data. *Remote Sens.* **2022**, *14*, 3380. [CrossRef]
32. Neuenschwander, A.; Guenther, E.; White, J.C.; Duncanson, L.; Montesano, P. Validation of ICESat-2 Terrain and Canopy Heights in Boreal Forests. *Remote Sens. Environ.* **2020**, *251*, 112110. [CrossRef]

remote sensing

Article

Deep Learning-Based Landslide Recognition Incorporating Deformation Characteristics

Zhihai Li [1,2], Anchi Shi [1,2], Xinran Li [3], Jie Dou [4], Sijia Li [1,2], Tingxuan Chen [1,2] and Tao Chen [3,*]

1 Power China Huadong Engineering Corporation Limited, Hangzhou 311122, China; li_zh5@hdec.com (Z.L.); shi_ac@hdec.com (A.S.); li_sj2@hdec.com (S.L.); chen_tx@hdec.com (T.C.)
2 Zhejiang Huadong Geotechnical Investigation & Design Institute Corporation Limited, Hangzhou 310004, China
3 School of Geophysics and Geomatics, China University of Geosciences, Wuhan 430074, China; xinranli@cug.edu.cn
4 Badong National Observation and Research Station of Geohazards, China University of Geosciences, Wuhan 430074, China; doujie@cug.edu.cn
* Correspondence: taochen@cug.edu.cn

Abstract: Landslide disasters pose a significant threat, with their highly destructive nature underscoring the critical importance of timely and accurate recognition for effective early warning systems and emergency response efforts. In recent years, substantial advancements have been made in the realm of landslide recognition (LR) based on remote sensing data, leveraging deep learning techniques. However, the intricate and varied environments in which landslides occur often present challenges in detecting subtle changes, especially when relying solely on optical remote sensing images. InSAR (Interferometric Synthetic Aperture Radar) technology emerges as a valuable tool for LR, providing more detailed ground deformation data and enhancing the theoretical foundation. To harness the slow deformation characteristics of landslides, we developed the FCADenseNet model. This model is designed to learn features and patterns within ground deformation data, with a specific focus on improving LR. A noteworthy aspect of our model is the integration of an attention mechanism, which considers various monitoring factors. This holistic approach enables the comprehensive detection of landslide disasters across entire watersheds, providing valuable information on landslide hazards. Our experimental results demonstrate the effectiveness of the FCADenseNet model, with an F1-score of 0.7611, which is 9.53% higher than that of FC_DenseNet. This study substantiates the feasibility and efficacy of combining InSAR with deep learning methods for LR. The insights gained from this research contribute to the advancement of regional landslide geological hazard monitoring, identification, and prevention strategies.

Keywords: InSAR; surface deformation rate; deep learning; landslide recognition

Citation: Li, Z.; Shi, A.; Li, X.; Dou, J.; Li, S.; Chen, T.; Chen, T. Deep Learning-Based Landslide Recognition Incorporating Deformation Characteristics. *Remote Sens.* **2024**, *16*, 992. https://doi.org/10.3390/rs16060992

Academic Editor: Francesca Ardizzone

Received: 10 January 2024
Revised: 27 February 2024
Accepted: 8 March 2024
Published: 12 March 2024

1. Introduction

Landslide is a geological disaster characterized by sudden, rapid, and irreversible movement. Typically occurring in terrain such as mountain slopes, river banks, or geotechnical slopes, landslides involve the downward sliding of soil and rocks due to the influence of gravity or other factors. Given China's mountainous landscape and distinctive energy gradient, coupled with human engineering activities, the country is prone to frequent geological disasters, including landslides [1]. In 2022, China experienced a total of 5659 geological disasters, with landslide geological disasters accounting for a significant 70%, totaling 3919 incidents. These events resulted in substantial life and economic losses [2]. Landslides, influenced by external factors such as intense or sustained precipitation, earthquakes, and human activities, as well as internal factors such as loose rock and soil structures and geological structural damage, are marked by their strong destructiveness and high risk [3]. Landslides pose a significant threat to the lives of society and individuals. Such disasters often lead to road destruction and disruptions in traffic, hindering investigators' swift access to the disaster site for immediate assessments. Therefore, the rapid and timely acquisition of landslide disaster information and the establishment of landslide inventory maps are crucial for effective subsequent responses to landslide risks [4,5].

In recent years, extensive studies have focused on landslide recognition (LR). The initial methods for LR involved manual visual interpretation. This method usually requires interpreters to have rich experiences, and is subjectively influenced and time-consuming, making it difficult to meet the needs of processing a large amount of information [6]. Subsequently, computer-aided methods were introduced. However, due to the increasingly complex spectral features, the classification efficacy remained unsatisfactory. As computer algorithms continued to evolve, researchers progressively integrated mainstream machine learning algorithms into LR, encompassing artificial neural networks [7], support vector machines [8], random forests [9], and decision trees [10]. Despite enhancing performance, these algorithms encountered difficulties in processing extensive remote sensing (RS) data and extracting deep-dimensional information from various image datasets. Deep learning algorithms (DLAs) can automatically extract meaningful feature representations from images [11]. Through the construction of deep neural networks, DLAs can acquire sophisticated and abstract features from raw data. Research on LR based on DLAs can be broadly categorized into three categories: classification, semantic segmentation, and object detection. Mainstream models based on CNNs (Convolutional Neural Networks), such as VGGNet [12], ResNet [13], DenseNet [14], have achieved notable success in remote sensing landslide recognition (RSLR). While object segmentation methods are effective, the accumulation of errors stemming from their reliance on predefined object features has prompted the rapid development of pixel-based methods. These pixel-based approaches prioritize labeling each pixel, thus capturing fine-grained information.

The semantic segmentation network initiated with a Fully Convolutional Network (FCN) replaces the traditional fully connected layer with a convolutional layer for the first time, and achieves end-to-end pixel-level classification [15]. Moreover, the semantic segmentation encoding–decoding architecture, as exemplified by U-Net, lays the groundwork for subsequent improvements [16]. For instance, Lei et al. used FCN and U-Net frameworks to extract landslide features, combining them with the Pyramid Pooling module to consider the mutual constraints between convolutional receptive fields and contextual information, thereby improving landslide mapping [17]. Liu et al. introduced a reconstruction U-Net model using hierarchical feature extraction for automatic extraction of post-earthquake landslides, surpassing traditional U-Net models by 13.8% [18]. Soares et al. delved into the effectiveness of a fully convolutional deep learning model based on U-Net in mountain landslide inventory maps, conducting experiments with patch pixels of varying sizes [19]. Qi et al. innovatively utilized residual learning blocks to replace each convolutional layer of the U-Net baseline model's encoding path. By integrating environmental information, they generated disaster maps, effectively circumventing model degradation resulting from limited landslide samples training [20]. The continual refinement and exploration of these methods contribute to the provision of more accurate and efficient solutions for LR.

Landslide recognition, as a pivotal task in geological disaster monitoring, plays a crucial role for early warning and disaster prevention. Currently, the main method is to combine optical remote sensing images and influencing factors. This combination, considering the spectral and spatial characteristics of landslide, is employed to monitor landslide-prone areas and effectively identify them [21]. However, the environments where landslides occur are often complex, marked by significant variations in landforms, vegetation, and land use types. Furthermore, adverse weather conditions, particularly cloud cover, can impede the quality and availability of optical RS images, rendering it challenging to capture changes on the surface.

To overcome these limitations, researchers have turned their attention to other RS data and technologies. InSAR (Interferometric Synthetic Aperture Radar) stands out as a RS technology based on Synthetic Aperture Radar data. By analyzing the phase difference of radar signals, InSAR observes subtle changes in surface deformation, reflecting alterations in underground geological structure and surface motion. This approach offers crucial insights into landslide activity and proves to be a potent and effective technical tool. Currently, applications like D-InSAR (Differential Interferometric Synthetic Aperture Radar) [22], PS-InSAR (Persistent Scatterer Interferometric Synthetic Aperture Radar) [23], SBAS-InSAR (Small Baseline Subset Interferometric Synthetic Aperture Radar) [24], developed based on InSAR methods, are widely used. In terms of geological disaster research, InSAR technology is mainly used for long-term deformation monitoring and characterization. Liu extracted two-dimensional deformation features of landslides using LOS/ALSAR-1 and ALOS/ALSAR-2 images under a new offset tracking method [25]. Dai used SBAS InSAR to monitor the surface deformation area of Zhouqu County, analyzed its deformation characteristics and triggering reasons, and identified 23 active landslide deformation characteristics [26]. Lattari combined with long short-term memory to model InSAR time series and achieve continuous monitoring of landslides [27]. In addition, Cai proposed an integrated algorithm for landslide multi-source displacement optimization estimation based on the Kalman filter, which integrates displacement observation results from multiple platforms into a unified time series to achieve high temporal resolution monitoring of landslide motion [28]. Liu used the DS-InSAR method to monitor and analyze the Woda landslide area, and the density of monitoring points in the ascending and descending orbit data increased by 25.1% and 22.9%, respectively, providing more accurate monitoring of landslide deformation [29].

Landslide recognition requires the support from multiple sources of data and methods. The comprehensive utilization of multiple data sources and technical means can significantly improve the landslide identification ability and reliability of results [30]. The factor systems commonly utilized in numerous studies primarily focus on external triggering factors and internal geological factors. However, these factor systems may not adequately capture the intricate evolution and deformation processes of landslides themselves. InSAR technology can capture small surface deformation information, and deep learning methods can obtain the intrinsic feature representation of landslides. Therefore, this study combines InSAR technology and deep learning methods, utilizing the surface deformation information obtained by InSAR technology as an internal indicator for landslide analysis. Combining RS images and geological factor data, regional landslide sample data are established to explore the coupling effect between it and other breeding factors; Based on deep learning methods, a dense connected semantic segmentation network based on full convolution is proposed, and attention mechanism is introduced to integrate various monitoring factors for landslide recognition and analysis. This comprehensive method can fully utilize the advantages of different data sources and technical means, provide comprehensive landslide analysis and evaluation, and provide strong support for landslide prevention and disaster management.

2. Study Area and Data

2.1. Study Area

The Baihetan Reservoir Area (BRA) is located in the border area between Sichuan Province and Yunnan Province in China, adjacent to five counties (districts): Huidong County, Ningnan County, Qiaojia County, Huize County, and Dongchuan District (Figure 1). This region lies between the Wumeng Mountains and the Lunan Mountains within the Hengduan Mountains, positioned on the southeastern edge of the Qinghai Tibet Plateau. It falls within the high mountain and plateau geomorphic units of southwestern Sichuan and northeastern Yunnan. The terrain exhibits significant variations, marked by complexity, with most mountain peaks surpassing an elevation of 2000 m. The BRA is situated in the Jinsha River Basin, encompassing a broad distribution of water systems, including tributaries such as Xiaojiang River and Pudu River, along with well-developed gullies. This area experiences a typical subtropical humid and alpine climate due to its undulating terrain, high mountains, deep valleys, and abundant precipitation. The combination of these factors makes it susceptible to elevated river water levels, impacting steep terrain and unstable slope. The dynamic geological environment and intricate water system contribute to the BRA being a high-risk area for landslide disasters [31]. Consequently, effective identification and analysis of landslides in this area are paramount for proactive prevention and response measures.

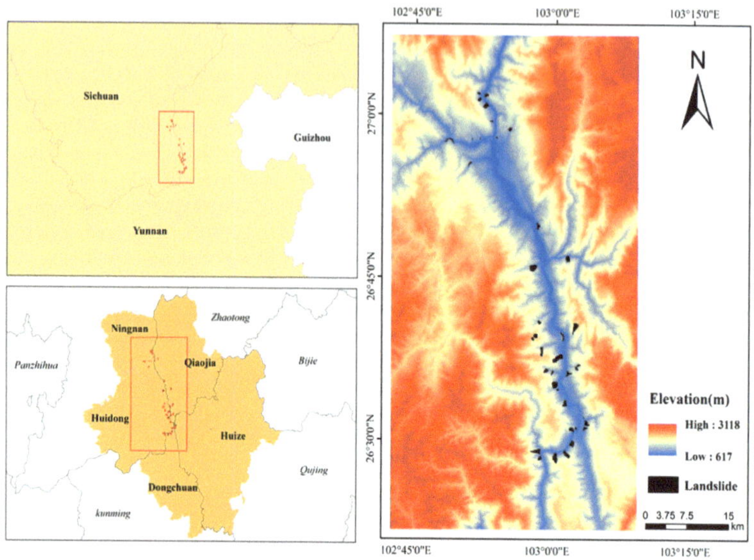

Figure 1. Location of the study area.

2.2. Data and Preprocessing

In order to detect landslides in the BRA, this study mainly collected Sentinel-1 images to offer surface deformation information, Sentinel-2 images to provide surface feature details. Digital elevation model (DEM) and regional geological maps, as well as derived factors were incorporated to furnish geological structure information. Table 1 outlines the main data sources and information utilized in this study.

Table 1. Data information.

Data	Information	Number/Cloud Coverage	Resolution	Data Source
Sentinel-1	9 January 2021–29 May 2023	103 scenes	5 m × 20 m	European Space Agency
Sentinel-2	20230529T034541_T48RTR	0.7637%	10 m	
	20230529T034541_T48RUR	1.2485%		
	20230521T033539_T48RTQ	0.0565%		
	20230521T033539_T48RUQ	0.4099%		
DEM	/		12.5 m	Alaska Satellite Facility
Lithology	/		/	Geologic Map

The deformation data in this study are derived from Sentinel-1 image provided by the European Space Agency. Sentinel-1A possesses all-weather observation capabilities, making it widely used in surface monitoring [32]. SBAS-InSAR, employed in this study, is an InSAR time series analysis technique based on multiple main images [33]. Its principle involves calculating the time and spatial baselines of multi-scene SAR images in different time periods within the coverage area. Subsequently, differential interference processing and phase unwrapping are performed on the selected interference pairs. A series of subsets are generated based on the time and spatial baselines, and the least squares method is used to estimate deformation parameters. The final step involves inversion calculation and geocoding to obtain deformation results. To ensure the data source base for SBAS-InSAR technology processing and the accuracy of obtaining surface deformation information in the study area, 103 Sentinel-1A images from 9 January 2021 to 29 May 2023, were selected. The imaging mode was interferometric wideband imaging (IW) with a resolution of 5 m × 20 m, a central incidence angle of 39.17° and VV single polarization mode. In the SBAS-InSAR method, the threshold for time baseline and spatial baseline is 60 days and 0–2%, respectively. With these limitations, a total of 310 interferograms are generated from the data, revealing a deformation rate in the study area ranging from −136.52 to 44.14 mm/year, as shown in Figure 2a. Based on the surface deformation results obtained from the above process, we selected the areas with obvious deformation and combined optical and Google Earth images for verification, and delineated landslide samples.

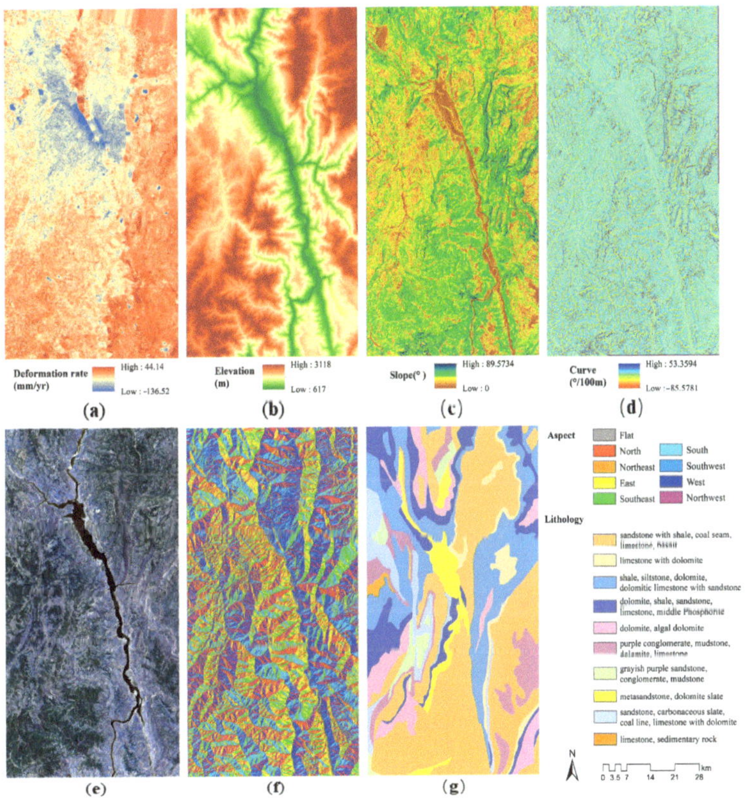

Figure 2. Landslide causative factors: (**a**) deformation rate, (**b**) elevation, (**c**) slope, (**d**) curvature, (**e**)optical image, (**f**) aspect, and (**g**) lithology.

The formation of landslides is a complex outcome involving the interaction of multiple factors. Exploring the relationship between internal and external factors and the occurrence of landslides enhances the accuracy of LR [34]. Therefore, optical images and geological factors were also chosen as landslide causative factors (LCFs), conducting LR together with deformation rate.

Sentinel-2 images are also sourced from the European Space Agency, encompassing a total of 4 scenes across the entire study area, with imaging dates of 21 May 2023 and 29 May 2023. Following preprocessing steps such as radiometric calibration and atmospheric correction, bands 2, 3, 4, and 8 with a resolution of 10 m were selected for band fusion, cropping, and embedding to enhance interpretability. Additionally, the DEM used in this study was obtained from the Alaska Satellite Facility, featuring a spatial resolution of 12.5 m. Given the close relationship between the occurrence of landslide geological disasters and LCFs, this study selected DEM-derived factors such as slope, aspect, curvature, and geological lithology maps as LCFs for LR. All LCFs were underwent processing based on ENVI5.3, SARscape platform and ArcGIS10.7, and the results were visualized using ArcGIS10.7 (Figure 2).

To ensure pixel position matching, the spatial resolution of all factors was resampled to 12.5 m. To standardize the data distribution across different dimensions and expedite the convergence speed of the model, the Z-score method was used to normalize each channel of input data [35]. Given the relatively small proportion of landslide samples, a data augmentation strategy was adopted to generate more training samples. This experiment utilized a sliding window size of 128 × 128 to scan and generate non-overlapping sample blocks. Divide landslide and non-landslide samples at a ratio of 7:3, and then combine 70% of the selected samples into the training dataset to learn landslide characteristics.

3. Method

This study is organized into three main components: data collection and processing, model training, predictive evaluation (Figure 3). The first part is data collection and processing, mainly including radar satellite data collection and processing in the research area, image and deformation data acquisition, sample annotation, and geological factor extraction. The second part is the model construction, a new FCADenseNet model is established based on semantic segmentation, incorporating attention modules. Additionally, three commonly used semantic segmentation networks are selected for comparison. The third part involves predictive evaluation, wherein the trained model is employed to predict the entire research area. Five evaluation metrics (EMs) for assessment of the performance of the LR models, and the LR map are generated based on the predictions.

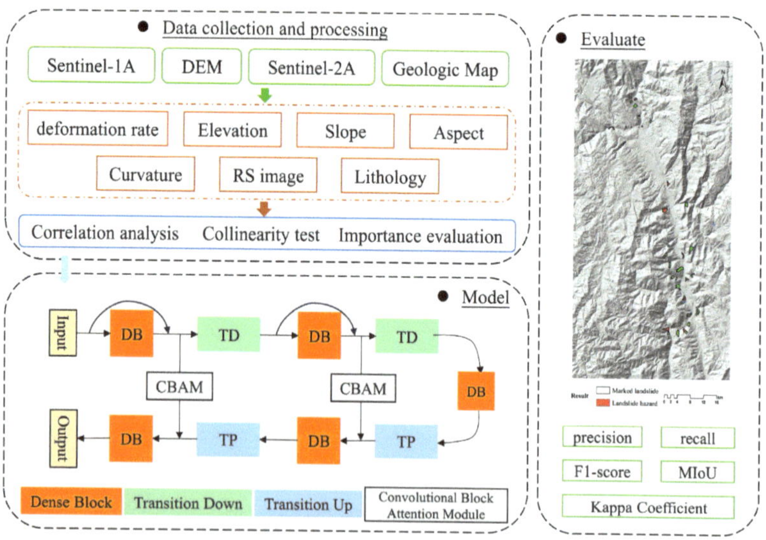

Figure 3. Flowchart of this study.

3.1. Selection of LCFs

The selection of LCFs has an important impact on the overall performance of the model. When factors exhibit high correlation, indicating a strong linear or non-linear relationship between them, it not only burdens the model with excessive data but also interferes with the weight allocation of different features during the feature learning process [36].

Consequently, this study conducted correlation analysis, collinearity test, and importance evaluation on the LCFs.

(1) Correlation analysis: Pearson correlation coefficient, also known as Pearson product moment correlation coefficient (PPMCC), serves as a metric for quantifying the linear correlation between two variables, denoted as X and Y. The correlation values range from -1 and 1, indicating the strength and direction of the relationship between the variables [37]. This coefficient is determined by evaluating the covariance and standard deviation of the two sets of data, as illustrated in Formula (1):

$$PPMCC = \frac{cov(X,Y)}{\sigma_X \sigma_Y} = \frac{\sum_{i=1}^{n}\left(X_i - \overline{X}\right)\left(Y_i - \overline{Y}\right)}{\sqrt{\sum_{i=1}^{n}\left(X_i - \overline{X}\right)^2}\sqrt{\sum_{i=1}^{n}\left(Y_i - \overline{Y}\right)^2}} \quad (1)$$

where n is the total number of samples, and $cov(X,Y)$ represents the covariance between variables X and Y. Additionally, σ_X, σ_Y signify the standard deviation of all samples. $\overline{X}$ and $\overline{Y}$ denote the average values of all samples.

(2) Collinearity test: Multicollinearity refers to the linear correlation among independent variables [38]. It involves examining tolerance (TOL) and variance inflation factor (VIT) between selected factors, which exhibit a reciprocal relationship. The calculation method of VIF is shown in Formula (2). The closer the VIF value is to 1, the weaker the correlation among the data, while a higher value indicates a more pronounced correlation.

$$VIF = \frac{1}{1 - R_i^2} \quad (2)$$

where R_i^2 represents the i-th variable and obtains the corresponding decision coefficient in the linear regression model.

(3) Importance evaluation: Factors have different contributions to the performance of the model, and the importance of LCFs is ranked using the Gini Index in the random forest model [39]. If a feature is frequently selected for node partitioning in the construction of random forests and can improve model performance after partitioning, then this feature is likely to be an important feature. Conversely, if a feature is rarely selected in the construction of a random forest or has little impact on model performance after partitioning, then this feature is likely to be an unimportant feature. By employing this approach, we can rank the importance of features.

3.2. FCADenseNet

The model initially incorporates fundamental concepts from DenseNet and ResNet, establishing connections between feature maps (FMs) at different levels through dense connections. This design enables each layer to directly access information from all previous layers. Simultaneously, the incorporation of residual connections aims to address the gradient dispersion problem that may arise with deepening the network. The architecture of the FCN encoder–decoder is subsequently retained, transforming the conventional DenseNet into a model well suited for image segmentation tasks. This adaptation allows for the generation of pixel-level segmentation predictions, effectively fusing spatial characteristics [40].

However, the dense feature reuse and connections may result in the loss of original information. Moreover, the smoothing and interpolation operations during the up-sampling process can cause the blurring of edge information. Although the model's emphasis on learning deep semantic information, it is crucial to preserve detailed information. Consequently, an attention module is introduced to both the decoding and encoder skip connection sections, with the weight adjusted to learn without unduly increasing the model's complexity. This strategic enhancement aims to better leverage low-level fine information and high-level semantic features. Figure 4 shows the overall model structure.

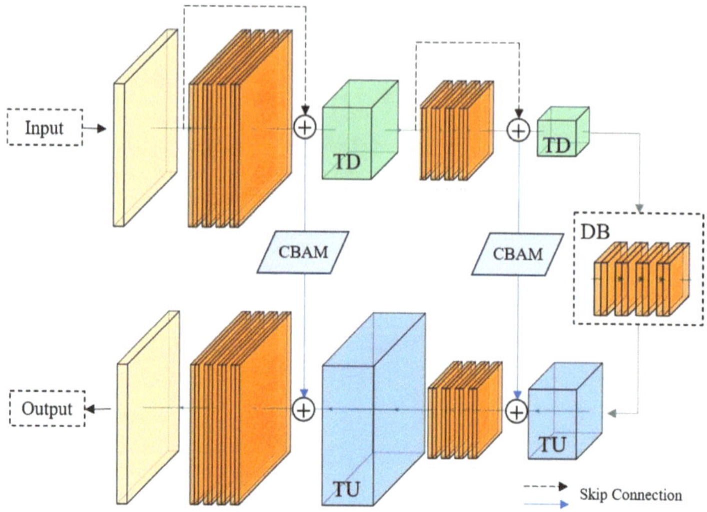

Figure 4. FCADenseNet network architecture.

The FCADenseNet model mainly consists of four modules: Dense Block (DB), Transition Down (TD), Transition Up (TU), and attention mechanism. In the DB, the FMs of each layer are densely connected to the FMs of all previous layers, to facilitating optimal information transmission (Figure 5). This dense connection method significantly augments the number of connections in the network, allowing each layer to directly access information from all previous layers. Leveraging features from previous layers enhances the expression ability of each layer, resulting in the extraction of richer semantic information. The formula governing this process is as follows:

$$X_L = F_l([X_0, X_1, \ldots, X_{L-1}]) \quad (3)$$

where $[X_0, X_1, \cdots, X_{L-1}]$ refers to the dense connections of the FMs of layers 0,. . ., $L-1$. F_l represents a combination of non-linear transformation functions: three operations are batch normalization (BN), ReLU activation function, and 3×3 Convolution convolutions [41,42].

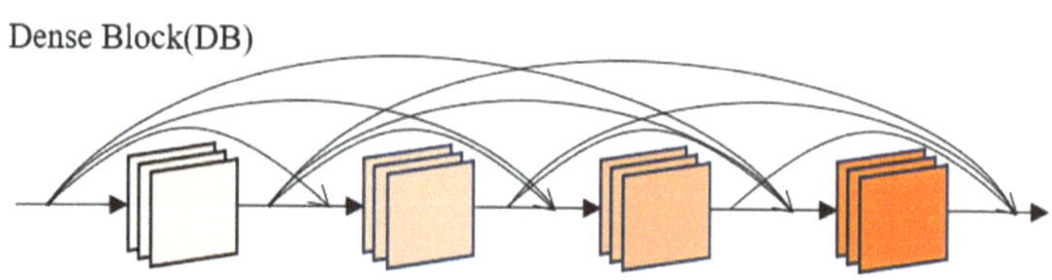

Figure 5. Dense Block block structure.

Following the dense connections, each layer F_l produces k FMs, referred to as growth rates, which are hyperparameters. Generally, opting for a smaller k tends to yield superior performance. To mitigate an upsurge in feature computation due to redundant learning, an internal 1×1 convolutional bottleneck layer is incorporated for feature compression.

The TD module employs a convolution operation akin to the DB module. It is distinguished by reducing the number of channels in the FM through a 1×1 convolution layer and a max-pooling operation, effectively diminishing reducing network parameters. In the decoder, TD and DB modules are arranged alternately, facilitating the extraction of global information from the input data. The decoder part is a combination of TU and DB modules. The TU module uses 3×3 convolutional layers and 2×2 up-samplingup-sampling layers to systematically restore low resolution FMs to their original resolution. The overall structure is shown in Figure 6.

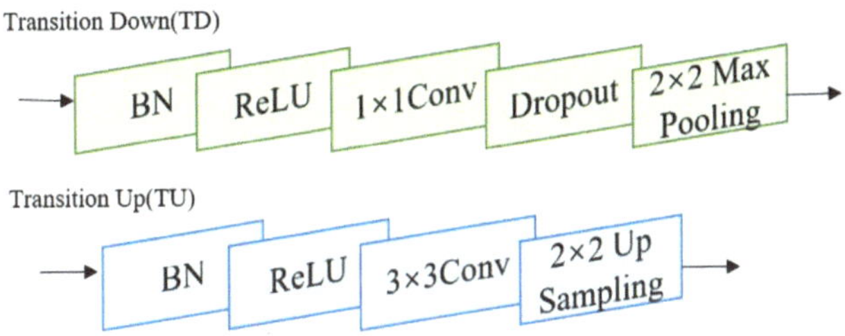

Figure 6. Transition Down, Transition Up structure.

Spectral and spatial features contained in different terrain characteristics in RS data are represented differently in FMs. The recognition ability will be greatly improved by pinpointing significant features of landslide objects in specific bands and spatial dimensions. To optimize the utilization of both low-level fine information and high-level semantic features, the Convolutional Block Attention Module (CBAM) [43] has been introduced to the shallow FMs in the skip connection section after the TU module. It is an efficient module suitable for feedforward neural networks, where the given feature map infers the attention map along two independent dimensions, and then multiplies the attention map with the input feature map for adaptive feature optimization. This is achieved through the learning of weights or attention distribution, facilitating the extraction and accentuation of key landmark features [44]. The structure of this attention module is shown in Figure 7. The channel attention component dynamically adjusts the importance level of channels in the FMs, while the spatial attention part underscores the significance of spatial positions in the FMs, and the pooling operation can generate information feedback for each pixel. By focusing on these key features, the information contained in the shallow FMs is emphasized, effectively mitigating information loss caused by the up-sampling process in the FM and improving overall model performance. Ultimately, the class distribution of each pixel is derived through sigmoid non-linear mapping.

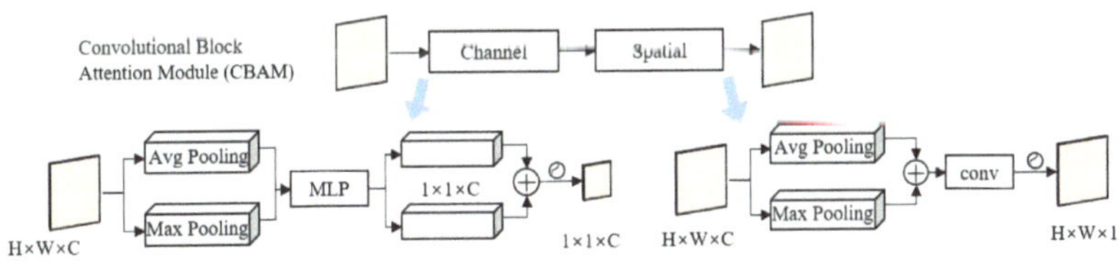

Figure 7. CBAM module structure.

3.3. Evaluation Metrics

Five commonly used evaluation metrics (EMs) were chosen to quantitatively evaluate the performance of the proposed LR method. These metrics include the precision (Pre), recall (Rec), F1-score (F1), Kappa coefficient (Kappa), and mean intersection over union (MIoU). These five EMs are defined as follows.

Pre refers to the proportion of correctly predicted positive samples to the total number of predicted positive samples. Rec refers to the proportion of correctly predicted positive samples to all positive samples. Usually, in order to comprehensively evaluate the performance of classification models, F1 is used to evaluate the model. When *F*1 is high, the model performs well [45]. The calculation formulas of these three EMs are as follows:

$$Pre = \frac{TP}{TP + FP} \quad (4)$$

$$Rec = \frac{TP}{TP + FN} \tag{5}$$

$$F1 = 2 \times \frac{Pre \times Rec}{Pre + Rec} \tag{6}$$

where TP is the number of correctly predicted positive samples, TN is the number of correctly predicted negative samples, FN is the number of predicted negative samples, and FP is the number of predicted positive negative samples.

Kappa coefficient is a method of evaluating consistency, ranging from [−1, 1]. The higher the value of the coefficient, the higher the classification accuracy achieved by the model [46]. Kappa coefficient and overall accuracy are mutually verified, and the results are objectively evaluated. The formula is as follows:

$$kappa = \frac{p_o - p_e}{1 - p_e} \tag{7}$$

$$p_o = \frac{TP + TN}{TP + TN + FP + FN} \tag{8}$$

$$p_e = \frac{(TP + FN)(TP + FP) + (FP + TN)(FN + TN)}{n^2} \tag{9}$$

MIoU calculates the ratio of the intersection and union of pixels in each category, and averages the results of all categories to obtain the final MIoU value [47]. When the true and predicted values are the same, the MIoU value is 1, indicating a perfect match in an ideal situation. The formula is as follows:

$$MIoU = \frac{1}{2} \times \left(\frac{TP}{TP + FN + FP} + \frac{TN}{TN + FN + FP}\right) \tag{10}$$

3.4. Experiment Setting

The construction of the experimental platform mainly includes three key aspects: hardware configuration, running environment setup, and parameter settings. The experimental code was written using the Python programming language, executed on the Pycharm2021 software, and built upon the Keras framework within TensorFlow. Firstly, the network weights are randomly initialized. The classifier of the network uses the sigmoid function, which is suitable for binary classification problems. The binary cross-entropy function is chosen as the loss function for LR. Using Adam algorithm as the optimization algorithm, the learning rate is 0.0001, the batch size is 8, and the network weights are fine tuned. Additionally, a Dropout layer with a value of 0.2 is incorporated during training to streamline the model and mitigate gradient disappearance. Finally, the ModelCheckpoint function is selected to monitor changes in the loss function and automatically save the model.

4. Results and Analysis

4.1. Analysis of LCFs

In this study, PyCharm 2020 was employed to calculate correlations, VIF, and TOL among the selected factors. Figure 8 shows the PPMCC visualization thermal map of six LCFs in the study area, including deformation rate. Positive and negative values represent the positive and negative correlation between the data, while the depth of the color indicates the strength of these correlations. Specifically, the deformation rate is positively correlated with elevation and slope features, with correlation coefficients of 0.19 and 0.14, respectively. It indicates that with the increase in height and slope in the region, the rate of surface deformation also rises. Relatively high elevations and steep slopes have a certain correlation with the occurrence of landslide events, which is reflected in surface deformation. Consequently, selecting elevation and slope features as discriminant factors proves beneficial for LR.

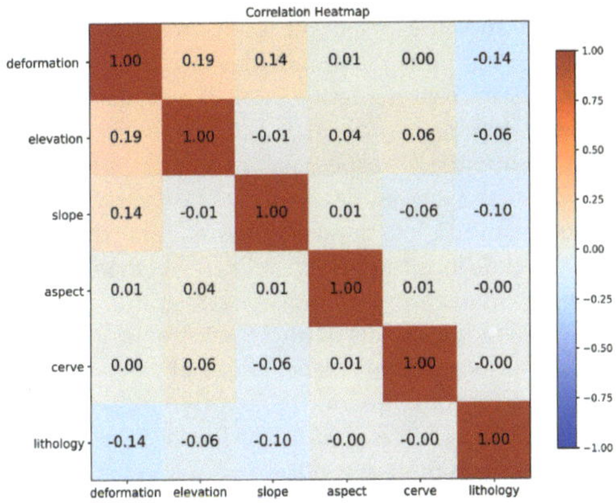

Figure 8. Spearman correlation coefficients of factors.

Table 2 shows the VIF and TOL values for the six LCFs. The results indicate a weak positive correlation among the selected LCFs. The VIF and TOL values for each LCF approach 1, indicating that the correlation among the selected LCFs in this study is not strong. There is no need for removal through multicollinearity testing, affirming the robustness of the chosen factors in the analysis.

Table 2. Multicollinearity test results of factors.

	Deformation Rate	Elevation	Slope	Aspect	Curve	Lithology
VIF	1.074	1.044	1.031	1.002	1.007	1.028
TOL	0.931	0.957	0.970	0.998	0.993	0.973

The importance ranking distribution of each LCFs in the study area is shown in Figure 9, with the highest importance value for surface deformation rate, followed by elevation and curvature. A high deformation rate indicates the acceleration or rapid subsidence of geological bodies, and a high probability indicates the occurrence of landslides; Elevation and curvature are also important factors in the occurrence of landslides. Stress concentration at high curvature and steep slopes at high altitudes are more prone to landslides, so their importance is also high. The importance of lithology is the lowest, as there are many kinds of lithology and complex distribution in the region.

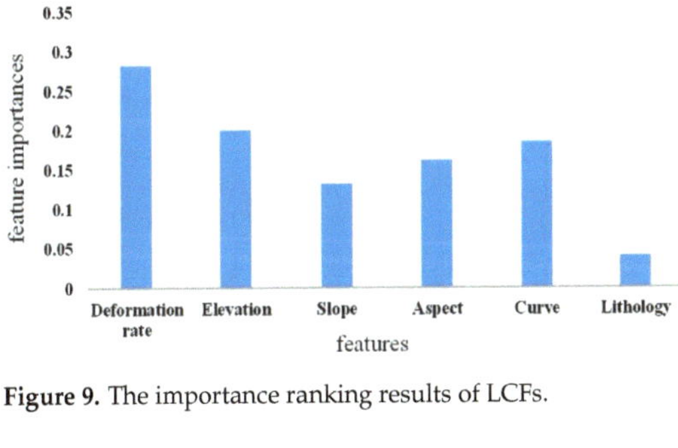

Figure 9. The importance ranking results of LCFs.

4.2. LR Results

In this section, the performance of LR result of the proposed FCADenseNet model are compared with three LR model, which are FCN, U-Net [48], and FC_DenseNet [40] network.

Since the landslide area in the study area occupies a small area, we chose to zoom in on the local area to view the model identification results. As shown in Figures 10 and 11, they belong to two representative areas of medium landslide (most landslides) and small landslide development in the region, respectively. According to the results, for medium landslide, Figure 10a,b, respectively, show the results of the FCN and U-Net models respectively. The FCN model seems to be overly regularized in identifying landslide boundaries, resulting in a noticeable rectangular shape. The U-Net model demonstrates stronger land-slide localization capabilities. But there are some difficulties in distinguishing similar backgrounds around landslides, resulting in incorrect predictions. Figure 10c,d, respectively, show the results of the FC_DenseNet and FCADenseNet models respectively. Considering the localization of landslides and the recovery of boundary features, the results are more accurate compared to the first two models, but there is still some noise in the results of the FC_DenseNet model. The landslide range identified by FCADenseNet in the four model results is more complete, and due to the introduction of attention, it effectively recovers the loss of edge information during the up-sampling process. It is worth noting that for the recognition results of small landslides, the range of landslides annotated based on prior knowledge is very small, and there are obvious differences between the results of models. Figure 11a,b show the identification results of FCN and U-Net models, respectively. FCN and U-Net models are difficult to identify such small landslides. Figure 11c,d show the results of the FC_DenseNet and FCADenseNet models, both of which accurately identify the range of small landslides. In comparison, FCADenseNet has fewer errors in identifying the surrounding areas.

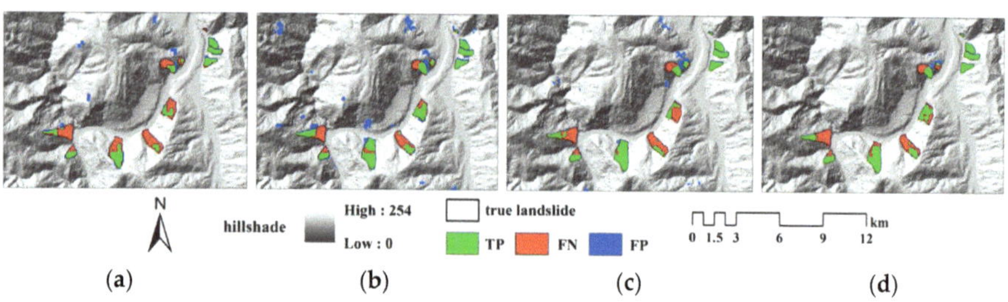

Figure 10. Recognition results of local medium landslide: (**a**) FCN, (**b**) U-Net, (**c**) FC_DenseNet, and (**d**) FCADenseNet.

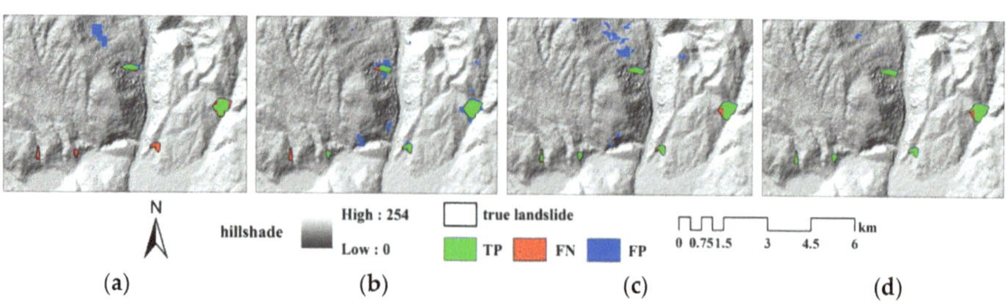

Figure 11. Recognition results of local small landslide: (**a**) FCN, (**b**) U-Net, (**c**) FC_DenseNet, and (**d**) FCADenseNet.

Furthermore, based on the visual results, we quantitatively analyzed the performance of the model using the five EMs. In the experiment, we opted not to use Overall Accuracy (OA) as one EM because the substantial imbalance between non-landslide and landslide pixels in the study area, with a ratio of 224:1. The proportion of non-landslide pixels is very high, which leads to a small change between the OA values of the model. From Table 3, the five EMs of FCADenseNet are superior to the other three models, achieving commendable LR results. The FCADenseNet model, augmented with an attention module, achieves the Pre of 0.8381, the Rec of 0 6972, the F1 of 0.7611, the kappa of 0.7602, and the MIoU of 0.8062. It is noteworthy that the results of the proposed FCADenseNet obtained the highest values of the five EMs, except for Rec. Compared to the FC_DenseNet model, the F1, Kappa, and MIoU demonstrate improvements of 14.3%, 15.1%, and 7.8%, respectively. This enhancement stems from our introduced method, adjusting the attention weight of the model to preserve shallow information on landslide boundary features, effectively recovering the loss of edge information during the up-sampling process. This adjustment is the main reason for improving model accuracy.

Table 3. The EMs of different models (the bold and underlined value means the highest EMs of the col).

	Pre	Rec	F1	Kappa	MIoU
FCN	0.6870	0.5354	0.6018	0.6002	0.7136
U-Net	0.4653	0.6908	0.5561	0.5543	0.6901
FC_DenseNet	0.6312	**0.7043**	0.6658	0.6599	0.7479
FCADenseNet	**0.8381**	0.6972	**0.7611**	**0.7602**	**0.8062**

4.3. Improvement of FCADenseNet Performance by Deformation Rate

To verify the impact of deformation rate on the performance of the model, we compared the LR results using the FCADenseNet model with or without deformation rate. Two sets of training datasets were created: RS (devoid of deformation rate) and RS-DR (incorporating deformation rate). The proposed FCADenseNet model was used to evaluate both datasets.

Similarly, we focused on a local area for a closer examination of the model's LR results. As shown in Figure 12a,b, they represent the LR outcomes of small landslide areas using RS and RS-DR training data, respectively. With sufficient learning from the FCADenseNet model, the recognition of small landslides is relatively accurate, but there is no pronounced increase in error recognition for non-landslide areas. Figure 12c,d represent the recognition results of the local landslide area using RS and RS-DR training data, respectively. The results of this area indicate that the model has some difficulties in identifying the overall range of landslides in this area, but with the addition of deformation rate features, partial positioning can be carried out within the landslide area, and the recognition results have been improved.

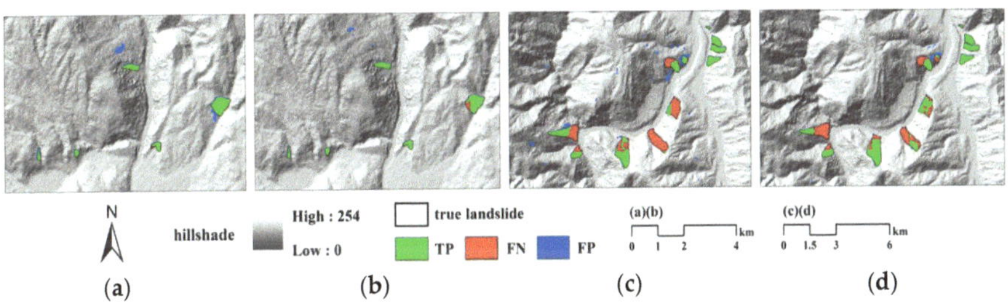

Figure 12. Recognition results: (**a**) local small landslides using RS, (**b**) local small landslides using RS-DR, (**c**) local medium landslide using RS, and (**d**) local medium landslide using RS-DR.

Table 4 presents the quantitative analysis results for the two datasets. When using the RS-DR dataset, the model achieved an F1, a kappa, and a MIoU of 0.7611, 0.7602, and 0.8062, respectively. Compared with the RS dataset, the model exhibited increases in 12.9%, 12.9%, and 7.1% in F1, kappa, and MIoU, respectively, highlighting the practicality of deformation rate in effectively distinguishing landslides. Notably, in the results from both datasets, the Rec value did not exhibit significant changes, while the Pre value demonstrated a notable improvement. Combining the results of Figure 12, it becomes evident that the RS-DR dataset provides a more detailed depiction of the morphology of localized landslide, resulting in clearer LR results. This indicates that adding deformation rate as a landslide feature has a positive effect on LR.

Table 4. The EMs of FCADenseNet for two training data (the bold and underlined value means the highest EMs of the col).

	Pre	Rec	F1	Kappa	MIoU
RS	0.6614	0.6878	0.6743	0.6729	0.7529
RS-DR	**0.8381**	**0.6972**	**0.7611**	**0.7602**	**0.8062**

5. Discussion

5.1. Comparisions of the Model's Effect

Currently, LR methods based on semantic segmentation usually adopt an encoding–decoding structure. In this structure, the stacking of convolutional layers can extract landslide-related information well. However, during the up-sampling process, the gradual recovery of feature maps may lead to the loss of information. To solve this problem, the FCADenseNet proposed in this article adds an attention mechanism module based on the original model and applies it to skip connections. This attention mechanism module focuses on the information contained in the shallow feature map, effectively reducing the information loss that may occur during the sampling process on the feature map. According to comparative experiments with the original FC_DenseNet, U-Net, and FCN networks, the FCADenseNet model has a more complete landslide boundary in terms of landslide recognition effect, and has stronger recognition capabilities for some small landslides. This adjustment effectively maintains the integrity of the original features during the information recovery process and is the main factor in improving model accuracy.

5.2. Influence of Deformation Rate on LR

In the BRA, the water level of the river undergoes constant changes due to factors such as rainfall and reservoir water storage. Additionally, the slope of the reservoir area experiences continuous fluctuations. To account for these variations, we incorporate a time series deformation change value, integrating it into LCFs to train the model for LR. Given that the FCADenseNet model has already demonstrated excellent LR results, we conducted a comparative analysis to assess the impact of including deformation rate features. The results indicate that, when deformation rate is combined, the model can more accurately identify landslide. However, some challenges remain, particularly in achieving precise landslide positioning. There are regional landslides in the results that have not been recognized by all models in this study, but there are in the sample library, which may be due to incorrect labeling of samples based on deformation results and images; and the "false detection" areas in the identification results may have some impact on the quantitative evaluation results because these areas have characteristics of landslide hazards but are not in our original sample library. Addressing this issue requires ongoing attention to surface deformation conditions, with on-site verification being crucial for subsequent research endeavors. In addition, it is difficult to accurately analyze the motion characteristics of some deep landslides solely through InSAR methods. The surface deformation rate characteristics have a certain improvement effect on the recognition of these shallow landslides, but the impact on deep landslides still needs to be explored.

5.3. Limitations and Future Work

This study attempts to combine landslide surface deformation with DLAs for landslide disaster recognition However, practical applications still face challenges in timely LR. Firstly, the orbital data of the Sentinel-1A satellite can be used to provide surface deformation data for LR. However, effectively analyzing orbit and descent data, leveraging richer SAR data resources such as interferograms, and fusing InSAR data into factor data in a more appropriate manner is still an unresolved issue. This is crucial for monitoring surface deformation with higher accuracy and obtaining more precise landslide cataloging data. Secondly, the utilization of DLAs requires a substantial amount of sample data to ensure accuracy, rendering it more suitable for ongoing landslides that require long-term monitoring. Research efforts should focus on establishing dynamic databases that can be updated in real time, thereby enhancing the generalization ability of existing models.

6. Conclusions

In this study, we aimed to validate the feasibility of combining InSAR technology with deep learning in LR. We obtained the annual deformation rate of the BRA as the main factor using SBAS-InSAR technology. This information, combined with the study area image and geological factors, formed a new dataset for LR. We constructed a newly Dense Connected Fully Convolutional Network model, incorporating an attention mechanism, FCADenseNet, for LR. To assess the performance of the model, comparisons were made with FCN, U-Net, and FC_DenseNet networks. The results demonstrate that the proposed FCADenseNet achieves the F1-score of 0.7611 and the kappa of 0.7602, which is more effective in this study and has certain advantages for small landslides. We also analyzed the impact of surface deformation rate on LR. The results indicate that the increase in surface deformation characteristics has a positive impact on LR. In the future, the integration of InSAR technology with the existing system will provide more timely information for the monitoring and rapid identification of landslide disasters. The establishment of a long-term dynamic database will play a proactive role in regional disaster monitoring and recognition.

Author Contributions: Conceptualization, T.C. (Tao Chen) and A.S.; methodology, T.C. (Tao Chen) and X.L.; validation, X.L., S.L. and T.C. (Tingxuan Chen); formal analysis, J.D.; investigation, Z.L. and A.S.; resources, T.C. (Tao Chen); data curation, X.L., S.L. and T.C. (Tingxuan Chen); writing—original draft preparation, Z.L. and X.L.; writing—review and editing, T.C. (Tao Chen), J.D. and A.S.; project administration, Z.L. and A.S. All authors have read and agreed to the published version of the manuscript.

Funding: This research was funded by the Key Science and Technology Plan Project of PowerChina Huadong Engineering Corporation Limited (KY2021–ZD–03).

Data Availability Statement: This study cannot publicly share the dataset due to third-party restrictions.

Conflicts of Interest: The authors Zhihai Li, Anchi Shi, Sijia Li and Tingxuan Chen are employed by China Power Construction Corporation East China Survey and Design Institute Co., Ltd. The remaining authors declare that the research was conducted in the absence of any commercial or financial relationships that could be construed as a potential conflict of interest.

References

1. Casagli, N.; Intrieri, E.; Tofani, V.; Gigli, G.; Raspini, F. Landslide Detection, Monitoring and Prediction with Remote-Sensing Techniques. *Nat. Rev. Earth Environ.* **2023**, *4*, 51–64. [CrossRef]
2. Lissak, C.; Bartsch, A.; De Michele, M.; Gomez, C.; Maquaire, O.; Raucoules, D.; Roulland, T. Remote Sensing for Assessing Landslides and Associated Hazards. *Surv. Geophys.* **2020**, *41*, 1391–1435. [CrossRef]
3. Gao, J.; Sang, Y. Identification and Estimation of Landslide-Debris Flow Disaster Risk in Primary and Middle School Campuses in a Mountainous Area of Southwest China. *Int. J. Disaster Risk Reduct.* **2017**, *25*, 60–71. [CrossRef]
4. Catani, F. Landslide detection by deep learning of non-nadiral and crowdsourced optical images. *Landslides* **2021**, *18*, 1025–1044. [CrossRef]
5. Mohan, A.; Singh, A.K.; Kumar, B.; Dwivedi, R. Review on remote sensing methods for landslide detection using machine and deep learning. *Trans. Emerg. Telecommun. Technol.* **2021**, *32*, e3998. [CrossRef]
6. Du, B.; Zhang, W. Research on object-oriented high resolution remote sensing image classification technology. *West. Resour.* **2016**, *5*, 135–138.
7. Long, L.; He, F.; Liu, H. The use of remote sensing satellite using deep learning in emergency monitoring of high-level landslides disaster in Jinsha River. *J. Supercomput.* **2021**, *77*, 8728–8744. [CrossRef]

8. Das, R.; Wegmann, K.W. Evaluation of Machine Learning-Based Algorithms for Landslide Detection across Satellite Sensors for the 2019 Cyclone Idai Event, Chimanimani District, Zimbabwe. *Landslides* **2022**, *19*, 2965–2981. [CrossRef]
9. Chen, F.; Yu, B.; Li, B. A Practical Trial of Landslide Detection from Single-Temporal Landsat8 Images Using Contour-Based Proposals and Random Forest: A Case Study of National Nepal. *Landslides* **2018**, *15*, 453–464. [CrossRef]
10. Krawczyk, B.; Woźniak, M.; Schaefer, G. Cost-Sensitive Decision Tree Ensembles for Effective Imbalanced Classification. *Appl. Soft Comput.* **2014**, *14*, 554–562. [CrossRef]
11. Ding, A.; Zhang, Q.; Zhou, X.; Dai, B. Automatic recognition of landslide based on CNN and texture change detection. In Proceedings of the 2016 31st Youth Academic Annual Conference of Chinese Association of Automation (YAC), Wuhan, China, 11–13 November 2016; pp. 444–448.
12. Simonyan, K.; Zisserman, A. Very Deep Convolutional Networks for Large-Scale Image Recognition. In Proceedings of the 2015 IEEE Conference on Computer Vision and Pattern Recognition (CVPR), Boston, MA, USA, 7–12 June 2015; pp. 509–518.
13. He, K.; Zhang, X.; Ren, S.; Sun, J. Deep Residual Learning for Image Recognition. In Proceedings of the 2016 IEEE Conference on Computer Vision and Pattern Recognition (CVPR), Las Vegas, NV, USA, 27–30 June 2016; pp. 770–778.
14. Chen, T.; Gao, X.; Liu, G.; Wang, C.; Zhao, Z.; Dou, J.; Niu, R.Q.; Plaza, A. BisDeNet: A New Lightweight Deep Learning-based Framework for Efficient Landslide Detection. *IEEE J. Sel. Top. Appl. Earth Obs. Remote Sens.* **2024**, *17*, 3648–3663. [CrossRef]
15. Shelhamer, E.; Long, J.; Darrell, T. Fully Convolutional Networks for Semantic Segmentation. *IEEE Trans. Pattern Anal. Mach. Intell.* **2017**, *39*, 640–651. [CrossRef] [PubMed]
16. Ronneberger, O.; Fischer, P.; Brox, T. U-Net: Convolutional Networks for Biomedical Image Segmentation. In *Medical Image Computing and Computer-Assisted Intervention—MICCAI 2015*; Navab, N., Hornegger, J., Wells, W.M., Frangi, A.F., Eds.; Lecture Notes in Computer Science; Springer International Publishing: Cham, Switzerland, 2015; Volume 9351, pp. 234–241.
17. Lei, T.; Zhang, Y.; Lv, Z.; Li, S.; Liu, S.; Nandi, A.K. Landslide Inventory Mapping from Bitemporal Images Using Deep Convolutional Neural Networks. *IEEE Geosci. Remote Sens. Lett.* **2019**, *16*, 982–986. [CrossRef]
18. Liu, P.; Wei, Y.; Wang, Q.; Chen, Y.; Xie, J. Research on Post-Earthquake Landslide Extraction Algorithm Based on Improved U-Net Model. *Remote Sens.* **2020**, *12*, 894. [CrossRef]
19. Soares, L.P.; Dias, H.C.; Grohmann, C.H. Landslide Segmentation with U-Net: Evaluating Different Sampling Methods and Patch Sizes. *arXiv* **2020**, arXiv:2007.06672.
20. Qi, W.; Wei, M.; Yang, W.; Xu, C.; Ma, C. Automatic Mapping of Landslides by the ResU-Net. *Remote Sens.* **2020**, *12*, 2487. [CrossRef]
21. Gorsevski, P.V.; Brown, M.K.; Panter, K.; Onasch, C.M.; Simic, A.; Snyder, J. Landslide Detection and Susceptibility Mapping Using LiDAR and an Artificial Neural Network Approach: A Case Study in the Cuyahoga Valley National Park, Ohio. *Landslides* **2016**, *13*, 467–484. [CrossRef]
22. Peduto, D.; Santoro, M.; Aceto, L.; Borrelli, L.; Gullà, G. Full Integration of Geomorphological, Geotechnical, A-DInSAR and Damage Data for Detailed Geometric-Kinematic Features of a Slow-Moving Landslide in Urban Area. *Landslides* **2021**, *18*, 807–825. [CrossRef]
23. Rosi, A.; Tofani, V.; Tanteri, L.; Stefanelli, C.T.; Agostini, A.; Catani, F.; Casagli, N. The new landslide inventory of Tuscany (Italy) updated with PS-InSAR: Geomorphological features and landslide distribution. *Landslides* **2018**, *15*, 5–19. [CrossRef]
24. Zhao, F.; Meng, X.; Zhang, Y.; Chen, G.; Su, X.; Yue, D. Landslide susceptibility mapping of karakorum highway combined with the application of SBAS-InSAR technology. *Sensors* **2019**, *19*, 2685. [CrossRef] [PubMed]
25. Liu, X.; Zhao, C.; Zhang, Q.; Lu, Z.; Li, Z. Deformation of the Baige Landslide, Tibet, China, Revealed Through the Integration of Cross-Platform ALOS/PALSAR-1 and ALOS/PALSAR-2 SAR Observations. *Geophys. Res. Lett.* **2020**, *47*, e2019GL086142. [CrossRef]
26. Dai, C.; Li, W.; Wang, D.; Lu, H.; Xu, Q.; Jian, J. Active Landslide Detection Based on Sentinel-1 Data and InSAR Technology in Zhouqu County, Gansu Province, Northwest China. *J. Earth Sci.* **2021**, *32*, 1092–1103. [CrossRef]
27. Lattari, F.; Rucci, A.; Matteucci, M. A Deep Learning Approach for Change Points Detection in InSAR Time Series. *IEEE Trans. Geosci. Remote Sens.* **2022**, *60*, 1–16. [CrossRef]
28. Cai, J.; Liu, G.; Jia, H.; Zhang, B.; Wu, R.; Fu, Y.; Xiang, W.; Mao, W.; Wang, X.; Zhang, R. A New Algorithm for Landslide Dynamic Monitoring with High Temporal Resolution by Kalman Filter Integration of Multiplatform Time-Series InSAR Processing. *Int. J. Appl. Earth Obs. Geoinf.* **2022**, *110*, 102812. [CrossRef]
29. Liu, Y.; Yang, H.; Wang, S.; Xu, L.; Peng, J. Monitoring and Stability Analysis of the Deformation in the Woda Landslide Area in Tibet, China by the DS-InSAR Method. *Remote Sens.* **2022**, *14*, 532. [CrossRef]
30. Zheng, X.; He, G.; Wang, S.; Wang, Y.; Wang, G.; Yang, Z.; Yu, J.; Wang, N. Comparison of Machine Learning Methods for Potential Active Landslide Hazards Identification with Multi-Source Data. *ISPRS Int. J. Geo-Inf.* **2021**, *10*, 253. [CrossRef]
31. Liu, M.; Yang, Z.; Xi, W.; Guo, J.; Yang, H. InSAR-Based Method for Deformation Monitoring of Landslide Source Area in Baihetan Reservoir, China. *Front. Earth Sci.* **2023**, *11*, 1253272. [CrossRef]
32. Oludare, O.; Kazeem, R.; Adebayo, A.; Awonusi, A.; Dare, A.; Ikumapayi, O.; Adaramola, B. An Assessment of Earthquake-Induced Landslides Distribution in Nepal Using Open-Source Applications on Sentinel-1 Tops SAR Imagery. *Int. J. Des. Nat. Ecodyn.* **2023**, *18*, 237–249. [CrossRef]
33. Berardino, P.; Fornaro, G.; Lanari, R.; Sansosti, E. A new algorithm for surface deformation monitoring based on small baseline differential SAR interferograms. *IEEE Trans. Geosci. Remote Sens.* **2002**, *40*, 2375–2383. [CrossRef]

34. Liu, T.; Chen, T.; Niu, R.; Plaza, A. Landslide Detection Mapping Employing CNN, ResNet, and DenseNet in the Three Gorges Reservoir, China. *IEEE J. Sel. Top. Appl. Earth Obs. Remote Sens.* **2021**, *14*, 11417–11428. [CrossRef]
35. Lv, Z.; Liu, T.; Shi, C.; Benediktsson, J.A. Local Histogram-Based Analysis for Detecting Land Cover Change Using VHR Remote Sensing Images. *IEEE Geosci. Remote Sens. Lett.* **2021**, *18*, 1284–1287. [CrossRef]
36. Varmuza, K.; Filzmoser, P. *Introduction to Multivariate Statistical Analysis in Chemometrics*; CRC Press: Boca Raton, FL, USA, 2016.
37. Gulick, S.P.S.; Barton, P.J.; Christeson, G.L.; Morgan, J.V.; McDonald, M.; Mendoza-Cervantes, K.; Pearson, Z.F.; Surendra, A.; Urrutia-Fucugauchi, J.; Vermeesch, P.M.; et al. Importance of Pre-Impact Crustal Structure for the Asymmetry of the Chicxulub Impact Crater. *Nat. Geosci.* **2008**, *1*, 131–135. [CrossRef]
38. Berk, K.N. Tolerance and Condition in Regression Computations. *J. Am. Stat. Assoc.* **1977**, *72*, 863–866.
39. Boulesteix, A.-L.; Bender, A.; Lorenzo Bermejo, J.; Strobl, C. Random Forest Gini Importance Favours SNPs with Large Minor Allele Frequency: Impact, Sources and Recommendations. *Brief. Bioinform.* **2012**, *13*, 292–304. [CrossRef] [PubMed]
40. Gao, X.; Chen, T.; Niu, R.; Plaza, A. Recognition and Mapping of Landslide Using a Fully Convolutional DenseNet and Influencing Factors. *IEEE J. Sel. Top. Appl. Earth Obs. Remote Sens.* **2021**, *14*, 7881–7894. [CrossRef]
41. Ghorbanzadeh, O.; Blaschke, T.; Gholamnia, K.; Meena, S.; Tiede, D.; Aryal, J. Evaluation of Different Machine Learning Methods and Deep-Learning Convolutional Neural Networks for Landslide Detection. *Remote Sens.* **2019**, *11*, 196. [CrossRef]
42. Xiao, X.; Lian, S.; Luo, Z.; Li, S. Weighted Res-UNet for High-Quality Retina Vessel Segmentation. In Proceedings of the 2018 9th International Conference on Information Technology in Medicine and Education (ITME), Hangzhou, China, 19–21 October 2018; pp. 327–331.
43. Woo, S.; Park, J.; Lee, J.-Y.; Kweon, I.S. CBAM: Convolutional Block Attention Module. *Eur. Conf. Comput. Vis.* **2018**, *11211*, 3–19.
44. Babak, M.; Dominik, S.; Andreas, U. *PRNU-Based Finger Vein Sensor Identification in the Presence of Presentation Attack Data*; Verlag der Technischen Universität Graz: Graz, Austria, 2019.
45. Ghorbanzadeh, O.; Crivellari, A.; Ghamisi, P.; Shahabi, H.; Blaschke, T. A Comprehensive Transferability Evaluation of U-Net and ResU-Net for Landslide Detection from Sentinel-2 Data (Case Study Areas from Taiwan, China, and Japan). *Sci. Rep.* **2021**, *11*, 14629. [CrossRef]
46. Othman, A.A.; Gloaguen, R.; Andreani, L.; Rahnama, M. Improving Landslide Susceptibility Mapping Using Morphometric Features in the Mawat Area, Kurdistan Region, NE Iraq: Comparison of Different Statistical Models. *Geomorphology* **2018**, *319*, 147–160. [CrossRef]
47. Sameen, M.I.; Pradhan, B. Landslide Detection Using Residual Networks and the Fusion of Spectral and Topographic Information. *IEEE Access* **2019**, *7*, 114363–114373. [CrossRef]
48. 48Li, C.; Yi, B.; Gao, P.; Li, H.; Sun, J.; Chen, X.; Zhong, C. Valuable Clues for DCNN-Based Landslide Detection from a Comparative Assessment in the Wenchuan Earthquake Area. *Sensors* **2021**, *21*, 5191.

remote sensing

MDPI

Article

Improving Seismic Fault Recognition with Self-Supervised Pre-Training: A Study of 3D Transformer-Based with Multi-Scale Decoding and Fusion

Zeren Zhang †, Ran Chen † and Jinwen Ma *

Department of Information and Computational Sciences, School of Mathematical Sciences, Peking University, Beijing 100871, China; eric_zhang@stu.pku.edu.cn (Z.Z.); chenran@stu.pku.edu.cn (R.C.)
* Correspondence: jwma@math.pku.edu.cn
† These authors contributed equally to this work.

Abstract: Seismic fault interpretation holds great significance in the fields of geophysics and geology. However, conventional methods of seismic fault recognition encounter various issues. For example, models trained on synthetic data often exhibit inadequate generalization when applied to field seismic data, and supervised learning is heavily dependent on the quantity and quality of annotated data, being susceptible to the subjectivity of interpreters. To address these challenges, we propose applying self-supervised pre-training methods to seismic fault recognition, exploring the transfer of 3D Transformer-based backbone networks and different pre-training methods on fault recognition tasks, thereby enabling the model to learn more powerful feature representations from extensive unlabeled datasets. Additionally, we propose an innovative pre-training strategy for the entire segmentation network based on the characteristics of seismic data and introduce a multi-scale decoding and fusion module that significantly improves recognition accuracy. Specifically, during the pre-training stage, we compare various self-supervision methods, like MAE, SimMIM, SimCLR, and a joint self-supervised learning approach. We adopt multi-scale decoding step-by-step fitting expansion targets during the fine-tuning stage. Ultimately merging features to refine fault edges, the model displays superior adaptability when handling narrow, elongated, and unevenly distributed fault annotations. Experiments demonstrate that our proposed method achieves state-of-the-art performance on Thebe, the currently largest publicly annotated dataset in this field.

Keywords: seismic faults detection; 3D segmentation; self-supervised; pre-training

Citation: Zhang, Z.; Chen, R.; Ma, J. Improving Seismic Fault Recognition with Self-Supervised Pre-Training: A Study of 3D Transformer-Based with Multi-Scale Decoding and Fusion. *Remote Sens.* **2024**, *16*, 922. https://doi.org/10.3390/rs16050922

Academic Editor: Shuying Li

Received: 4 February 2024
Revised: 28 February 2024
Accepted: 4 March 2024
Published: 6 March 2024

1. Introduction

1.1. Background

Seismic data obtained through reflection surveys represent an indirect depiction of the intricate physical structures within the Earth's subsurface [1]. Three-dimensional seismic data offer a more detailed imaging and complex mapping of sub-surface structures than 2D because of denser sampling [2]. In 3D seismic images, geological strata create consistent reflections, whereas faults, due to brittle deformation, cause reflection discontinuities [3,4]. This contrast enables identifying and extracting geological features like strata and faults from the data. Seismic fault interpretation is crucial in geophysics and geology for tasks like stratigraphic analysis, examining ancient landscapes, reservoir characterization, well planning, resource assessment, and geological hazard risk mitigation [4–8].

Traditionally, manual and experiential seismic fault interpretation is susceptible to the influence of data quality and interpreter subjectivity [1,9]. Moreover, in large 3D seismic images, identifying stratum and fault mapping is a time-consuming and labor-intensive process [3]. Various automated fault recognition methods are proposed in response to these limitations, including phase unwrapping [10,11], ant-tracking [12,13], waveform classification [14], slope techniques [15], and Hough transform [16]. However, most feature-based methods impose high computational costs and display susceptibility to noise [1]. These methods often require manual adjustment and heuristic parameter tuning to adapt to different datasets.

The emergence of machine learning [17] has spurred the advancement of intelligent seismic fault recognition algorithms [18,19]. End-to-end fault recognition methods based on convolutional neural networks (CNNs) frame fault recognition as a classification or segmentation task [20–22]. Wu et al. [22] employed synthetic data to train an end-to-end 3D-Unet network named FaultSeg3D for fault recognition tasks. Furthermore, An et al. [23] utilized a 2D segmentation network, Mobile DeepLabV3+, for per-pixel segmentation to extract faults and introduced Thebe, the largest currently publicly available field seismic fault dataset. However, existing models trained on small synthetic datasets demonstrate performance degradation and limited generalization when applied to field seismic data. This is due to the heavy reliance on supervised deep learning on accurate annotations. Consequently, these models need improvement in dealing with data with significant feature variations, quality issues, high annotation costs, and subjectivity.

1.2. Related Work

Transformer [24] models, which rely on attention mechanisms, have advantages over CNNs in capturing global context information and obtaining full-scale features. Self-supervised learning methods using the Transformer structure, such as Vision Transformer (ViT) [25], show excellent performance across various visual tasks. Additionally, the Swin-Transformer [26], optimized through hierarchical and sliding window strategies, further enhances the performance of the ViT. On the other hand, masked autoencoders (MAEs) [27] demonstrate commendable image understanding capabilities through self-supervised learning. The simplified mask image modeling (SimMIM) [28], tailored specifically for the Swin-Transformer, uses a straightforward masked image modeling (MIM) approach, a lightweight decoder, and a more adaptable token processing mechanism, making it well-suited for downstream tasks requiring multi-scale modeling. As a self-supervised technique that learns similarities in data through contrastive learning, a simple framework for contrastive learning of representations (SimCLR) [29] also demonstrates significant competitiveness in image understanding tasks.

Seismic data processing tasks are now increasingly utilizing self-supervised learning techniques, such as seismic velocity inversion, stratigraphic phase semantic segmentation, seismic data denoising, etc. [30–32], which demonstrates the applicability of self-supervised pre-training methods in seismic data feature extraction. However, only a few studies apply the Transformer framework to fault recognition tasks. Yang et al. [33] presented a multitask learning network capable of simultaneous stratigraphy extraction and fault detection. They utilized Transformer architecture for stratigraphy estimation, fortifying the model's robustness with expert geological interpretation. Tang et al. [34] proposed a novel method that amalgamates the Unet architecture with a Transformer encoder for 2.5D fault detection. The model, trained on synthetic datasets, demonstrates superior performance on the Netherlands F3 dataset [35] compared to the complete 3D Unet model. The Dual Unet with Transformer model proposed by Wang et al. [36] combines the traditional U-Net with the Transformer U-Net, and they found that the binary cross-entropy loss (BCE) performed best after comparing six different loss functions. Finally, FaultSSL [37] is a semi-supervised fault recognition framework. Its supervised learning component uses synthetic data and a small number of 2D label data, while the unsupervised learning component relies on two proxy tasks—PaNning Consistency (PNC) and PaTching Consistency (PTC). The approach reduces the reliance on synthetic data and purely supervised learning methods but still needs a certain amount of labeled data support. In summary, recent research has made progress in applying the Transformer architecture to solve fault recognition problems. However, training on synthetic datasets limits the generalization of field datasets.

Although the potential advantages of semi-supervised learning have been preliminarily explored, more quantitative evaluations are still needed.

1.3. Motivation

As previously discussed, seismic fault recognition technology has evolved significantly, transitioning from traditional manual interpretation to advanced deep learning models. Despite these advancements, supervised learning models, heavily dependent on accurate labeling, exhibit limitations when processing field seismic data. In response to these limitations, self-supervised learning has become a powerful solution to compensate for the limitations of supervised learning. However, applying the Transformer framework in fault recognition tasks and using self-supervised learning methods within this context are still largely undeveloped, and the resulting performance has yet to reach the expected level. Consequently, this study explores the potential gains of using self-supervised pre-training methods in seismic fault recognition. We adopt a 3D Transformer-based backbone network, and during the pre-training phase, we conduct a detailed exploration of various self-supervised pre-training techniques, including MAE, SimMIM, and SimCLR, allowing the model to learn more robust feature representations from large-scale unlabeled data. Our findings reveal that combining the Swin UNEt TRansformer (Swin-UNETR) [38] backbone with the SimMIM pre-training task significantly enhances seismic fault recognition capabilities. We specially design our model architecture to handle the complexity inherent in seismic fault recognition patterns. Considering the sparse distribution features of seismic fault data, we innovatively improve the Swin-UNETR architecture, realizing multi-scale decoding and fusion. Furthermore, our findings reveal that the Swin-UNETR model's decoder has significantly more parameters than its backbone network, which suggests that pre-training the entire segmentation model can further enhance fault detection accuracy. Our method, referred to as FaultSeg Swin-UNETR, demonstrates excellent adaptability and precision in detecting seismic faults.

2. Datasets

In their comprehensive 2023 review on fault recognition [39], An and colleagues synthesized information from 73 seismic datasets, of which only three field datasets and four synthetic datasets open-sourced seismic data and labels, providing a public baseline for research. Only two open-sourced datasets with annotations are 3D, including the synthetic dataset FaultSeg [22] created by Wu and the field dataset Thebe [23] collected by An's team. FaultSeg simulates seismic signal patterns, providing detailed explanations and opening its dataset and code. Although synthetic datasets can partially reduce the reliance on expert annotations, the quality difference of synthetic data significantly influences the model's performance when dealing with practical data. When choosing the dataset type, researchers must consider that the ultimate purpose of all models is to solve real earthquake data-related problems. Hence, testing the model's performance through field earthquake datasets is a straightforward approach to gauging its performance. Currently, the Thebe dataset is the largest publicly available earthquake fault dataset, providing many detailed pixel-level expert annotations, allowing researchers to compare the performance of different models more accurately and identify their advantages and limitations.

2.1. Datasets Employed

Earthquake fault recognition models trained with data from actual work areas can fully understand and handle the complexity and uniqueness of field earthquake data. Such models have better generalization abilities than models trained on simulated datasets. They can adapt to varying conditions and maintain high accuracy even when encountering previously unseen data. Therefore, considering the practical applicability of seismic fault recognition models, our approach primarily relies on seismic data obtained from actual working areas. Specifically, we use a large number of unlabeled private data for pre-training.

Then, we fine-tune it on the labeled field dataset Thebe and verify the model's effectiveness on both the publicly available synthetic dataset FaultSeg and the field dataset Thebe.

- Synthetic dataset
  A synthetic dataset was used in the earliest work using deep learning methods for fault recognition [22]. This dataset consists of 220 seismic volumes of 128 × 128 × 128 and corresponding fault labels. The training set slice images are depicted in Figure 1.

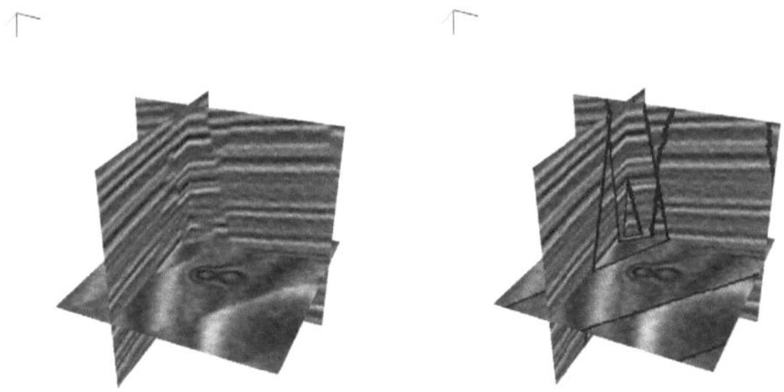

Figure 1. Three-dimensional visual examples of the synthetic dataset. A synthetic seismic image of the synthetic dataset and the corresponding true fault image (with labeling ones on faults and zeros elsewhere) is overlaid with the seismic image.

- Field dataset
  Most relevant works on fault recognition are trained on this synthetic dataset and then qualitatively analyzed on field fault datasets. However, deep learning methods typically require much training data to mitigate overfitting. Deep networks are prone to learning patterns specific to the synthetic data, which may not apply to real-world scenarios. Therefore, during the fine-tuning process, we conducted significant experiments using Thebe [40], the currently largest publicly available seismic faults dataset. The dataset, originally from a seismic survey called Thebe Gas Field in the Exmouth Plateau of the Carnarvan Basin on the NW shelf of Australia, is represented in Python Numpy format. The Thebe dataset has a size of 1803 × 1537 × 3174. Because of the high correlation between adjacent slices, random partitioning along the crossline or inline direction was not a reasonable choice. Following the approach proposed by An et al. [40], we divided the data along the crossline direction. The first 900 slices were used for training, the next 200 for validation, and the remaining 703 for testing. The 640 × 640 × 640 cube segmented from the Thebe dataset is depicted in Figure 2. A comparison between Figures 1 and 2 reveals that the field seismic data are comparatively more complex than the synthetic seismic data. The field seismic dataset exhibits a more intricate fault distribution and finer fault annotation lines, making transferring the model trained on synthetic data to field data challenging.
- Pre-training dataset
  In addition, as self-supervised pre-training does not require annotations, we collected private datasets from 15 different working areas for pre-training tasks. These datasets are diverse in their geological features and size, and notably, they all lack fault annotations. Our goal was for the network to independently discover and learn the intrinsic characteristics of fault data across various contexts from various working areas during the self-supervised learning phase. We visualize some of the data used for pre-training in Figure 3.

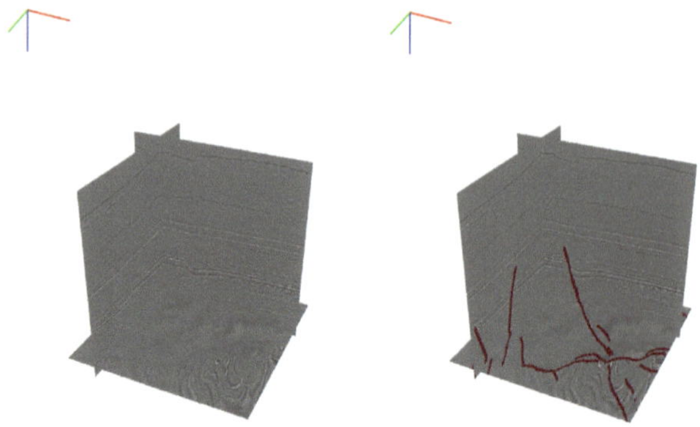

Figure 2. Three-dimensional visual examples of the Thebe dataset. A synthetic seismic image is cropped from the Thebe dataset, and the corresponding true fault image (with labeling ones on faults and zeros elsewhere) is overlaid with the cropped seismic image.

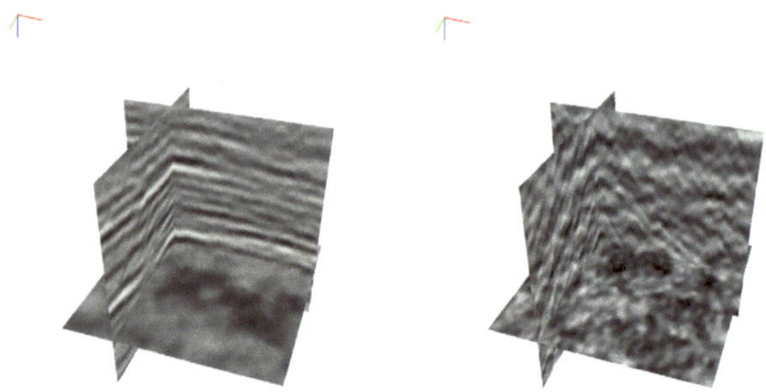

Figure 3. Three-dimensional visual examples of the pre-training dataset. The figure illustrates 128 × 128 × 128 segmentations from the pre-training datasets of various working areas.

2.2. Data Processing

To ensure consistency during training and comparability during predictions of seismic data, we initially performed z-score standardization on the data. Assuming S represents the original three-dimensional seismic data and μ and σ respectively denote its mean and standard deviation, we utilized Equation (1) to standardize the original data.

$$S_{\text{standardized}} = \frac{S-\mu}{\sigma} \tag{1}$$

These preprocessing steps prevent data from being overly concentrated or dispersed, mitigating the influence of varying feature scales. This enhances the stability and convergence speed of the model, facilitating a more straightforward acquisition of fault distribution characteristics. The distributions of standardized fault data and annotated data are illustrated in Figure 4. The Thebe dataset is publicly available. Expert interpreters from the Fault Analysis Group at University College Dublin manually labeled the faults in the seismic data. Because of the slanted nature of the work area, preprocessing involved creating a bounding rectangle to form a 3D cube and filling in blank areas with zeros. This resulted in a higher kurtosis for the Thebe dataset than synthetic and field data typically used for pre-training, which usually resemble a normal distribution. The field seismic cube shows skewness, with many zeros and near-zero values, making its distribution notably different from the synthetic and pre-training data. Therefore, fine-tuning post-pre-training on unlabeled datasets was crucial to addressing these distributional discrepancies.

Figure 5a is a histogram showing the statistical distribution of 0–1 fault labels for 10,000 data points randomly sampled from the Thebe and synthetic datasets; on the x-axis,

False represents non-fault points, i.e., points with a label of 0, while True represents fault points, i.e., points with a label of 1. The y-axis represents the total number of voxel points with a label of 0 or 1 in different datasets. It reveals substantial label imbalance, where fault annotations are significantly fewer than non-fault annotations. The field dataset, in particular, exhibits a more pronounced imbalance, making fault interpretation tasks more challenging. Consequently, models trained on synthetic datasets demonstrate limited feature extraction capabilities when applied to field datasets. Figure 5b displays the distribution of fault and non-fault points after merging the Thebe and synthetic datasets. The distinct difference in fault distribution between the synthetic and the Thebe datasets highlights the crucial significance of the model's simultaneous learning to identify seismic fault and non-fault features.

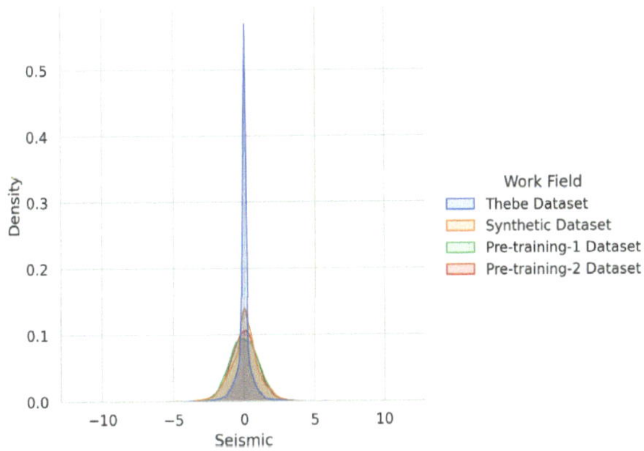

Figure 4. The amplitude distribution of seismic datasets after z-score standardization.

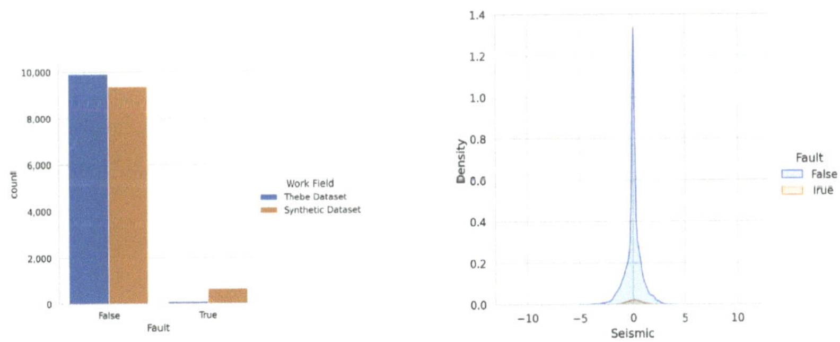

(**a**) (**b**)

Figure 5. (**a**) The distribution of labels in synthetic seismic data and field seismic data exhibiting a 0–1 distribution. (**b**) Labels distribution of faults and non-faults after merging synthetic and field datasets.

Figure 6 is a boxplot of the data distribution. The values on the vertical axis represent the seismic data values after z-score normalization. The False part on the left shows the distribution of non-fault data for Thebe and synthetic data, and the True part on the right shows fault data. The distribution of non-fault data in Thebe on the far left even loses the shape of the box because of the over-concentration of zero values and includes many outliers, in which the maximum outlier even reaches a positive 12.5. However, the synthetic data values are evenly distributed on both sides of the 0 axis after standardization. The annotation of fault data is relatively uniform, with both Thebe, in blue, and the synthetic data, in orange, being standard and symmetrical. The boxplot vividly illustrates the highly concentrated distribution of non-fault points within the Thebe dataset and the considerable presence of outliers. This challenges the model, potentially leading to biases toward specific features or regions during training, neglecting other crucial information. Such biases can impede the model's comprehensive understanding of geological structures.

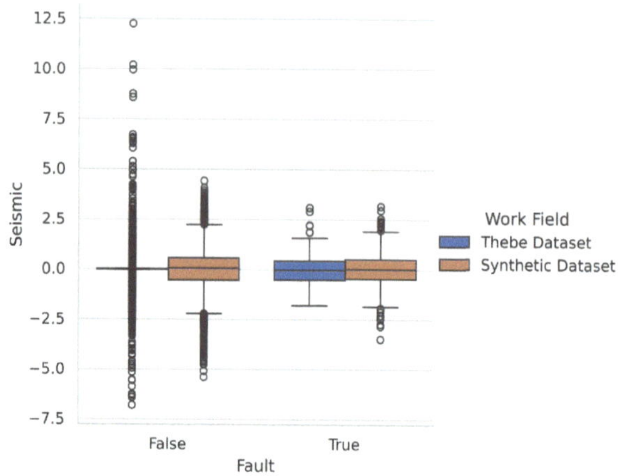

Figure 6. The boxplots of fault and non-fault data in the Thebe dataset and synthetic dataset.

Additionally, random flipping, 90-degree random rotations, and random cropping are common data augmentation techniques in the computer vision domain. They expand the training dataset, enabling the model to learn diverse features better. Random cropping effectively extracts feature information of different scales and positions, bolstering the model's generalization ability. Random flipping and 90-degree random rotations aid the model in learning features from various orientations, enhancing its robustness. This combination of techniques augments the model's capability to extract seismic data features and generalizability, bolstering its effectiveness and stability in real-world applications.

3. Seismic Fault Recognition with Self-Supervised Pre-Training

Inspired by successful self-supervised pre-training schemes in image processing, we explored various pre-training tasks on a large volume of unlabeled seismic field data using 3D Transformer-based backbones. These backbones include U-Net TRansformers (UNETR) [41] and Swin-UNETR, with pre-training strategies comprising MAE, SimMIM, SimCLR, as well as three joint self-supervised tasks used in [38]: masked volume inpainting, image rotation, and contrastive coding. Following this pre-training phase, the network learned rich features within the seismic domain. We fine-tuned and tested it on field datasets, discovering that the strategy combining Swin-UNETR with SimMIM yields the best results. Building on this insight, to address challenges such as the scarcity of seismic fault data and the narrow and uneven distribution of seismic faults, we further refined the Swin-UNETR model. We propose a full-network pre-training scheme named FaultSeg Swin-UNETR specifically for seismic fault segmentation tasks. We observe a notable disparity between the size of parameters in the backbone network and the decoder head of Swin-UNETR. To address this, we employed full-network self-supervised pre-training with SimMIM, which achieved better outcomes. Additionally, we found that introducing multi-scale decoding and fusion modules significantly enhanced identification precision. Section 3.1 introduces the backbone network and pre-training schemes, while Section 3.2 discusses the full-network pre-training of FaultSeg Swin-UNETR.

3.1. Backbone and Pre-Training Methods

3.1.1. 3D Transformer-Based Backbone

Our research uses a Transformer-based model, different from many past studies that used CNNs [23,42], for fault recognition in seismic images. It is adept at capturing multi-scale contexts and managing tasks that require understanding long-distance relationships, which is crucial for segmenting images by pixel relationships. Because of the difficulty of annotating seismic faults, it is common for researchers to annotate faults slice by slice along a specific direction (inline or crossline) within a particular working area. Therefore, previous studies often used 2D segmentation networks for fault recognition, but these could not capture the full spatial context of seismic faults. While 2.5D data have been employed to address this partially, proper integration of 3D spatial information necessitates a 3D segmentation network. Therefore, our model uses a 3D Transformer backbone as the main network.

UNETR and Swin-UNETR [38], as classic 3D Transformer-based frameworks, redefine the task of 3D cube fault recognition as a sequence-to-sequence prediction problem. These architectures divide the input 3D seismic image into volumetric patches. Each patch is linearly embedded and then processed through a series of Transformer or Swin-Transformer blocks, enabling the models to learn rich, context-aware data representations. UNETR combines the architecture of U-Net with the advantages of Transformers and typically features an encoder–decoder structure: the encoder utilizes Transformer blocks to capture global context information, while the decoder employs conventional convolutional layers to reconstruct the segmentation output of seismic faults intricately. Swin-UNETR builds upon the UNETR concept but incorporates the Swin-Transformer as its backbone. "Swin" in Swin-Transformer stands for "shifted window", a novel attention mechanism that adopts a window-based approach to handle local information in images better. As shown in Figure 7, the Swin-Transformer consists of multiple Swin blocks. In Swin-UNETR, the Swin-Transformer's shifted windowing approach is adapted to handle 3D images, allowing the model to dynamically adjust its focus between smaller, local areas and the broader global context. This mechanism provides a more nuanced understanding of the spatial relationships within volumetric data, which is crucial for accurate segmentation and analysis in seismic imaging. For more details about Swin-Transformer, please refer to Appendix A.

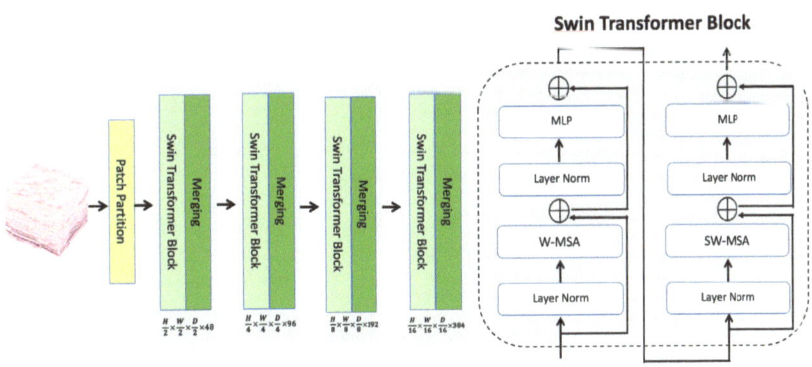

Figure 7. Architecture of 3D Swin-Transformer backbone.

3.1.2. Self-Supervised Pre-Training of Seismic Data

We collected many unlabeled field seismic data, covering many scenarios. In order to explore the potential information within seismic data, we compare several mainstream self-supervised pre-training methods, including MAE, SimMIM, SimCLR, and the masked volume inpainting, image rotation, and contrastive coding implemented within the Swin-UNETR architecture. MAE and SimMIM focus on improving the model's prediction capabilities for masked parts within the data, thereby strengthening the model's overall understanding of seismic data. Meanwhile, SimCLR enhances the model's ability to discriminate between different seismic features via contrastive learning. Masked volume inpainting trains the model on how to infer the overall structure from partially visible seismic data; image rotation compels the model to learn the direction invariance of seismic images; contrastive coding encourages the model to identify and understand the differences and similarities within different parts of the images. Our goal was to comprehensively evaluate the applicability of each method in the field of earthquakes to determine which method performs better on seismic datasets.

3.2. *FaultSeg Swin-UNETR*

This research introduces FaultSeg Swin-UNETR, a self-supervised pre-training model for seismic fault segmentation based on Transformer. This model innovatively combines the Swin-UNETR architecture with the SimMIM pre-training method, optimizes the uneven distribution of parameters through full-network pre-training, and incorporates multi-scale decoding fusion modules to adapt to the peculiarities of seismic data.

3.2.1. Full-Network Pre-Training

Our findings highlight that the Swin-UNETR backbone, particularly when combined with the SimMIM pre-training strategy, outperforms traditional tasks like masked volume inpainting, image rotation, and contrastive coding in terms of earthquake fault recognition capability. This empirical evidence led us to adopt the Swin-UNETR backbone in conjunction with the SimMIM pre-training approach in our proposed FaultSeg Swin-UNETR method.

Figure 8 illustrates the workflow of using SimMIM in seismic data. Given a fixed size ($128 \times 128 \times 128$) seismic dataset, it was first divided into 262,144 non-overlapping patches of size $2 \times 2 \times 2$. After specifying a mask ratio (usually 75%), a random portion of the seismic data was masked from these 262,144 patches to serve as the input for the backbone network, Swin-Transformer. However, it is not reasonable to randomly mask overly dense patches, as it would make the reconstruction task more similar to a super-resolution task. Therefore, we adopted the strategy of simulating larger patch sizes in ViT, where during random masking, we simulated the effect of $16 \times 16 \times 16$ large patches on $2 \times 2 \times 2$ small patches. In the specific implementation, after the image was divided into patches, it could go through a tokenization module to convert it into tokens inputted to the Transformer. The masking operation was then performed on these tokens (Figure 8 optional branch). Consider the 3D seismic dataset represented by S. The formulation for the masking operation can be elegantly expressed as Equation (2).

$$\text{MaskedInput} = \text{PE}(\text{Tokenization}(S) \odot \mathcal{M} + \text{Repeat}(\text{MaskToken}) \odot (1 - \mathcal{M})) \quad (2)$$

Herein, Tokenization$(\cdot)$ represents the operator that segments the input seismic data into blocks and then projects them into tokens that can be accepted by the Swin-Transformer. PE$(\cdot)$ denotes the application of positional embeddings to introduce spatial awareness. The term $\mathcal{M}$ represents a masking matrix, which is designed to obscure a random 75% of the input tokens, thereby facilitating a robust learning process. MaskToken, a tunable parameter, serves to supplant the elements obscured by the mask, thus preserving the integrity of the data's structural composition. After inputting the masked tokens into the Transformer, they went through its internal self-attention mechanism. Afterward, we obtained a feature map with a smaller spatial size at the output. We needed a simple upsampling network (composed of transpose convolutions) to upsample it to the same size as the input. Then, through the same patch-based masking operation, we calculated the l_1 loss between the masked regions and the original input. Despite the difficulty of annotating earthquake data, collecting it was relatively easier. This made it possible to perform SimMIM on many unlabeled seismic data. Compared to using pre-trained backbone networks for supervised tasks such as classification, segmentation, and object detection on natural scene data (such as Imagenet, COCO), directly using earthquake data for self-supervised learning could learn the essential features of the seismic domain, making it easier and more accurate in the downstream fine-tuning process.

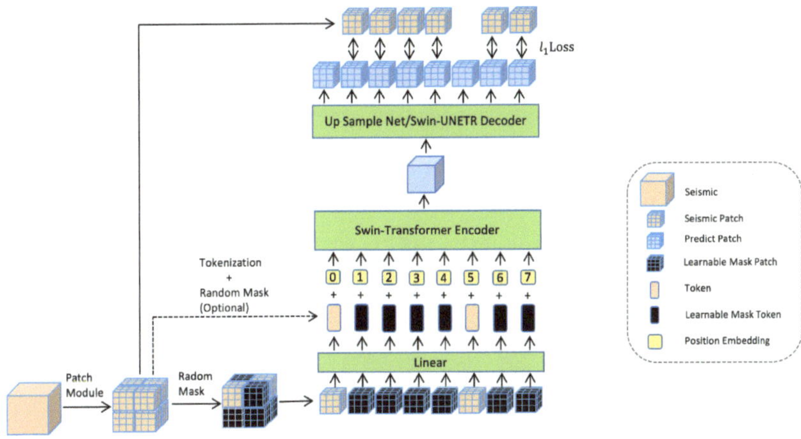

Figure 8. Overview of simplified mask image modeling.

Furthermore, we found a significant disparity in parameter proportion between the backbone network and the segmentation decoder network in the Swin-UNETR network used for seismic fault segmentation. This mainly arose from using the self-attention mechanism in the Swin-Transformer, resulting in far fewer parameters than the 3D convolutional layers in the segmentation decoder. Specifically, the parameter size of the Swin-Transformer backbone network is only one-eighth of the total parameter size of the segmentation network (see Table 1). This design contradicts the ideas of many self-supervised pre-training models. Generally, in self-supervised pre-training frameworks, the parameter size of the backbone network is much larger than that of the decoding heads of downstream tasks. The backbone network mainly accomplishes the learning of input data features, and it is preferable to have a simple design for the decoding heads. In light of these observations, we propose an innovative shift in strategy. Rather than confining the self-supervised learning task, such as SimMIM, to merely a network component, we advocate for utilizing the entire Swin-UNETR structure, including both the backbone and the segmentation head, for self-supervised pre-training. This method is particularly suitable for the inherently image-to-image nature of most seismic-related downstream tasks. To some extent, it can solve the problem of unreasonable parameter settings when Swin-UNETR integrates the Unit architecture. The original self-supervised pre-training method of Swin-UNETR only focuses on the model's backbone and needs to fully explore the potential of unlabeled data. Our training strategy not only allows the decoder to learn prior information from a large number of unlabeled data during the pre-training stage but also ensures that this pre-training information can be fully utilized in FaultSeg Swin-UNETR through the parameter replication of the three decoder heads downstream tasks.

Consequently, we engaged in extensive pre-training of the complete Swin-UNETR network on a vast corpus of seismic fault data. In response to this situation, we made certain modifications to the input processing. We performed patching and masking operations directly on the seismic data at the input stage, in contrast to the typical approach of manipulating tokens post the patch embedding layer. This reimagined strategy addresses the parameter distribution challenge and aligns more congruently with the intrinsic requirements of seismic fault recognition tasks. This direct masking operation on the input seismic

S can be described by Equation (3), where $F_{\text{patch}}(\cdot)$ denotes the operation of segmenting the input into patches and $\mathcal{M}$ represents the mask matrix.

$$\text{MaskedInput} = F_{\text{patch}}(S) \odot \mathcal{M} + \text{Repeat}(\text{MaskPatch}) \odot (1 - \mathcal{M}). \tag{3}$$

Additionally, the dimension of MaskPatch is also smaller than that of MaskToken. This approach is more in line with the original design of SimMIM. It allows the backbone network and decoder to learn some essential features from unlabeled data during the pre-training phase. The goal of the SimMIM pre-training task is to reconstruct masked seismic data. The learning objective for the task is

$$\mathcal{L}_{SimMIM} = \|\mathcal{M} \odot (S_{reconstruct} - S)\|_1. \tag{4}$$

During the pre-training phase, the network starts learning from weights initialized randomly. The chosen optimizer is AdamW, and a cosine annealing learning rate decay strategy is employed.

Table 1. Parameters of the Swin-Transformer-based model.

Model Name	Backbone	Decode Head	Total Params
Swin-Transformer + Up Sample Net	7.8 M	3.1 M	10.9 M
Swin-UNETR	7.8 M	54.1 M	61.9 M
FaultSeg Swin-UNETR	7.8 M	93.3 M	101.1 M

3.2.2. Multi-Scale Decoding and Fusion Module

Figure 9 illustrates the segmentation network FaultSeg Swin-UNETR that we designed for fault recognition. Based on our findings, earthquake faults are typically narrow and account for a deficient proportion, resulting in a serious class imbalance issue in training targets. In order to alleviate this phenomenon, we use multiple segmentation decoders to learn from fault segments that have been expanded at different levels. Specifically, let us assume our annotated samples are $\{S_j, F_j\}_{j=1}^{N}$, where $S_j \in R^{128\times128\times128}$ represents the input seismic and $F_j \in \{0,1\}^{128\times128\times128}$ represents the corresponding labeled faults. The Swin-Transformer encodes the input seismic and then decodes it separately by three segmentation decoders. These three decoders need to learn seismic faults that have been expanded by $1x$, $3x$, and $5x$, respectively. Our main loss function is defined as Equation (5), where $\text{Dilate}(F_j, 2i-1)$ represents the j-th labeled faults F_j after being dilated $2i-1$ times using the dilation operator, and $\text{Decode}_{i,j} = \text{Decoder}_{\text{i}}(\text{Backbone}(S_j))$ represents the predicted dilated fault result of seismic sample S_i after passing through the j-th decoder of FaultSeg Swin-UNETR. We conceptualize fault identification as a per-pixel binary classification problem for 3D seismic data, thereby configuring the loss function as binary cross-entropy (BCE) loss. The design of the main loss function is employed to supervise the three decoding branches of our FaultSeg Swin-UNETR, enabling them to independently learn the annotated fault information after being dilated at three different scales. Afterward, to restore the original fault body, we introduce a fusion module that re-calibrates the fault bodies learned from the three branches to their original thickness. The fusion module consists of a series of 3D convolutions. The three predictions of dilated faults $\{\text{Decode}_{i,j}\}_{i=1}^{3}$ are concatenated along the channel dimension and passed through the fusion module. We define the fusion loss function as Equation (6), where $\text{Fusion}(\cdot)$ denotes the fusion layer we designed, and $\text{Cat}(\cdot)$ represents the concatenation operator. Given that the decoded outputs $\text{Decode}_{i,j} \in \mathbb{R}^{H\times W\times D}$ are three-dimensional, we extend an additional dimension to the predicted results $\{\text{Decode}_{i,j}\}_{i=1}^{3}$ from the three decoding heads and concatenate them along this new dimension, resulting in $\text{Cat}(Decode_{1,j}, Decode_{2,j}, Decode_{3,j}) \in \mathbb{R}^{H\times W\times D\times 3}$. Since the fusion module is designed solely to learn how to restore the expanded fault body during gradient backpropagation, we disconnect it from the main segmentation network. The fusion loss function $\mathcal{L}_{fusion}$ does not participate in the updates of the main network. The total loss is composed of a weighted combination of the losses from two parts (Equation (7)), where λ is used to adjust the strength of the supervisory signal between the backbone network of FaultSeg Swin-UNETR and its fusion module.

$$\mathcal{L}_{main} = \frac{1}{3N}\sum_{i=1}^{3}\sum_{j=1}^{N} \mathrm{BCE}(\mathrm{Decode}_{i,j}, \mathrm{Dilate}(F_j, 2i-1)) \tag{5}$$

$$\mathcal{L}_{fusion} = \frac{1}{N}\sum_{j=1}^{N} \mathrm{BCE}(\mathrm{Fusion}(\mathrm{Cat}(Decode_{1,j}, Decode_{2,j}, Decode_{3,j})), F_j) \tag{6}$$

$$\mathcal{L}_{total} = \mathcal{L}_{main} + \lambda\mathcal{L}_{fusion} \tag{7}$$

The redesigned FaultSeg Swin-UNETR network can learn fault features more easily. It is worth noting that after pre-training with Swin-UNETR on SimMIM, we can create three copies of the parameters of the decoder part to initialize the weights of FaultSeg Swin-UNETR. This allows us to leverage the prior knowledge obtained from pre-training in the seismic fault domain, speeding up network convergence and improving recognition accuracy. In the finetuning phase, we start from the weights obtained during self-supervised pre-training. We continue to use the AdamW optimizer and employ a cosine annealing learning rate decay strategy to gradually decrease the learning rate, ensuring the convergence of the network.

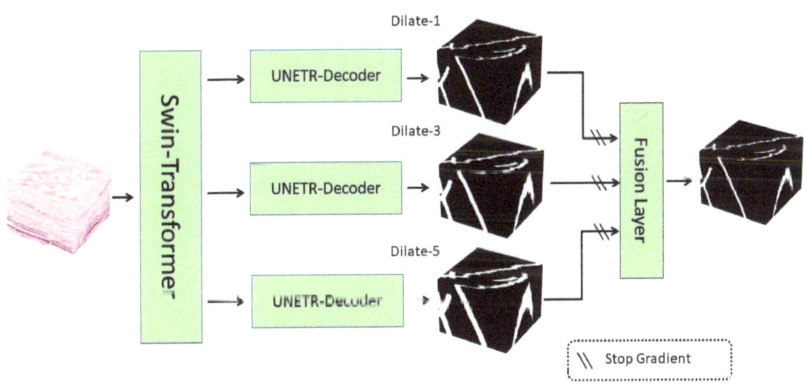

Figure 9. The architecture of the FaultSeg Swin-UNETR network.

4. Experimental Results and Analysis

This section introduced the experiments conducted and the results obtained. Section 4.1 details the evaluation metrics used for fault recognition tasks. Section 4.2 describes the training details of our algorithm. Finally, Sections 4.3 and 4.4 provide a detailed discussion of the results and an ablation study.

4.1. Evaluation Metrics

Fault recognition is a typical binary segmentation task; therefore, classic classification task metrics such as precision, recall, and F1 score can be used to measure the model's performance. However, evaluating the model solely based on precision or recall is not reasonable because of the low proportion of real earthquake faults. Therefore, following the work of An et al. [23], we use the per-image best threshold (OIS) and fixed contour threshold (ODS) metrics, calculated based on the F1 score, to assess the performance of the model. In the case of deep segmentation networks, the output is a 3D score volume with logits of the same size as the input seismic data. After applying the sigmoid activation function, the output is a 3D score volume ranging from 0 to 1. To calculate the OIS and ODS, the seismic faults in the test set are considered as multiple images along the crossline, $i = 1, 2, ..., N$. The F1 score is computed for the network's output score volume at 100 different thresholds. The best threshold is then selected either per image or for the entire volume to evaluate the model's performance. In this context, let us denote the model's output score as Y_{pred} and the ground truth labels for the faults as GT. Both Y_{pred} and GT are represented as 3D volumes with dimensions $H \times W \times D$. The evaluation metrics, OIS and ODS, are defined by the following equations (refer to Equation (8)):

$$
\begin{aligned}
OIS &= \frac{1}{N}\sum_{i=1}^{N} \max\left(\left\{ \mathrm{F1}\left(Y_{pred}[i,:,:] > \frac{j}{100}, GT[i,:,:]\right)\right\}_{j=1}^{100}\right) \\
ODS &= \max\left(\left\{ \frac{1}{N}\sum_{i=1}^{N} \mathrm{F1}\left(Y_{pred}[i,:,:] > \frac{j}{100}, GT[i,:,:]\right)\right\}_{j=1}^{100}\right)
\end{aligned} \tag{8}
$$

The formulations reveal that ODS, which determines the optimal threshold across the entire dataset, typically yields a value less than or equal to OIS. The latter metric computes the optimal threshold for each individual slice. Consequently, ODS offers a more comprehensive assessment of the model's efficacy across the whole seismic dataset.

4.2. Training and Validation

Regarding the datasets from Thebe and our collection, we divided the entire dataset into smaller chunks because of its large size. During the self-supervised stage, we used a patch size of $128 \times 128 \times 128$ and a stride of $64 \times 64 \times 64$ for chunking. Since we applied random masking to the input to the network each time, we did not perform excessive data augmentation. Instead, we only normalized each chunk of data by adjusting its mean to 0 and standard deviation to 1, making it easier for the neural network to process. During the fine-tuning stage, we divided the Thebe data into blocks of size $256 \times 256 \times 256$ with a corresponding stride of $128 \times 128 \times 128$. Before inputting each data block into the segmentation network, we applied random flips, random rotations by 90 degrees, and random cropping of seismic data from the blocks of size $256 \times 256 \times 256$ to $128 \times 128 \times 128$.

Additionally, we performed data normalization and other data augmentation operations. We utilized the slide window inference technique during the inference stage to generate predictions for the entire work area. This involved predicting a volume of size $128 \times 128 \times 128$ at a time, with a sliding stride of $64 \times 64 \times 64$. We calculated the average values in overlapping regions to address inconsistent predicted results at the boundaries. This choice of data augmentation method helped effectively mitigate the issue of inconsistent predictions at the junctions. The code for the entire pre-training and fine-tuning tasks was implemented using PyTorch 1.12.1 and accelerated using 4×4 NVIDIA A100 GPUs supported by the High-performance Computing Platform of Peking University. Because of memory limitations, we set the batch size for all tasks to two. The SimMIM self-supervised pre-training for the backbone network Swin-Transformer and the whole segmentation network Swin-UNETR was trained for 300 epochs. During the fine-tuning stage, all Transformer-based models ran for 100 epochs.

4.3. Results

Table 2 shows the OIS and ODS performance indicators of our FaultSeg Swin-UNETR model on the Thebe dataset. We primarily compare our results with An's work [42]. In that study, they found that the DeeplabV3 segmentation network pre-trained on ImageNet with ResNet-101 performed the best and even surpassed the foundational work of FaultSeg in 3D fault detection. Our experimental results show that our end-to-end SimMIM pre-training FaultSeg Swin-UNETR surpasses previous methods in both indicators, improving by at least 0.25 points compared to An's state-of-the-art work. In this section, we delve into the motivations behind the design of FaultSeg Swin-UNETR, validate the effectiveness of various pre-training strategies through a series of experiments, and confirm that every technique we propose significantly improves the detection accuracy of seismic faults.

Our research used UNETR and Swin-UNETR based on the 3D Transformer architecture as backbone networks. We conducted experiments on several self-supervised pre-training tasks, including MAE, SimMIM, SimCLR, and the masked volume inpainting, image rotation, and contrastive coding implemented within the Swin-UNETR. The results show that even in the absence of self-supervised pre-training, the performance of UNETR and Swin-UNETR is superior to the traditional 3D convolutional network-based FaultSeg3D, emphasizing the effectiveness of the combined Transformer and U-Net architecture for seismic fault segmentation tasks. However, the performance of the Swin-UNETR model without pre-training is similar to that of UNETR and needs improvement. Both backbone and entire segmentation networks can achieve better results than UNETR or Swin-UNETR under self-supervised pre-training. This indicates that self-supervised pre-training further enhances the ability of these models to extract seismic fault features. Despite this, we observe that after self-supervised pre-training, the performance improvement of UNETR is different than expected. This may be because, compared to CNN, the Transformer architecture requires more training samples to achieve optimal performance. Although we provided a large number of unsupervised data, the fault annotation data in the Thebe dataset still needs to be improved for the standard version of the Transformer. Relative to this, the Swin-Transformer structure with fewer parameters is more suitable for this task.

We improved the multi-task self-supervised pre-training method proposed by Tang et al. for Swin-UNETR in the field of medical imaging [38]. The previously designed tasks (masked volume inpainting, image rotation, and contrastive coding) have limited pre-training effects on seismic data, as shown in the 12th row of Table 2. Although these multi-task self-supervised pre-training tasks have a role in improving model performance, they are not as effective as the SimMIM method in seismic fault recognition tasks. Likewise, the effect of SimMIM is superior to MAE and SimCLR, indicating that the simple SimMIM is more suitable for fault recognition tasks. The original three-task pre trained task of Swin-UNETR may increase the complexity and difficulty of model training. Contrastive learning in SimCLR focuses more on global features, which may not be sufficient to capture tiny changes in seismic data. MAE is more suitable for ViT structures. In comparison, SimMIM, designed for Swin structures, provides a more straightforward and efficient method, focusing on the most critical learning tasks. Predicting the masked part encourages the model to pay attention to the subtle differences in the image, making it more effective in seismic fault detection.

Pre-training the entire segmentation network with SimMIM can yield the most significant improvements in downstream segmentation tasks. We attribute this phenomenon to the fact that all network components gain substantial prior knowledge during the pre-training stage and are not just limited to the Swin-Transformer backbone with fewer parameters. However, this may also be due to the irrational design of Swin-UNETR, whose backbone parameters are much smaller than those of the segmentation decoding head, only one-eighth the size. The design goal of most self-supervised pre-training networks is to make the backbone as complex and parameter-rich as possible, while smaller parameter networks usually handle downstream tasks. Nevertheless, pre-training the entire segmentation network can significantly alleviate this problem. As shown in the lower half of Table 2, both Swin-UNETR and FaultSeg Swin-UNETR, which incorporates the multi-scale decoding and fusion module, have greatly improved performance because of full-network pre-training.

Ultimately, our FaultSeg Swin-UNETR model is optimized explicitly for seismic faults' sparseness and slender characteristics. This model can restore the original fault structure more accurately by introducing multi-scale decoding objectives and integrated fusion modules. We deeply analyze the design of FaultSeg Swin-UNETR to demonstrate its rationality. Even without pre-training support, FaultSeg Swin-UNETR significantly outperforms the standard Swin-UNETR in detection performance. This enhancement allows us to further improve the performance of models based on the Swin-UNETR structure, making it reach the industry-leading level on the Thebe dataset.

Table 2. Performance on the Thebe dataset using OIS and ODS metrics, evaluating FaultSeg Swin-UNETR multi-scale decoding and fusion module design and SimMIM pre-training strategy for the entire segmentation network.

Model	Pre-Trained Method	OIS	ODS
UNet	-	0.769	0.766
HED	-	0.811	0.806
RCF	-	0.806	0.800
DeeplabM	-	0.759	0.756
DeeplabV3	Imagenet-pre-trained	0.849	0.845
FaultSeg3D	-	0.840	0.836
UNETR	-	0.845	0.841
UNETR	SimCLR	0.849	0.845
UNETR	MAE	0.847	0.844
UNETR	SimMIM	0.846	0.843
Swin-UNETR	-	0.844	0.840
Swin-UNETR	Three-task-pre-trained	0.852	0.845
Swin-UNETR	SimCLR	0.855	0.852
Swin-UNETR	MAE	0.859	0.857
Swin-UNETR	SimMIM	0.861	0.857
Swin-UNETR	Overall SimMIM	0.872	0.868
FaultSeg Swin-UNETR	-	0.851	0.847
FaultSeg Swin-UNETR	SimMIM	0.866	0.862
FaultSeg Swin-UNETR	Overall SimMIM	**0.875**	**0.870**

4.4. Ablation Study

Although neural networks based on the Transformer model require a large number of training data to achieve outstanding detection performance, we conducted fine-tuning experiments on a small-scale synthesized dataset to demonstrate that self-supervised pre-training can still achieve specific results under such stringent conditions. For our experiments, we used a synthetic dataset of 220 samples of size $128 \times 128 \times 128$, as used by Wu et al. [21] in the FaultSeg3D model. We only used 200 samples to train the model, and the model selection was based on the performance of 20 validation samples from the synthetic dataset. After completing the fine-tuning, we directly evaluated the performance of the trained model on the test set of Thebe using the OIS and ODS metrics (Table 3). Because of the smaller scale and lower quality of the synthetic dataset compared to field seismic data, the model's performance showed a significant drop. However, compared to the FaultSeg3D model, the Swin-UNETR model performed significantly better, and further improvements can be achieved through the SimMIM self-supervised pre-training. Similar to previous conclusions, using the FaultSeg Swin-UNETR model combined with self-supervised pre-training yielded the best results. Moreover, the model could still achieve acceptable predictive performance on a relatively more minor annotated dataset, which indicates that our method can maintain decent performance in transferring between different work regions even with limited annotated data. This finding provides a baseline data volume reference for new work regions with less data. In addition, it also suggests our method provides an efficient pre-trained model. This pre-trained model offers an optimal starting point for fault prediction in new work regions. This experiment also significantly improves deep network performance by introducing large-scale annotated data. Though the Thebe dataset is currently the most extensive earthquake fault annotation dataset, it still needs more data compared to medical imaging to fully demonstrate the advantages of deep models.

Table 3. Testing results on the Thebe dataset using our proposed method fine-tuned on a small synthetic dataset.

Model	Pre-Trained Method	OIS	ODS
FaultSeg3D	-	0.441	0.431
Swin-UNETR	Three-task-pre-trained	0.482	0.473
Swin-UNETR	SimMIM	0.503	0.495
FaultSeg Swin-UNETR	Overall SimMIM	0.525	0.510

Because of the sizeable seismic data volume, it is often necessary to divide it into blocks for training data in the deep network. However, because of the limitation of GPU memory, the size of the input data given to the segmentation network is only 128 × 128 × 128. We used the sliding window method during prediction to infer the entire test data volume. To improve the consistency between sliding windows, we usually chose half of the window size as the sliding stride and took the average value of logits at the overlapping areas. In our deep network setup, the window size was 128 × 128 × 128, and the sliding stride was 64 × 64 × 64. However, the choice of block size during data partitioning was also worth considering. Since random cropping during the training process can serve as a powerful data augmentation technique and help the model learn fracture information at different positions in the data volume, it was necessary to set the block size larger than the input size that the network could accept. Therefore, we explored the block size of the data volume based on the Swin-UNETR baseline model. From Table 4, it can be seen that if the block size is set to 128 × 128 × 128, which means no random cropping during the training process, the performance is the worst in terms of OIS and ODS indicators. As we gradually increase the block size, both evaluation metrics significantly improve. However, the block size should not be too large, as this can lead to a significant difference in the randomly cropped 128-sized inputs encountered by each epoch when fitting the same data block, making it difficult for the network to converge. In this baseline model, we found that the OIS and ODS scores were highest when the block size was 256. Therefore, we used this setting in all experiments, as shown in Table 2.

Table 4. Comparison of different data chunking sizes.

Model	Chunk Size	OIS	ODS
Swin-UNETR	128 × 128 × 128	0.830	0.827
	192 × 192 × 192	0.841	0.835
	256 × 256 × 256	0.844	0.840
	512 × 512 × 512	0.832	0.829

The proportion of positive samples in the fault labels is too low (Figure 5). Therefore, it is easy to consider increasing the weight of positive samples in the BCE loss to enhance the model's ability to detect faults. However, adjusting the weight of positive samples only changes the balance between precision and recall and does not significantly affect the OIS and ODS, which are evaluated using the F1 score. This is also why we do not use the average precision (AP) as an evaluation metric. We can increase the precision of the model by adjusting the weight of positive samples in the BCE loss. However, this comes at the cost of reducing recall. The experimental results in Table 5 also confirm our observation. Therefore, we used the regular BCE loss without any special weighting for positive samples in the remaining experiments.

Table 5. The impact of BCE positive weight.

Model	Positive Weight	OIS	ODS
Swin-UNETR	1.0	0.844	0.840
	5.0	0.845	0.839
	10.0	0.843	0.841

5. Discussion

This section presents test cases for the 2D slices and 3D cubes of the Thebe dataset. The 2D showcase is shown in Figure 10, and the 3D showcase is shown in Figure 11.

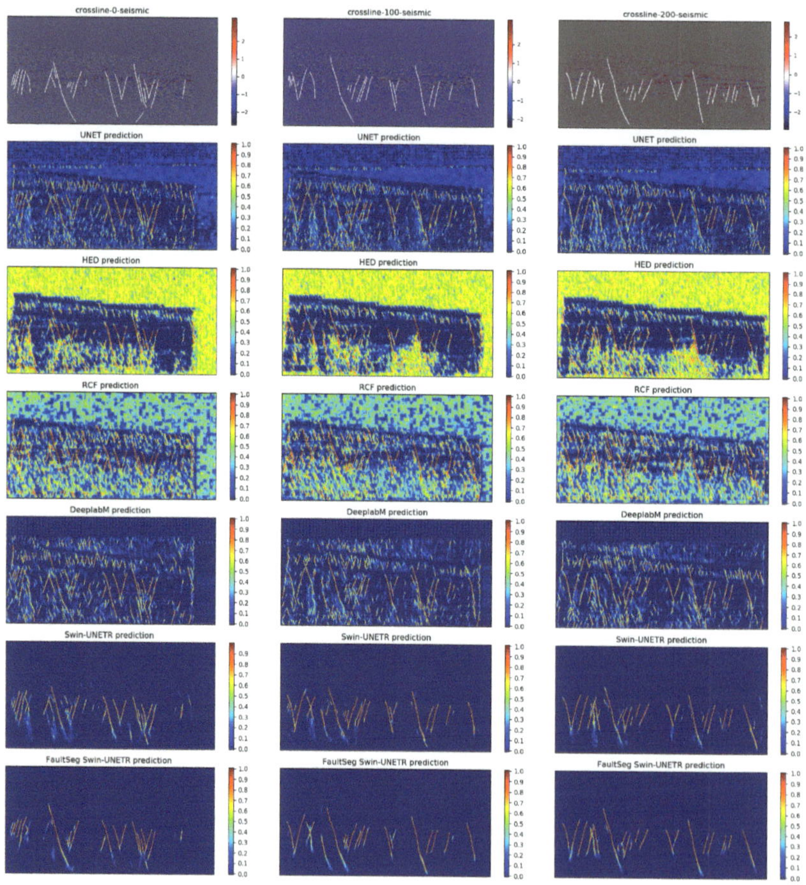

Figure 10. Comparison of different deep learning models on the Thebe test sets' 0th, 100th, and 200th crosslines. The first row represents expert annotations, followed by the second through fifth rows, displaying the inferences from the models proposed by An et al. [23] for comparison. The sixth and seventh rows present the visual probability maps predicted by Swin-UNETR and our FaultSeg Swin-UNETR, respectively.

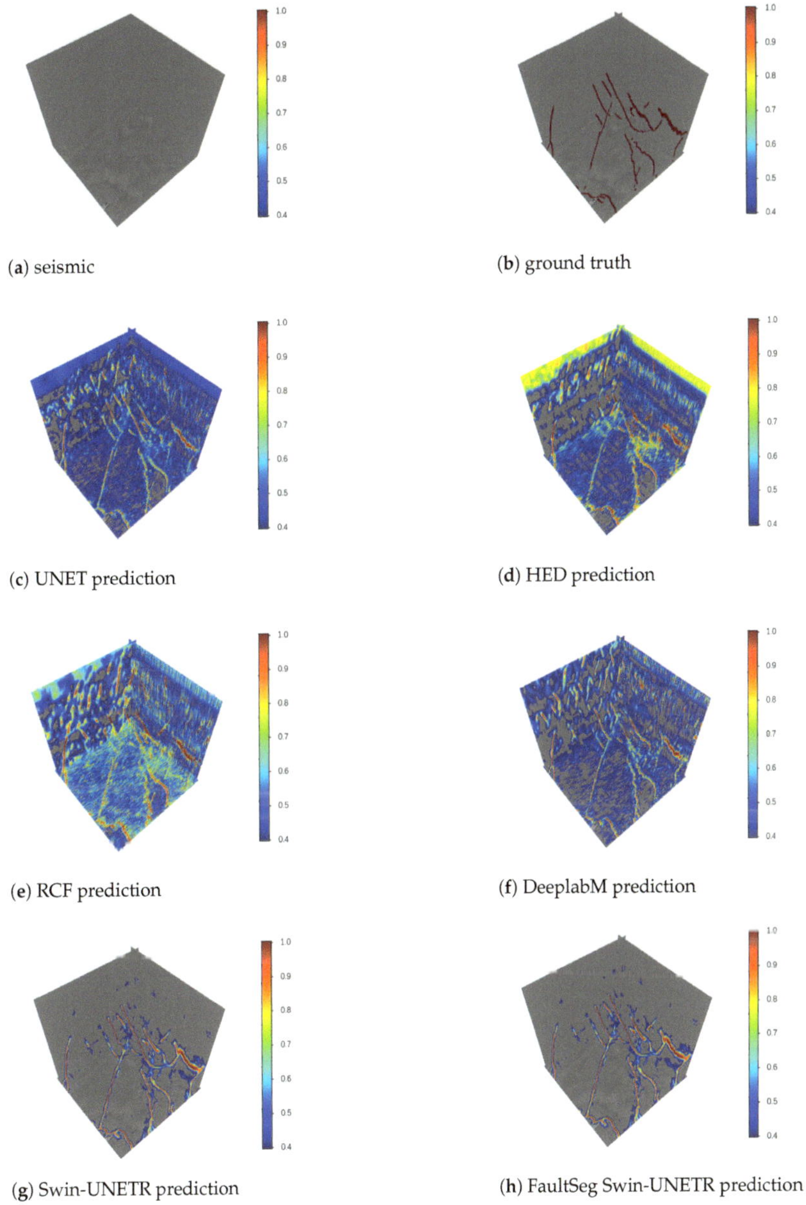

(**a**) seismic (**b**) ground truth

(**c**) UNET prediction (**d**) HED prediction

(**e**) RCF prediction (**f**) DeeplabM prediction

(**g**) Swin-UNETR prediction (**h**) FaultSeg Swin-UNETR prediction

Figure 11. The 3D images are 640 × 640 × 640 cubes extracted from the testing subset of the Thebe dataset. We superimposed various deep learning models' predictive probabilities onto profiles in different directions, denoted as (**c**–**h**) for UNET, HED, RCF, DeeplabM, Swin-UNETR, and FaultSeg Swin-UNETR, respectively. Notably, FaultSeg Swin-UNETR accurately delineated the majority of fault lines.

5.1. Performance of 2D Slices

We tested the performance of several fault recognition methods on parts of the Thebe seismic data test set. Figure 10 displays the crossline of 0, 100, and 200 traces. The white lines in the first row represent the real faults of the strata on the seismic images as interpreted by experts for comparison with the performance of each automatic fault recognition model. We use fault likelihood to represent the prediction results of each model. The range of each pixel in the figure is zero to one. Values close to one (red) represent a higher probability of faults, and values close to zero (blue) represent a higher probability of non-faults. As expected, as shown in the quantitative evaluation results in Table 2, compared to the prediction results of the four models proposed by An et al. [23]—UNET (second row), HED (third row), RCF (fourth row), DeeplabM (fifth row)—Swin-UNTER achieves significant performance improvement, and our model achieves more accurate interpretations. Continuous fault prediction for the Thebe dataset means better results [23]. Specifically, the four models in Figure 10 (second through fifth rows) are susceptible to noise. In the probability map, many minor points and line segments appear in the non-fault part and are misjudged as faults, confirming the work [23], which said, "All four models struggle with the migration noise within the deeper layers, which the models do not see during the training phase". In comparison, UNet and DeeplabM are more accessible to identify non-faults than HED and RCF. At the same time, the HED and RCF models with weighted cross-entropy loss functions pay more attention to the fault part, so it is easier to misjudge non-fault pixels as faults. In particular, model RCF goes to the extreme of labeling non-fault pixels or noise as faults. Additionally, when calculating OIS and ODS values, these four models only extract the concentrated distribution of faults in the image. However, we lack prior knowledge of the concentrated fault areas in practical applications. The significant misjudgment in non-fault areas challenges the models when applied in real scenarios. Encouragingly, the Swin-UNETR segmentation model shows good prediction performance. Its prediction outcomes align more closely with expert interpretations than the previous four models, effectively identifying noise and non-fault areas, indicating this model's successful extraction of noise and non-fault features. The model structure and self-supervised pre-training tasks effectively improve the fault recognition capabilities. However, Swin-UNETR also has some redundant false detections and artifacts, and some faults are disconnected. This is due to the sparse and narrow earthquake faults, resulting in highly imbalanced data annotation. Therefore, the FaultSeg Swin-UNETR we proposed solves this problem to a certain extent through a multi-scale fusion decoder. As shown in the last row of Figure 10, our method is improved based on Swin-UNETR. Most fault points have a very high probability (close to one), the noise points are minimal, the predicted faults are complete, and they have more excellent continuity and refinement, almost comparable to expert explanations.

5.2. Performance of 3D Volumes

To further verify our model's fault recognition performance on 3D seismic volumes, we visualize the 3D inferences of Swin-UNETR, FaultSeg Swin-UNETR, and four other fault recognition methods on the Thebe test set in Figure 11. Figure 11c–f display six fault recognition probability maps computed by UNET, HED, RCF, DeeplabM, Swin-UNETR, and our FaultSeg Swin-UNETR methods. Figure 11a exhibits the extracted seismic volume, while the deep red annotations in Figure 11b represent seismic faults interpreted by experts.

Consistent with the 2D scenario, FaultSeg Swin-UNETR performs best in calculation accuracy, fault clarity, and completeness compared to the former four methods. It showcases more apparent fault characteristics and more continuous fault tracking in the 3D visualization. To observe multiple sets of fault features in different directions distinctly, we specifically visualize fault hard labels in Figure 12 using the following method:

$$\begin{aligned} \text{label}[Y_{pred} < 0.7] &= 0 \\ \text{label}[Y_{pred} \geq 0.7] &= 1 \end{aligned} \tag{9}$$

where Y_{pred} is the model output score, i.e., the probability of that pixel point being predicted as a fault. We regard voxels with a prediction probability greater than 0.7 as faults and the others as non-faults. The faults obtained by the former four methods are noisier, with numerous misidentified non-fault regions, discontinuities at connection points, and a lack of coherent structures within complex fault areas. However, our method provides clear and accurate fault segmentation results with good continuity at the intersection of three faces, aligning closely with real fault annotations. Dou et al. [43] explained that larger angles between faults and slices lead to poorer predictive outcomes. The skewed distribution of faults in the Thebe dataset poses substantial learning challenges for models and likely contributes to misjudgments in the former four models. Our model adeptly extracts 3D features, significantly enhancing predictive consistency. Compared to Swin-UNETR, FaultSeg Swin-UNETR demonstrates finer fault segmentation results with fewer misidentified fault points, indicating the effective enhancement of the model's ability to characterize faults through the multi-scale fusion module.

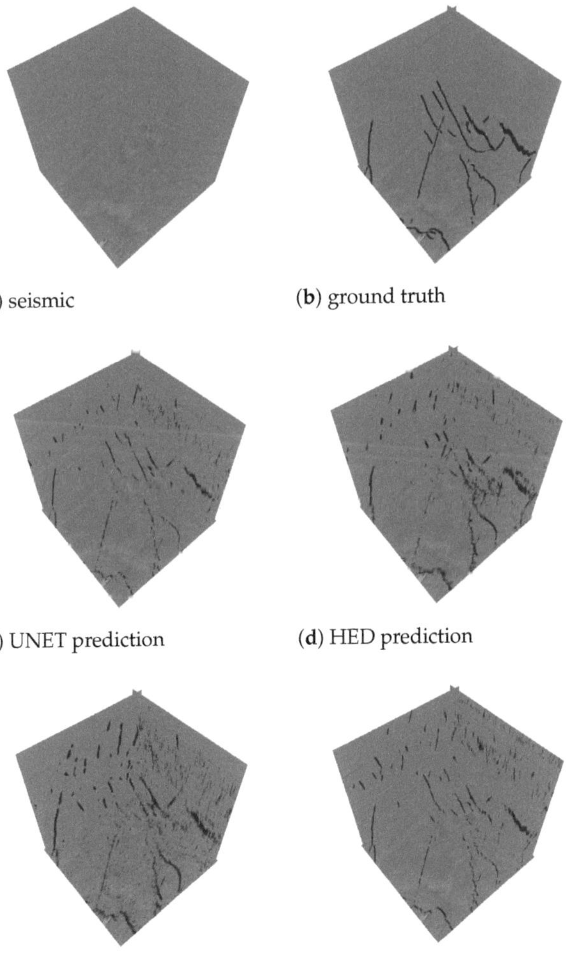

(**a**) seismic (**b**) ground truth

(**c**) UNET prediction (**d**) HED prediction

(**e**) RCF prediction (**f**) DeeplabM prediction

Figure 12. *Cont.*

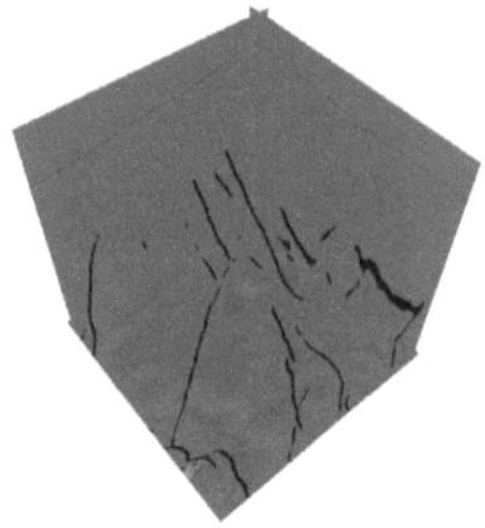

(g) Swin-UNETR prediction

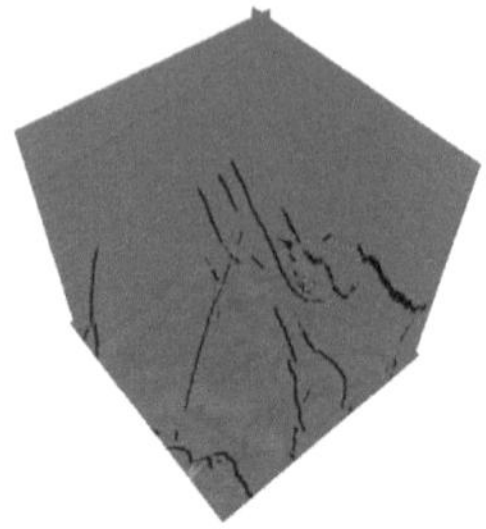

(h) FaultSeg Swin-UNETR prediction

Figure 12. Visualizations of the hard scores predicted by the six models on the Thebe dataset are presented in a 3D cube. The deep red indicates that the probability of a pixel being predicted as a fault is equal to or greater than 70%; otherwise, it is considered a non-fault.

6. Conclusions

In our work, we explore the application of the most popular self-supervised pre-training in seismic fault recognition in recent years.

In summary, the contributions of this article are as follows:

- Utilizing the 3D Swin-Transformer backbone network, we investigated diverse pre-training methods with a substantial volume of field 3D seismic data. The integration of SimMIM's pre-training method with the enhanced Swin-UNETR model markedly improved performance. Consequently, we introduced FaultSeg Swin-UNETR, a method meticulously crafted for the unique characteristics of seismic data.
- We improved the Swin-UNETR model structure to adapt to the sparse distribution of seismic fault data and the narrow line profile characteristics in the inline or cross-line directions, promoting multi-scale decoding and fusion, thereby making fault detection easier.
- Furthermore, upon recognizing the significant imbalance in the number of parameters between the decoder and the backbone network in the Swin-UNETR model, we proposed a strategy for pre-training the complete segmentation model, which further improves fault detection accuracy.
- In the end, our proposed method achieves state-of-the-art performances on the Thebe dataset according to the standard metrics of the optimal image scale (OIS) and optimal dataset scale (ODS) metrics.
- Our research significantly advances the precision and efficiency of seismic fault recognition, overcoming the constraints associated with reliance on annotated datasets. This breakthrough paves the way for developing more robust and generalizable models capable of addressing the inherent complexities of field seismic data.

Our research robustly demonstrates the substantial potential of self-supervised learning in interpreting seismic data. Using self-supervised pre-training on extensive datasets, we find that models can generalize and significantly enhance identification accuracy, even when only a minimal number of labeled data are available in new work zones. This is particularly valuable considering the often complex and costly nature of annotating seismic datasets. The models can identify and segment faults, thus aiding geologists and engineers in precisely understanding subsurface structures, which have potential application values in earthquake risk assessment, research on seismic resilience of infrastructure and buildings [44,45], etc. However, our current focus has been mainly on fault recognition. Future work can include a broader range of seismic image-to-image tasks, such as stratigraphic layer segmentation. Implementing multi-objective training can improve the model's feature extraction and generalization capabilities.

Furthermore, our validation work has been confined to the two publicly available datasets we could access because of the scarcity of publicly available labeled datasets. Given the opportunity, we would conduct extensive testing across various geological conditions. In summary, seismic analysis needs a universally applicable pre-trained model akin to ImageNet [46] in computer vision or the GPT [47] series in natural language processing. In addition, there needs to be more high-quality, open-source, large-scale, and diverse seismic datasets within the field. Given these challenges, it is imperative to forge a data-driven, unified model akin to SAM [48]. A model fortified with semi-automated annotation capabilities will significantly enlarge our fault dataset, propelling us toward creating a universally applicable fault recognition AI.

Author Contributions: Conceptualization, J.M. and Z.Z.; methodology, Z.Z. and R.C.; software, Z.Z.; validation, Z.Z. and R.C.; formal analysis, R.C.; investigation, Z.Z. and R.C.; writing—original draft preparation, Z.Z. and R.C.; writing—review and editing, J.M. and R.C.; visualization, Z.Z. and R.C.; supervision, J.M.; project administration, J.M. All authors have read and agreed to the published version of the manuscript.

Funding: This research was funded by the National Key Research and Development Program of China under Grant 2022YFB2703002.

Data Availability Statement: In this study, the pre-training dataset cannot be disclosed because of privacy constraints. Nevertheless, to ensure transparency and reproducibility, we have shared the checkpoint of the pre-training network on private data. Experts can use this checkpoint to replicate our methods. This selection serves the research objectives while minimizing potential impacts on privacy. The Thebe datasets we used in this paper are available in Harvard Dataverse: https://dataverse.harvard.edu/dataset.xhtml?persistentId=doi:10.7910/DVN/YBYGBK (accessed on 6 May 2023).

Acknowledgments: Supported by the High-performance Computing Platform of Peking University. Thanks for the private unlabeled seismic data provided by BGP Inc., China National Petroleum Corporation.

Conflicts of Interest: The authors declare no conflicts of interest.

Appendix A

Our study incorporates the Swin-Transformer encoder, a sophisticated feature extraction network tailored for analyzing 3D data. This network processes three dimensional seismic data, denoted as $S \in \mathbb{R}^{H \times W \times D}$, where H, W, and D represent the dimensions of inline, crossline, and timeline, respectively. The initial step involves a patch partitioning of the input 3D seismic data into non-overlapping volumetric blocks, each of size $2 \times 2 \times 2$. This partitioning results in a total of $\frac{H}{2} \times \frac{W}{2} \times \frac{D}{2}$ blocks, which are subsequently flattened and mapped into a 48-dimensional embedding space through a linear projection layer, forming the inputs for the Swin-Transformer.

The architecture of the Swin-Transformer is divided into four distinct stages, each meticulously designed to extract information from the 3D data volumes in a hierarchical manner. Within each stage, a reduction of the spatial dimensions by half along each axis leads to a significant decrease in the number of tokens, specifically to one-eighth of their original count, while simultaneously doubling the channel count. This process is visually represented in the left half of Figure 7, which annotates the dimensions of the feature maps fed into each stage. Comprising Swin-Transformer blocks and a merging module, each stage is pivotal in transforming the feature maps, with the merging module solely responsible for all such deformations.

The detailed structure of a Swin-Transformer block is showcased in the right half of Figure 7, emphasizing the employment of two window-based self-attention mechanisms: window multi-head self-attention (W-MSA) and shifted-window multi-head self-attention (SW-MSA). These mechanisms are instrumental in calculating self-attention scores among the input tokens. Considering $s^{in} \in \mathbb{R}^{H' \times W' \times D' \times C'}$ as the input to a Swin-Transformer block, where it is perceived as tokens of dimension C' arranged across a grid of

$H' \times W' \times D'$, the forward pass through this module can be articulated through the following set of equations. It is noteworthy that the dimensions of the feature maps are preserved throughout this process:

$$s' = \text{W-MSA}(\text{LN}(s^{in})) + s^{in},$$
$$s'' = \text{MLP}(\text{LN}(s')) + s',$$
$$s''' = \text{SW-MSA}(\text{LN}(s'')) + s'',$$
$$s^{out} = \text{MLP}(\text{LN}(s''')) + s'''.$$

In this context, LN denotes the layer normalization layer, strategically placed before each sub-module within the network to address the internal covariate shift dilemma encountered during training phases. The MLP, or multi-layer perceptron, along with W-MSA and SW-MSA, the dual window-based self-attention mechanisms devised for the Swin-Transformer, play crucial roles. These mechanisms partition the input tokens into groups confined by a predetermined window size of $M \times M \times M$, leading to $\frac{H'}{M} \times \frac{W'}{M} \times \frac{D'}{M}$ groups, within which self-attention scores are computed as follows:

$$\text{Attention}(Q, K, V) = \text{Softmax}\left(\frac{QK^T}{\sqrt{d_k}}\right)V$$

Here, $Q = W^q s$, $K = W^k s$, and $V = W^v s$ represent the query, key, and value components within the self-attention framework, respectively, with d_k reflecting their dimensionalities. The matrices W^q, W^k, and W^v facilitate the transformation of the input s into Q, K, and V, respectivelyx. In the W-MSA approach, tokens are partitioned in a manner that allocates equal window sizes across the board. Conversely, the SW-MSA strategy involves segmenting the input tokens by offsetting the windows along the three spatial dimensions by $\frac{M}{2}$ each. Figure A1 shows the visual results of these two window partitioning mechanisms. This partitioning scheme effectively balances the computational demands of the self-attention mechanism with the interactions among tokens across different window partitions, ensuring a harmonious integration of computational efficiency and inter-token dynamics.

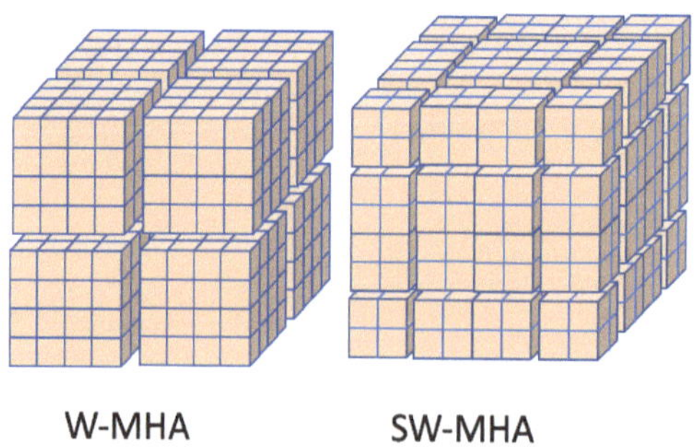

Figure A1. W-MHA and SW-MHA window partitioning mechanism.

References

1. Alcalde, J.; Bond, C.E.; Johnson, G.; Kloppenburg, A.; Ferrer, O.; Bell, R.; Ayarza, P. Fault interpretation in seismic reflection data: An experiment analysing the impact of conceptual model anchoring and vertical exaggeration. *Solid Earth* **2019**, *10*, 1651–1662. [CrossRef]
2. Bouvier, J.; Kaars-Sijpesteijn, C.; Kluesner, D.; Onyejekwe, C.; Van der Pal, R. Three-dimensional seismic interpretation and fault sealing investigations, Nun River Field, Nigeria. *AAPG Bull.* **1989**, *73*, 1397–1414.
3. Wu, X.; Geng, Z.; Shi, Y.; Pham, N.; Fomel, S.; Caumon, G. Building realistic structure models to train convolutional neural networks for seismic structural interpretation. *Geophysics* **2020**, *85*, WA27–WA39. [CrossRef]
4. Fossen, H. *Structural Geology*; Cambridge University Press: New York, NY, USA, 2010; pp. 205–207.
5. Posamentier, H.W.; Davies, R.J.; Cartwright, J.A.; Wood, L. *Seismic Geomorphology—An Overview*; Special Publications; Geological Society: London, UK, 2007.

6. Knipe, R.J.; Jones, G.; Fisher, Q. *Faulting, Fault Sealing and Fluid Flow in Hydrocarbon Reservoirs: An Introduction*; Special Publications; Geological Society: London, UK, 1998.
7. Ottesen Ellevset, S.; Knipe, R.; Svava Olsen, T.; Fisher, Q.; Jones, G. *Fault Controlled Communication in the Sleipner Vest Field, Norwegian Continental Shelf: Detailed, Quantitative Input for Reservoir Simulation and Well Planning*; Special Publications; Geological Society: London, UK, 1998; Volume 147. pp. 283–297.
8. Richards, F.L.; Richardson, N.J.; Bond, C.E.; Cowgill, M. *Interpretational Variability of Structural Traps: Implications for Exploration Risk and Volume Uncertainty*; Geological Society: London, UK, 2015; Volume 421, pp. 7–27.
9. Hale, D. Fault surfaces and fault throws from 3D seismic images. In Proceedings of the 2012 SEG Annual Meeting, Las Vegas, NV, USA, 4–9 November 2012.
10. Stark, T.J. Unwrapping instantaneous phase to generate a relative geologic time volume. In Proceedings of the 2003 SEG Annual Meeting, Dallas, TX, USA, 26–31 October 2003.
11. Wu, X.; Zhong, G. Generating a relative geologic time volume by 3D graph-cut phase unwrapping method with horizon and unconformity constraints. *Geophysics* **2012**, *77*, O21–O34. [CrossRef]
12. Silva, C.C.; Marcolino, C.S.; Lima, F.D. Automatic fault extraction using ant tracking algorithm in the Marlim South Field, Campos Basin. In Proceedings of the 2005 SEG Annual Meeting, Houston, TX, USA, 6–11 November 2005.
13. Pedersen, S.I.; Randen, T.; Sønneland, L.; Steen, Ø. Automatic fault extraction using artificial ants. In Proceedings of the 2002 SEG Annual Meeting, Salt Lake City, UT, USA, 6–9 October 2002.
14. Figueiredo, A.M.; Gattass, M.; Szenberg, F. Seismic horizon mapping across faults with growing neural gas. In Proceedings of the 10th International Congress of the Brazilian Geophysical Society, Rio de Janeiro, Brazil, 19–23 November 2007.
15. Zinck, G.; Donias, M.; Daniel, J.; Guillon, S.; Lavialle, O. Fast seismic horizon reconstruction based on local dip transformation. *J. Appl. Geophys.* **2013**, *96*, 11–18. [CrossRef]
16. Wang, Z.; AlRegib, G. Automatic fault surface detection by using 3D Hough transform. In Proceedings of the 2014 SEG Annual Meeting, Denver, CO, USA, 26–31 October 2014.
17. Bishop, C.M.; Nasrabadi, N.M. *Pattern Recognition and Machine Learning*; Springer: New York, NY, USA, 2006; Volume 4.
18. Kortström, J.; Uski, M.; Tiira, T. Automatic classification of seismic events within a regional seismograph network. *Comput. Geosci.* **2016**, *87*, 22–30. [CrossRef]
19. Guitton, A.; Wang, H.; Trainor-Guitton, W. Statistical imaging of faults in 3D seismic volumes using a machine learning approach. In *SEG Technical Program Expanded Abstracts 2017*; Society of Exploration Geophysicists: Houston, TX, USA, 2017; pp. 2045–2049.
20. Zhao, T.; Mukhopadhyay, P. A fault detection workflow using deep learning and image processing. In Proceedings of the 2018 SEG International Exposition and Annual Meeting, Anaheim, CA, USA, 14–19 October 2018.
21. Wu, X.; Shi, Y.; Fomel, S.; Liang, L. Convolutional neural networks for fault interpretation in seismic images. In Proceedings of the 2018 SEG International Exposition and Annual Meeting, Anaheim, CA, USA, 14–19 October 2018.
22. Wu, X.; Liang, L.; Shi, Y.; Fomel, S. FaultSeg3D: Using synthetic data sets to train an end-to-end convolutional neural network for 3D seismic fault segmentation. *Geophysics* **2019**, *84*, IM35–IM45. [CrossRef]
23. An, Y.; Guo, J.; Ye, Q.; Childs, C.; Walsh, J.; Dong, R. Deep convolutional neural network for automatic fault recognition from 3D seismic datasets. *Comput. Geosci.* **2021**, *153*, 104776. [CrossRef]
24. Vaswani, A.; Shazeer, N.; Parmar, N.; Uszkoreit, J.; Jones, L.; Gomez, A.N.; Kaiser, Ł.; Polosukhin, I. Attention is all you need. In Proceedings of the 31st Conference on Neural Information Processing Systems (NIPS 2017), Long Beach, CA, USA, 4–9 December 2017, Volume 30.
25. Dosovitskiy, A.; Beyer, L.; Kolesnikov, A.; Weissenborn, D.; Zhai, X.; Unterthiner, T.; Dehghani, M.; Minderer, M.; Heigold, G.; Gelly, S.; et al. An image is worth 16 × 16 words: Transformers for image recognition at scale. *arXiv* **2020**, arXiv:2010.11929.
26. Liu, Z.; Lin, Y.; Cao, Y.; Hu, H.; Wei, Y.; Zhang, Z.; Lin, S.; Guo, B. Swin transformer: Hierarchical vision transformer using shifted windows. In Proceedings of the IEEE/CVF International Conference on Computer Vision, Montreal, BC, Canada, 11–17 October 2021; pp. 10012–10022.
27. He, K.; Chen, X.; Xie, S.; Li, Y.; Dollár, P.; Girshick, R. Masked autoencoders are scalable vision learners. In Proceedings of the IEEE/CVF Conference on Computer Vision and Pattern Recognition, New Orleans, LA, USA, 18–24 June 2022; pp. 16000–16009.
28. Xie, Z.; Zhang, Z.; Cao, Y.; Lin, Y.; Bao, J.; Yao, Z.; Dai, Q.; Hu, H. Simmim: A simple framework for masked image modeling. In Proceedings of the IEEE/CVF Conference on Computer Vision and Pattern Recognition, New Orleans, LA, USA, 18–24 June 2022; pp. 9653–9663.
29. Chen, T.; Kornblith, S.; Norouzi, M.; Hinton, G. A simple framework for contrastive learning of visual representations. *PMLR* **2020**, *119*, 1597–1607.
30. Liu, B.; Jiang, P.; Wang, Q.; Ren, Y.; Yang, S.; Cohn, A.G. Physics-driven self-supervised learning system for seismic velocity inversion. *Geophysics* **2023**, *88*, R145–R161. [CrossRef]
31. Monteiro, B.A.; Oliveira, H.; dos Santos, J.A. Self-Supervised Learning for Seismic Image Segmentation From Few-Labeled Samples. *IEEE Geosci. Remote Sens. Lett.* **2022**, *19*, 8028805. [CrossRef]
32. Xu, Z.; Luo, Y.; Wu, B.; Meng, D.; Chen, Y. Deep Nonlocal Regularizer: A Self-Supervised Learning Method for 3D Seismic Denoising. *IEEE Trans. Geosci. Remote. Sens.* **2023**, *61*, 5921517. [CrossRef]
33. Yang, J.; Wu, X.; Bi, Z.; Geng, Z. A multi-task learning method for relative geologic time, horizons, and faults with prior information and transformer. *IEEE Trans. Geosci. Remote. Sens.* **2023**, *61*, 5907720. [CrossRef]

34. Tang, Z.; Wu, B.; Wu, W.; Ma, D. Fault Detection via 2.5 D Transformer U-Net with Seismic Data Pre-Processing. *Remote Sens.* **2023**, *15*, 1039. [CrossRef]
35. Silva, R.M.; Baroni, L.; Ferreira, R.S.; Civitarese, D.; Szwarcman, D.; Brazil, E.V. Netherlands dataset: A new public dataset for machine learning in seismic interpretation. *arXiv* **2019**, arXiv:1904.00770.
36. Wang, Z.; You, J.; Liu, W.; Wang, X. Transformer assisted dual U-net for seismic fault detection. *Front. Earth Sci.* **2023**, *11*, 1047626. [CrossRef]
37. Dou, Y.; Dong, M.; Li, K.; Xiao, Y. FaultSSL: Seismic Fault Detection via Semi-supervised learning. *arXiv* **2023**, arXiv:2309.02930.
38. Tang, Y.; Yang, D.; Li, W.; Roth, H.R.; Landman, B.; Xu, D.; Nath, V.; Hatamizadeh, A. Self-supervised pre-training of swin transformers for 3d medical image analysis. In Proceedings of the IEEE/CVF Conference on Computer Vision and Pattern Recognition, New Orleans, LA, USA, 18–24 June 2022; pp. 20730–20740.
39. An, Y.; Du, H.; Ma, S.; Niu, Y.; Liu, D.; Wang, J.; Du, Y.; Childs, C.; Walsh, J.; Dong, R. Current state and future directions for deep learning based automatic seismic fault interpretation: A systematic review. *Earth-Sci. Rev.* **2023**, *243*, 104509. [CrossRef]
40. An, Y.; Guo, J.; Ye, Q.; Childs, C.; Walsh, J.; Dong, R. A gigabyte interpreted seismic dataset for automatic fault recognition. *Data Brief* **2021**, *37*, 107219. [CrossRef]
41. Hatamizadeh, A.; Tang, Y.; Nath, V.; Yang, D.; Myronenko, A.; Landman, B.; Roth, H.R.; Xu, D. Unetr: Transformers for 3d medical image segmentation. In Proceedings of the IEEE/CVF Conference on Computer Vision and Pattern Recognition, New Orleans, LA, USA, 18–24 June 2022; pp. 574–584.
42. An, Y.; Dong, R. Understanding the Effect of Different Prior Knowledge on CNN Fault Interpreter. *IEEE Access* **2023**, *11*, 15058–15068. [CrossRef]
43. Dou, Y.; Li, K.; Zhu, J.; Li, T.; Tan, S.; Huang, Z. MD loss: Efficient training of 3-D seismic fault segmentation network under sparse labels by weakening anomaly annotation. *IEEE Trans. Geosci. Remote Sens.* **2022**, *60*, 5919014. [CrossRef]
44. Kourehpaz, P.; Molina Hutt, C. Machine learning for enhanced regional seismic risk assessments. *J. Struct. Eng.* **2022**, *148*, 04022126. [CrossRef]
45. Forcellini, D. An expeditious framework for assessing the seismic resilience (SR) of structural configurations. *Structures* **2023**, *56*, 105015. [CrossRef]
46. Deng, J.; Dong, W.; Socher, R.; Li, L.J.; Li, K.; Fei-Fei, L. Imagenet: A large-scale hierarchical image database. In Proceedings of the 2009 IEEE Conference on Computer Vision and Pattern Recognition, Miami, FL, USA, 20–25 June 2009; pp. 248–255.
47. Floridi, L.; Chiriatti, M. GPT-3: Its nature, scope, limits, and consequences. *Minds Mach.* **2020**, *30*, 681–694. [CrossRef]
48. Kirillov, A.; Mintun, E.; Ravi, N.; Mao, H.; Rolland, C.; Gustafson, L.; Xiao, T.; Whitehead, S.; Berg, A.C.; Lo, W.Y.; et al. Segment anything. *arXiv* **2023**, arXiv:2304.02643.

remote sensing

Article

Tracking the 2D/3D Morphological Changes of Tidal Flats Using Time Series Remote Sensing Data in Northern China

Zhiquan Gan [1], Shurong Guo [1], Chunpeng Chen [2,3], Hanjie Zheng [1], Yuekai Hu [2], Hua Su [1] and Wenting Wu [1,2,*]

1 Key Laboratory of Spatial Data Mining and Information Sharing of Ministry of Education, National & Local Joint Engineering Research Center of Satellite Geospatial Information Technology, Fuzhou University, Fuzhou 350108, China; 215527035@fzu.edu.cn (Z.G.); 235527017@fzu.edu.cn (S.G.); 215527033@fzu.edu.cn (H.Z.); suhua@fzu.edu.cn (H.S.)
2 State Key Laboratory of Estuarine and Coastal Research, East China Normal University, Shanghai 200050, China; cpchen@stu.ecnu.edu.cn (C.C.); ykhu.sklec@stu.ecnu.edu.cn (Y.H.)
3 Lancaster Environment Centre, Lancaster University, Lancaster LA1 4YQ, UK
* Correspondence: wuwt@fzu.edu.cn

Abstract: Tidal flats in northern China are essential parts of the East Asian-Australasian Flyway, the densest pathway for migratory waterbirds, and are of great ecological and economic importance. They are threatened by human activities and climate change, raising the urgency surrounding tracking the spatiotemporal dynamics of tidal flats. However, there is no cost-effective way to map morphological changes on a large spatial scale due to the inaccessibility of the mudflats. In this study, we proposed a pixel-based multi-indices tidal flat mapping algorithm that precisely characterizes 2D/3D morphological changes in tidal flats in northern China using time-series remote sensing data. An overall accuracy of 0.95 in delineating tidal flats to a 2D extent was achieved, with 11,716 verification points. Our results demonstrate that the reduction in sediment discharge from rivers along the coastlines of the Yellow and Bohai Seas has resulted in an overall decline in the area of tidal flats, from 4856.40 km^2 to 4778.32 km^2. Specifically, 3D analysis showed that significant losses were observed in the mid-to-high-tidal flat zones, while low-elevation tidal flats experienced an increase in area due to the transformations in mid-to-high-tidal flats. Our results indicate that the sediment inputs from rivers and the succession of native vegetation are the primary drivers leading to 2D/3D morphological changes of tidal flats following the cessation of extensive land reclamation in northern China.

Keywords: tidal flats; 2D/3D morphological changes; remote sensing; time series; Yellow and Bohai Seas

Citation: Gan, Z.; Guo, S.; Chen, C.; Zheng, H.; Hu, Y.; Su, H.; Wu, W. Tracking the 2D/3D Morphological Changes of Tidal Flats Using Time Series Remote Sensing Data in Northern China. *Remote Sens.* **2024**, *16*, 886. https://doi.org/10.3390/rs16050886

Academic Editor: Jorge Vazquez

Received: 15 January 2024
Revised: 21 February 2024
Accepted: 29 February 2024
Published: 2 March 2024

1. Introduction

Tidal flats, situated between the average high tide and average low tide lines, are integral components of coastal zones [1]. They act as transitional zones linking terrestrial and marine ecosystems [2] and provide a number of ecological services, including storm surge protection and carbon sequestration [3,4]. Tidal flats also provide habitats for a diverse range of wildlife, including migratory birds [5], fish, and other marine organisms [6,7]. Moreover, they play a crucial role in supporting economic activities such as port facilities, tourism, and fisheries [3,8]. Global tidal flats have decreased in area by 16% (>20,000 km^2) [9]. Approximately 10% of the global population resides in coastal regions with elevations below 10 m [10]. However, tidal flats face increasing threats from human activities, such as land reclamation [11,12] and aquaculture [13], which amplifies the exposure of tidal flats to coastal erosion [14] and rising sea levels driven by climate change [15,16]. It is crucial to carefully consider the balance between the preservation and utilization of tidal flats to ensure the future sustainable development of coastal areas.

Therefore, accurate monitoring and understanding of the spatial distribution and temporal changes in tidal flat morphology are of paramount importance and necessity.

The tidal flats in northern China, along with the shorelines of the Yellow Sea and Bohai Sea (YBS), have exhibited tremendous losses in tidal flat area due to extensive land reclamation over the past decades. These wetlands constitute a core component of the East Asian-Australasian Flyway, one of the densest pathways for migratory waterbirds. The area of tidal flats in the YBS has been drastically reduced by 39.2% from 1981 to 2016 due to human activities [17]. The Chinese government has implemented stringent bans on marine utilization since 2018 in an attempt to cease the continuous loss of tidal flats over the past decades. However, the fate of tidal flats in the YBS is still ambiguous as they are exposed to risks in the form of sediment starvation and sea level rise [18,19]. Therefore, large-scale mapping with timely updates is needed to track the 2D/3D morphological changes of tidal flats in the YBS after the bans.

Poor accessibility and highly dynamic challenges pose great obstacles to the morphological mapping of tidal flats [20], resulting in a morphological data gap in these areas. Various techniques have been employed to reconstruct the morphology of tidal flats. Conventional in situ measurement methods, such as use of the Real-Time Kinematic Global Navigation Satellite System and the total station [21], are time-consuming and labor intensive and can only collect limited data in a small region with sparse samples. The rise of unmanned aerial vehicles (UAV) [22] and airborne LiDAR [23] has enabled the acquisition of precise digital elevation models with high spatial resolution for localized, small patches of observed areas. Nevertheless, the high cost of UAV and airborne LiDAR limits timely updates of morphological data in space and time. Since the first launch of the Earth observation satellite in the 1970s, remote sensing (RS) has provided the potential for rapid monitoring of the surface of Earth on a large spatiotemporal scale. Using remote sensing data to map morphological changes in tidal flats, with large spatial coverage and frequent updates, has emerged as a widely adopted and cost-effective approach.

The waterline method [24] is a commonly employed approach to derive tidal flats morphology using satellite images and has been applied in various regions worldwide [25–28]. It entails extracting waterlines from time remote sensing images and assigning measured or simulated tidal elevation information to these waterlines, then reconstructing the three-dimensional morphology of tidal flats by interpolating the assigned tidal values. However, there are uncertainties in morphological data on a large scale derived from the waterline method due to the spatial inconsistency in the elevation of waterlines [29]. Recently, a pixel-based morphology mapping method was adopted to reconstruct the three-dimensional morphology of tidal flats [30,31], which links the inundation frequency to the tidal flat elevation of pixels to mitigate the spatial inconsistency of waterline height on a large scale. This method is easy to reproduce in other regions since it relies only on a simple remote sensing index [31,32]. The key to reconstructing high-precision morphology of tidal flats is to classify the water and land area accurately in each image. However, in the context of large-scale tidal flats morphology mapping, variations in sediment input from rivers lead to substantial disparities in the concentration of suspended sediment in coastal waters, which introduces great uncertainty to morphological mapping. Therefore, water-land separation should employ different water indices based on coastal water bodies. Thus, two primary research questions needed to be well addressed in our study:

(1) How can we accurately extract the water surface area, at different degrees of turbidity, of coastal water over a large spatial scale?
(2) How do tidal flats change in 2D/3D scale after the cessation of coastal reclamation, and what are the main drivers?

To address the aforementioned issues, this study aims to develop a pixel-based, multi-indices tidal flats mapping algorithm (PMITMA) to accurately delineate 2D/3D morphological changes in tidal flats using multiple sources of time-series remote sensing imagery on the Google Earth Engine (GEE) platform. Based on the application of tidal flats in the YBS, we will have a better understanding of spatiotemporal and morphological changes in tidal flats in the YBS after institution of the marine utilization bans and the influence of continuous human activities.

2. Material and Methods

2.1. Study Area

The study region extends from the Yalu River Estuary to the Yangtze River Delta region in China (30°38′–41°03′N, 117°29′–122°03′E) (Figure 1a). The region north of 34°20′N is characterized by a temperate monsoon climate, whereas the southern portion has a subtropical monsoon climate. The mean annual temperature ranges from 5 to 20 °C, and the average annual precipitation varies from 400 to 1300 mm [33]. The coastline stretches approximately 7978 km and passes through the provincial districts of Liaoning, Hebei, Tianjin, Shandong, Jiangsu, and Shanghai, with around 60 million people inhabiting the low-lying coastal regions. The staggering population pressure is driving the rapid expansion of cities, industries, and agriculture, which in turn encroaches upon the already limited space of tidal flats [34].

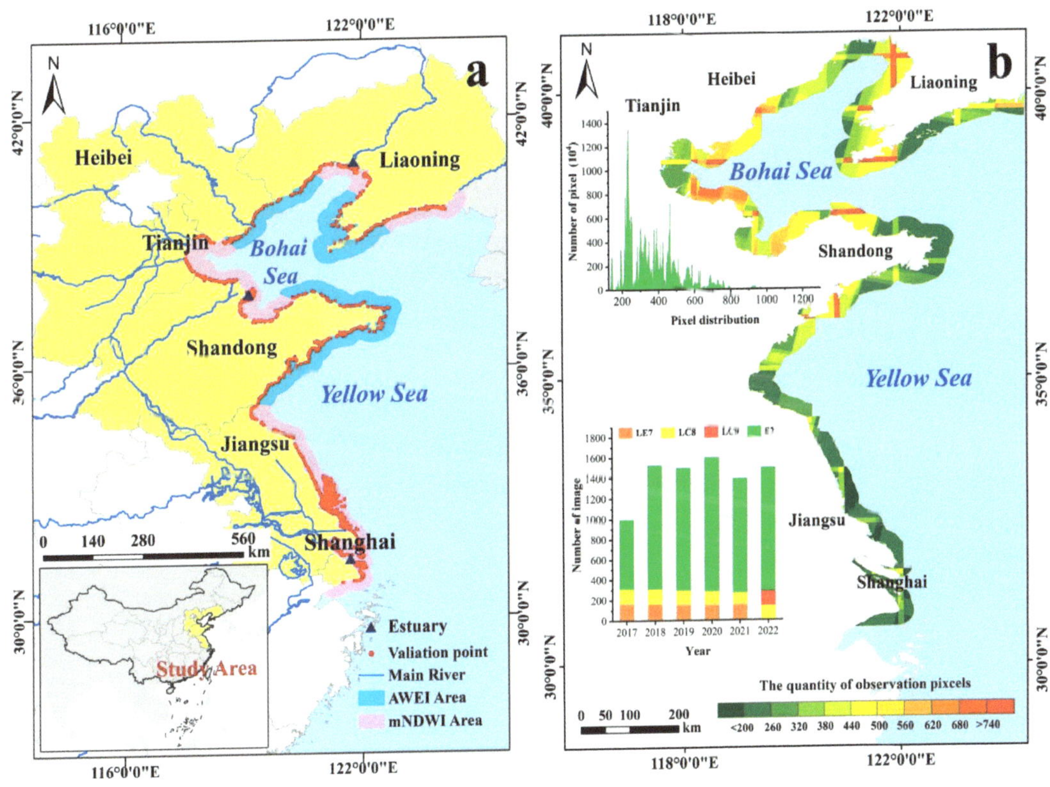

Figure 1. Geographical setting of the study area. (**a**) Location of the study area and water indices used on different shorelines. (**b**) The number of satellite images and the distribution of observation pixels.

In the past half-century, sediment in the Yellow and Bohai Seas has primarily originated from three major rivers, namely the Yellow River, Yangtze River, and Liao River [35], including underwater sand from the abandoned mouth of the Yellow River [36]. During the Ming and Qing dynasties, the Yellow River discharged a significant amount of sediment into the Yellow Sea through the Huai River, contributing to sedimentation in northern Jiangsu. However, the diversion of the Yellow River into the Bohai Sea in 1855 led to the gradual abandonment of the Yellow River Delta in northern Jiangsu, resulting in the formation of extensive underwater sandbars. Over the past 22 years, the annual average sediment input from the Yellow River and Yangtze River has significantly decreased by 292 million and 371 million tons, respectively, representing a reduction of 81.33% and 60.80% compared to the previous decades of the last century. The average concentration of suspended sediment in the nearshore waters of these areas ranges from approximately 40–80 mg/L, while ranging from 20–40 mg/L along the coast on either side of the river mouths [37]. These sediments, nurtured by the combined forces of ocean currents, have contributed to the formation of significant tidal flat wetlands, including the Yellow River Delta, Yangtze River Delta, and Liao River Delta, as well as radiating sandbars in central Jiangsu. These tidal flat wetlands provide ecosystem services worth approximately $30 billion annually [34]. The width of the major tidal flats can reach up to 25 km in the Bohai and Yellow Seas. Therefore, the study area is delimited to within a 25 km buffer of the coastline.

2.2. Data Sources

In this study, we utilized two widely used and freely accessible satellite datasets, the Landsat series and Sentinel-2A/B. The Landsat 7/8/9 satellites cover a broad range of wavelengths, spanning from visible to thermal infrared, making them suitable for identifying coastal water bodies. The bands of Landsat series images have a spatial resolution of 30 m, and the satellites revisit the same area every 16 days. The Sentinel-2, equipped with the Multi-Spectral Imager (MSI), encompasses similar spectral bands as the Landsat series. Furthermore, the Sentinel-2 satellites offer a higher spatial resolution. The blue, green, red, and near-infrared band channels have a resolution of 10 m. Moreover, the Sentinel-2 satellites revisit the same area every 5 days.

The integrated utilization of these two satellites enables more frequent observations of the tidal flats, enhancing the temporal resolution of remote sensing imagery and providing detailed and accurate information for monitoring surface changes and environmental dynamics. We utilized a total of 1776 images from the Landsat7/8/9 satellite and 6272 images from the Sentinel-2 spanning from 1 January 2017 to 31 December 2022 (Figure 1b). These image data are divided into three periods, including 2017–2018, 2019–2020, and 2021–2022, in order to conduct the time-series analysis for tidal flats mapping of time stages in the YBS. To mitigate the influence of cloud coverage on earth observation, we selected images with a cloud cover of less than 30% and utilized the bitmask band from Sentinel-2 images and Landsat images to mask the pixels covered by opaque and cirrus clouds. Figure 1b illustrates the distribution of observation pixels across the entire study area after masking.

2.3. Pixel-Based Multi-Indices Tidal Flat Mapping Algorithm

A pixel-based multi-indices tidal flat mapping algorithm was developed to accurately map the morphology of tidal flats in the YBS using the GEE platform. It encompasses three key components: (1) water frequency generation from time-series images, (2) tidal flats extraction, and (3) accuracy assessment (Figure 2).

2.3.1. Assessing Temporal Changes in Coastline

The extensive and meandering coastline of the Yellow and Bohai Seas exhibits varying suspended sediment content in its nearshore water bodies [37,38]. Various water indices exhibit different levels of effectiveness in differentiating between tidal flats and water bodies. Hence, the selection of a suitable water index should be based on the sediment content of these bodies. The most commonly used water indices [39] for water-land separation are the Normalized Difference Water Index (NDWI) [40], the modified Normalized Difference Water Index (mNDWI) [41], and the Automated Water Extraction Index (AWEI) [42] (Table 1). NDWI, the initially proposed water index, calculates the reflectance of green and near-infrared bands to effectively differentiate between clear water bodies and land. The mNDWI, an enhanced variant of NDWI that utilizes the shortwave infrared band instead of the near-infrared band, improves its sensitivity [43] and more accurately extracts water bodies. AWEI was originally developed to mask water bodies influenced by shadows cast by clouds, buildings, and mountains, and it consistently exhibits excellent performance across diverse water types and quality conditions [44–47]. To compare the dynamic characteristics of different water indices on tidal flats and water bodies, we selected 24 images (one image per month) from the high turbid regions of the Yellow River Estuary and the Yangtze River Estuary, captured in 2021. A total of 1440 sample points were collected by selecting 30 sample points each from water bodies and tidal flats along the water-land boundary in every image. The values of three water indices for the sample points were obtained. According to the results (Figure 3), NDWI did not effectively separate water bodies (Figure 3(a1,b1)) from tidal flats in January, March, August, September, and November. In contrast, AWEI (Figure 3(a3,b3)) successfully separated water bodies in all months except January, while mNDWI (Figure 3(a2,b2)) demonstrated the highest degree of separation among all the indices. Consistent with previous research findings, mNDWI is often suitable for areas with relatively turbid water, but it may lead to over-extraction in some clear water areas [48]. Accordingly, a few points in the figure showing tidal flats were misclassified as water bodies by mNDWI. AWEI demonstrated higher accuracy than NDWI in extracting clear water bodies [49,50] and also exhibited better extraction capabilities in slightly turbid water areas (Figure 3). Based on thorough consideration, we have determined that mNDWI will be applied in the Yalu River Basin, Liaohe Basin, Yellow River Basin, and Yangtze River Basin, whereas AWEI will be employed in the remaining regions (Figure 1a).

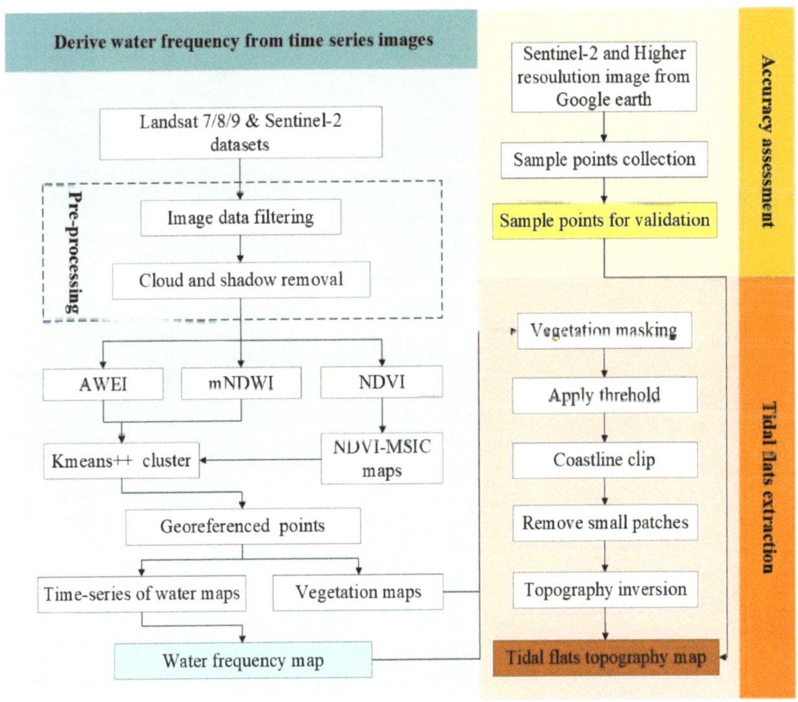

Figure 2. Workflow for mapping tidal flats.

Table 1. The spectral indices used in this study.

Index	Tidal Height (cm)
Normalized Difference Water Index (NDWI)	$\frac{\rho_{Green}-\rho_{NIR}}{\rho_{Green}+\rho_{NIR}}$
Modified Normalized Difference Water Index (mNDWI)	$\frac{\rho_{Green}-\rho_{SWIR}}{\rho_{Green}+\rho_{SWIR}}$
Automated Water Extraction Index (AWEI)	$\rho_{Blue}+2.5*\rho_{Green}-1.5(\rho_{NIR}+\rho_{SWIR1})-0.25\rho_{SWIR2}$
Normalized Difference Vegetation Index (NDVI)	$\frac{\rho_{NIR}-\rho_{Red}}{\rho_{NIR}+\rho_{Red}}$

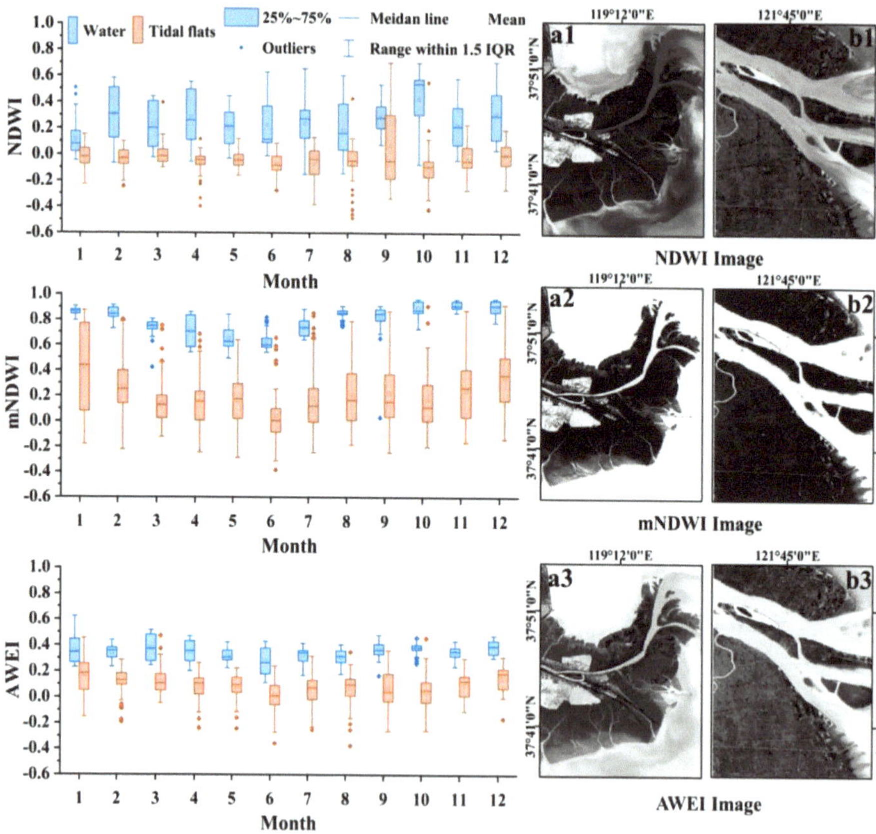

Figure 3. Comparison of water index characteristics between tidal flat and water body. (**a1**) NDWI in Yangtze River Estuary, (**a2**) mNDWI in Yangtze River Estuary, (**a3**) AWEI in Yangtze River Estuary, (**b1**) NDWI in Yellow River Estuary, (**b2**) mNDWI in Yellow River Estuary, (**b3**) AWEI in Yellow River Estuary.

2.3.2. Water Frequency Generation from Time Series Images

Some studies directly employ threshold segmentation for water-land separation. This is a simple pixel-level comparison method. However, for large-scale image data and large-scale regions, there are significant differences in grayscale values between different images and areas. It can be challenging to determine a universal threshold that can adapt to the classification requirements of the most complex regions [51,52], especially in turbid water areas. Thus, a semi-automated method integrated with the K-Means++ algorithm [53], which is an unsupervised classifier, along with georeferenced points, was used to separate water and land pixels in the time series water index images (Figure 4). Herein, the utilization of the K-means++ initialization method enhanced the clustering results by ensuring a more even distribution of the initial centroids, decreasing the likelihood of suboptimal solutions that can arise from random initialization in the traditional K-means algorithm. Through our multiple experiments (set cluster numbers ranging from two to 10) with visual comparison, we found that the K-mean++ algorithm achieves optimal performance for land-water separation when the number of clusters is set to six (Figure 4c). By incorporating additional georeferenced points (Figure 4c), the points intersecting with the six clusters were classified as water, while the rest were classified as land (Figure 4d). Thus, the binary image of land and water area was generated from every single image of the time-series (Figure 4d). The water frequency is calculated on a pixel-by-pixel basis using the binary time-series images over each study period (Figure 4e). The calculation formula is as follows:

$$f(x,y) = \frac{\sum_{i=1}^{n} B_k(x,y)}{n} \quad (1)$$

where (x,y) represents the spatial coordinates of each pixel, $f(x,y)$ is the water frequency of each pixel, n is the number of binary images, and $B_k(x,y)$ is the binary image of the k-th scene.

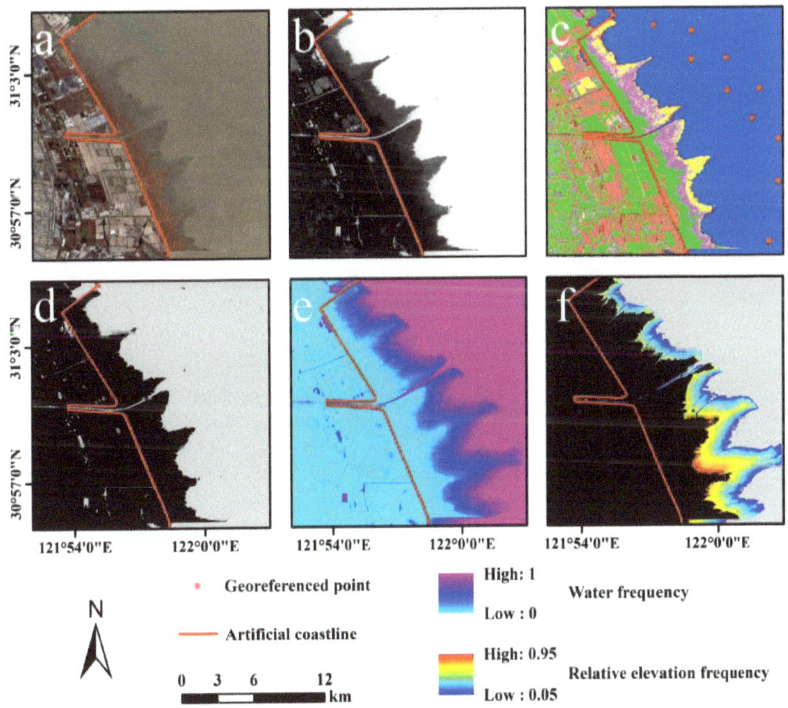

Figure 4. Illustration of workflow with case site in the Nanhui tidal flats of the Yangtze Estuary. (**a**) Satellite images, (**b**) mNDWI Water Index, (**c**) K-means++ classification and geographic landmarks, (**d**) water-land binary image, (**e**) water frequency, and (**f**) relative elevation.

2.3.3. Tidal Flat Extraction

Tidal flats can be more effectively identified in a water frequency map. We adopted thresholds of 0.05 and 0.95 to determine the upper and lower boundaries of tidal wetlands, as it has been empirically proven effective [54]. It is worth noting that the presence of vegetation on tidal flats significantly influences the accuracy of tidal flats terrain mapping, as their height exceeds that of the tidal flats. Therefore, we adopted the Maximum Spectral Index Composite (MSIC) method to generate an NDVI-MSIC map to determine the largest potential area of vegetation presence. Subsequently, the K-means++ method described in the previous section was applied to construct a binary vegetation map for masking vegetation.

Nonetheless, certain features present in both land and sea, such as reservoirs, fish farms, fish rafts, scattered clouds, and other objects with high fluctuation, can affect the accurate extraction of tidal flats. Consequently, to address this issue, an artificial coastline was employed to mask inland areas, and the "connectedPixelCount" function in GEE was utilized to remove pixel clusters containing fewer than 120 units. A negative correlation exists between the elevation of the intertidal zone and the frequency of tidal inundation [30]. Tidal flats with higher elevations exhibit a low frequency of water submergence, approaching zero, whereas tidal flats with lower elevations experience a high frequency of water submergence, approaching one. Therefore, we calculated the one-water frequency for each grid of the obtained water frequency map in ArcGIS to represent the relative elevation frequency of tidal flats (Figure 4f). The relative elevation values range from 0.05 to 0.95, with 0.95 representing the highest point and 0.05 representing the lowest point of the tidal flats.

2.3.4. Accuracy Assessment

To validate the accuracy of the tidal flats map, two methods were employed to assess the accuracy. A total of 21,171 validation points were collected from the lowest-tide images (Figure 1a) from Sentinel-2 from 2017 to 2022 and high-resolution imagery from Google Earth and used for the accuracy assessment of RS-derived tidal flats. Strategic selection of tidal gauge stations situated along the nearshore areas of the Yellow and Bohai Seas was performed, and 23 in total were chosen in order to guarantee at least one tidal gauge station in each path/row of the satellite images. These points were categorized into tidal flats and non-tidal flats, e.g., water, vegetation, and land. The seaward edge detection of tidal flats is key to verifying the accuracy of RS-derived tidal flats products, and 11,716 validation points were located at the seaward edges of the tidal flats (Figure 1a). A confusion matrix and the F1 score were employed to evaluate the accuracy of tidal flats with the following equation:

$$F1_{Score} = \frac{2TP}{2TP + FN + FP} \tag{2}$$

where TP represents the count of correctly classified tidal flats samples, FP represents the count of incorrectly classified non-tidal flat samples as tidal flats, and FN represents the count of incorrectly classified tidal flats samples as non-tidal flats. Secondly, we compared the results by overlaying the lowest tide images. The aforementioned experiments were conducted on GEE.

3. Results

3.1. Accuracy Assessment Result

The accuracy of coastal land cover maps was assessed for three periods (Table 2). Results showed that the year 2020 achieved the highest accuracy with an overall precision of 95.83%. The validation results indicate that the tidal flat extent delineated by PMTMA aligns closely with the validation sites. Detailed accuracy evaluation results for each period are provided in Table 2.

We tested the relative elevation with ICESat-2 derived elevation of tidal flats, revealing a p-value of less than 0.01 and R^2 (Figure 5), indicating a significant positive correlation between relative elevation and ICESat-2 elevation.

Table 2. Accuracy of tidal flats map.

Period	Class	Tidal Flats	Non-Tidal Flat	UA (%)	OA (%)
2018	Tidal flats	3607	95	97.43	95.77
	Non-tidal flat	197	3012	93.86	
	PA (%)	94.82	96.94		
	F1_Score	96.10	95.37		
2020	Tidal flats	3755	110	97.15	95.83
	Non-tidal flat	185	3025	94.23	
	PA (%)	95.30	96.80		
	F1_Score	96.21	95.49		
2022	Tidal flats	3737	81	97.87	95.60
	Non-tidal flat	235	3132	93.02	
	PA (%)	94.08	97.47		
	F1_Score	95.93	95.19		

Note: PA is producer's accuracy; UA is user's accuracy; OA is overall accuracy.

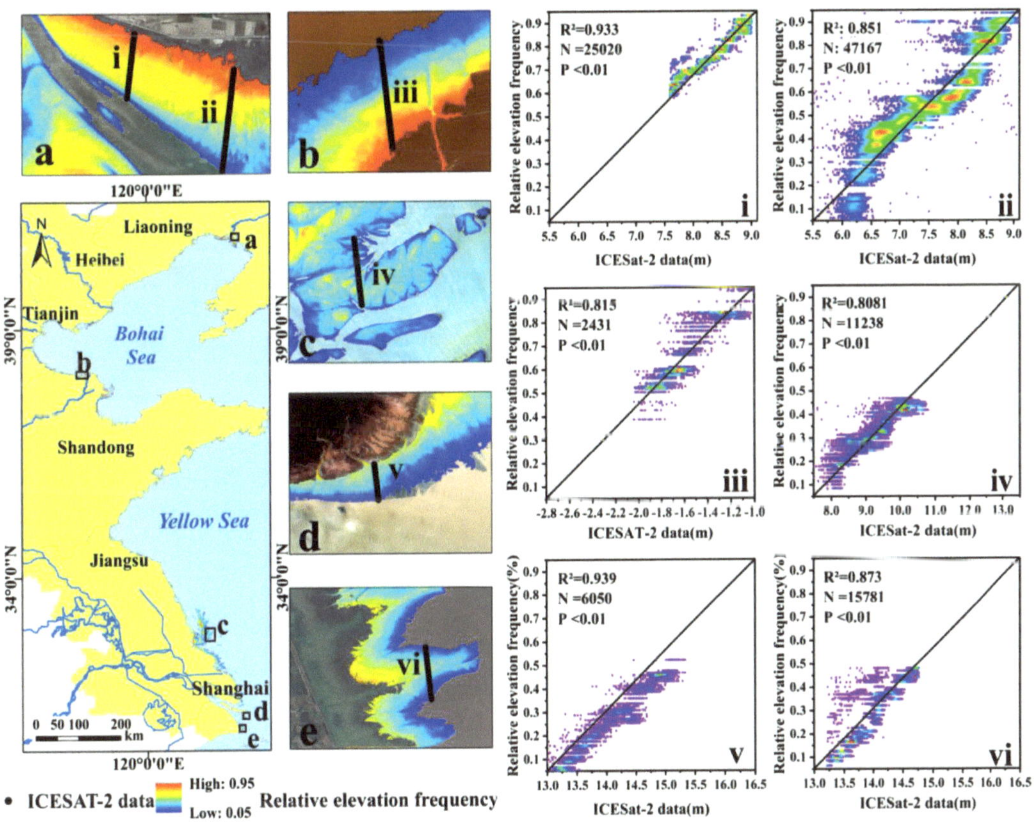

Figure 5. Correlation between the relative elevation frequency of tidal flats and ICESAT-2 data. (**a**) Liaohe Estuary, (**b**) Diaokou River Estuary, (**c**) radial sand ridges (RSRS) in Jiangsu middle coast, (**d**) Jiuduan sands, and (**e**) Nanhui Beach.

3.2. 2D Changes in Tidal Flats

The total area of the tidal flats of the YBS was 4778.36 km^2 in 2022 and is of great spatial variation. Tidal flats were primarily distributed at the edge of the large river deltas, such as the Yangtze River Delta and Yellow River Delta, etc. Jiangsu Province had the most tidal flats among the provinces with a total area of 2279.71 km^2, while Tianjin had the least, with only 58.94 km^2. The high, middle, and low-tidal flats were 640.08 km^2, 1374.28 km^2, and 2763 km^2 throughout the YBS in 2022, respectively (Figure 6).

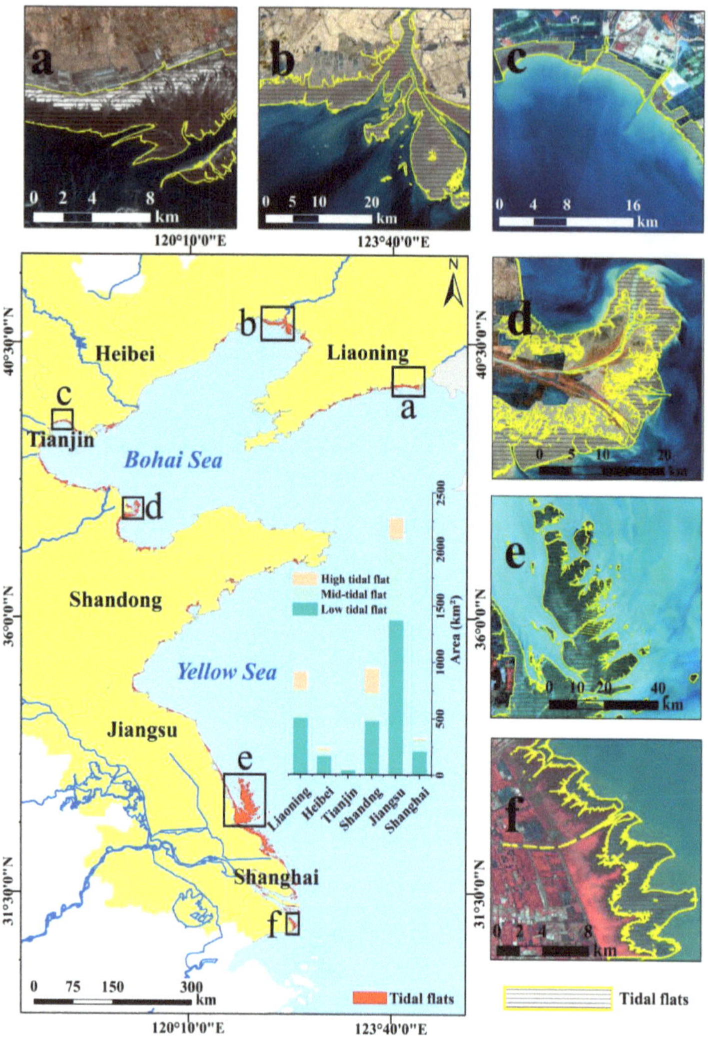

Figure 6. The spatial distribution coverage and areas of the three types of tidal flats along the coast of the Bohai and Yellow Seas in 2022. Subfigures show tidal flats in (**a**) the Jiangsu middle coast radial sand, (**b**) Liao River Estuary, (**c**) Bohai Bay, (**d**) Yellow River Estuary, (**e**) the Jiangsu middle coast, and (**f**) Yangtze River Estuary.

Over the period from 2016 to 2022, tidal flats suffered an overall loss in area across nine regions (Figure 7), resulting in a net loss of approximately 70.52 km^2, representing 1.79% of the total. The gain and loss in tidal flats area from 2018 to 2022 were 105.25 km^2 and 175.78 km^2, respectively. The losses in tidal flats area from 2018 to 2022 and from 2020 to 2022 were 10.62 km^2 and 59.91 km^2, respectively. Figure 7 provides an overview of the changes in tidal flat areas in different regions, revealing the following trends: (1) Regions L2, H2, and C2 exhibited gaining trends. Region C2 experienced the greatest gain over the entire study period, while region L2 had the smallest gain, with gains of 111.25 km^2 and 17.09 km^2, respectively. (2) In contrast, regions L1, L3, H1, H3, C1, and C3 showed a losing trend in tidal flat areas. Region C1 experienced the highest loss of an area of 115.61 km^2, followed by region H1, which lost 62.40 km^2. Region C3 experienced the smallest loss, with a loss of only 16.08 km^2. Notably, apart from regions C3 and C2, most of the tidal flats near the estuary experienced gain, while the bays and coastal tidal flats further away from the estuary were subjected to loss.

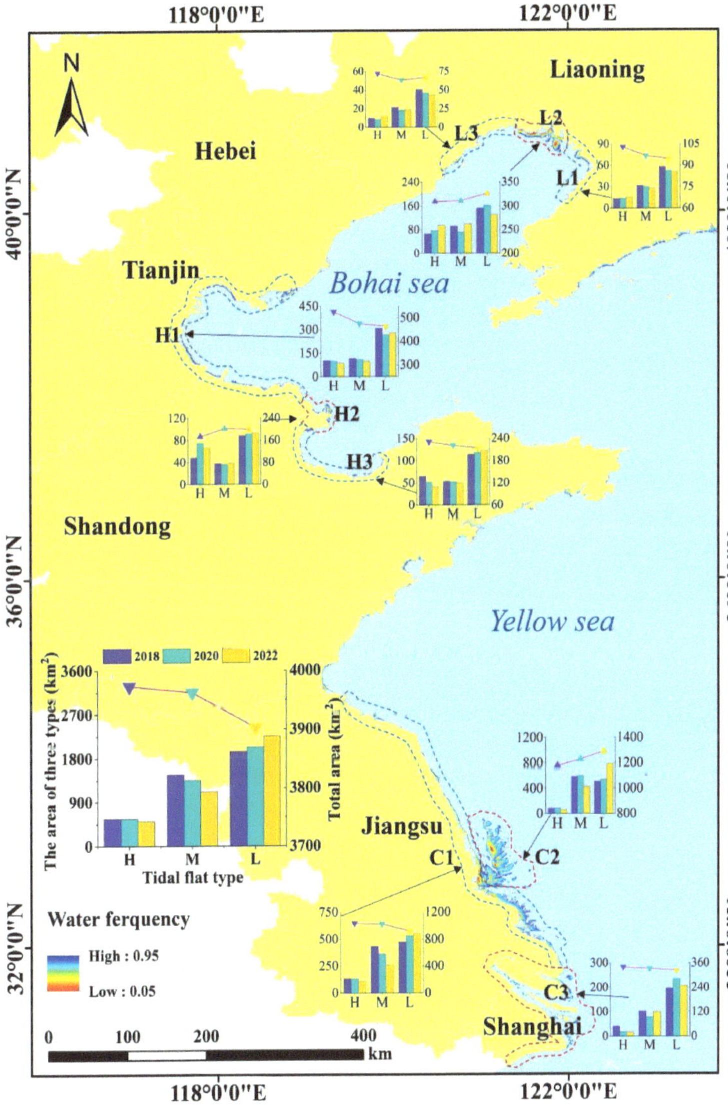

Figure 7. The tidal flat locations of different regions. Starting from the northeastern coastline and heading south, the sections are identified as L1, L2, L3, H1, H2, H3, C1, C2, and C3.

3.3. 3D Changes in Tidal Flats

Considering the high level of variation in tidal ranges in the study area, the 3D morphology of tidal flats was analyzed using the changes in inundation probability to evaluate their vulnerability to the rising sea level. A classification system for tidal flats was defined in our study, including high-tidal flats, mid-tidal flats, and low-tidal flats, with inundation probability from 0.05 to 0.35, 0.35 to 0.65, and 0.65 to 0.95, respectively.

There is an overall erosion trend in the 3D morphology of tidal flats. During the period from 2018 to 2022, the areas of high-tidal flats and mid-tidal flats experienced reductions of 50.24 km^2 and 344.13 km^2, respectively. In contrast, the area of low-tidal flats increased by 323.85 km^2. Among them, the C3 and C2 regions witnessed the most significant decreases in the areas of high-tidal flats and mid-tidal flats, with reductions of 24.37 km^2 and 154.23 km^2, respectively. Moreover, there were localized areas where the high-tidal flats and mid-tidal flats exhibited increases. The L2 region experienced the most substantial increases, of 29.68 km^2 and 8.16 km^2, respectively, while H2 had increases of 18.652 and 1.46 km^2. Conversely, the low-tidal flat areas exhibited a general growth trend. Among these, the C2 region experienced the most substantial increase, reaching 283.69 km^2, followed by C1 and C3 with 76.55 km^2 and 11.27 km^2, respectively. Some individual areas experienced decreases, such as the H1 region, which decreased by 28.78 km^2.

Herein, we also conducted a transect analysis of tidal flats for a detailed illustration of morphological changes (Figure 8). The evolution patterns of tidal flats can be classified into four forms: (1) rapid erosion, where the shoreline retreated more than 500 m inland, and experienced a relative elevation frequency decrease exceeding 20%; (2) slow erosion, where the shoreline receded less than 500 m inland and witnessed a relative elevation frequency decrease of less than 20%; (3) rapid accretion, where the shoreline expanded more than 500 m towards the sea and demonstrated a relative elevation frequency increase exceeding 20%, and (4) gradual accretion, where the shoreline expanded less than 500 m towards the sea and exhibited a relative elevation frequency increase of less than 20%.

The mid- and low-tidal flats decreased simultaneously (Figure 9f,h,i), with a noticeable landward retreat of the lower boundary of tidal flats, indicating rapid erosion. We also found that the mid-tidal flats eroded and transformed into low-tidal flats (Figures 9b–d,g,j,k and 10a), while the low-tidal flats experienced a slight erosion or even slight accretion, with the less pronounced retreat of the tidal flats inward. In Figure 9e, a significant expansion of tidal flats can be observed, while Figures 9a and 10b,d demonstrate a slow accretion in tidal flats. Combining the previous results of the 2D morphology of tidal flats, it can be inferred that the L1, L3, and H1 regions are primarily undergoing rapid erosion, while H3, C1, and C3 are in a state of slow erosion. The pattern of gradual accretion with slow growth dominates the L2 tidal flats. The H2 tidal flats exhibit complex changes, influenced by all four patterns, transitioning from a period of rapid accretion to a period of slow erosion. C2 is mainly characterized by slow erosion. In light of the above analysis, it can be inferred that the elevation of tidal flats in the nine regions is primarily undergoing erosion, with a minority experiencing accretion.

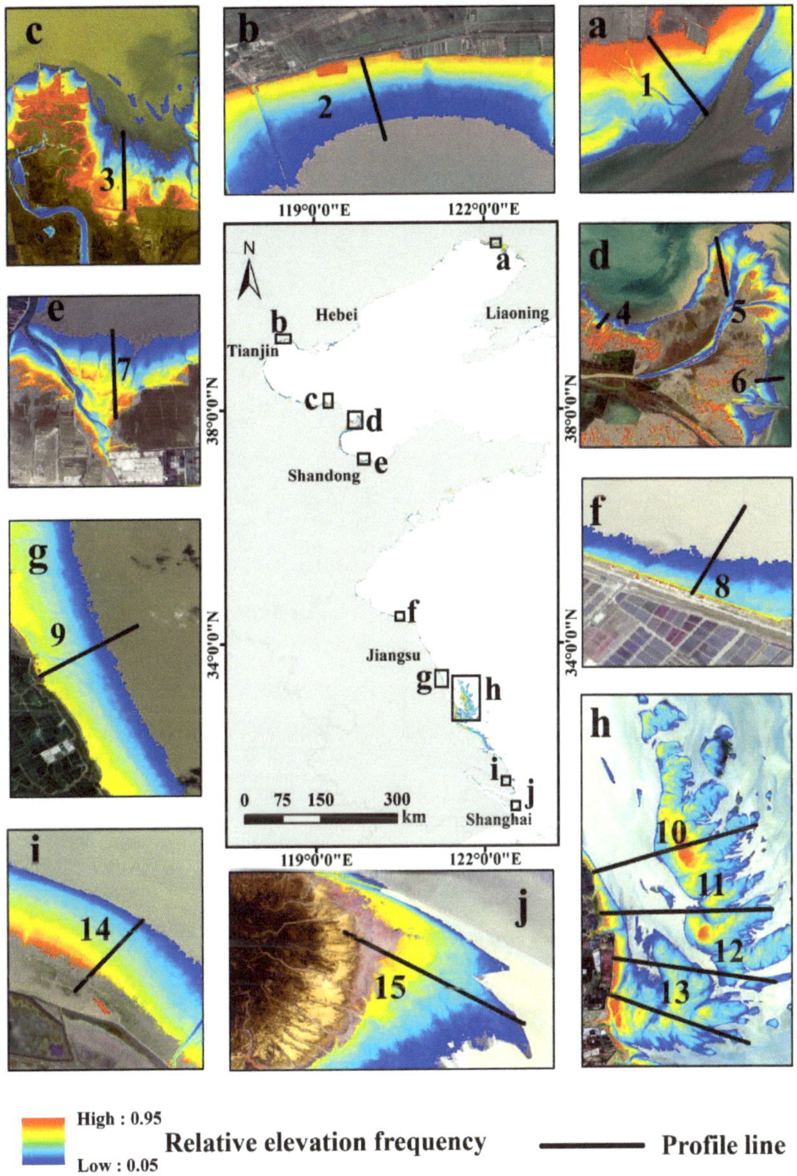

Figure 8. The profile of different regions. (**a**) Profile lines in the Liao River Estuary, (**b**,**c**) profile lines in the Bohai Bay, (**d**) profiles lines in the Yellow River Estuary, (**e**) profile line in the Laizhou Bay, (**f**–**h**) profile lines in the central-northern coastal section of Jiangsu, (**i**) profile line in the Chongming east beach, and (**j**) profile line in the Jiuduansha sandbank.

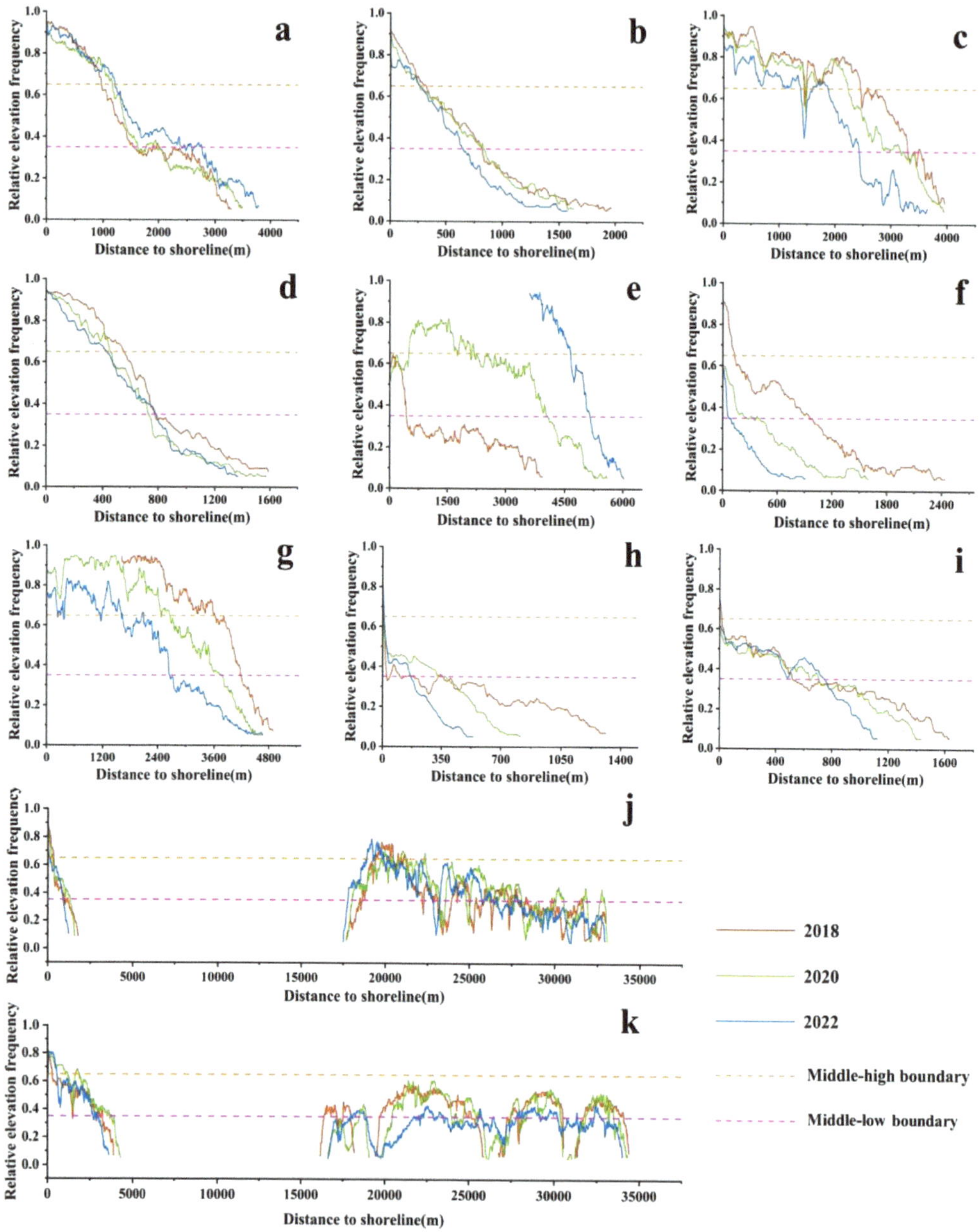

Figure 9. The morphological changes in transects over the study period. (**a**) Represents the profile line of "1" in Figure 8, (**b**) represents the profile line of "2" in Figure 8, (**c**) represents the profile line of "3" in Figure 8, (**d**) represents the profile line of "4" in Figure 8, (**e**) represents the profile line of "5" in Figure 8, (**f**) represents the profile line of "6" in Figure 8, (**g**) represents the profile line of "7" in Figure 8, (**h**) represents the profile line of "8" in Figure 8, (**i**) represents the profile line of "9" in Figure 8, and (**j**) represents the profile line of "10" in Figure 8, (**k**) represents the profile line of "11" in Figure 8.

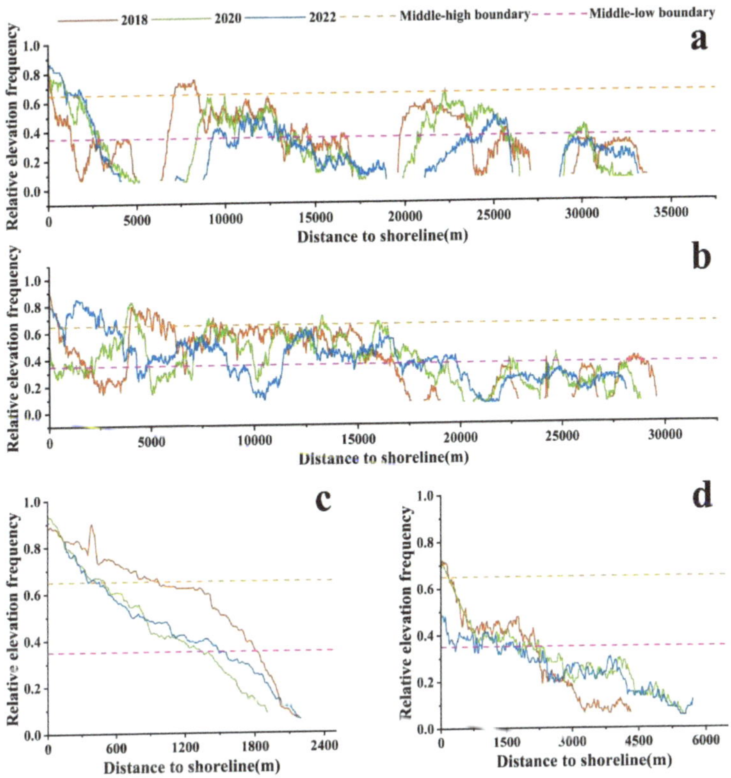

Figure 10. The profile plots of RSRS are drawn based on the positions of the profile lines. (**a**) Represents the profile line of "12" in Figure 8, (**b**) represents the profile line of "13" in Figure 8, (**c**) represents the profile line of "14" in Figure 8, and (**d**) represents the profile line of "15" in Figure 8.

4. Discussion

4.1. Robustness and Uncertainties

In this study, we extensively utilized Landsat and Sentinel-2 images from 2017 to 2022 through the GEE platform. To account for the dynamic changes in tidal inundation, we improved a specialized extraction model for tidal flat wetlands, yielding satisfactory classification results. This can be mainly attributed to three factors. Firstly, the combination of freely available Landsat with Sentinel-2 data has improved the revisit period of the satellites (greater than 2–5 days) and provided higher spatial resolution (10 m). The higher temporal resolution facilitates more frequent capturing of tidal flat information during low-tide and high-tide periods, while the improved spatial resolution of Sentinel-2 enables finer mapping of tidal flats. This is crucial for monitoring the dynamic intertidal wetland environment [55]. Secondly, pixel-based image synthesis reduces the uncertainty caused by tidal variations within the scenes and low-quality observation pixels. Thirdly, the use of appropriate water indices based on the sediment content of different nearshore water bodies has greatly reduced the misclassification and omission of tidal flats. Finally, the utilization of the GEE platform allows for rapid access and parallel processing of thousands of satellite images. This efficient platform facilitates the handling of large volumes of data, enabling timely and effective analysis. Our proposed PITMA method is applicable to other tidal flat areas. Firstly, the method is grounded in principles of physics and geography. It calculates the inundation probability of each pixel in the tidal flat based on the elevation of various regions and the tidal range during the same time period, which are intrinsic attributes of tidal flats. Secondly, our extensive tidal flat dataset spans 7978 km, capturing various tidal flat types and complex water bodies worldwide, and its accuracy has been rigorously validated. Therefore, our method can be easily applied to other tidal flat areas with remote sensing data.

We compared our study with several other datasets of tidal flats, namely the 2018 Chinese Coastal Wetland Map (referred to as FUDAN/OU) [54], the 2020 Chinese Tidal Flat Map (referred to as CTF) [48], and the East Asia Tidal Wetland Map circa 2020 (referred to as MTWN) [56], in terms of area (Figure 11) and spatial distribution (Figure 12). Due to disparities in data sources, methods, and time spans, variations exist among these tidal flat maps. Our tidal flat map covers a larger overall area compared to FUDAN/OU, but a smaller area compared to MTWN. In some regions, CTF shows slightly more extensive tidal flats compared to our study, except in Jiangsu, where it shows smaller extents.

These differences are expected since it relies on only one year of Landsat data. FUDAN/OU misclassifies nearshore tidal flats as seawater [54], leading to an incomplete extraction of tidal flats in their map. In contrast, our study utilizes a combination of two years of Sentinel-2 imagery and Landsat series imagery and uses two indices to classify the land and water with higher accuracy, enabling us to capture finer and more extensive tidal flat features. The slightly larger overall tidal flat extent in CTF is mainly attributed to the use of different spectral indices. CTF utilized NDVI for comprehensive analysis in determining the minimum water area, whereas we employed mNDWI and AWEI to accurately differentiate between tidal flats and water bodies of different turbidity levels. NDVI-MSIC incorrectly classifies some sediment-laden areas as tidal flats (Figure 12(a2)), and misclassifies vegetation as tidal flats (Figure 12(d2)), whereas the more water-sensitive mNDWI accurately distinguishes suspended sediments from tidal flats. Furthermore, in the low-latitude regions, we extracted a larger tidal flats area than CTF (Figure 12(b2,c2)). The utilization of only one and a half years of Sentinel-2 imagery in CTF failed to capture more tidal flat areas. These factors contribute to the discrepancies between our results and CTF. Regarding MTWN, the primary inconsistency lies in the methodology employed. We differentiated tidal flats using a 0.05–0.95 water frequency threshold based on time-series images. In contrast, MTWN employed a random forest method to identify tidal flats and water bodies, and due to spectral similarities, it may have extracted some intertidal mudflats or sandy beaches located above the high tide line (Figure 12(a3,c3)). However, this aspect is not reflected in our results, resulting in smaller extents compared to MTWN.

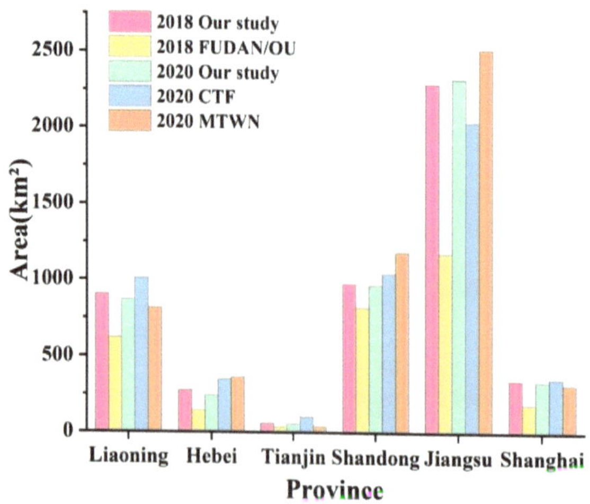

Figure 11. Comparisons of tidal flat areas with other results, including MTWN, CTF, and FUDAN/OU.

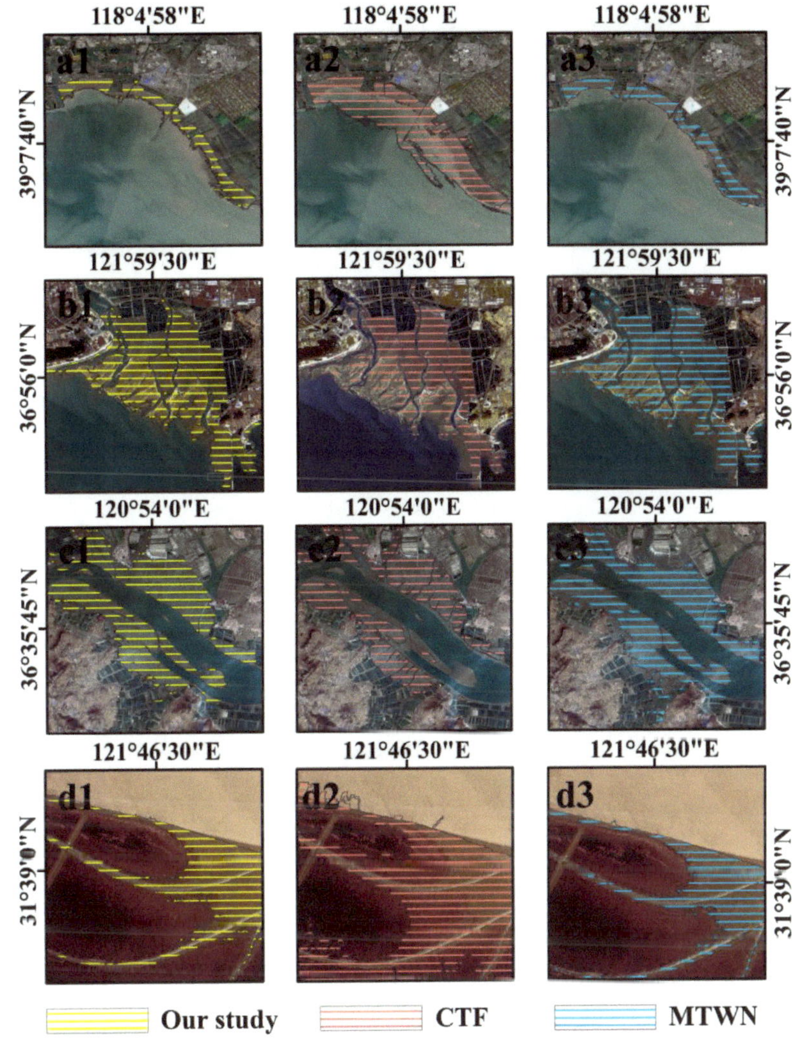

Figure 12. Comparison of the tidal flat extent in our study (No. 1), Jia's study (CTF) (No. 2), and Zhang's study (MTWN) (No. 3) in (**a1–a3**) Bohai Bay, (**b1–b3**) Wulei Island Bay, (**c1–c3**) Dingzi Port, and (**d1–d3**) Chongming east tidal flat.

However, certain objective factors can influence the accuracy of the extraction outcomes. Firstly, the study area has a significant north-south span, with the southern part characterized by a subtropical monsoon climate and frequent cloud cover, which makes it challenging to obtain an adequate quantity of images. Secondly, our research is limited to the exposed tidal flats observed in satellite imagery, and it cannot be determined whether the actual highest and lowest tides can be observed. Lastly, despite employing a frequency-based approach to effectively utilize data from both satellite sources, there may still be a minor presence of cloud pixels on the images even after applying quality assurance (QA) bands for cloud removal [57]. These remaining cloud pixels can potentially affect the precision of our results.

4.2. Drivers of Tidal Flat Dynamics in YBS

4.2.1. Tidal Flat Changes Induced by Sediment

The sediment supply and the distance from the estuary directly or indirectly influence erosion and deposition of tidal flats. Among the rivers that transport freshwater and sediment to the Bohai Sea, the Yellow River exhibits the highest runoff and sediment load, constituting 77.96% and 96.69% of the total, respectively [58]. The dominant current on the western side of the Yellow Sea is the Yellow Sea Coastal Current (YSCC) [59,60]. It originates from the southern part of the Bohai Sea and flows eastward along the northern coast of the Shandong Peninsula via the Dengzhou Waterway (Figure 13). After bypassing Chengshantou, it turns southward and reaches the coast of Jiangsu. On the eastern side, the primary current is the Yellow Sea Warm Current (YSWC). The Yellow Sea Warm Current enters from the northern part of the Bohai Strait and divides into two branches: the northern branch flows eastward along the western coast of the Bohai Sea, generating a cyclonic circulation in Liaodong Bay, and the southern branch enters the Bohai Bay and follows the coast until it exits through the southern part of the Bohai Strait [61,62]. The sediments are transported by these ocean currents to the estuary and its surrounding coastline. However, the construction of numerous reservoirs and water diversion projects has led to a significant decrease in the amount of sediment transported by rivers [19]. The average annual sediment transport of the Yangtze River and Yellow River significantly decreased around the turn of the 21st century. As the transportation of sediment by ocean currents to tidal flats located far from the estuary continues, this will lead to tidal flat erosion closer to the estuary mouth, as sediment import becomes smaller than export (Figure 14a). Our research findings revealed that the loss of tidal flats has occurred in distinct stages over time, encompassing both severe and mild erosion. For instance, in the H1 and H3 regions, sediment from the Yellow River is primarily influenced by the counterclockwise circulation of the southern branch of the Bohai Sea, resulting in its predominant transportation to the H1 area. The H1 region experienced mild erosion due to its greater distance from the river mouth, reduced sediment transport, and wave erosion. The loss of tidal flat wetlands is associated with a gradual reduction in the area of mid-high-tidal flats, which transition to low-tidal flats. In contrast, the H3 region encounters even lower sediment transport, posing challenges for the transportation of sediment from the Yellow River mouth to the northern Bohai Bay area. The loss of tidal flat wetlands in this region is consistent with the overall decrease in the frequency of relative elevation. Profile change diagrams also illustrate a significant reduction in the width of the entire tidal flat, indicating a noticeable retreat inland. Importantly, if the supply of sediment remains inadequate, mild erosion may progress towards severe erosion, resulting in the eventual disappearance of low-tidal flats due to ongoing wave erosion. From an alternative standpoint, tidal flat erosion can be classified into two scenarios: managed coastlines (Figure 14a) and natural coastlines (Figure 14b). In managed coastlines, human activities maintain the high-tidal flats, while the mid- and low-tidal flats shift landward due to reduced sediment supply, resulting in an overall reduction in tidal flat width. Conversely, in natural coastlines, the high and low water lines move inland due to reduced sediment transport and wave action. However, the overall area of the tidal flats does not undergo significant changes.

Although the tidal flat area has generally decreased in the aforementioned regions, there have been periods of increase in tidal flat wetland areas throughout the time series. In the case of H2, the diversion of the Yellow River northward in 1996 resulted in the loss of its primary sediment supply to the southern region of the estuary [63]. Consequently, the tidal flats in the region have experienced a gradual erosion process. However, their proximity to the river mouth and the impact of tidal currents have resulted in reduced erosion rates over time, as sediment transport to the southern delta has played a mitigating role. The tidal flats in the northern region have exhibited a trend of northward expansion, albeit at a decelerated growth rate, attributed to the prolonged decline in sediment supply. Sediment has experienced a decline in recent years, increasing from an average of 152 million tons (2017–2018) to 292 million tons (2019–2020), and then decreasing to 184 million tons

(Figure 15b). Throughout this entire period, the overall area of tidal flats has displayed a substantial increase, transitioning to a slightly diminished state, consequently leading to a reduction in the extent of high-tidal flats. Similarly, in the C2 region, the RSRS has experienced an overall expansion in the total area of tidal flats. However, the area of mid- to high-tidal flats has consistently declined, while only the low-tidal flats have shown an increase. This situation presents a relatively complex scenario. The sediments primarily originate from the submerged delta of the former Yellow River, which was redirected northward to Shandong and emptied into the sea in 1855, in addition to the Huai and Yangtze Rivers' discharge [59]. The YSCC (Figure 13) transports sediments from the sandbar located in the abandoned Yellow River delta to the south along the coastline [64] and acts as a protective barrier against erosion for the radiating sand shoal and the northwest coast. However, over time (from 1855 to the present), the former Yellow River delta has been gradually diminishing, leading to a decrease in suspended sediment [65]. The YSCC acts as a powerful force, propelling the radiating sand shoal from the northwest to the southeast, resulting in the gradual erosion of mid-high-tidal flats and even low-tidal flats in the northern part of C2. The remaining water currents disperse across the nearshore tidal channels and small sand ridges, generating residual currents and eddies that persistently transport sediment toward the southeast [66]. This process leads to the erosion of internal sediment and the conversion of mid-high-tidal flats. The tidal flats undergo a gradual process of flattening, giving the impression of sedimentation in terms of their area. However, in reality, they may be undergoing erosion.

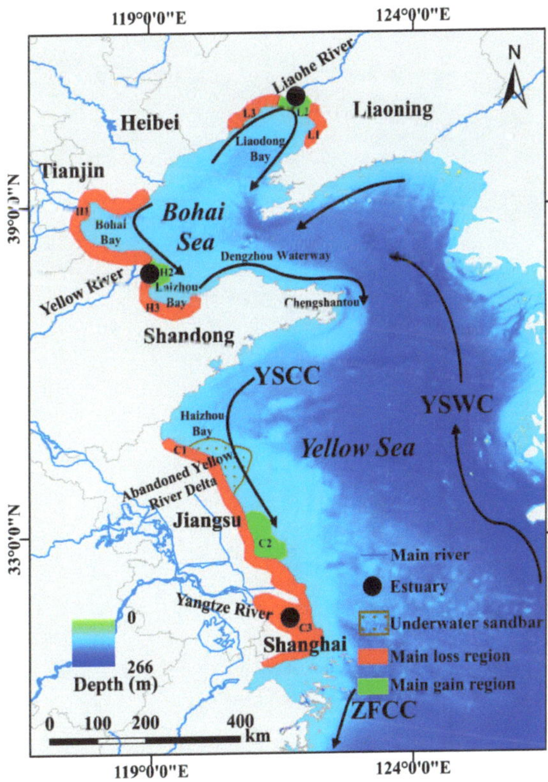

Figure 13. Submarine bathymetry and circulation distribution in the Yellow and Bohai Seas. (YSWC) Yellow Sea Warm Current, (YSCC) Yellow Sea Coastal Current, (ZFCC) Zhejiang and Fujian Coastal Current. The bathymetry is from NOAA National Centers for Environmental Information. 2022: ETOPO 2022 15 Arc-Second Global Relief Model.

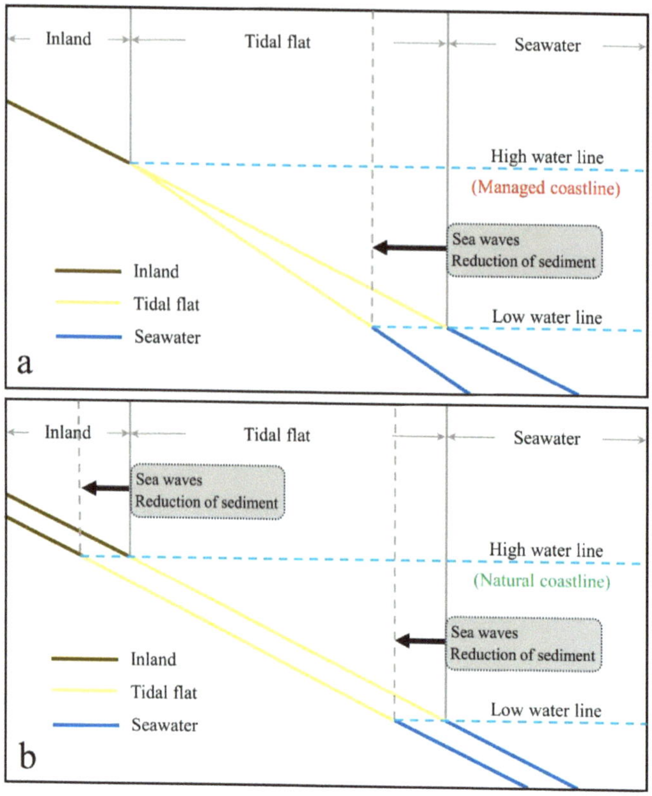

Figure 14. Two forms of tidal flat erosion. (**a**) The form of managed coastline erosion, and (**b**) the form of natural coastline erosion.

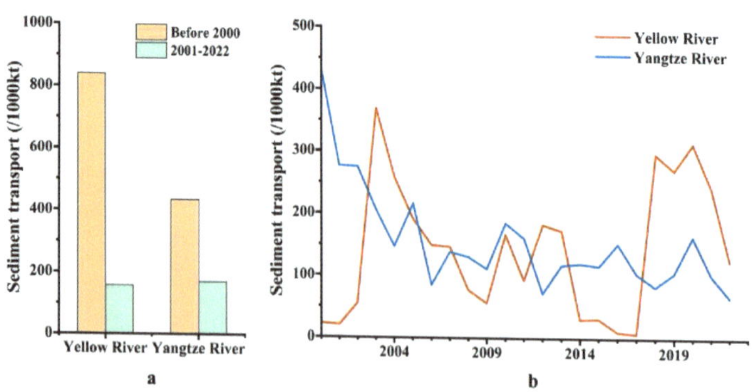

Figure 15. Sediment transport of major rivers. (**a**) The average sediment transport of rivers during the early 21st century, and (**b**) sediment transport of major rivers in the past twenty years.

4.2.2. Tidal Flat Changes Induced by Vegetation

Coastal vegetation expansion plays a significant role in specific areas. Our six-year time series analysis of relative elevation frequencies of tidal flats has revealed the direct impact of vegetation growth on the expansion or reduction of tidal flats in specific wetland systems. The energy transferred to high-tidal flats by waves is limited, creating favorable conditions for the ongoing growth of new vegetation and sediment stabilization, which establishes suitable boundary conditions for vegetation survival. Observation of C3, situated at the estuary, reveals continuous sediment accumulation in this region, resulting in the outward expansion of tidal flat boundaries, despite the overall area exhibiting a declining trend. This expansion is attributed to vegetation encroaching directly upon the survival space of the high-tidal flats [67], covering an area of 88.60 km^2 (Figure 16). There seems to be a correlation between the vegetation increase and sediment transport. Over the three periods of 2018, 2020, and 2022, the total sediment transport increased from 187.1 million tons to 269 million tons and then decreased to 168.5 million tons (Figure 15b), while the increase in vegetation area decreased from 62.89 km^2 to 25.71 km^2. Thus, while the expansion of vegetation encroaches directly upon the tidal flat area, the reduction in sediment transport hinders the rapid accumulation of sediment in the high-tidal flats to meet the conditions for vegetation survival, thereby limiting further vegetation expansion.

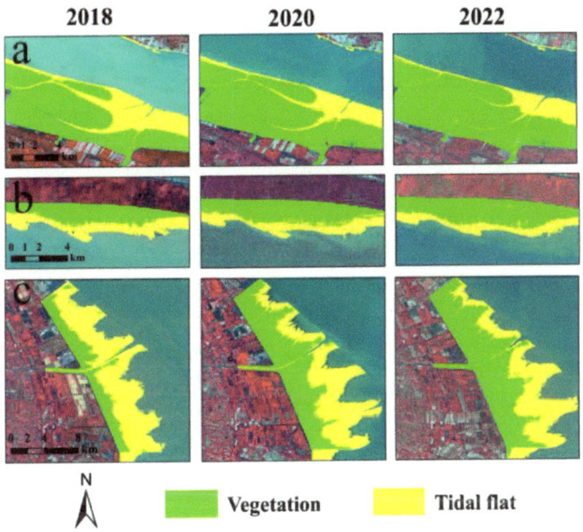

Figure 16. The changes in tidal flats and vegetation in the Yangtze River Estuary from 2018 to 2022. (**a**) The northern portion of Chongming East Beach, (**b**) the southern portion of Hengduan Sands, and (**c**) the eastern portion of Nanhui Beach.

5. Conclusions

Our study proposed a PMITMA algorithm for mapping coastal tidal flats, which integrates multi-source time-series satellite imagery and utilizes muti-indices for water index calculation, along with the K-Means++ method for water-land separation. This model effectively addresses challenges related to tides, waves, and cloud contamination. Consequently, it generates highly detailed maps of tidal flat coverage along the coastlines of the Yellow and Bohai Seas from 2017 to 2022. These maps offer robust data support for the scientific management and utilization of coastal tidal flat resources while facilitating high-quality development in the coastal zone.

Our research findings indicate a shift in the causes of tidal flat changes, from previous human reclamation to sediment transport and vegetation growth. In specific regions, such as the Yangtze River delta, the expansion of vegetation covers an area of 88.60 km^2, while sediment from the Yellow River and Yangtze River decreased by 81.33% and 60.80%, respectively, during the transition from the twentieth to the twenty-first century. Over the study period, the tidal flat area in the YBS decreased by 78 km^2, resulting in a current total area of 4778.32 km^2. The primary change pattern entailed a conversion from mid-high-tidal flats to low-tidal flats, along with the loss of pre-existing low-tidal flats. The most affected areas by these losses were the bays and coastal segments distant from river mouths.

Author Contributions: Conception, W.W.; Methodology, Z.G. and W.W.; Validation, Z.G.; Data Curation, Z.G. and S.G.; Writing—original draft, Z.G.; Writing—review and editing, C.C., H.Z., H.S., Y.H. and W.W.; Funding Acquisition, W.W. All authors have read and agreed to the published version of the manuscript.

Funding: This research was funded by the National Natural Science Foundation of China (Grant No. 42301540), the Open Research Fund of State Key Laboratory of Estuarine and Coastal Research (Grant number SKLEC-KF202307), the Natural Science Foundation of Fujian Province (Grant No. 2022J05024), the Education Department of Fujian Province (Grant No. JAT210027), and the Natural Science Foundation for Distinguished Young Scholars of Fujian Province, China (Grant No. 2021J06014).

Data Availability Statement: The data presented in this study are available on request from the corresponding author.

Acknowledgments: We appreciate the critical and constructive comments and suggestions from the reviewers that helped improve the quality of this manuscript.

Conflicts of Interest: The authors declare no conflicts of interest.

References

1. Dyer, K.R.; Christie, M.C.; Wright, E.W. The classification of intertidal mudflats. *Cont. Shelf Res.* **2000**, *20*, 1039–1060. [CrossRef]
2. Murray, N.J.; Phinn, S.R.; Clemens, R.S.; Roelfsema, C.M.; Fuller, R.A. Continental Scale Mapping of Tidal Flats across East Asia Using the Landsat Archive. *Remote Sens.* **2012**, *4*, 3417–3426. [CrossRef]
3. Ma, Z.; Melville, D.S.; Liu, J.; Chen, Y.; Yang, H.; Ren, W.; Zhang, Z.; Piersma, T.; Li, B. Rethinking China's new great wall. *Science* **2014**, *346*, 912–914. [CrossRef]
4. Solomon, S.; Plattner, G.-K.; Knutti, R.; Friedlingstein, P. Irreversible climate change due to carbon dioxide emissions. *Proc. Natl. Acad. Sci. USA* **2009**, *106*, 1704–1709. [CrossRef]
5. Rodriguez, J.F.; Saco, P.M.; Sandi, S.; Saintilan, N.; Riccardi, G. Potential increase in coastal wetland vulnerability to sea-level rise suggested by considering hydrodynamic attenuation effects. *Nat. Commun.* **2017**, *8*, 16094. [CrossRef]
6. Dhanjal-Adams, K.L.; Hanson, J.O.; Murray, N.J.; Phinn, S.R.; Wingate, V.R.; Mustin, K.; Lee, J.R.; Allan, J.R.; Cappadonna, J.L.; Studds, C.E.; et al. The distribution and protection of intertidal habitats in Australia. *Emu Austral Ornithol.* **2016**, *116*, 208–214. [CrossRef]
7. Phinn, S.R.; Kovacs, E.M.; Roelfsema, C.M.; Canto, R.F.; Collier, C.J.; McKenzie, L.J. Assessing the potential for satellite image monitoring of seagrass thermal dynamics: For inter- and shallow sub-tidal seagrasses in the inshore Great Barrier Reef World Heritage Area, Australia. *Int. J. Digit. Earth* **2018**, *11*, 803–824. [CrossRef]
8. Ghosh, S.; Mishra, D.R.; Gitelson, A.A. Long-term monitoring of biophysical characteristics of tidal wetlands in the northern Gulf of Mexico—A methodological approach using MODIS. *Remote Sens. Environ.* **2016**, *173*, 39–58. [CrossRef]
9. Murray, N.J.; Phinn, S.R.; DeWitt, M.; Ferrari, R.; Johnston, R.; Lyons, M.B.; Clinton, N.; Thau, D.; Fuller, R.A. The global distribution and trajectory of tidal flats. *Nature* **2019**, *565*, 222–225. [CrossRef]
10. McGranahan, G.; Balk, D.; Anderson, B. The rising tide: Assessing the risks of climate change and human settlements in low elevation coastal zones. *Environ. Urban.* **2007**, *19*, 17–37. [CrossRef]
11. Sengupta, D.; Chen, R.; Meadows, M.E.; Choi, Y.R.; Banerjee, A.; Zilong, X. Mapping Trajectories of Coastal Land Reclamation in Nine Deltaic Megacities using Google Earth Engine. *Remote Sens.* **2019**, *11*, 2621. [CrossRef]
12. Wu, W.; Gao, Y.; Chen, C.; Sun, Y.; Su, H. A Framework for Assessing the Dynamic Coastlines Induced by Urbanization Using Remote Sensing Data: A Case Study in Fujian, China. *Remote Sens.* **2022**, *14*, 2911. [CrossRef]
13. Ren, C.; Wang, Z.; Zhang, Y.; Zhang, B.; Chen, L.; Xi, Y.; Xiao, X.; Doughty, R.B.; Liu, M.; Jia, M.; et al. Rapid expansion of coastal aquaculture ponds in China from Landsat observations during 1984–2016. *Int. J. Appl. Earth Obs. Geoinf.* **2019**, *82*, 101902. [CrossRef]
14. Jung, P.S.; Seo, H.-J.; Min, P.S.; Park, S.-H.; Ahn, I.J.; Seo, K.S. The Analysis of Coastal Erosion and Erosion Impact Assessment in the East Coast. *J. Korean Soc. Coast. Ocean Eng.* **2021**, *33*, 246–256. [CrossRef]
15. Beeharry, Y.D.; Bekaroo, G.; Bokhoree, C.; Phillips, M.R. Impacts of sea-level rise on coastal zones of Mauritius: Insights following calculation of a coastal vulnerability index. *Nat. Hazards* **2022**, *114*, 27–55. [CrossRef]
16. Lovelock, C.E.; Feller, I.C.; Reef, R.; Hickey, S.; Ball, M.C. Mangrove dieback during fluctuating sea levels. *Sci. Rep.* **2017**, *7*, 1680. [CrossRef]
17. Yim, J.; Kwon, B.-O.; Nam, J.; Hwang, J.H.; Choi, K.; Khim, J.S. Analysis of forty years long changes in coastal land use and land cover of the Yellow Sea: The gains or losses in ecosystem services. *Environ. Pollut.* **2018**, *241*, 74–84. [CrossRef]
18. Mentaschi, L.; Vousdoukas, M.I.; Pekel, J.-F.; Voukouvalas, E.; Feyen, L.J.S.R. Global long-term observations of coastal erosion and accretion. *Sci. Rep.* **2018**, *8*, 12876. [CrossRef]
19. Dethier, E.N.; Renshaw, C.E.; Magilligan, F.J.J.S. Rapid changes to global river suspended sediment flux by humans. *Science* **2022**, *376*, 1447–1452. [CrossRef]

20. Wu, W.; Zhi, C.; Gao, Y.; Chen, C.; Chen, Z.; Su, H.; Lu, W.; Tian, B. Increasing fragmentation and squeezing of coastal wetlands: Status, drivers, and sustainable protection from the perspective of remote sensing. *Sci. Total Environ.* **2022**, *811*, 152339. [CrossRef]
21. Pari, Y.; Murthy, M.R.; Subramanian, B.; Ramachandran, S. Morphological changes at Vellar estuary, India—Impact of the December 2004 tsunami. *J. Environ. Manag.* **2008**, *89*, 45–57. [CrossRef]
22. Chen, B.; Yang, Y.; Wen, H.; Ruan, H.; Zhou, Z.; Luo, K.; Zhong, F. High-resolution monitoring of beach topography and its change using unmanned aerial vehicle imagery. *Ocean Coast. Manag.* **2018**, *160*, 103–116. [CrossRef]
23. Zhang, H.; Wang, L.; Zhao, Y.; Cao, J.; Xu, M. Application of Airborne LiDAR Measurements to the Topographic Survey of the Tidal Flats of the Northern Jiangsu Radial Sand Ridges in the Southern Yellow Sea. *Front. Mar. Sci.* **2022**, *9*, 871156. [CrossRef]
24. Mason, D.C.; Davenport, I.J.; Robinson, G.J.H.; Flather, R.A.; McCartney, B.S. Construction of an Inter-Tidal Digital Elevation Model by the Water-Line Method. *Geophys. Res. Lett.* **1995**, *22*, 3187–3190. [CrossRef]
25. Salameh, E.; Frappart, F.; Turki, I.; Laignel, B. Intertidal topography mapping using the waterline method from Sentinel-1 &-2 images: The examples of Arcachon and Veys Bays in France. *ISPRS J. Photogramm. Remote Sens.* **2020**, *163*, 98–120. [CrossRef]
26. Zhang, S.; Xu, Q.; Wang, H.; Kang, Y.; Li, X. Automatic Waterline Extraction and Topographic Mapping of Tidal Flats From SAR Images Based on Deep Learning. *Geophys. Res. Lett.* **2022**, *49*, e2021GL096007. [CrossRef]
27. Ryu, J.-H.; Kim, C.-H.; Lee, Y.-K.; Won, J.-S.; Chun, S.-S.; Lee, S. Detecting the intertidal morphologic change using satellite data. *Estuar. Coast. Shelf Sci.* **2008**, *78*, 623–632. [CrossRef]
28. Chen, L.C.; Rau, J.Y. Detection of shoreline changes for tideland areas using multi-temporal satellite images. *Int. J. Remote Sens.* **1998**, *19*, 3383–3397. [CrossRef]
29. Chang, M.; Li, P.; Li, Z.; Wang, H. Mapping Tidal Flats of the Bohai and Yellow Seas Using Time Series Sentinel-2 Images and Google Earth Engine. *Remote Sens.* **2022**, *14*, 1789. [CrossRef]
30. Chen, C.; Zhang, C.; Tian, B.; Wu, W.; Zhou, Y. Tide2Topo: A new method for mapping intertidal topography accurately in complex estuaries and bays with time-series Sentinel-2 images. *ISPRS J. Photogramm. Remote Sens.* **2023**, *200*, 55–72. [CrossRef]
31. Wu, W.; Zhang, M.; Chen, C.; Chen, Z.; Yang, H.; Su, H. Coastal reclamation shaped narrower and steeper tidal flats in Fujian, China: Evidence from time-series satellite data. *Ocean Coast. Manag.* **2024**, *247*, 106933. [CrossRef]
32. Sagar, S.; Roberts, D.; Bala, B.; Lymburner, L. Extracting the intertidal extent and topography of the Australian coastline from a 28 year time series of Landsat observations. *Remote Sens. Environ.* **2017**, *195*, 153–169. [CrossRef]
33. Mao, D.; Liu, M.; Wang, Z.; Li, L.; Man, W.; Jia, M.; Zhang, Y. Rapid Invasion of Spartina Alterniflora in the Coastal Zone of Mainland China: Spatiotemporal Patterns and Human Prevention. *Sensors* **2019**, *19*, 2308. [CrossRef]
34. MacKinnon, J.; Verkuil, Y.I.; Murray, N. *IUCN Situation Analysis on East and Southeast Asian Intertidal Habitats, with Particular Reference to the Yellow Sea (Including the Bohai Sea)*; Occasional Paper of the IUCN Species Survival Commission no. 47; IUCN: Gland, Switzerland, 2012.
35. Wei, Y.; Jiao, J.; Zhao, G.; Zhao, H.; He, Z.; Mu, X.J.C. Spatial–temporal variation and periodic change in streamflow and suspended sediment discharge along the mainstream of the Yellow River during 1950–2013. *Catena* **2016**, *140*, 105–115. [CrossRef]
36. Qiao, S.; Shi, X.; Wang, G.; Zhou, L.; Hu, B.; Hu, L.; Yang, G.; Liu, Y.; Yao, Z.; Liu, S. Sediment accumulation and budget in the Bohai sea, Yellow Sea and East China sea. *Mar. Geol.* **2017**, *390*, 270–281. [CrossRef]
37. Zhou, Z.; Bian, C.; Chen, S.; Li, Z.; Jiang, W.; Wang, T.; Bi, R. Sediment concentration variations in the East China Seas over multiple timescales indicated by satellite observations. *J. Mar. Syst.* **2020**, *212*, 103430. [CrossRef]
38. Wang, S.; Mao, Y.; Zheng, L.; Qiu, Z.; Bilal, M.; Sun, D. Remote sensing of water turbidity in the Eastern China Seas from Geostationary Ocean Colour Imager. *Int. J. Remote Sens.* **2020**, *41*, 4080–4101. [CrossRef]
39. Xu, H. Development of remote sensing water indices:a review. *J. Fuzhou Univ. Nat. Sci. Ed.* **2021**, *49*, 613–625.
40. McFeeters, S.K. The use of the normalized difference water index (NDWI) in the delineation of open water features. *Int. J. Remote Sens.* **1996**, *17*, 1425–1432. [CrossRef]
41. Xu, H. Modification of normalised difference water index (NDWI) to enhance open water features in remotely sensed imagery. *Int. J. Remote Sens.* **2006**, *27*, 3025–3033. [CrossRef]
42. Feyisa, G.L.; Meilby, H.; Fensholt, R.; Proud, S.R. Automated Water Extraction Index: A new technique for surface water mapping using Landsat imagery. *Remote Sens. Environ.* **2014**, *140*, 23–35. [CrossRef]
43. Singh, K.V.; Setia, R.; Sahoo, S.; Prasad, A.; Pateriya, B. Evaluation of NDWI and MNDWI for assessment of waterlogging by integrating digital elevation model and groundwater level. *Geocarto Int.* **2015**, *30*, 650–661. [CrossRef]
44. Acharya, T.D.; Subedi, A.; Lee, D.H. Evaluation of Water Indices for Surface Water Extraction in a Landsat 8 Scene of Nepal. *Sensors* **2018**, *18*, 2580. [CrossRef]
45. Fisher, A.; Flood, N.; Danaher, T. Comparing Landsat water index methods for automated water classification in eastern Australia. *Remote Sens. Environ.* **2016**, *175*, 167–182. [CrossRef]
46. Nguyen, U.N.T.; Pham, L.T.H.; Dang, T.D. an automatic water detection approach using Landsat 8 OLI and Google earth engine cloud computing to map lakes and reservoirs in New Zealand. *Environ. Monit. Assess.* **2020**, *192*, 616. [CrossRef]
47. Yang, Y.; Liu, Y.; Zhou, M.; Zhang, S.; Zhan, W.; Sun, C.; Duan, Y. Landsat 8 OLI image based terrestrial water extraction from heterogeneous backgrounds using a reflectance homogenization approach. *Remote Sens. Environ.* **2015**, *171*, 14–32. [CrossRef]
48. Jia, M.; Wang, Z.; Mao, D.; Ren, C.; Wang, C.; Wang, Y. Rapid, robust, and automated mapping of tidal flats in China using time series Sentinel-2 images and Google Earth Engine. *Remote Sens. Environ.* **2021**, *255*, 112285. [CrossRef]

49. Sunder, S.; Ramsankaran, R.; Ramakrishnan, B. Inter-comparison of remote sensing sensing-based shoreline mapping techniques at different coastal stretches of India. *Environ. Monit. Assess.* **2017**, *189*, 290. [CrossRef]
50. Le Hung, T.; Thi Giang, L.; Van Hoan, K.; Thi Minh Ly, T.; Thi Thu Nga, N. Application of remote sensing technique for shoreline change detection in Ninh Binh and Nam Dinh provinces (Vietnam) during the period 1988 to 2018 based on water indices. *Russ. J. Earth Sci.* **2020**, *20*, ES2004. [CrossRef]
51. Huang, C.; Chen, Y.; Zhang, S.; Wu, J. Detecting, Extracting, and Monitoring Surface Water From Space Using Optical Sensors: A Review. *Rev. Geophys.* **2018**, *56*, 333–360. [CrossRef]
52. Li, D.; Wu, B.; Chen, B.; Xue, Y.; Zhang, Y. Review of water body information extraction based on satellite remote sensing. *J. Tsinghua Univ. Sci. Technol.* **2020**, *60*, 147–161. [CrossRef]
53. Arthur, D.; Vassilvitskii, S. k-means++: The advantages of careful seeding. In Proceedings of the Eighteenth Annual ACM-SIAM Symposium on Discrete Algorithms, New Orleans, LO, USA, 7–9 January 2007; pp. 1027–1035.
54. Wang, X.; Xiao, X.; Zou, Z.; Hou, L.; Qin, Y.; Dong, J.; Doughty, R.B.; Chen, B.; Zhang, X.; Chen, Y.; et al. Mapping coastal wetlands of China using time series Landsat images in 2018 and Google Earth Engine. *ISPRS J. Photogramm. Remote Sens.* **2020**, *163*, 312–326. [CrossRef]
55. Wu, W.; Zhi, C.; Chen, C.; Tian, B.; Chen, Z.; Su, H. Detecting annual anthropogenic encroachment on intertidal vegetation using full Landsat time-series in Fujian, China. *GIScience Remote Sens.* **2022**, *59*, 2266–2282. [CrossRef]
56. Zhang, Z.; Xu, N.; Li, Y.; Li, Y. Sub-continental-scale mapping of tidal wetland composition for East Asia: A novel algorithm integrating satellite tide-level and phenological features. *Remote Sens. Environ.* **2022**, *269*, 112799. [CrossRef]
57. Dong, J.; Xiao, X.; Menarguez, M.A.; Zhang, G.; Qin, Y.; Thau, D.; Biradar, C.; Moore, B., III. Mapping paddy rice planting area in northeastern Asia with Landsat 8 images, phenology-based algorithm and Google Earth Engine. *Remote Sens. Environ.* **2016**, *185*, 142–154. [CrossRef]
58. Wang, H.; Wang, A.; Bi, N.; Zeng, X.; Xiao, H. Seasonal distribution of suspended sediment in the Bohai Sea, China. *Cont. Shelf Res.* **2014**, *90*, 17–32. [CrossRef]
59. Su, M.; Yao, P.; Wang, Z.B.; Zhang, C.K.; Stive, M.J.F. Exploratory morphodynamic modeling of the evolution of the Jiangsu coast, China, since 1855: Contributions of old Yellow River-derived sediment. *Mar. Geol.* **2017**, *390*, 306–320. [CrossRef]
60. Liu, J.; Saito, Y.; Kong, X.; Wang, H.; Xiang, L.; Wen, C.; Nakashima, R. Sedimentary record of environmental evolution off the Yangtze River estuary, East China Sea, during the last similar to 13,000 years, with special reference to the influence of the Yellow River on the Yangtze River delta during the last 600 years. *Quat. Sci. Rev.* **2010**, *29*, 2424–2438. [CrossRef]
61. Zheng, Q.; Fang, G.; Song, Y.T. Introduction to special section: Dynamics and circulation of the Yellow, East, and South China Seas. *J. Geophys. Res.* **2006**, *111*, C11S01. [CrossRef]
62. Wang, A.; Wang, H.; Bi, N.; Wu, X. Sediment transport and dispersal pattern from the Bohai Sea to the Yellow Sea. *J. Coast. Res.* **2016**, *74*, 104–116. [CrossRef]
63. Xu, H.; Jia, A.; Song, X.; Bai, Y. Extraction and spatiotemporal evolution analysis of tidal flats in the Bohai Rim during 1984-2019 based on remote sensing. *J. Geogr. Sci.* **2023**, *33*, 76–98. [CrossRef]
64. Kim, W.; Doh, S.-J.; Yu, Y.; Lee, Y.I. Magnetic evaluation of sediment provenance in the northern East China Sea using fuzzy c-means cluster analysis. *Mar. Geol.* **2013**, *337*, 9–19. [CrossRef]
65. Wang, Y.; Liu, Y.; Jin, S.; Sun, C.; Wei, X. Evolution of the topography of tidal flats and sandbanks along the Jiangsu coast from 1973 to 2016 observed from satellites. *ISPRS J. Photogramm. Remote Sens.* **2019**, *150*, 27–43. [CrossRef]
66. Ni, W.; Wang, Y.; Zou, X.; Zhang, J.; Gao, J. Sediment dynamics in an offshore tidal channel in the southern Yellow Sea. *Int. J. Sediment Res.* **2014**, *29*, 246–259. [CrossRef]
67. Wu, W.-T.; Zhou, Y.-X.; Tian, B.J.O.; Management, C. Coastal wetlands facing climate change and anthropogenic activities: A remote sensing analysis and modelling application. *Ocean. Coast. Manag.* **2017**, *138*, 1–10. [CrossRef]

remote sensing

MDPI

Article

Accurate and Rapid Extraction of Aquatic Vegetation in the China Side of the Amur River Basin Based on Landsat Imagery

Mengna Chen [1,2], Rong Zhang [2,*], Mingming Jia [2], Lina Cheng [2,3], Chuanpeng Zhao [2], Huiying Li [4] and Zongming Wang [2]

1 School of Land Engineering, Chang'an University, Xi'an 710054, China; 2021135007@chd.edu.cn
2 Key Laboratory of Wetland Ecology and Environment, Northeast Institute of Geography and Agroecology, Chinese Academy of Sciences, Changchun 130102, China; jiamingming@iga.ac.cn (M.J.); chengln23@mails.jlu.edu.cn (L.C.); zhaochuanpeng@iga.ac.cn (C.Z.); zongmingwang@iga.ac.cn (Z.W.)
3 College of Earth Sciences, Jilin University, Changchun 130061, China
4 School of Environmental and Municipal Engineering, Qingdao University of Technology, Qingdao 266520, China; lihy@qut.edu.cn
* Correspondence: zhangrong@iga.ac.cn

Abstract: Since the early 1950s, the development of human settlements and over-exploitation of agriculture in the China side of the Amur River Basin (CARB) have had a major impact on the water environment of the surrounding lakes, resulting in a decrease of aquatic vegetation. According to the United Nations Sustainable Development Goals, a comprehensive understanding of the extent and variability of aquatic vegetation is crucial for preserving the structure and functionality of stable aquatic ecosystems. Currently, there is a deficiency in the CARB long-sequence dataset of aquatic vegetation distribution in China. This shortage hampers effective support for actual management. Therefore, the development of a fast, robust, and automatic method for accurate extraction of aquatic vegetation becomes crucial for large-scale applications. Our objective is to gather information on the spatial and temporal distribution as well as changes in aquatic vegetation within the CARB. Utilizing a hybrid approach that combines the maximum spectral index composite and Otsu algorithm, along with the integration of convolutional neural networks (CNN) and random forest, we applied this methodology to obtain an annual dataset of aquatic vegetation spanning from 1985 to 2020 using Landsat series imagery. The accuracy of this method was validated through both field investigations and Google Images. Upon assessing the confusion matrix spanning from 1985 to 2020, the producer accuracy for aquatic vegetation classification consistently exceeded 87%. Further quantitative analysis unveiled a discernible decreasing trend in both the water and vegetation areas of lakes larger than 20 km^2 within the CARB over the past 36 years. Specifically, the total water area decreased from 3575 km^2 to 3412 km^2, while the vegetation area decreased from 745 km^2 to 687 km^2. These changes may be attributed to a combination of climate change and human activities. These quantitative data hold significant practical implications for establishing a scientific restoration path for lake aquatic vegetation. They are particularly valuable for constructing the historical background and reference indices of aquatic vegetation.

Keywords: aquatic vegetation; spatial distribution; CNN; random forest

Citation: Chen, M.; Zhang, R.; Jia, M.; Cheng, L.; Zhao, C.; Li, H.; Wang, Z. Accurate and Rapid Extraction of Aquatic Vegetation in the China Side of the Amur River Basin Based on Landsat Imagery. *Remote Sens.* **2024**, *16*, 654. https://doi.org/10.3390/rs16040654

Academic Editors: Yanni Dong, Tao Chen and Chao Chen

Received: 5 January 2024
Revised: 31 January 2024
Accepted: 7 February 2024
Published: 9 February 2024

1. Introduction

Lakes serve as crucial links in the interaction of various elements within terrestrial ecosystems [1–3]. They fulfill diverse ecological functions, including the regulation of river runoff and the breeding of aquatic organisms, and play a significant strategic role in safeguarding national social and economic interests, and ecological security [4–6]. Aquatic vegetation is a crucial component of lake ecosystems, and its degradation and loss serve as important indicators of ecosystem decline in lakes [7,8]. It also plays a crucial role in maintaining the balance of lake ecosystems by stabilizing sediments, sequestering carbon, improving water quality, facilitating nutrient cycling, and providing food and habitat for a variety of aquatic organisms [9–12]. Aquatic vegetation can be categorized into three types based on its position relative to the water surface: emergent vegetation, floating-leaved vegetation, and submerged vegetation [13].

The Amur River Basin, spanning Russia, China, and Mongolia, ranks among the ten largest river basins in the world [14]. The Chinese part of the basin constitutes 48% of the entire basin and hosts the majority of freshwater lakes in Northeast China. It stands as the region with the most abundant water resources and densest concentration of lakes in Northeast China. Additionally, it boasts the largest natural forest area and serves as a crucial commercial grain base in China [15,16]. More than 93% of the population of the entire basin lives on the China side of the Amur River Basin (CARB) [17]. Since the early 1950s, the development of human settlements and agriculture in the CARB has significantly impacted the region's natural wetlands [18,19]. The excessive use of groundwater, fertilizers, pesticides, and other agricultural inputs, along with the overdevelopment of agriculture, has resulted in the degradation of cultivated land and the serious pollution of wetlands. These practices have had a significant impact on the aquatic environment of the surrounding lakes [20]. This has led to the decline of aquatic vegetation in the lakes, significantly affecting the living environment quality of the surrounding residents. Moreover, it causes negative impacts on the effective utilization of water supply resources and hampers social and economic development. A comprehensive understanding of temporal and spatial changes in aquatic vegetation within the basin is indispensable to assessing ecosystem services and supporting the sustainable conservation of the region's wetlands.

Traditional methods of monitoring aquatic vegetation rely on labor-intensive field surveys, making it challenging to obtain continuous spatial distribution information. As a non-contact detection method, remote sensing has unique advantages in the monitoring and extraction of aquatic vegetation information [21]. In general, satellite remote sensing data with varying resolutions can achieve an overall accuracy of aquatic vegetation group classification above 80% [22]. However, due to the influence of water and other environmental factors, the radiation transfer mechanism of the aquatic vegetation canopy becomes highly complex, thereby increasing the difficulty of its accurate identification. In recent years, various classification methods have been applied to monitor changes in aquatic vegetation. For example, Jing et al. used micro unmanned aerial vehicle images to generate visible vegetation indices, establish decision tree rules, and automatically classify the ecological area of the Xin Cheng Waterside Forest Park in Beijing, achieving an overall accuracy of 91.7% [23]. Chen et al. proposed a new concave–convex decision function for detecting submerged aquatic vegetation (SAV) in Lake Ulansuhai, with classification accuracies of 92.17% and 91.79% for July and August, respectively [24]. Wang et al. first proposed an enhanced aquatic vegetation index based on the difference in two short-wave infrared (SWIR) bands to extract the aquatic vegetation above the water surface. Subsequently, using Taihu Lake as an example, they accurately extracted seasonal and annual distributions of aquatic vegetation [25]. Ashford described a field investigation method and image processing workflow based on unmanned aerial systems (UAS), which was used to map local, exotic, and invasive aquatic vegetation, indicating that small UAS can investigate, map, and detect features of interest in a large area [26]. Additionally, machine learning methods, including random forest (RF) classification and deep learning, have also been employed [27–29], demonstrating significant potential in the field of aquatic vegetation extraction. Aquatic vegetation extraction based on deep learning achieves precise identification by learning features from a large amount of remote sensing image data. Therefore, deep learning technology provides new insights and approaches for aquatic vegetation extraction. The results of Ji et al. indicated that the integration of CNN and RF classifiers slightly outperforms independent RF classification and CNN classification methods in terms of classification accuracy [30]. However, there is currently a lack of research on integrating CNN and RF classifiers for aquatic vegetation information extraction.

The specific objectives of this study are as follows: (1) to propose a novel and robust classification method by integrating image composite algorithms, CNN, and RF classifiers; (2) to apply the proposed method to generate annual maps of aquatic vegetation, water, and other land cover types in the CARB from 1985 to 2020; (3) to analyze the spatiotemporal changes of aquatic vegetation, water, and other land cover types during the period of climate change from 1985 to 2020. This study extends the spatial transferability of the model and provides strong methodological and theoretical support for future extraction and monitoring of aquatic vegetation.

2. Materials and Methods

2.1. Study Area

As shown in Figure 1, the CARB (41°45′–53°33′N, 115°13′–135°05′E) encompasses most of Heilongjiang and parts of Jilin and Inner Mongolia. We selected the China side of the Amur River Basin with lakes larger than 20 km^2 as the study area. The basin has a monsoon climate, with winds alternating between continental and maritime with the seasons. The summer is short, warm, and humid, with temperatures ranging from 16–18 °C in July. The winter is long (November to March), dry, and cold, and January is the coldest month, with temperatures ranging from −31 to −15 °C. Nearly two-thirds of the basin's precipitation occurs from June to August. The peak period of vegetation biomass in the study area occurs from July to August, while the ice cover period lasts from November to March the following year. The average annual temperature ranges from −8 to 6 °C, and the average annual precipitation ranges from 250 to 800 mm. The area exhibits abundant water resources and diverse flora and fauna, making it a significant ecological hotspot. The aquatic vegetation types of the CARB are mainly reed communities, wild barley communities, *Potamogeton wrightii* Morong communities, *Calamagrostis epigeios* (L.) Roth communities, *Suaeda salsa* communities, *Zizania latifolia* (Griseb.) Turcz. ex Stapf communities, Typha orientalis Presl communities, and so on. Moreover, over the past three decades, it has thrived as an agricultural region, particularly renowned for its prominent role in rice cultivation.

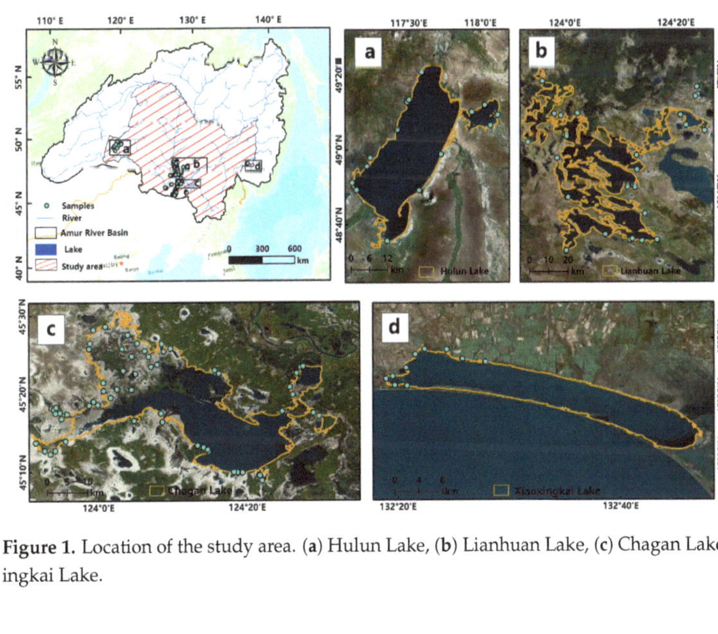

Figure 1. Location of the study area. (**a**) Hulun Lake, (**b**) Lianhuan Lake, (**c**) Chagan Lake, (**d**) Xiaoxingkai Lake.

2.2. Remote Sensing Data

The Landsat series, initially launched in 1972, has a 16-day revisit period and a spatial resolution of 30 m. Due to the limited availability of Landsat-4 TM satellite imagery in China, this study utilized surface reflectance data from Landsat-8 OLI, Landsat-7 ETM+, and Landsat-5 TM. The Scan Line Corrector (SLC) of the Landsat-7 ETM+ sensor failed after 2003, resulting in striping gaps in Landsat-7 image data. This study utilized Landsat-7 ETM+ data from 1999 to 2002. For 2012, only Landsat-7 ETM+ satellite imagery was available, which required banding correction before use. A total of 6175 Landsat surface reflectance images were selected from the Google Earth Engine (GEE) data catalog for the period between 1985 and 1 June 2020, with a focus on the dates from 15 May to 15 October. The temporal distribution of the Landsat satellite series used in this study and the effective observation quantity of Landsat-8 satellite in 2020 are shown in Figure 2.

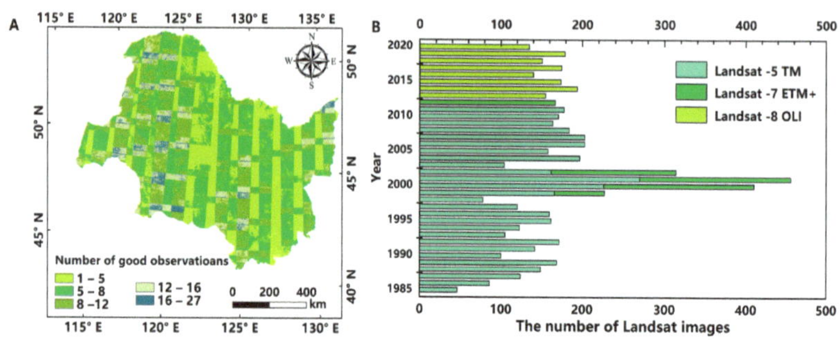

Figure 2. Spatial distribution and the number of good observations across the CARB in 2020 by Landsat-8 OLI (**A**) and number of Landsat images used in the study (**B**).

2.3. Ground Sample Data

The validation data used in this study consisted of 286 field survey samples, high-resolution images obtained from Google Earth, and high-resolution images from national geographic monitoring. The labelme tool was employed for manual visual interpretation to create a labeled dataset of 1200 images with pixels of 128 × 128, which included water surfaces, aquatic vegetation (emergent vegetation, floating-leaved vegetation, and submerged vegetation), and other surface features such as muddy areas in lake transition zones and terrestrial vegetation. In this dataset, 70% of the labels were used for training, while 30% of the samples were reserved for validating the accuracy of the classification results.

2.4. Methodologies

2.4.1. Basic Idea

The mapping workflow for aquatic vegetation, water, and other land cover types (including inland vegetation, algal blooms, agricultural fields, bare soil, built-up areas, etc.) in the CARB is illustrated in Figure 3. Firstly, we utilized the Landsat imagery along with image composite (the specified band value of each pixel at all pixel positions in the image stack is sorted pixel by pixel, maximum value, mean value, or median value, etc., of the specified band are extracted, and finally, all these pixels are formed into an image) and Otsu algorithm (Otsu) to delineate the maximum spatial extent of water surfaces of large lakes (larger than 20 km^2) in the CARB. Subsequently, we developed a novel and robust classification method by combining the image composite, CNN, and RF classifier. This classification approach consists of three steps: (1) obtaining a synthesized image collection of the CARB lakes from high-quality observational data; (2) extracting features from the image collection using a CNN model; (3) classifying the extracted CNN features using an RF classifier.

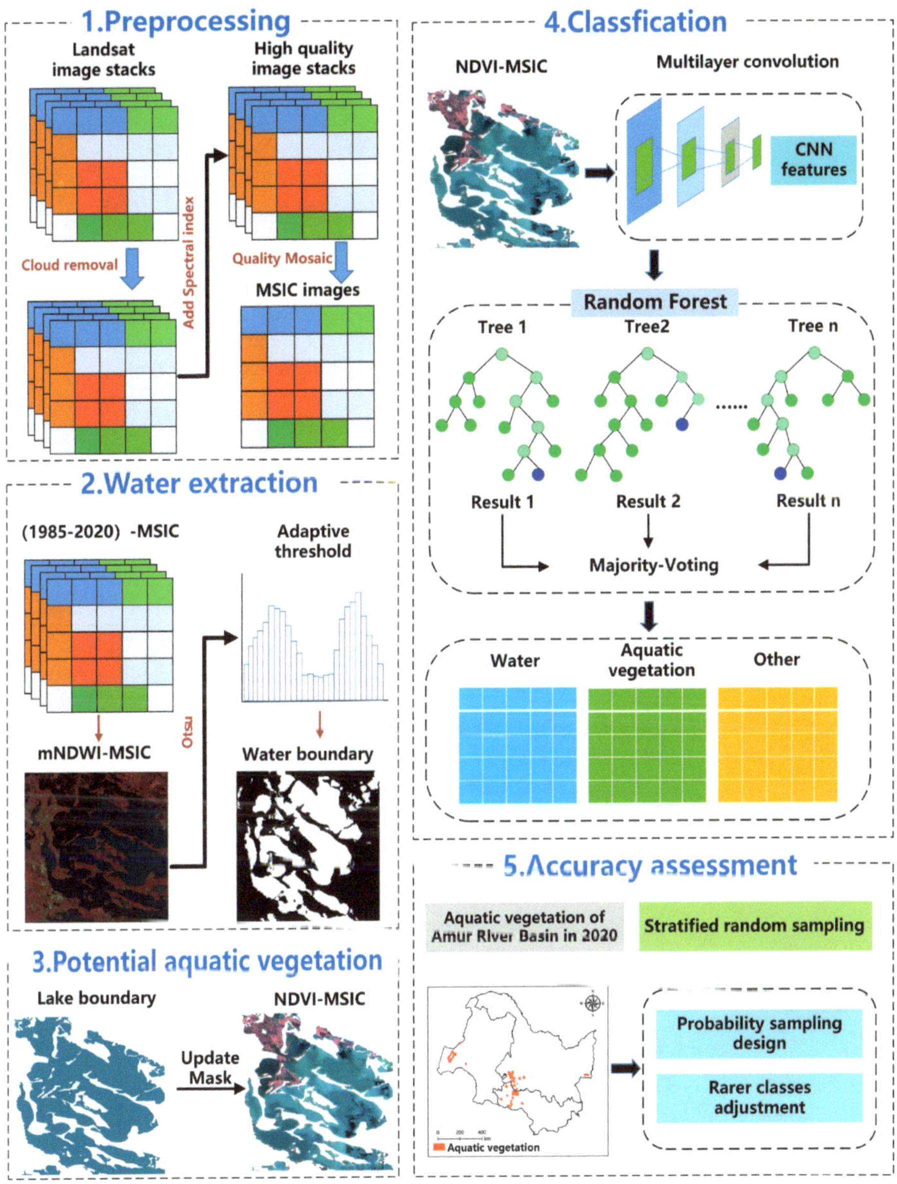

Figure 3. Technical roadmap of aquatic vegetation extraction.

2.4.2. The Maximum Spectral Index Composite (MSIC) and Otsu

The GEE API offers the "imageCollection.qualityMosaic" function, which utilizes a maximum spectral index composite (MSIC). The MSIC algorithm selects a specified band as a sorting criterion on a pixel-by-pixel basis. It then computes the maximum value after sorting the pixels based on this criterion. The quality measure band is used as the primary source for each pixel in the final composite image, where the pixel values are obtained by overlaying the specified band data. This approach ensures that the resulting composite image showcases the highest quality pixels, which are defined as the pixels with the maximum spectral index selected at each pixel position in the time series image stack after cloud removal, based on the selected band. The Otsu, also referred to as the maximum inter-class variance method, was proposed by Otsu in 1975 [31] and has been further studied [32]. This algorithm divides an image into two parts, distinguishing between the background and foreground based on the grayscale characteristics of the image. The optimal threshold is determined by selecting the threshold that maximizes the inter-class variance and minimizes the in-class variance which is then applied for automatic binary classification of the image, enabling efficient segmentation.

2.4.3. Spectral Feature Construction

This study relies on the GEE platform to construct spectral features and enhance the extraction of aquatic vegetation information. These feature variables have been carefully selected and integrated to provide comprehensive and reliable analysis of aquatic vegetation. The spectral index formulas used in this study are shown in Table 1.

Table 1. Formulas of the spectral indices used in this study.

Spectral Name	Expression	Source
NDVI (Normalized Differential Vegetation Index)	$NDVI = \frac{NIR-Red}{NIR+Red}$	[33]
mNDWI (Modified Normalized Difference Water Index)	$mNDWI = \frac{Green-SWIR}{Green+SWIR}$	[34,35]
FAI (Floating Algae Index)	$FAI = (NIR - Red - (SWIR - Red) \times \frac{(\lambda NIR-\lambda Red)}{(\lambda SWIR-\lambda Red)}$	[36]

NIR represents Near-Infrared band Reflectance; Red represents infrared band reflectance; *SWIR* represents Short-Wave Infrared band reflectance; Green represents green band reflectance; λNIR, λRed, and $\lambda SWIR$ are the center wavelengths of the near-infrared, red, and shortwave infrared bands, respectively.

2.4.4. Delineating Spatial Extents of Lakes in the CARB

According to research conducted by Jia et al. [35], the mNDWI (Modified Normalized Difference Water Index) effectively enhances the distinction between artificial shorelines, such as embankments, roads, and pond edges, and the surrounding water. It outperforms the NDWI (Normalized Difference Water Index) in extracting intertidal zones and inland surface water coverage. Therefore, this study selected mNDWI for the extraction of lake water. In this study, the MSIC synthetic image was constructed annually using the mNDWI, and the maximum water surface image was generated. The OA was applied to select the threshold for mNDWI, facilitating the determination of the maximum water surface range on a yearly basis. Subsequently, these ranges were merged to obtain the maximum coverage of lake water surfaces within the study area from 1985 to 2020.

2.4.5. Convolutional Neural Network (CNN)

CNN, proposed by Lecun et al. in the late 1980s [37], is a deep learning model primarily used for image recognition and computer vision tasks. The principle consists of the following key concepts: (1) Convolutional Layer: The convolutional layer is a core component of CNN. It extracts the features of the image by convolution operations on the input image using a series of learnable filters, or convolution nuclei. Each filter slides over the image and calculates the inner product of its corresponding local receptive field, generating a Feature Map. Through the combination of multiple filters, the convolution layer can learn different features, such as edges, textures, etc. (2) Pooling Layer: the pooling layer is used to downsample the feature map, reduce the size and number of parameters of the feature map, and retain the main features. Common pooling operations include Max Pooling and Average Pooling, which select the maximum value or average value of the experience field, respectively, as the pooled value. (3) Activation Function: The activation function introduces nonlinear transformations that enable the CNN to learn nonlinear features. Common activation functions include Rectified Linear Unit (ReLU), Sigmoid, and Tanh. (4) Fully Connected Layer: The fully connected layer flattens the output of the convolutional layer and the pooling layer and fully connects it with the previous layer to convert the extracted feature map into the output of classification or regression. Fully connected layers typically use a Multi-Layer Perceptron (MLP) structure. (5) Back propagation: Forward propagation and back propagation are interdependent processes. Forward propagation calculates and stores the results of each layer of the neural network from the input layer to the output layer in turn. On the other hand, back propagation uses chain rule to calculate the gradient of loss function relative to the weight of each layer, so as to update the weight to minimize the loss function. Loss function plays a key role in the training process by measuring the difference between the predicted results of the model and the actual tags. Through back propagation, CNN can gradually extract the abstract features of the image and automatically learn more advanced feature representations during the training process.

The CNN structure adopted in this study is shown in Figure 4. The network structure of DeepPlabv 3+ is used to extract features [38]. The model is trained with labeled data sets. The main body of its encoder is the DCNN feature extraction network with atrous convolution, and the Xception_65 network structure and Xception_71 network structure are selected. DeepPlabv3+ introduces atrous spatial pyramid pooling (ASPP) to fuse context information at different scales. The ASPP module includes several parallel branches of atrous convolution. Each branch has a different dilation rate. In this way, rich contextual information can be captured at different scales and fused into the segmentation results. The decoder fuses the features of the bottom layer and the top layer by techniques such as upsampling and skip connections, and the extracted CNN features are used as the input of the random forest classifier.

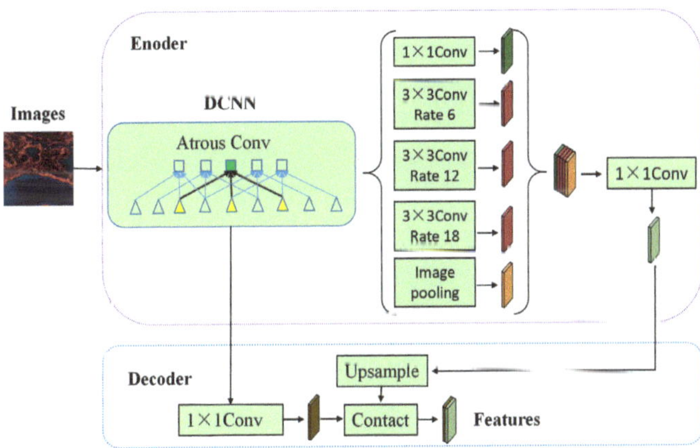

Figure 4. Network structure of CNN feature extraction.

2.4.6. Random Forest (RF) Classification Algorithm

RF is an ensemble learning algorithm known for its advantages, including insensitivity to multivariate collinearity, no requirement for data normalization, and good tolerance for missing and unbalanced data [39]. RF is an integrated method consisting of classification trees and regression trees. The final classification decision is determined by averaging the probabilities of class assignments computed across all generated trees [40]. There are two random processes in RF. First, a training set is created for each tree by sampling replacements from the original training data set. Second, when the nodes of the tree are split, random features are selected from the total features without replacement. RF has been widely adopted because it can improve the robustness and performance of classification based on bootstrap sampling and random feature combination strategies [41]. In addition, it can process multiple variables, rank them, assess variable importance based on performance, and determine the computationally optimal number of trees through algorithm testing. Moreover, RF can overcome the overfitting problem of decision trees and exhibits strong resilience to noise and outliers.

2.4.7. Accuracy Assessment

In this study, a confusion matrix was constructed to obtain the Kappa coefficient, Overall Accuracy (OA), Producer Accuracy (PA), and User Accuracy (UA). OA expresses the overall degree of agreement in the matrix. UA represents the likelihood that a classified object matches the ground situation. PA shows the percentage of an object type, which was correctly classified. Kappa coefficient states how well the classification results agree with the reference data. The formulas are shown in Table 2.

Table 2. Formulas of overall accuracy, Kappa coefficient, producer accuracy, and user accuracy.

Evaluation Index	Equation	Description
Overall Accuracy	$OA = \frac{\sum_{i=1}^{k} X_i}{N}$	The ratio of correctly classified category pixels to total category pixels
Kappa coefficient	$Kappa = \frac{OA-P}{1-P}$ $P = \frac{1}{N^2}\left(\sum_{i=1}^{m} X_i \bullet \sum_{i=1}^{r} X_i\right)$	Used to evaluate the consistency of classification results
Producer Accuracy	$PA = \frac{X_i}{\sum_{i=1}^{m} X_i}$	The ratio of the number of correctly classified pixels of a class to the total number of true reference pixels of that class
User Accuracy	$UA = \frac{X_i}{\sum_{i=1}^{r} X_i}$	The ratio of the number of correctly classified pixels of a class to the total number of pixels of that class

3. Results

3.1. Accuracy Assessment

Table 3 presents the comprehensive confusion matrix for the classification results in 2020. The OA of the land cover map reached 90.6%. The UA for aquatic vegetation and water was 91.07% and 90.91%, respectively, while the UA for other land cover types was 89.29%. Notably, the OA for aquatic vegetation and other land cover types fell below 90%, although the highest PA for water reached 93.75%. When assessing the confusion matrix from 1985 to 2020, the PA of the land cover maps consistently exceeded 87%. The evaluation, conducted through confusion matrix calculations, indicated strong agreement between the obtained maps and ground measurement points.

Table 3. Classification confusion matrix and precision analysis in 2020.

Class	Aquatic Vegetation	Water	Other	Total	UA
Aquatic vegetation	51	2	3	56	91.07%
Water	3	30	0	33	90.91%
Other	3	0	25	28	89.29%
Total	57	32	28	117	
PA	89.47%	93.75%	89.29%		
OA	90.60%	Kappa	85.13%		

3.2. Analysis of the Spatial–Temporal Changes of the Overall Water and Aquatic Vegetation in the Study Area

The overall lake changes are shown in Figure 5. Aquatic vegetation fluctuated greatly during 1985–2001, showing an overall increasing trend, rising from 745 km^2 to 851 km^2. Aquatic vegetation showed a decreasing trend from 2001 to 2003, reducing from 851 km^2 to 596 km^2. The change in aquatic vegetation remained relatively stable from 2004 to 2019, with a significant increase observed in 2020. The overall water surface of the CARB shows a shrinking trend. Water surface showed a slight increasing trend from 1985 to 1987, rising from 3575 km^2 to 3775 km^2, a slight decreasing trend from 1988 to 1989, from 3671 km^2 to 3619 km^2, and a small increasing trend from 1990. By 1997, water surface was in a relatively stable state. Due to artificial intervention, water surface showed an overall decreasing trend from 1998 to 2012, reducing from 3795 km^2 to 2902 km^2. To promote the integrated protection and restoration of mountains, rivers, forests, fields, lakes, grasslands, and deserts, a series of fundamental, pioneering, and long-term efforts has been undertaken. Water surface showed a small decline from 2013 to 2018, gradually recovering after 2019. From 1997 to 2011, other land use classes increased sharply from 349 km^2 to 1135 km^2. With population growth and economic development, the demand for food, agricultural products, and construction also increased, causing people to sacrifice part of the water surface and aquatic vegetation to meet the needs of development. By 2013, it dropped sharply to 612 km^2. From 2014 to 2018, the area of other landincreased. However, by 2020 other land area reduced to 522.4 km^2.

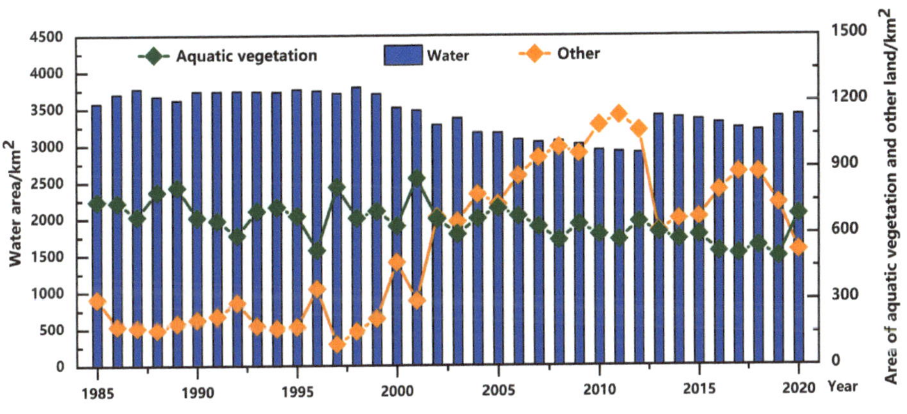

Figure 5. Changes in total aquatic vegetation and water area of the whole study from 1985 to 2020.

The changes in the main lakes (Hulun Lake, Lianhuan Lake, Chagan Lake and Xiaoxingkai Lake) are shown in Figure 6. Aquatic vegetation showed a slight decreasing trend from 1985 to 1987, from 486 km^2 to 485 km^2, and began to increase in 1988. In 1989, the aquatic vegetation was relatively stable, with an area of 549 km^2, and then it showed a decreasing trend. In 1992, aquatic vegetation area was 332 km^2, and during 1992–1996 it showed a fluctuating trend. In 1996, the aquatic vegetation area was 315 km^2, and this increased to 477 km^2 in 1997. From 1997 to 2006, aquatic vegetation fluctuated in a small range. In 2006, the aquatic vegetation area was 340 km^2, and during 2006–2020, the area of aquatic vegetation was relatively stable. In 2020, the area of aquatic vegetation was 428 km^2. Overall, the change trend of water surface and other land in these four typical lakes is close to that in all the lakes as a whole.

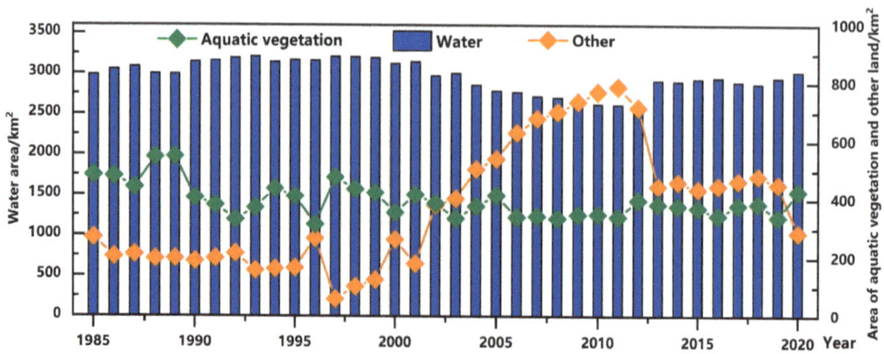

Figure 6. Changes in aquatic vegetation and water area of Hulun Lake, Lianhuan Lake, Chagan Lake, and Xiaoxingkai Lake from 1985 to 2020.

3.2.1. Temporal and Spatial Changes of Aquatic Vegetation in Hulun Lake

As depicted in Figure 7, aquatic vegetation in Hulun Lake is primarily concentrated in the south, southwest, and northeast, typically growing at the lake's edge. From 1985 to 1989, there was a continuous increase in aquatic vegetation, while the lake area remained relatively stable. In 1990, the water area started to increase, whereas aquatic vegetation began to decrease, indicating a decline in the water quality of the lake. The area of water and aquatic vegetation was relatively stable from 1990 to 2001, steadily decreased from 2002 to 2010, remained stable from 2011 to 2013, and showed a steady increase from 2014 to 2020. The overall water changes in Hulun Lake were predominantly concentrated in the eastern, northeastern, southern, and southeastern parts of the lake, while other locations remained relatively stable. The most dramatic changes occurred in the northeast part of the lake, where the lake area expanded significantly in 1990 and remained stable until 2001. However, the water surface in the region experienced a dramatic shrinkage in 2005 and complete drying between 2006 and 2019. Nevertheless, in 2020, water began to reappear in this area, leading to a significant increase in aquatic vegetation. From 2002 to 2020, the overall aquatic vegetation in Hulun Lake exhibited slight fluctuations but demonstrated an overall increasing trend.

3.2.2. Temporal and Spatial Changes of Aquatic Vegetation in Lianhuan Lake

As depicted in Figure 8, aquatic vegetation around the lake is primarily concentrated in the northern part, with a small amount distributed along the lake boundary. From 1985 to 1989, the water surface of the lake remained relatively stable. It exhibited an increasing trend in 1990, gradually decreased from 1991 to 1995, and then increased again in 1996. The area of the water surface initially decreased from 1996 to 2001 and then increased. However, in 2002, the water surface significantly shrank. Subsequently, from 2003 to 2013, the water surface increased steadily, followed by a period of decline from 2014 to 2018. Notably, there was an upward trend from 2017 to 2020. The most significant changes occurred in the northeastern part of the lake. From 1985 to 1989, the surface area of aquatic vegetation increased slightly. From 1989 to 1996, it first decreased slightly and then increased. In 1997, the surface area of aquatic vegetation increased significantly. In 1998, there was a slight increase, but in 1999, the surface area of aquatic vegetation began to decrease sharply. It maintained a relatively stable state in 2001, steadily increased from 2002 to 2005, and then gradually decreased from 2006 to 2013. From 2014 to 2020, the water vegetation area exhibited a pattern of initially increasing and then decreasing. Generally speaking, the aquatic vegetation area showed a decline from 1985 to 2020. Strengthening the protection and restoration of aquatic vegetation in lakes is essential.

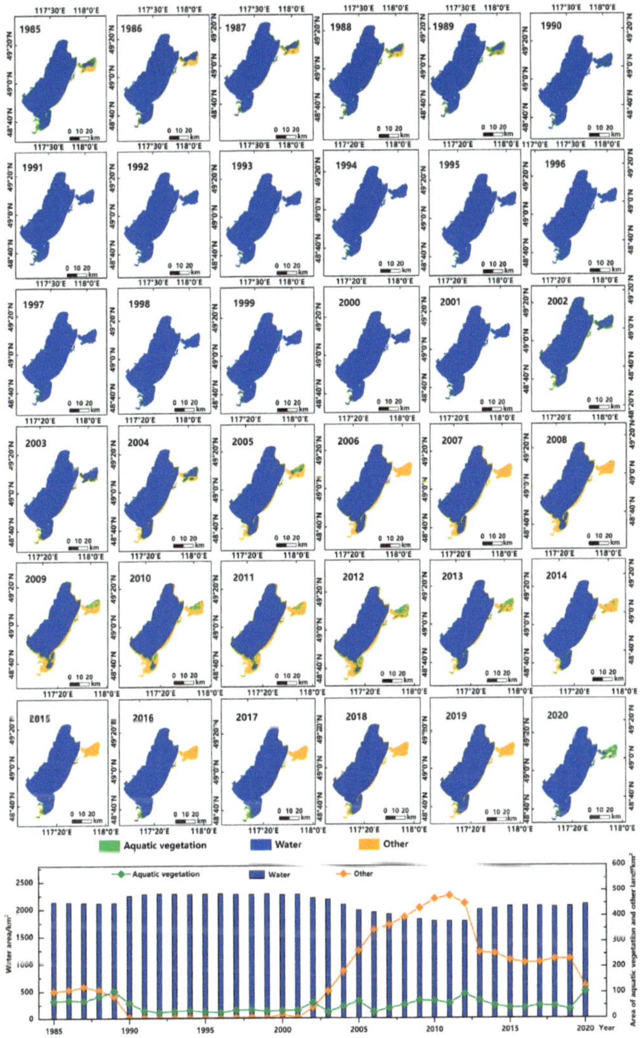

Figure 7. Spatial and temporal distribution and area changes of aquatic vegetation and water of Hulun Lake from 1985 to 2020.

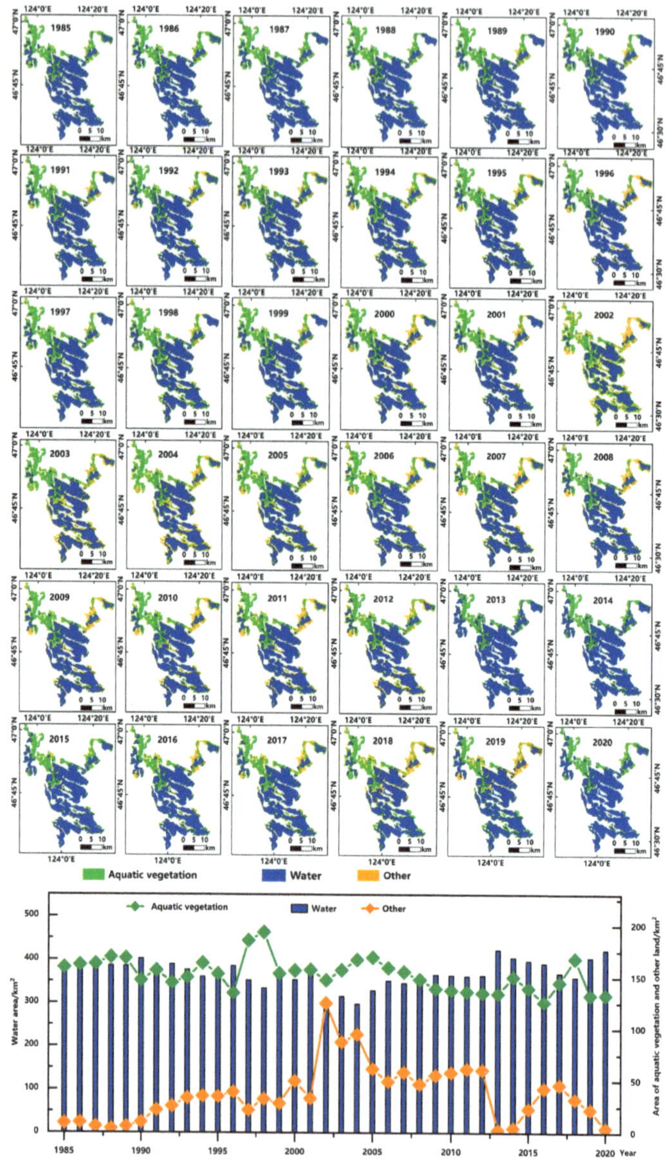

Figure 8. Spatial and temporal distribution and area changes of aquatic vegetation and water of Lianhuan Lake from 1985 to 2020.

3.2.3. Temporal and Spatial Changes of Aquatic Vegetation in Chagan Lake

As shown in Figure 9, aquatic vegetation in Chagan Lake is primarily concentrated in the northwestern and northeastern regions. The water surface of Chagan Lake increased significantly from 1985 to 1987 and decreased from 1987 to 1990. The lake surface area increased from 1988 to 1993, started to decrease from 1994 to 1996, and experienced a sharp increase in 1997–1998. From 1999 to 2002, the lake area steadily decreased, and from 2003 to 2020, it exhibited little fluctuation, remaining essentially stable. The primary reason for the change in the overall water surface of Chagan Lake is the transformation occurring in the northwest part of the lake, while other areas remain relatively stable. From 1985 to 2020, the aquatic vegetation area of Chagan Lake exhibited significant fluctuations. It decreased from 1985 to 1987, followed by a notable increase in 1988. The aquatic vegetation area decreased from 1988 to 1992, increased from 1992 to 1994, and remained stable from 1994 to 1995. There was a significant decrease in the area of aquatic vegetation in 1996. Subsequently, the area fluctuated greatly from 1996 to 2001 and showed a downward trend from 2001 to 2008. However, the aquatic vegetation area increased in 2005 and continued to rise from 2008 to 2020.

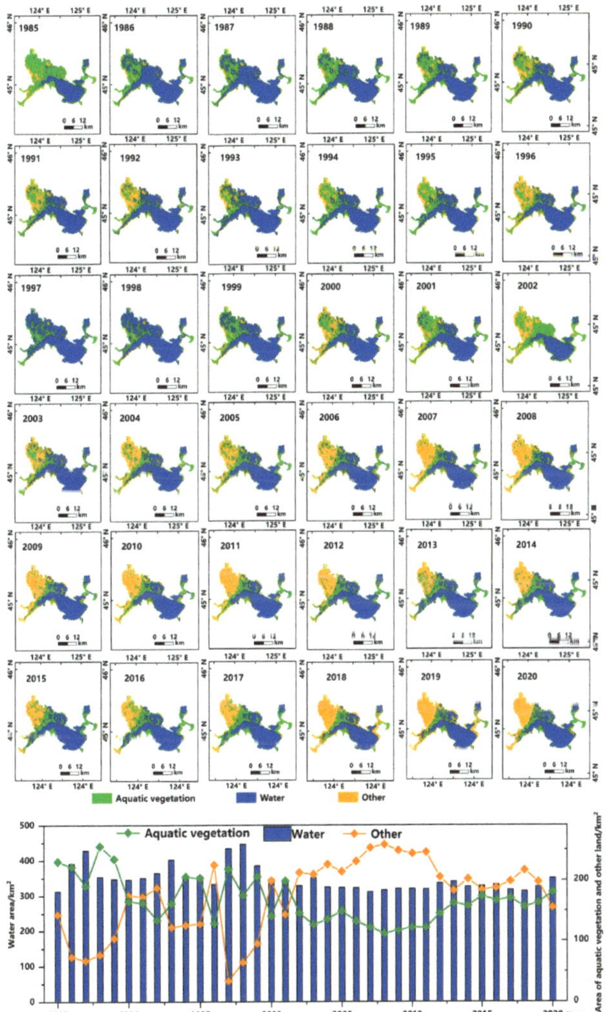

Figure 9. Spatial and temporal distribution and area changes of aquatic vegetation and water of Chagan Lake from 1985 to 2020.

3.2.4. Temporal and Spatial Changes of Aquatic Vegetation in Xiaoxingkai Lake

As shown in Figure 10, aquatic vegetation in Xiaoxingkai Lake is primarily concentrated in the eastern and southeastern parts, with a smaller amount distributed near the lake boundary. According to the statistical chart depicting the water and aquatic vegetation area of Xiaoxingkai Lake from 1985 to 2020, it is evident that the surface of Xiaoxingkai Lake exhibited an overall growth trend during this period. Specifically, from 1985 to 2002, the surface experienced slight fluctuations but remained relatively stable. Subsequently, from 2002 to 2020, the lake surface area showed a steady and gradual increase. In terms of the aquatic vegetation area, significant fluctuations were observed in Xiaoxingkai Lake from 1985 to 2020. Initially, there was a decrease in the aquatic vegetation area from 1985 to 1987, followed by a substantial increase from 1988 to 1990. The aquatic vegetation area experienced a continuous decrease from 1990 to 1992. However, from 1992 to 1997, the area exhibited a pattern of first increasing, then decreasing, and then increasing again. Finally, from 1997 to 2020, there was a consistent trend of decreasing aquatic vegetation area.

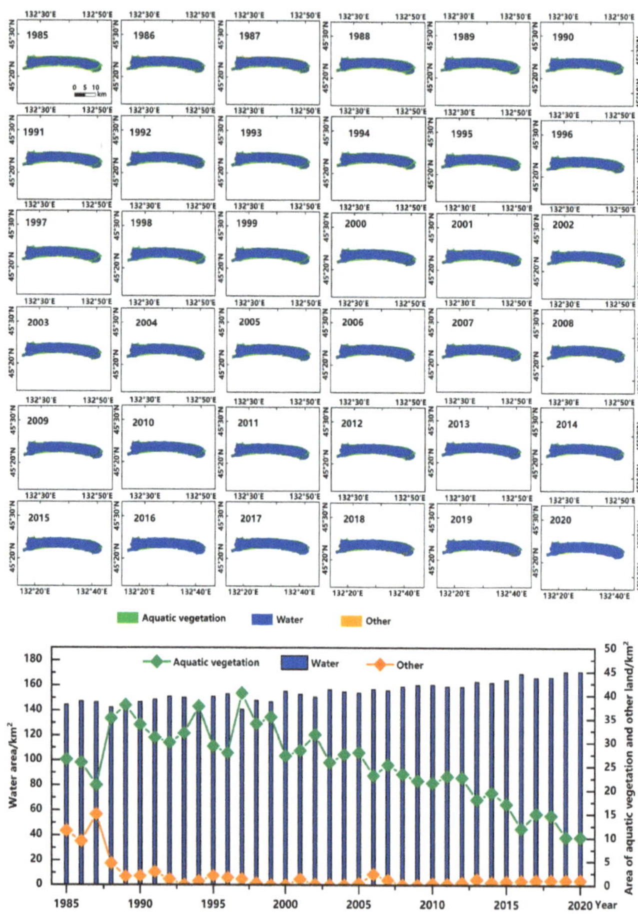

Figure 10. Spatial and temporal distribution and area changes of aquatic vegetation and water of Xiaoxingkai Lake from 1985 to 2020.

4. Discussion

4.1. The Advantages of the Model in Large-Scale Extraction of Aquatic Vegetation

In this study, a MSIC-Otsu and CNN-RF classifier method were developed to accurately extract aquatic vegetation. The maximum water surface boundary of the lake and the maximum coverage area of aquatic vegetation were determined by the time series Landsat images on the GEE platform and MSIC-Otsu. The CNN-RF classifier was used to extract the distribution of aquatic vegetation, water, and other land covers. The successful implementation of this study can be mainly attributed to three factors: the availability of a large number of free Landsat images, the powerful parallel computing capability of the GEE platform, and the robust MSIC-Otsu and CNN-RF methods. GEE can rapidly access and process a large volume of Landsat imagery without the need for additional masking or extensive pre-processing and post-processing steps. In CNN, the classifier consists of a fully connected layer, and the final decision criterion is based on calculation accumulation and activation. The advantage of CNN lies in its ability to directly extract more representative and abstract features from the original data [42–44]. This includes extracting texture information of different types of land use in remote sensing images, shape features such as boundary shape, geometric shape, and contour shape, spectral features encompassing spectral reflectance, spectral intensity, and spectral features, context features referring to the spatial relationship of surrounding pixels to infer the features of the target, and multi-scale features obtained through multi-layer convolution and pooling operations to capture information at different scales in remote sensing images. Li et al. used the open source dataset DeepSat to build and test the CNN model. The results proved that the CNN model was effective in land use classification and CNN did have the learning ability to extract the most critical and effective information from the training dataset [45]. Ståhl et al. used deep CNN to identify wetland areas in historical maps using deep convolutional neural networks. The presented deep CNN performs well [46]. As a more complex and reliable classifier, the RF classifier has been proven to achieve superior classification performance through strategies such as boosting and bagging [47,48]. Aiming at the problem of image classification and recognition, Xi proposed a hybrid model based on CNN, and input the features extracted by CNN into RF for classification. The results show that the classification accuracy of the hybrid model and the generalization ability is also improved [49]. Therefore, by leveraging the GEE platform and combining the strengths of both CNN and RF, we can fully capitalize on their advantages. In addition, MSIC-Otsu is a fully automatic met hod that can effectively extract the maximum water boundary of the lake to a certain extent, identify the potential area of aquatic vegetation distribution, thereby reducing the likelihood of missed measurements. Moreover, in the mapping process, there is no need for human intervention, training samples, or auxiliary information. Practice has proven that this method can efficiently and directly extract aquatic vegetation, overcoming the limitations of traditional methods such as fuzzy water boundaries, spectral mixing, and classification accuracy. In addition, during the vegetation extraction process, NDVI was less affected by turbid water signals, offering certain advantages in extracting submerged vegetation information. This improvement contributes to the accuracy and robustness of aquatic vegetation extraction. Therefore, the fusion of CNN and RF can effectively address the challenge of extracting aquatic vegetation, resulting in more reliable and accurate classification results.

In order to verify the effectiveness of CNN-RF, we compared it with the results extracted by Otsu, CNN, and RF. The classification was realized on the GEE platform, Otsu used an adaptive threshold to extract the results, and the decision tree value of RF was set to 100. We used PyTorch as a deep learning framework, CNN-RF and CNN had trained 200 epochs with a learning rate of 0.007, and the optimizer was sgd. The weight attenuation value was 0.0004, with a total of 1100 training samples and 100 test samples. The classification accuracy is shown in Table 4.

It is worth mentioning that CNN-RF outperforms other methods in classification accuracy. The Kappa coefficient of CNN-RF is 0.85, and OA is 0.91, which is higher than other classification methods. Then, taking Chagan Lake in 2020 as an example, we compare the enlarged images of the extraction results of different classification methods. As shown in Figure 11, CNN-RF is good at distinguishing aquatic vegetation/inland vegetation near water areas, providing clearer land cover boundaries and more concentrated distribution results. In contrast, RF needs further improvement in accurately separating aquatic vegetation from inland vegetation. The extraction result of CNN is inferior to that of CNN-RF, and the accuracy of OA in extracting water is higher, but there is a lot of misclassification in the process of extracting aquatic vegetation and inland vegetation.

Table 4. Accuracy of different classification methods.

Land Use	Accuracy	Otsu	RF	CNN	CNN-RF
Aquatic vegetation	UA	0.75	0.88	0.89	0.91
	PA	0.82	0.89	0.91	0.89
Water	UA	0.91	0.91	0.91	0.91
	PA	0.94	0.91	0.91	0.94
Other	UA	0.75	0.86	0.89	0.90
	PA	0.62	0.83	0.86	0.90
	Kappa	0.68	0.81	0.84	0.85
	OA	0.80	0.88	0.90	0.91

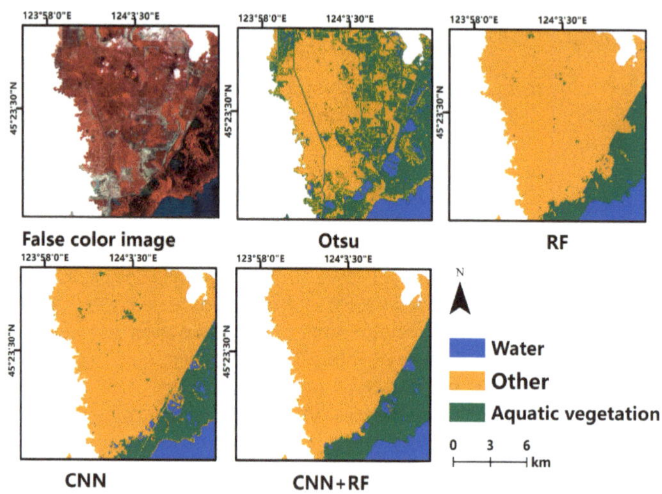

Figure 11. Classification maps of different classification methods.

4.2. Uncertainties in Aquatic Vegetation Mapping

In this study, we monitored the areal extents of aquatic vegetation over the past 36 years. However, due to the limited spatial resolution of Landsat series satellite data, there are challenges in identifying patches less than 30 m. Additionally, Landsat satellites have a relatively extended revisit period of 16 days, which may not capture short-term fluctuations in aquatic vegetation, especially during seasons characterized by rapid growth or decline. Moreover, we cannot guarantee that each pixel contains enough cloud-free observations to acquire key information about land cover after cloud masking. In situations where both algae blooms and aquatic vegetation are present in water, especially if the density of the algae bloom is low and the vegetation signal is weak, the NDVI value may become negative and omit the vegetation signal during the extraction process, potentially leading to misidentification with water. For example, from August to September 2017, there was a weak algal bloom in the middle and lower part of Hulun Lake, and the weak algal bloom did not produce obvious color change or texture characteristics, but this area was divided into water areas in this paper. Despite these uncertainties, the extraction of aquatic vegetation relies on consistent data sources and methods, encompassing long-term (36 years) multi-temporal information on aquatic vegetation and dynamics in the CARB.

5. Conclusions

In this study, we proposed a high-precision and robust classification method for extracting aquatic vegetation The CNN-RF classification method used in this paper has higher accuracy and can quickly process a large amount of data at one time, achieving efficient classification and greater robustness compared to CNN and RF alone. The basic idea involves employing an image composite to construct spatially and spectrally uniform images. We composed the maximum mNDWI index and extracted the water surface area using Otsu binary segmentation. Only lakes with an area larger than 20 km^2 were considered in this paper. The mask extracts the maximum NDVI composite image to identify potential aquatic vegetation areas. The resulting maximum NDVI composite image is then inputted into the CNN model to extract features. Subsequently, these features are fed into the RF classifier for accurate classification, enabling the precise mapping of the spatial and temporal distribution of aquatic vegetation from 1985 to 2020. According to the accuracy assessment, all the maps in the dataset exhibit high producer and user accuracy. Water surface and aquatic vegetation areas larger than 20 km^2 in the CARB show an overall decreasing trend from 1985 to 2020, reducing from 3575 km^2 and 745 km^2 to 3412 km^2 and 687 km^2, respectively. Among them, the water area of Hulun Lake exhibits a decreasing trend, while the water areas of Chagan Lake and Xiaoxingkai Lake are relatively stable. The water area of Lianhuan Lake exhibits an initial decrease and then increases. In terms of aquatic vegetation, the area in Lianhuan Lake remains relatively stable. However, in Hulun Lake, the aquatic vegetation area shows an initial increase, then decreases significantly, and then fluctuates and rises. Chagan Lake's aquatic vegetation area fluctuates and then rises slowly, and Xiaoxingkai Lake's aquatic vegetation area fluctuates and then continues to decline. The proposed method for aquatic vegetation information extraction in this paper not only extends the spatial portability of the model but also provides a powerful method and theoretical support for future extraction and monitoring of aquatic vegetation. The dataset of aquatic vegetation is invaluable for creating historical background maps of aquatic vegetation and evaluating the habitats of aquatic organisms, including their applicability to various species. In this study, an intriguing question emerges: can we further optimize the RF model to enhance interpretability when extracting aquatic vegetation from lakes using the CNN-RF approach? Additionally, future work will involve the development of a Landsat calibration model using high-resolution images. This calibration model aims to refine Landsat satellite-derived aquatic vegetation images, enabling a long-term series of fine classifications for various aquatic vegetation groups.

Author Contributions: Conceptualization, R.Z. and M.J.; Methodology, M.C. and L.C.; Software, M.C., L.C. and C.Z.; Validation, R.Z., M.J., H.L. and Z.W.; Data curation, L.C.; Writing—original draft, M.C.; Writing—review & editing, R.Z. and M.J.; Visualization, M.C., L.C. and C.Z.; Supervision, R.Z., C.Z., H.L. and Z.W.; Funding acquisition, M.J., H.L. and Z.W. All authors have read and agreed to the published version of the manuscript.

Funding: This research was jointly funded by the National Key Research and Development Program of China (2019YFA0607101); the National Natural Science Foundation of China (42171379 and 42101372); the Youth Innovation Promotion Association of Chinese Academy of Sciences (No. 2021227); the National Natural Science Youth Foundation of China (No. 42001383); the National Natural Science Foundation of Shandong Province (No. ZR2020QD020).

Data Availability Statement: The data presented in this study are available on request from the corresponding author. The data are not publicly available due to group policy or confidentiality agreement.

Acknowledgments: We express our gratitude to everyone who helped us to successfully complete this research.

Conflicts of Interest: The authors declare no conflicts of interest.

References

1. Yang, G.; Ma, R.; Zhang, L.; Jiang, J.; Yao, S.; Zhang, M.; Zeng, H.D. Lake Status, Major Problems and Protection Strategy in China. *J. Lake Sci.* **2010**, *22*, 799–810.
2. Parkes, M. Personal Commentaries on "Ecosystems and Human Well-Being: Health Synthesis—A Report of the Millennium Ecosystem Assessment". *EcoHealth* **2006**, *3*, 136–140. [CrossRef]
3. Inácio, M.; Barceló, D.; Zhao, W.; Pereira, P. Mapping Lake Ecosystem Services: A Systematic Review. *Sci. Total Environ.* **2022**, *847*, 157561. [CrossRef] [PubMed]
4. Jeppesen, E.; Søndergaard, M.; Søndergaard, M.; Christoffersen, K. *The Structuring Role of Submerged Macrophytes in Lakes*; Springer Science & Business Media: Berlin/Heidelberg, Germany, 2012; Volume 131, ISBN 1-4612-0695-2.
5. Horppila, J.; Nurminen, L. Effects of Submerged Macrophytes on Sediment Resuspension and Internal Phosphorus Loading in Lake Hiidenvesi (Southern Finland). *Water Res.* **2003**, *37*, 4468–4474. [CrossRef] [PubMed]
6. Jia, M.; Mao, D.; Wang, Z.; Ren, C.; Zhu, Q.; Li, X.; Zhang, Y. Tracking Long-Term Floodplain Wetland Changes: A Case Study in the China Side of the Amur River Basin. *Int. J. Appl. Earth Obs. Geoinf.* **2020**, *92*, 102185. [CrossRef]
7. Murphy, F.; Schmieder, K.; Baastrup-Spohr, L.; Pedersen, O.; Sand-Jensen, K. Five Decades of Dramatic Changes in Submerged Vegetation in Lake Constance. *Aquat. Bot.* **2018**, *144*, 31–37. [CrossRef]
8. Zhang, Y.; Liu, X.; Qin, B.; Shi, K.; Deng, J.; Zhou, Y. Aquatic Vegetation in Response to Increased Eutrophication and Degraded Light Climate in Eastern Lake Taihu: Implications for Lake Ecological Restoration. *Sci. Rep.* **2016**, *6*, 23867. [CrossRef] [PubMed]
9. Fourqurean, J.W.; Duarte, C.M.; Kennedy, H.; Marbà, N.; Holmer, M.; Mateo, M.A.; Apostolaki, E.T.; Kendrick, G.A.; Krause-Jensen, D.; McGlathery, K.J. Seagrass Ecosystems as a Globally Significant Carbon Stock. *Nat. Geosci.* **2012**, *5*, 505–509. [CrossRef]
10. Massicotte, P.; Bertolo, A.; Brodeur, P.; Hudon, C.; Mingelbier, M.; Magnan, P. Influence of the Aquatic Vegetation Landscape on Larval Fish Abundance. *J. Great Lakes Res.* **2015**, *41*, 873–880. [CrossRef]
11. Hilt, S.; Gross, E.M.; Hupfer, M.; Morscheid, H.; Mählmann, J.; Melzer, A.; Poltz, J.; Sandrock, S.; Scharf, E.-M.; Schneider, S. Restoration of Submerged Vegetation in Shallow Eutrophic Lakes–A Guideline and State of the Art in Germany. *Limnologica* **2006**, *36*, 155–171. [CrossRef]
12. Zeng, L.; He, F.; Dai, Z.; Xu, D.; Liu, B.; Zhou, Q.; Wu, Z. Effect of Submerged Macrophyte Restoration on Improving Aquatic Ecosystem in a Subtropical, Shallow Lake. *Ecol. Eng.* **2017**, *106*, 578–587. [CrossRef]
13. Luo, J.H.; Yang, J.Z.C.; Duan, H.T.; Lu, L.R.; Sun, Z.; Xin, Y.H. Research Progress of Aquatic Vegetation Remote Sensing in Shallow Lakes. *Natl. Remote Sens. Bull.* **2022**, *26*, 68–76.
14. Wen, K.; Gao, B.; Li, M. Quantifying the Impact of Future Climate Change on Runoff in the Amur River Basin Using a Distributed Hydrological Model and CMIP6 GCM Projections. *Atmosphere* **2021**, *12*, 1560. [CrossRef]
15. Wang, Z.; Song, K.; Ma, W.; Ren, C.; Zhang, B.; Liu, D.; Chen, J.M.; Song, C. Loss and Fragmentation of Marshes in the Sanjiang Plain, Northeast China, 1954–2005. *Wetlands* **2011**, *31*, 945–954. [CrossRef]
16. Mao, D.; Tian, Y.; Wang, Z.; Jia, M.; Du, J.; Song, C. Wetland Changes in the Amur River Basin: Differing Trends and Proximate Causes on the Chinese and Russian Sides. *J. Environ. Manag.* **2021**, *280*, 111670. [CrossRef] [PubMed]
17. Sokolova, G.V.; Verkhoturov, A.L.; Korolev, S.P. Impact of Deforestation on Streamflow in the Amur River Basin. *Geosciences* **2019**, *9*, 262. [CrossRef]
18. Han, D.; Gao, C.; Liu, H.; Yu, X.; Li, Y.; Cong, J.; Wang, G. Vegetation Dynamics and Its Response to Climate Change during the Past 2000 Years along the Amur River Basin, Northeast China. *Ecol. Indic.* **2020**, *117*, 106577. [CrossRef]
19. Zou, Y.; Wang, L.; Xue, Z.; Mingju, E.; Jiang, M.; Lu, X.; Yang, S.; Shen, X.; Liu, Z.; Sun, G. Impacts of Agricultural and Reclamation Practices on Wetlands in the Amur River Basin, Northeastern China. *Wetlands* **2018**, *38*, 383–389. [CrossRef]
20. Zhang, J.; Ma, K.; Fu, B. Wetland Loss under the Impact of Agricultural Development in the Sanjiang Plain, NE China. *Environ. Monit. Assess.* **2010**, *166*, 139–148. [CrossRef]
21. Bolpagni, R.; Bresciani, M.; Laini, A.; Pinardi, M.; Matta, E.; Ampe, E.M.; Giardino, C.; Viaroli, P.; Bartoli, M. Remote Sensing of Phytoplankton-Macrophyte Coexistence in Shallow Hypereutrophic Fluvial Lakes. *Hydrobiologia* **2014**, *737*, 67–76. [CrossRef]
22. Luo, J.; Ma, R.; Duan, H.; Hu, W.; Zhu, J.; Huang, W.; Lin, C. A New Method for Modifying Thresholds in the Classification of Tree Models for Mapping Aquatic Vegetation in Taihu Lake with Satellite Images. *Remote Sens.* **2014**, *6*, 7442–7462. [CrossRef]
23. Jing, R.; Deng, L.; Zhao, W.J.; Gong, Z.N. Object-Oriented Aquatic Vegetation Extracting Approach Based on Visible Vegetation Indices. *Ying Yong Sheng Tai Xue Bao J. Appl. Ecol.* **2016**, *27*, 1427–1436.
24. Chen, Q.; Yu, R.; Hao, Y.; Wu, L.; Zhang, W.; Zhang, Q.; Bu, X. A New Method for Mapping Aquatic Vegetation Especially Underwater Vegetation in Lake Ulansuhai Using GF-1 Satellite Data. *Remote Sens.* **2018**, *10*, 1279. [CrossRef]
25. Wang, H.; Li, Y.; Zeng, S.; Cai, X.; Bi, S.; Liu, H.; Mu, M.; Dong, X.; Li, J.; Xu, J. Recognition of Aquatic Vegetation above Water Using Shortwave Infrared Baseline and Phenological Features. *Ecol. Indic.* **2022**, *136*, 108607. [CrossRef]
26. Ashford, P. Field Methods and Photogrammetric Techniques for Mapping Aquatic Vegetation Using Unmanned Aerial Systems. Master's Thesis, University of Georgia, Athens, GA, USA, 2019.
27. Yuan, Q.; Shen, H.; Li, T.; Li, Z.; Li, S.; Jiang, Y.; Xu, H.; Tan, W.; Yang, Q.; Wang, J. Deep Learning in Environmental Remote Sensing: Achievements and Challenges. *Remote Sens. Environ.* **2020**, *241*, 111716. [CrossRef]
28. Li, B.; Yang, G.; Wan, R.; Dai, X.; Zhang, Y. Comparison of Random Forests and Other Statistical Methods for the Prediction of Lake Water Level: A Case Study of the Poyang Lake in China. *Hydrol. Res.* **2016**, *47*, 69–83. [CrossRef]

29. Zhao, C.; Jia, M.; Wang, Z.; Mao, D.; Wang, Y. Toward a Better Understanding of Coastal Salt Marsh Mapping: A Case from China Using Dual-Temporal Images. *Remote Sens. Environ.* **2023**, *295*, 113664. [CrossRef]
30. Ji, M.; Liu, L.; Du, R.; Buchroithner, M.F. A Comparative Study of Texture and Convolutional Neural Network Features for Detecting Collapsed Buildings after Earthquakes Using Pre-and Post-Event Satellite Imagery. *Remote Sens.* **2019**, *11*, 1202. [CrossRef]
31. Otsu, N. A Threshold Selection Method from Gray-Level Histograms. *IEEE Trans. Syst. Man Cybern.* **1979**, *9*, 62–66. [CrossRef]
32. Wang, Z.H.; Xin, C.L.; Sun, Z.; Luo, J.H.; Ma, R.H. Automatic Extraction Method of Aquatic Vegetation Types in Small Shallow Lakes Based on Sentinel-2 Data: A Case Study of Cuiping Lake. *Remote Sens. Inf* **2019**, *34*, 132–141.
33. Rouse, J.W., Jr.; Haas, R.H.; Deering, D.W.; Schell, J.A.; Harlan, J.C. *Monitoring the Vernal Advancement and Retrogradation (Green Wave Effect) of Natural Vegetation*; NASA: Washington, DC, USA, 1974.
34. Duan, Y.; Li, X.; Zhang, L.; Chen, D.; Ji, H. Mapping National-Scale Aquaculture Ponds Based on the Google Earth Engine in the Chinese Coastal Zone. *Aquaculture* **2020**, *520*, 734666. [CrossRef]
35. Jia, M.; Wang, Z.; Mao, D.; Ren, C.; Wang, C.; Wang, Y. Rapid, Robust, and Automated Mapping of Tidal Flats in China Using Time Series Sentinel-2 Images and Google Earth Engine. *Remote Sens. Environ.* **2021**, *255*, 112285. [CrossRef]
36. Simis, S.G.; Peters, S.W.; Gons, H.J. Remote Sensing of the Cyanobacterial Pigment Phycocyanin in Turbid Inland Water. *Limnol. Oceanogr.* **2005**, *50*, 237–245. [CrossRef]
37. LeCun, Y.; Bottou, L.; Bengio, Y.; Haffner, P. Gradient-Based Learning Applied to Document Recognition. *Proc. IEEE* **1998**, *86*, 2278–2324. [CrossRef]
38. Chen, L.-C.; Zhu, Y.; Papandreou, G.; Schroff, F.; Adam, H. Encoder-Decoder with Atrous Separable Convolution for Semantic Image Segmentation. *arXiv* **2018**, arXiv:1802.02611.
39. Rodriguez-Galiano, V.F.; Ghimire, B.; Rogan, J.; Chica-Olmo, M.; Rigol-Sanchez, J.P. An Assessment of the Effectiveness of a Random Forest Classifier for Land-Cover Classification. *ISPRS J. Photogramm. Remote Sens.* **2012**, *67*, 93–104. [CrossRef]
40. Parmar, A.; Katariya, R.; Patel, V. A Review on Random Forest: An Ensemble Classifier. In Proceedings of the International Conference on Intelligent Data Communication Technologies and Internet of Things (ICICI) 2018, Coimbatore, India, 7–8 August 2018; Springer: Cham, Switzerland, 2019; pp. 758–763.
41. Liaw, A.; Wiener, M. Classification and Regression by randomForest. *R News* **2002**, *2*, 18–22.
42. Sun, T.; Liu, M.; Ye, H.; Yeung, D.-Y. Point-Cloud-Based Place Recognition Using CNN Feature Extraction. *IEEE Sens. J.* **2019**, *19*, 12175–12186. [CrossRef]
43. Zhang, M.; Li, W.; Du, Q. Diverse Region-Based CNN for Hyperspectral Image Classification. *IEEE Trans. Image Process.* **2018**, *27*, 2623–2634. [CrossRef]
44. Zeiler, M.D.; Fergus, R. Visualizing and Understanding Convolutional Networks. In Proceedings of the Computer Vision–ECCV 2014: 13th European Conference, Zurich, Switzerland, 6–12 September 2014; Proceedings, Part I 13. Springer: Cham, Switzerland, 2014; pp. 818–833.
45. Li, M.; Wang, L.; Wang, J.; Li, X.; She, J. Comparison of Land Use Classification Based on Convolutional Neural Network. *J. Appl. Remote Sens.* **2020**, *14*, 016501. [CrossRef]
46. Ståhl, N.; Weimann, L. Identifying Wetland Areas in Historical Maps Using Deep Convolutional Neural Networks. *Ecol. Inform.* **2022**, *68*, 101557. [CrossRef]
47. Freund, Y.; Schapire, R.E. A Decision-Theoretic Generalization of on-Line Learning and an Application to Boosting. *J. Comput. Syst. Sci.* **1997**, *55*, 119–139. [CrossRef]
48. Breiman, L. Bagging Predictors. *Mach. Learn.* **1996**, *24*, 123–140. [CrossRef]
49. Xi, E. Image Classification and Recognition Based on Deep Learning and Random Forest Algorithm. *Wirel. Commun. Mob. Comput.* **2022**, *2022*, 2013181. [CrossRef]

remote sensing

MDPI

Article

A High-Precision Target Geolocation Algorithm for a Spaceborne Bistatic Interferometric Synthetic Aperture Radar System Based on an Improved Range–Doppler Model

Chao Xing [1], Zhenfang Li [1], Fanyi Tang [1,*], Feng Tian [2] and Zhiyong Suo [1]

[1] National Key Laboratory of Radar Signal Processing, Xidian University, Xi'an 710071, China; xingchao@xidian.edu.cn (C.X.); lzf@xidian.edu.cn (Z.L.); zysuo@xidian.edu.cn (Z.S.)

[2] Nanjing Research Institute of Electronics Technology, Nanjing 210039, China; feng_tian@stu.xidian.edu.cn

* Correspondence: 18021210100@stu.xidian.edu.cn; Tel.: +86-137-2073-0517

Abstract: A trend in the development of spaceborne Synthetic Aperture Radar (SAR) technology is the shift from a single-satellite repeated observation mode to a multi-satellite collaborative observation mode. However, current multi-satellite collaborative geolocation algorithms face challenges, such as geometric model mismatch and poor baseline estimation accuracy, arising from highly dynamic changes among multi-satellites. This paper introduces a high-precision and efficient geolocation algorithm for a spaceborne bistatic interferometric SAR (BiInSAR) system based on an improved range–Doppler (IRD) model. The proposed algorithm encompasses three key contributions. Firstly, a comprehensive description of the spatial baseline geometric model unique to the bistatic configuration is provided, with a specific focus on deriving the perpendicular baseline expression. Secondly, IRD geolocation functions are established to meet the specific requirements of the bistatic configuration. Then, a novel BiInSAR geolocation algorithm based on the IRD's functions is proposed, which can significantly improve the target geolocation accuracy by modifying the range–Doppler equation to suit the bistatic configuration. Meanwhile, a low-coupling parallel calculation method is proposed, which can improve the calculation speed by two to three times. Finally, the accuracy and efficiency of the algorithm are demonstrated using experimental data acquired by the TH-2 satellite, which is China's first spaceborne BiInSAR system. The experimental results prove that the IRD algorithm exhibits geolocation accuracy with an average error of less than 1 m and a standard deviation of less than 2.5 m while maintaining computational efficiency at a calculation speed of 1,429,678 pixels per second.

Keywords: interferometric synthetic aperture radar; InSAR; bistatic configuration; improved range–Doppler; IRD; geolocation algorithm; TH-2 satellite

Citation: Xing, C.; Li, Z.; Tang, F.; Tian, F.; Suo, Z. A High-Precision Target Geolocation Algorithm for a Spaceborne Bistatic Interferometric Synthetic Aperture Radar System Based on an Improved Range–Doppler Model. *Remote Sens.* **2024**, *16*, 532. https://doi.org/10.3390/rs16030532

Academic Editor: Piotr Samczynski

Received: 5 December 2023
Revised: 12 January 2024
Accepted: 29 January 2024
Published: 30 January 2024

1. Introduction

Spaceborne Interferometric Synthetic Aperture Radar (InSAR) has proven to be a highly effective method for acquiring high-resolution and high-precision digital elevation model (DEM) data [1–5]. The emergence of bistatic InSAR (BiInSAR) configurations, wherein two formation-flying satellites act as a single-pass radar interferometer, offers a novel and flexible approach to radar imaging, with distinct advantages over traditional repeat-pass interferometry [6,7]. Notable examples include the TerraSAR-X/TanDEM-X system developed by the German Aerospace Center [8–10], the TH-2 system developed by China [7,11], and the TwinSAR-L system developed by China [12–14].

However, the dynamic nature of the baseline in BiInSAR systems, owing to the rapid relative motion between satellites, presents a significant challenge. Existing baseline calibration models struggle to adequately describe the evolving baseline over time, necessitating innovative solutions. One such solution is the introduction of a pixel-related baseline model based on geometrical shifts [12], which accurately captures satellite position changes.

Furthermore, adapting repeat-pass InSAR geolocation algorithms to BiInSAR systems is non-trivial. A bistatic-to-monostatic equivalent imaging algorithm has been proposed to address this issue [15], simplifying the geolocation model by reducing the number of antenna phase centers (APCs) from two to one. However, this method assumes that the transmitter and receiver platforms follow a rectilinear path with constant and equal velocities, a condition not met by all BiInSAR systems. Researchers have consequently explored direct bistatic imaging algorithms capable of broader applications [16–19]. For instance, a three-dimensional localization approach based on the numerical range–Doppler algorithm and entropy minimization principle was introduced in [20], enhancing image quality and localization accuracy. Nonetheless, this approach does not fully exploit phase information within the images and overlooks the impact of system errors such as atmospheric delays and baseline measurement errors.

While bistatic-to-monostatic equivalent algorithms simplify the bistatic geolocation model, their applicability is constrained [15,21]. Direct bistatic geolocation algorithms, although more accurate, often demand time-consuming iterative operations [20,22]. Therefore, this paper concentrates on enhancing both the operational efficiency and accuracy of direct bistatic geolocation.

First, an improved range–Doppler (IRD) model is proposed to solve the geometric model mismatch problem in the BiInSAR system. The IRD model fully considers the different propagation paths between the transmitting antenna, receiving antenna, and ground targets. Then, an accurate target geolocation equation based on the IRD model is established. It can overcome the slant-range measurement error caused by APC equivalence and the Doppler frequency error caused by velocity equivalence in conventional geolocation algorithms. Apart from the efforts to improve geolocation accuracy, the challenge of accelerating calculation efficiency is another issue that researchers are interested in [23]. In this paper, a low-coupling parallel calculation method is proposed, which can improve the calculation speed by two to three times.

Second, a bistatic interferometric baseline calibration method based on an unwrapped interferogram and ground control points (GCPs) is introduced. The interferometric baseline vector is dissected into two orthogonal components: the parallel baseline, aligned with the line of sight (LoS), and the perpendicular baseline, orthogonal to the LoS. The measurement error of the parallel baseline can be estimated according to the linear relationship between the elevation measurement error and the parallel baseline measurement error. The perpendicular baseline can be determined via the least-squares fitting method. When constructing the baseline error-solving equation, the differences between the BiInSAR geometry and monostatic InSAR (MoInSAR) geometry are considered, and the information contained in the interference phase is fully utilized.

The rest of this article is organized as follows. Section 2 presents the geometry model of the bistatic interferometric baseline. The IRD model used for geolocation is presented, and its influencing factors are analyzed in Section 3. Section 4 introduces the implementation process of the algorithm, including the ICESat-2 data filtering method, baseline calibration method, IRD geolocation method, and parallel processing method. The accuracy and efficiency of the proposed algorithm are verified through two groups of experiments in Section 5. Finally, Section 6 summarizes the conclusions.

2. Interferometric Baseline Calibration

At the beginning of this section, the geometry model of the BiInSAR baseline configuration is described. Then, the principle of the baseline projection method from a three-dimensional vector to a two-dimensional vector is explained, and the mathematical expression of the baseline is deduced.

2.1. Geometry Model of BiInSAR Baseline Configuration

As shown in Figure 1a, two satellites are flying in approximately parallel trajectories. The trajectories of the active satellite S_{AT} and passive satellite S_{PR} are represented by the red dotted line and the blue dotted line, respectively. An italicized S_{AT}/S_{PR} denotes the satellite's name, whereas a bold $\mathbf{S_{AT}}/\mathbf{S_{PR}}$ represents the satellite's position vectors. The red and blue arrows represent the flight directions of S_{AT} and S_{PR}, respectively. The yellow line between $\mathbf{S}_{AT}$ and $\mathbf{S}_{PR}$ represents the instantaneous baseline (ISB). In Figure 1b, the green line between $\mathbf{S}_{AT}$ and $\mathbf{S}_{PR}$ represents the interferometric baseline (ITB). The ISB in Figure 1a reflects the relative position relationship between the two satellites at the same time, whereas the ITB in Figure 1b represents the relative position of S_{AT} and S_{PR} at different moments but with the same view. The yellow triangle P_T at the bottom of Figure 1b symbolizes an arbitrary target on Earth's surface. In Figure 1b, the red dashed line from S_{AT} to P_T indicates the signal transmission propagation path R_{AT}, whereas the blue dashed line from P_T to S_{PR} signifies the signal reception propagation path R_{PR}. Similarly, S_{AR}'s receiving signal propagation path R_{AR} is represented by a red dashed line extending from P_T to S_{AR}.

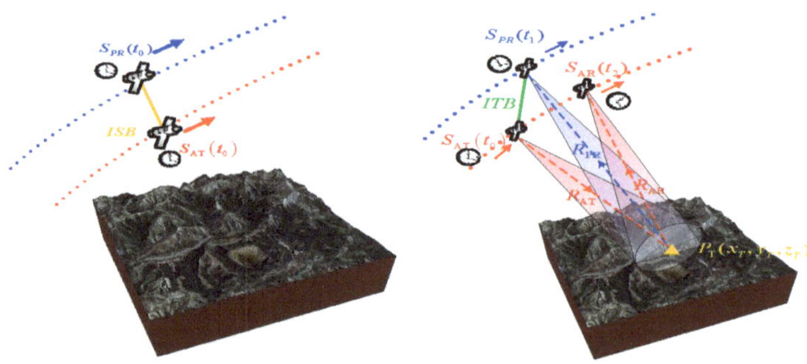

(a) (b)

Figure 1. Geometry model of BiInSAR baseline configuration. (**a**) ISB at the same moment. (**b**) ITB at different moments with the same view.

The ITB cannot be measured directly since the differential GPS device can only measure the relative position relationship at the same moment, and SAR image products do not contain prior information about when S_{AT} and S_{PR} have the same view. Fortunately, the moments of the same view can be estimated from BiInSAR imaging geometry. Firstly, a ground target with three-dimensional coordinates is extracted from reference the DEM according to the latitude range and longitude range of the observed scene. The symbol $\mathbf{P}_T$=(x_T,y_T,z_T) represents the position of a ground target. Then, the zero-Doppler moments of S_{AT} and S_{PR} can be searched according to Equations (1) and (2).

$$f_{d_{PR}}(t_0;t_1) = -\frac{\mathbf{V}_{AT}(t_0)\cdot(\mathbf{S}_{AT}(t_0)-\mathbf{P}_T)}{\lambda\|\mathbf{S}_{AT}(t_0)-\mathbf{P}_T\|} - \frac{\mathbf{V}_{PR}(t_1)\cdot(\mathbf{S}_{PR}(t_1)-\mathbf{P}_T)}{\lambda\|\mathbf{S}_{PR}(t_1)-\mathbf{P}_T\|} \tag{1}$$

$$f_{d_{AT}}(t_0;t_2) = -\frac{\mathbf{V}_{AT}(t_0)\cdot(\mathbf{S}_{AT}(t_0)-\mathbf{P}_T)}{\lambda\|\mathbf{S}_{AT}(t_0)-\mathbf{P}_T\|} - \frac{\mathbf{V}_{AT}(t_2)\cdot(\mathbf{S}_{AT}(t_2)-\mathbf{P}_T)}{\lambda\|\mathbf{S}_{AT}(t_2)-\mathbf{P}_T\|} \tag{2}$$

where $f_{d_{PR}}$ and $f_{d_{AT}}$ represent the Doppler frequency of a dual-satellite cooperative operation and the Doppler frequency of a single-satellite independent operation, respectively. The symbol λ represents the carrier wavelength of the radar signal.

Geometric registration methods are usually used to calculate when two satellites have the same view [24], and the obtained time is called the Doppler center time (DCT). Then, the APC positions of satellite S_{AT} and satellite S_{PR} can be acquired using interpolation functions based on the original orbit information and DCT.

2.2. Projection Principle of BiInSAR Baseline

This subsection introduces the projection principle of the **ITB** from a three-dimensional vector to a two-dimensional vector. As depicted in Figure 2a, the **ITB** vector is pointing from S_{AT} at time t_0 to S_{PR} at time t_1. The mathematical expression of the **ITB** can be expressed by Equation (3).

$$\mathbf{ITB}(t_0, t_1) = \mathbf{S}_{AT}(t_0) - \mathbf{S}_{PR}(t_1) \tag{3}$$

The **ITB** can be dissected into two orthogonal components: the across-track baseline **ACB** and the along-track baseline **ALB**. The **ALB** is parallel to satellite velocity $\mathbf{V}_{AT}$, and the **ACB** is perpendicular to $\mathbf{V}_{AT}$. The **ALB** is mainly used for ground moving-target velocity estimation, and the **ACB** is mainly used for topography mapping and ground deformation detection applications. The focus of this paper lies in the field of topography mapping, thus necessitating a meticulous estimation of the **ACB**. The green plane represents the zero-Doppler plane (ZDP) of a ground target P_T. Because both the **ACB** and ZDP are perpendicular to $\mathbf{V}_{AT}$, the **ACB** can be deemed as the projection of the **ITB** in a two-dimensional plane. The red point S_{AP} is the projection of S_{AT} in the ZDP. Similarly, the blue point S_{PP} is the projection of S_{PR} in the ZDP. The existing BiInSAR baseline calibration algorithms are all defined in a two-dimensional plane, called the ZDP, as shown in Figure 2b. The mathematical formulation presented in this study establishes a conversion relationship between the **ITB** and **ACB**.

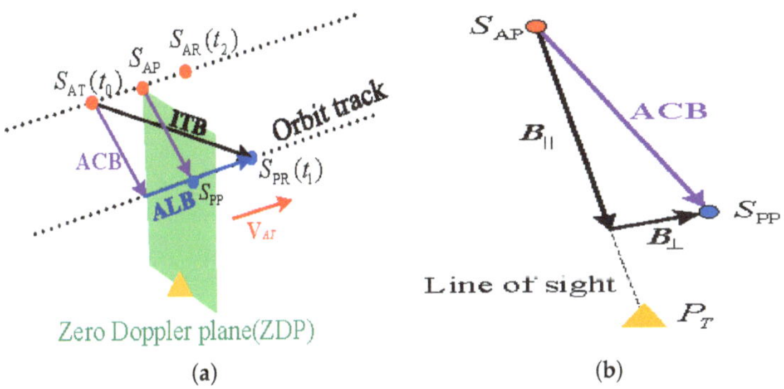

Figure 2. Projection principle of BiInSAR baseline. (**a**) Three-dimensional baseline vector. (**b**) Two-dimensional baseline vector.

It can be seen that the **ACB** is perpendicular to $\mathbf{V}_{AT}$, but this is not enough to determine its specific direction in the ZDP. The **ALB** vector is computed first, followed by the **ACB** based on the **ALB**. The **ALB** is parallel to $\mathbf{V}_{AT}$, and the modulus of the **ALB** can be calculated through a vector dot operation, as shown in Equation (4).

$$\|\mathbf{ALB}(t_0, t_1)\| = \mathbf{ITB}(t_0, t_1) \cdot \mathbf{V}_{AT}(t_0) / \|\mathbf{V}_{AT}(t_0)\| \tag{4}$$

In Equation (4), the symbol $\|\cdot\|$ represents the modulus of a vector. $\mathbf{V}_{AT}(t_0)/\|\mathbf{V}_{AT}(t_0)\|$ represents the unit vector along $\mathbf{V}_{AT}$ at time t_0. Then, the **ACB** can be calculated through a vector subtraction operation, as shown in Equation (5).

$$\mathbf{ACB}(t_0, t_1) = \mathbf{ITB}(t_0, t_1) - \mathbf{ALB}(t_0, t_1) \tag{5}$$

In Figure 2b, the $\mathbf{B}_{\|}$ component is pointing from $\mathbf{S}_{AP}$ to $\mathbf{P}_T$ along the radar line of sight (LoS). $\mathbf{B}_{\perp}$ is pointing from $\mathbf{S}_{PP}$ to the LoS, and it is perpendicular to the LoS. The direction of the LoS varies from near range to far range, so the directions of $\mathbf{B}_{\|}$ and $\mathbf{B}_{\perp}$ are also dynamic.

3. The IRD Model and Analysis of Influencing Factors

In this section, the IRD model is proposed, which can effectively reflect the geometric characteristics in the bistatic configuration, and the influencing factors of geolocation accuracy are analyzed.

3.1. Introduction to the Proposed IRD Model

This subsection introduces the proposed IRD geolocation model. First, an equation system for target position $\mathbf{P}_T$ is given according to the slant-range information and Doppler information, as shown in Equations (6)–(8). Equation (6) expresses the slant-range relationship between $\mathbf{P}_T$ and $\mathbf{S}_{AT}/\mathbf{S}_{AR}$. $\mathbf{P}_T$ is distributed over an ellipse, with $\mathbf{S}_{AT}$ and $\mathbf{S}_{AR}$ as the focal points, as shown in Figure 3. The solid red line represents the slant range between the active satellite and target, containing R_{AT} and R_{AR}. R_{AT} represents the distance from $\mathbf{S}_{AT}$ to $\mathbf{P}_T$, and R_{AR} represents the distance from $\mathbf{P}_T$ to $\mathbf{S}_{AR}$. In Equation (7), R_{PR} represents the distance from $\mathbf{P}_T$ to $\mathbf{S}_{PR}$. There are two methods used to calculate R_{PR}. One is based on the delay time from signal transmission to signal reception and is usually used in stereo SAR geolocation applications. The other is based on the interferometry phase (ITP), which is adopted in the InSAR geolocation algorithm. In Equation (8), f_{dc} represents the Doppler center frequencies of $\mathbf{S}_{AT}$ and $\mathbf{S}_{AR}$ at target $\mathbf{P}_T$. The conversion relationship between the ITP and R_{PR} can be expressed by Equation (9). In Equation (9), ϕ represents the ITP.

$$f_{RA}(\mathbf{S}_{AT}(t_0), \mathbf{S}_{AR}(t_2)) = \|\mathbf{S}_{AT}(t_0) - \mathbf{P}_T\| + \|\mathbf{S}_{AR}(t_2) - \mathbf{P}_T\| - R_{AT} - R_{AR} = 0 \quad (6)$$

$$f_{RP}(\mathbf{S}_{AT}(t_0), \mathbf{S}_{PR}(t_1)) = \|\mathbf{S}_{AT}(t_0) - \mathbf{P}_T\| + \|\mathbf{S}_{PR}(t_1) - \mathbf{P}_T\| - R_{AT} - R_{PR} = 0 \quad (7)$$

$$f_{DA}(\mathbf{S}_{AT}(t_0), \mathbf{S}_{AR}(t_2), \mathbf{P}_T) = -\frac{\mathbf{V}_{AT}(t_0) \cdot (\mathbf{S}_{AT}(t_0) - \mathbf{P}_T)}{\lambda \cdot \|\mathbf{S}_{AT}(t_0) - \mathbf{P}_T\|} - \frac{\mathbf{V}_{AR}(t_2) \cdot (\mathbf{S}_{AR}(t_2) - \mathbf{P}_T)}{\lambda \cdot \|\mathbf{S}_{AR}(t_2) - \mathbf{P}_T\|} - f_{dc} = 0 \quad (8)$$

$$R_{PR} = R_{AR} + \frac{\phi}{2\pi}\lambda \quad (9)$$

where $\mathbf{S}_{AT} = (x_{AT}, y_{AT}, z_{AT})$, $\mathbf{S}_{AR} = (x_{AR}, y_{AR}, z_{AR})$, $\mathbf{S}_{PR} = (x_{PR}, y_{PR}, z_{PR})$, $\mathbf{V}_{AT} = (v_{xt}, v_{yt}, v_{zt})$, and $\mathbf{V}_{AR} = (v_{xr}, v_{yr}, v_{zr})$.

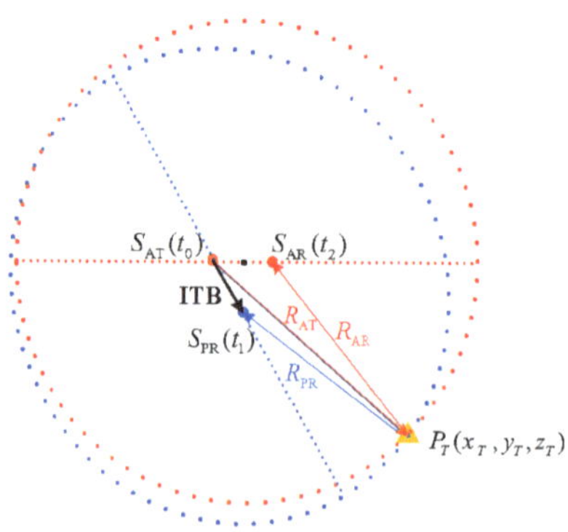

Figure 3. The proposed IRD model.

Considering there is an addition operation of two square roots in Equations (6) and (7), it is impossible to calculate the coordinate value of $\mathbf{P}_T$ through a closed-form expression. The Newton iteration method is a commonly used method to solve such nonlinear equations. Subsequently, a comprehensive explanation of the process for deriving the formula is provided. In Equation (10), to obtain the target geolocation sensitivity of the f_{RA} range equation in the x-axis, the partial derivative of f_{RA} with respect to x_T is calculated.

Then, the partial derivatives of f_{RA} with respect to y_T and z_T are calculated to obtain the sensitivity of the range equation in the y-axis and z-axis, as seen in Equations (11) and (12).

$$\frac{\partial f_{RA}}{\partial x_T} = -\frac{x_{AT} - x_T}{R_{AT}} - \frac{x_{AR} - x_T}{R_{AR}} \tag{10}$$

$$\frac{\partial f_{RA}}{\partial y_T} = -\frac{y_{AT} - y_T}{R_{AT}} - \frac{y_{AR} - y_T}{R_{AR}} \tag{11}$$

$$\frac{\partial f_{RA}}{\partial z_T} = -\frac{z_{AT} - z_T}{R_{AT}} - \frac{z_{AR} - z_T}{R_{AR}} \tag{12}$$

Similarly, the target geolocation sensitivity of the f_{RP} range equation in three axes can be obtained by performing a partial derivative operation, as shown in Equations (13)–(15).

$$\frac{\partial f_{RP}}{\partial x_T} = -\frac{x_{AT} - x_T}{R_{AT}} - \frac{x_{PR} - x_T}{R_{PR}} \tag{13}$$

$$\frac{\partial f_{RP}}{\partial y_T} = -\frac{y_{AT} - y_T}{R_{AT}} - \frac{y_{PR} - y_T}{R_{PR}} \tag{14}$$

$$\frac{\partial f_{RP}}{\partial z_T} = -\frac{z_{AT} - z_T}{R_{AT}} - \frac{z_{PR} - z_T}{R_{PR}} \tag{15}$$

To solve the IRD equation system, it is also necessary to perform a partial derivative operation of f_{DA} with respect to x_T, y_T, and z_T, as shown in Equations (16)–(18).

$$\begin{aligned}\frac{\partial f_{DA}}{\partial x_T} &= \frac{-v_{xt}\cdot(x_T - x_{AT})R_{AT}^2 + (x_T - x_{AT})\cdot\mathbf{V}_{\mathrm{AT}}\cdot(\mathbf{P}_T - \mathbf{S}_{\mathrm{AT}})}{\lambda R_{AT}^3} \\ &+ \frac{-v_{xr}\cdot(x_T - x_{AR})R_{AR}^2 + (x_T - x_{AR})\cdot\mathbf{V}_{\mathrm{AR}}\cdot(\mathbf{P}_T - \mathbf{S}_{\mathrm{AR}})}{\lambda R_{AR}^3}\end{aligned} \tag{16}$$

$$\begin{aligned}\frac{\partial f_{DA}}{\partial y_T} &= \frac{-v_{yt}\cdot(y_T - x_{AT})R_{AT}^2 + (y_T - y_{AT})\cdot\mathbf{V}_{\mathrm{AT}}\cdot(\mathbf{P}_T - \mathbf{S}_{\mathrm{AT}})}{\lambda R_{AT}^3} \\ &+ \frac{-v_{yr}\cdot(y_T - y_{AR})R_{AR}^2 + (y_T - y_{AR})\cdot\mathbf{V}_{\mathrm{AR}}\cdot(\mathbf{P}_T - \mathbf{S}_{\mathrm{AR}})}{\lambda R_{AR}^3}\end{aligned} \tag{17}$$

$$\begin{aligned}\frac{\partial f_{DA}}{\partial z_T} &= \frac{-v_{zt}\cdot(z_T - z_{AT})R_{AT}^2 + (z_T - z_{AT})\cdot\mathbf{V}_{\mathrm{AT}}\cdot(\mathbf{P}_T - \mathbf{S}_{\mathrm{AT}})}{\lambda R_{AT}^3} \\ &+ \frac{-v_{zr}\cdot(z_T - z_{AR})R_{AR}^2 + (z_T - z_{AR})\cdot\mathbf{V}_{\mathrm{AR}}\cdot(\mathbf{P}_T - \mathbf{S}_{\mathrm{AR}})}{\lambda R_{AR}^3}\end{aligned} \tag{18}$$

By combining Equations (10)–(18), the iterative solution formula of target coordinates $(x_T, y_T, z_T)^T$ is obtained.

$$\begin{bmatrix} \frac{\partial f_{RA}}{\partial x_T}\Big|_{x_T = x_T^k} & \frac{\partial f_{RA}}{\partial y_T}\Big|_{y_T = y_T^k} & \frac{\partial f_{RA}}{\partial z_T}\Big|_{z_T = z_T^k} \\ \frac{\partial f_{RP}}{\partial x_T}\Big|_{x_T = x_T^k} & \frac{\partial f_{RP}}{\partial y_T}\Big|_{y_T = y_T^k} & \frac{\partial f_{RP}}{\partial z_T}\Big|_{z_T = z_T^k} \\ \frac{\partial f_{DA}}{\partial x_T}\Big|_{x_T = x_T^k} & \frac{\partial f_{DA}}{\partial y_T}\Big|_{y_T = y_T^k} & \frac{\partial f_{DA}}{\partial z_T}\Big|_{z_T = z_T^k} \end{bmatrix} \cdot \begin{bmatrix} \Delta x_T^{k+1} \\ \Delta y_T^{k+1} \\ \Delta z_T^{k+1} \end{bmatrix} = \begin{bmatrix} \Delta R_A \\ \Delta R_P \\ \Delta f_d \end{bmatrix} \tag{19}$$

where $P_T^k = (x_T^k, y_T^k, z_T^k)^T$ represents the current iteration value, and $P_T^{k+1} = (x_T^{k+1}, y_T^{k+1}, z_T^{k+1})^T$ represents the next iteration value. When $k = 0$, P_T^k represents the initial value of the iteration. $\mathbf{e} = (\Delta R_A, \Delta R_P, \Delta f_D)^T$ represents the geolocation error estimation value of the current observation vector. $\mathbf{u} = \left(\Delta x_T^{k+1}, \Delta y_T^{k+1}, \Delta z_T^{k+1}\right)^T$ represents the iterative correction value. Equation (19) can be expressed in matrix form as follows:

$$\mathbf{A} \cdot \mathbf{u} = \mathbf{e} \tag{20}$$

3.2. Influencing Factors of Geolocation Accuracy

There are two types of factors influencing geolocation accuracy: the measurement errors of radar parameters and the errors introduced by InSAR signal processing [25]. As illustrated in Figure 4, the measurement errors include the satellite timing error, APC position vector measurement error, and APC velocity vector measurement error [26]. The errors introduced by InSAR signal processing include the interferometric baseline estimation error [27], atmospheric delay estimation error [28], and interferogram phase unwrapping error [29].

The satellite timing error affects the records of image start and end times, thus indirectly influencing the interpolation result of satellite orbital state vectors [26]. This timing error is a kind of system error, and it can be estimated and corrected by inter-satellite time synchronization signals. The measurement values of APC position and velocity vectors are extracted from satellite ephemeris data, and different kinds of ephemeris data have different accuracies. Ephemeris data can be classified into three levels: precise ephemeris data, fast ephemeris data, and real-time ephemeris data [30]. Precise ephemeris data are the most precise but are not available until 12 days after the SAR image is acquired. Fast ephemeris data are moderately accurate and are available 2 days after the SAR image is acquired. Real-time ephemeris data are less accurate but are available as soon as the SAR image is acquired. So, the measurement errors of APC position and velocity vectors can be reduced using high-precision ephemeris data.

The interferometric baseline error is a kind of system error, and it influences the position vector accuracy of $\mathbf{S}_{PR}$. The GCPs or ICESat-2 laser altimetry data can serve as a valuable reference for calibrating the interferometric baseline error. The atmospheric delay error is a kind of random error, and it is closely related to meteorological conditions such as cloud thickness and water vapor content [28]. The atmospheric delay error can be estimated using statistical model methods or meteorological data methods [31]. The interferometric phase unwrapping error is a kind of random error, and it influences the range accuracy of $\hat{R}_{ATP}$. The interferometric phase unwrapping error can be reduced using a multi-baseline phase unwrapping algorithm.

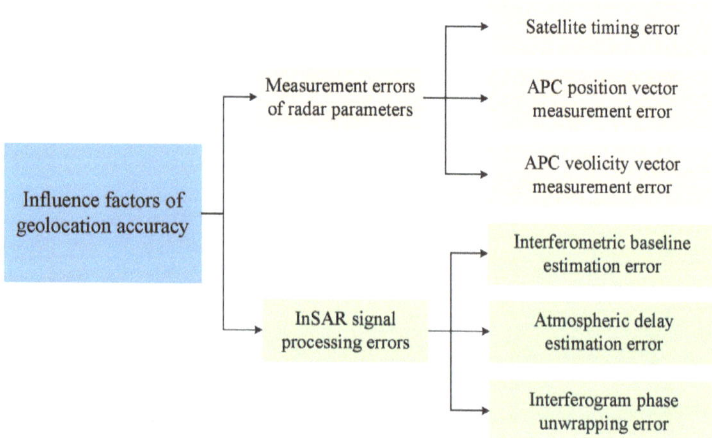

Figure 4. Influencing factors of geolocation accuracy.

4. Methodology

In this section, descriptions of the methods used in this paper are presented, including the ICESat-2 data filtering method, the baseline calibration method, the IRD geolocation method, and the low-coupling parallel calculation method.

4.1. ICESat-2 Data Filtering Method

The ICESat-2 data download file includes detailed parameter specifications, which record the quality information of the data, including meteorological information such as cloud thickness and terrain slope angle, etc. This quality information can be used as an important reference when conducting data filtering. If the cloud thickness is more than 3 m or the terrain slope angle is greater than 20 degrees, the data measurement accuracy of that day is determined to be low, and all sample points in that area are filtered out.

The ICESat-2 data are further filtered utilizing the AW3D DSM and World DEM datasets as reference datasets. The AW3D DSM dataset provides a publicly available digital surface model (DSM) with a spatial resolution of 30 m, similar to the ICESat-2 data. The elevation measurements of the AW3D DSM dataset are better than 7 m in areas with terrain slopes less than 30 degrees and better than 12 m in areas with terrain slopes greater than 30 degrees. The World DEM dataset also provides a publicly available digital elevation model (DEM) with a spatial resolution of 30 m. This DEM exhibits a measurement accuracy of better than 2 m in areas characterized by terrain slopes below 30 degrees and better than 5 m in regions with terrain slopes greater than or equal to 30 degrees.

The above two reference datasets can be used as the basis for screening ICESat-2 data. If the elevation measurement of ICESat-2 data differs from the two reference elevation data by more than 10 m (the threshold can be adjusted according to the specific situation) at the same latitude and longitude coordinates, it is considered an abnormal measurement sample point.

The data filtering process of ICESat-2 data is illustrated by the flowchart in Figure 5. In the flowchart, H_{ALT} represents the elevation of the ICESat-2 samples, H_{AW3D} represents the elevation of the AW3D DSM samples, and H_{DEM} represents the elevation of the World DEM samples.

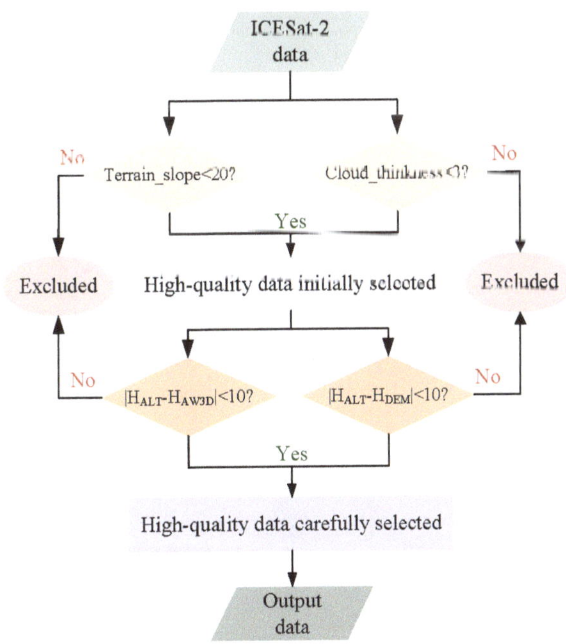

Figure 5. The flowchart for filtering ICESat-2 altimetry data.

4.2. Baseline Calibration Method

Step 1: Coregistration of SAR image data and ICESat-2/GCP data

The ICESat-2/GCP data record the geographic coordinates of each sample point T_k, and its correspondence with the pixel coordinates ($line_k$,$sample_k$) in a SAR image can be determined by applying Equation (21)–(23).

$$\frac{(\mathbf{T}_k - \mathbf{S}_{AT}(t_k)) \cdot \mathbf{V}_{AT}(t_k)}{\lambda \|\mathbf{T}_k - \mathbf{S}_{AT}(t_k)\|} + \frac{(\mathbf{T}_k - \mathbf{S}_{AR}(t_k + \Delta t)) \cdot \mathbf{V}_{AR}(t_k + \Delta t)}{\lambda \|\mathbf{T}_k - \mathbf{S}_{AR}(t_k + \Delta t)\|} = 0 \tag{21}$$

where t_k represents the zero-Doppler moment of the sample point, and there is a correspondence between t_k and the SAR image start time t_{start}, as shown in Equation (22). The symbol Δt in Equation (21) represents the propagation time of radar signals from transmission to reception, which can be calculated using Equation (23).

$$t_k = t_{start} + line_k \cdot PRT \tag{22}$$

PRT in Equation (22) represents the pulse repetition time interval. The values of PRT and t_{start} are recorded in the meta.xml file accompanying the SAR images.

$$\Delta t = \frac{R_{near} + PixelSpace \cdot sample_k}{c} \tag{23}$$

where R_{near} denotes the slant range of the first pixel in each line, and $PixelSpace$ represents the sampling interval between adjacent pixels along the range dimension. Both parameters are measured in meters and are recorded in detail in the meta.xml file accompanying the SAR image. The symbol c represents the propagation velocity of electromagnetic waves.

Step 2: Calculation of elevation measurement error

The calculation method of the elevation measurement error can be divided into two cases based on the type of reference data. If the reference data are the GCP data with a similar resolution to the SAR image, the elevation measurement error can be directly obtained by subtracting the DEM elevation of the matching point from the GCP elevation, as shown in Equation (24).

$$h_{err} = h_{GCP} - h_{DEM} \tag{24}$$

For the second case, the correlation between the ICESat-2 reference data and experimental data is complex. The SAR image resolution is 3m× 3m, whereas the ICESat-2 laser footprint diameter is about 17.5m. According to the resolution ratio relationship, an ICESat-2 laser footprint contains about 25 SAR image pixels. The average elevation of these 25 SAR image pixels is calculated and subsequently compared with the ICESat-2 laser altimetry data to derive the elevation measurement error [32], as shown in Equation (25).

$$h_{err} = h_{ICESat-2} - \frac{\sum_{k=1}^{M} h_{DEM\_k}}{M} \tag{25}$$

where M represents the number of SAR image pixels within the ICESat-2 laser footprint, and h_{DEM_k}represents the elevation measurement value of the k-th pixel.

Figure 6 exhibits a model diagram, where the red circle depicts the contour line of the ICESat-2 laser footprint, and the red dot in the center represents the center of the footprint. The solid green circles represent the locations of SAR image pixels within the footprint.

Step 3: Estimation of parallel baseline error

The parallel baseline error $B_{\|err}$ is parallel to the line of sight (LoS) and exhibits a linear relationship with the elevation measurement error h_{err}, as shown in Equation (26).

$$B_{\|err} = h_{err} \cdot \frac{B_{\perp}}{R_{AT} \sin(\theta_i)} \tag{26}$$

where $B_{\perp}$ represents the vertical baseline length, R_{AT} represents the distance between the ICESat-2/GCP point and the satellite S_{AT}, and θ_i represents the incidence angle of the corresponding pixel in the SAR image.

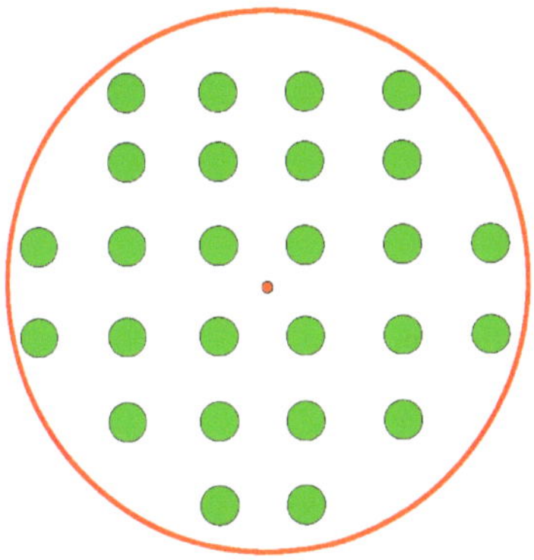

Figure 6. The correlation between the ICESat data and experimental data.

Step 4: Estimation of perpendicular baseline error

The interferometric baseline error is a kind of systematic error that varies slowly over time. At any given moment, the baseline error vector can be decomposed into two orthogonal components: the parallel baseline error $\mathbf{B}_{\|err}$ and the perpendicular baseline error $\mathbf{B}_{\perp err}$. $\mathbf{B}_{\|err}$ is parallel to the LoS, and $\mathbf{B}_{\perp err}$ is perpendicular to the LoS. The direction of the LoS varies from near range to far range, so the directions of $\mathbf{B}_{\|err}$ and $\mathbf{B}_{\perp err}$ are also dynamic for different range samples. In Figure 7, the decomposition result of the baseline error vector $\Delta\mathbf{B}_s$ at a near-range sample comprises $\mathbf{B}_{\|err1}$ and $\mathbf{B}_{\perp err1}$, whereas the corresponding result at a far-range sample consists of $\mathbf{B}_{\|err2}$ and $\mathbf{B}_{\perp err2}$.

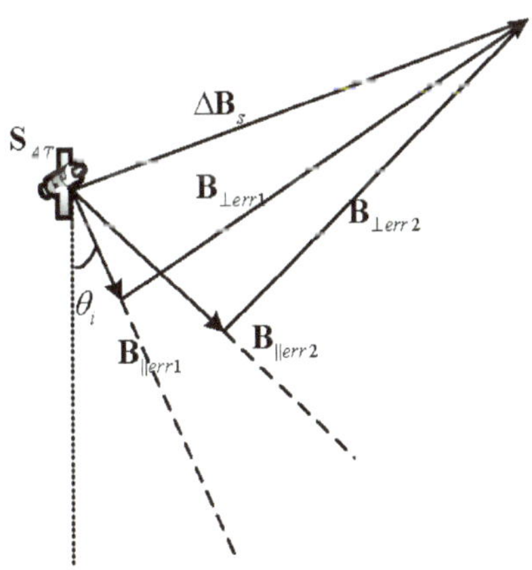

Figure 7. Interferometric baseline error vector decomposition diagram.

Based on the equality of the vector synthesis results obtained from these two decomposition methods, an equation can be established, as shown in Equation (27).

$$\mathbf{B}_{\|err1} + \mathbf{B}_{\perp err1} = \mathbf{B}_{\|err2} + \mathbf{B}_{\perp err2} = \Delta\mathbf{B}_s \tag{27}$$

Then, the estimation of the perpendicular baseline error can be determined via the least-squares fitting method.

Step 5: Baseline error coordinate system conversion

According to the relationship between the baseline error and radar incidence angle, the expression of the baseline error in the satellite's local coordinate system can be deduced, as shown in Equations (28)–(30).

$$x_s = B_{\perp}\cos(\theta_i) + B_{\|}\sin(\theta_i) \tag{28}$$

$$z_s = B_{\|}\cos(\theta_i) - B_{\perp}\sin(\theta_i) \tag{29}$$

$$\Delta\mathbf{B}_s = (x_s, 0, z_s)^T \tag{30}$$

Step 6: Baseline error correction

The baseline measurement error estimated using the baseline calibration method is typically defined in the satellite's local coordinate system, whereas the satellite orbital data (including position and velocity vectors) is defined in the Earth-centered Earth-fixed (ECEF) coordinate system. Therefore, it is necessary to convert the baseline error estimation from the satellite's local coordinate system into the ECEF coordinate system and subsequently integrate it with the original satellite orbital data. The satellite position vector $\mathbf{S_{AT}}$ and velocity vector $\mathbf{V_{AT}}$ in the coordinate transformation formula are associated with the azimuth time of pixels, so it is impossible to carry out baseline correction for the entire scene based on the baseline error expression of reference points. The formula for coordinate transformation from the satellite's local coordinate system to the ECEF coordinate system is shown in Equations (31)–(33).

$$\mathbf{T}_x = -\mathbf{S}_{AT}(t_{azi}) \times \mathbf{V}_{AT}(t_{azi}) \tag{31}$$

where $\mathbf{T}_x$ represents the transformation expression of the x-axis from the satellite's local coordinate system to the ECEF coordinate system, where the symbol "×" represents the cross-product of two vectors. $\mathbf{S_{AT}}(\mathbf{t_{azi}})$ and $\mathbf{V_{AT}}(\mathbf{t_{azi}})$ represent the satellite's position vector and velocity at time t_{azi}, respectively.

$$\mathbf{T}_z = \mathbf{T}_x \times \mathbf{V}_{AT}(t_{azi}) \tag{32}$$

where $\mathbf{T}_z$ represents the transformation expression of the z-axis from the satellite's local coordinate system to the ECEF coordinate system.

$$\Delta\mathbf{B}_E = x_s \cdot \frac{\mathbf{T}_x}{||\mathbf{T}_x||} + z_s \cdot \frac{\mathbf{T}_z}{||\mathbf{T}_z||} \tag{33}$$

where $\Delta\mathbf{B}_E$ represents the baseline error vector in the ECEF coordinate system, and $\Delta\mathbf{B}_S = (x_s, 0, z_s)^T$ represents the baseline error vector in the satellite's local coordinate system.

4.3. IRD Geolocation Method

This subsection introduces the flowchart of the proposed IRD geolocation algorithm, as illustrated in Figure 8. The whole process can be divided into four steps. Here, the algorithm processing steps are described in detail.

Step 1: Data pre-processing

This step includes orbital interpolation processing and baseline calibration. The original orbital sampling points are sparse, usually one record per second. However, the two adjacent pixels in SAR images are sampled more densely than this. The acquisition of precise measurements of the orbital position and velocity at the moment of ground target focusing necessitates the implementation of orbital interpolation processing. Additionally, baseline calibration is also completed in this step. It is worth noting that the estimated baseline calibration value is a constant number in the local coordinate system, but it becomes a variable number after projection into the Earth-centered fixed coordinate system.

Therefore, a pixel-wise correction of the satellite's position is imperative during the orbital interpolation process.

Step 2: Establishment of geolocation equation system

The geolocation equation system comprises two range equations and one Doppler equation. The input parameters of the range equation include the satellite transmitter position $\mathbf{S}_{AT}$ and receiver position $\mathbf{S}_{AR}/\mathbf{S}_{PR}$ corresponding to the zero-Doppler moment of ground target $\mathbf{P}_T$. These can be obtained through the orbital interpolation process in Step 1. Additionally, the slant range between the satellite and ground target can be estimated based on the delay time from signal transmitting to receiving. Assuming that the satellite $\mathbf{S}_{AT}$ sends a signal at time t_0 and receives it at time t_1, the propagation time of the electromagnetic wave is $t_1 - t_0$. The propagation path of electromagnetic waves during this period can be divided into two distinct segments: the transmission distance R_{AT} and the reception distance R_{AR}. The sum range $\hat{R}_{ATA}$ consists of R_{AT} and R_{AR}, which can be estimated using Equation (34).

$$\hat{R}_{ATA} = R_{AT} + R_{AR} = c \cdot (t_1 - t_0) \tag{34}$$

where c represents the propagation speed of electromagnetic waves in a vacuum; however, the propagation speed of electromagnetic waves in the atmosphere is slower than that in a vacuum. The estimated value of $\hat{R}_{ATA}$ according to Equation (34) is larger than the true value. Thus, range calibration is needed. In the other range equation, the sum range of R_{AT} and R_{PR} is calculated based on the interferometric phase φ and slant range $\hat{R}_{ATA}$, as shown in Equation (35).

$$\hat{R}_{ATP} = R_{AT} + R_{PR} = \hat{R}_{ATA} + \frac{\varphi}{2\pi}\lambda \tag{35}$$

The relative values of $\hat{R}_{ATP}$ and $\hat{R}_{ATA}$ in Equation (35) are determined with exceptional precision, because the measurement sensitivity of the interference phase method is higher than that of the time delay method.

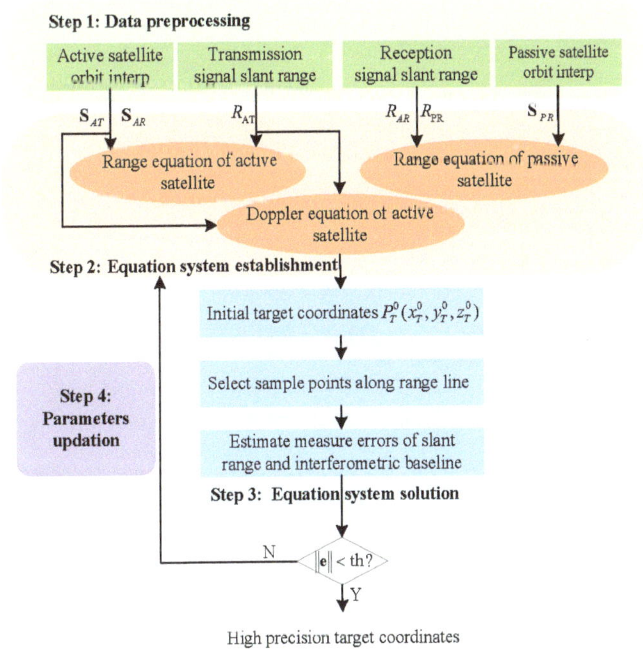

Figure 8. Flowchart of BiInSAR target geolocation algorithm.

Step 3: Equation system solution

The Newton iteration method is utilized to solve the nonlinear geolocation equations, aiming to minimize the standard deviation of the errors. First, an initial value of $P_T^0 = (x_T^0, y_T^0, z_T^0)^T$ is estimated based on the pixel coordinates (azi, rgi) of P_T and the radar parameters. Then, the initial value P_T^0 is substituted into Equation (19) to calculate the iterative correction vector $\mathbf{u}$. The relative error is calculated by selecting sample points along the range line to ensure the consistency of all ground targets. The ICESat-2 laser points are matched to the experimental SAR data, enabling the accurate estimation of range errors caused by atmospheric propagation delays. These range errors seriously reduce the consistency of target geolocation accuracy.

Step 4: Accuracy evaluation and parameter updating

The geolocation deviation vector $\mathbf{e}$ and iterative correction vector $\mathbf{u}$ can be obtained according to Equation (19). The vector $\mathbf{e}$ contains evaluation information about the range deviation and Doppler center frequency deviation. If both the range deviation and Doppler center frequency deviation are below the threshold, the target geolocation equation convergence and we obtain the output coordinate $\hat{\mathbf{P}}_T^k = (x_T^k, y_T^k, z_T^k)^T$. Otherwise, if the range deviation or the Doppler center frequency deviation is higher than the threshold, the estimation of $\hat{\mathbf{P}}_T^k$ is not accurate enough. A new estimation value is given by $\hat{\mathbf{P}}_T^{k+1} = \hat{\mathbf{P}}_T^k + \mathbf{u}$. If the iterations reach the upper limit and do not converge, it means that the input parameters of Equation (19) have significant errors. Therefore, it is necessary to correct errors such as baseline vector, atmosphere delay, etc.

4.4. Low-Coupling Parallel Calculation Method

The low-coupling parallel calculation method aims to decompose the time-consuming computational tasks in the algorithm and assign these tasks to different threads. Each thread independently acquires the necessary data from shared memory, performs data processing, and ultimately returns the processed results to the main thread. The entire processing procedure consists of three fundamental stages: data preparation, task allocation, and parallel processing, as illustrated in Figure 9.

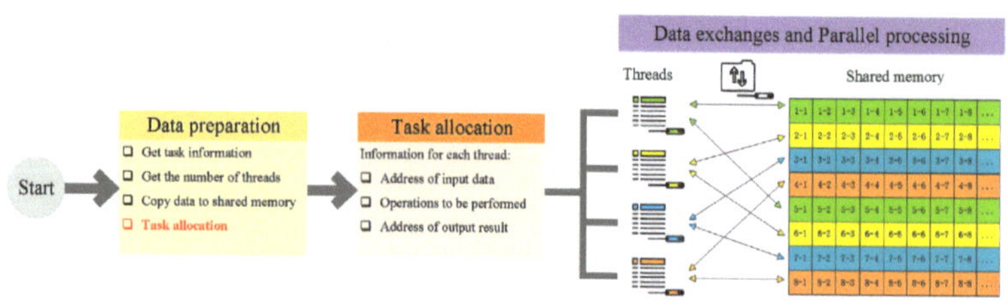

Figure 9. A flowchart of the parallel processing method.

Step 1: Data preparation

Data preparation mainly involves obtaining task information, determining the number of threads available for the task, copying input data into shared memory, and assigning tasks according to the number of threads.

Step 2: Task allocation

Task allocation includes assigning the input and output data addresses for each thread, as well as specifying the operations to be performed.

Step 3: Parallel processing

After receiving the task, each thread engages in data interaction with the shared memory and parallel processing. In Figure 9, four distinct color-coded thread symbols (green, yellow, blue, and orange) are employed to represent four independent computing units. The rectangular block on the right signifies the data stored in shared memory. The data correlation between each thread and memory is achieved by employing a consistent color scheme. The data required by each thread for processing can be distributed in the shared memory either continuously or discretely, depending on the actual demand.

5. Experimental Design and Analysis of Results

In this section, two groups of experiments are designed to verify the proposed IRD geolocation algorithm. It is proven that the proposed algorithm overcomes the problem of poor precision in monostatic equivalent (MoE) algorithms [21] and exhibits a faster iteration speed compared to the existing entropy minimize principle (EMP) algorithm.

5.1. Group 1: Digital Simulation Experiment

This subsection presents a digital simulation experiment, and the simulation parameters are listed in Table 1. The simulated scene comprises nine precisely georeferenced point targets. The effectiveness of the proposed algorithm is verified through three distinct test conditions. The first test takes place under ideal conditions, devoid of any measurement errors, and yields geolocation accuracy on the order of 0.001 m. In contrast, the second test takes place under challenging conditions, marked by multiple measurement errors and the absence of calibration. In this scenario, geolocation accuracy is on the order of 5 m. For the third test, multiple measurement errors are introduced, but calibration processes are incorporated, resulting in geolocation accuracy on the order of 0.1 m. These experiments offer valuable insights into the algorithm's performance under varying conditions, from ideal to real-world scenarios, providing a comprehensive assessment of its capabilities.

Table 1. Simulation experiment parameters of BiInSAR strip imaging mode.

Satellite Parameter	Value
Semi-major axis	6913.140 km
Orbital altitude	535.00 km
Inclination	97.54°
Eccentricity	0
Off-nadir angle	41.19°
Radar frequency	9.2 GHz
Transmitted bandwidth	150 MHz
Sampling rate	180 MHz
Pulse repetition frequency	6,763 Hz
Transmitter velocity	7681.69 m/s
Antenna size (azimuth × range)	5 m × 3 m

The nine point targets are distributed at approximately equal intervals in the range and azimuth dimensions, as displayed in Figure 10. The objective of this distribution design is to explore the potential correlation between the geolocation error distributions and pixel coordinates, encompassing azimuth and range coordinates. The baseline measurement errors include the parallel baseline error and perpendicular baseline error. In this experiment, the baseline measurement error $\Delta\vec{B}_e = \left(b_1 \cdot \vec{B}_{\parallel} + b_2 \cdot \vec{B}_{\perp}\right)$ is a constant value, and the values of b_1 and b_2 are listed in Table 2. The value of the slant-range measurement error ΔR_e is affected by cloud thickness and terrain height. In this experiment, the value of ΔR_e is set according to the exponential model and terrain height, as shown in Equation (36).

$$\Delta R_e(h, \theta_{inc}) = \frac{ZPD}{\cos(\theta_{inc})} \cdot e^{-\frac{h}{H}} \tag{36}$$

where the Zenith Path Delay (ZPD) is an empirical parameter, fixed at a constant 2.3 m for an X-band signal. The reference height H is set to 6000 m, and h is the scene average height, which can be extracted from a coarse DEM. The types and values of measurement errors are listed in Table 2.

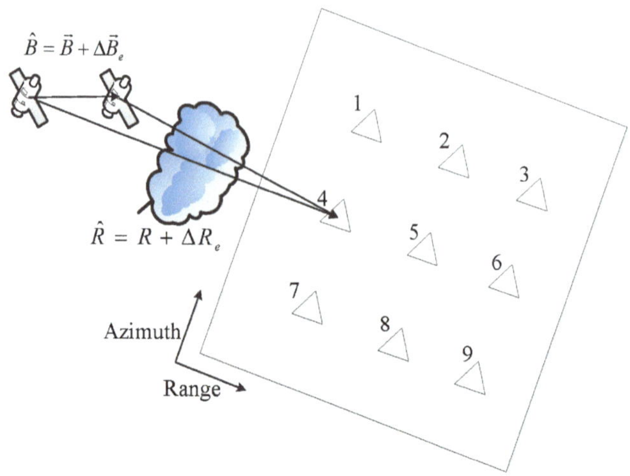

Figure 10. Design of simulation experiment.

Table 2. Types and values of measurement errors in the simulation experiment.

Parameter Type	Parameter Name	Measurement Error Value
Radar parameters	Satellite timing error	0.0015 s
	APC position vector measurement error	0.5 m
	APC velocity vector measurement error	0.01 m/s
Signal processing parameters	Parallel baseline error	0.002 m
	Perpendicular baseline error	0.002 m
	Atmospheric delay estimation error	0.2–0.8 m
	Interferometric phase unwrapping error	3°

The geolocation accuracies of point targets under different test conditions are depicted in Figure 11. The red line represents the geolocation error trend of the nine point targets under ideal conditions (no measurement errors). The blue line represents the geolocation error trend under bad conditions with the baseline measurement error. The orange line represents the geolocation error trend under worse conditions with both the baseline and range errors. The purple line represents the geolocation error trend under good conditions with the baseline measurement error and carrying out parameter calibration. The pink line represents the geolocation error trend under good conditions with multiple types of measurement errors and carrying out parameter calibration. A partially enlarged image of the curves at the bottom of Figure 11 is provided on the right of the original image.

Through this series of comparative experiments, it was found that the geolocation accuracy was mainly affected by two factors. One factor was the number of error types and the magnitude of measurement errors, and the other was whether parameter calibration was carried out. The measurement accuracy of the slant range based on the time delay method was around 1 m. In our method, the measurement accuracy can reach the same order as the carrier wavelength, usually at the decimeter or even centimeter level. The root-mean-square error (RMSE) is used to reflect the accuracy of target geolocation. It can be seen that even if there are measurement errors, as long as the magnitude of errors does not exceed the threshold, the proposed method can effectively estimate the values of the measurement errors. By carrying out parameter calibration, the geolocation accuracy can reach a similar level to the accuracy under ideal conditions, as exhibited in Table 3.

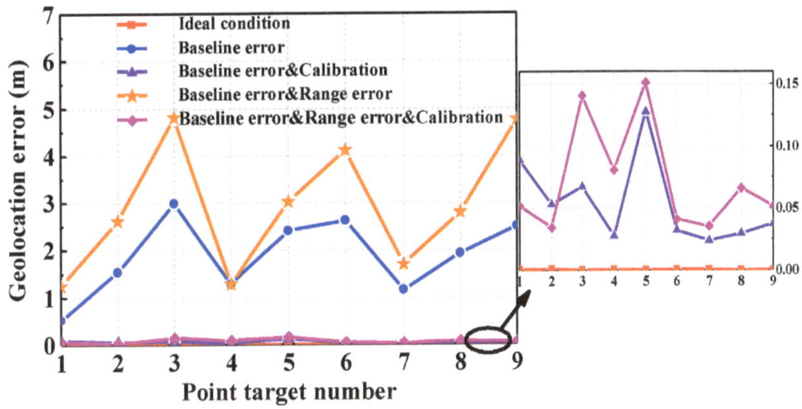

Figure 11. Geolocation accuracy of point targets under different test conditions.

Table 3. RMSEs of target geolocation under different test conditions.

Conditions	Number of Error Types	Parameter Calibration	RMSE of Target Geolocation
Ideal	Zero	-	0.001 m
Bad	One	No	2.45 m
Worse	Two or more	No	3.21 m
Good	One	Yes	0.08 m
Good	Two or more	Yes	0.11 m

The geolocation errors of the point targets were evaluated, and the comparison results of the different algorithms are exhibited in Figure 12. The proposed IRD algorithm exhibits a lower average error and smaller standard deviation compared to the MoE and EMP algorithms. The experimental results depicted in Figure 12 demonstrate the absence of a significant correlation between geolocation errors and pixel coordinates.

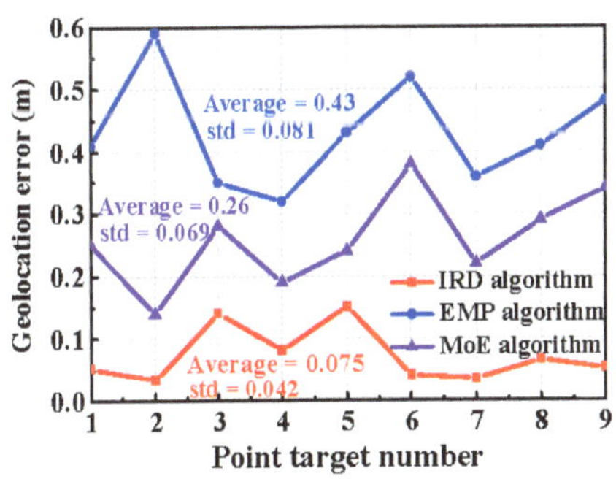

Figure 12. A comparison of geolocation errors across different algorithms.

Based on the existing experiments, a group of incremental baseline errors with a step of 1cm was added to the experimental data. Subsequently, an estimation value for the baseline error was obtained using the proposed algorithm and evaluated based on the residual error. The test results are exhibited in Figure 13, where the horizontal axis represents the baseline error introduced in the experiment, and the vertical axis represents the residual error after conducting baseline calibration. The curve in Figure 13 demonstrates that the residual baseline errors exhibit a fluctuation trend centered around zero means, with amplitudes smaller than 0.5 mm under a boundary condition where the baseline measurement error does not exceed 20 cm.

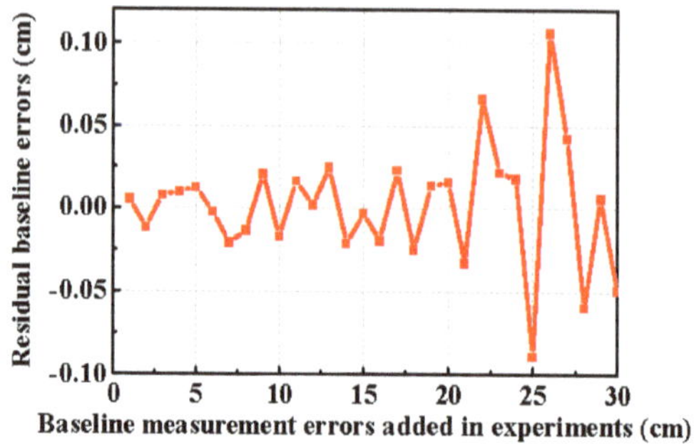

Figure 13. The boundary condition experiment for the baseline error.

5.2. Group 2: Real SAR Data Experiment

In this subsection, two typical kinds of bistatic SAR satellite data are used as examples to verify the efficiency of the proposed geolocation algorithm in DEM generation applications.

The first test data were acquired by the TH2-01A and TH2-01B satellites on 26 September 2019. The key parameters of this test data are listed in Table 4. The TH-2 satellites and the whole radar system were designed and built in China and launched on 30 April 2019. The imaging mode of the test data is a strip map, and both the azimuth resolution and range resolution are 3 m. The amplitude image and coherence image of the TH-2 test data are exhibited in Figure 14.

Table 4. Key parameters of TH-2 test data.

Parameter Name	Parameter Value
Satellite name	TH2-01A, TH2-01B
Orbital height	580 Km
Incidence angle	42.1°
Nearest range	600 Km
Resolution	3 m
Perpendicular baseline length	280 m
Height of ambiguity	21 m
Average coherence	0.91

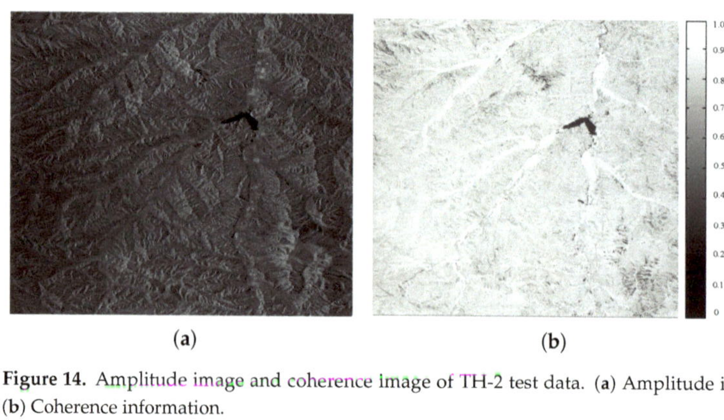

Figure 14. Amplitude image and coherence image of TH-2 test data. (**a**) Amplitude information. (**b**) Coherence information.

The size of the original SAR image is 23,716 × 23,596 pixels. After conducting the necessary pre-processing tasks for the SAR images, the efficiency of the proposed IRD geolocation algorithm can be verified. Data pre-processing includes image coregistration, interferometric phase extraction, phase filtering, and phase unwrapping. These pre-processing jobs can be performed using various commercial software, such as Gamma, ENVI SARscape, RDSpace, PIE, SAR studio, etc. These data pre-processing tasks are applicable to both monostatic satellite data and bistatic satellite data. The primary focus of this paper is on geolocation processing.

There are 10 trihedron reflectors in the test image, and their geographical coordinates can be obtained via a high-precision GPS device. The GPS device exhibits a measurement accuracy at the level of 0.5 m, rendering its measurement values suitable as reference benchmarks for evaluating the precision of diverse geolocation algorithms. The distribution of the 10 GCPs is illustrated in Figure 15. The red squares in Figure 15 represent the pixel coordinates of the GCPs in the SAR image, and at the end of each red connecting line is a partially enlarged image. Figure 15 is defined in the slant-range coordinate system, where the horizontal axis represents the slant-range dimension, and the vertical axis represents the azimuth dimension.

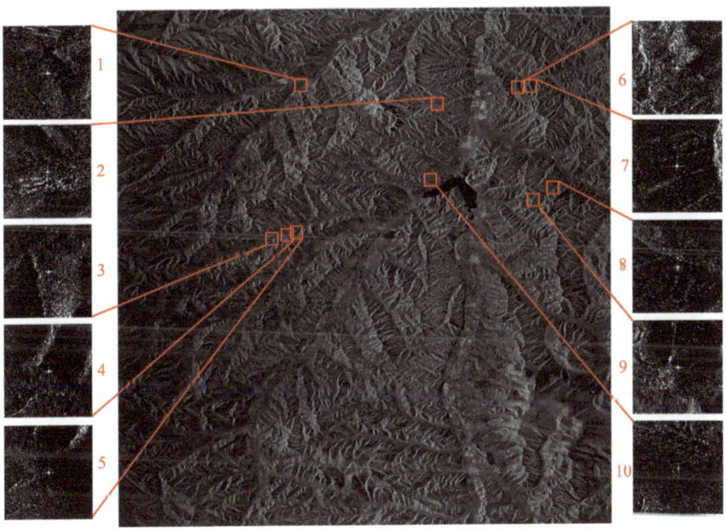

Figure 15. Pixel coordinates of GCPs and their partially enlarged images.

The proposed IRD geolocation algorithm was utilized to compute geographical coordinates for all pixels within the SAR image, subsequently enabling the generation of a three-dimensional topographic map of the observed scene, as depicted in Figure 16. The red dots represent the positions of the GCPs, and the numbers near the red dots represent the order of the GCPs. Figure 16 is defined in the geographical coordinate system, where the horizontal axis represents the longitude coordinate, and the vertical axis represents the latitude coordinate. Since the distributions of the GCPs are discrete and sparse, ICESat-2 laser altimetry data are provided as reference data to evaluate the geolocation accuracy of the proposed method. The distribution of ICESat-2 laser altimetry data is illustrated in Figure 17, where the white dots represent the geographic positions of laser points.

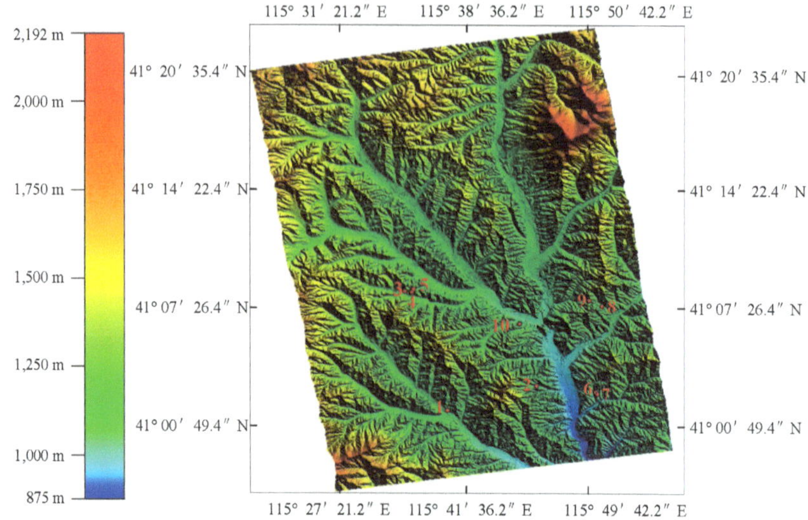

Figure 16. Three- dimensional topographic map of the geolocation results.

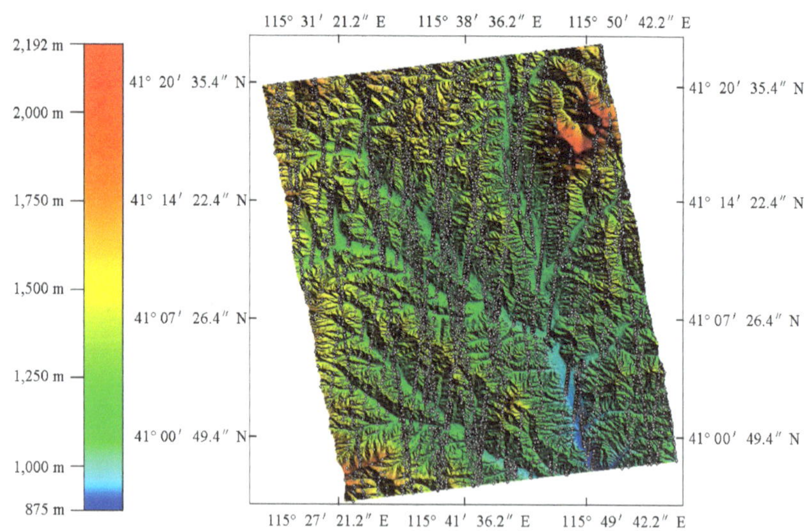

Figure 17. ICESat-2 laser altimetry data distribution.

The results depicted in Figure 18 demonstrate that the proposed IRD algorithm exhibits geolocation errors smaller than those observed with the EMP and MoE algorithms, as evidenced by both the GCP reference data and the ICESat-2 laser altimetry data. In Figure 18a, the horizontal axis represents the order of the GCPs, and the vertical axis represents the geolocation error of each point. In Figure 18b, the horizontal axis represents the statistics intervals of elevation error, with a range of [−15 m, 15 m] and an interval of 0.5 m. The vertical axis represents the number of pixels in each interval. It can be seen that the average elevation error of the IRD algorithm is closer to zero and its standard deviation(std) is smaller than the other two algorithms.

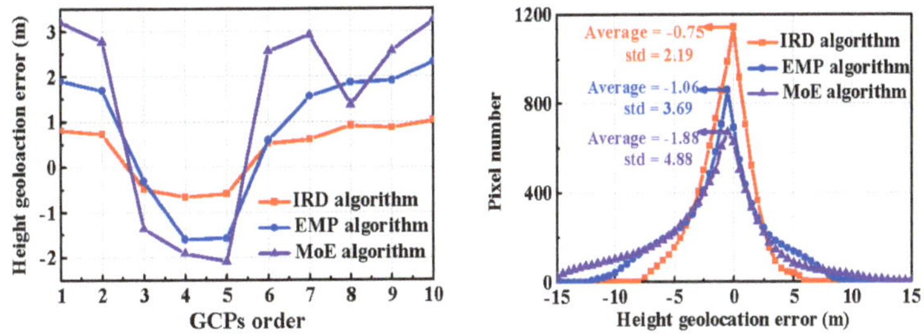

(a) (b)

Figure 18. Evaluation of height geolocation accuracy. (**a**) Evaluation results based on GCPs. (**b**) Evaluation results based on ICESat-2 laser altimetry data.

In Figure 18, we can see that both the GCP reference data and the ICESat-2 laser altimetry data prove that the geolocation errors of the proposed IRD algorithm are smaller than those of the EMP and MoE algorithms. The evaluated RMSE of the IRD algorithm based on the GCP reference data is 0.743 m, whereas the RMSEs of the EMP and MoE algorithms are 0.778 m and 2.49 m, respectively. And, the RMSEs of the three algorithms based on the ICESat laser altimetry data are 3.190 m, 3.439 m, and 5.234 m, respectively. The reason why the RMSEs evaluated based on ICESat-2 laser altimetry data are larger than the RMSEs evaluated based on the GCP reference data is that the ICESat-2 laser altimetry measurement values represent the average height values for each footprint of the laser, whereas the GCP measurement values represent specific point heights. The TH-2 test image coverage is in a mountainous area, and the average height (ICESat-2 data) does not exactly match the geolocation product, so the evaluation result is larger than the real geolocation error.

In Figure 18a, the geolocation error curves show depressions at control points 3, 4, and 5. The local terrain slope angle of this area is statistically analyzed to assist in investigating the cause. The local terrain slope map of the generated digital elevation model (DEM) is presented in Figure 19a, with the positions of the control points denoted by red dots. The color bar on the right side of Figure 19a represents the degrees of the terrain slope angle. Additionally, Figure 19b illustrates a one-dimensional terrain slope profile along the longitude direction for three selected control points. It can be seen that the variation trends for the terrain slope angle around control points 3, 4, and 5 are at critical points of rising and falling. Due to the unique topography, these three control points exhibit distinct geolocation errors compared to the remaining control points. The geolocation errors demonstrate a correlation with the pixel coordinates, which arises from the differences in the terrain height and/or slope angle of the GCPs.

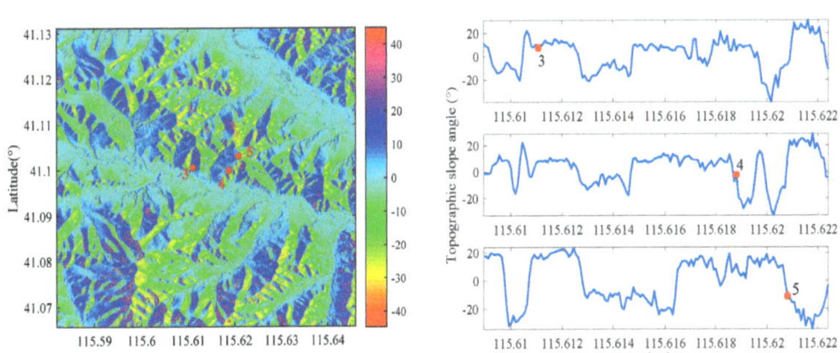

(a) (b)

Figure 19. The local terrain slope angle of DEM. (**a**) Two-dimensional view. (**b**) One-dimensional view.

5.3. Algorithm Efficiency Improvement

In this subsection, the efficiency of the proposed low-coupling parallel calculation method is verified. It can improve the calculation speed by two to three times. The core idea is to reduce the coupling between computing units and improve computing efficiency via the CPU-based parallel computing strategy. The creation of a shared memory block enables universal access by all threads. Subsequently, individual data processing is performed in parallel by each thread, culminating in the submission of the processed results to the main thread. When designing parallel computing algorithms, special consideration should be given to ensuring that the memory addresses of the local variables defined by each thread are unique and that there are no memory conflicts. For example, the absolute phase matrix and the phase-to-elevation conversion matrix are defined as shared memory that can be accessed by all threads, whereas the satellite orbit position vector, velocity vector, and slant-range variables are defined as local variables. Algorithm designers allocate computing tasks reasonably among various computing units, thereby improving computational efficiency.

Table 5 shows the processing times for different geographic positioning algorithms under the same experimental conditions. The computer used in our experiment has an Intel (R) Core (TM) i7-4900 processor running @ 3.2 GHz, with eight threads and 32 GB of memory. The experimental data include two sets of TH-2 data and two sets of TerraSAR-X data. The average size of the experimental data is 18848 × 26045 pixels, covering an area of around 30 km^2. It can be seen that regardless of the geolocation accuracy, the IRD algorithm processes around 1,429,678 pixels per second, whereas the EMP algorithm processes about 674,157 pixels per second. This proves that the proposed IRD algorithm requires less time than the EMP algorithm to process the same test data.

Table 5. The processing times of different geolocation algorithms.

Information on Test Data		Data Processing Times	
Test Data ID	**Test Data Size**	**EMP Algorithm**	**IRD Algorithm**
TH2-01BA-InSAR-20190926	21,096 × 23,584 pixels	12.3 min	**5.8 min**
TH2-01AB-InSAR-20191015	23,716 × 23,596 pixels	14.25 min	**6.65 min**
TDM1-SAR-BIST-SM-20180223	13,206 × 28,796 pixels	9.3 min	**4.13 min**
TDM1-SAR-BIST-SM-20130101	17,374 × 28,204 pixels	11.2 min	**5.3 min**

6. Conclusions

This paper aims to improve geolocation accuracy and calculation efficiency in target geolocation. Firstly, a bistatic interferometric baseline calibration model and an improved range–Doppler (IRD) model are proposed. These two models effectively reflect the geometric characteristics in a bistatic configuration. Then, the calculation steps and derivation process of the baseline calibration method and the IRD target geolocation method are elaborated in detail. A processing flowchart is provided to introduce how the parallel processing method proposed in this paper improves the efficiency of the algorithm. The proposed algorithm is verified using a pair of bistatic spaceborne SAR data acquired by the TH-2 satellite, which is China's first spaceborne BiInSAR system. Compared with the MoE and EMP algorithms, the accuracies of the proposed IRD algorithm can be improved by around 80 percent and 40 percent, respectively. Meanwhile, compared with the EMP algorithm, the calculation efficiency is improved by two to three times by reducing the coupling between computation units and designing a parallel calculation strategy. In conclusion, the proposed BiInSAR geolocation algorithm based on the IRD model has been successfully employed in the DEM generation and topographic mapping mission of the TH-2 satellite.

Author Contributions: Conceptualization, C.X. and F.T. (Fanyi Tang); methodology, F.T. (Feng Tian); software, C.X.; validation, C.X.; investigation, C.X.; writing—original draft preparation, C.X.; writing—review and editing, Z.L., F.T. (Fanyi Tang), F.T. (Feng Tian), and Z.S.; funding acquisition, Z.L. and Z.S. All authors have read and agreed to the published version of the manuscript.

Funding: This research was funded by the National Natural Science Foundation of China (grant no. 62031005).

Data Availability Statement: Not applicable.

Acknowledgments: The authors would like to thank the anonymous reviewers for their helpful comments and suggestions.

Conflicts of Interest: The authors declare no conflicts of interest.

References

1. Prats-Iraola, P.; Pinheiro, M.; Rodriguez-Cassola, M.; Scheiber, R.; Lopez-Dekker, P. Bistatic SAR image formation and interferometric processing for the stereoid Earth explorer 10 candidate mission. In Proceedings of the IGARSS 2019–2019 IEEE International Geoscience and Remote Sensing Symposium (IGARSS), Yokohama, Japan, 28 July–2 August 2019; pp. 106–109.
2. Sun, Z.; Wu, J.; Pei, J.; Li, Z.; Huang, Y.; Yang, J. Inclined geosynchronous spaceborne–airborne bistatic SAR: Performance analysis and mission design. *IEEE Trans. Geosci. Remote Sens.* **2016**, *54*, 343–357. [CrossRef]
3. Wang, R. First bistatic demonstration of digital beamforming in elevation with TerraSAR-X as an illuminator. *IEEE Trans. Geosci. Remote Sens.* **2016**, *54*, 842–849. [CrossRef]
4. Liu, F.; Fan, X.; Zhang, T.; Liu, Q. GNSS-based SAR interferometry for 3-D deformation retrieval: Algorithms and feasibility study. *IEEE Trans. Geosci. Remote Sens.* **2018**, *56*, 5736–5748. [CrossRef]
5. Tian, F.; Lu, Z.; Wang, Y.; Suo, Z.; Li, Z.; Guo, H. Fast Geolocation Solution and Accuracy Analysis for Bistatic InSAR Configuration of Geostationary Transmitter With LEO Receivers. *Geosci. Remote Sens. Lett.* **2022**, *19*, 4014305. [CrossRef]
6. Krieger, G.; Moreira, A.; Fiedler, H.; Hajnsek, I.; Werner, M.; Younis, M.; Zink, M. TanDEM-X: A satellite formation for high-resolution SAR interferometry. *IEEE Trans. Geosci. Remote Sens.* **2007**, *45*, 3317–3341. [CrossRef]
7. Lou, L.; Liu, Z.; Zhang, H. TH-2 satellite engineering design and implementation. *Acta Geod. Cartogr. Sin.* **2020**, *49*, 1252-1264.
8. Martone, M.; Braeutigam, B.; Rizzoli, P. Coherence evaluation of TanDEM-X interferometric data. *ISPRS J. Photogramm. Remote Sens.* **2012**, *73*, 21–29. [CrossRef]
9. Wang, R.; Lv, X.; Chai, H.; Zhang, L. A Three-Dimensional Block Adjustment Method for Spaceborne InSAR Based on the Range-Doppler-Phase Model. *Remote Sens.* **2023**, *15*, 1046. [CrossRef]
10. Fritz, T.; Rossi, C.; Yague-Martinez, N.; Rodriguez-Gonzalez, F.; Lachaise, M.; Breit, H. Interferometric processing of TanDEM-X data. In Proceedings of the 2011 IEEE International Geoscience and Remote Sensing Symposium, Vancouver, BC, Canada, 24–29 July 2011; pp. 2428–2431.
11. Zhang, H.; Gu, D.; Ju, B.; Shao, K.; Yi, B.; Duan, X.; Huang, Z. Precise Orbit Determination and Maneuver Assessment for TH-2 Satellites Using Spaceborne GPS and BDS2 Observations. *Remote Sens.* **2021**, *13*, 5002. [CrossRef]
12. Mou, J.; Wang, Y.; Hong, J.; Wang, Y.; Wang, A. Baseline Calibration of L-Band Spaceborne Bistatic SAR TwinSAR-L for DEM Generation. *Remote Sens.* **2023**, *15*, 3024. [CrossRef]
13. Mou, J.; Wang, Y.; Hong, J.; Wang, Y.; Wang, A.; Sun, S.; Liu, G. First Assessment of Bistatic Geometric Calibration and Geolocation Accuracy of Innovative Spaceborne Synthetic Aperture Radar LuTan-1. *Remote Sens.* **2023**, *15*, 5280. [CrossRef]
14. Mou, J.; Wang, Y.; Hong, J.; Wang, Y.; Wang, A. Geometric Calibration of Spaceborne Bistatic SAR LT-1 for Generation of High-Accuracy DEM. In Proceedings of the 2023 8th International Conference on Signal and Image Processing (ICSIP), Wuxi, China, 8–10 July 2023; pp. 914–918.
15. Ding, J.; Zhang, Z.; Xing, M.; Bao, Z. A New Look at the Bistatic-to-Monostatic Conversion for Tandem SAR Image Formation. *IEEE Trans. Geosci. Remote Sens.* **2008**, *5*, 392–395. [CrossRef]
16. Raines, E.; Park, J.; Johnson, J.T.; Burkholder, R.J. A Comparison of Bistatic and Monostatic Radar Images of 1-D Perfectly Conducting Rough Surfaces. *Geosci. Remote Sens. Lett.* **2022**, *19*, 1–5. [CrossRef]
17. Zhou, X.-k.; Chen, J.; Wang, P.-b.; Zeng, H.-c.; Fang, Y.; Men, Z.-r.; Liu, W. An Efficient Imaging Algorithm for GNSS-R Bi-Static SAR. *Remote Sens.* **2019**, *11*, 2945. [CrossRef]
18. Zhang, H.; Tang, J.; Wang, R.; Deng, Y.; Wang, W.; Li, N. An Accelerated Backprojection Algorithm for Monostatic and Bistatic SAR Processing. *Remote Sens.* **2018**, *10*, 140. [CrossRef]
19. Li, T.; Chen, K.-S.; Jin, M. Analysis and Simulation on Imaging Performance of Backward and Forward Bistatic Synthetic Aperture Radar. *Remote Sens.* **2018**, *10*, 1676. [CrossRef]
20. Wu, J.; Xu, Y.; Zhong, X.; Sun, Z.; Yang, J. A Three-Dimensional Localization Method for Multistatic SAR Based on Numerical Range-Doppler Algorithm and Entropy Minimization. *Remote Sens.* **2017**, *9*, 470. [CrossRef]
21. Li, J.; Yang, Q.; Li, Z.; Wu, J.; Xia, W.; Yang, J. A Blind Localization Method Based on Monostatic Equivalent for Bistatic SAR. In Proceedings of the IGARSS 2022–2022 IEEE International Geoscience and Remote Sensing Symposium, Kuala Lumpur, Malaysia, 17–22 July 2022; pp. 1836–1839.

22. Wang, Y.; Lu, Z.; Tian, F.; Suo, Z.; Li, Z.; Zhu, Y. Spaceborne Bistatic InSAR Geolocation with Geostationary Transmitter. In Proceedings of the 2019 IEEE International Conference on Signal, Information and Data Processing (ICSIDP), Chongqing, China, 11–13 December 2019; pp. 1–4.
23. Gao, C.; Li, Y.; Yang, J.; Zhang, Y. A Location Recall Strategy for Improving Efficiency of User-Generated Short Text Geolocalization. *IEEE Trans. Comput.* **2022**, *9*, 1419–1431. [CrossRef]
24. Sansosti, E.; Berardino, P.; Manunta, M.; Serafino, F.; Fornaro, G. Geometrical SAR image registration. *IEEE Trans. Geosci. Remote Sens.* **2006**, *44*, 2861–2870. [CrossRef]
25. Xu, H.; Kang, C. Equivalence Analysis of Accuracy of Geolocation Models for Spaceborne InSAR. *IEEE Trans. Geosci. Remote Sens.* **2010**, *48*, 480–490.
26. Gisinger, C.; Balss, U.; Pail, R.; Zhu, X.X.; Montazeri, S.; Gernhardt, S.; Eineder, M. Precise Three-Dimensional Stereo Localization of Corner Reflectors and Persistent Scatterers with TerraSAR-X. *IEEE Trans. Geosci. Remote Sens.* **2015**, *53*, 1782–1802. [CrossRef]
27. Gruber A.; Wessel B.; Huber M.; Roth, A. Operational TanDEM-X DEM calibration and first validation results. *ISPRS J. Photogramm. Remote Sens.* **2012**, *73*, 39–49. [CrossRef]
28. Fornaro, G.; D'Agostino, N.; Giuliani, R.; Noviello, C.; Reale, D.; Verde, S. Assimilation of GPS-Derived Atmospheric Propagation Delay in DInSAR Data Processing. *IEEE J. Sel. Top. Appl. Earth Obs. Remote Sens.* **2015**, *8*, 784–799. [CrossRef]
29. Pepe, A. The Correction of Phase Unwrapping Errors in Sequences of Multi-Temporal Differential SAR Interferograms. In Proceedings of the IGARSc2020–2020 IEEE International Geoscience and Remote Sensing Symposium, Waikoloa, HI, USA, 26 September–2 October 2020. **2015**, *8*, 784–799.
30. Wang, K.; Liu, J.; Su, H.; El-Mowafy, A.; Yang, X. Real-Time LEO Satellite Orbits Based on Batch Least-Squares Orbit Determination with Short-Term Orbit Prediction. *Remote Sens.* **2023**, *15*, 133. [CrossRef]
31. Liu, Q.; Zeng, Q.; Zhang, Z. Evaluation of InSAR Tropospheric Correction by Using Efficient WRF Simulation with ERA5 for Initialization. *Remote Sens.* **2023**, *15*, 273. [CrossRef]
32. González, J.H.; Antony, J.M.W.; Bachmann, M.; Krieger, G.; Zink, M.; Schrank, D.; Schwerdt, M. Bistatic system and baseline calibration in TanDEM-X to ensure the global digital elevation model quality. *ISPRS J. Photogramm. Remote Sens.* **2012**, *73*, 3–11. [CrossRef]

remote sensing

Article

Enhancing Hyperspectral Anomaly Detection with a Novel Differential Network Approach for Precision and Robust Background Suppression

Jiajia Zhang [1,2], Pei Xiang [1], Xiang Teng [1], Dong Zhao [1,3], Huan Li [1,*], Jiangluqi Song [1], Huixin Zhou [1] and Wei Tan [4]

1 School of Physics, Xidian University, Xi'an 710071, China; jiajizhang2@student.unimelb.edu.au (J.Z.); pxiang@stu.xidian.edu.cn (P.X.); 21051110101@stu.xidian.edu.cn (X.T.); dzhao@cwxu.edu.cn (D.Z.); jlqsong@xidian.edu.cn (J.S.); hxzhou@mail.xidian.edu.cn (H.Z.)
2 School of Mathematics and Statistics, The University of Melbourne, Parkville, VIC 3010, Australia
3 School of Electronics and Information Engineering, Wuxi University, Wuxi 214105, China
4 Science and Technology on Complex System Control and Intelligent Agent Cooperation Laboratory, Beijing Mechano-Electronic Engineering Institute, Beijing 100074, China; wtan@stu.xidian.edu.cn
* Correspondence: huanli@xidian.edu.cn; Tel.: +86-029-88202553

Abstract: The existing deep-learning-based hyperspectral anomaly detection methods detect anomalies by reconstructing a clean background. However, these methods model the background of the hyperspectral image (HSI) through global features, neglecting local features. In complex background scenarios, these methods struggle to obtain accurate background priors for training constraints, thereby limiting the anomaly detection performance. To enhance the capability of the network in extracting local features and improve anomaly detection performance, a hyperspectral anomaly detection method based on differential network is proposed. First, we posit that anomalous pixels are challenging to be reconstructed through the features of surrounding pixels. A differential convolution method is introduced to extract local punctured neighborhood features in the HSI. The differential convolution contains two types of kernels with different receptive fields. These kernels are adopted to obtain the outer window features and inner window features. Second, to improve the feature extraction capability of the network, a local detail attention and a local Transformer attention are proposed. These attention modules enhance the inner window features. Third, the obtained inner window features are subtracted from the outer window features to derive differential features, which encapsulate local punctured neighborhood characteristics. The obtained differential features are employed to reconstruct the background of the HSI. Finally, the anomaly detection results are extracted from the difference between the input HSI and the reconstructed background of the HSI. In the proposed method, for each receptive field kernel, the optimization objective is to reconstruct the input HSI rather than the background HSI. This way circumvents problems where the background constraint biases might affect detection performance. The proposed method offers researchers a new and effective approach for applying deep learning in a local area to the field of hyperspectral anomaly detection. The experiments are conducted with multiple metrics on five real-world datasets. The proposed method outperforms eight state-of-the-art methods in both subjective and objective evaluations.

Keywords: hyperspectral anomaly detection; differential convolution; local detail attention; local transformer attention

Citation: Zhang, J.; Xiang, P.; Teng, X.; Zhao, D.; Li, H.; Song, J.; Zhou, H.; Tan, W. Enhancing Hyperspectral Anomaly Detection with a Novel Differential Network Approach for Precision and Robust Background Suppression. *Remote Sens.* **2024**, *16*, 434. https://doi.org/10.3390/rs16030434

Academic Editors: Yanni Dong, Tao Chen and Chao Chen

Received: 22 December 2023
Revised: 17 January 2024
Accepted: 18 January 2024
Published: 23 January 2024

1. Introduction

Hyperspectral images (HSIs) are 3D image matrices captured by remote sensing imaging systems [1–3]. They encompass numerous narrow and approximately continuous spectral bands, covering the ultraviolet, visible, and infrared ranges [4–6]. The abundance of spectral information is advantageous for discriminating targets of interest with distinct spectral characteristics [7–9]. As a result, HSI has been applied in the field of target detection [10–12].

Hyperspectral anomaly detection refers to the unsupervised detection of targets that exhibit differences in both spatial and spectral characteristics from the surrounding background, without any prior information [13,14]. This technique has been applied in different areas, such as environmental monitoring, resource exploration, and precision agriculture [15,16]. Despite the unique advantages of hyperspectral anomaly detection, it faces various challenges, including the absence of prior information, diverse background types, and noise interference [17,18].

In past decades, plenty of hyperspectral anomaly detection methods have emerged to solve various problems. The existing hyperspectral anomaly detection methods can be categorized into traditional methods and deep-learning-based methods [19,20]. The traditional methods can be further classified into statistical-based methods and data-representation-based methods [21,22]. The statistical-based methods aim to model the background distribution of the HSI and identify target locations through hypothesis testing [23,24]. In the statistical-based methods, the Reed–Xiaoli (RX) method [25] is recognized as one of the most representative methods. This method assumes that the background of HSI obeys a multivariate Gaussian distribution. The anomalous targets are detected by calculating the Mahalanobis distance between the test pixels and the background. However, in the real HSI, the mean and covariance used by the RX method are susceptible to interference from anomalies and noise when the background information is complex. This will impact detection performance. To reduce the impact of anomalies and noise on detection results, methods like local RX [26] and kernel RX [27,28] are proposed. To precisely model complex backgrounds in situations with multipixel anomalies, the two-step generalized likelihood ratio test (2S-GLRT) [29] has been introduced. This method is a generalization of the RX method. It employs Gaussian hypothesis testing to detect anomalous targets by modeling multipixel anomalies and background. Data-representation-based methods posit that background pixels can be reconstructed by some similar background pixels [30]. Among them, the collaborative representation detector (CRD) [31] assumes that each background pixel can be approximated by other pixels in its spatial neighborhood while anomalous pixels cannot. This method uses background reconstruction errors to extract anomalies. Detectors based on low-rank sparse representation [32,33] consider that the background is low-rank and anomalies are sparse. These methods achieve low-rank and sparse representation through the constructed background dictionary. For instance, Zhang et al. [34] propose a hyperspectral anomaly detection method based on low-rank sparse matrix decomposition and Mahalanobis distance. This method utilizes the constructed background dictionary for low-rank and sparse decomposition and employs Mahalanobis distance for anomaly detection. Wang et al. [35] introduce a hyperspectral anomaly detection method based on principal component analysis and tensor low-rank and sparse representation (PCA-TLRSR). In this method, a vector low-rank and sparse representation method is used instead of the traditional 2D representation to preserve the 3D information of the HSI. A comprehensive 3D background dictionary is constructed for low-rank and sparse representation. The traditional methods have been successfully applied in practical applications due to the efficient execution [36]. However, for complex HSI scenarios, statistical-based methods struggle to acquire accurate background statistical features, which may affect detection performance. The data-representation-based methods face challenges in establishing precise background dictionaries, which reduces the accuracy of the obtained low-rank and sparse matrices.

Deep-learning-based methods can extract latent deep features from the HSI to achieve precise mapping of complex scenes through outstanding nonlinear fitting capabilities. The deep-learning-based methods can be categorized into supervised methods and unsupervised methods depending on the availability of labels [37,38]. The supervised methods train classifiers under a plethora of labels to facilitate anomaly detection. However, due to the absence of ground truth labels, these methods use other methods to generate pseudo-labels [39]. For instance, Li et al. [40] utilize reference data with labels to train a convolutional neural network (CNN), extracting differentiating features between spectra. Then, the trained CNN is employed for anomaly detection. Rao et al. [41] propose a hyperspectral anomaly detection method based on a Siamese Transformer network. This method utilizes a small set of prior spectral signatures and spectral unmixing to construct pseudo-labels for training data. The Siamese network is then optimized for anomaly target detection. The unsupervised methods extract deep features from the HSI in the absence of labels. Currently, generative models have been applied in the field of hyperspectral anomaly detection [38]. For example, Xiang et al. [42] introduce a method based on a guided autoencoder network (GAED). This method acquires the guided image through the spectral similarity. Then, the guided image is used to guide the training process of the autoencoder. An anomaly detection result is obtained from the reconstruction error. Wang et al. [43] propose an anomaly detection method based on the autonomous anomaly detection network (Auto-AD). The Auto-AD method reconstructed the background of the HSI by training on random noise and utilized residuals for anomaly detection. Jiang et al. [44] introduce a low-rank embedded network (LREN). In this method, a low-rank prior is introduced into the autoencoder network to guide the network optimization towards the lowest rank for background reconstruction. The anomaly detection result is extracted from the residual image. Fu et al. [45] present a plug-and-play denoising regularization anomaly detection method named DeCNNAD. This method introduces an effective CNN denoiser into low-rank representations to reconstruct a clean background. The above deep-learning-based methods globally model the HSI background as low-rank [8]. These methods have yielded satisfactory results. However, these methods overlook the local features around the anomalies, which are important features in hyperspectral anomaly detection. Training only with global constraints makes it difficult to acquire a clean reconstructed background. This limitation constrains the effectiveness of these methods in anomaly detection.

As analyzed above, the detection performance of the existing traditional methods is limited by the finite feature descriptors and imprecise background statistical models. Moreover, deep-learning-based methods are susceptible to neglecting the surrounding local features of anomalies, which can lead to overfitting of the training. The reconstructed background is then interfered with by anomalies. To solve these problems, motivated by the CRD [31], a local prior is introduced into the deep leaning method. We extend the assumption of the CRD from pixels to features. We posit that anomalous features in HSI are difficult to be represented by local features in the surrounding neighborhoods. So, the neighborhood features make it easy to reconstruct a clean background. To extract the local punctured features, To solve these problems, a hyperspectral anomaly detection method based on differential network is proposed. In the differential network, a novel differential convolution is designed through the difference among two types of kernels. For each type of kernel, the objective is to reconstruct the input HSI instead of the background HSI. Background HSI is estimated through the difference operation. This way avoids the problem that the reconstructed background HSI receives interference from anomalies. Specifically, in the proposed method, first, 5×5 and 3×3 convolutional kernels are employed to extract outer window features and inner window features, respectively. Second, to enhance the expression capability of internal window features for details and crucial information, outer window features are utilized to guide inner window features, and two local guidance attention modules are proposed. Among them, local detail attention strengthens the ability of inner window features to extract edge details, while local Transformer attention enhances the focus ability of inner window features on salient information. Third, the obtained inner window features are subtracted from the outer window features to derive differential features. These differential features can be used to reconstruct a clean background of the HSI. Fourth, by concatenating all differential features along the channel dimension and applying a 1×1 convolution for feature fusion, the reconstructed background HSI is obtained. Finally, the anomaly detection result is derived from the residuals between the input HSI and the reconstructed background HSI.

The main contributions of this work are as follows:

(1) A differential convolutional network is proposed to extract the local punctured neighborhood features for background reconstruction.
(2) To enhance the capability of the network to extract details, a local detail attention module is proposed.
(3) To enhance the focusing ability of the network on important features, a local Transformer attention module is proposed.

This paper is organized as follows. In Section 2, the proposed method is introduced in detail. In Section 3, the experimental details and results are shown, and the effectiveness of the proposed method is discussed. In Section 4, we conclude this paper.

2. Proposed Method

The framework of the proposed method is shown in Figure 1. The 5×5 and 3×3 convolution are used to extract outer window features and inner window features. The proposed local detail attention and local Transformer attention are added to the network for enhancing the inner window features. The differential features are obtained by the difference between the outer window features and inner window features. All the differential features are concatenated and fused to obtain the background HSI. The anomaly detection result is obtained through the residual of the input HSI and background HSI.

In the following sections, we first describe the differential convolutional neural network in detail in Section 2.1. Then, to improve the feature representation capabilities, two local guidance attention modules named local detail attention and local Transformer attention are introduced in Sections 2.2 and 2.3. In Section 2.4, the training loss of the proposed method is described. In Section 2.5, the network prediction and anomaly target extraction are explicated.

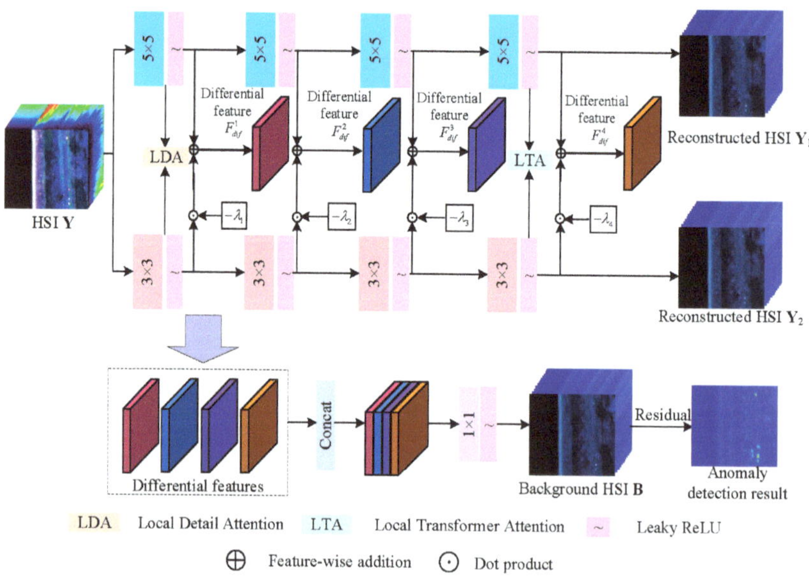

Figure 1. The whole structure of the proposed DifferNet.

2.1. *Differential Convolutional Neural Network*

Recently, many anomaly detection methods based on CNNs and autoencoders have been proposed to obtain satisfactory results. These methods reconstruct a clean background and subsequently detect anomalies from residual images. CNN-based methods utilize simple convolutional kernels to extract the spatial–spectral features of the target. The global low-rank constraints are adopted during network training. However, local features around the anomalies are ignored. As training progresses, there is a possibility that the network indiscriminately learns anomalous features, resulting in the contamination of reconstruction background by anomalies. As such, this training process is uncontrollable. On the other hand, the methods based on autoencoders usually employ autoencoder networks for spectral reconstruction and add spatial features through other strategies. This, to some extent, severs the spatial–spectral relationship in the HSI. To further exploit the local features of anomalies, inspired by CRD, an anomaly detection method based on a differential convolutional neural network (DifferNet) is proposed.

We represent the input HSI as $\mathbf{Y} = (y_1, y_2, \ldots, y_{M \cdot N}) \in \mathbb{R}^{M \times N \times B}$, where M and N denote the spatial dimensions of the HSI, and B represents the spectral dimension. In the input HSI, background pixels can be represented by other pixels within their local neighborhood, while anomalous pixels cannot. Thus, in a local window, the objective function can be expressed as

$$\arg\min_{\theta} \|y_i - f(X_s(y_i);\theta)\|_2^2 + \alpha\|\theta\|_2^2, \quad i = 1, 2, \ldots, M \cdot N \tag{1}$$

where $X_s(y_i)$ represents the set of local punctured neighborhood of pixel y_i, and X_s is the set of local punctured neighborhoods at all the pixels. θ denotes the network parameters to be optimized, $f(\cdot)$ is the mapping function from network input to output, and α is the regularization parameter.

To extract the local punctured neighborhood pixels of the pixel to be tested, a differential convolution method is proposed. The general form of the proposed convolution can be expressed

$$F = w \circledast X_s + b \tag{2}$$

where w is the convolution weights defined on the set X_s, b is the bias defined on the set X_s, $\circledast$ denotes the convolution operation, and F denotes the convolution output feature. X_s can be represented as

$$X_s = X_{out} \ominus X_{in} \tag{3}$$

where X_{out} and X_{in} are the sets of pixels in the local outer window and local inner window centered by pixel y_i, respectively. $\ominus$ denotes the set difference operation. It is worth noting that the convolution weights w and bias b operate on the punctured neighborhood. Therefore, their values are zero on the local inner window X_{in}. Thus, Equation (2) can be expressed as

$$\begin{aligned} F &= w \circledast (X_{out} \ominus X_{in}) + b \\ &= w_{out} \circledast X_{out} - w_{in} \circledast X_{in} + b \\ &= F' - F'' \end{aligned} \tag{4}$$

$$w = w_{out} - w_{in} \tag{5}$$

$$F' = w_{out} \circledast X_{out} + b_{out} \tag{6}$$

$$F'' = w_{in} \circledast X_{in} + b_{in} \tag{7}$$

where w_{out} and w_{in} represent the weights of the convolution defined on X_{out} and X_{in}, respectively. The weights w_{out} and w_{in} have the same values on set X_{in}. Similarly, b_{out} and b_{in} denote the biases of the convolution defined on X_{out} and X_{in}, respectively. Biases b_{out} and b_{in} have the same values on set X_{in}. F' and F'' represent the convolution output features on X_{out} and X_{in}, respectively.

Equation (4) indicates that we can obtain the estimated feature of the test pixel y_i in the local punctured neighborhood X_s by taking the difference of convolution features with different receptive field sizes. To implement such convolution operations, the most direct approach is to extract smaller convolution kernels within a larger receptive field. However, this approach completely neglects inner window pixels in the convolution operation, potentially introducing artificial biases in the results. Therefore, we relax the constraints on convolution kernels w_{out} and w_{in}, allowing for different weights and biases on the set X_{in}. Correspondingly, to enhance the effectiveness of the difference, learnable weighting parameters λ are introduced to the differential operation. Thus, the proposed differential convolution can be expressed as

$$\begin{aligned} g(X_{out}, X_{in}) &= F' - \lambda \times F'' \\ &= w_{out} \circledast X_{out} + b_{out} - \lambda \times (w_{in} \circledast X_{in} + b_{in}) \end{aligned} \quad (8)$$

Specifically, the proposed DifferNet is structured as illustrated in Figure 1. We employ convolution kernels of sizes 5×5 and 3×3 to construct the differential convolution. For the j-th layer, with the outer window input features F_{out}^{j-1} and the inner window input features F_{in}^{j-1}, the differential convolution produces three output features, including outer window output feature F_{out}^{j}, inner window output feature F_{in}^{j}, and differential output feature F_{dif}^{j}. This process can be represented as

$$Z_{out}^{j} = w_{out}^{j} \circledast F_{out}^{j-1} + b_{out}^{j}, j = 1, 2, \ldots. \quad (9)$$

$$F_{out}^{j} = \max\left(\alpha_{slope} \times Z_{out}^{j}, Z_{out}^{j}\right), j = 1, 2, \ldots. \quad (10)$$

$$Z_{in}^{j} = w_{in}^{j} \circledast F_{in}^{j-1} + b_{in}^{j}, j = 1, 2, \ldots. \quad (11)$$

$$F_{in}^{j} = \max\left(\alpha_{slope} \times Z_{in}^{j}, Z_{in}^{j}\right), j = 1, 2, \ldots. \quad (12)$$

$$F_{dif}^{j} = F_{out}^{j} - \lambda_j \times F_{in}^{j}, j = 1, 2, \ldots. \quad (13)$$

where w_{out}^{j} and b_{out}^{j} represent the weights and biases of the j-th 5×5 convolutional layer, respectively. Similarly, w_{in}^{j} and b_{in}^{j} denote the weights and biases of the j-th 3×3 convolutional layer, respectively. Z_{out}^{j} and Z_{in}^{j} are the outputs of the 5×5 convolutional layer and the 3×3 convolutional layer, respectively. λ_j is the weighting coefficient for the j-th differential convolutional layer. Following each convolutional layer, the activation function is the LeakyReLU function, as shown in Equation (10), where α_{slope} is the slope of the negative axis. The input features F_{out}^{0} and F_{in}^{0} are derived from the input HSI $\mathbf{Y}$. The outer window output feature and inner window output feature at the final level contribute to the reconstruction of the HSI.

After all the differential features are obtained, they are concatenated along the channel dimension to obtain the combined differential feature $F_{combine}$. Subsequently, a 1×1 convolutional layer and LeakyReLU activation are employed to fuse all differential features. The predicted background HSI $\mathbf{B}$ is obtained. This process can be described as

$$F_{combine} = f_{Concat}(F_{dif}^{1}, F_{dif}^{2}, F_{dif}^{3}, F_{dif}^{4}) \quad (14)$$

$$\mathbf{B} = f_{LeakyReLU}(f_{Conv}(F_{combine})) \quad (15)$$

where $f_{Concat}(\cdot)$ denotes the channel concatenation operation, $f_{Conv}(\cdot)$ represents the 1×1 convolution operation, and $f_{LeakyReLU}(\cdot)$ is the LeakyReLU activation function.

Due to the differences in receptive fields in the proposed differential convolution, the 5×5 branch and the 3×3 branch exhibit variations in their reconstructive capabilities for HSI. Compared to the 3×3 convolutional layer, the 5×5 convolutional layer can capture information over a broader range, leading to better perception and localization of edge details. Therefore, to enhance the expressive capability of the 3×3 convolutional layer, we propose two guidance attention modules, namely the local detail attention (LDA) and the local Transformer attention (LTA). These modules leverage the features from the 5×5 convolutional layer to guide the features of the 3×3 convolutional layer. Then, the finely detailed fusion features can be obtained. In the following two sections, we will provide a detailed explanation of these two guidance attention modules.

2.2. Local Detail Attention

The proposed structure of differential network successfully divides the traditional background reconstruction into two steps: comprehensive feature representation of input HSI and background features extraction through difference operation. During the feature representation step, the more comprehensive the features extracted by convolutional networks, the more beneficial it is for reconstructing the background. To enhance the capability of network to extract details and acquire comprehensive feature representations of the input HSI, a local detail attention is designed. Traditional convolutional operations typically utilize the same convolution kernel across the entire image. However, this convolutional approach lacks adaptive handling of variations in local details, resulting in a degree of information loss. To effectively capture local details in the HSI, inspired by the work [46], a local detail attention module is proposed.

The proposed local detail attention module is illustrated in Figure 2. Given the input inner window feature $F_{in} \in \mathbb{R}^{M \times N \times C}$ and the input outer window feature $F_{out} \in \mathbb{R}^{M \times N \times C}$, where C is the number of feature channels, the local detail attention learns a set of base convolutional kernels $\mathbf{W} = \{W_i | i = 1, 2, \ldots, n_{base}\} \in \mathbb{R}^{n_{base} \times C \times c_{group} \times k \times k}$ from the input outer window features F_{out}. These base convolutional kernels are associated with fusion coefficients at different positions in the image. Utilizing these base convolutional kernels and fusion coefficients, a fused convolutional kernel is calculated and applied to the image for convolution. Then, the detail attention features are obtained. The obtained detail attention features are added to the input inner window feature F_{in}, yielding the output inner window feature F_{inda}. F_{inda} is used to replace F_{in} as the input for the next-level differential convolution, introducing more detail attention into the inner window features.

The base convolutional kernels $\mathbf{W}$ consist of n_{base} group convolution kernels, each with a size of $C \times c_{group} \times k \times k$. c_{group} represents the number of channels per group in group convolution, and the number of groups is $\frac{C}{c_{group}}$. k denotes the spatial dimension of the convolutional kernel. The group convolution can balance the efficiency and effectiveness of the module.

Specifically, first, to obtain the fusion coefficients for the base convolutional kernels, a fusion attention module is introduced. This module integrates local information through a 3×3 convolution. Subsequently, a SimpleGate activation function [47] is employed for further feature transformation. Following that, a 1×1 convolution is used for feature fusion and channel mapping to n_{base}. A residual connection containing a 1×1 convolution is utilized to supplement additional detail features.

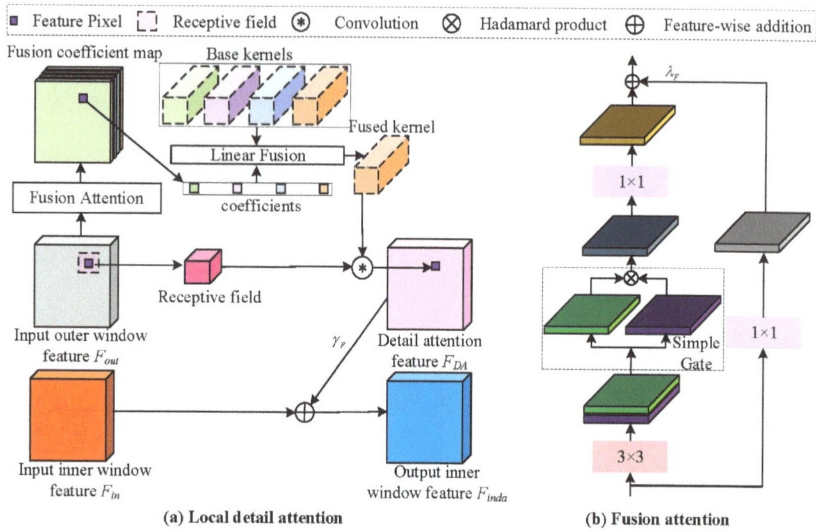

Figure 2. The proposed local detail attention.

Second, after the fusion attention module, the fusion coefficient map $D \in \mathbb{R}^{M\times N\times n_{base}}$ are obtained. Combining with the base convolutional kernels $\mathbf{W}$, the fused convolutional kernel $K(i,j) \in \mathbb{R}^{C\times c_{group}\times k\times k}$ at position (i,j) can be calculated by

$$K(i,j) = \sum_{t=1}^{n_{base}} (D(i,j,t) \times W_t) \tag{16}$$

where $W_t \in \mathbb{R}^{C\times c_{group}\times k\times k}$ represents the t-th convolutional kernel in the base convolutional kernels $\mathbf{W}$, and $D(i,j,t)$ denotes the t-th fusion coefficient at position (i,j) in the fusion coefficient map D.

Third, the obtained fused convolutional kernel $K(i,j)$ is utilized to compute the feature $F_{DA}(i,j) \in \mathbb{R}^{1\times C}$ at position (i,j) in the detail attention feature $F_{DA} \in \mathbb{R}^{M\times N\times C}$.

$$F_{DA}(i,j) = f_{GroupConv}(F_{out}(i,j), K(i,j)) \tag{17}$$

where $f_{GroupConv}(\cdot)$ denotes the group convolution operation, and $F_{out}(i,j)$ represents the features in the receptive field at position (i,j) in the input outer window feature F_{out}.

Finally, the obtained detail attention features F_{DA} are weighted summed with the input inner window feature F_{in} to obtain the output inner window feature F_{inda}.

$$F_{inda} = F_{in} + \gamma_F \times F_{DA} \tag{18}$$

where γ_F represents the learnable weighting parameter. The resulting F_{inda} is used to replace F_{in} as the input for the next-level differential convolution.

The pseudo-code table for the proposed local detail attention is shown in Algorithm 1.

Algorithm 1 Pseudo-code of the proposed local detail attention.

Input:
F_{out}, input outer window feature with shape [1, C, M, N]
F_{in}, input inner window feature with shape [1, C, M, N]

Output: F_{inda}, output inner window feature with shape [1, C, M, N]

Hyperparameters: n_{base} the number of base convolutional kernels

Operators:
FA, fusion attention model
Conv, Convolution

```
Fusion coefficients C = FA(F_out) with shape [1, n_base, M, N]
for i = 1 to M
  for j = 1 to N
    W = zeros([1, C, 5, 5])
    for k = 1 to n_base do
      W = W + C[i, j, k] × W[K]
    end for
    F_DA[i, j, :] = Conv(F_out[i − 2:i + 2, j − 2:j + 2, :], W)
  end for
end for
F_inda = F_in + γ_F × F_DA
```

2.3. Local Transformer Attention

In recent years, Transformer module [48–50] has gained widespread attention and research due to its outstanding performance. It is a global attention module capable of establishing long-range dependencies across images, exhibiting strong feature extraction capabilities. However, the computation of attention weights in Transformer involves matrix multiplication over global pixels, resulting in significant computational and memory requirements. This hinders its application in hyperspectral anomaly detection. Considering that anomalous targets in hyperspectral anomaly detection typically occupy a few pixels in spatial dimension, the establishment of global dependencies for feature extraction is unnecessary. Compared to global information, nonlocal information is more useful for anomaly detection. Therefore, to focus on nonlocal essential information, we modify the way of calculating the attention coefficients and propose a local Transformer attention (LTA) module. Through the LTA model, the network is capable of focusing on salient regions in the scene, enhancing its ability to represent features in the input HSI. This contributes to the comprehensiveness of feature representation, promoting the reconstruction of background HSI.

The proposed local Transformer attention module is illustrated in Figure 3. First, given the input inner window feature $F_{in} \in \mathbb{R}^{M \times N \times C}$ and input outer window feature $F_{out} \in \mathbb{R}^{M \times N \times C}$, a query map $Q \in \mathbb{R}^{M \times N \times C}$ is obtained by performing a 1×1 convolution on F_{out}. Second, two convolutional layers are employed to approximate the matrix multiplication in Transformer, yielding the correlation map $F_{qv} \in \mathbb{R}^{M \times N \times r^2}$. Here, an $r \times r$ convolutional layer is used to capture the nonlocal receptive field and map the feature channels to r^2, where r denotes the spatial size of the nonlocal receptive field. The LeakyReLU function enhances the nonlinear representation capacity. Then, a 1×1 convolutional layer are employed to merge channel features, establishing the correlation of attention coefficients within the receptive field. Third, the correlation coefficients $F_{qv}(i,j)$ at position (i,j) is extracted and reshaped into $r \times r$. Fourth, the Transformer attention feature $F_{TA} \in \mathbb{R}^{M \times N \times C}$ at position (i,j) can be calculated by

$$F_{TA}(i,j) = Softmax(F_{qv}(i,j)) \circledast F_{out}^{r}(i,j) \quad (19)$$

where $F_{out}^{r}(i,j)$ represents the nonlocal region of size r within F_{out} at position (i,j).

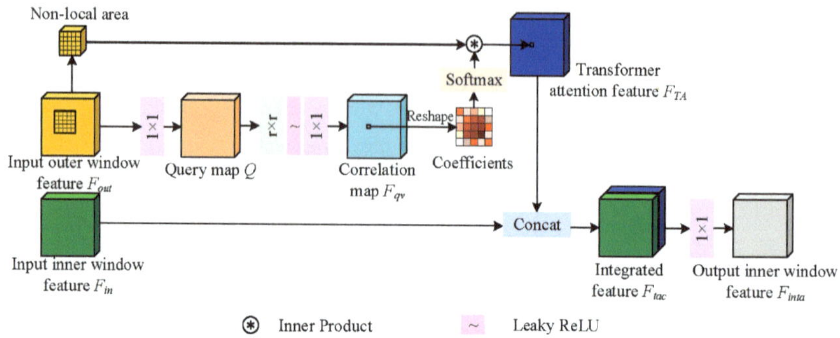

Figure 3. The proposed local Transformer attention.

Fifth, the obtained Transformer attention feature is concatenated with F_{in} along the channel dimension. Then, the integrated feature F_{tac} is obtained. It can be represented by

$$F_{tac} = f_{Concat}(F_{in}, F_{TA}) \tag{20}$$

Finally, through a 1×1 convolutional layer for feature fusion, the output inner window feature $F_{inta} \in \mathbb{R}^{M \times N \times C}$ is obtained. The F_{inta} is used to replace F_{in} as the input for the next-level differential convolution.

The pseudo-code table for the proposed local Transformer attention is shown in Algorithm 2.

Algorithm 2 Pseudo-code of the proposed local Transformer attention.

Input:
- F_{out}, input outer window feature with shape [1, C, M, N]
- F_{in}, input inner window feature with shape [1, C, M, N]

Output: F_{inta}, output inner window feature with shape [1, C, M, N]

Hyperparameters: c_{group} the number of group channels

Operators:
- **Conv**(F, r), feature convolution on feature F with kernels size r
- **act**, Leaky ReLU activation **Cat**, Concatenation

Generate query map Q = **Conv**(F_{out}, 1) with output shape [1, C, M, N]
Extract nonlocal features F_{nl} = **act**(**Conv**(Q, r)) with output shape [1, r^2, M, N]
Map to coefficients F_{qv} = **Conv**(F_{nl}, 1) with output shape [1, r^2, M, N]
for i = 1 to M
  for j = 1 to N
    F_{TA}[i, j, :] = F_{out}[i−r:i+r, j−r:j+r, :]·**Softmax**(F_{qv}[i, j, :])
  end for
end for
F_{tac} = **Cat**(F_{in}, F_{TA})
F_{inta} = **Conv**(F_{tac}, 1)

2.4. Loss Function

In the proposed DifferNet, it is imperative to fully explore features in the HSI, encompassing both anomalous and background features. This approach differs from the existing CNN-based methods, which focus on extracting background features while ignoring anomalous features. The proposed method avoids the problem of background contamination caused by anomalies in traditional networks.

For the outputs of the 5×5 branch $\mathbf{Y}_1 \in \mathbb{R}^{M \times N \times B}$ and the 3×3 branch $\mathbf{Y}_2 \in \mathbb{R}^{M \times N \times B}$, the loss function is defined as follows

$$L_{reconstruct} = SmoothL1(\mathbf{Y}_1, \mathbf{Y}) + SmoothL1(\mathbf{Y}_2, \mathbf{Y}) \tag{21}$$

where $SmoothL1(\cdot)$ is the smooth L1 function [51]. It is calculated by

$$f_{SmoothL1}(x, y) = \begin{cases} 0.5(x-y)^2, & |x-y| < 1 \\ |x-y| - 0.5, & |x-y| > 1 \end{cases} \tag{22}$$

For the predicted background HSI $\mathbf{B}$ of the differential network, as anomalies occupy a small portion in the HSI, we choose the input HSI $\mathbf{Y}$ as the optimization target in consideration of computational efficiency. The loss function of the background is formulated as

$$L_{background} = SmoothL1(\mathbf{B}, \mathbf{Y}) \tag{23}$$

Therefore, the total loss function is

$$L = \frac{1}{2}(L_{reconstruct} + L_{background}) \tag{24}$$

2.5. Anomaly Target Extraction

After the training completion of the proposed DifferNet, the HSI $\mathbf{Y}$ is fed into the network to obtain the predicted background HSI $\mathbf{B} = (b_1, b_2, \ldots, b_{M \cdot N})$. The difference between the input HSI and the predicted background HSI is calculated to obtain the anomaly detection result. The anomaly detection result E can be represented as

$$E_i = \sum_{j-1}^{B}(|y_i(j) - b_i(j)|) \underset{\mathbf{B}}{\overset{\mathbf{T}}{\gtrless}} \xi, i = 1, 2, \ldots, M \cdot N \tag{25}$$

where $y_i(j)$ is the j-th element of the input pixel y_i, $b_i(j)$ is the j-th element of the predicted background pixel b_i, and E_i represents the value of the anomaly detection result E at position i, ξ is the detection threshold to separate the targets $\mathbf{T}$ from the background $\mathbf{B}$.

The pseudo-code table for the proposed method is shown in Algorithm 3.

Algorithm 3 Pseudo-code of the proposed method.

Input: $\mathbf{Y}$, hyperspectral image with shape [1, B, M, N]

Output: E, anomaly detection result

Hyperparameters:
- n_{base} the number of base convolutional kernels
- c_{group}, the number of group channels
- p_{ld}, the position of local detail attention
- p_{lt}, the position of local Transformer attention
- l_r, learning rate
- β_{max}, maximum training epochs
- β_{step}, the number of epochs at which learning rate decays
- σ_{decay}, factor by which learning rate decays

Operators:
- **DN**, differential network with local detail attention and local Transformer attention
- **L**, Loss function
- **norm1**, Vector L1 norm

Algorithm 3 *cont.*

Construct the network using parameters n_{base}, c_{group}, p_{ld}, and p_{lt}.
for i = 1 to β_{max} **do**
  Predicting process $[\mathbf{Y}_1, \mathbf{Y}_2, \mathbf{Y}_{differ}] = \mathbf{DN}(\mathbf{Y})$
  Total loss is calculated by $t_{loss} = \mathbf{L}(\mathbf{Y}_1, \mathbf{Y}_2, \mathbf{Y}_{differ}; \mathbf{Y})$
  Weight gradients of back propagation $G = \frac{\partial t_{loss}}{\partial W}$
  if i **mod** β_{step} is 0 **do**
    $l_r = \sigma_{decay} \times l_r$
  end if
  Weights update by $W = W - l_r \times G$
end for
Background prediction $[\_, \_, \mathbf{B}] = \mathbf{DN}(\mathbf{Y})$
Anomaly detection E = **norm1**($\mathbf{Y}-\mathbf{B}$, dim=1)

3. Results and Disscussion

In this section, different experiments are conducted to prove the superiority of the proposed method. In the experiments, we use eight state-of-the-art methods as comparison methods. The comparison methods include the RX [25], the CRD [31], the 2S-GLRT [29], the PCA-TLRSR [35], the GAED [42], the Auto-AD [43], the LREN [44], and the DeCNNAD [45] methods. Firstly, we introduce the experimental datasets and the evaluation metrics. Then, the parameters of the proposed method and the comparison methods are analyzed. Next, the experimental results of different methods are performed on five real-world datasets. Finally, the structural effectiveness and the module effectiveness are discussed.

3.1. Experimental Datasets

(1) Bay Champagne: This dataset was collected at Bay Champagne, France, by the Airborne Visible/Infrared Imaging Spectrometer (AVIRIS) sensor [52] on 4 July 2010. This HSI is 100 × 100 in size with spatial resolution of 4.4 m, as shown in Figure 4a. The HSI data have 188 bands. All the bands are used in the experiments. The ship in the scene is regarded as an anomaly target.
(2) Pavia: This dataset was collected at Pavia, Italy, by Reflective Optics System Imaging Spectrometer (ROSIS-03) sensor [52]. This HSI is 150 × 150 in size with spatial resolution of 1.3 m, as shown in Figure 4b. The HSI has 102 bands. The cars on the bridge are treated as anomaly targets.
(3) SpecTIR: This dataset was obtained from the SpecTIR hyperspectral aircraft Rochester experiment [53]. This HSI is 180 × 180 in size with spatial resolution of 1 m and 120 in bands with spectral resolution of 5 nm. In the experiments, we select a 100 × 100 area with 120 bands as the experimental dataset, as shown in Figure 4c. The artificial-colored square fabrics are regarded as the anomaly targets.
(4) WHU-Hi-River: This dataset was collected at a long river bank in Honghu, Hubei Province of China on 21 March 2018 [54]. This HSI is 105 × 168 in size with spatial resolution of 6 cm, as shown in Figure 4d. The HSI has 135 bands ranging from 0.4 μm to 1 μm. Two plastic plates and two gray panels are treated as anomaly targets.
(5) MUUFLGulfport: This dataset was collected at University of Southern Mississippi Gulf Park Campus, Long Beach, Mississippi, in November 2010 [55,56]. This HSI is 325 × 220 in size and 72 in bands with 10 nm spectral resolution ranging from 375 nm to 1050 nm, as shown in Figure 5. In the experiments, 64 bands are selected by removing the noise bands. Four cloth panels in the scene are regarded as anomaly targets.

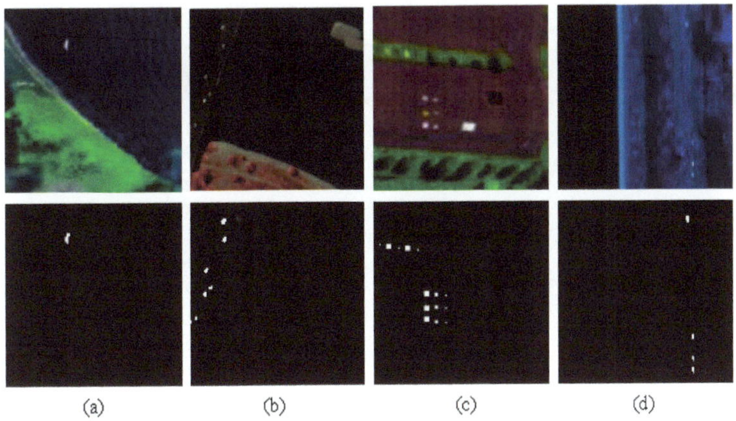

Figure 4. Pseudo-color image and the ground truth of four datasets: (a) Bay Champagne; (b) Pavia; (c) SpecTIR; (d) WHU-Hi-River.

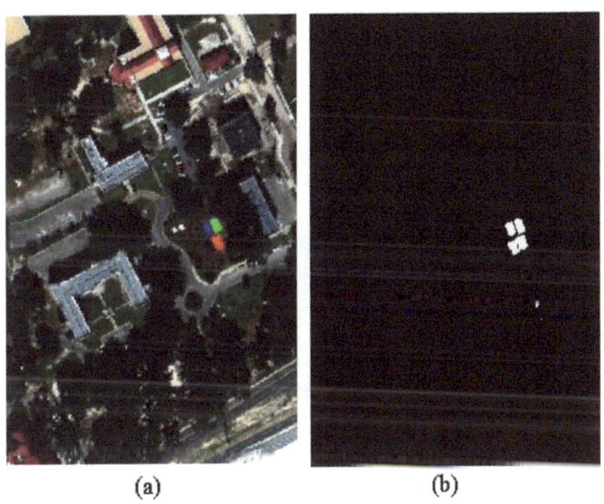

Figure 5. Pseudo-color image and the ground truth of the MUUFLGulfport dataset: (a) pseudo-color image; (b) ground truth.

3.2. Evaluation Metrics

In the experiments, three evaluation metrics are adopted, including three-dimensional receiver operating characteristic curve (3D ROC) [57], two-dimensional ROC curve (2D ROC), and area under the 2D ROC curve (AUC). The 3D ROC curve can illustrate the relationships between the probability of detection (P_D), the probability of false alarm (P_F), and the threshold (τ). The 3D ROC can be decomposed into three 2D curves: (P_D, P_F), (P_D,τ), and (P_F,τ). They can show the detection performance and evaluate separation between anomalies and the background. The AUC values of these three 2D curves are donated by $AUC_{(D,F)}$, $AUC_{(D,\tau)}$, and $AUC_{(F,\tau)}$, respectively. The $AUC_{(D,F)}$ is used to evaluate the overall performance of the detectors. The $AUC_{(D,\tau)}$ is used to evaluate the target detection performance of detectors. The $AUC_{(F,\tau)}$ is used to evaluate the background suppression performance of detectors. The larger the $AUC_{(D,F)}$ and $AUC_{(D,\tau)}$, the better the performance. The smaller the value of $AUC_{(F,\tau)}$, the better the performance.

3.3. Parameter Analysis and Experimental Setup

In this section, the experimental parameters of the proposed method and the comparison methods are performed. The parameters of the eight state-of-the-art comparison methods are set according to the suggestions of the authors or the performance in the

$AUC_{(D,F)}$. Specifically, in the RX method, there are no additional parameters that need to be set. In the CRD method, the sizes of the double window (w_{out}, w_{in}) are set to (7, 5), (7, 5), (25, 21), (9, 5), and (17, 5) on the five datasets of Bay Champagne, Pavia, MUUFLGulfport, SpecTIR, and WHU-Hi-River, respectively. In the 2S-GLRT method, the double window sizes are set to (17, 15), (21, 5), (25, 21), (9, 5), and (11, 7) on the five datasets. In the PCA-TLRSR method, the numbers of principle components are set to 15, 9, 17, 11, and 13 on the five datasets. In the GAED method, the weights and bias are randomly initialized. As suggested by the authors, on all five datasets, the window size c is set to 7, the learning rate l_a is set to 0.4, the penalty coefficient β is set to 1, the number of iterations is set to 300, and the dimension of the hidden layer is set to 25. In the Auto-AD methods, there are no uncertain parameters that need to be set. In the LREN method, the weights and bias are randomly initialized. The number of clusters and the size of hidden layer are set to 7 and 9 on all the datasets as suggested. The regularization parameters are set to 0.01, 0.01, 0.01, 1.0, and 1.0 on the five datasets, respectively. In the DeCNNAD method, the number of clusters is set to 9, 8, 8, 13, and 6 on the five datasets, respectively. The regularization parameters (β, λ) are set to (0.01, 0.01), (0.01, 0.01), (0.0001, 0.0001), (0.001, 0.01), and (0.01, 0.01) on the five datasets, respectively.

For the proposed method, before our experiments, the datasets are first normalized. The weights and bias are randomly initialized. The hyperparameters include learning rate l_r, maximum training epochs β_{max}, the number of epochs at which learning rate decays β_{step}, and the factor by which learning rate decays σ_{decay}. In our experiments, on all five datasets, $l_r = 0.001$, $\beta_{max} = 4000$, $\beta_{step} = 1000$, $\sigma_{decay} = 0.5$. The uncertain parameters include the number of base convolutional kernels n_{base} and the number of group channels in the group convolution c_{group}. In addition, the position of the local detail attention and local Transformer attention also influence the performance of the proposed method. Hence, we analyze the positions p_{ld} and p_{lt} of these two modules, where p_{ld} represents the position of the local detail attention and p_{lt} represents the position of the local Transformer attention. To analyze the stability of these parameters, some parameters are conducted by varying the parameters. Because the weights and bias are randomly initialized, the performance of different experiments will randomly change. So, to reduce the influence of random initialization, we fix the random seed on different datasets during the parameter analysis experiments. This ensures that the performance of the proposed method with different parameters is comparable.

We vary the value of the n_{base} in 2, 4, 8, 16, 32, 64, and 128. The value of the parameter c_{group} is varied in 1, 2, 4, and 8. The optimal results of the n_{base} and c_{group} are shown in Figure 6. To constrain the size of the network, only one module is adopted for both the local detail attention and the local Transformer attention. We vary the values of the p_{ld} and p_{lt} in 1, 2, 3, which means that the position is after the first, second, or third convolution layer in the network, respectively. The optimal results of the p_{ld} and p_{lt} are shown in Figure 7.

As shown in Figure 6, the performance on the datasets of MUUFLGulfport and WHU-Hi-River is stable while changing the n_{base} and c_{group}. The $AUC_{D,F}$ are concentrated above 0.9976 on these two datasets. The range of the $AUC_{D,F}$ is about 0.0008 and 0.0006 on the datasets of MUUFLGulfport and WHU-Hi-River, respectively. On the datasets of Bay Champagne, Pavia, and WHU-Hi-River, the performance is unstable while changing the n_{base} and c_{group}. The range of the $AUC_{D,F}$ is 0.0094, 0.0042, and 0.0110 on these three datasets, respectively. However, most $AUC_{D,F}$ values of these three datasets are concentrated above 009935. According to the $AUC_{D,F}$ performance, the parameters (n_{base}, c_{group}) are set to (8, 1), (32, 8), (32, 1), (8, 4), and (64, 8) on the datasets of Bay Champagne, Pavia, MUUFLGulfport, SpecTIR, and WHU-Hi-River, respectively.

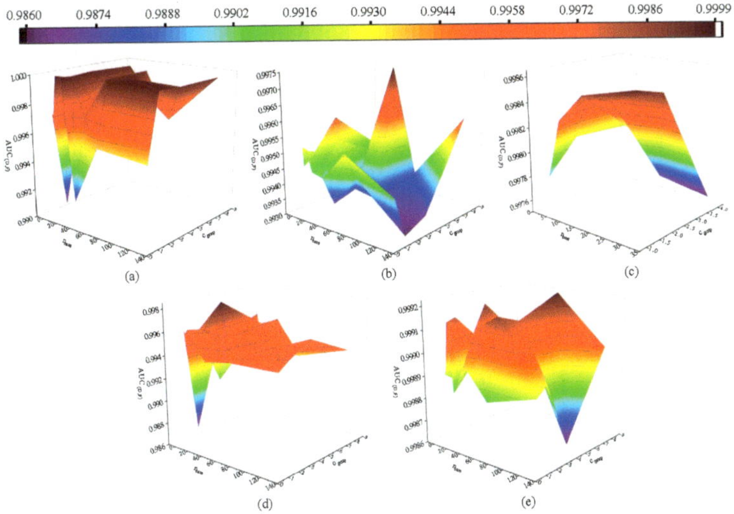

Figure 6. Effects of the number of base convolutional kernels n_{base} and the number of group channels c_{group} on the performance of the proposed method for different datasets: (**a**) Bay Champagne; (**b**) Pavia; (**c**) MUUFLGulfport; (**d**) SpecTIR; (**e**) WHU-Hi-River.

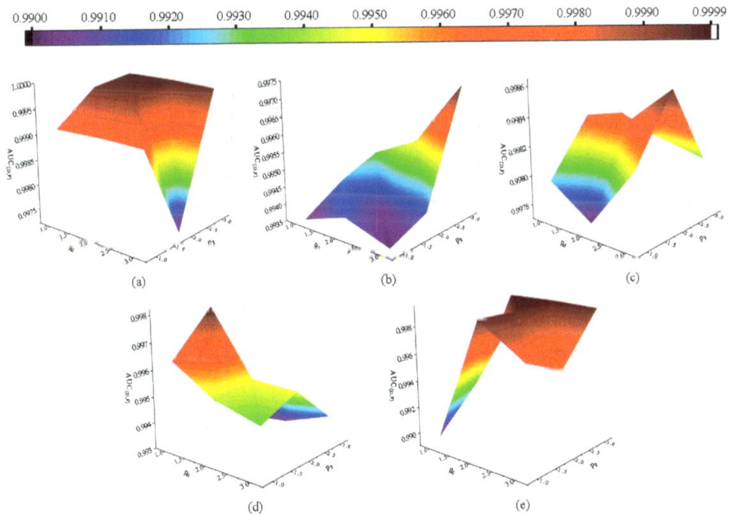

Figure 7. Effects of the local detail attention position p_{ld} and the local Transformer position p_{lt} on the performance of the proposed method for different datasets: (**a**) Bay Champagne; (**b**) Pavia; (**c**) MUUFLGulfport; (**d**) SpecTIR; (**e**) WHU-Hi-River.

As shown in Figure 7, the performance on the five datasets is stable while changing the p_{ld} and p_{lt}. The $AUC_{D,F}$ are concentrated above 0.9930. The range of the $AUC_{D,F}$ is 0.0027, 0.0038, 0.0008, 0.0048, and 0.0096 on the five datasets, respectively. According to the $AUC_{D,F}$ performance, the parameters (p_{ld}, p_{lt}) are set to (1, 3), (3, 3), (3, 2), (1, 2), and (1, 3) on the datasets of Bay Champagne, Pavia, MUUFLGulfport, SpecTIR, and WHU-Hi-River, respectively.

3.4. Experimental Results

In this section, the detection performance of the proposed method is analyzed compared with eight state-of-the-art methods, including the RX, the CRD, the 2S-GLRT, the PCA-TLRSR, the GAED, the Auto-AD, the LREN, and the DeCNNAD. The detection results of these methods are shown in Figure 8.

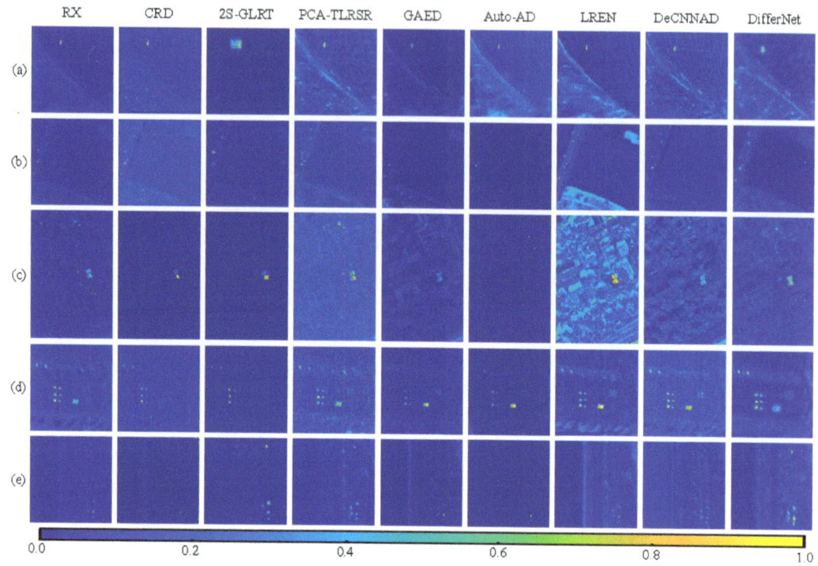

Figure 8. Detection performance of different methods on five datasets: (**a**) Bay Champagne; (**b**) Pavia; (**c**) MUUFLGulfport; (**d**) SpecTIR; (**e**) WHU-Hi-River.

As shown in Figure 8, in the results of the RX method, the background is relatively clean, but the anomaly targets are not completely detected. This is due to the common occurrence of spectral mixing in real datasets, resulting in spectral interference from background pixels around the targets. Such interference can impact the overall detection performance. Similar detection performance also exists in the results of CRD, GAED, and Auto-AD. In the results of 2S-GLRT, the contour information of the anomalies is lost. This is attributed to the fact that the features in the local inner window change slowly as the window slides across the HSI, causing block effects. In the results of the PCA-TLRSR method, there are some noises in the detection results. This is because that the low-rank and sparse decomposition theory is adopted in the PCA-TLRSR. The sparse matrix contains both anomalies and noises. As a result, it is hard for the PCA-TLRSR to distinguish the targets and noises. In the results of the LREN method, the background is clear. This is because this method constructs a global lowest rank dictionary. When the background is complex, the constructed dictionary is not comprehensive enough for the background, resulting in the inability to fully represent background components. This leads to insufficient background suppression. In the results of the DeCNNAD method, the background is clear and there are some noises in the results. This is because a denoiser is adopted in this method to obtain the background dictionary. The quality of the constructed dictionary depends on the performance of the denoiser. When the denoiser removes certain edges in the scene, the reconstructed background also loses these edges, leaving some background edges in the detection results. Furthermore, noises and anomalies are not distinguished in this method. This leads to some noises existing in the detection results.

Specifically, for the Bay Champagne dataset, the anomaly targets in the detection results of RX, GAED, Auto-AD, and LREN are not fully detected. The contours of the anomaly targets in the detection result of 2S-GLRT are lost. The background in the result of

PCA-TLRSR is clear. The detection results of CRD and proposed DifferNet are satisfactory. For the Pavia dataset, the anomaly targets are not well separated from the background in the results of RX, CRD, GAED, Auto-AD, and DeCNNAD. The shape of the targets is lost in the result of 2S-GLRT. The distinctiveness between the anomalies and background is not significant in the result of PCA-TLRSR. The background in the result of LREN is clear. In the result of the proposed DifferNet, all the anomaly targets are highlighted compared with the background. For the dataset of MUUFLGulfport, the Auto-AD method fails to detect the anomaly targets. The background is suppressed in the results of the CRD and 2S-GLRT methods, but the anomaly targets are partially suppressed as well. In the results of the RX and PCA-TLRSR methods, the roof at the top of the scene is enhanced incorrectly. In the results of the GAED and DeCNNAD methods, the anomaly targets are not well distinguished from the background. The background is clear in the result of LREN. In the result of the proposed DifferNet method, the anomaly targets are well separated from the background. For the dataset of SpecTIR, the small targets are submerged in the background in the results of RX, CRD, 2S-GLRT, PCA-TLRSR, LREN, and DeCNNAD. There are some noises in the results of PCA-TLRSR and DeCNNAD. In the results of PCA-TLRSR, GAED, Auto-AD, LREN, and DeCNNAD, a square background in the middle of the scene is not suppressed. In the result of the proposed method, the detection performance is satisfactory. For the dataset of WHU-Hi-River, the anomaly targets are not well detected in the results of CRD, GAED, and Auto-AD. In the results of RX, LREN, and DeCNNAD, the distinction between the targets and background is not significant. The background is clear in the results of PCA-TLRSR, LREN, and DeCNNAD. The shape of the anomaly targets is changed in the result of 2S-GLRT. The proposed DifferNet obtains a satisfactory detection result. Overall, the proposed method obtains superior results on the five datasets compared with other state-of-the-art methods.

From the objective evaluation perspective, 3D ROC curves and AUC values are adopted to evaluate the performance of the methods. The 3D ROC curves (P_D, P_F, τ) on different datasets are shown in Figure 9. The curves of the proposed method marked in red are higher than those of the other methods on the datasets of Bay Champagne and WHU-Hi-River. On the other datasets, the curves of the proposed DifferNet are close to those of the other methods.

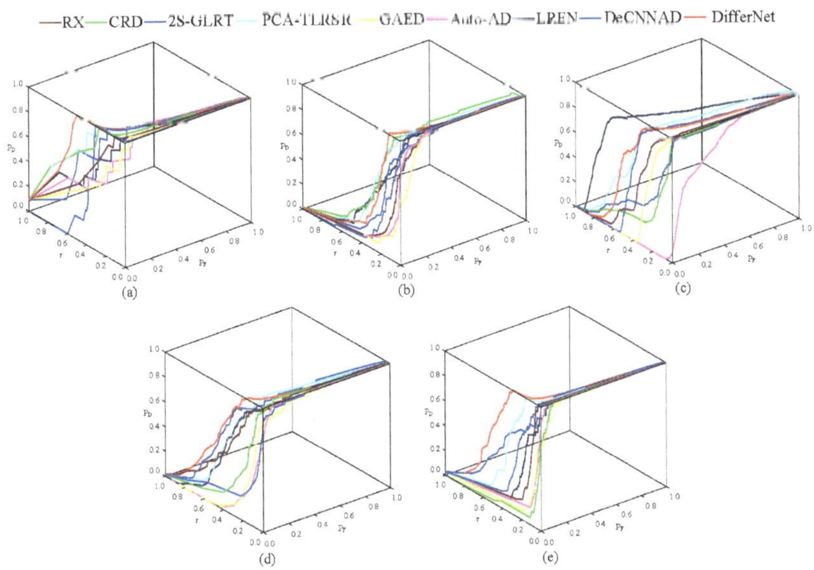

Figure 9. 3D ROC curves (P_D, P_F, τ) on the five datasets: (**a**) Bay Champagne; (**b**) Pavia; (**c**) MUUFLGulfport; (**d**) SpecTIR; (**e**) WHU-Hi-River.

The 2D ROC curves (P_D, P_F) on different datasets are shown in Figure 10. The corresponding $AUC_{(D,F)}$ values are shown in Table 1. As shown in Figure 10, the (P_D, P_F) curves of the proposed DifferNet marked in red are higher than those of the other methods on the datasets of Bay Champagne, Pavia, SpecTIR, and WHU-Hi-River. On the dataset of MUUFLGulfport, the curves of the proposed method and comparison methods are mixed, but the curve of the proposed method is at the upper part of all the curves. In Table 1, the best $AUC_{(D,F)}$ values are in bold. The proposed method achieves the best $AUC_{(D,F)}$ on all the datasets. This means that the proposed method achieves a high detection rate at a low false alarm rate. Compared with the second-best values, the $AUC_{(D,F)}$ values of the proposed method are higher by 0.0001, 0.0105, 0.0006, 0.0110, and 0.0005 on the five datasets, respectively. Overall, the performance of the proposed method is excellent.

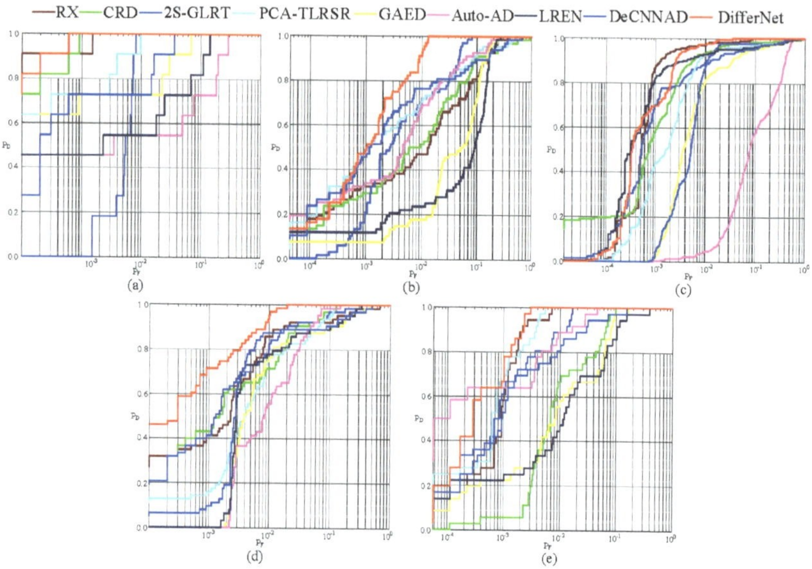

Figure 10. 2D ROC curves (P_D, P_F) on the five datasets: (**a**) Bay Champagne; (**b**) Pavia; (**c**) MUUFLGulfport; (**d**) SpecTIR; (**e**) WHU-Hi-River.

Table 1. The values of $AUC_{(D,F)}$ for different methods on the five datasets, the best values are in blod.

Datasets	Methods								
	RX	CRD	2S-GLRT	PCA-TLRSR	GAED	Auto-AD	LREN	DeCNND	DifferNet
Bay Champagne	0.9998	0.9998	0.9946	0.9985	0.9889	0.9276	0.9669	0.9938	**0.9999**
Pavia	0.9538	0.9453	0.9868	0.9664	0.9348	0.9767	0.9102	0.9601	**0.9973**
MUUFLGulfport	0.9980	0.9886	0.9878	0.9844	0.9760	0.8453	0.9849	0.9819	**0.9986**
SpecTIR	0.9748	0.9870	0.9844	0.9824	0.9664	0.9812	0.9716	0.9710	**0.9980**
WHU-Hi-River	0.9988	0.9802	0.9971	0.9988	0.9716	0.9954	0.9599	0.9899	**0.9993**

The 2D ROC curves (P_D, τ) on different datasets are shown in Figure 11. The corresponding $AUC_{(D,\tau)}$ values are shown in Table 2. As shown in Figure 11, the (P_D, τ) curves of the proposed DifferNet marked in red are higher than those of the other methods on the datasets of Bay Champagne, SpecTIR, and WHU-Hi-River. On the datasets of Pavia and MUUFLGulfport, the curves of the proposed method are not at the top of all the curves. However, the curves of the proposed method are at the upper part of all the curves. In Table 2, the best $AUC_{(D,\tau)}$ values are in bold. The proposed method achieves the best $AUC_{(D,\tau)}$ values on the datasets of Bay Champagne, SpecTIR, and WHU-Hi-River, which are higher by 0.0108, 0.0425, and 0.1588 compared with the second-best values, respectively.

On the datasets of Pavia and MUUFLGulfport, the CRD method and the LREN method achieve the best values, respectively. The best values on the two datasets are higher by 0.0603 and 0.2393 than that of the proposed method, respectively. As a result, the proposed DifferNet has a good detection rate compared with the other methods.

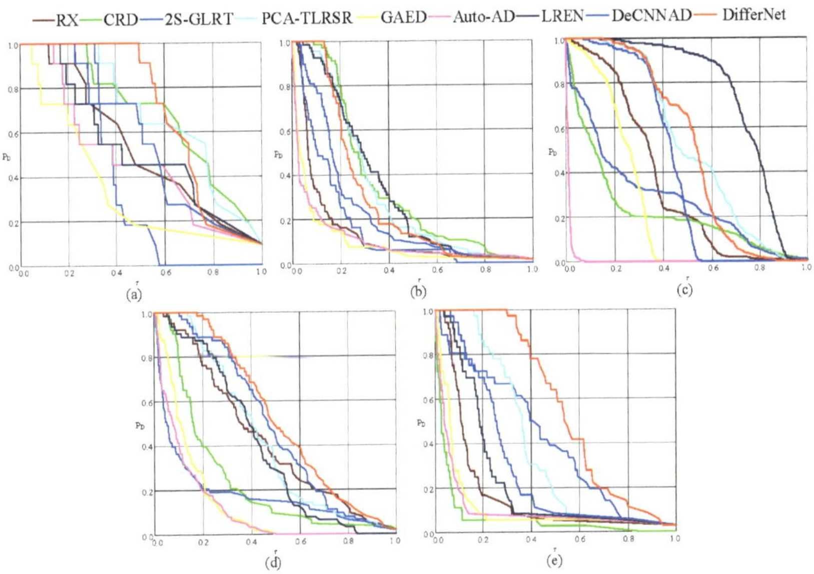

Figure 11. 2D ROC curves (P_D, τ) on the five datasets: (**a**) Bay Champagne; (**b**) Pavia; (**c**) MUUFLGulfport; (**d**) SpecTIR; (**e**) WHU-Hi-River.

Table 2. The values of $AUC_{(D,\tau)}$ for different methods on the five datasets, the best values are in bold.

Datasets	Methods								
	RX	CRD	2S-GLRT	PCA-TLRSR	GAED	Auto-AD	LREN	DeCNND	DifferNet
Bay Champagne	0.5314	0.6854	0.3967	0.6848	0.3422	0.4743	0.5265	0.5642	**0.6962**
Pavia	0.1343	**0.3479**	0.1607	0.3197	0.0924	0.1018	0.3284	0.2254	0.2876
MUUFLGulfport	0.3481	0.2198	0.2804	0.5253	0.2403	0.0110	**0.7499**	0.4175	0.5106
SpecTIR	0.4263	0.2432	0.1815	0.4278	0.1374	0.1149	0.3928	0.4879	**0.5304**
WHU-Hi-River	0.1698	0.0642	0.4098	0.3862	0.1133	0.1049	0.2329	0.2894	**0.5686**

The 2D ROC curves (P_F, τ) on different datasets are shown in Figure 12. The corresponding $AUC_{(F,\tau)}$ values are shown in Table 3. As shown in Figure 12, the (P_F, τ) curves of the proposed DifferNet marked in red are mixed with those of the comparison methods on the five datasets, which are neither the highest nor the lowest. On the datasets of Bay Champagne and MUUFLGulfport, the curve of the 2S-GLRT is the lowest one among all the curves. On the datasets of Pavia and SpecTIR, the curve of the Auto-AD is lower than the other 2D ROC curves. On the dataset of WHU-Hi-River, the curves of CRD and Auto-AD are relatively lower than those of the other methods. In Table 3, the best $AUC_{(F,\tau)}$ values are in bold. On the datasets of Bay Champagne, MUUFLGulfport, and SpecTIR, the 2S-GLRT achieves the minimum $AUC_{(F,\tau)}$ values of 0.0080, 0.0011, and 0.0052, which are lower by 0.0839, 0.0706, and 0.0477 than those of the proposed method. On the datasets of Pavia and WHU-Hi-River, the Auto-AD method achieves the minimum $AUC_{(F,\tau)}$ values of 0.0013 and 0.0008, which are lower by 0.0326 and 0.0496 than those of the proposed method. The $AUC_{(F,\tau)}$ values of the proposed method on the five datasets are between the maximum value and the minimum value, respectively. This means that the proposed method has a certain background suppression capability. Overall, the proposed DifferNet method obtains satisfactory and advanced detection results.

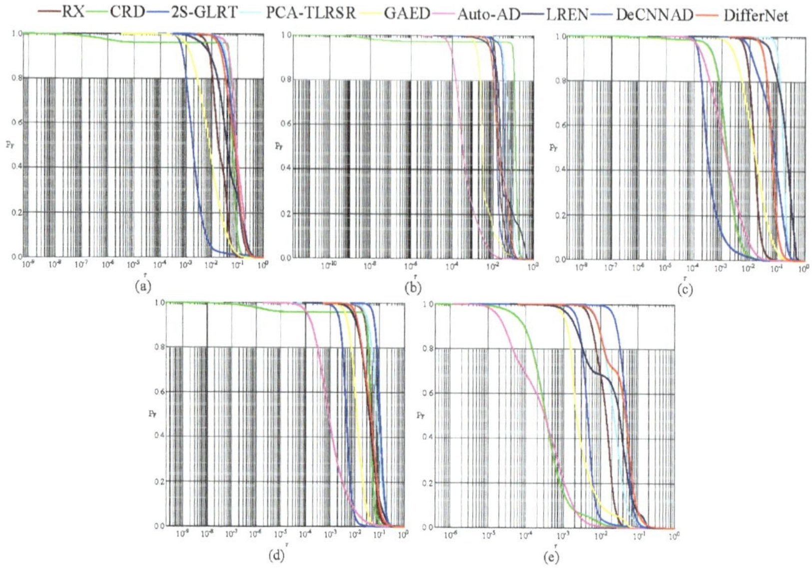

Figure 12. 2D ROC curves (P_F, τ) on the five datasets: (**a**) Bay Champagne; (**b**) Pavia; (**c**) MUUFLGulfport; (**d**) SpecTIR; (**e**) WHU-Hi-River.

Table 3. The values of $AUC_{(F,\tau)}$ for different methods on the five datasets, the best values are in bold.

Datasets	Methods								
	RX	**CRD**	**2S-GLRT**	**PCA-TLRSR**	**GAED**	**Auto-AD**	**LREN**	**DeCNND**	**DifferNet**
Bay Champagne	0.0259	0.0685	**0.0080**	0.0955	0.0160	0.1141	0.0754	0.0900	0.0919
Pavia	0.0233	0.1338	0.0186	0.0748	0.0085	**0.0013**	0.0750	0.0390	0.0339
MUUFLGulfport	0.0172	0.0025	**0.0011**	0.1665	0.0269	0.0034	0.2290	0.0959	0.0717
SpecTIR	0.0555	0.0453	**0.0052**	0.0915	0.0167	0.0053	0.0550	0.1140	0.0529
WHU-Hi-River	0.0150	0.0019	0.0064	0.0286	0.0061	**0.0008**	0.0388	0.0508	0.0504

To evaluate the execution efficiency of all the methods, the time consumption of all the methods on the five datasets is evaluated. It is worth noting that Auto-AD, LREN, and the proposed method are implemented by Python 3. The rest of the methods are conducted using Matlab 2022b. All the methods are executed on a computer with an Intel Core^{TM} i9-12900H produced by Intel the United States, 16 GB RAM produced by Samsung South Korea, and NVDIA GeForce RTX 3060 Laptop GPU produced by Lenovo China. We measure the speed of the methods both with and without GPU acceleration separately. The time consumption results are shown in Table 4, and the best values are in bold. The results show that the RX method has the fastest processing speed without GPU acceleration. With the acceleration of GPU, the DeCNNAD method achieves the best time performance on the dataset of MUUFLGulfport, and the Auto-AD method achieves the best time performance on the rest of the datasets. The time consumption is relatively high compared with other methods without GPU acceleration, and it has an acceptable running speed with GPU acceleration. In the future, we need to further optimize the DifferNet to improve efficiency and detection performance.

Table 4. Time consumptions of different methods on the five datasets, the best values are in bold.

Datasets	GPU	Time(s)								
		RX	CRD	2S-GLRT	PCA-TLRSR	GAED	Auto-AD	LREN	DeCNND	DifferNet
Bay Champagne	✗	**0.0437**	2.7322	173.9472	6.4567	449.0402	58.2303	118.0271	-	626.4889
	✓	-	-	-	-	-	**8.3890**	56.2896	21.5019	36.6918
Pavia	✗	**0.0387**	5.5085	33.5038	8.2466	502.2639	291.2721	231.4080	-	1913.1071
	✓	-	-	-	-	-	**23.0904**	116.5415	26.7872	91.4145
MUUFLGulfport	✗	**0.1687**	366.6264	4602.1543	40.7141	163.3412	1201.6475	638.6995	-	3479.0125
	✓	-	-	-	-	-	61.2890	315.9486	**51.3641**	251.3833
SpecTIR	✗	**0.0344**	4.3991	16.0232	3.8254	270.6453	41.9761	109.0282	-	704.3134
	✓	-	-	-	-	-	**2.9343**	52.4552	23.9450	31.9733
WHU-Hi-River	✗	**0.0780**	117.1494	38.8183	11.0424	346.9338	110.9527	189.1533	-	1606.3941
	✓	-	-	-	-	-	**8.1995**	91.7981	25.3940	78.3617

The initialization of weights and bias in the CNN directly affects the detection performance. To investigate the stability of the proposed method under different initialization conditions, we remove the fixed random seed and conduct 20 repeated experiments. We record the range of $AUC_{(D,F)}$ values obtained, and the statistical results are shown in Table 5.

Table 5. Stability of the proposed method with random initialization.

Datasets	$AUC_{(D,F)}$				
	Maxmum	Minimum	Mean	Variance	Range
Bay Champagne	0.9999	0.9961	0.9990	1.00×10^{-6}	3.80×10^{-3}
Pavia	0.9973	0.9928	0.9944	9.26×10^{-7}	4.50×10^{-3}
MUUFLGulfport	0.9986	0.9949	0.9979	6.11×10^{-7}	3.70×10^{-3}
SpecTIR	0.9980	0.9908	0.9965	2.55×10^{-6}	7.20×10^{-3}
WHU-Hi-River	0.9993	0.9838	0.9981	1.21×10^{-5}	1.56×10^{-2}

As shown in Table 5, the proposed method exhibits relatively low variance and range across 20 experiments. This means that the proposed method has good parameter stability.

To demonstrate the noise robustness of the proposed method, we conduct experiments by adding various types of noise to the original dataset. The added noises are the Gaussian noise with mean 0 and variance 0.01, salt and pepper noise with density 0.01, and uniform multiplicative noise with mean 0 and variance 0.01. The detection results are shown in Figure 13 and Table 6. To further illustrate the false alarm rates of these methods, we also calculate the false positive rate (FPR), while the detection rate is 0.9. The results are shown in Table 7.

Table 6. The values of $AUC_{(D,F)}$ on the five datasets with different noises, the best values are in bold.

Noises	Datasets				
	Bay Champagne	Pavia	MUUFLGulfport	SpecTIR	WHU-Hi-River
no noise	**0.9999**	**0.9973**	**0.9986**	**0.9980**	**0.9993**
Gaussian	0.9535	0.9530	0.9774	0.8907	0.9889
Salt & pepper	0.9943	0.9648	0.9867	0.9439	0.9889
Multiplicative noise	0.9925	0.9280	0.9914	0.9866	0.9921

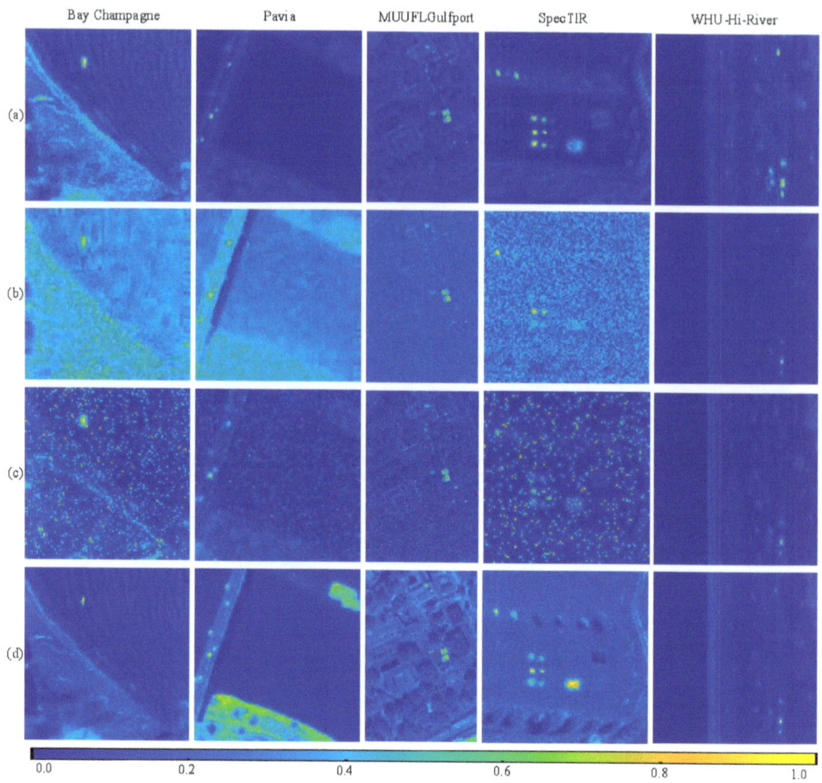

Figure 13. Detection performance on five datasets with different noises: (**a**) without noise; (**b**) with Gaussian noise; (**c**) with salt and pepper noise; (**d**) with uniform multiplicative noise.

Table 7. The values of FPR on the five datasets with different noises, the best values are in bold.

Noises	Datasets				
	Bay Champagne	**Pavia**	**MUUFLGulfport**	**SpecTIR**	**WHU-Hi-River**
no noise	2.00×10^{-4}	**0.0089**	**0.0026**	**0.0086**	**0.0023**
Gaussian	0.0162	0.2046	0.0332	0.3867	0.0293
Salt & pepper	0.0159	0.0638	0.0352	0.0759	0.0336
Multiplicative noise	0.0206	0.1479	0.0129	0.0263	0.0318

As shown in Figure 13, the results of input without noises are the best detection results. When the Gaussian noises are added to the input datasets, the detection performance decreases. When the salt and pepper noises are added to the inputs, the detection results are slightly better than those of inputs with Gaussian noises. When the multiplicative noises are added to the inputs, the detection results are better than those of inputs with salt and pepper noises. Despite the interference of noises with the detection results, the anomaly targets can still be detected. In Table 6, on the datasets of Bay Champagne, Pavia, and WHU-Hi-River, the performance of inputs with salt and pepper noises is better than inputs with the rest of the noises. On the datasets of MUUFLGulfport and SpecTIR, the performance of inputs with multiplicative noises is better than inputs with rest of the noises. The Gaussian noises have the greatest impact on the detection performance. In Table 7, the FPR values of inputs without noise are low. With Gaussian noises, the FPR values increase significantly compared with other noises. The impact of salt and pepper noises and multiplicative noises on the FPR values is similar. As a result, Gaussian noise has a significant impact on the performance of the proposed method, while salt and pepper noise and multiplicative noise have relatively minor effects on the proposed method. The noise robustness of the proposed method is acceptable.

3.5. Ablation Analysis

In this section, the attention module effectiveness and the structure effectiveness are discussed.

To investigate the effectiveness of the attention module, the DifferNet without any attention and DifferNet with only one type of attention in different positions are compared with the proposed DifferNet. The $AUC_{(D,F)}$ is adopted to evaluate the detection performance. The detection results of these methods are shown in Figure 14. In Figure 14, $LTA_i, i = 1,2,3$ denotes that the local Transformer attention (LTA) is added after the i-th convolutional layer, and $LDA_i, i = 1,2,3$ denotes that the local detail attention (LDA) is added after the i-th convolutional layer. Although the results of these methods are close to each other, background edges still remain in the results of DifferNet without attention and DifferNet with LTA. This means that the LDA module can preserve the background edges well so that the residual map contains few background edges. In addition, the anomaly targets are not complete and prominent in the results of DifferNet with LDA. This demonstrates that the LTA is capable of effectively focusing on the primary information within the local window. It can enhance the expressive ability of CNN, making our DifferNet model the background accurately. The corresponding $AUC_{(D,F)}$ values are shown in Table 8. The DifferNet with both LTA and LDA achieves the best values on the five datasets. The experimental results show that the LTA can improve the expressive ability of CNN and facilitate precise modeling of the background. The LDA can enhance the ability of the network to represent background edges. Both attention modules are effective in anomaly detection.

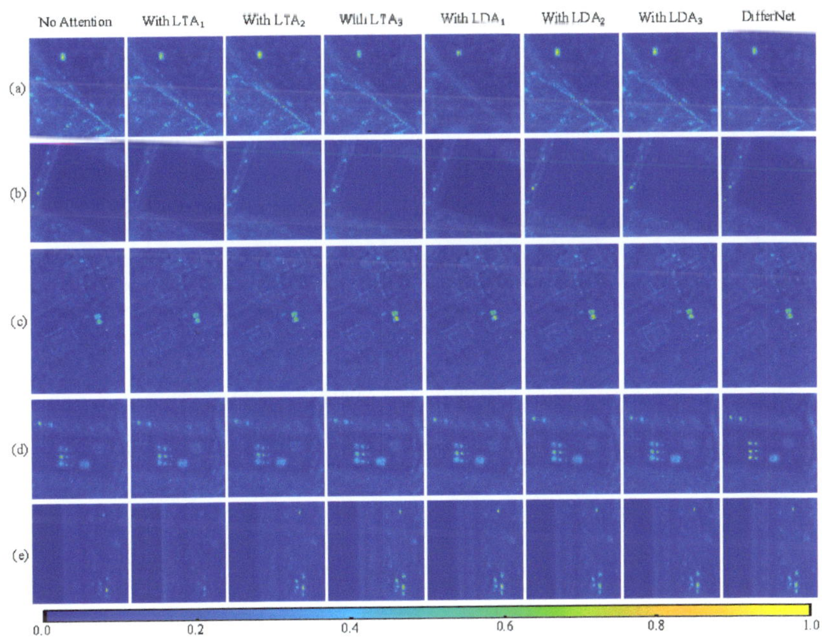

Figure 14. Detection results of the proposed method with different attentions on the five datasets: (**a**) Bay Champagne; (**b**) Pavia; (**c**) MUUFLGulfport; (**d**) SpecTIR; (**e**) WHU-Hi-River.

Table 8. $AUC_{(D,F)}$ values of the proposed method with different attentions on the five datasets, the best values are in bold.

Versions	Datasets				
	Bay Champagne	Pavia	MUUFLGulfport	SpecTIR	WHU-Hi-River
No Attention	0.9991	0.9943	0.9977	0.9959	0.9990
With LTA1	0.9989	0.9944	0.9982	0.9946	0.9923
With LTA2	0.9996	0.9936	0.9982	0.9947	0.9990
With LTA3	0.9995	0.9943	0.9980	0.9942	0.9988
With LDA1	0.9958	0.9930	0.9982	0.9960	0.9990
With LDA2	0.9995	0.9941	0.9983	0.9960	0.9990
With LDA3	0.9994	0.9943	0.9982	0.9959	0.9992
DifferNet	**0.9999**	**0.9973**	**0.9986**	**0.9980**	**0.9993**

To further demonstrate the efficiency of the proposed LTA module, we assess the time consumption and storage occupancy in comparison to the traditional Transformer model. The evaluation is conducted on inputs with spatial sizes of 100×100 and 150×150. The results are shown in Table 9. The results show that the proposed LTA exhibits significant advantages over traditional Transformer in terms of both time consumption and spatial occupancy.

Table 9. Time consumption and storage occupancy of LTA and Transformer, the best values are in bold.

Input Size	Methods	Storage (Byte)	Time(s)	
			CPU	GPU
100 × 100	Transformer	1549.28 M	0.2829	0.0012
	LTA	**372.52 M**	**0.0427**	**0.0009**
150 × 150	Transformer	3919.28 M	1.4296	0.2428
	LTA	**2488.64 M**	**0.1513**	**0.0012**

To investigate the effectiveness of the differential structure of the DifferNet, the network only with 3×3 convolution kernels and the network only with 5×5 convolution kernels are compared with the DifferNet without attention. The loss functions of these methods include a smooth L1 function. The $AUC_{(D,F)}$ is adopted as the evaluation indicator. The detection results are shown in Figure 15. The anomaly targets in the results of 3×3 convolution and 5×5 convolution are not prominent compared with the differential convolution. This means that the differential convolution is effective in preserving anomaly information. The $AUC_{(D,F)}$ values are shown in Table 10. The differential convolution achieves the best detection performance compared with standard 3×3 and 5×5 convolution. As such, the framework of the proposed method is effective and excellent in the anomaly detection field.

Table 10. $AUC_{(D,F)}$ values of the differential convolution and standard convolution on the five datasets, the best values are in bold.

Datasets	Versions		
	5 × 5 Conv	3 × 3 Conv	Differ Conv
Bay Champagne	0.9962	0.9981	**0.9991**
Pavia	0.9918	0.9923	**0.9943**
MUUFLGulfport	0.9899	0.9946	**0.9977**
SpecTIR	0.9877	0.9884	**0.9959**
WHU-Hi-River	0.9976	0.9980	**0.9990**

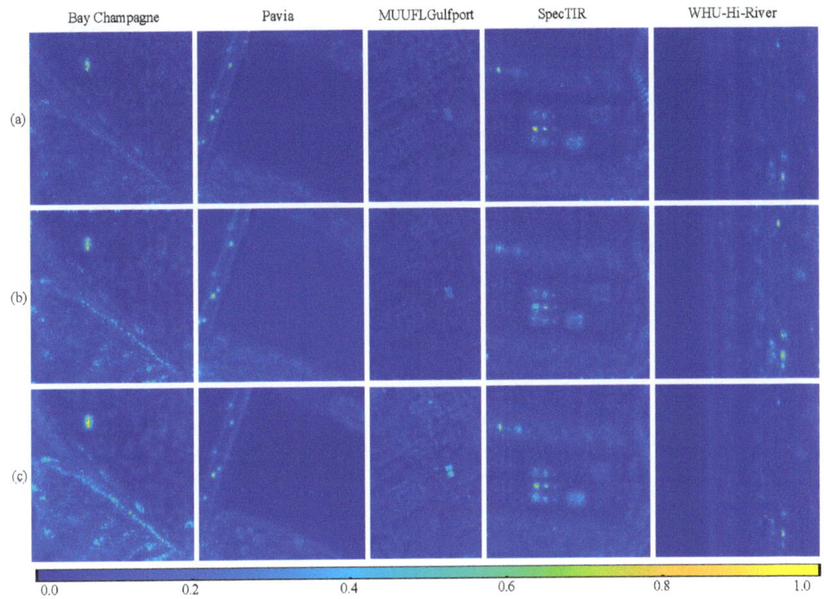

Figure 15. Detection results of the differential convolution and standard convolution on the five datasets: (**a**) 5 × 5 convolution; (**b**) 3 × 3 convolution; (**c**) differential convolution.

To further illustrate the effectiveness of differential convolution for anomaly detection, we apply the differential convolution (DC) to the Auto-AD method. Specifically, all the 3 × 3 convolutions are replaced by 3 × 3 and 5 × 5 convolutions, and the difference operation is adopted in the decoder part of the Auto-AD. The experimental results are shown in Figure 16. The corresponding $AUC_{(D,F)}$ values are shown in Table 11.

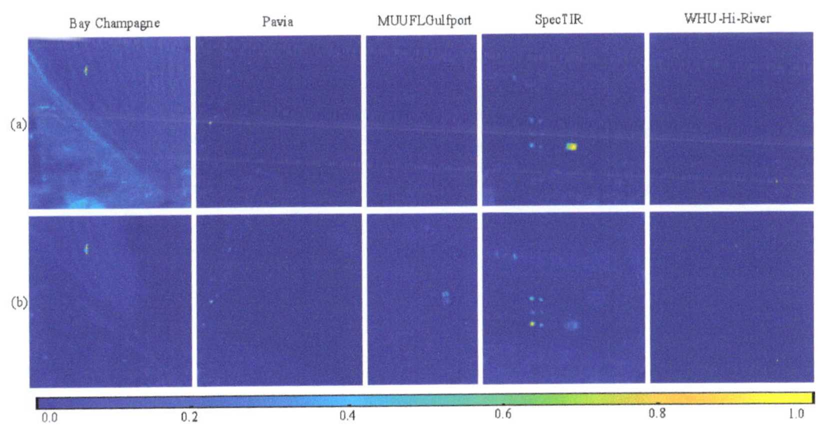

Figure 16. Detection results of the Auto-AD with and without differential convolution on the five datasets: (**a**) results of Auto-AD; (**b**) results of Auto-AD with differential convolution.

As shown in Figure 16, the background in the results of Auto-AD with DC is suppressed. The anomaly targets are highlighted on the five datasets. The performance of the Auto-AD with DC is better than that of Auto-AD. The AUC values in Table 11 show that the differential convolution can improve the performance of Auto-AD on most of the datasets. These experiments demonstrate that the differential convolution is effective.

Table 11. $AUC_{(D,F)}$ values of the Auto-AD with and without differential convolution on the five datasets, the best values are in bold.

Auto-AD	Datasets				
	Bay Champagne	Pavia	MUUFLGulfport	SpecTIR	WHU-Hi-River
with DC	0.9276	0.9767	0.8453	0.9812	**0.9954**
without DC	**0.9997**	**0.9873**	**0.9980**	**0.9934**	0.9911

4. Conclusions

In this work, a hyperspectral anomaly detection method based on differential network is proposed. Firstly, 5×5 and 3×3 convolutional kernels are employed separately to extract outer window features and inner window features. Secondly, to enhance the expression capability of internal window features for details and important information, outer window features are utilized to guide inner window features, and two local guidance attention modules are introduced. Specifically, the local detail attention module enhances the extraction capability of inner window features for edge details, while the local Transformer attention module boosts the focusing ability of inner window features on salient information. Thirdly, weighted difference is applied to the obtained outer and inner window features to derive differential features. These differential features represent local punctured neighborhood characteristics in the HSI, facilitating the reconstruction of a clean background. Fourthly, by concatenating all differential features along the channel dimension and applying a 1×1 convolution for feature fusion, the background HSI is obtained. Finally, anomaly detection results are obtained from the residuals between the input HSI and the background HSI. Comparative experiments with eight state-of-the-art methods are conducted on five real-world datasets. The experimental results demonstrate that the proposed method enhances anomalies while suppressing background. The proposed method achieves superior performance in terms of the detection results, including ROC curves and AUC values. Ablation studies indicate that each part of the proposed method contributes positively to detection performance. The proposed method provides a novel differential structure and decomposes the background reconstruction task into two steps for the first time, including comprehensive feature representation and background feature extraction. It can avoid the problem that anomaly features interfere with the background reconstruction. Furthermore, the two proposed guidance attention modules can ensure the completeness of feature extraction. These methods can provide researchers with inspiring ideas and drive the development in the field of hyperspectral anomaly detection. However, the proposed method performs poorly in terms of execution efficiency, affecting its practical applicability in engineering. In our future work, we plan to conduct further research on network architecture, streamline the structure, and enhance the efficiency of the method while maintaining detection effectiveness.

Author Contributions: Conceptualization, J.Z., P.X. and X.T.; methodology, J.Z.; software, J.Z.; validation, X.T., J.Z. and W.T.; formal analysis, J.Z., P.X. and X.T.; investigation, J.Z.; resources, X.T.; data curation, J.Z. and P.X.; writing—original draft preparation, J.Z.; writing—review and editing, P.X., D.Z. and H.L.; visualization, J.Z. and P.X.; supervision, H.Z.; project administration, H.L., J.S. and H.Z.; funding acquisition, D.Z., H.L., J.S. and H.Z. All authors have read and agreed to the published version of the manuscript.

Funding: This research was funded by the 111 Project (B17035), Aeronautical Science Foundation of China (201901081002), Youth Foundation of Shaanxi Province (2021JQ-182), the Natural Science Foundation of Jiangsu Province (BK20210064), the Wuxi Innovation and Entrepreneurship Fund "Taihu Light" Science and Technology (Fundamental Research) Project (K20221046), the Start-up Fund for Introducing Talent of Wuxi University (2021r007), National Natural Science Foundation of China (62001443, 62105258), Natural Science Foundation of Shandong Province (ZR2020QE294), the Fundamental Research Funds for the Central Universities (QTZX23059, QTZX23009), the Basic

Research Plan of Natural Science in Shaanxi Province (2023-JC-YB-062), Research Scholarship Fund of Xidian University.

Data Availability Statement: Data are available from the corresponding author upon reasonable request or from the data publishers as shown in the paper.

Conflicts of Interest: The authors declare no conflicts of interest.

References

1. Xiang, P.; Zhou, H.; Li, H.; Song, S.; Tan, W.; Song, J.; Gu, L. Hyperspectral anomaly detection by local joint subspace process and support vector machine. *Int. J. Remote Sens.* **2020**, *41*, 3798–3819. [CrossRef]
2. Xiao, Q.; Zhao, L.; Chen, S.; Li, X. Robust Tensor Low-Rank Sparse Representation with Saliency Prior for Hyperspectral Anomaly Detection. *IEEE Trans. Geosci. Remote Sens.* **2023**. [CrossRef]
3. Guan, J.; Lai, R.; Li, H.; Yang, Y.; Gu, L. DnRCNN: Deep recurrent convolutional neural network for HSI destriping. *IEEE Trans. Neural Netw. Learn. Syst.* **2022**, *34*, 3255–3268. [CrossRef]
4. Xiang, P.; Song, J.; Li, H.; Gu, L.; Zhou, H. Hyperspectral anomaly detection with harmonic analysis and low-rank decomposition. *Remote Sens.* **2019**, *11*, 3028. [CrossRef]
5. Zhang, Z.; Hu, B.; Wang, M.; Arun, P.V.; Zhao, D.; Zhu, X.; Hu, J.; Li, H.; Zhou, H.; Qian, K. Hyperspectral Video Tracker Based on Spectral Deviation Reduction and a Double Siamese Network. *Remote Sens.* **2023**, *15*, 1579. [CrossRef]
6. Zhao, D.; Zhu, X.; Zhang, Z.; Arun, P.V.; Cao, J.; Wang, Q.; Zhou, H.; Jiang, H.; Hu, J.; Qian, K. Hyperspectral video target tracking based on pixel-wise spectral matching reduction and deep spectral cascading texture features. *Signal Process.* **2023**, *209*, 109033. [CrossRef]
7. Zhang, Z.; Zhu, X.; Zhao, D.; Arun, P.V.; Zhou, H.; Qian, K.; Hu, J. Hyperspectral Video Target Tracking Based on Deep Features with Spectral Matching Reduction and Adaptive Scale 3D Hog Features. *Remote Sens.* **2022**, *14*, 5958. [CrossRef]
8. Lin, S.; Zhang, M.; Cheng, X.; Shi, L.; Gamba, P.; Wang, H. Dynamic Low-Rank and Sparse Priors Constrained Deep Autoencoders for Hyperspectral Anomaly Detection. *IEEE Trans. Instrum. Meas.* **2023**, *73*, 2500518. [CrossRef]
9. Zhao, D.; Cao, J.; Zhu, X.; Zhang, Z.; Arun, P.V.; Guo, Y.; Qian, K.; Zhang, L.; Zhou, H.; Hu, J. Hyperspectral Video Target Tracking Based on Deep Edge Convolution Feature and Improved Context Filter. *Remote Sens.* **2022**, *14*, 6219. [CrossRef]
10. Xu, K.; Zhang, H.; Li, Y.; Zhang, Y.; Lai, R.; Liu, Y. An ultra-low power tinyml system for real-time visual processing at edge. *IEEE Trans. Circ. Syst. II-Express Briefs* **2023**, *70*, 2640–2644. [CrossRef]
11. Wang, L.; Wang, Y.; Li, Z.; Wu, C.; Xu, M.; Shao, M. Eliminating spatial correlations of anomaly: Corner-Visible Network for Unsupervised Hyperspectral Anomaly Detection. *IEEE Trans. Geosci. Remote Sens.* **2023**, *61*, 5529114. [CrossRef]
12. Zhu, X.; Zhang, H.; Hu, B.; Huang, K.; Arun, P.V.; Jia, X.; Zhao, D.; Wang, Q.; Zhou, H.; Yang, S. DSP-Net: A dynamic spectral-spatial joint perception network for hyperspectral target tracking. *IEEE Geosci. Remote Sens. Lett.* **2023**, *20*, 5510905. [CrossRef]
13. Xiang, P.; Li, H.; Song, J.; Wang, D.; Zhang, J.; Zhou, H. Spectral Spatial complementary decision fusion for hyperspectral anomaly detection. *Remote Sens.* **2022**, *14*, 943. [CrossRef]
14. Wu, Z.; Wang, B. Background Reconstruction via 3D-Transformer Network for Hyperspectral Anomaly Detection. *Remote Sens.* **2023**, *15*, 4592. [CrossRef]
15. Wang, N.; Shi, Y.; Li, H.; Zhang, G.; Li, S.; Liu, X. Multi-Prior Graph Autoencoder with Ranking-Based Band Selection for Hyperspectral Anomaly Detection. *Remote Sens.* **2023**, *15*, 4430. [CrossRef]
16. Li, Y.; Jiang, K.; Xie, W.; Lei, J.; Zhang, X.; Du, Q. A Model-Driven Deep Mixture Network for Robust Hyperspectral Anomaly Detection. *IEEE Trans. Geosci. Remote Sens.* **2023**, *61*, 5522916. [CrossRef]
17. Wang, X.; Wang, Y.; Mu, Z.; Wang, M. FCAE-AD: Full Convolutional Autoencoder Based on Attention Gate for Hyperspectral Anomaly Detection. *Remote Sens.* **2023**, *15*, 4263. [CrossRef]
18. Yang, X.; Tu, B.; Li, Q.; Li, J.; Plaza, A. Graph Evolution-Based Vertex Extraction for Hyperspectral Anomaly Detection. *IIEEE Trans. Neural Netw. Learn. Syst.* **2023**, 1–15. [CrossRef]
19. Su, H.; Wu, Z.; Zhang, H.; Du, Q. Hyperspectral anomaly detection: A survey. *IEEE Geosci. Remote Sens. Mag.* **2021**, *10*, 64–90. [CrossRef]
20. Tu, B.; Yang, X.; Zhou, C.; He, D.; Plaza, A. Hyperspectral anomaly detection using dual window density. *IEEE Trans. Geosci. Remote Sens.* **2020**, *58*, 8503–8517. [CrossRef]
21. Chang, C.I. Hyperspectral anomaly detection: A dual theory of hyperspectral target detection. *IEEE Trans. Geosci. Remote Sens.* **2021**, *60*, 5511720. [CrossRef]
22. Li, S.; Zhang, K.; Duan, P.; Kang, X. Hyperspectral anomaly detection with kernel isolation forest. *IEEE Trans. Geosci. Remote Sens.* **2019**, *58*, 319–329. [CrossRef]
23. Matteoli, S.; Diani, M.; Corsini, G. A tutorial overview of anomaly detection in hyperspectral images. *IEEE Aerosp. Electron. Syst. Mag.* **2010**, *25*, 5–28. [CrossRef]
24. Salem, M.B.; Ettabaa, K.S.; Hamdi, M.A. Anomaly detection in hyperspectral imagery: an overview. In Proceedings of the International Image Processing, Applications and Systems Conference, Sfax, Tunisia, 5–7 November 2014; pp. 1–6.

25. Reed, I.S.; Yu, X. Adaptive multiple-band CFAR detection of an optical pattern with unknown spectral distribution. *IEEE Trans. Acoust. Speech Signal Process.* **1990**, *38*, 1760–1770. [CrossRef]
26. Molero, J.M.; Garzon, E.M.; Garcia, I.; Plaza, A. Analysis and optimizations of global and local versions of the RX algorithm for anomaly detection in hyperspectral data. *IEEE J. Sel. Top. Appl. Earth Observ. Remote Sens.* **2013**, *6*, 801–814. [CrossRef]
27. Kwon, H.; Nasrabadi, N.M. Kernel RX-algorithm: A nonlinear anomaly detector for hyperspectral imagery. *IEEE Trans. Geosci. Remote Sens.* **2005**, *43*, 388–397. [CrossRef]
28. Kwon, H.; Nasrabadi, N.M. Hyperspectral anomaly detection using kernel RX-algorithm. In Proceedings of the 2004 International Conference on Image Processing, 2004, ICIP '04, Singapore, 24–27 October 2004; Volume 5, pp. 3331–3334.
29. Liu, J.; Hou, Z.; Li, W.; Tao, R.; Orlando, D.; Li, H. Multipixel anomaly detection with unknown patterns for hyperspectral imagery. *IEEE Trans. Neural Netw. Learn. Syst.* **2021**, *33*, 5557–5567. [CrossRef]
30. Wang, S.; Hu, X.; Sun, J.; Liu, J. Hyperspectral anomaly detection using ensemble and robust collaborative representation. *Inf. Sci.* **2023**, *624*, 748–760. [CrossRef]
31. Li, W.; Du, Q. Collaborative representation for hyperspectral anomaly detection. *IEEE Trans. Geosci. Remote Sens.* **2014**, *53*, 1463–1474. [CrossRef]
32. Peng, J.; Sun, W.; Li, H.C.; Li, W.; Meng, X.; Ge, C.; Du, Q. Low-rank and sparse representation for hyperspectral image processing: A review. *IEEE Geosci. Remote Sens. Mag.* **2021**, *10*, 10–43. [CrossRef]
33. Su, H.; Wu, Z.; Zhu, A.X.; Du, Q. Low rank and collaborative representation for hyperspectral anomaly detection via robust dictionary construction. *ISPRS-J. Photogramm. Remote Sens.* **2020**, *169*, 195–211. [CrossRef]
34. Zhang, Y.; Du, B.; Zhang, L.; Wang, S. A low-rank and sparse matrix decomposition-based Mahalanobis distance method for hyperspectral anomaly detection. *IEEE Trans. Geosci. Remote Sens.* **2015**, *54*, 1376–1389. [CrossRef]
35. Wang, M.; Wang, Q.; Hong, D.; Roy, S.K.; Chanussot, J. Learning tensor low-rank representation for hyperspectral anomaly detection. *IEEE Trans. Cybern.* **2022**, *53*, 679–691. [CrossRef] [PubMed]
36. Zhang, L.; Ma, J.; Cheng, B.; Lin, F. Fractional fourier transform-based tensor RX for hyperspectral anomaly detection. *Remote Sens.* **2022**, *14*, 797. [CrossRef]
37. Xu, Y.; Zhang, L.; Du, B.; Zhang, L. Hyperspectral anomaly detection based on machine learning: An overview. *IEEE J. Sel. Top. Appl. Earth Observ. Remote Sens.* **2022**, *15*, 3351–3364. [CrossRef]
38. Hu, X.; Xie, C.; Fan, Z.; Duan, Q.; Zhang, D.; Jiang, L.; Wei, X.; Hong, D.; Li, G.; Zeng, X.; et al. Hyperspectral anomaly detection using deep learning: A review. *Remote Sens.* **2022**, *14*, 1973. [CrossRef]
39. Xie, W.; Zhang, X.; Li, Y.; Lei, J.; Li, J.; Du, Q. Weakly supervised low-rank representation for hyperspectral anomaly detection. *IEEE Trans. Cybern.* **2021**, *51*, 3889–3900. [CrossRef]
40. Li, W.; Wu, G.; Du, Q. Transferred deep learning for anomaly detection in hyperspectral imagery. *IEEE Geosci. Remote Sens. Lett.* **2017**, *14*, 597–601. [CrossRef]
41. Rao, W.; Gao, L.; Qu, Y.; Sun, X.; Zhang, B.; Chanussot, J. Siamese transformer network for hyperspectral image target detection. *IEEE Trans. Geosci. Remote Sens.* **2022**, *60*, 5526419. [CrossRef]
42. Xiang, P.; Ali, S.; Jung, S.K.; Zhou, H. Hyperspectral anomaly detection with guided autoencoder. *IEEE Trans. Geosci. Remote Sens.* **2022**, *60*, 5538818. [CrossRef]
43. Wang, S.; Wang, X.; Zhang, L.; Zhong, Y. Auto-AD: Autonomous hyperspectral anomaly detection network based on fully convolutional autoencoder. *IEEE Trans. Geosci. Remote Sens.* **2021**, *60*, 5503314. [CrossRef]
44. Jiang, K.; Xie, W.; Lei, J.; Jiang, T.; Li, Y. LREN: Low-rank embedded network for sample-free hyperspectral anomaly detection. *AAAI Conf. Artif. Intell.* **2021**, *35*, 4139–4146. [CrossRef]
45. Fu, X.; Jia, S.; Zhuang, L.; Xu, M.; Zhou, J.; Li, Q. Hyperspectral anomaly detection via deep plug-and-play denoising CNN regularization. *IEEE Trans. Geosci. Remote Sens.* **2021**, *59*, 9553–9568. [CrossRef]
46. Zhang, Y.; Li, D.; Shi, X.; He, D.; Song, K.; Wang, X.; Qin, H.; Li, H. Kbnet: Kernel basis network for image restoration. *arXiv* **2023**, arXiv:2303.02881.
47. Chen, L.; Chu, X.; Zhang, X.; Sun, J. Simple baselines for image restoration. In *European Conference on Computer Vision*; Springer: Cham, Switzerland, 2022; pp. 17–33.
48. Dosovitskiy, A.; Beyer, L.; Kolesnikov, A.; Weissenborn, D.; Zhai, X.; Unterthiner, T.; Dehghani, M.; Minderer, M.; Heigold, G.; Gelly, S.; et al. An image is worth 16 $\times$ 16 words: Transformers for image recognition at scale. *arXiv* **2020**, arXiv:2010.11929.
49. Vaswani, A.; Shazeer, N.; Parmar, N.; Uszkoreit, J.; Jones, L.; Gomez, A.N.; Kaiser, Ł.; Polosukhin, I. Attention is all you need. *Adv. Neural Inf. Process. Syst.* **2017**, *30*, 5999–6009.
50. Mao, X.; Qi, G.; Chen, Y.; Li, X.; Duan, R.; Ye, S.; He, Y.; Xue, H. Towards robust vision transformer. In Proceedings of the IEEE/CVF Conference on Computer Vision and Pattern Recognition (CVPR), New Orleans, LA, USA, 18–24 June 2022; pp. 12042–12051.
51. Schmidt, M.; Fung, G.; Rosales, R. Fast optimization methods for l1 regularization: A comparative study and two new approaches. In Proceedings of the Machine Learning: ECML 2007: 18th European Conference on Machine Learning, Warsaw, Poland, 17–21 September 2007; Proceedings 18; Springer: Berlin/Heidelberg, Germany, 2007; pp. 286–297.
52. Kang, X.; Zhang, X.; Li, S.; Li, K.; Li, J.; Benediktsson, J.A. Hyperspectral anomaly detection with attribute and edge-preserving filters. *IEEE Trans. Geosci. Remote Sens.* **2017**, *55*, 5600–5611. [CrossRef]

53. Herwegab, J.A.; Kerekesa, J.P.; Weatherbeec, O.; Messingera, D.; van Aardta, J.; Ientiluccia, E.; Ninkova, Z.; Faulringa, J.; Raquenoa, N.; Meolad, J. Spectir hyperspectral airborne rochester experiment data collection campaign. In Proceedings of the SPIE, Bellingham, WA, USA, 27 December 2012.
54. Wang, S.; Wang, X.; Zhong, Y.; Zhang, L. Hyperspectral anomaly detection via locally enhanced low-rank prior. *IEEE Trans. Geosci. Remote Sens.* **2020**, *58*, 6995–7009. [CrossRef]
55. Gader, P.; Zare, A.; Close, R.; Aitken, J.; Tuell, G. *Muufl Gulfport Hyperspectral and Lidar Airborne Data Set*; Tech. Rep. REP-2013-570; University Florida: Gainesville, FL, USA, 2013.
56. Du, X.; Zare, A. *Technical Report: Scene Label Ground Truth Map for MUUFL Gulfport Data Set*; University of Florida: Gainesville, FL, USA, 2017.
57. Chang, C.I. An effective evaluation tool for hyperspectral target detection: 3D receiver operating characteristic curve analysis. *IEEE Trans. Geosci. Remote Sens.* **2020**, *59*, 5131–5153. [CrossRef]

remote sensing

Article

Optimizing Terrain Classification Methods for the Determination of Bedrock Depth and the Average Shear Wave Velocity of Soil

Inhyeok Choi and Dongyoup Kwak *

Department of Civil and Environmental Engineering, Hanyang University, ERICA, 55 Hanyangdaehak-ro, Sangnok-gu, Ansan 15588, Republic of Korea; ssonagbi11@hanyang.ac.kr
* Correspondence: dkwak@hanyang.ac.kr; Tel.: +82-31-400-5141

Abstract: The advancement of remote sensing has enabled the creation of high-resolution Digital Elevation Models (DEMs). Topographic features such as slope gradient (SG), local convexity (LC), and surface texture (ST), derived from DEMs, are related to subsurface geological conditions. In South Korea, bedrock depth ($D_{bedrock}$) and the average shear wave velocity of soil (V_{Ssoil}) serve as metrics for determining the site class, which represents the degree of site amplification in seismic design criteria. These metrics, typically measured through geotechnical and geophysical investigations, require predictive methods for preliminary estimation over large areas. Previous studies developed an automatic terrain classification (AC) scheme using SG, LC, and ST, and subsequent research revealed that terrain classification effectively represents subsurface conditions such as $D_{bedrcok}$ and average shear wave velocity down to 30 m depth. However, AC intrinsically depends on the regional features of DEMs, dividing regions based on nested means of topographic features (SG, LC, and ST). In this study, we developed two terrain classification methods to determine the thresholds of class divisions, aiming to optimize $D_{bedrock}$ and V_{Ssoil} predictions: Sequentially Optimized Classification (SOC) and Non-Sequentially Optimized Classification (NOC). Through the study of the sensitivity of terrain classification methods, smoothing levels, and threshold levels for terrain class generation, we identified the best classification method by comparing it with the geological and mountainous region distribution. Subsequently, we developed DEM-dependent regression models for each class to enhance the accuracy of predicting $D_{bedrock}$ and V_{Ssoil}. The main findings of this study are: (1) the terrain class map suggested in this study represents the distribution of alluvial plane and mountainous regions well, and (2) the DEM calibration for each class provides increased accuracy of $D_{bedrock}$ and V_{Ssoil} predictions in South Korea. We anticipate that the terrain class map, along with $D_{bedrock}$ and V_{Ssoil} maps, will be effectively utilized in geological interpretations and land-use planning for seismic design.

Keywords: DEM; slope; convexity; texture; bedrock depth; average shear wave velocity; terrain classification

Citation: Choi, I.; Kwak, D. Optimizing Terrain Classification Methods for the Determination of Bedrock Depth and the Average Shear Wave Velocity of Soil. *Remote Sens.* **2024**, *16*, 233. https://doi.org/10.3390/rs16020233

Academic Editors: Yanni Dong, Chao Chen and Tao Chen

Received: 25 November 2023
Revised: 27 December 2023
Accepted: 3 January 2024
Published: 6 January 2024

1. Introduction

South Korea's national seismic design standard classifies soil using bedrock depth ($D_{bedrock}$) and the average shear wave velocity of soil layers (V_{Ssoil}) [1]. Ground investigations at new construction sites typically involve drilling surveys to determine $D_{bedrock}$ and geophysical methods such as downhole and suspension P-S logging tests for V_{Ssoil}. However, these ground surveys are often impractical for existing structures and numerous buildings which lack prior data. A preliminary approach that estimates $D_{bedrock}$ and V_{Ssoil} using available topographic data (e.g., elevation, slope) can offer valuable insights for broader area site classifications. Representatively, Wald and Allen [2] and Allen and Wald [3] developed a relationship model for V_{S30} (i.e., time-averaged shear wave velocity down to 30 m depth) using the slope calculated from a Digital Elevation Model (DEM).

In addition, there are studies using machine learning techniques with DEM data to develop a $D_{bedrock}$ map in China [4], and using linear regression analysis based on DEM and slope to develop a V_{S30} model in South Korea [5]. Moreover, there are studies that predict the soil–rock interface and rockhead elevation using machine learning techniques with borehole data [6,7]. As noted, DEM has been the main resource used in the prediction of subsurface conditions.

The advancement of remote sensing has significantly progressed the development of DEM [8]. Traditionally, DEMs were derived from stereo pairs of satellite images, providing global coverage with a moderate resolution [9]. In the 2000s, the Shuttle Radar Topography Mission (SRTM) marked a significant shift by using radar interferometry to achieve a 30 m resolution mapping of the Earth [10]. Further enhancement came with aerial Light Detection and Ranging (LiDAR) systems, which brought DEM accuracy down to centimeter-level precision, proving to be particularly advantageous in forested areas due to their capability to penetrate vegetation [11]. In South Korea, the National Geographic Information Institute (NGII) provides a DEM with a 90 m resolution for the Korean Peninsula and one with a 5 m resolution for South Korea derived from aerial photogrammetry using stereo pairs of satellite images and topography maps [12]. In addition, the advent of modern LiDAR technology has further refined its accuracy, which enables it to provide a 1 m resolution DEM for city regions. As a result, South Korea's DEMs are available in 1 m, 5 m, and 90 m resolutions, with the 90 m resolution DEM being accessible via the National Spatial Data Infrastructure (NSDI) portal [13]. However, the 1 m and 5 m resolution DEMs remain restricted from public access.

As the DEM provides the elevation of the surface, surface characteristics such as slope gradient (SG), local convexity (LC), and surface texture (ST) can be generated from the DEM [14]. These topographic features are instrumental in analyzing surface soil characteristics and the underlying geology. Iwahashi and Pike [14] (IP07) introduced a terrain classification scheme based on the SG, LC, and ST topographic features. They established an automatic terrain classification method using these features to distinguish 16 terrain classes and validated their approach by comparing the classifications with known geology and landforms. Table 1 lists dominant landforms and lithologic units per terrain class in Japan [14]. Low-number classes (1–8) are related to mountains and volcanic landforms, while high-number classes (9–16) are related to soil deposits. These terrain classes effectively represent the geomorphic and geologic features of the surface, but as noted in IP07, the classification's effectiveness might vary depending on regional differences, resolution, and vertical intervals in the DEMs used.

Table 1. Dominant landforms and lithologic units per terrain class in Japan (adapted from Iwahashi and Pike [14]).

Class	Landforms and Lithology	Class	Landforms and Lithology
1	Mountain. Cretaceous accretionary complexes (plutonic rocks)	9	Volcanic hill. Holocene pyroclastic flow deposits
2	Volcano. Holocene mafic volcanic rocks	10	Volcanic footslope. Pleistocene volcanic debris
3	Mountain footslope. Chert (exotic blocks)	11	Valley bottom plain. Pliocene marine sedimentary rocks
4	Mountain footslope. Holocene mafic volcanic rocks	12	Alluvial fan. Holocene sediments
5	Volcanic hill. Pleistocene pyroclastic flow deposits	13	Terrace covered with volcanic ash soil. Pleistocene sediments
6	Volcanic footslope. Pleistocene volcanic debris	14	Alluvial fan. Pleistocene sediments
7	Mountain footslope. Pliocene mafic volcanic rocks	15	Sand dunes. Holocene sediments
8	Mountain footslope. Pleistocene volcanic debris	16	Natural levee. Holocene sediments

Terrain classes from IP07 have been applied in various studies worldwide to deduce geotechnical properties such as the average shear wave velocity in the top 30 m (V_{S30}) and bedrock depth ($D_{bedrock}$). For instance, Vilanova et al. [15] developed a V_{S30} model for Portugal that incorporated terrain classifications to predict seismic site behaviors. Similarly, other studies have utilized DEM-based terrain classifications to map seismic risks and site conditions in diverse regions like Iran [16], Indonesia [17], North Korea [18], and California [19]. Moreover, Furze et al. [20] demonstrated the use of terrain attributes in mapping $D_{bedrock}$ with a random forest algorithm.

This study aims to find terrain classes which correspond to geologic units in South Korea, and infer $D_{bedrock}$ and V_{Ssoil} values based on terrain classes and DEM. We implemented three methods for terrain classification: (1) Automated Classification (AC) aligned with the IP07 approach, (2) Sequentially Optimized Classification (SOC) that defines thresholds to minimize $D_{bedrock}$ standard deviation at each division stage, and (3) Non-Sequentially Optimized Classification (NOC) that employs a Monte-Carlo simulation to identify thresholds, minimizing $D_{bedrock}$ standard deviation across all classes. We utilize a 90 m resolution DEM from the NSDI portal. The 1 m DEM only covers city regions, and the 5 m DEM is not open to public access. The 30 m DEM from the SRTM project [10] is available, but we chose the local DEM to capture the local topographical variations. Using the 90 m DEM, we generated multiple SG, LC, and ST layers for sensitivity analysis of the smoothing window and threshold effects. We further calculated $D_{bedrock}$ by fitting a linear model to the elevation of the DEM at each class. Finally, the resulting terrain classes are compared with geological layers and the prediction performance of $D_{bedrock}$ and V_{Ssoil} are evaluated.

2. Data

This section discusses the ground investigation data utilized in this study and the methods employed for deriving SG, LC, and ST. Ground investigation provides $D_{bedrock}$ and V_{Ssoil}, while SG, LC, and ST are prediction variables used for estimating terrain classes.

2.1. Ground Investigations

The ground investigation data encompass borehole logs with geological strata and Standard Penetration Test (SPT) results, as well as V_S profiles. These elements determine the $D_{bedrock}$ and V_{Ssoil} at a given site. $D_{bedrock}$ is ascertained by identifying the depth to a soft rock layer or a stiffer layer, while V_{Ssoil} is an average of the shear wave velocities of soil layers calculated from V_S travel times. The Geotechnical Information Database System (GeoInfo) in South Korea [21] is a crucial repository for this data, containing 349,869 borehole records as of October 2023. Among these, 115,608 borehole logs reach the bedrock for $D_{bedrock}$ calculations, but only 1891 logs include V_S profiles derived from geophysical investigations. To overcome the deficit in V_S profiles, we have leveraged SPT N-values, which are widely available, correlating them to V_S using the N-V_S model [22]. In the absence of N-values, a median V_S for each soil layer, as defined in Heo and Kwak [22], is utilized to fill in the missing V_S data. Additionally, we filtered out sites that are poorly located due to the potential errors from the coordination system when the raw data are supplied. Ultimately, we produced $D_{bedrock}$ and V_{Ssoil} data for 113,208 sites. Figure 1 presents the distribution of these sites in South Korea.

2.2. Topographic Features

The terrain classification in this study leverages DEM-related topographic features: SG, LC, and ST. These features' calculations are influenced by kernel window size and the thresholds of binary classification. Figure 2 shows the study area, the DEM used in this study, and three example topographic features (SG, LC, and ST) derived from the DEM. The derivation of SG, LC, and ST is described in the following sessions.

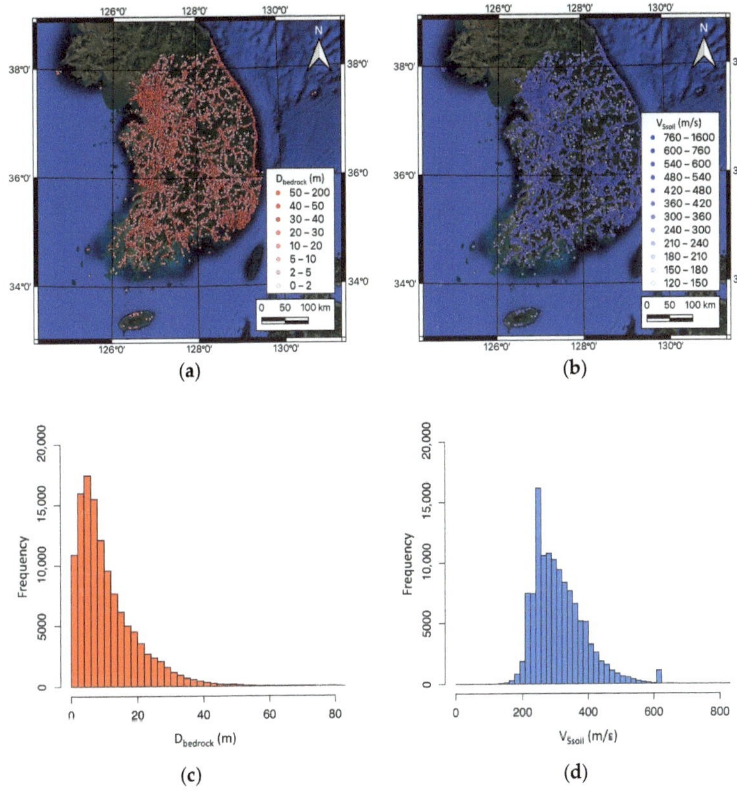

Figure 1. Distribution maps of borehole locations where (**a**) $D_{bedrock}$ and (**b**) V_{Ssoil} are available, and histograms of (**c**) $D_{bedrock}$ and (**d**) V_{Ssoil}.

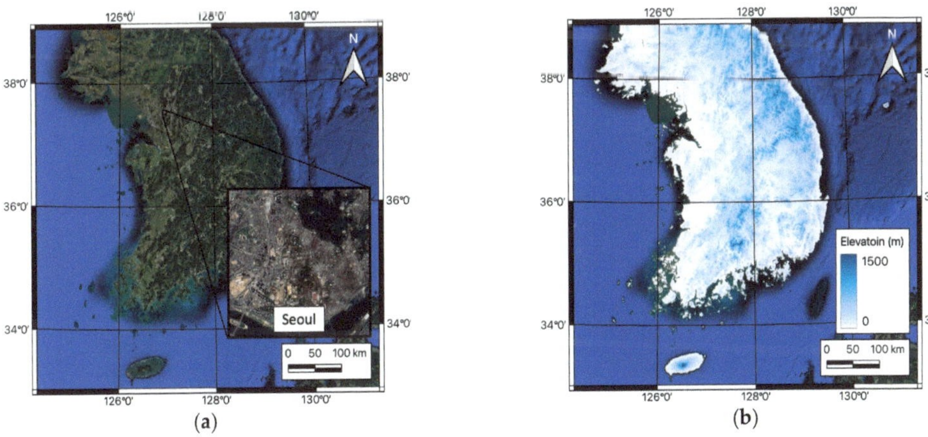

Figure 2. *Cont.*

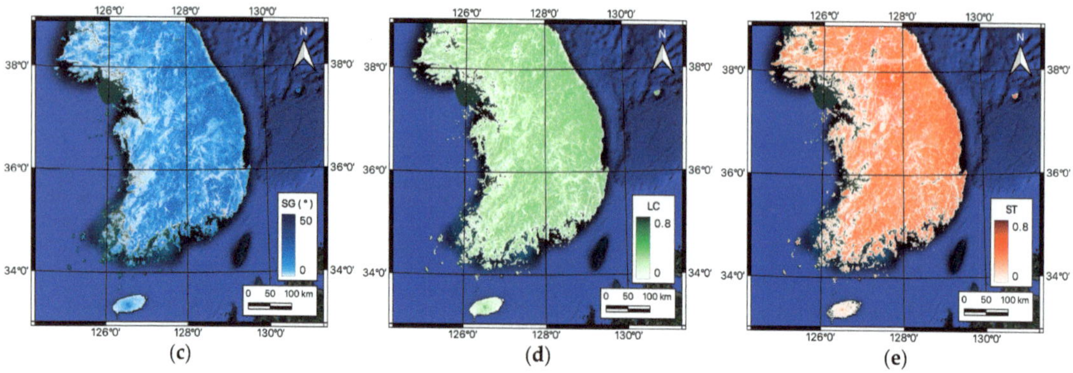

Figure 2. (**a**) Map of South Korea and study area (Seoul), (**b**) DEM used in this study, (**c**) slope gradient (SG), (**d**) local convexity (LC), and (**e**) surface texture (ST) derived from DEM in South Korea.

2.2.1. Slope Gradient (SG)

SG quantifies the change in elevation over a distance. To compute SG, a 3 × 3 kernel matrix, where the sum of each value is zero, is first applied to the same dimension of the elevation matrix to derive dx and dy components as per Equations (1)–(4) [23]:

$$kernel_{SG,x} = \begin{bmatrix} 1 & 0 & -1 \\ 2 & 0 & -2 \\ 1 & 0 & -1 \end{bmatrix} \tag{1}$$

$$kernel_{SG,y} = \begin{bmatrix} 1 & 2 & 1 \\ 0 & 0 & 0 \\ -1 & -2 & -1 \end{bmatrix} \tag{2}$$

$$dx = \frac{[elev] : kernel_{SG,x}}{sum(abs(kernel_{SG,x})) \times res_{DEM}} \tag{3}$$

$$dy = \frac{[elev] : kernel_{SG,y}}{sum(abs(kernel_{SG,y})) \times res_{DEM}} \tag{4}$$

where $kernel_{SG,x}$ and $kernel_{SG,y}$ are kernel matrices for the x and y directions, respectively, and $[elev]$ is a matrix with DEM values (i.e., elevations), and : indicates the inner product of the two matrices. The summation of the absolute values of the kernel matrix and the resolution of DEM (res_{DEM}) are divided into Equations (3) and (4) for normalization. Therefore, slope in radian (slp) is the arctangent of the square root of the sum of the squares of dx and dy (Equation (5)).

$$slp = \tan^{-1}\left(\sqrt{dx^2 + dy^2}\right) \tag{5}$$

SG is the value in degree converted from slp. If the size of kernel window increases, it smooths the SG values. Figure 3 compares SG calculations over different kernel sizes for the Seoul region in South Korea. Among various smooth levels, the SG with a 3 × 3 kernel window is adopted for use in the terrain classification carried out in this study because the SG with a 3 × 3 kernel window does not strongly emphasize the noise from the DEM, as noted in Iwahashi et al. [24].

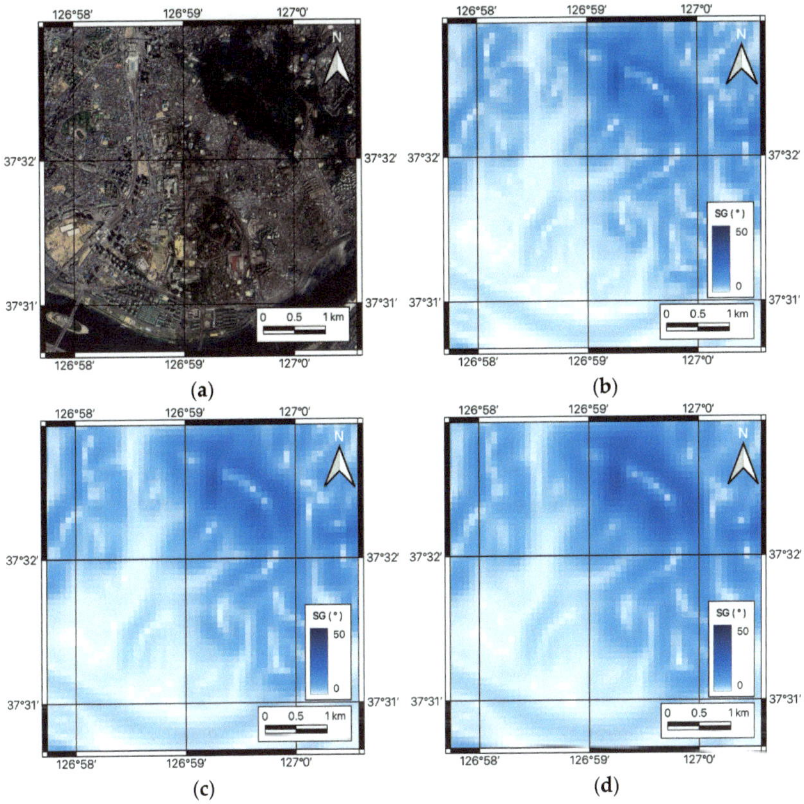

Figure 3. (**a**) Satellite image for the Seoul region and slope gradient maps with (**b**) 3 × 3 kernel window, (**c**) 5 × 5 kernel window, and (**d**) 7 × 7 kernel window.

2.2.2. Local Convexity (LC)

LC measures surface curvature. By applying various kernel matrices to the DEM, LC quantifies the convexity or concavity of the terrain:

$$conv = [elev] : kernel_{LC} \tag{6}$$

where $kernel_{LC}$ is a kernel matrix and $conv$ is the resulting convexity. Positive values of $conv$ indicate a convex shape, while negative values indicate a concave shape. In IP07, rather than using raw $conv$ values, binary values, which are zero if $conv$ is less than a certain threshold and one if $conv$ is greater than the threshold, are averaged within a certain radius to calculate LC. Iwahashi et al. [24] calculated LC using thresholds greater than 0 for the determination of concave shape. In addition, the window size of $kernel_{LC}$ can be varied. For the 3 × 3 window size, the Laplacian kernel is used, and for the 5 × 5 and 7 × 7 sizes, Laplacian of Gaussian kernels are used:

$$kernel_{LC,3\times3} = \begin{bmatrix} 0 & -1 & 0 \\ -1 & 4 & -1 \\ 0 & -1 & 0 \end{bmatrix} \tag{7}$$

$$kernel_{LC,5\times5} = \begin{bmatrix} 0 & 0 & -1 & 0 & 0 \\ 0 & -1 & -2 & -1 & 0 \\ -1 & -2 & 16 & -2 & -1 \\ 0 & -1 & -2 & -1 & 0 \\ 0 & 0 & -1 & 0 & 0 \end{bmatrix} \quad (8)$$

$$kernel_{LC,7\times7} = \begin{bmatrix} 0 & 0 & -1 & -1 & -1 & 0 & 0 \\ 0 & -1 & -3 & -3 & -3 & -1 & 0 \\ -1 & -3 & 0 & 7 & 0 & -3 & -1 \\ -1 & -3 & 7 & 24 & 7 & -3 & -1 \\ -1 & -3 & 0 & 7 & 0 & -3 & -1 \\ 0 & -1 & -3 & -3 & -3 & -1 & 0 \\ 0 & 0 & -1 & -1 & -1 & 0 & 0 \end{bmatrix} \quad (9)$$

Figure 4 shows LCs for the Seoul region with different kernel sizes and thresholds (1, 2, and 3 m). The radius of circular range is determined based on the kernel window sizes (10 grids for 3 × 3, 15 grids for 5 × 5, and 20 grids for 7 × 7 window size). Window sizes larger than the 3 × 3 size (Figure 4d–i) smooth LC too much so that the resolution level is significantly decreased regardless of threshold variation. The higher threshold ignores the low-level concave shape so that it distinguishes flat and concave areas more sharply (e.g., Figure 4c comparing to Figure 4a). Thus, we selected the 3 × 3 window size with 1 to 3 m threshold values for the sensitivity study in the subsequent section.

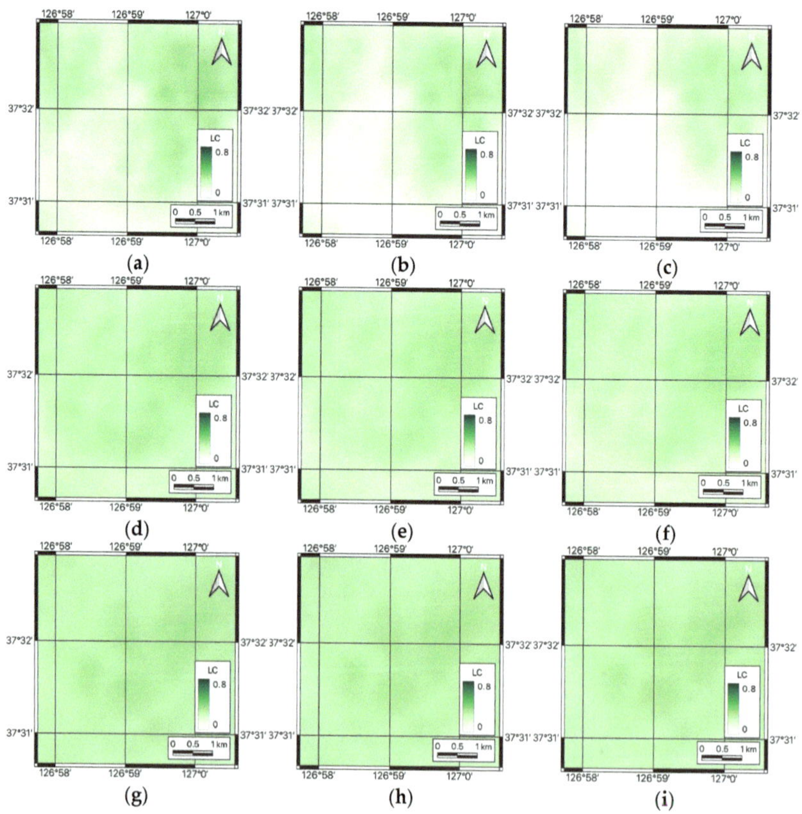

Figure 4. Maps of local convexity for Seoul region with (**a–c**) 3 × 3 kernel windows, (**d–f**) 5 × 5 kernel windows, and (**g–i**) 7 × 7 kernel windows with 1 m (**a**,**d**,**g**), 2 m (**b**,**e**,**h**), and 3 m (**c**,**f**,**i**) thresholds.

2.2.3. Surface Texture (ST)

ST assesses terrain roughness. In IP07, ST is calculated as the average number of either pits or peaks within a certain circular range. If a DEM value is higher or lower than the median within a window, one is assigned to the cell; otherwise, zero is assigned. The average of these binary values within a circular range is determined to be the ST value. An ST close to one indicates a highly textured area, while an ST close to zero indicates flat terrain. This method is highly dependent on the vertical interval of the DEM; if the vertical interval is at the millimeter level, almost every cell would be assigned one, whereas if the vertical interval is large (e.g., 10 m), most cells would be assigned zero. To address this issue, Iwahashi et al. [24] introduced a threshold; the absolute value of the difference between a DEM value and the median must be greater than this threshold for a cell to be assigned as one.

Figure 5 displays the STs for the Seoul region using 3 × 3, 5 × 5, and 7 × 7 window sizes with three different thresholds (1, 2, and 3 m). Similar to the LC calculation, the radius of the circular range is dependent on window sizes (10 grids for 3 × 3, 15 grids for 5 × 5, and 20 grids for 7 × 7 window size). The level of smoothing in the ST correlates with the selected window size, ranging from light smoothing (3 × 3) to heavy smoothing (7 × 7), and the threshold level more clearly distinguishes between flat and textured areas (e.g., Figure 5c comparing to Figure 5a).

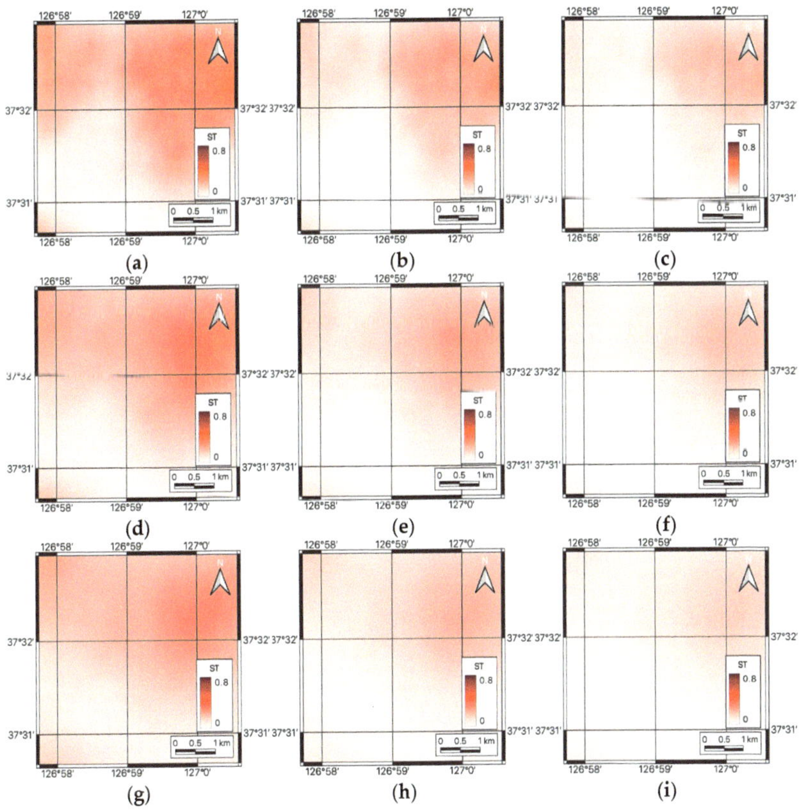

Figure 5. Maps of surface texture for Seoul region with (**a–c**) 3 × 3 kernel windows, (**d–f**) 5 × 5 kernel windows, and (**g–i**) 7 × 7 kernel windows with 1 m (**a**,**d**,**g**), 2 m (**b**,**e**,**h**), and 3 m (**c**,**f**,**i**) thresholds.

2.3. Geological Features

The 1:50,000 geologic maps, collected from the Korean Institute of Geoscience and Mineral Resources (KIGAM) [25], provide a detailed representation of geological formations and alluvial distribution on the surface of the earth. Furthermore, in conjunction with maps of mountainous regions from the national cadastral map collection of the NSDI [13], we found that geological formations in mountainous regions are from an old geologic era, and Quaternary alluvium is distributed around the outside of the mountainous region. Comparing the terrain classification results with these geological maps helps in assessing the accuracy of the topographic features validated in IP07 [14].

Figure 6a shows the distribution of surface geology in the Seoul region. Quaternary alluvium dominates near the river, and a region older than Quaternary (Precambrian and Jurassic) can be seen in the hilly and mountainous areas. The mountainous region map delineates the extent of mountainous terrains (Figure 6b). This map highlights the areas of high elevation, ruggedness, and slope steepness, which are relevant to terrain classification.

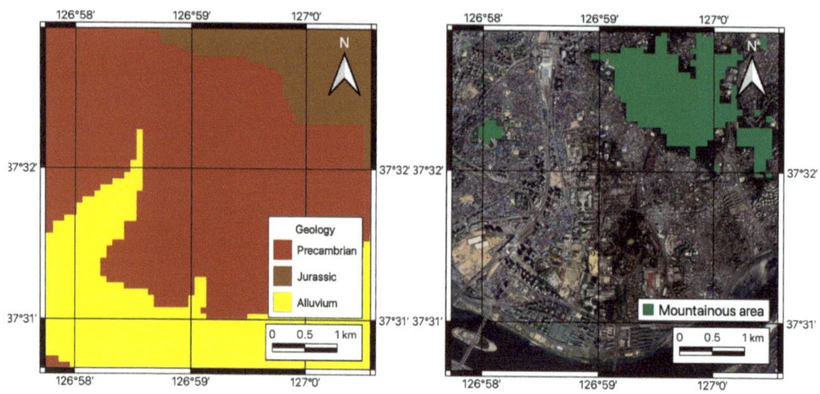

(a) (b)

Figure 6. Maps of (**a**) surface geology from 1:50,000 resolution geology map and (**b**) mountainous region obtained from national cadastral map for the Seoul region.

3. Methods of Terrain Classification

3.1. Analysis Cases

SG, LC, and ST are parameters which are influenced by the size of the smoothing window and threshold values, which yield diverse values and, consequently, different terrain classifications. Lighter smoothing typically results in highly variable SG, LC, and ST values over small areas. The thresholds that determine a positive value for LC and ST influence the level of consideration given to convexity and texture. Therefore, it is crucial to determine which combination of smoothing level and threshold produces terrain classes that best correlate with geologic formation and $D_{bedrock}$ and V_{Ssoil} within the region of South Korea. For SG analysis, a 3 × 3 window size is consistently used. In the case of LC, the kernel window size is fixed at 3 × 3, but the thresholds for determining a positive value vary from 1 to 3 m. For ST, the window sizes range from 3 × 3 to 7 × 7, with corresponding thresholds which also range from 1 to 3 m. This approach results in a total of nine cases, as detailed in Table 2. It is noted that the DEM provided by NSDI has a base resolution of 90 m grid size with a vertical interval measured in micrometers.

Table 2. Analysis of cases with varying filter sizes and vertical intervals for the calculation of slope gradient (SG), local convexity (LC), and surface texture (ST).

Case	Threshold for LC (m)	Window Size [1] and Calculation Radius for ST (grids)	Thresholds for ST (m)
1	1	$3 \times 3, 10$	1
2	1	$3 \times 3, 10$	2
3	1	$3 \times 3, 10$	3
4	2	$5 \times 5, 15$	1
5	2	$5 \times 5, 15$	2
6	2	$5 \times 5, 15$	3
7	3	$7 \times 7, 20$	1
8	3	$7 \times 7, 20$	2
9	3	$7 \times 7, 20$	3

[1] 1 window size = 90 m grid.

3.2. Automated Classification (AC)

Iwahashi and Pike [14] proposed an Automated Classification (AC) method which partitions a study area into 8, 12, or 16 terrain classes based on SG, LC, and ST. These classes, though derived from topographic data, correlate strongly with the landforms and geological structures in the region [14]. This study adopts the 16-class system to accommodate all combinations of SG, LC, and ST values. This system is suitable for a wide area with diverse topographical features [14]. These 16 classes are defined based on three threshold values for SG, LC, and ST, respectively, which are determined through nested means at each classification phase. Figure 7 shows the criteria defining the 16 classes. SG, LC, and ST values are divided sequentially at each phase according to their 1st (SG_{TH1}, LC_{TH1}, ST_{TH1}), 2nd (SG_{TH2}, LC_{TH2}, ST_{TH2}), and 3rd (SG_{TH3}, LC_{TH3}, ST_{TH3}) phase thresholds, forming 16 distinct classes. In the 1st phase, classes 1 to 4 are defined by following steps:

(1) Divide the SG into two groups based on the SG_{TH1}, which is the mean of the whole SG;
(2) Divide the nested LC, where the SG is steeper than SG_{TH1} (SG1), into two groups based on the LC_{TH1}, which is the mean of the nested LC (i.e., LC1 that is greater than LC_{TH1} and LC2 that is lower than LC_{TH1});
(3) Divide the nested ST, where the SG is steeper than SG_{TH1} (SG1), into two groups based on the ST_{TH1}, which is the mean of the nested ST (i.e., ST1 that is greater than ST_{TH1} and ST2 that is lower than ST_{TH1});
(4) Denote regions with [SG1, LC1, and ST1] as Class 1, with [SG1, LC1, ST2] as Class 2, with [SG1, LC2, ST1] as Class 3, and with [SG1, LC2, ST2] as Class 4.

The 2nd phase defines classes 5 to 8 similarly to the 1st phase but using an SG that is gentler than SG_{TH1} and steeper than SG_{TH2} (SG2) with LC_{TH2} and ST_{TH2}. The 3rd phase defines classes 9 to 16, where classes 9 to 12 are defined using SG3, which is an SG gentler than SG_{TH2} and steeper than SG_{TH3} and classes 13 to 16 are defined using SG4, which is an SG gentler than SG_{TH3}. For the 3rd phase, LC_{TH3} and ST_{TH3} are applied to classes 9 to 16.

Figure 8 presents the terrain classification map of the Seoul region, generated using the AC for each case outlined in Table 2. The terrain classifications exhibit minimal variation across the cases. Within the same window size category (Cases 1–3, 4–6, and 7–9), the regions classified as plains (high number classes) extend into areas previously categorized as mountainous or hilly edges (low number classes) with an increasing number of thresholds. Modifying the LC threshold (e.g., Cases 1, 4, and 7) does not substantially alter the extent of the classes; however, there is a slight increase in class numbers in mountainous regions.

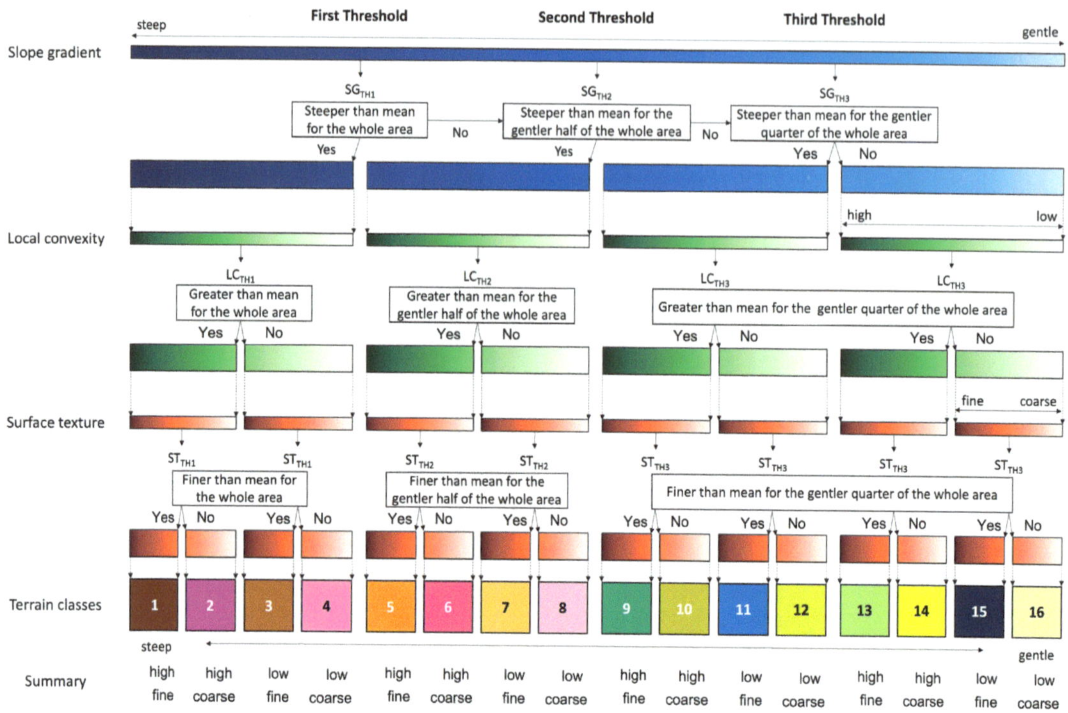

Figure 7. Schematic drawing showing partitioning slope gradients, local convexity, and surface texture based on nested means suggested by Iwahashi and Pike [14].

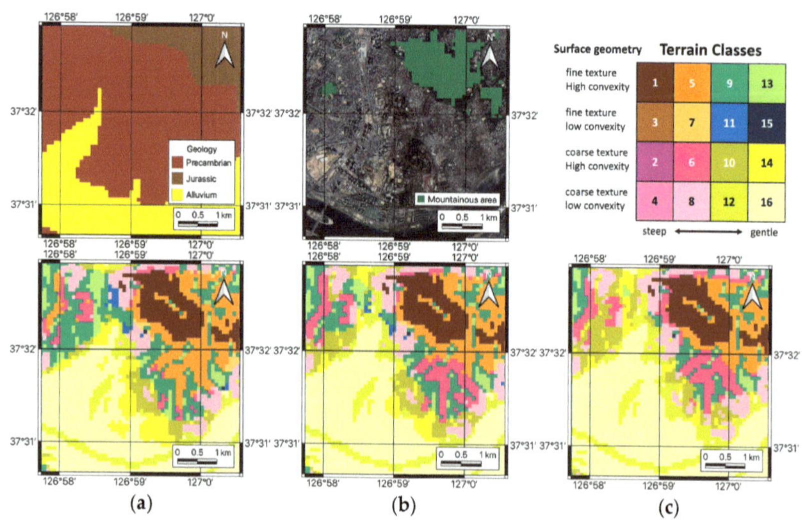

Figure 8. *Cont.*

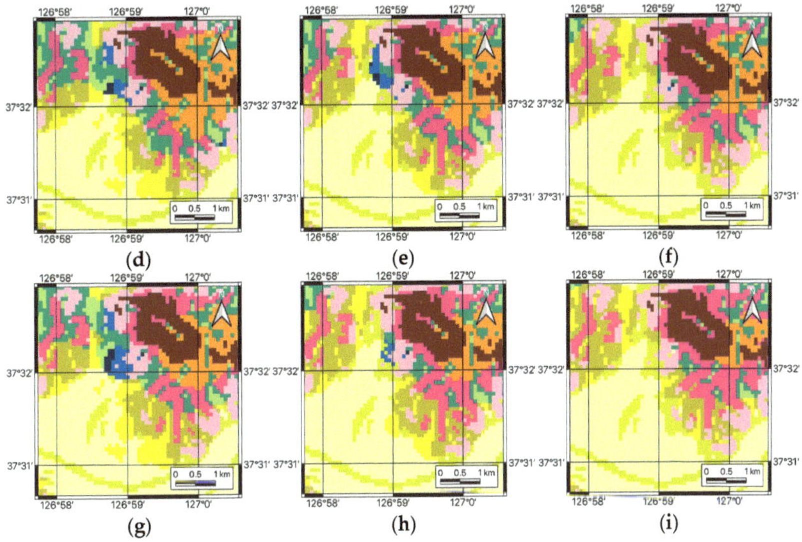

Figure 8. Maps of terrain classes of Seoul region classified using automatic terrain classification scheme proposed by Iwahashi and Pike [14] for Case 1 to Case 9 which correspond to (**a**–**i**).

3.3. Sequentially Optimized Classification (SOC)

The AC method primarily categorizes major terrain types based on SG. Typically, mountainous regions, characterized by steep SG values, are associated with shallow bedrock depth and rapid shear-wave soil velocity. Conversely, areas with gentle SG are often composed of sedimentary layers and landfills, where deeper $D_{bedrock}$ and slower V_{Ssoil} are expected. Consequently, employing SG as a primary classification criterion is justified. However, the AC's approach of utilizing the nested mean as the threshold in South Korea, results in a high SG threshold because most of the area is mountainous, and so the average SG is high. As a result, AC's approach cannot distinguish flatter regions into various classes.

To address this, we introduced Sequentially Optimized Classification (SOC), which determines thresholds for SG, LC, and ST by minimizing the standard deviation in the $D_{bedrock}$ residuals at each classification phase. Residuals are calculated as the difference between the actual $D_{bedrock}$ and the median $D_{bedrock}$ of a terrain group, expressed in natural logarithm. For instance, in the first phase, the optimal SG threshold (SG_{TH1}) is identified by finding the value that minimizes the standard deviation of $D_{bedrock}$ residuals for the two groups—divided into low and high SG_{TH1}. $D_{bedrock}$ predictions for each group are determined by the median value within that group. This procedure is replicated to define optimal thresholds for SG, LC, and ST in subsequent phases.

Figure 9 presents the terrain classification maps for the Seoul region utilizing SOC for each case. Compared to AC, there is a reduction in the terrain class numbers across all regions. The regions with high class number (greater than 9) alluvial deposits are greatly shrunk, and mountainous regions are mostly composed of class numbers 1 to 3 for all cases. The influence of smoothing levels and threshold values on terrain classification is observed to be similar to that in AC-derived classes.

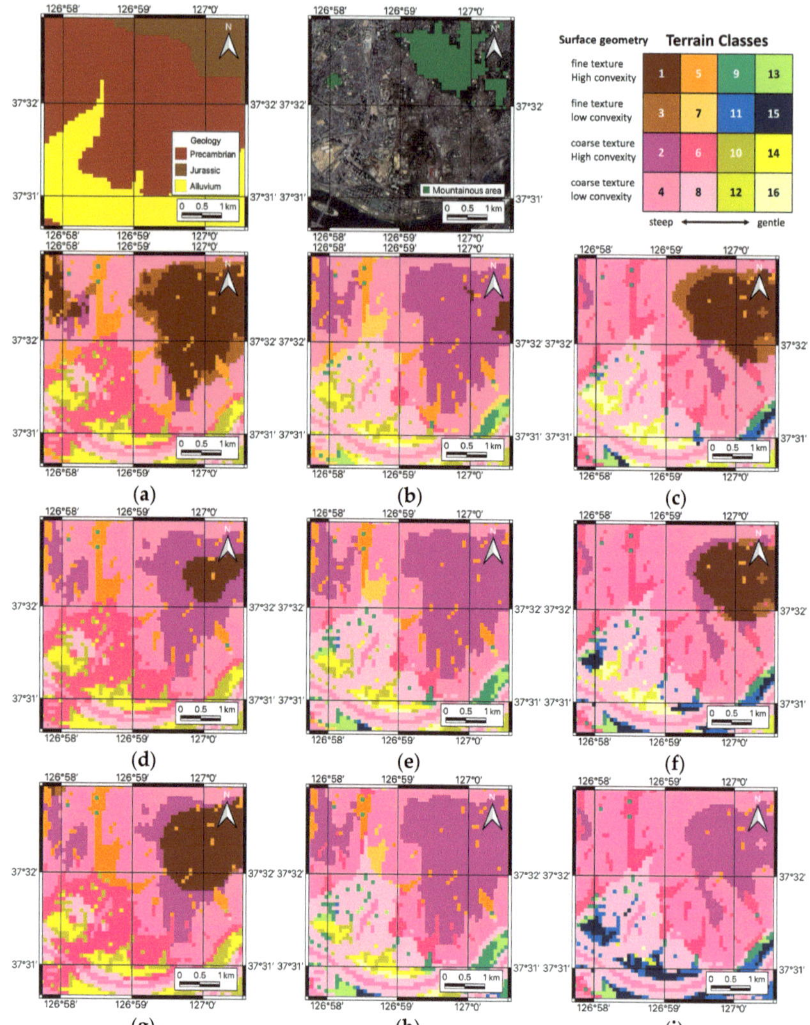

Figure 9. Maps of terrain classes in the Seoul region classified using the Sequentially Optimized Classification scheme for $D_{bedrock}$ prediction in Case 1 to Case 9 which are correspond to (**a**–**i**).

3.4. Non-Sequentially Optimized Classification (NOC)

While AC and SOC begin their classification with SG, followed by LC and ST, implying a dependency of the latter classifications on the initial SG division, the Non-Sequentially Optimized Classification (NOC) approach differs by simultaneously optimizing thresholds for SG, LC, and ST. NOC employs random number generation to minimize the overall standard deviation of $D_{bedrock}$ residuals. To this end, we generated 100,000 sets of uniformly distributed random numbers within the range of 0 to 20 degrees for SG, and 0 to 0.6 for both LC and ST, subsequently identifying the optimized threshold set. This method allows for the determination of SG, LC, and ST thresholds that are not predicated on their 'parent' classification group, unlike the sequential dependencies in AC and SOC.

Figure 10 shows the terrain class maps generated using NOC in the Seoul region for each case. Comparing these to the terrain class maps generated by AC and SOC, the NOC maps show a better match with the actual alluvial plain distribution and mountainous region distribution. In particular, Case 2 (Figure 10b) shows the best match; the top-right area has the lowest class number, which corresponds to the mountainous region, and high class numbers (yellowish colors), indicating flat regions, correspond to Quaternary alluvium. Case 1 (Figure 10a) has a limited area of high class numbers in the flat region. Cases 3 and 7 have similar class distribution to Case 2, but their representation of mountainous regions and alluvial deposits is worse than Case 2. Cases 4 to 6 and 7 to 8 do not represent the geologic distribution well either.

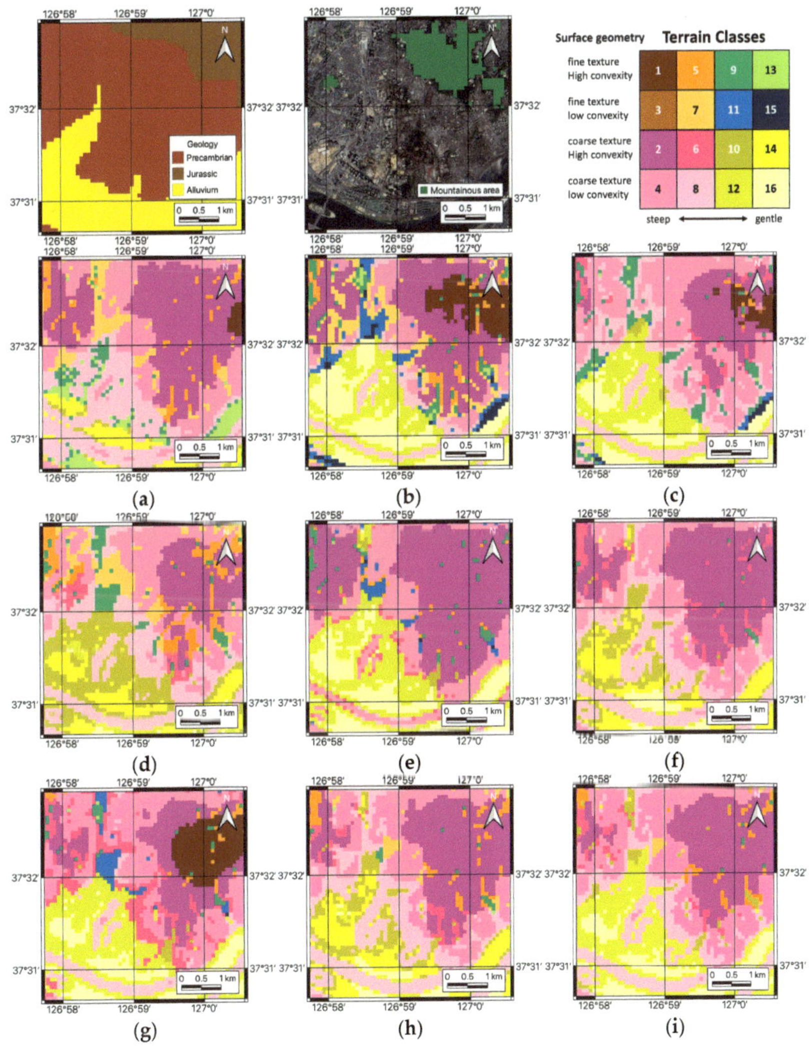

Figure 10. Maps of terrain classes in the Seoul region classified using the Non-Sequentially Optimized Classification scheme for $D_{bedrock}$ prediction in Case 1 to Case 9, which correspond to (**a**–**i**).

4. Results

4.1. Prediction of Bedrock Depth and Average V_S of Soil Layers per Terrain Class

Upon classifying a region, we obtain the mean $D_{bedrock}$ and V_{Ssoil} for each class, calculate the residuals, and assess prediction accuracy by estimating the standard deviation of residuals. SOC follows the classification scheme suggested by IP07 (i.e., AC): it classifies

SG, LC, and ST in order. The difference between SOC and AC is that SOC does not use the class division thresholds as the mean of each topographic feature, but instead uses thresholds which minimize the standard deviations. On the other hand, NOC randomly selects class division thresholds for SG, LC, and ST, and finds the set which minimizes the standard deviations. The main difference between SOC and NOC is the independence of determining the division thresholds. In SOC, a selected division threshold is dependent on the division threshold selected in the previous phase. For example, the division threshold for LC is dependent on the division threshold for SG, and the division threshold for ST is dependent on the division threshold for LC. On the other hand, in NOC, the division thresholds for SG, LC, and ST are not dependent each other because all the thresholds are randomly selected.

To determine the most effective case and method among the nine cases and three methods, we analyzed the standard deviations in $D_{bedrock}$ and V_{Ssoil} residuals, assigning equal weight to each class across all cases with AC, SOC, and NOC, as detailed in Table 3. As a result, Case 1 in the NOC scheme stands out across all methods, but this is due to the very limited number of data points within the classes (only two points for class 11 and six points for class 15) causing low standard deviations. We selected Case 2 (3 $\times$ 3 grid, 2 m threshold) from NOC method as the optimal case due to following reasons; 1) number of data points are at least greater than seven for all the classes, and 2) the distribution of terrain classes corresponds well to the alluvial plane (high number class) and mountainous regions (low number class).

Table 3. Standard deviation (σ_{ln}) in natural log unit of $D_{bedrock}$ and V_{Ssoil} for terrain classes classified using AC, SOC, and NOC schemes for all cases.

Case	AC		SOC		NOC	
	σ_{ln} ($D_{bedrock}$)	σ_{ln} (V_{Ssoil})	σ_{ln} ($D_{bedrock}$)	σ_{ln} (V_{Ssoil})	σ_{ln} ($D_{bedrock}$)	σ_{ln} (V_{Ssoil})
1	0.871	0.234	0.819	0.211	0.721	0.19
2	0.865	0.235	0.807	0.21	0.831	0.229
3	0.869	0.235	0.781	0.211	0.833	0.225
4	0.868	0.234	0.8	0.212	0.833	0.221
5	0.866	0.234	0.791	0.208	0.784	0.232
6	0.864	0.235	0.773	0.209	0.832	0.211
7	0.867	0.234	0.776	0.219	0.833	0.228
8	0.868	0.235	0.793	0.206	0.824	0.218
9	0.864	0.236	0.775	0.21	0.819	0.22

Table 4 presents the thresholds defined by AC, SOC, and NOC for Case 2. AC's SG thresholds are significantly higher than those of the other methods. Given the predominantly mountainous terrain of the Korean Peninsula, this leads to excessively high SG thresholds in AC, failing to accurately represent the relatively flat terrains, and consequently resulting in a greater standard deviation for $D_{bedrock}$ and V_{Ssoil} (refer to Table 3). On the other hand, the SG threshold in SOC's second phase is too low, so the flatter area is occupied with low class numbers. The SG thresholds in NOC are well distributed. Thus, even though the SOC scheme provides a lower standard deviation for $D_{bedrock}$ and V_{Ssoil} in general, we selected NOC to match the geological representation.

Table 4. Thresholds classifying slope gradient (SG), local convexity (LC), and surface texture (ST) of terrain classes using the Automated (AC), Sequentially Optimized (SOC), and Non-Sequentially Optimized Classification (NOC) schemes for Case 2.

Scheme/Case	Phase	SG (deg)	LC	ST
AC/Case 2	1	11.08	0.254	0.234
	2	4.67	0.205	0.174
	3	1.86	0.152	0.109
SOC/Case 2	1	2.1	0.214	0.263
	2	0.32	0.154	0.046
	3	0.12	0.071	0.030
NOC/Case 2	1	3.77	0.188	0.232
	2	1.88	0.242	0.079
	3	0.58	0.197	0.048

Figure 11 displays boxplots for $D_{bedrock}$ and V_{Ssoil} and Table 5 lists the number of data points and the median and standard deviation of $D_{bedrock}$ and V_{Ssoil} for each terrain class under NOC for Case 2. The medians for $D_{bedrock}$ and V_{Ssoil} are calculated by first removing outliers, then computing the median for each group. This outlier elimination helps to exclude anomalous $D_{bedrock}$ and V_{Ssoil} data that were potentially caused by coordinate errors in the raw dataset. The median $D_{bedrock}$ increases with the higher number of terrain classes which correspond to the flat region. For V_{Ssoil}, the median values become slower for high number of classes, but the difference between the steep region (low number classes) and the flat region (high number classes) is not significant.

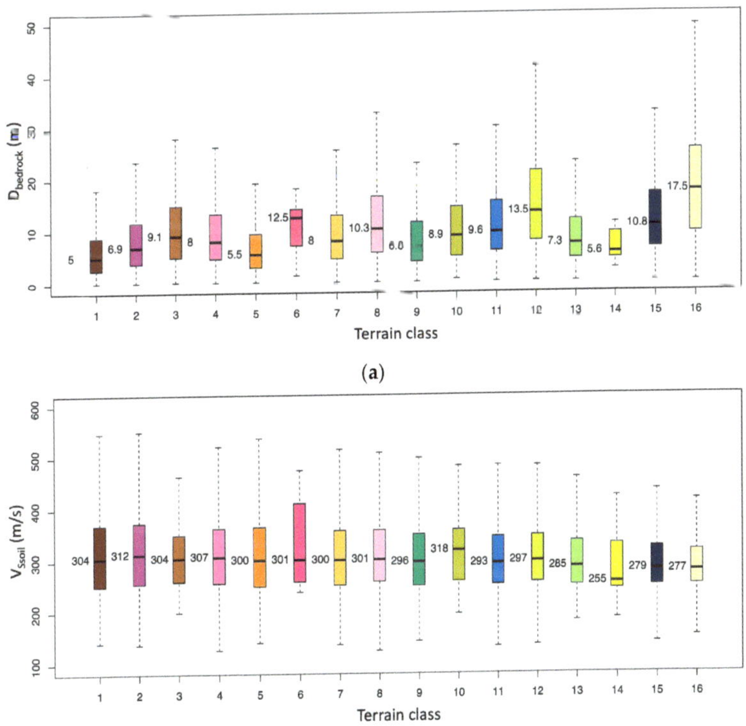

(a)

(b)

Figure 11. Boxplots of (**a**) $D_{bedrock}$ and (**b**) V_{Ssoil} for each terrain class using NOC with Case 2.

Table 5. Median (x_m) and standard deviation (σ_{ln}) of natural log unit of $D_{bedrock}$ and V_{Ssoil} for terrain classes classified using NOC with Case 2.

Terrain Class	Number of Data Points	$D_{bedrock}$		V_{Ssoil}	
		x_m (m)	σ_{ln}	x_m (m/s)	σ_{ln}
1	21,597	5.0	0.953	304	0.254
2	18,071	6.9	0.925	312	0.252
3	63	9.1	0.879	304	0.201
4	10,487	8.0	0.885	307	0.237
5	4154	5.5	0.927	300	0.258
6	8	12.5	0.878	301	0.266
7	12,954	8.0	0.855	300	0.241
8	7121	10.3	0.868	301	0.233
9	5032	6.8	0.872	296	0.236
10	126	8.9	0.727	318	0.225
11	12,280	9.6	0.781	293	0.220
12	11,062	13.5	0.822	297	0.215
13	883	7.3	0.751	285	0.228
14	17	5.6	0.644	255	0.216
15	5250	10.8	0.768	279	0.201
16	13,404	17.5	0.757	277	0.187

4.2. Regression with DEM

While elevation is a key predictor for $D_{bedrock}$ and V_{Ssoil}, the terrain classes do not inherently convey elevation values. The alluvial deposits become thicker (increase in $D_{bedrock}$) and the soil particle size becomes finer (decrease in V_{Ssoil}) along a decreasing elevation. To refine the prediction of $D_{bedrock}$ and V_{Ssoil} considering the soil deposit environment, we employ a power model based on DEM elevation for each class. Figure 12 plots elevation against $D_{bedrock}$ and V_{Ssoil} by class. In steep terrain (low number classes, 1 to 7), $D_{bedrock}$ are not greatly influenced by elevation, whereas in flatter terrains (high number classes, 8–16), there is a discernible correlation between elevation and $D_{bedrock}$. For V_{Ssoil}, the elevation dependence is not as strong as $D_{bedrock}$, but there is slight increase in V_{Ssoil} with an increase in elevation. This suggests that incorporating elevation into a regression model could enhance the precision of $D_{bedrock}$ and V_{Ssoil} predictions beyond terrain classification.

To account for elevation dependency, the following power models for $D_{bedrock}$ and V_{Ssoil} are applied:

$$D_{bedrock} = c_0 \times elev^{c_1} \tag{10}$$

$$V_{Ssoil} = d_0 \times elev^{d_1} \tag{11}$$

where $[c_0, c_1]$ and $[d_0, d_1]$ are regression coefficients for each class, and *elev* represents the elevation in meters. Note that the resultant $D_{bedrock}$ is in meters and V_{Ssoil} is in meters/second. Table 6 provides the regression coefficients and their significance level (i.e., p-value) as well as their standard deviations for each class. For $D_{bedrock}$, classes of [3,6,14] do not markedly improve prediction accuracy with model adjustments (p-value > 0.1%) due to limited number of data points and flat model prediction. In addition, class 10 has the most limited range of available elevations even though the p-value is less than 0.1%. Note that we used a p-value of 0.1% as the significance level. For V_{Ssoil}, classes [5,6,10,13,14] have highest p-values, and class 3 also has the most limited range of elevation. For these classes (i.e., 3, 6, 10, 14 for $D_{bedrock}$ and 3, 5, 6, 10, 13, 14 for V_{Ssoil}), use of the median is recommended rather than using fitted models. Utilizing elevation models for each class reduces the total standard deviation from [0.865, 0.233], which solely uses median estimation, to [0.844, 0.231] for $D_{bedrock}$ and V_{Ssoil}, respectively. Note that we limited the elevation range to [0.1, 1000 m] when applying it to models as not to have excessive $D_{bedrock}$ and V_{Ssoil}.

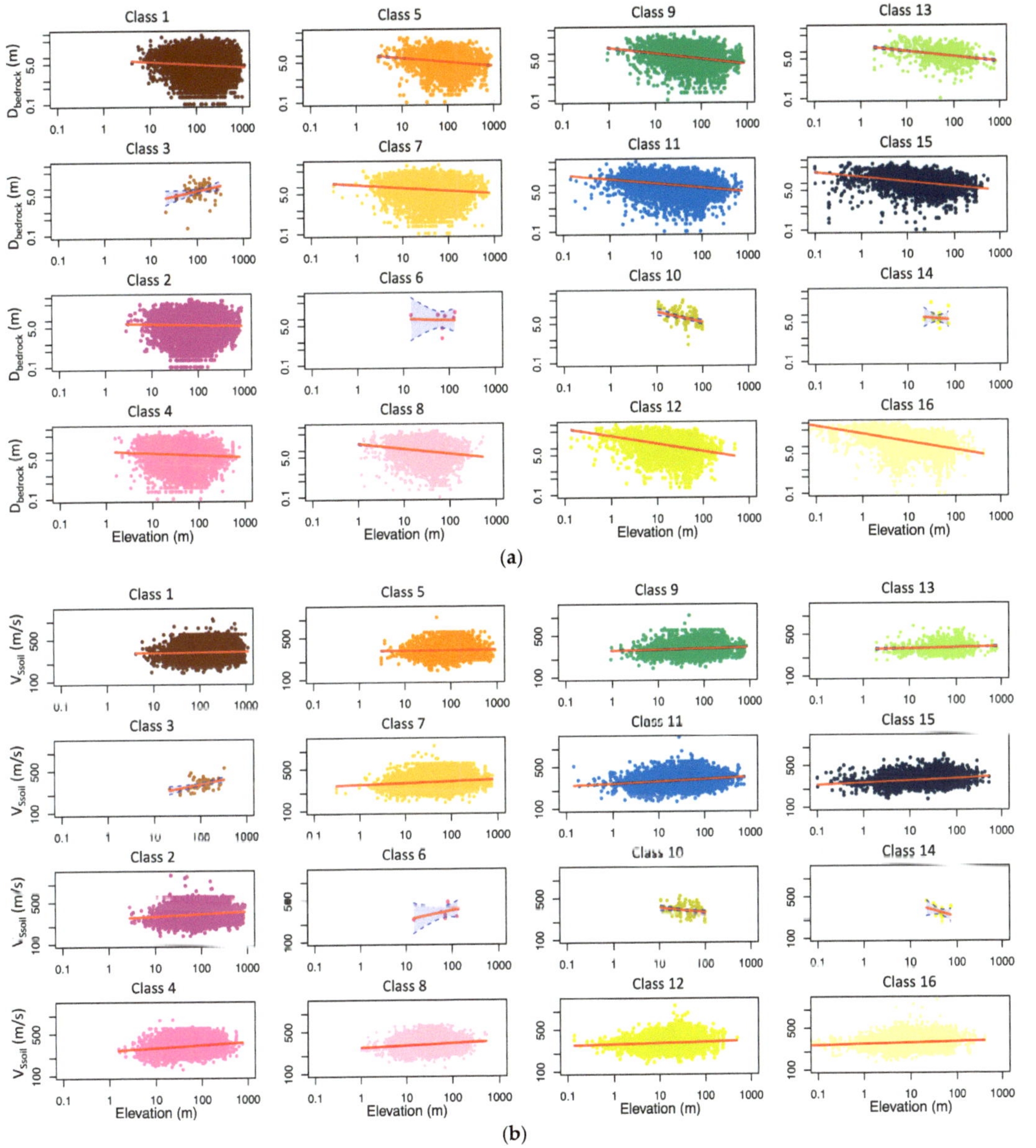

Figure 12. (**a**) $D_{bedrock}$ versus elevation and (**b**) V_{Ssoil} versus elevation for all classes. A fit model at each class is shown as red line.

Table 6. Regression coefficients for $D_{bedrock}$ and V_{Ssoil} prediction and standard deviation of residuals at each terrain class.

Class	$D_{bedrock}$				V_{Ssoil}			
	c_0	c_1	σ_{ln}	*p*-Value	d_0	d_1	σ_{ln}	*p*-Value
1	7.88	−0.111	0.95	<0.1%	296.6	0.009	0.254	<0.1%
2	7.38	−0.041	0.925	<0.1%	265.4	0.038	0.25	<0.1%
3	1.02	0.434	0.849	4.1%	148.1	0.152	0.185	<0.1%
4	10.31	−0.084	0.882	<0.1%	257.0	0.047	0.234	<0.1%
5	11.43	−0.18	0.914	<0.1%	302.5	0.004	0.258	46.3%
6	11.09	−0.059	0.877	91.4%	163.4	0.16	0.243	30.9%
7	11.98	−0.125	0.847	<0.1%	273.1	0.028	0.239	<0.1%
8	18.54	−0.212	0.852	<0.1%	270.5	0.036	0.232	<0.1%
9	16.72	−0.23	0.839	<0.1%	279.2	0.017	0.235	<0.1%
10	42.07	−0.433	0.69	<0.1%	410.9	−0.075	0.221	4.6%
11	16.97	−0.179	0.758	<0.1%	261.7	0.037	0.216	<0.1%
12	32.00	−0.321	0.77	<0.1%	283.6	0.019	0.214	<0.1%
13	16.93	−0.218	0.713	<0.1%	277.1	0.015	0.227	3.4%
14	12.84	−0.166	0.642	78.3%	647.4	−0.229	0.206	24.8%
15	18.60	−0.202	0.724	<0.1%	256.8	0.034	0.196	<0.1%
16	34.87	−0.341	0.656	<0.1%	270.2	0.019	0.186	<0.1%

Figure 13 presents maps of $D_{bedrock}$ and V_{Ssoil} across the Korean Peninsula using models as recommended in Table 6. The most regions have a $D_{bedrock}$ of <20 m and a V_{Ssoil} of >300 m/s, but some alluvium regions (e.g., Busan, Pohang, or Goheung) have deep $D_{bedrock}$ (20 to 75 m) and soft V_{Ssoil} (250 to 280 m/s).

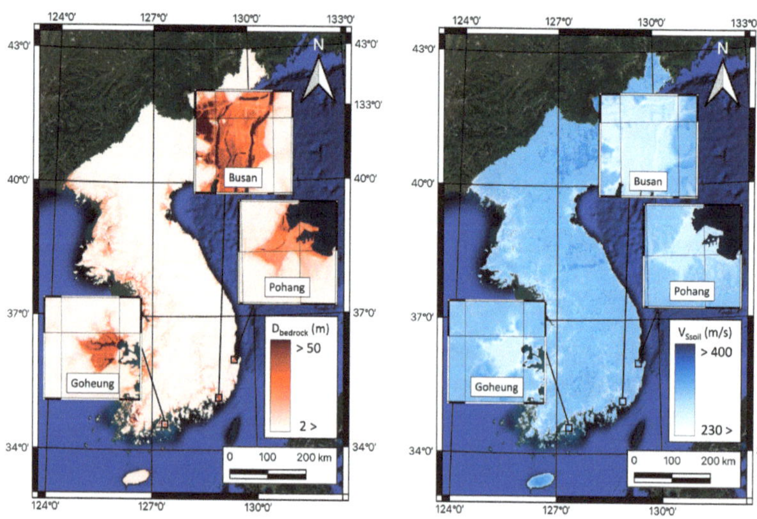

(a) (b)

Figure 13. Maps of (**a**) $D_{bedrock}$ (red color gradient) and (**b**) V_{Ssoil} (blue color gradient) for the Korean Peninsula.

5. Discussion

5.1. $D_{bedrock}$ and V_{Ssoil} Prediction Performance

This study suggests two ways of predicting $D_{bedrock}$ and V_{Ssoil}: (1) using the median values of $D_{bedrock}$ and V_{Ssoil} at each class, and (2) using the elevation model (Equations (10) and (11)), for which the coefficients are listed in Table 6. As noted above, for classes [3,6,10,14] for $D_{bedrock}$ and [3,5,6,10,13,14] for V_{Ssoil}, median values were used instead of the elevation model. We evaluated the prediction performance of these two methods by comparing the standard deviation of the residuals using the natural log unit (σ_{ln}) for $D_{bedrock}$ and V_{Ssoil}. Figure 14 shows the σ_{ln} at each class and the total data points. In the case of $D_{bedrock}$, the reduction in σ_{ln} is high for high class numbers (9–16), where the maximum reduction was seen in class 16 (13.3% reduction). In the case of V_{Ssoil}, the reduction in σ_{ln} is not significant; the maximum reduction was 2.3% for class 15. Again, we found that the elevation has more of an impact on the prediction of soil thicknesses in flatter regions, but the V_S of a soil deposit is not related to elevation.

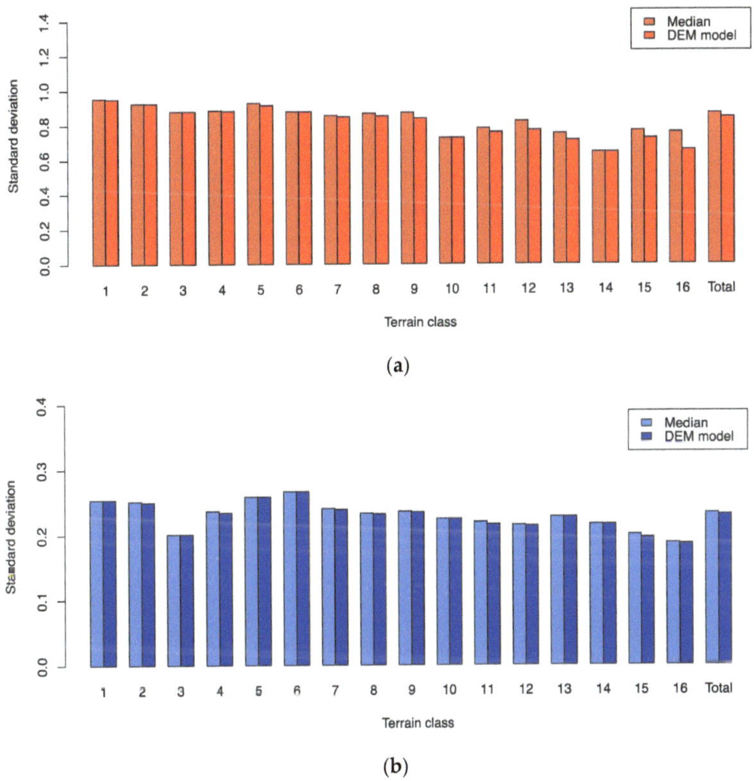

Figure 14. Comparison between standard deviations of residuals for each class using median predictions and elevation models for (**a**) $D_{bedrock}$ and (**b**) V_{Ssoil}.

5.2. Comparison with Other Thresholds

Previous studies revealed that terrain classification is related to geological formations [6,16]. Comparing terrain class maps for nine cases from three methods with geological and mountainous maps of the Seoul region (Figures 8–10), we decided that Case 2 of the NOC method is the best-matched case. To further validate the applicability of the terrain class map developed in this study, we compared the terrain class maps for Case 2 with the 1:50,000 scale geological and mountainous maps of different regions (Busan, Pohang, and Goheung) in South Korea, as shown in Figure 15. The comparison revealed a good correspondence between terrain classes and geologic divisions for the NOC method. For instance, classes with low class numbers showed a significant overlap with mountainous terrains.

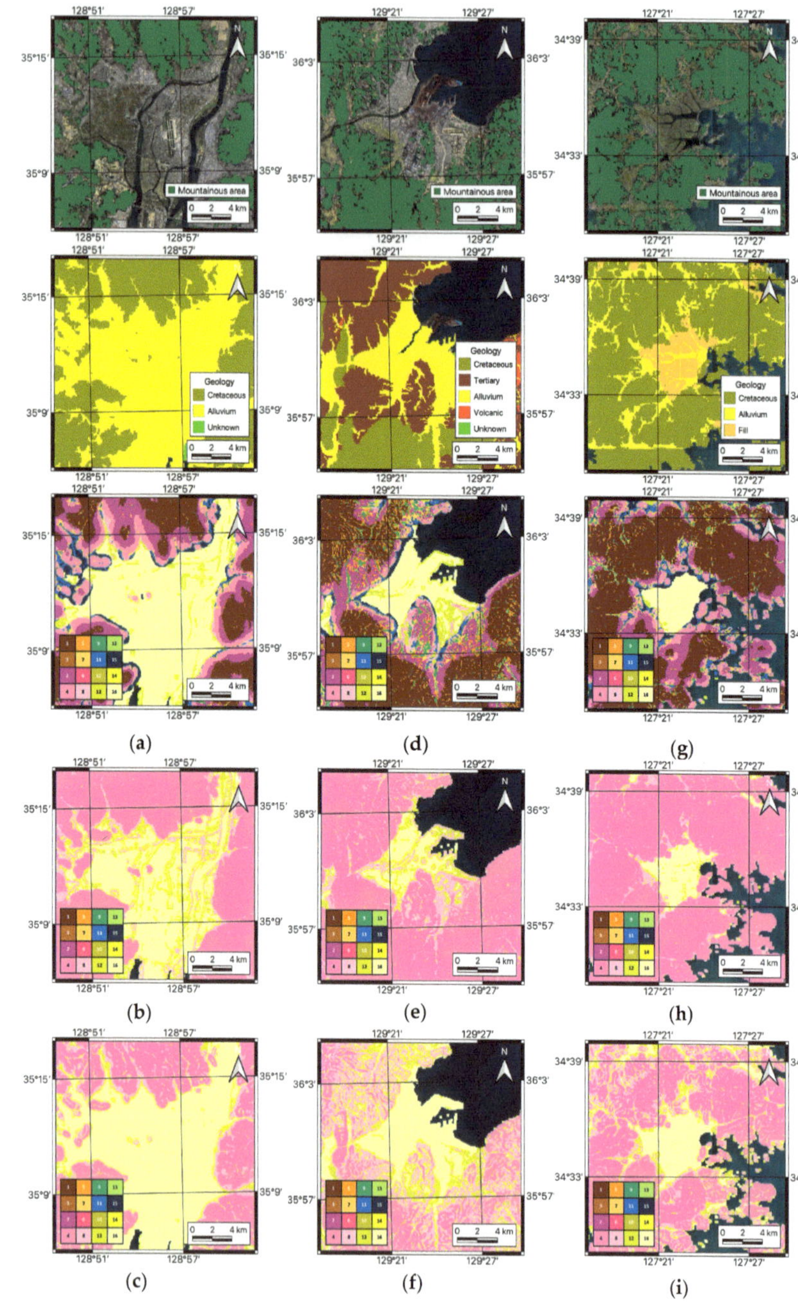

Figure 15. Maps of mountainous and surface geology, and the classification of terrain class maps for the Busan (**a**–**c**), Pohang (**d**–**f**), and Goheung (**g**–**i**) regions based on this study (**a**,**d**,**g**), the world (**b**,**e**,**h**), and Japan (**c**,**f**,**i**).

In addition, the thresholds determined in this study were specifically optimized for predicting $D_{bedrock}$ and V_{Ssoil}. However, it is crucial to examine how these thresholds are juxtaposed with those from other studies, which may have been developed for different regions or under varying geological conditions. To establish the robustness of our optimized thresholds, we compared the terrain classes generated using our methodology with classes defined by thresholds from other research efforts. Table 7 compares the thresholds for SG, LC, and ST chosen in this study with thresholds created for the world and Japan [14]. The world contains all terrain regions on the Earth so the median SG is low, while for Japan, the median SG is high because Japan consists of volcanic islands. The SG thresholds in this study are in between the world and Japan cases, and the LC and ST thresholds are lower than other cases. Figure 15 shows terrain class distributions defined using the world and Japanese cases for the Busan and Pohang regions. The world case predicts a narrower range of alluvium for class number 16 and cannot distinguish the mountainous regions well. The Japanese case, which has high SG thresholds, assigned class number 16 to too large an extent, and the resolution of classification is low for mountainous regions. This suggests that a region-specific calibration of the thresholds is imperative for achieving meaningful terrain classification.

Table 7. Thresholds classifying slope gradient (SG), local convexity (LC), and surface texture (ST) to terrain classes by this study (NOC, Case 2) and other studies.

Region	Phase	SG (deg)	LC	ST
Korea (90 m grid; NOC, Case 2)	1	3.77	0.188	0.232
	2	1.88	0.242	0.079
	3	0.58	0.197	0.048
World (1 km grid; AC)	1	1.76	0.456	0.669
	2	0.48	0.454	0.639
	3	0.20	0.450	0.590
Japan (270 m grid; AC)	1	8.05	0.462	0.650
	2	3.26	0.451	0.606
	3	1.30	0.439	0.539

6. Conclusions

This study presented a comprehensive approach for classifying terrain in South Korea using DEMs, focusing on estimating $D_{bedrock}$ and V_{Ssoil}. By applying three distinct classification methods—Automated Classification (AC), Sequentially Optimized Classification (SOC), and Non-Sequentially Optimized Classification (NOC)—to DEM-derived topographic features (slope gradient, local convexity, and surface texture), we effectively correlated terrain classes with subsurface geotechnical properties and geologic formation. Our analysis revealed that the NOC method, particularly for Case 2 (3 × 3 window size, 2 m threshold), provided the most accurate terrain classification when being correlated with geological and mountainous maps in addition to minimizing the standard deviation of errors in $D_{bedrock}$ and V_{Ssoil}. The main findings of this study are listed below.

(1) The results of this study indicate the capability of the NOC method to not only predict subsurface conditions but also reflect the geologic formation of the landscape. The $D_{bedrock}$ and V_{Ssoil} are the result of geologic formations, so the terrain classes derived from our study can be applied to geological interpretations.
(2) The incorporation of a regression model based on DEM elevation significantly enhanced the prediction accuracy for $D_{bedrock}$, and showed a moderate enhancement in predicting V_{Ssoil}.
(3) This study highlights the importance of considering regional specificity when setting thresholds for terrain classification, as evidenced by the varying effectiveness of world and Japan cases applied to South Korean regions.

Our classification scheme and $D_{bedrock}$ and V_{Ssoil} prediction models have the potential to serve as a preliminary guide for geologic explorations and land-use planning as well as site assessments, particularly in areas where detailed geologic maps and ground investigations are unavailable.

Although the suggested method can be applied to any region, the limitation of the suggested terrain class divisions is that they are dependent on the classification method, smoothing, and the thresholds. Thus, the use of other DEMs (e.g., high resolution DEM) or the application of our methods to other regions require further sensitivity analyses. In addition, the values of $D_{bedrock}$ and V_{Ssoil} are from a limited number of sites, so the suggested results can not represent the values for entire region. Therefore, a further sensitivity study is required if a different DEM is used or the method is applied to other regions, and a validation process is essential for data points that are not used in the model development.

Future work could extend the scope of this methodology by incorporating a broader array of geospatial data to refine terrain classification and enhance the prediction of subsurface properties. While the current study utilized SG, LC, and ST for terrain classification and elevation data for $D_{bedrock}$ and V_{Ssoil} refinement, integrating additional topographic attributes such as relative elevation and curvature could potentially yield more precise results. Moreover, the employment of machine learning techniques with an increased number of topographic attributes targeting geologic, geomorphic, and subsurface conditions is a promising avenue for the classification of regions.

Author Contributions: Conceptualization, I.C. and D.K.; methodology, I.C. and D.K.; software, I.C.; validation, I.C.; formal analysis, I.C.; investigation, D.K.; resources, I.C.; data curation, I.C.; writing—original draft preparation, I.C. and D.K.; writing—review and editing, I.C. and D.K.; visualization, I.C.; supervision, D.K.; funding acquisition, D.K. All authors have read and agreed to the published version of the manuscript.

Funding: This work was supported by the Ministry of Land, Infrastructure and Transport with the grant number RS-2022-00143584 and the Basic Research Program through the National Research Foundation of Korea funded by the Ministry of Science and ICT with the grant number RS-2023-00220751.

Data Availability Statement: The data presented in this study are available in following resources: boring log data at GeoInfo (https://www.geoinfo.or.kr accessed on 15 August 2023), DEM and cadastral map at NSDI portal (http://www.nsdi.go.kr accessed on 15 August 2023), and geological map at Geo Big Data Open Platform (https://data.kigam.re.kr accessed on 15 August 2023).

Acknowledgments: We are grateful to NSDI for providing the DEM and cadastral maps, MOLIT for operating and maintaining GeoInfo geotechnical database, and KIGAM for providing geological map.

Conflicts of Interest: The authors declare no conflicts of interest. The funders had no role in the design of the study; in the collection, analyses, or interpretation of data; in the writing of the manuscript; or in the decision to publish the results.

References

1. MOLIT. *General Seismic Design. KDS 17 10 00*; Ministry of Land, Infrastructure and Transport: Sejong, Republic of Korea, 2018.
2. Wald, D.J.; Allen, T.I. Topographic slope as a proxy for seismic site conditions and amplification. *Bull. Seismol. Soc. Am.* **2009**, *97*, 1379–1395. [CrossRef]
3. Allen, T.I.; Wald, D.J. On the use of high-resolution topographic data as a proxy for seismic site conditions (V_{S30}). *Bull. Seismol. Soc. Am.* **2009**, *99*, 935–943. [CrossRef]
4. Yang, F.; Shangguan, W.; Zhang, J.; Hu, B. Depth-to-bedrock map of China at a spatial resolution of 100 meters. *Sci. Data* **2020**, *7*, 2.
5. Choi, I.; Yoo, B.; Kwak, D. Development of Korean Peninsula VS30 map based on proxy using linear regression analysis. *KSCE J. Civ. Environ. Eng.* **2022**, *42*, 35–44. (In Korean)
6. Yang, H.Q.; Chu, J.; Qi, X.; Wu, S.; Chiam, K. Bayesian evidential learning of soil-rock interface identification using boreholes. *Comput. Geotech.* **2023**, *162*, 105638. [CrossRef]
7. Wang, Z.Z.; Hu, Y.; Guo, X.; He, X.; Kek, H.Y.; Ku, T.; Goh, S.H.; Leung, C.F. Predicting geological interfaces using stacking ensemble learning with multi-scale features. *Can. Geotech. J.* **2023**, *60*, 1036–1054. [CrossRef]
8. Yang, L.; Meng, X.; Zhang, X. SRTM DEM and its application advances. *Remote Sens.* **2011**, *32*, 3875–3896. [CrossRef]
9. Toutin, T. Three-dimensional topographic mapping with ASTER stereo data in rugged topography. *IEEE Trans. Geosci. Remote Sens.* **2002**, *40*, 2241–2247. [CrossRef]

10. Farr, T.G.; Rosen, P.A.; Caro, E.; Crippen, R.; Duren, R.; Hensley, S.; Kobrick, M.; Paller, M.; Rodriguez, E.; Roth, L.; et al. The Shuttle Radar Topography Mission. *Rev. Geophys.* **2007**, *45*, 2005RG000183. [CrossRef]
11. Hofton, M.A.; Minster, J.B.; Blair, J.B. Decomposition of laser altimeter waveforms. *IEEE Trans. Geosci. Remote Sens.* **2000**, *38*, 1989–1996. [CrossRef]
12. National Geographic Information Institute (NGII). Available online: http://www.ngii.go.kr (accessed on 18 July 2023).
13. National Spatial Data Infrastructure Portal (NSDI). Available online: http://www.nsdi.go.kr (accessed on 15 August 2023).
14. Iwahashi, J.; Pike, R.J. Automated classifications of topography from DEMs by an unsupervised nested-means algorithm and a three-part geometric signature. *Geomorphology* **2007**, *86*, 409–440. [CrossRef]
15. Vilanova, S.P.; Narciso, J.; Carvalho, J.P.; Lopes, I.; Quinta0Ferreira, M.; Pinto, C.C.; Borges, J.; Nemser, E.S. Developing a Geologically-Based VS30 Site-Conditions Model for Portugal: Methodology and Assessment of the Performance of Proxies. *Bull. Seismol. Soc. Am.* **2018**, *108*, 322–337. [CrossRef]
16. Karimzadeh, S.; Feizizadeh, B.; Matsuoka, M. DEM-based Vs30 map and terrain surface classification in nationwide scale—A case study in Iran. *ISPRS Int. J. Geo-Inf.* **2019**, *8*, 537. [CrossRef]
17. Irsyam, M.; Asrurifak, M.; Mikhail, R.; Wahdiny, I.I.; Rustiani, S. Development of Nationwide Vs30 Map and Calibrated Conversion Table for Indonesia using Automated Topographical Classification. *J. Eng. Technol. Sci.* **2017**, *49*, 457–471. [CrossRef]
18. Kim, H.S.; Sun, C.G.; Lee, M.G.; Cho, H.I. Terrain Proxy-Based Site Classification for Seismic Zonation in North Korea within a Geospatial Data-Driven Workflow. *Remote Sens.* **2021**, *13*, 1844. [CrossRef]
19. Yong, A.; Hough, S.E.; Iwahashi, J.; Braverman, A. A terrain-based site-conditions map of California with implications for the contiguous United States. *Bull. Seismol. Soc. Am.* **2012**, *102*, 114–128. [CrossRef]
20. Furze, S.; O'Sullivan, A.M.; Allard, S.; Pronk, T.; Curry, R.A. A high-resolution, random forest approach to mapping depth-to-bedrock across shallow overburden and post-glacial terrain. *Remote Sens.* **2021**, *13*, 4210. [CrossRef]
21. Ministry of Land, Infrastructure and Transport (MOLIT). Available online: https://www.geoinfo.or.kr (accessed on 15 August 2023). (In Korean)
22. Heo, G.S.; Kwak, D.Y. V_S prediction model using SPT-N values and soil layers in South Korea. *J. Korean Geotech. Soc.* **2022**, 53–66. (In Korean)
23. Horn, B.K. Hill shading and the reflectance map. *Proc. IEEE* **1981**, *61*, 14–47. [CrossRef]
24. Iwahashi, J.; Yamazaki, D.; Nakano, T.; Endo, R. Classification of topography for ground vulnerability assessment of alluvial plains and mountains of Japan using 30 m DEM. *Prog. Earth Planet. Sci.* **2021**, *8*, 3. [CrossRef]
25. Korean Institute of Geoscience and Mineral Resources (KIGAM). Available online: https://data.kigam.re.kr (accessed on 1 February 2023). (In Korean).

remote sensing

MDPI

Article

Spatiotemporal Analysis and Prediction of Carbon Emissions from Energy Consumption in China through Nighttime Light Remote Sensing

Zhaoxu Zhang [1,2,3], Shihong Fu [1], Jiayi Li [1], Yuchen Qiu [1], Zhenwei Shi [4] and Yuanheng Sun [5,*]

1 School of Environmental Science and Engineering, Tiangong University, Tianjin 300387, China; zhangzhaoxu@tiangong.edu.cn (Z.Z.); 2113620105@tiangong.edu.cn (S.F.); 2213620136@tiangong.edu.cn (J.L.)
2 The Eighth Geological Brigade, Hebei Bureau of Geology and Mineral Resources Exploration, Qinhuangdao 066000, China
3 Marine Ecological Restoration and Smart Ocean Engineering Research Center of Hebei Province, Qinhuangdao 066000, China
4 Key Laboratory of Technology in Geo-Spatial Information Processing and Application System, Chinese Academy of Sciences, Beijing 100190, China; shizw@aircas.ac.cn
5 Environmental Information Institute, Navigation College, Dalian Maritime University, Dalian 116026, China
* Correspondence: yhsun@dlmu.edu.cn

Abstract: With burgeoning economic development, a surging influx of greenhouse gases, notably carbon dioxide (CO_2), has precipitated global warming, thus accentuating the critical imperatives of monitoring and predicting carbon emissions. Conventional approaches employed in the examination of carbon emissions predominantly rely on energy statistics procured from the National Bureau of Statistics and local statistical bureaus. However, these conventional data sources, often encapsulated in statistical yearbooks, exclusively furnish insights into energy consumption at the national and provincial levels, so the assessment at a more granular scale, such as the municipal and county levels, poses a formidable challenge. This study, using nighttime light data and statistics records spanning from 2000 to 2019, undertook a comparative analysis, scrutinizing various modeling methodologies, encompassing linear, exponential, and logarithmic models, with the aim of assessing carbon emissions across diverse spatial scales. A multifaceted analysis unfolded, delving into the key attributes of China's carbon emissions, spanning total carbon emissions, per capita carbon emissions, and carbon emission intensity. Spatial considerations were also paramount, encompassing an examination of carbon emissions across provincial, municipal, and county scales, as well as an intricate exploration of spatial patterns, including the displacement of the center of gravity and the application of trend analyses. These multifaceted analyses collectively contributed to the endeavor of predicting China's future carbon emission trajectory. The findings of the study revealed that at the national scale, total carbon emissions exhibited an annual increment throughout the period spanning 2000 to 2019. Secondly, upon an in-depth evaluation of model fitting, it was evident that the logarithmic model emerged as the most adept in terms of fitting, presenting a mean R^2 value of 0.83. Thirdly, the gravity center of carbon emissions in China was situated within Henan Province, and there was a discernible overall shift towards the southwest. In 2025 and 2030, it is anticipated that the average quantum of China's carbon emissions will reach 7.82×10^2 million and 25.61×10^2 million metric tons, with Shandong Province emerging as the foremost contributor. In summary, this research serves as a robust factual underpinning and an indispensable reference point for advancing the scientific underpinnings of China's transition to a low-carbon economy and the judicious formulation of policies governing carbon emissions.

Keywords: nighttime light; carbon emission; spatiotemporal analysis; scale; prediction

Citation: Zhang, Z.; Fu, S.; Li, J.; Qiu, Y.; Shi, Z.; Sun, Y. Spatiotemporal Analysis and Prediction of Carbon Emissions from Energy Consumption in China through Nighttime Light Remote Sensing. *Remote Sens.* **2024**, *16*, 23. https://doi.org/10.3390/rs16010023

Academic Editors: Yanni Dong, Tao Chen and Chao Chen

Received: 31 October 2023
Revised: 13 December 2023
Accepted: 19 December 2023
Published: 20 December 2023

1. Introduction

Today's society faces various challenges and threats, such as resource constraints, energy depletion, reduced food production, environmental pollution, overpopulation, and global warming [1–3]. Amongst them, global warming has emerged as the most critical issue, directly impacting the natural environment and the sustainable development of human society and economy [4]. The factors that affect global warming are natural [5], such as solar radiation, atmospheric radiation, and atmospheric circulation, and anthropogenic [6,7]. The predominant contributor to the rise in global temperatures and global warming is large-scale human activities that emit substantial amounts of greenhouse gases, such as CO_2, into the atmosphere [8,9]. These greenhouse gases are highly transmissive of solar radiation and highly absorbent of longwave radiation reflected from the Earth's surface, leading to an increase in global temperature, a phenomenon known as the greenhouse effect [10,11]. The latest assessment by the Intergovernmental Panel on Climate Change shows that the global average temperature increased by 1.09 °C from the second half of the 19th century to the first decades of the 21st century [12]. The main reason is the increased concentration of CO_2. Carbon emissions have become a focus of interest for all countries, and the study of carbon emissions has garnered global attention, progressively becoming a topic of international concern [13].

As China experiences rapid industrialization and urbanization, and with its economic development heavily reliant on fossil energy, the country has generated significant carbon emissions [14]. China has become the country with the highest CO_2 emission in the world, and its share of CO_2 emissions has increased from 10.9% in 1990 to 28.61% in 2018 [15]. To control energy consumption and reduce CO_2 emissions, China has established corresponding targets and policies. Notably, the emission reduction targets for 2030 stipulate that carbon emissions will peak by that year, and non-fossil energy will constitute 20% of total energy consumption [16]. Concurrently, the Chinese government plans to peak carbon emissions around 2030 [17]. Against this backdrop, China has proposed the dual carbon goals of achieving carbon peaking and carbon neutrality, aligning with the demands of the current era [18].

Against the backdrop of frequent global warming and environmental issues, remote sensing technology offers several advantages, including easy accessibility, rapid updates, and extensive coverage [19]. Consequently, remote sensing plays a pivotal role in monitoring greenhouse gases and atmospheric pollutants [20]. (1) Direct calculation is based on greenhouse gas remote sensing satellites and data [21]. The monitoring technology of these satellites quantifies greenhouse gas concentrations through non-contact methods involving various technologies such as Fourier transform infrared spectroscopy (FTIR), differential absorption lidar (DIAL), laser heterodyne spectroscopy (LHS), and spatial heterodyne spectroscopy (SHS) [22]. (2) Remote sensing observations are used for indirect accounting and inversion of greenhouse gases [23]. Currently, optimal estimation methods and differential optical absorption spectroscopy (DOAS)-related algorithms are the mainstream approaches for estimating non-CO_2 gas concentrations from satellite observations. The column concentration accuracy can reach 1% for CH_4, 10% for NO_2, 1% for N_2O, and 1% for O_3. (3) In the monitoring of atmospheric pollution particles based on remote sensing data [24], a three-dimensional monitoring network for urban air pollution is built based on ground environmental air automatic monitoring stations, and the environmental quality of the entire region is simulated and analyzed using measured data from the stations [25]. This network facilitates the simulation and analysis of regional environmental quality using measured data from stations, including inhalable particulate matter, fine particulate matter, nitrogen dioxide, sulfur dioxide, carbon monoxide, and ozone. In summary, remote sensing technology plays a crucial role in various aspects of atmospheric environment monitoring such as trace gases, atmospheric particulate matter, volatile organic compounds (VOCs), and hazardous substances [26].

In China, CO_2 emissions are generally quantified using energy statistics [13]. The country's vast territory and regional disparities make the collection of energy statistics burdensome and complex [27]. Meanwhile, China's research on carbon emissions is mainly based on energy statistics from the National Bureau of Statistics and local statistical bureaus. However, statistical yearbooks only provide energy consumption data at the national and provincial levels. Data on the energy consumption of the country's prefectural cities at county and township levels and even smaller scales are unavailable, so research cannot be conducted on carbon emissions at small scales [28]. Furthermore, the quality of data statistics varies across different regions of China, and energy consumption statistics in national and local energy statistical yearbooks are lagging [29].

With the advancement of remote sensing technology, the use of satellite data to measure carbon emissions is gaining momentum [30]. Remote sensing technology enables the periodic and rapid acquisition of macro-level information with minimal data constraints [31,32]. Therefore, this technology is widely used in related fields, such as agriculture and forestry monitoring, weather and meteorological forecasting, hydrological water monitoring, environmental monitoring, mapping and surveying, military reconnaissance, and geology and minerals [33–35]. Compared to traditional methods like statistical approaches, remote sensing technology overcomes the spatial and temporal limitations associated with data collection [32,36]. Thus, remote sensing technology provides data support for studying carbon emissions at various scales [37]. Nighttime light imagery, a special kind of remote sensing imagery, has long light sequences that effectively record light information related to human activities [38]. Consequently, nighttime light data have become a crucial source for carbon emission studies [39].

This research aims to control the total amount of carbon emissions in China and explore the patterns and development trends of energy carbon emissions. The study focuses on three main objectives: (1) Using statistical and nighttime light data from 2000 to 2019, a model for fitting carbon emissions and nighttime lighting is constructed to estimate carbon emissions at different scales (province, city, and county). This task compensates for the limitations of statistical data. (2) The basic characteristics of total carbon emissions, per capita carbon emissions, and carbon emission intensity at different scales (province, city, and county) in 2000, 2005, 2010, 2015, and 2019 are examined. Carbon emission characteristics are analyzed using spatial center of gravity migration, trend changes, and spatial aggregation methods. (3) Based on the grey prediction model GM (1, 1), provincial-scale carbon emissions for 2025 and 2030 are forecasted using carbon emission data from 2000 to 2019. This research aims to provide a reference for the scientific and sustainable development of a low-carbon economy in China and for the formulation of corresponding carbon emission reduction policies by regional governments.

2. Materials

2.1. Study Area

As the world's most populous country, China is undergoing rapidly growing urbanization as it enters the 21st century [40]. By the end of 2019, mainland China comprised 360 prefecture-level cities and 2774 county-level cities [41]. This research excludes Tibet, Hong Kong, Macau, and Taiwan due to regional differences and the unavailability of statistics in these areas (Figure 1). To align with available official statistics and to reflect the continuous changes in China's carbon emissions, the study covers a period of 20 years, from 2000 to 2019.

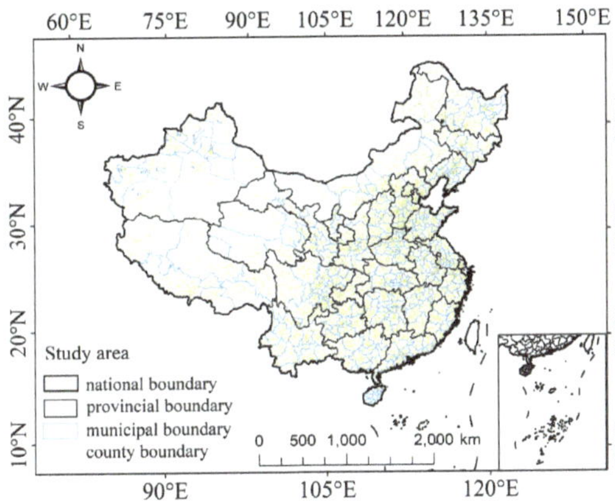

Figure 1. Research area.

2.2. Study Data

2.2.1. Nighttime Light Data

Data were obtained from the Long Time Series Nighttime Light Dataset of China (2000–2020) published by Zhong Xiaoya in the Global Change Research Data Publishing and Repository [42]. This dataset for China was based on the calibration and fusion of DMSP/OLS annual moonlight image data and NPP/VIIRS monthly image data, with a spatial resolution of 1 km. Both the oversaturation issue of DMSP/OLS data and the noise problem in NPP/VIIRS data had been addressed. The continuity and availability of the data were conducive to the extraction of information on cities and towns across the country over an extended period.

The original DMSP/OLS and NPP/VIIRS data present challenges due to variations in sensors, spatiotemporal resolutions, data processing methods, and pixel interpretations, which pose difficulties in integrating long-term nighttime lighting data. These data issues significantly constrain the potential applications of nighttime light data and may also impact the effectiveness of research relying on such data. Therefore, researchers conducted a series of data processing steps on DMSP/OLS images from 2000 to 2013 and monthly NPP/VIIRS images from 2012 to 2020. Through preprocessing, mutual correction, and saturation correction techniques applied to DMSP/OLS images, annual DMSP/OLS datasets were obtained. Simultaneously, preprocessing steps including outlier processing and median-based annual synthesis on NPP/VIIRS images were performed to obtain annual NPP/VIIRS datasets. Finally, by fusing the annual EANTLI dataset with the NPP/VIIRS dataset through spatial resolution integration, fitting function construction and continuous corrections were made resulting in long-term nighttime light datasets [42].

2.2.2. Statistical Data

To calculate carbon emissions from energy consumption at different scales (provinces, municipalities, and districts), annual statistics were obtained from the National Bureau of Statistics of the People's Republic of China. Data on 11 energy sources were collected based on the China Energy Statistical Yearbook from 2000 to 2019, as well as from provincial and local energy balances (including autonomous regions and municipalities directly under the central government, but excluding Tibet, Taiwan, Hong Kong, and Macao) from the relevant city statistics. Additionally, the carbon emission factors for different energy sources were derived from the IPCC Guidelines for National Greenhouse Gas Inventories.

3. Methods

Figure 2 illustrates the technical route of this study, which encompasses the following: (1) data collection and processing; (2) carbon emission model construction and comparison; (3) application of the results, which mainly includes carbon emission analysis at different scales, pattern analysis, and center of gravity analysis; and (4) prediction of carbon emissions for the years 2025 and 2030 using the grey model.

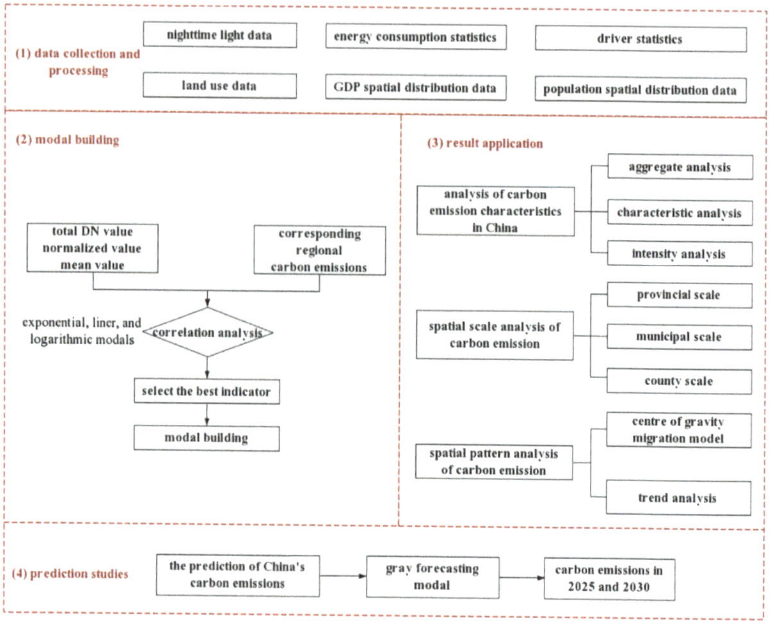

Figure 2. The flowchart of this study.

3.1. Carbon Emission Modeling

3.1.1. Accounting for Energy Carbon Emissions in China

To calculate China's carbon emissions from energy consumption, the most authoritative international carbon emission calculation method was used in this study. To calculate the total amount of carbon emissions from each variety of energy sources, this method was to convert the total consumption of various energy sources into the consumption of standard coal by folding the standard coal coefficient, and then use the standard coal multiplied by the respective carbon emission coefficients [43]. The CO_2 emissions are calculated using the following formula:

$$E = \sum_{i=1}^{j} EN_i \times EF_i \times \frac{44}{12} \quad (1)$$

where E is the emission of CO_2 from j energy sources ($\times 10^4$ t), EN_i is the energy consumption of the energy source category ($\times 10^4$ t), EF_i is the carbon emission factor of the energy source category, i is the type of energy source, and j is the number of energy sources.

The energy types, reference coefficient of standard coal, and carbon emission coefficients are shown in Table 1.

Table 1. The energy types, reference coefficient of standard coal, and carbon emission coefficients.

Type of Energy	Reference Coefficient of Standard Coal (kJ/kg)	Carbon Emission Coefficients
raw coal	0.7143	0.7559
refined coal	0.7143	0.9000
coke (processed coal used in blast furnace)	0.9714	0.8550
coke oven gas	0.6143	0.3548
crude oil	1.4286	0.5857
petrol	1.4714	0.5538
diesel	1.4714	0.5714
diesel oil	1.4571	0.5921
fuel oil	1.4286	0.6185
liquefied petroleum gas	1.7143	0.5042
refinery dry gas	1.5714	0.4602
petroleum	1.3300	0.4483

3.1.2. Extraction of DN Values

Extensive research has shown that the total nighttime light intensity value (T) and carbon emissions have a good correlation [44]. T is the sum of DN values in a region. In this study, the indicator of total T in the study area was selected to fit the energy carbon emissions. The image element values of EANTLI nighttime lighting data were extracted using the provincial administrative boundaries from 2000 to 2019. The T-value of the study area was calculated subsequently. The T-value is calculated using the following formula:

$$T = \sum_{i=0}^{n} DN_i \tag{2}$$

where DN_i is the value of the image element in region i and n is the number of image elements.

3.1.3. Model Calculations at Different Scales

In this study, exponential, linear, and logarithmic models were constructed. The best model was selected to achieve the calculation of carbon emissions at different scales using the coefficient of determination (R^2). In the fitting comparison, the correlation coefficients of the exponential, linear, and logarithmic models were very high (R^2 were all more than 0.7). Among them, the logarithmic fit was the best, and the mean value of R^2 was 0.83. To calculate and convert the sum of carbon dioxide emissions and DN values, the logarithmic model was selected in this study. The constructed model was calculated using the following formula:

The provincial conversion model is as follows:

$$E_{ij} = K \times \ln\left(T_{ij}\right) - B \tag{3}$$

The downscaled municipal and county conversion model is as follows:

$$E_{ij} = \frac{T_{kj}}{T_{ij}} K \times \ln(T_{ij}) - B \tag{4}$$

where E_{ij} is the carbon emissions of province i in year j, T_{kj} is the total value of city and county k in year j, T_{ij} is the total value of nighttime light, K is the provincial simulation coefficient, and B is the simulation intercept.

3.2. Correlation Analysis

The Pearson correlation coefficient (R) can reflect the degree and direction of correlation between two variables [45]. In this study, the correlation coefficient between the total intensity value of nighttime lighting and carbon emissions from energy consumption from 2000 to 2019 was calculated. The correlation coefficient calculation formula is as follows:

$$R = \frac{\sum_{i=1}^{n}(X_i - \overline{X})(Y_i - \overline{Y})}{\sqrt{\sum_{i=1}^{n}(X_i - \overline{X})^2 (Y_i - \overline{Y})^2}} \quad (5)$$

where R is the correlation coefficient, n is the length of the time series, X is the monthly average total intensity value of night lights, and Y is the carbon emission value of energy consumption.

3.3. Center of Gravity Migration Model

The concept of the center of gravity migration model comes from mechanics in physics and represents a mathematical weighted average. Its application to geography began with calculating the population center of gravity, expressed through latitude and longitude coordinates. By using a migration model based on the center of gravity, we could simulate temporal changes in carbon emissions' center of gravity, providing a more intuitive representation of their spatial and temporal dynamics within the study area [46]. To investigate China's changing carbon emissions patterns, this model was used to calculate the carbon emissions' center of gravity between 2000 and 2019. The center of gravity migration model is the following formulas:

$$\overline{X}_t = \frac{\sum_{i=1}^{n} E_{it} x_{it}}{\sum_{i=1}^{n} E_{it}} \quad (6)$$

$$\overline{Y}_t = \frac{\sum_{i=1}^{n} E_{it} y_{it}}{\sum_{i=1}^{n} E_{it}} \quad (7)$$

where $\overline{X}_t$ is the latitude of the center of gravity of China's carbon emissions in year t, $\overline{Y}_t$ is the longitude of the center of gravity of China's carbon emissions in year t, E_{it} is the carbon emissions on grid i in year t, x_{it} is the latitude coordinate of the center of grid i in year t, and y_{it} is the longitude coordinate of the center of grid i in year t.

3.4. Trend Analysis

Trend analysis is a method of predicting the trend of a variable over time by performing a linear regression analysis [45]. To analyze the growth trend of carbon emissions in different regions, the following formula is used to calculate the trend (slope) of carbon emissions:

$$\text{slope} = \frac{n \times \sum_{i=1}^{n} x_i t_i - \left(\sum_{i=1}^{n} x_i\right)\left(\sum_{i=1}^{n} t_i\right)}{n \times \sum_{i=1}^{n} {t_i}^2 - \left(\sum_{i=1}^{n} t_i\right)^2} \quad (8)$$

where slope is the trend in carbon emissions, x_i is the scaled exponential value, n is the length of the study period, and t_i is the year from 2000 to 2019. Positive slope values indicate an increase in carbon emissions, while negative slope values indicate a decrease. Therefore, the slope value can display the trend of carbon emissions during the study period. To further classify carbon emissions, trends were categorized into more refined levels including negative growth type, slow growth, slower growth, medium growth, faster growth, and rapid growth. The specific classification standards are shown in Table 2. x is the average value of the slope, and s is the standard deviation.

Table 2. Criteria for classifying the type of growth in carbon emissions.

Type of Carbon Growth	Delineation Criterion
negative growth	slope < 0
slow growth	0 ≤ slope < x~0.5 s
slower growth	x~0.5 s ≤ slope < x + 0.5 s
medium growth	x + 0.5 s ≤ slope < x + 1.5 s
faster growth	x + 1.5 s ≤ slope < x + 2.5 s
rapid growth	slope ≥ x + 2.5 s

3.5. Predictive Modeling

Grey theory posits that all random variables are grey quantities and grey processes that change within a certain range and over a certain period. Data processing does not aim to establish the statistical law and probability distribution of data; it processes the original data so that they become regular time series data, on the basis of which a mathematical model is established. The GM series model is the basic model of grey prediction theory. In particular, the GM (1, 1) model is widely used in the field of predicting carbon emissions [47]. The GM (1, 1) is as follows:

Suppose $X^{(0)}$ as a time sequence:

$$X^{(0)} = (x^{(0)}(1), x^{(0)}(2), \cdots, x^{(0)}(n)) \tag{9}$$

where $X^{(0)}$ is a nonnegative series

$$X^{(0)} = (x^{(0)}(1), x^{(0)}(2), \cdots, x^{(0)}(n)) \tag{10}$$

where $x^{(0)}(k) \geq 0,\ k = 1, 2, \cdots, n$.

$X^{(1)}$ can be obtained as 1-AGO series of $X^{(0)}$:

$$X^{(1)} = (x^{(1)}(1), x^{(1)}(2), \cdots, x^{(1)}(n)) \tag{11}$$

where $x^{(1)}(k) = \sum_{i=1}^{k} x^{(0)}(i),\ k = 1, 2, \cdots, n$

$Z^{(1)}$ is the adjacent mean value generation sequence of $X^{(1)}$, also called the background value:

$$Z^{(1)} = (z^{(1)}(2), z^{(1)}(3), \cdots, z^{(1)}(n)) \tag{12}$$

where $z^{(1)}(k) = 0.5x^{(1)}(k) + 0.5x^{(1)}(k-1),\ k = 2, 3, \cdots, n$.

The basic form of the GM (1, 1) model can be defined as

$$x^{(0)}(k) + az^{(1)}(k) = b \tag{13}$$

The grey model can be written as the matrix form:

$$Y = B\hat{a} \tag{14}$$

Applying the least squares method, the estimated coefficients $[a, b]^T$ can be calculated as

$$Y = \begin{bmatrix} x^{(0)}(2) \\ x^{(0)}(3) \\ \vdots \\ x^{(0)}(n) \end{bmatrix} \quad B = \begin{bmatrix} -z^{(1)}(2) & 1 \\ -z^{(1)}(3) & 1 \\ \vdots & \vdots \\ -z^{(1)}(n) & 1 \end{bmatrix} \quad \hat{a} = \begin{bmatrix} a \\ b \end{bmatrix} \tag{15}$$

The whitening equation is defined as

$$\frac{dx^{(1)}}{dt} + ax^{(1)} = b \tag{16}$$

With $x^{(1)}(1) = x^{(0)}(1)$, the predicted value $\hat{x}^{(1)}(k+1)$ is calculated as

$$\hat{x}^{(1)}(k+1) = \left(x^{(1)}(0) - \frac{b}{a}\right)e^{-ak} + \frac{b}{a} \quad (k = 1, 2, \cdots, n) \tag{17}$$

4. Results

This section primarily presents the results and application analysis of carbon emission calculation, aiming to explore the characteristics of China's carbon emissions from 2000 to 2019. Firstly, we conducted calculations for total carbon emissions, per capita carbon emissions, and carbon emission intensity to analyze their respective characteristics. Secondly, models were developed to calculate total carbon emissions at various scales, including provinces, cities, and counties. Additionally, per capita carbon emissions and provincial-level carbon emission intensity were calculated to conduct a spatial scale analysis of China's carbon emissions. Lastly, through the utilization of a shift-of-focus model along with trend analysis and prediction techniques for carbon emission data, the spatial pattern of China's carbon emissions was analyzed while its spatiotemporal dynamic changes were also studied.

4.1. Carbon Emission Calculation Based on Statistical Data

4.1.1. National-Scale Carbon Emission Calculation

A. National-scale carbon emissions

The total amount of carbon emissions from energy emissions in China is depicted in Figure 3. From 2000 to 2019, with the exception of 2015, there was a consistent increase in total carbon emissions, escalating from 4.42 Gt in 2000 to 15.83 Gt in 2019. The average annual growth rate during this period was 7.08%. Notably, the growth rate was more rapid between 2003 and 2006, and again in 2013, with an average annual growth rate of approximately 14.56%. The highest growth rate recorded was observed in the year 2004, reaching 17.36%. Subsequently, carbon emissions exhibited a slower pace of growth from 2014 to 2019, with an average annual growth rate of 1.44%. In 2015, there was a notable decline, with a negative growth rate of −1.06%.

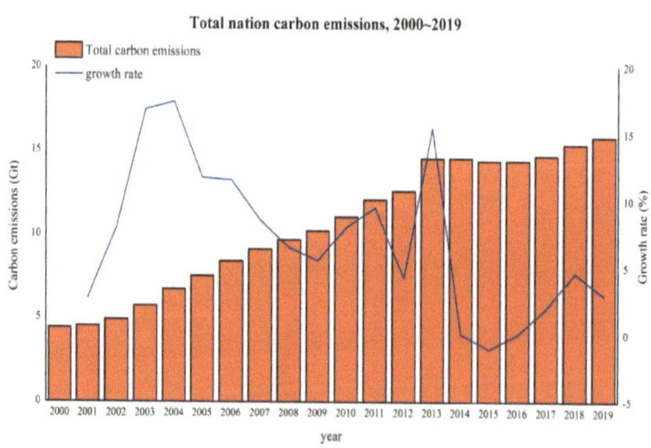

Figure 3. Total national carbon emissions and growth rate, 2000 to 2019.

B. National-scale per capita carbon emissions

China's per capita carbon emissions are presented in Figure 4. Per capita carbon emissions indicated an increasing trend from 2000 to 2019. In 2000, the per capita carbon emissions were 3.49 tons per person, which rose to 11.23 tons per person by 2019. The average annual growth rate during this period was 6.48%. Notably, positive growth rates were recorded from 2000 to 2013 and from 2017 to 2020, with an average annual growth rate of approximately 7.87%. The highest growth rate occurred in the year 2004 (16.67%). Conversely, negative growth rates were observed from 2014 to 2016, resulting in an average annual decline of −0.92%. The largest negative growth rate was registered in the year 2015 (−1.55%).

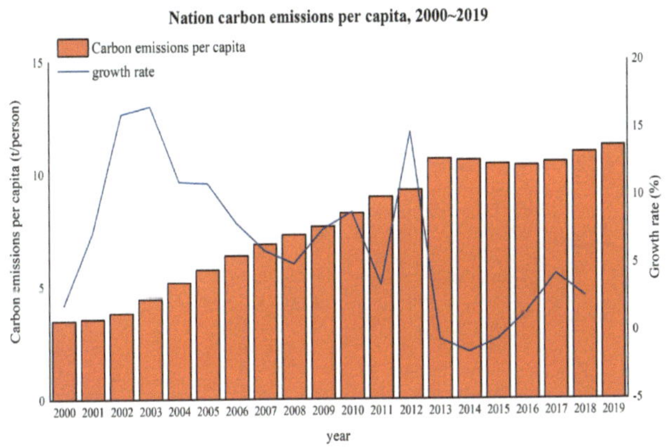

Figure 4. Total national per capita carbon emissions and growth rates, 2000 to 2019.

C. National-scale carbon emission intensity

The term "carbon emission intensity" is defined as the amount of carbon emitted per unit of economic output, where higher carbon intensity indicates less efficient energy use and greater utilization of energy and other resources for the same level of output. In practical research, carbon emissions per unit of gross national product are often utilized to calculate intensity values. Figure 5 illustrates China's national carbon emission intensity and its growth rate from 2000 to 2019. During this period, China experienced a downward trend in carbon emission intensity, with a decrease in total emissions from 4.41 tons per CNY 10,000 to 1.60 tons per CNY 10,000. This corresponded to an average annual reduction rate of 5.09%. These findings signify a notable decline in China's carbon emission intensity over the study period, largely attributable to its economic development.

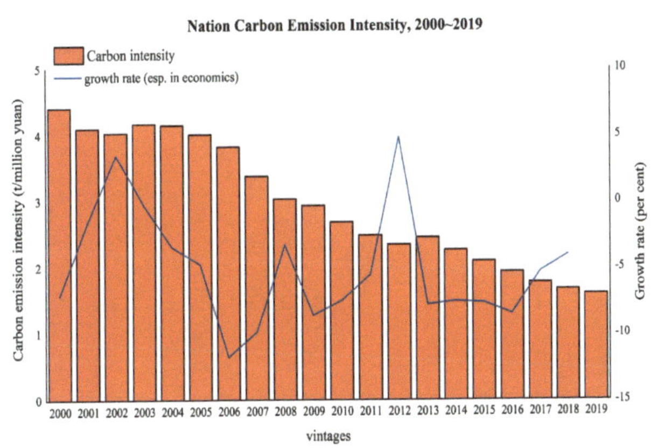

Figure 5. National carbon emission intensity and growth rate, 2000 to 2019.

4.1.2. Provincial-Scale Carbon Emission Calculation

A. Correlation analysis between nighttime lighting and carbon emissions at the provincial scale

To choose a more effective indicator from night light data to fit carbon emissions, this study used the total T-value for correlation with the calculated carbon emission data of each administrative district. The results of this correlation, measured by the Pearson correlation coefficient (R), between the total T-value and carbon emissions are presented in Table 3.

Table 3. Pearson's correlation coefficients between total T-value and carbon emissions.

Provinces	R	Provinces	R
Beijing	0.875	Henan	0.824
Tianjin	0.925	Hubei	0.835
Hebei	0.919	Hunan	0.779
Shanxi	0.856	Guangdong	0.955
Inner Mongolia	0.777	Guangxi	0.949
Liaoning	0.916	Hainan	0.927
Jilin	0.838	Chongqing	0.798
Heilongjiang	0.403	Sichuan	0.771
Shanghai	0.885	Guizhou	0.832
Jiangsu	0.980	Yunnan	0.803
Zhejiang	0.910	Shanxi	0.992
Anhui	0.965	Gansu	0.977
Fujian	0.931	Qinghai	0.966
Jiangxi	0.950	Ningxia	0.950
Shandong	0.879	Xinjiang	0.962
average value	0.878		
variance	0.012		

The correlation between the total T-value and carbon emissions at the province level not only passed the significance level test but also had a high average R-value of 0.878 with a minor variation of 0.012. Except for Heilongjiang, every province had an R-value greater than 0.7. The regions with R-values above 0.9 were Tianjin, Hebei, Liaoning, Jiangsu, Zhejiang, Anhui, Guangdong, Guangxi, Hainan, Shaanxi, Gansu, Qinghai, Ningxia, and Xinjiang. These results indicate a robust connection between the T-value and overall carbon emissions, suggesting the feasibility of estimating China's carbon emissions from its nighttime lighting data.

Owing to the large size of China and the pronounced regional variations in resources, climate, and economics between provinces, three models (logarithmic, exponential, and linear models) were compared to determine the best fit. Ultimately, DN values and the overall carbon dioxide emissions at the province level were computed and fitted. Figure 6 illustrates the fitting impacts of various function models.

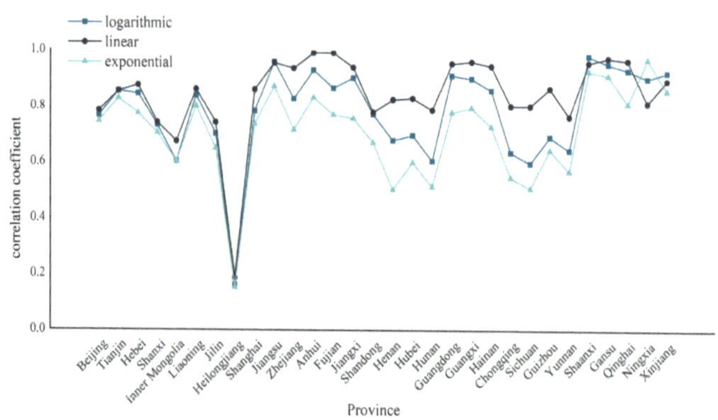

Figure 6. Correlation coefficients fitted by different models (logarithmic, linear, exponential).

In the fitting comparison, the correlation coefficients of the exponential, linear, and logarithmic models were relatively high, with the mean value of their R^2 being more than 0.7. The logarithmic model had the best fitting effect, and the mean value of R^2 was 0.83. Therefore, the logarithmic model was chosen as the calculation model for carbon emissions at the provincial-, municipal-, and county-level scales. The parameters of the logarithmic model for each province are shown in Table 4.

Table 4. Results of logarithmic fitted equations by province.

Province	K	B	R^2	Province	K	B	R^2
Beijing	10,323	−123,246	0.784	Henan	44,159	−549,038	0.824
Tianjin	18,796	−229,368	0.855	Hubei	20,609	−234,537	0.829
Hebei	96,166	−1,000,000	0.876	Hunan	15,023	−162,156	0.788
Shanxi	97,614	−1,000,000	0.742	Guangdong	97,592	−1,000,000	0.955
Inner Mongolia	68,284	−849,388	0.675	Guangxi	15,718	−182,095	0.962
Liaoning	69,561	−891,158	0.861	Hainan	4930	−52,555	0.946
Jilin	22,061	−264,133	0.742	Chongqing	6937.1	−70,012	0.803
Heilongjiang	15,572	−176,999	0.186	Sichuan	13,922	−149,693	0.803
Shanghai	14,669	−169,436	0.861	Guizhou	7679.2	−67,410	0.865
Jiangsu	56,584	−744,330	0.957	Yunnan	12,616	−141,700	0.765
Zhejiang	27,813	−343,871	0.938	Shaanxi	35,147	−431,815	0.959
Anhui	19,060	−218,990	0.992	Gansu	13,832	−155,499	0.976
Fujian	19,154	−231,076	0.992	Qinghai	5176.1	−53,851	0.967
Jiangxi	12,202	−133,406	0.941	Ningxia	16,676	−184,892	0.814
Shandong	160,087	−2,000,000	0.779	Xinjiang	58,191	−748,947	0.896

B. Provincial-scale carbon emissions

In this study, spatial distribution maps of provincial carbon emissions in China were generated based on the logarithmic model (Figure 7). From 2000 to 2019, China's provinces had an overall trend of rising carbon emissions. China's total carbon emissions in 2000 were less than 2.0×10^2 million tons in 22 provinces. By 2005, there were only 11 provinces left. By 2010, only five provinces in China—Beijing, Chongqing, Hainan, Ningxia, and Qinghai—had carbon emissions below 2.0×10^2 million tons. The provinces of Beijing, Hainan, and Qinghai had less than 2.0×10^2 million tons by 2015. The provinces of Beijing, Chongqing, Hainan, and Qinghai fell short of this threshold by 2019. In 2010, the total carbon emissions of Shandong, Hebei, and Jiangsu amounted to 1.24×10^2, 9.40×10^2, and 8.31×10^2 million tons, respectively. The number of provinces surpassing the threshold of 8.00×10^2 million tons reached five in 2015 and increased to seven provinces by 2019. By 2019, Shandong Province exhibited the highest carbon emissions at a level of 1.77×10^2 million tons while all other provinces remained below the threshold of 1.30×10^2 million tons.

By further calculating the annual average of the total carbon emissions of each province, a bar chart depicting the average carbon emissions was generated (Figure 8). The average value of the total carbon emissions of each province was 3.84×10^2 million tons from 2000 to 2019. Shandong had the highest annual average of carbon emissions at 11.1×10^2 million tons, while Hainan had the lowest at 53.7 million tons. Zhejiang, Henan, Inner Mongolia, Guangdong, Liaoning, Shanxi, Jiangsu, Hebei, and Shandong all exceeded the average level of carbon emissions.

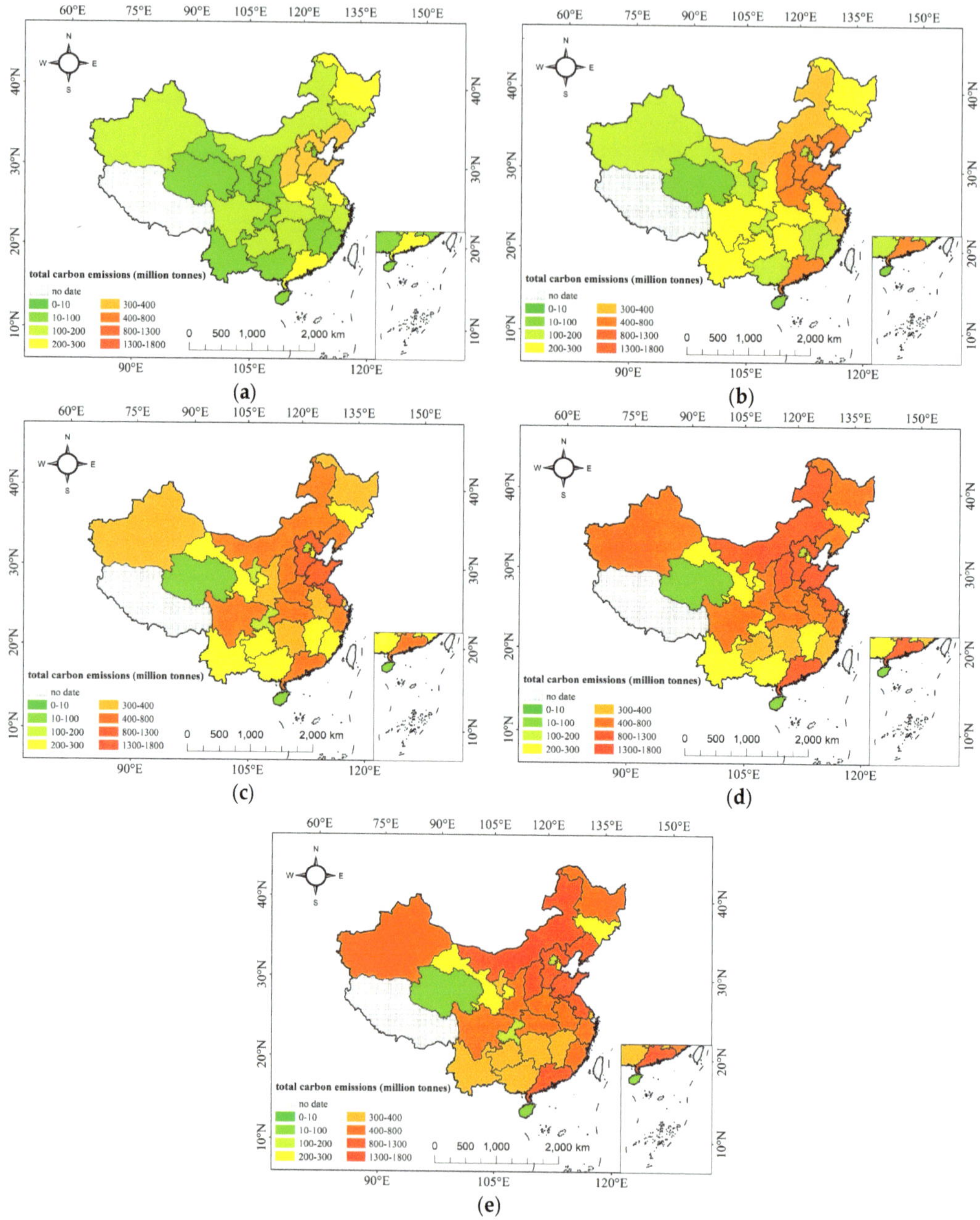

Figure 7. Provincial carbon emissions (2000, 2005, 2010, 2015, 2019). (**a**) Provincial carbon emissions in 2000. (**b**) Provincial carbon emissions in 2005. (**c**) Provincial carbon emissions in 2010. (**d**) Provincial carbon emissions in 2015. (**e**) Provincial carbon emissions in 2019.

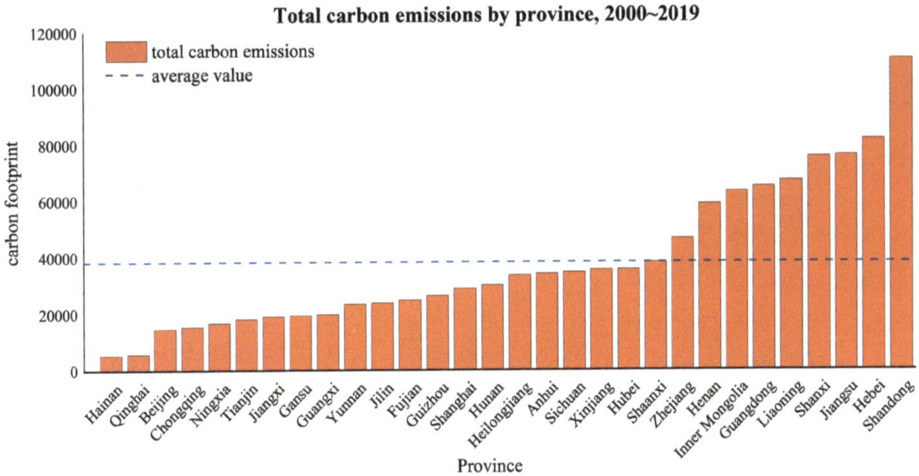

Figure 8. Average provincial carbon emissions in China (2000~2019).

C. Provincial carbon emissions per capita

The provincial per capita carbon emissions were calculated from 2000 to 2019 (Figure 9). Overall, there was an increasing trend in provincial per capita carbon emissions during this period. In 2000, only eight provinces, namely Beijing, Tianjin, Shanghai, Heilongjiang, Jilin, Liaoning, Inner Mongolia, and Shanxi, had per capita carbon emissions of more than 5 t/person. The remaining 22 provinces had less than that amount. Shanxi had the highest per capita carbon emission at 9.86 t/person. By 2005, five provinces—Shanghai, Inner Mongolia, Shanxi, Ningxia, and Liaoning—had carbon emissions within the range of 10–20 t/person, while 11 southern regions (excluding Guizhou) still emitted less than 5 t/person. Other provinces fell between these two ranges. In 2010, the number of provinces with average per capita carbon emissions of less than 5 t/person was reduced to only two provinces, Guangxi and Jiangxi, and the per capita carbon emissions of Inner Mongolia, Ningxia, and Shanxi were greater than 20 t/person. In 2015, the per capita carbon emissions of all provinces were greater than 5 t/person, with Ningxia topping 40 t/person. The per capita carbon emissions of two provinces, Shanxi and Xinjiang, resided in the range of 30–40 t/person. The per capita carbon emissions of three provinces, namely Liaoning, Tianjin, and Shanghai, ranged between 20 and 30 t/person. Additionally, the number of provinces with per capita carbon emissions below 20 t/person decreased to 14. In 2019, the per capita carbon emissions in coastal areas exceeded 10 t/person, while the number of provinces with per capita emissions ranging from 5 to 10 t/person decreased to 10. Notably, Inner Mongolia and Ningxia exhibited significantly higher levels at 51.56 t/person and 64.39 t/person, respectively.

The histogram of the average value of per capita carbon emissions in 30 provinces from 2000 to 2019 was obtained by further calculation (Figure 10). The mean value of per capita carbon emissions across all provinces over a span of two decades was 10.29 t/person. Notably, Guangxi had the lowest per capita carbon emission at 4.17 t/person, while Ningxia recorded the highest at 27.87 t/person. Eight provinces, namely Ningxia, Inner Mongolia, Shanxi, Xinjiang, Tianjin, Liaoning, Shanghai, Shandong, Hebei, and Qinghai, had per capita carbon emissions exceeding the national average.

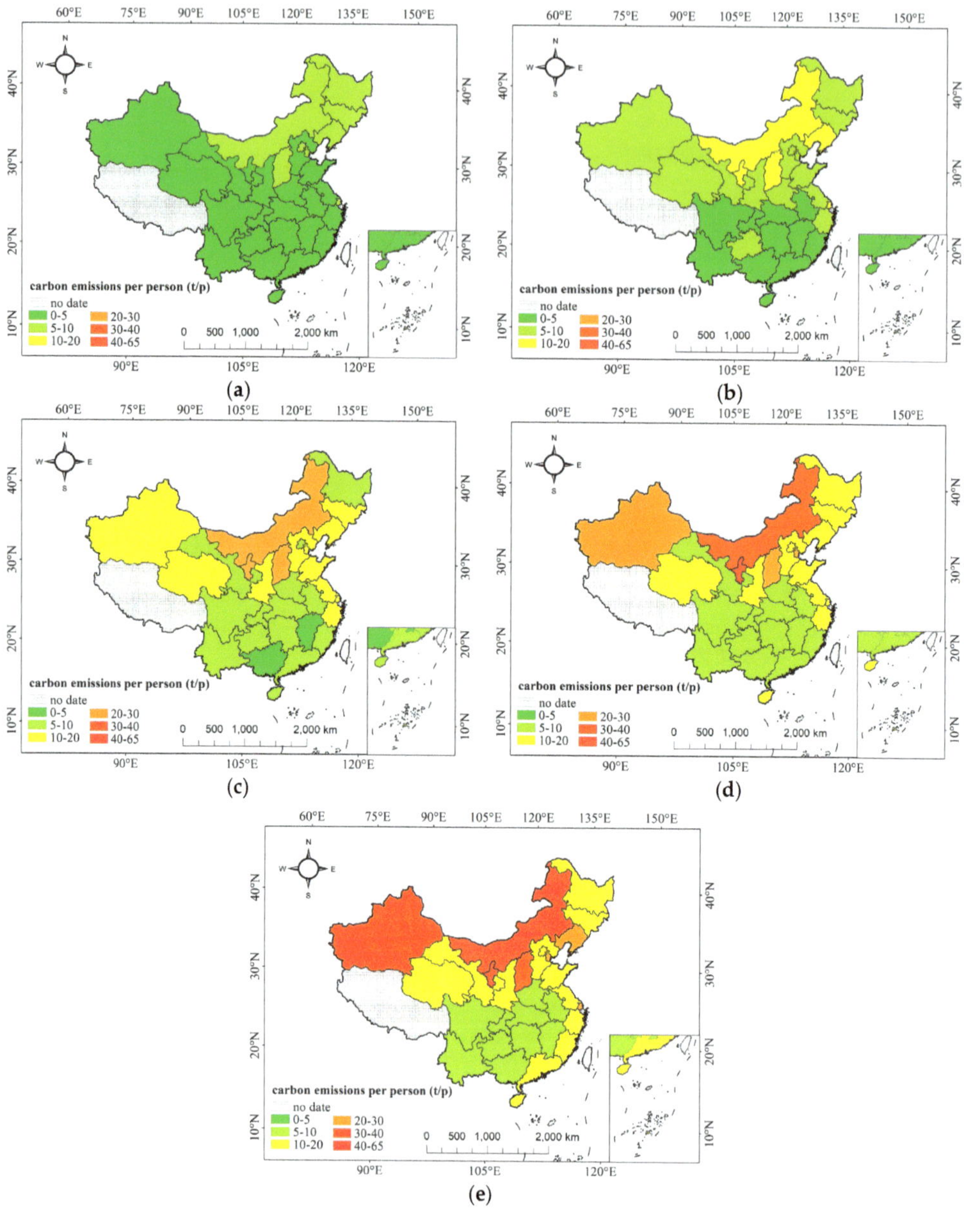

Figure 9. Carbon emissions per capita at the provincial level in China (2000, 2005, 2010, 2015, 2019). (**a**) Provincial carbon emissions per capita in 2000. (**b**) Provincial carbon emissions per capita in 2005. (**c**) Provincial carbon emissions per capita in 2010. (**d**) Provincial carbon emissions per capita in 2015. (**e**) Provincial carbon emissions per capita in 2019.

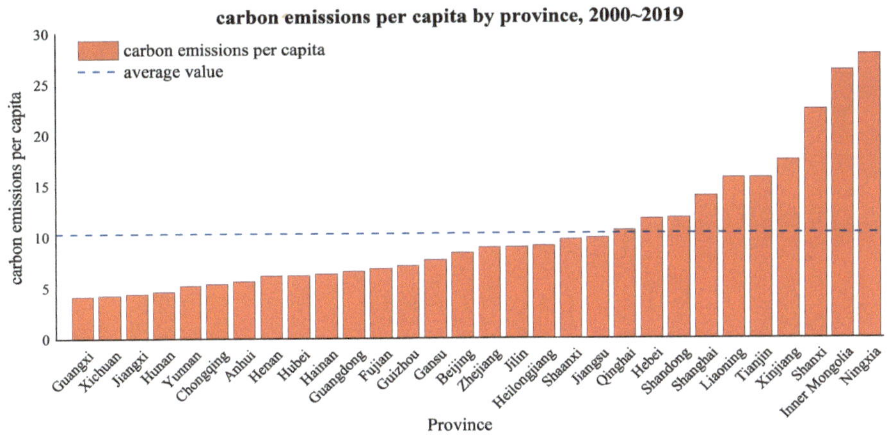

Figure 10. Average provincial per capita carbon emissions in China (2000~2019).

D. Provincial carbon intensity

Figure 11 shows the provincial carbon emission intensity in China from 2000 to 2019. The carbon emission intensity of China's provinces exhibited a declining trend from 2000 to 2019. In 2000, Ningxia exhibited the highest carbon emission intensity at 120.91 t/CNY 10,000. The carbon emission intensity of four provinces, Shanxi, Inner Mongolia, Ningxia, and Xinjiang, was located in the range of 40–80 t/CNY 10,000. There were seven provinces located in the range of 20–40 t/CNY 10,000, namely Hebei, Liaoning, Shandong, Guizhou, Shaanxi, Gansu, and Qinghai. Three provinces, Beijing, Shanghai, and Guangdong, had carbon emission intensities below 10 t/CNY 10,000. In 2005, Ningxia still had the highest carbon emission intensity, at 41.19 t/CNY 10,000. The carbon emission intensity of three provinces, Shanxi, Inner Mongolia, and Xinjiang, decreased to 20–40 t/CNY 10,000. The provinces of Qinghai, Gansu, Shaanxi, Guizhou, Hebei, Liaoning, and Jilin exhibited a range of 10–20 t/CNY 10,000, while the remaining 19 provinces demonstrated values below 10 t/CNY 10,000. In 2010, seven provinces, Beijing, Jiangsu, Shanghai, Zhejiang, Fujian, Jiangxi, and Shandong, had a carbon emission intensity below 2 t/CNY 10,000. In 2015, 15 provinces had a carbon emission intensity of less than 1 t/CNY 10,000; 9 provinces had a carbon emission intensity between 1 and 2 t/CNY 10,000; and 6 provinces, Gansu, Ningxia, Inner Mongolia, Shanxi, Heilongjiang, and Liaoning, had a carbon emission intensity between 2 and 10 t/CNY 10,000. In 2019, only Shanxi exceeded 2 tons per CNY 10,000, while Heilongjiang, Jilin, Liaoning, and Gansu ranged from 1 to 2 tons. The other 25 provinces had intensities below 1 ton.

The average carbon emission intensity value of 30 provinces from 2000 to 2019 was obtained through further calculations (Figure 12). Over the course of 20 decades, each province exhibited an average carbon emission intensity value of 7.37 t/CNY 10,000. Among them, Beijing had the lowest carbon emission intensity of 1.60 t/CNY 10,000. Ningxia had the highest carbon emission intensity, at 28.46 t/CNY 10,000. Thirty percent of the provinces exceed the national average for all provinces, including four provinces (Ningxia, Inner Mongolia, Shanxi, and Xinjiang). The aforementioned phenomenon highlighted the necessity of enhancing energy efficiency in these provinces.

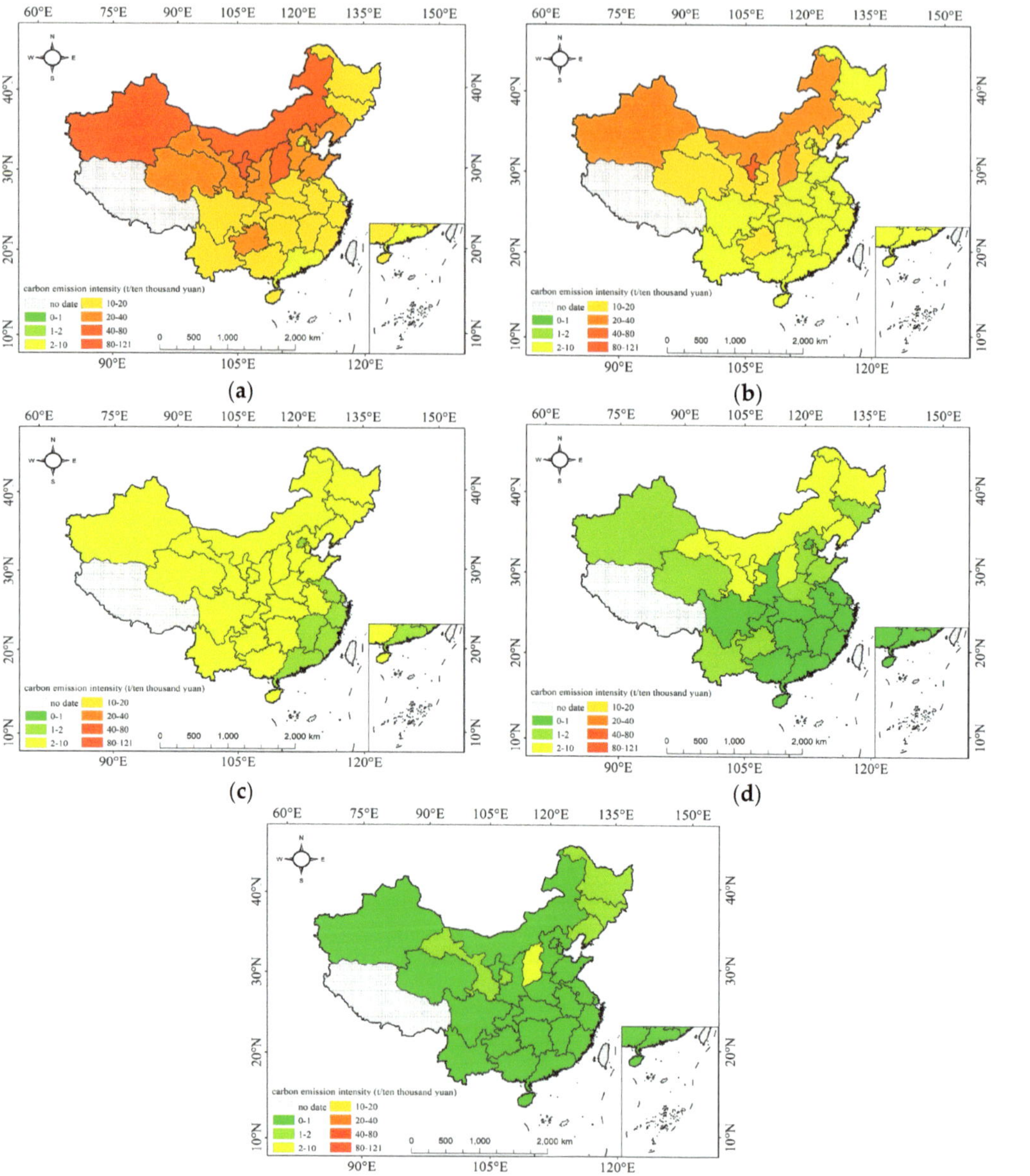

Figure 11. Provincial carbon emission intensity in China (2000, 2005, 2010, 2015, 2019). (**a**) Provincial carbon intensity, 2000. (**b**) Provincial carbon intensity, 2005. (**c**) Provincial carbon intensity in 2010. (**d**) Provincial carbon intensity, 2015. (**e**) Provincial carbon intensity in 2019.

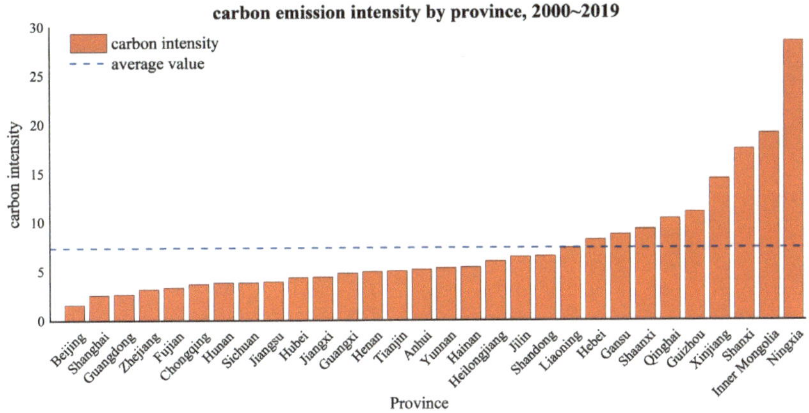

Figure 12. Provincial carbon emission intensity in China (2000~2019).

4.2. Carbon Emission Calculation Based on Remote Sensing Data of Nighttime Lighting

4.2.1. Municipal-Scale Carbon Emission Calculation

Municipal-scale carbon emission accounting was carried out based on the logarithmic model. Figure 13 displays the municipal-scale carbon emissions. The municipal-scale carbon emissions showed an increasing trend in 2000, 2005, 2010, 2015, and 2019, with mean values of 1.35×10^7 t, 2.38×10^7 t, 3.53×10^7 t, 4.15×10^7 t, and 4.71×10^7 t. Meanwhile, the carbon emissions of 30.00%, 29.17%, 30.56%, 32.50%, and 32.22% of the cities were higher than the average value. In 2000, carbon emissions below 1.5×10^7 t were observed in only 73.33% of cities, while the remaining two cities (Beijing and Shanghai) exhibited significantly higher emissions exceeding 1×10^8 t. In 2005, a decrease was noted as only 55.00% of cities had carbon emissions below 1.5×10^7 t, whereas the number of cities surpassing the threshold of 1×10^8 t rose to 13, with Shanghai accounting for emissions greater than 2×10^8 t. By 2010, less than half (40.56%) of China's cities demonstrated carbon emissions below the level of 1.5×10^7 t. Meanwhile, the number of cities with carbon emissions greater than 1×10^8 t increased rapidly to 27. Increasingly concerning trends persisted through to 2015 when approximately one-third (33.61%) of Chinese cities reported carbon emissions at or below the threshold value set at 1.5×10^7 t, while 37 cities exceeded an emission level greater than that specified by 1×10^8 t. In 2019, 29.72% of Chinese cities had carbon emissions less than 1.5×10^7 t, accounting for less than one-third, while the number of cities with carbon emissions greater than 1×10^8 t increased to 48.

Spatial analysis revealed that high-carbon-emission areas were distributed in five major regions in 2000, namely the Heilongjiang–Liaoning–Jilin, Beijing–Tianjin–Hebei, Shanghai–Nanjing–Hangzhou, Pearl River Delta, and Sichuan–Chongqing urban agglomerations. By 2005, the number of cities with high carbon emissions continued to rise within five major regions while also witnessing the emergence of high-value city clusters in areas such as Shandong Peninsula and Shanxi. In 2010, the number of high-value cities increased rapidly, with a notable concentration observed in seven specific regions: (1) Beijing–Tianjin–Hebei urban agglomerations; (2) Shandong Peninsula city cluster; (3) Chongqing and other urban agglomerations of Sichuan–Chongqing; (4) Shanghai–Nanjing–Hangzhou city cluster; (5) Ordos, Hohhot, Baotou, Yulin; (6) Changchun, Harbin, Dalian, and other urban agglomerations in Northeast China; (7) the Pearl River Delta. In 2015, some cities in Xinjiang began to show high carbon emissions, and the carbon emissions of cities in various regions of the country showed significant growth. In 2019, the number of cities with large carbon emissions continued to increase; in particular, the carbon emissions in Beijing–Tianjin–Hebei, Shandong Peninsula, Shanxi, Inner Mongolia, Erdos City, Jiangsu, and Shanghai increased more significantly.

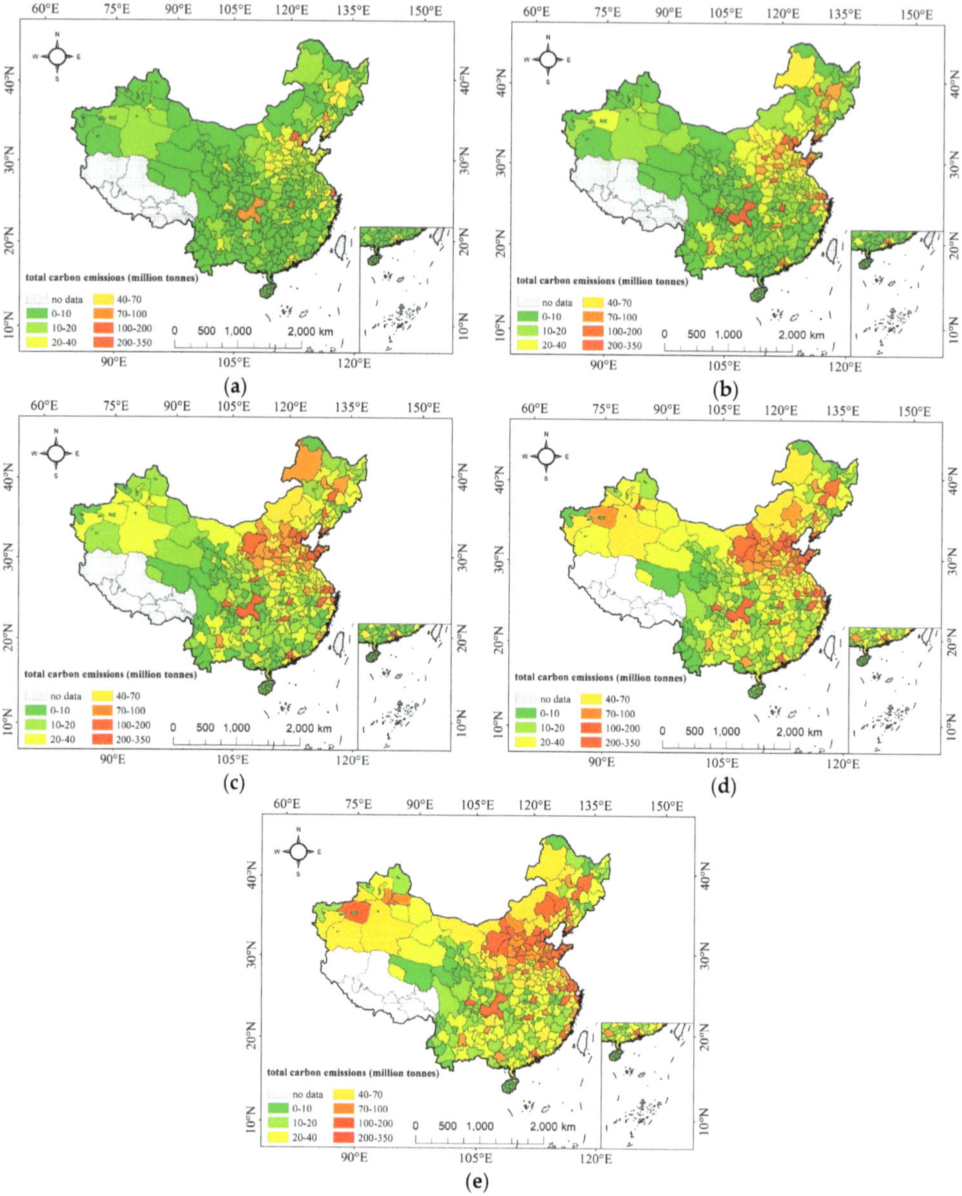

Figure 13. Municipal-scale carbon emissions (2000, 2005, 2010, 2015, 2019). (**a**) Municipal carbon emissions in 2000. (**b**) Municipal carbon emissions in 2005. (**c**) Municipal carbon emissions in 2010. (**d**) Municipal carbon emissions in 2015. (**e**) Municipal carbon emissions in 2019.

4.2.2. County-Scale Carbon Emission Calculation

County-scale carbon emission accounting was carried out based on the logarithmic model. Figure 14 shows the county-scale carbon emissions. China's county-scale carbon emissions increased in 2000, 2005, 2010, 2015, and 2019, with annual averages of 1.71×10^6 t, 3.03×10^6 t, 4.52×10^6 t, 5.32×10^6 t, and 6.04×10^6 t. In these years, the emissions of

70.21%, 69.82%, 68.73%, 69.06%, and 69.46% of counties (districts) were lower than the average value.

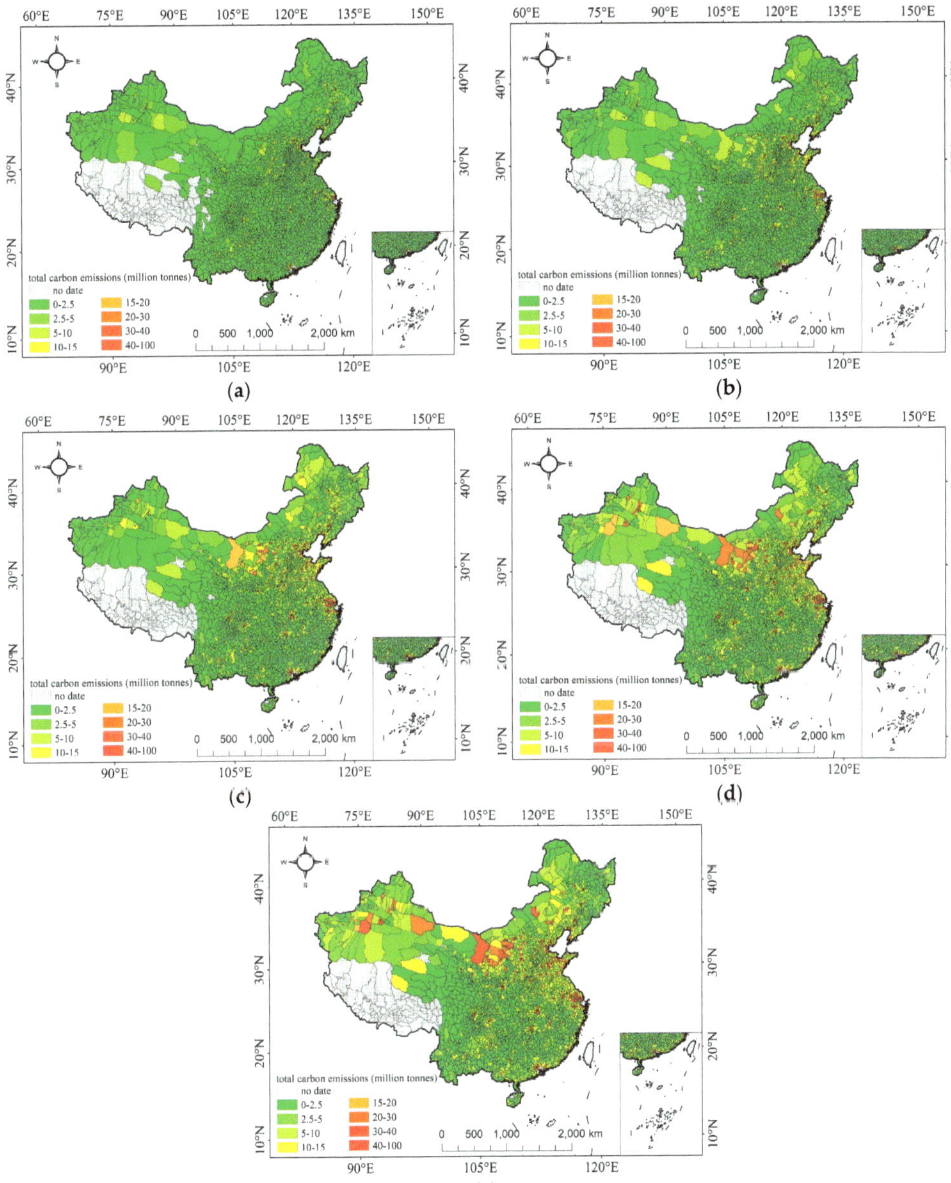

Figure 14. County-scale carbon emissions (2000, 2005, 2010, 2015, 2019). (**a**) County-level carbon emissions in 2000. (**b**) County-level carbon emissions in 2005. (**c**) County-level carbon emissions in 2010. (**d**) County-level carbon emissions in 2015. (**e**) County-level carbon emissions in 2019.

In 2000, 98.45% of the counties in China had carbon emissions lower than 1×10^7 t. Six counties had carbon emissions greater than 2×10^7 t, and only Pudong New Area had carbon emissions greater than 3×10^7 t. In 2005, carbon emissions lower than 1×10^7 t were observed in 93.98% of the counties. The number of counties with carbon emissions exceeding 2×10^7 t had risen to 25, while the number of counties with carbon emissions surpassing 3×10^7 t had increased to 7. In 2010, 87.85% of counties had carbon emissions less than 1×10^7 t, while the number of counties with carbon emissions greater than 2×10^7 t increased to 71, and the number of counties with carbon emissions greater than 3×10^7 t increased to 21. In 2015, 86.51% of counties had carbon emissions below 1×10^7 t, while the number of counties with emissions exceeding 2×10^7 t increased to 113, and the number of counties with emissions exceeding 3×10^7 t increased to 35. By contrast, in 2019, only 83.77% of counties had carbon emissions below the threshold of 1×10^7 t, while the number of counties with emissions greater than or equal to both thresholds rose significantly. There were 146 counties for the former and 61 for the latter.

The overall trend of county-level carbon emissions in China indicated a higher concentration in the eastern regions compared to the western regions, and a greater magnitude in the northern areas relative to the southern areas. In 2000, the number of high-carbon-emission areas at the county scale was low, and these areas had a dotted distribution. In 2005, high-carbon-emission areas increased, scattered in areas such as Beijing–Tianjin–Hebei, Shandong Peninsula, Jiangsu, Zhejiang, Shanghai, and the Pearl River Delta. In 2010, the number of high-carbon-emission areas increased rapidly, and these areas were characterized by a patchy distribution. The high values of carbon emissions were mainly concentrated in the following regions. (1) Tianjin, Shandong, Liaoning; (2) Shanghai; (3) Foshan, Zhongshan; (4) Ordos. Other high-value areas were sporadically located in the central and western parts of China. By 2015, high-carbon-emission areas continued to expand, especially the high-carbon-emission area in Ordos, which was centered around Alxa Left Banner, Etuoke Banner, and Dongsheng District. In 2019, urban areas such as Shaya County, Kuche County, Shihezi City, and Hami City in Xinjiang began to experience high carbon emissions, while the other major carbon emission centers significantly expanded. High-carbon-emission areas in the central and western regions still showed a dotted distribution, such as Banan District in Chongqing, Chenghua District in Chengdu, and Furong District in Changsha.

4.3. Analysis of the Spatial Dependence of China's Energy Carbon Emissions

In this study, the CO_2 emission data from 2000 to 2019 were selected to calculate the centroid of China's CO_2 emissions, incorporating carbon emission weights corresponding to administrative boundaries as a means of capturing spatial variations in China's carbon emissions. Utilizing the center of gravity model, several key outcomes were obtained: the latitude and longitude coordinates of China's CO_2 emissions center of gravity from 2000 to 2019 (Table 5), the migration distance and rate of this center of gravity (Table 6), and the migration trajectory of this center of gravity (Figure 15).

Table 5. Latitude and longitude coordinates of the center of gravity of carbon emissions.

Year	**2000**	**2001**	**2002**	**2003**	**2004**
Longitude Latitude	115.74 34.47	115.63 34.41	115.44 34.30	115.43 34.24	115.35 34.16
Year	2005	2006	2007	2008	2009
Longitude Latitude	115.50 34.20	115.41 34.137	115.33 34.06	115.38 34.08	115.33 34.00
Year	2010	2011	2012	2013	2014
Longitude Latitude	115.37 34.00	115.34 34.00	115.29 34.04	115.23 34.00	115.23 33.98
Year	2015	2016	2017	2018	2019
Longitude Latitude	115.31 33.96	115.34 33.94	115.32 33.91	115.30 33.93	115.31 33.97

Table 6. Distance and rate of carbon emission transport.

Year	2000–2005	2005–2010	2010–2015	2015–2019
Distance (km)	37.00	25.29	7.23	1.29
Speed (km/y)	7.40	5.06	1.45	0.32

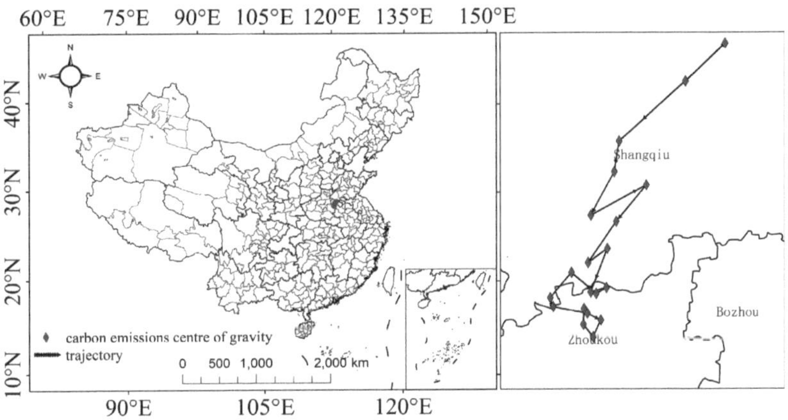

Figure 15. Carbon emission transport pathway in China.

As illustrated in Figure 15, the gravity center of carbon emission was located in Henan Province from 2000 to 2019. From 2000 to 2005, the gravity center of carbon emission shifted 37 km to the southwest, corresponding to 7.40 km/y. In the next 5 years, the same trend was observed but with a rate of 5.06 km/y. The move of this center further slowed down in 2010–2015 at a rate of 1.45 km/y to the southwest. However, this trend was slightly reversed from 2015 to 2019, with an average moving rate of 0.32 km/y to the northwest. The above analysis suggests that the gravity center of carbon emission shifted to the southwest, with a distance of 68.68 km, corresponding to 3.61 km/y.

4.4. *Trend Analysis of Carbon Emissions in China*

Figure 16 reflects the trend of carbon emissions growth at different scales in China from 2000 to 2019. The types of slow change (slow growth and slower growth) were more prevalent than rapid change (faster growth and rapid growth) and medium-speed change. Three provinces exhibited medium growth rates. The types of rapid changes were distributed in most parts of North China and the northern part of Central China. The type of slow change was distributed in the central and western regions and the northeast region. Among them, Shandong belonged to the category of rapid growth. Inner Mongolia and Jiangsu were faster-growing regions. Shanxi, Xinjiang, Hebei, and Guangdong were medium-growth models. The provinces exhibiting slower growth included Jilin, Liaoning, Yunnan, Guizhou, Sichuan, Hunan, Shaanxi, Zhejiang, and Fujian. Meanwhile, Heilongjiang, Hainan, Gansu, Qinghai, Beijing, and Tianjin were characterized by slow growth.

4.5. *Projections of Energy Carbon Emissions*

The forecast of China's carbon emissions based on the grey prediction model is depicted in Figure 17. In 2025, the average value of China's carbon emissions was 7.82 $\times$ 10^2 million tons. The provinces with emissions higher than the average value were Shandong (25.61 $\times$ 10^2 million tons), Jiangsu (16.39 $\times$ 10^2 million tons), Inner Mongolia (17.00 $\times$ 10^2 million tons), Hebei (15.54 $\times$ 10^2 million tons), Shanxi Province (14.16 $\times$ 10^2 million tons), Guangdong (12.70 $\times$ 10^2 million tons), Liaoning (11.54 $\times$ 10^2 million tons), Xinjiang (10.89 $\times$ 10^2 million tons), Henan (10.15 $\times$ 10^2 million tons), and Zhejiang (8.76 $\times$ 10^2 million tons). In 2030, the average value of China's carbon emissions was 9.44 $\times$ 10^2 million tons.

The provinces with emissions higher than the average value were Shandong (31.88 × 10^2 million tons), Jiangsu (20.10 × 10^2 million tons), Inner Mongolia (21.76 × 10^2 million tons), Hebei (18.50 × 10^2 million tons), Shanxi Province (16.77 × 10^2 million tons), Guangdong (15.22 × 10^2 million tons), Xinjiang (14.34 × 10^2 million tons), Liaoning (13.41 × 10^2 million tons), Henan (11.68 × 10^2 million tons), and Zhejiang (10.39 × 10^2 million tons). Spatial analysis indicates China's high-carbon-emission areas in 2025 and 2030 were mainly distributed in the northern part of the country.

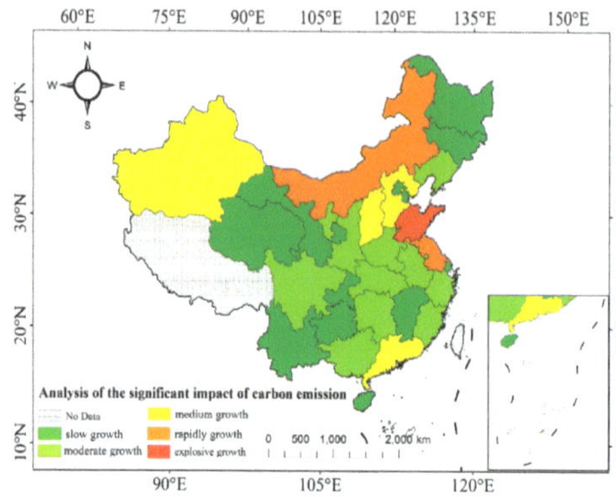

Figure 16. Trend of carbon emissions in China.

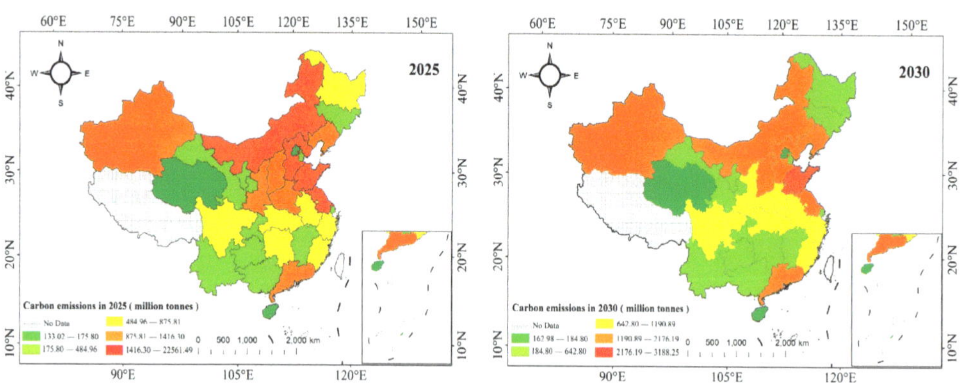

(a) (b)

Figure 17. Carbon emission projections (2025, 2030). (**a**) Carbon emission projections for 2025. (**b**) Carbon emission projections for 2030.

5. Discussion

With the rapid urbanization and economic development in China from 2000 to 2019, both national and provincial scales witnessed a continuous increase in total carbon emissions and per capita carbon emissions (Figures 3 and 4) [48]. However, advancements in science and technological innovation have led to improved energy utilization efficiency, resulting in a decline in the intensity of carbon emissions over time (Figure 5) [49]. To enhance our understanding of controlling total carbon emissions, this study investigated the relationship between carbon emissions and nighttime light brightness (Table 3). A model was constructed to assess carbon emissions at smaller scales (municipal and county levels) and analyze their spatial distribution characteristics (Figure 6).

China's carbon emissions displayed a discernible spatial pattern, characterized by higher levels in the northern regions compared to the southern regions and greater emission levels in the eastern areas relative to the western ones [50]. Shandong Province, as a prominent industrial hub with highly carbon-intensive energy structures, ranked highest among all provinces for its substantial contribution to national carbon emissions (Figure 7) [51]. The center of gravity for China's overall carbon emissions was located within Henan Province, but had been shifting towards the southwest due to various factors (Figure 15). Those included the attraction of industries and population resulting in increased energy demand. Moreover, national policies aimed at addressing regional imbalances had facilitated industrial structure adjustments along with the relocation of traditional heavy industries towards the southwest region, further contributing to this phenomenon [52].

In this study, carbon emissions were calculated not only at the national and provincial levels but also at the municipal and county levels (Figures 13 and 14). This finding compensated for the lack of data at the city and county scales, thereby facilitating precise focus by national and local governments on achieving carbon emission control. The investigation of China's spatial distribution characteristics of carbon emissions holds significant importance in elucidating their spatial patterns and historical dynamics. In light of China's "dual carbon" goal, this research predicted that by 2025 and 2030 [53], the average carbon emissions would be 7.82×10^2 million tons and 9.44×10^2 million tons, respectively. These predictions provide valuable data support for formulating energy conservation and emission reduction policies in China.

As a prominent economic hub with a substantial population, Shandong has experienced a consistent rise in carbon emissions attributed to its robust economic development. The province's electric power industry is highly developed, particularly its coal-based sector which significantly contributes to its overall carbon emissions [54]. Located on the eastern coast of China, Shandong plays a crucial role as an industrial center encompassing 41 major industrial categories. Shandong exhibits both the highest current level of carbon emissions and the highest predicted future increases for two reasons (Figure 17): (1) The dominance of heavy industry in Shandong's economy will persist, leading to significant combustion of fossil fuels and consequent escalation in CO_2 emissions. (2) Plans are underway for major manufacturing industries such as automobile manufacturing and chemical materials within different industrial parks or bases, which will further contribute to elevated levels of carbon emissions.

The research still has some limitations, and the following issues need to be addressed: (1) The calculation of carbon emissions only considers energy consumption factors, neglecting carbon emission factors associated with construction, population, industry, and other aspects. (2) This research primarily focuses on carbon sources. However, it is crucial to consider the impact of carbon sinks on carbon emissions. Since the Industrial Revolution, there has been a substantial decline in forests and grasslands. Therefore, exploring monitoring methods for carbon sinks is imperative. (3) While this study accounts for carbon emissions at the city and county levels, further exploration is needed regarding accounting methods for smaller scales such as towns and villages as well as specific point sources of carbon emissions.

6. Conclusions

Utilizing nighttime light remote sensing, energy statistics, population spatial distribution, GDP spatial distribution, and other data, this study measured the total CO_2 emissions at national and provincial levels over the past 20 years using national statistics data. Nighttime-class EANTLI nighttime light data from 2000 to 2019 were utilized; the sum of DN values (T) was extracted; and the total CO_2 emissions and DN values were adopted to conduct a correlation analysis and linear, exponential, and logarithmic fitting. The best function was selected for conversion model construction. The provincial conversion model was improved; the total CO_2 emissions at municipal and county levels where data are missing were measured; and the characteristics of China's carbon emissions (total carbon emissions, per capita carbon emissions, and carbon emission intensity), spatial scale of carbon emissions (national, provincial, municipal, and county scales), and spatial pattern of carbon emissions (center of gravity migration and trend analysis) were analyzed. Then, the carbon emissions in China were forecasted. The conclusions are as follows:

(1) Nationally, from 2000 to 2019, the total amount of carbon emissions generally increased annually. The total amount rose from 4.42 Gt to 15.83 Gt, with an average annual growth rate of 7.08%. Per capita carbon emissions showed an increasing trend, rising from 3.49 t/person to 11.23 t/person, with an average annual growth rate of 6.48%. The overall trend of China's carbon emission intensity during 2000~2019 displayed a decreasing trend, decreasing from 4.41 tons per CNY 1 million to 1.60 tons per CNY 1 million, with an average annual decrease of 5.09%.

(2) The fitting comparison indicated that the correlation coefficients of the exponential, linear, and logarithmic models were high, with their mean R^2 value exceeding 0.7. The logarithmic model exhibited the best fitting effect, with a mean R^2 value of 0.83. Therefore, a logarithmic model was chosen for the conversion calculation of energy carbon emissions and the sum of DN values.

(3) The center of gravity of carbon emissions was located within Henan Province in China, with a general tendency to move from west to south. The migration of the center of gravity of carbon emissions indicated an increasing proportion and higher growth rate of carbon emissions in the western and southern regions compared to other parts of the country.

(4) Based on the grey prediction model GM (1, 1), China's carbon emissions for 2025 and 2030 were predicted. In 2025, the average value of China's carbon emissions is predicted to be 7.82×10^2 million tons, with Shandong Province having the highest emissions at 225.61×10^2 million tons. In 2030, the average value is expected to be 9.44×10^2 million tons, with Shandong Province remaining the highest at 31.88×10^2 million tons.

As a result of an analysis of the spatial scale of carbon emissions, a carbon emission accounting model was constructed, which solved the problem of a large amount of missing statistical data at the city and county levels during carbon emission accounting. By using nighttime light data to measure carbon emissions, a more detailed scale of carbon emissions could be obtained. This achieved the clarity and targeting of China's total carbon emission control goals. To better achieve the goal of carbon neutrality, China needs to emphasize the need to comprehensively consider the impact of different environments, industrial structures, and socio-economic characteristics during formulation. The results of this study offer valuable insights for the scientific development of China's low-carbon economy and the strategic formulation of carbon emission policies.

Author Contributions: Conceptualization, Z.Z. and Y.S.; methodology, Z.Z. and S.F.; validation, Z.Z. and J.L.; formal analysis, Z.S.; investigation, Y.S., S.F. and J.L.; data curation, S.F., J.L. and Y.Q.; writing—original draft preparation, Z.Z. and Y.S.; writing—review and editing, Z.Z.; funding acquisition, Y.S. All authors have read and agreed to the published version of the manuscript.

Funding: This research was funded by the Open Foundation of Marine Ecological Restoration and Smart Ocean Engineering Research Center of Hebei Province (HBMESO2311) and the National Natural Science Foundation of China (42301391).

Data Availability Statement: The nighttime light remote sensing data are available from https://www.geodoi.ac.cn/WebCn/doi.aspx?Id=3100, accessed on 1 January 2023. The statistical data can be obtained from https://data.stats.gov.cn/, accessed on 1 May 2020.

Conflicts of Interest: The authors declare no conflict of interest.

References

1. Radočaj, D.; Obhođaš, J.; Jurišić, M.; Gašparović, M. Global Open Data Remote Sensing Satellite Missions for Land Monitoring and Conservation: A Review. *Land* **2020**, *9*, 402. [CrossRef]
2. Shiklomanov, A.N.; Bradley, B.A.; Dahlin, K.M.; Fox, A.M.; Gough, C.M.; Hoffman, F.M.; Middleton, E.M.; Serbin, S.P.; Smallman, L.; Smith, W.K. Enhancing Global Change Experiments through Integration of Remote-Sensing Techniques. *Front. Ecol. Environ.* **2019**, *17*, 215–224. [CrossRef]

3. Zhang, H.; Huang, M.; Qing, X.; Li, G.; Tian, C. Bibliometric Analysis of Global Remote Sensing Research during 2010–2015. *ISPRS Int. J. Geo-Inf.* **2017**, *6*, 332. [CrossRef]
4. Akbari, E.; Alavipanah, S.K.; Jeihouni, M.; Hajeb, M.; Haase, D.; Alavipanah, S. A Review of Ocean/Sea Subsurface Water Temperature Studies from Remote Sensing and Non-Remote Sensing Methods. *Water* **2017**, *9*, 936. [CrossRef]
5. Roy, I. Addressing on Abrupt Global Warming, Warming Trend Slowdown and Related Features in Recent Decades. *Front. Earth Sci.* **2018**, *6*, 136. [CrossRef]
6. Fang, J.; Zhu, J.; Wang, S.; Yue, C.; Shen, H. Global Warming, Human-Induced Carbon Emissions, and Their Uncertainties. *Sci. China Earth Sci.* **2011**, *54*, 1458–1468. [CrossRef]
7. Lenton, T.M.; Xu, C.; Abrams, J.F.; Ghadiali, A.; Loriani, S.; Sakschewski, B.; Zimm, C.; Ebi, K.L.; Dunn, R.R.; Svenning, J.-C.; et al. Quantifying the Human Cost of Global Warming. *Nat. Sustain.* **2023**, *6*, 1237–1247. [CrossRef]
8. Yoro, K.O.; Daramola, M.O. CO_2 Emission Sources, Greenhouse Gases, and the Global Warming Effect. In *Advances in Carbon Capture*; Elsevier: Amsterdam, The Netherlands, 2020; pp. 3–28.
9. Zhang, T.; Zhang, W.; Yang, R.; Liu, Y.; Jafari, M. CO_2 Capture and Storage Monitoring Based on Remote Sensing Techniques: A Review. *J. Clean. Prod.* **2021**, *281*, 124409. [CrossRef]
10. Jeffry, L.; Ong, M.Y.; Nomanbhay, S.; Mofijur, M.; Mubashir, M.; Show, P.L. Greenhouse Gases Utilization: A Review. *Fuel* **2021**, *301*, 121017. [CrossRef]
11. Mikhaylov, A.; Moiseev, N.; Aleshin, K.; Burkhardt, T. Global Climate Change and Greenhouse Effect. *Entrep. Sustain. Issues* **2020**, *7*, 2897. [CrossRef]
12. Lee, H.; Calvin, K.; Dasgupta, D.; Krinner, G.; Mukherji, A.; Thorne, P.; Trisos, C.; Romero, J.; Aldunce, P.; Barrett, K.; et al. *AR6 Synthesis Report: Climate Change 2023*; Summary for Policymakers; United Nations Intergovernmental Panel on Climate Change: Interlaken, Switzerland, 2023.
13. Chen, J.; Gao, M.; Huang, S.; Hou, W. Application of Remote Sensing Satellite Data for Carbon Emissions Reduction. *J. Chin. Econ. Bus. Stud.* **2021**, *19*, 109–117. [CrossRef]
14. Shang, M.; Ma, Z.; Su, Y.; Shaheen, F.; Khan, R.; Mohd Tahir, L.; Khalid Anser, M.; Zaman, K. Understanding the Importance of Sustainable Ecological Innovation in Reducing Carbon Emissions: Investigating the Green Energy Demand, Financial Development, Natural Resource Management, Industrialisation and Urbanisation Channels. *Econ. Res.* **2023**, *36*, 2137823. [CrossRef]
15. Gregg, J.S.; Andres, R.J.; Marland, G. China: Emissions Pattern of the World Leader in CO_2 Emissions from Fossil Fuel Consumption and Cement Production. *Geophys. Res. Lett.* **2008**, *35*, 1–5. [CrossRef]
16. Zhang, X.; Geng, Y.; Shao, S.; Wilson, J.; Song, X.; You, W. China's Non-Fossil Energy Development and Its 2030 CO_2 Reduction Targets: The Role of Urbanization. *Appl. Energy* **2020**, *261*, 114353. [CrossRef]
17. Zheng, J.; Mi, Z.; Coffman, D.; Milcheva, S.; Shan, Y.; Guan, D.; Wang, S. Regional Development and Carbon Emissions in China. *Energy Econ.* **2019**, *81*, 25–36. [CrossRef]
18. Wang, Y.; Guo, C.; Chen, X.; Jia, L.; Guo, X.; Chen, R.; Zhang, M.; Chen, Z.; Wang, H. Carbon Peak and Carbon Neutrality in China: Goals, Implementation Path and Prospects. *China Geol.* **2021**, *4*, 720–746. [CrossRef]
19. Obland, M.D.; Campbell, J.; Kooi, S.; Fan, T.F.; Carrion, W.; Hicks, J.; Lin, B.; Nehrir, A.R.; Browell, E.V.; Meadows, B.; et al. Technology Advancements for Active Remote Sensing of Carbon Dioxide from Space Using the Active Sensing of CO_2 Emissions over Nights, Days, and Seasons (ASCENDS) CarbonHawk Experiment Simulator. *EPJ Web Conf.* **2018**, *176*, 02018. [CrossRef]
20. Scheibenreif, L.M.; Mommert, M.; Borth, D. Estimation of Air Pollution with Remote Sensing Data: Revealing Greenhouse Gas Emissions from Space. *arXiv* **2021**, arXiv:2108.13902.
21. Penman, J.; Baltuck, M.; Green, C.; Olofsson, P.; Raison, J.; Woodcock, C. *Integrating Remote-Sensing and Ground-Based Observations for Estimation of Emissions and Removals of Greenhouse Gases in Forests: Methods and Guidance from Global Forest Observation Initiative*; GEO-GFOI MGD Document Version 1.0; Global Forest Observations Initiative: Geo Geneva, Switzerland, 2014.
22. Xiu, L.; Ling-Xue, W.; Wei-Qi, J.; Xia, W. The Development of Optical Remote Measurement for Hazardous Gas Leakage. *Infrared Technol.* **2009**, *31*, 562–563.
23. Global Forest Observations Initiative. *Integration of Remote-Sensing and Ground-Based Observations for Estimation of Emissions and Removals of Greenhouse Gases in Forests: Methods and Guidance from the Global Forest Observations Initiative*; Edition 2.0; Global Forest Observations Initiative: Geneva, Switzerland, 2016.
24. Ju, T.; Liang, Z.; Liu, W.; Li, B.; Huang, R.; Geng, T. Monitoring of Air Pollution by Remote Sensing in Lanzhou City from 2010 to 2019. *Water Air Soil Pollut.* **2022**, *233*, 359. [CrossRef]
25. Tanaka, K.; Hayami, H.; Saino, H.; Miura, K.; Itahashi, S.; Saito, S. Three-Dimensional Analysis of Haze by PM2.5 and Other Pollutants Based on Ground- and Elevated-Level Monitoring in Central Tokyo: Example in Early Part of December 2015. *J. Jpn. Soc. Atmos. Environ.* **2017**, *52*, 51–58.
26. Liu, Z.; Sun, Y.; Zeng, Y.; Guan, Y.; Huang, Y.; Chen, Y.; Li, D.; Mo, L.; Chen, S.; Mai, B. Semi-Volatile Organic Compounds in Fine Particulate Matter on a Tropical Island in the South China Sea. *J. Hazard. Mater.* **2022**, *426*, 128071. [CrossRef]
27. Gao, F.; Wu, J.; Xiao, J.; Li, X.; Liao, S.; Chen, W. Spatially Explicit Carbon Emissions by Remote Sensing and Social Sensing. *Environ. Res.* **2023**, *221*, 115257. [CrossRef]
28. Zhao, M.; Zhou, Y.; Li, X.; Cao, W.; He, C.; Yu, B.; Li, X.; Elvidge, C.D.; Cheng, W.; Zhou, C. Applications of Satellite Remote Sensing of Nighttime Light Observations: Advances, Challenges, and Perspectives. *Remote Sens.* **2019**, *11*, 1971. [CrossRef]

29. Liu, Z. Near-Real-Time Methodology for Assessing Global Carbon Emissions. *Chin. Sci. Bull. Chin.* **2023**, *68*, 830–840. [CrossRef]
30. De Sy, V.; Herold, M.; Achard, F.; Avitabile, V.; Baccini, A.; Carter, S.; Clevers, J.G.; Lindquist, E.; Pereira, M.; Verchot, L. Tropical Deforestation Drivers and Associated Carbon Emission Factors Derived from Remote Sensing Data. *Environ. Res. Lett.* **2019**, *14*, 094022. [CrossRef]
31. Pohl, C.; Van Genderen, J.L. Review Article Multisensor Image Fusion in Remote Sensing: Concepts, Methods and Applications. *Int. J. Remote Sens.* **1998**, *19*, 823–854. [CrossRef]
32. Zhang, Z.; Xu, W.; Qin, Q.; Long, Z. Downscaling Solar-Induced Chlorophyll Fluorescence Based on Convolutional Neural Network Method to Monitor Agricultural Drought. *IEEE Trans. Geosci. Remote Sens.* **2020**, *59*, 1012–1028. [CrossRef]
33. Shanmugapriya, P.; Rathika, S.; Ramesh, T.; Janaki, P. Applications of Remote Sensing in Agriculture-A Review. *Int. J. Curr. Microbiol. Appl. Sci* **2019**, *8*, 2270–2283. [CrossRef]
34. Zhang, Z.; Xu, W.; Shi, Z.; Qin, Q. Establishment of a Comprehensive Drought Monitoring Index Based on Multisource Remote Sensing Data and Agricultural Drought Monitoring. *IEEE J. Sel. Top. Appl. Earth Obs. Remote Sens.* **2021**, *14*, 2113–2126. [CrossRef]
35. Chen, C.; Liang, J.; Xie, F.; Hu, Z.; Sun, W.; Yang, G.; Yu, J.; Chen, L.; Wang, L.; Wang, L.; et al. Temporal and Spatial Variation of Coastline Using Remote Sensing Images for Zhoushan Archipelago, China. *Int. J. Appl. Earth Obs. Geoinf.* **2022**, *107*, 102711. [CrossRef]
36. Chen, C.; Liang, J.; Yang, G.; Sun, W. Spatio-Temporal Distribution of Harmful Algal Blooms and Their Correlations with Marine Hydrological Elements in Offshore Areas, China. *Ocean Coast. Manag.* **2023**, *238*, 106554. [CrossRef]
37. Wu, H.; Li, Z.-L. Scale Issues in Remote Sensing: A Review on Analysis, Processing and Modeling. *Sensors* **2009**, *9*, 1768–1793. [CrossRef]
38. Zhang, L.; Zhang, J.; Li, X.; Zhou, K.; Ye, J. The Impact of Urban Sprawl on Carbon Emissions from the Perspective of Nighttime Light Remote Sensing: A Case Study in Eastern China. *Sustainability* **2023**, *15*, 11940. [CrossRef]
39. Chen, J.; Gao, M.; Cheng, S.; Liu, X.; Hou, W.; Song, M.; Li, D.; Fan, W. China's City-Level Carbon Emissions during 1992–2017 Based on the Inter-Calibration of Nighttime Light Data. *Sci. Rep.* **2021**, *11*, 3323. [CrossRef]
40. Gu, C. Urbanization: Processes and Driving Forces. *Sci. China Earth Sci.* **2019**, *62*, 1351–1360. [CrossRef]
41. Zhou, Y.; Wang, H.; Qiu, H. Population Aging Reduces Carbon Emissions: Evidence from China's Latest Three Censuses. *Appl. Energy* **2023**, *351*, 121799. [CrossRef]
42. Zhong, X.; Yan, Q.; Li, G. Long Time Series Nighttime Light Dataset of China (2000–2020). *Digit. J. Glob. Chang. Data Repos* **2022**, *6*, 416–424.
43. Shan, Y.; Guan, D.; Liu, J.; Mi, Z.; Liu, Z.; Liu, J.; Schroeder, H.; Cai, B.; Chen, Y.; Shao, S.; et al. Methodology and Applications of City Level CO_2 Emission Accounts in China. *J. Clean. Prod.* **2017**, *161*, 1215–1225. [CrossRef]
44. Shi, K.; Chen, Y.; Yu, B.; Xu, T.; Chen, Z.; Liu, R.; Li, L.; Wu, J. Modeling Spatiotemporal CO_2 (Carbon Dioxide) Emission Dynamics in China from DMSP-OLS Nighttime Stable Light Data Using Panel Data Analysis. *Appl. Energy* **2016**, *168*, 523–533. [CrossRef]
45. Zhang, Z.; Xu, W.; Qin, Q.; Chen, Y. Monitoring and Assessment of Agricultural Drought Based on Solar-Induced Chlorophyll Fluorescence during Growing Season in North China Plain. *IEEE J. Sel. Top. Appl. Earth Obs. Remote Sens.* **2020**, *14*, 775–790. [CrossRef]
46. Wang, Y.; Xia, T.; Shataer, R.; Zhang, S.; Li, Z. Analysis of Characteristics and Driving Factors of Land-Use Changes in the Tarim River Basin from 1990 to 2018. *Sustainability* **2021**, *13*, 10263. [CrossRef]
47. Wang, Y.; Dang, Y.; Li, Y.; Liu, S. An Approach to Increase Prediction Precision of GM (1, 1) Model Based on Optimization of the Initial Condition. *Expert Syst. Appl.* **2010**, *37*, 5640–5644. [CrossRef]
48. Yang, T.; Liu, J.; Mi, H.; Cao, Z.; Wang, Y.; Han, H.; Luan, J.; Wang, Z. An Estimating Method for Carbon Emissions of China Based on Nighttime Lights Remote Sensing Satellite Images. *Sustainability* **2022**, *14*, 2269. [CrossRef]
49. Guan, D.; Meng, J.; Reiner, D.M.; Zhang, N.; Shan, Y.; Mi, Z.; Shao, S.; Liu, Z.; Zhang, Q.; Davis, S.J. Structural Decline in China's CO_2 Emissions through Transitions in Industry and Energy Systems. *Nat. Geosci.* **2018**, *11*, 551–555. [CrossRef]
50. Cai, B.; Wang, X.; Huang, G.; Wang, J.; Cao, D.; Baetz, B.W.; Liu, L.; Zhang, H.; Fenech, A.; Liu, Z. Spatiotemporal Changes of China's Carbon Emissions. *Geophys. Res. Lett.* **2018**, *45*, 8536–8546. [CrossRef]
51. Zhu, L.; Xing, H.; Hou, D. Analysis of Carbon Emissions from Land Cover Change during 2000 to 2020 in Shandong Province, China. *Sci. Rep.* **2022**, *12*, 8021. [CrossRef]
52. Xiao, D.; Niu, H.; Guo, J.; Zhao, S.; Fan, L. Carbon Storage Change Analysis and Emission Reduction Suggestions under Land Use Transition: A Case Study of Henan Province, China. *Int. J. Environ. Res. Public Health* **2021**, *18*, 1844. [CrossRef]
53. Wen, L.; Diao, P. Simulation Study on Carbon Emission of China's Electricity Supply and Demand under the Dual-Carbon Target. *J. Clean. Prod.* **2022**, *379*, 134654. [CrossRef]
54. Liu, K.; Wang, J.; Kang, X.; Liu, J.; Xia, Z.; Du, K.; Zhu, X. Spatio-Temporal Analysis of Population-Land-Economic Urbanization and Its Impact on Urban Carbon Emissions in Shandong Province, China. *Land* **2022**, *11*, 266. [CrossRef]

remote sensing

Article

High-Resolution Mapping of Mangrove Species Height in Fujian Zhangjiangkou National Mangrove Nature Reserve Combined GF-2, GF-3, and UAV-LiDAR

Ran Chen [1,2], Rong Zhang [1], Chuanpeng Zhao [1], Zongming Wang [1] and Mingming Jia [1,*]

1 Key Laboratory of Wetland Ecology and Environment, Northeast Institute of Geography and Agroecology, Chinese Academy of Sciences, Changchun 130102, China; chenran211@mails.ucas.ac.cn (R.C.); zhangrong@iga.ac.cn (R.Z.); zhaochuanpeng@iga.ac.cn (C.Z.); zongmingwang@iga.ac.cn (Z.W.)

2 University of Chinese Academy of Sciences, Beijing 101408, China

* Correspondence: jiamingming@iga.ac.cn

Abstract: Mangroves as an important blue carbon ecosystem have a unique ability to sequester and store large amounts of carbon. The height of mangrove forest is considered to be a critical factor in evaluating carbon sink capacity. However, considering the highly complicated nature of the mangrove system, accurate estimation of mangrove species height is challenging. Gaofen-2 (GF-2) panchromatic and multispectral sensor (PMS), Gaofen-3 (GF-3) SAR images, and unmanned aerial vehicle-light detection and ranging (UAV-LiDAR) data have the capability to capture detailed information about both the horizontal and vertical structures of mangrove forests, which offer a cost-effective and reliable approach to predict mangrove species height. To accurately estimate mangrove species height, this study obtained a variety of characteristic parameters from GF-2 PMS and GF-3 SAR data and utilized the canopy height model (CHM) derived from UAV-LiDAR data as the observed data of mangrove forest height. Based on these parameters and the random forest (RF) regression algorithm, the mangrove species height result had a root-mean-square error (RMSE) of 0.91 m and an R^2 of 0.71. The *Kandelia obovate* (KO) exhibited the tallest tree height, reaching a maximum of 9.6 m. The polarization features, HH, VV, and texture feature, mean_1 (calculated based on the mean value of blue band in GF-2 image), had a reasonable correlation with canopy height. Among them, the most significant factor in determining the height of mangrove forest was HH. In areas where it is difficult to conduct field surveys, the results provided an opportunity to update access to acquire forest structural attributes.

Keywords: tree height; GF-2 optical; GF-3 SAR; UAV-LiDAR; random forest

Citation: Chen, R.; Zhang, R.; Zhao, C.; Wang, Z.; Jia, M. High-Resolution Mapping of Mangrove Species Height in Fujian Zhangjiangkou National Mangrove Nature Reserve Combined GF-2, GF-3, and UAV-LiDAR. *Remote Sens.* **2023**, *15*, 5645. https://doi.org/10.3390/rs15245645

Academic Editors: Yanni Dong, Tao Chen and Chao Chen

Received: 8 November 2023
Revised: 30 November 2023
Accepted: 4 December 2023
Published: 6 December 2023

1. Introduction

Mangrove forests are plant communities consisting of evergreen trees or shrubs, thriving in the tropical and subtropical inter tidal coasts [1]. They play a crucial role in preventing coastal erosion, maintaining water quality, and preserving biodiversity [2–4]. Despite mangrove forests only constituting less than 1% of the total tropical forest area, they remarkably contribute to 11% of the total terrestrial carbon input into the ocean [5–7]. However, human activities like deforestation, coastal development, and pollutant discharge have significantly reduced the global mangrove forest area [8,9], endangering their ecological services and carbon storage. Therefore, mapping the spatial distribution of mangrove species and height can furnish a crucial database for the conservation and management of mangrove forests. In addition, accurately and promptly identified tree height provides a key data basis for estimating mangrove forests' carbon storage and carbon fluxes. However, accurately mapping mangrove forest height presents numerous challenges due to the absence of sensitive image features and the low accuracy of existing retrieval models.

Mangrove forest heights are influenced by complex factors such as tides and community structure [10]. Traditional surveys, while comprehensive, are time-consuming and labor-intensive, making it challenging to meet the demands of real-time, large-scale monitoring [11]. Recently, remote sensing technology has gained popularity as a tool for acquiring tree height information due to its extensive geographical coverage, real-time data acquisition, and multi-angle imaging capabilities. For example, Hyyppä et al. [12] used Spot PAN and Landsat TM data to retrieve forest mean height; Kayitakire et al. [13] assessed the ability of 1 m resolution IKONOS-2 images to estimate the tree height of common spruce. However, these methods often yield low estimation accuracies as optical imagery is not sensitive to the vertical information of trees [14]. The advent and rapid adoption of light detection and ranging (LiDAR) technologies, which provide precise three-dimensional information, have significantly improved the accuracy of tree height capture [15]. Simard et al. [16] produced a global wall-to-wall canopy height map based on the Shuttle Radar Topography Mission (SRTM) 30 m resolution global digital elevation model (DEM), and Geoscience Laser Altimeter System (GLAS) global Lidar altimetry products. A study by Lefsky et al. [17] indicated that stand structure and the mean height and height variability found in LiDAR data are closely associated. However, spaceborne LiDAR data is typically used only for obtaining large-scale tree height information due to its large laser beam width and low laser point density. Furthermore, the high cost of acquiring airborne LiDAR data for the same area is likely one of the biggest obstacles hindering research on large-area tree height extraction. As a result, an increasing number of studies are combining optical images and LiDAR data to accurately map tree heights across large areas. For example, Wang et al. [18] estimated the mangrove forests' overall height and above-ground biomass (AGB) throughout Hainan Island by combining field sample plots, UAV-LiDAR data and Sentinel-2 image. Zhu et al. [19] combined Landsat-8 images and Multiple Altimeter Beam Experimental Lidar (MABEL) data to create a map of tree height with a 30 m spatial resolution. However, optical images have limitations including saturation problems and the influence of weather conditions, atmosphere, and moisture [20].

With its all-weather, day-and-night operation and penetration capabilities, Synthetic Aperture Radar (SAR) is widely utilized for tree height retrieval [21,22]. Previous studies have demonstrated a strong correlation between tree height and polarization decomposition parameters, as well as radar vegetation indices extracted from polarization SAR [23–25]. Furthermore, SAR data can be integrated with optical imagery to capture three-dimensional structural information of forests [26]. Ghosh et al. [27] investigated the potential of Sentinel-2 optical and Sentinel-1 SAR data to develop a novel approach for canopy height estimation in the Bhitarkanika wildlife sanctuary. Existing research primarily integrated Sentinel-2 optical and Sentinel-1 SAR data to construct a feature set for tree height inversion, enabling regional tree height mapping. As the spatial resolution of images is improved, more abundant structural details regarding mangrove forests can be obtained. Compared with the most commonly used Sentinel-1 SAR and Sentinel-2 optical imagery, the high-spatial-resolution Gaofen-3 SAR and Gaofen-2 optical imagery could provide more spectral and structural information for mapping tree height. However, it is not yet clear the capacity of integrating GF-2 Optical and GF-3 SAR for estimating mangrove forest height.

For tree height retrieval, machine learning is a currently widely employed method, with the advantages of low sample consumption and few parameters [28]. In these studies, a common task is use a series of remote sensing observations to develop a predictive model, aiming to predict forest conditions or attributes in unfamiliar scenarios [29]. Among these methods, RF is the most popular and widely used [30]. In a study by Wilkes et al. [31], canopy height was predicted in a 2.9 million ha area of heterogeneous temperate forests by utilizing the RF algorithm, which linked LiDAR-derived canopy height with a combination of satellite imagery (Landsat and MODIS), resulting in an RMSE of 5.68 m. Wang et al. [18] combined field sample plots, UAV-LiDAR data and Sentinel-2 image to retrieve the height of Hainan Island's mangrove forest using the RF algorithm, resulting in an R^2 value of 0.67.

These studies demonstrated the ability of RF regression model to unveil intricate non-linear connections among variables and the simplicity in tuning the model parameters [30]. Currently, there is a lot of research focused on obtaining forest structure information using RF model [18,19], but there are few studies on mangrove forest height estimation based on GF-2, GF-3, and UAV-LiDAR data. On the other hand, the addition of the mangrove species information could improve the accuracy and stability of the RF regression model in retrieving mangrove forest height.

Based on the above, the purpose of this research is to examine the possibility of using the RF algorithm to estimate the height of mangrove species combining GF-2, GF-3, and UAV-LiDAR data. We extracted characteristic parameters of estimating mangrove species height based on GF-2 and GF-3 imagery. And the CHM derived from UAV-LiDAR data was used as the observed height data of mangrove forest for model training and validation. The specific objectives of this paper include (1) extracting mangrove extent and species information; (2) mining the characteristic parameters related to mangrove species height from GF-2 and GF-3 imagery; (3) mapping the height of the mangrove species by combining UAV-LiDAR, characteristic parameters and RF regression model in the study area; and (4) analyzing the validity parameters used to retrieve the mangrove species height and the spatial characteristics of the mangrove species height in our study area. The results of this study will contribute to a better understanding of mangrove ecosystems and their ecological dynamics, providing basic data for conservation and management efforts. Furthermore, the proposed methodology combining satellite imagery and UAV-LiDAR data with the RF algorithm can serve as a foundation for future research in vegetation height estimation and monitoring in other coastal regions.

2. Materials and Methods

2.1. Study Area

The Fujian Zhangjiangkou National Mangrove Nature Reserve (FZNNR) (117°24′~117°30′E, 23°53′~23°56′N) is situated in the Zhangjiang River Estuary in Yunxiao, Fujian Province, with a total area of 2360 hm^2. The reserve was established in 1992 and was designated as a National Nature Reserve in 2003. In 2008, it was recognized as a Wetlands of International Importance. The reserve was established in 1992, and in 2003, it received the designation of National Nature Reserve. It was recognized as an Internationally Significant Wetland in 2008. This region boasts the most extensive and thriving natural mangrove community located north of the Tropic of Cancer [32]. The mangroves of FZNNR provides an ideal habitat for more than 150 species of birds, 240 species of aquatic animals, and nearly 400 species of aquatic organisms. The majority of the mangrove forests are concentrated within the core zone of the FZNNR (Figure 1).

FZNNR is a warm humid temperate region with a subtropical maritime monsoon climate. The average yearly temperature is 21.2 °C, and the average yearly precipitation is 1714.5 mm, with the majority of the precipitation falling between April and September. The reserve is generally composed of estuarine waters, intertidal forest wetlands (mangrove swamps), mudflats (intertidal mudflats), and salt marshes, of which the most important type is mangrove forests. *Kandelia obovate* (KO), *Aegiceras corniculatum* (AC), and Avicennia marina (AM) are dominant species which account for 55% of the total wetland of FZNNR. Since 2014, *Spartina alterniflora* (SA) has been expanding explosively in the mudflats at the seaward mudflats. Nowadays, the invasion is still intensifying, Spartina alterniflora is competing with the mangroves at the fringe, and threatening the habitat of local mangroves.

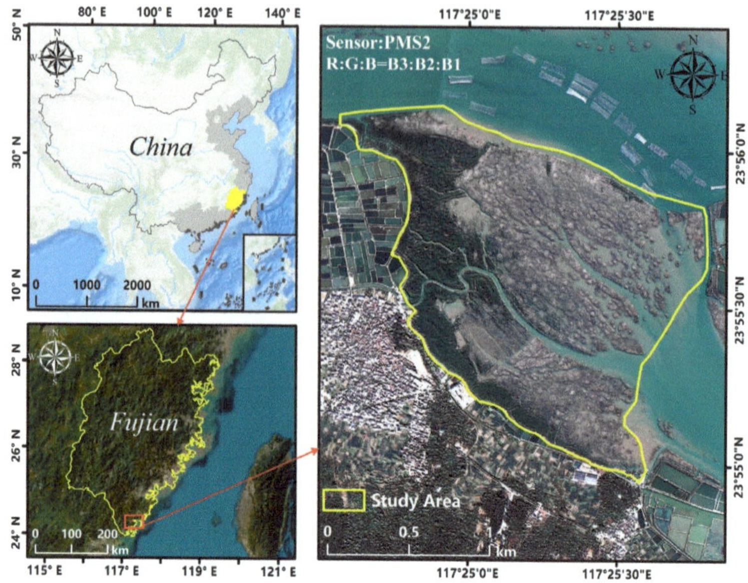

Figure 1. Location of the study area.

2.2. Data and Pre-Processing

2.2.1. UAV Multi-Spectral Data

The UAV multi-spectral image utilized within the scope of the research was acquired from a DJI Genie 4pro (DJI, Shenzhen, China) equipped with a 1-inch CMOS (Complementary Metal-Oxide-Semiconductor) image sensor. The data were collected on 12 November 2020 when weather was clear and sunny. The collection conditions were low tide period. The flight altitude was set to 200 m, and the course overlap and side overlap were 75% and 80%, respectively. Using Agisoft Metashape-pro 1.8.5 software, we carried out the pre-processing of the UAV image. The processed UAV multi-spectral image has a 0.05 m resolution with the near-infrared band, green band, red band, and blue band, and is used for mangrove extent extraction and species classification.

2.2.2. Gaofen-2 and Gaofen-3 Data

GF-2 PMS image and GF-3 polarimetric SAR data were used to calculate the characteristic indices for the retrieval of mangrove species height. GF-2 is the first civilian optical remote sensing satellite independently developed by China with a revisit period of 5 days. The images include four multispectral bands with 4 m resolution and one panchromatic band with 1 m resolution. The cloud-free GF-2 images were obtained from the website of the China Resources Satellite Application Center (http://www.cresda.com, accessed on 24 June 2023) during low-tide on 25 November 2020. The GF-2 images pre-processing include radiometric calibration, atmospheric correction, geometric correction, and image fusion. Atmospheric correction was carried out using FLAASH atmospheric correction module to remove the effects of atmospheric scattering and absorption. Multispectral bands and panchromatic band were fused using the nearest neighbor diffusion pan-sharpening algorithm. The pre-processing was conducted using the Chinese Domestic Satellite Support Tool in ENVI 5.3.

Developed by the China National Space Administration (CNSA), GF-3 SAR was launched in August 2016 and it is the first Chinese satellite to gather C-band SAR data with multiple polarizations. The spatial resolution of GF-3 data can range from 1 m to 500 m, and the imaging modes cover both single and full polarizations. GF-3 has the most imaging modes (12 imaging modes) covering the traditional strip imaging mode and scanning imaging mode, as well as wave imaging mode for ocean applications and global observation imaging mode. The China Resources Satellite Application Center website (http://www.cresda.com, accessed on 12 April 2023) provided the GF-3 polarimetric SAR data. We acquired two images covering the study area, with level 1A products of SLC (Single-Looking Complex Product) type. One FSI (Fine Striping I) mode image was taken on 29 December 2020, with dual polarization (VV and VH) at 2.6 m spatial resolution. The other FSII (Fine Striping II) mode image was taken on 4 June 2022, with dual polarization (HH and HV) at 4.8 m spatial resolution. The pre-processing was carried out making use of the SARscape module in ENVI 5.6, including data import and generation of SLC files, multi-view processing, filtering, geocoding, and radiometric calibration.

2.2.3. UAV-LiDAR Data

We collected the UAV-LiDAR data on 24 September 2022 during low-tide period by a HS40P sensor mounted on a DJI M600 PRO UAV (DJI, Shenzhen, China). Capable of producing up to 720,000 points per second, the laser sensor operates at a wavelength of 905 nm. In the study area, a total of four flights were conducted at a speed of 6 m/s and an altitude of 100 m. The main processing steps for UAV-LiDAR data include point cloud decomposition, denoising, and ground point classification. Based on base station and POS data, the point cloud decomposition was conducted to calculate accurate geographical positions using LiAcquire 3.3.0 software. The rest of data processing was conducted in LiDAR360 5.3.0 software, including point cloud denoising, and ground and non-ground point classification [33]. At a spatial resolution of 0.5 m, a digital surface model (DSM) and digital elevation model (DEM) were created. UAV-LiDAR data can obtain mangrove forest height with high accuracy [34]. The CHM was produced by deducting the DEM from the DSM. The CHM was used for training and verifying the height prediction model. A total of 340 sample points were randomly selected, and the values of LiDAR-derived height and 47 features were extracted for the training and validation of the height model.

2.2.4. Reference Data

We carried out field surveys in FZNNR from 12 to 15 November 2020. To record the species of mangrove forest and their respective coordinates, a Global Navigation Satellite System (GNSS) equipment was used. Finally, we determined 140 ground points within the scope of the research as a result of the complex growing environment of mangrove forests. According to the field surveys, KO, AC, and AM are the primary mangrove species found in the study area.

2.3. Methods

Figure 2 displays the overall workflow of this study. Firstly, we identified mangrove species based on object-based random forest (OBRF) method and the UAV multi-spectral images. This classification process encompassed multi-scale segmentation, feature extraction, classification, and accuracy assessment. Then, we built a mangrove species height model based on RF regression algorithm and GF-2 images, GF-3 images, and UAV-LiDAR to retrieve the mangrove species height of FZNNR. The retrieval step included feature extraction, model construction, retrieval, and evaluation. Finally, we analyzed the spatial distribution characteristics of mangrove species height and explored the key features for mangrove species height retrieval.

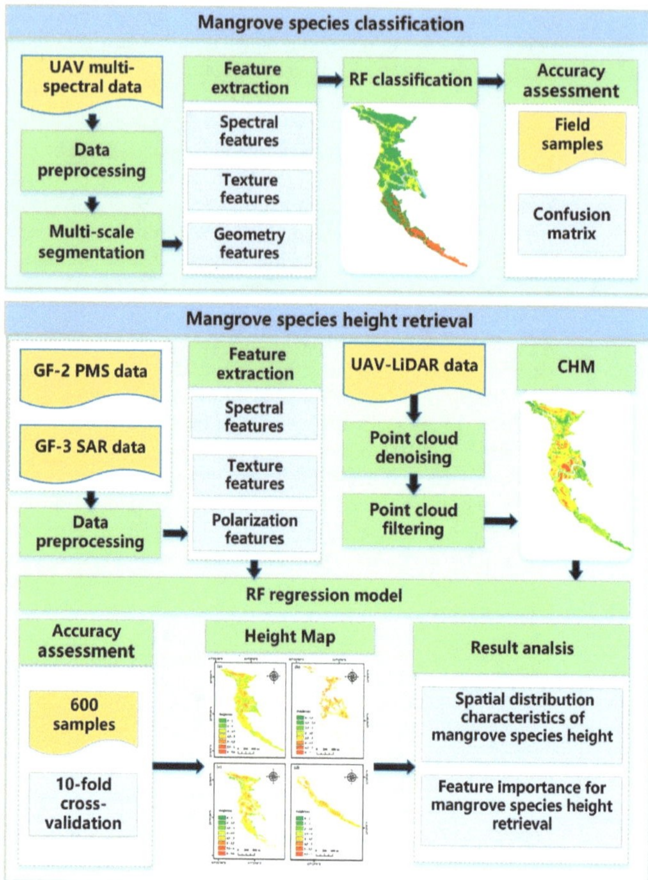

Figure 2. The workflow overview of mapping the mangrove species height.

2.3.1. Mangrove Species Classification

To extract mangrove species, we set six landscape types, AC, KO, AM, SA, water, and mudflat, based on the field investigation of landscape types in the study area. In the study, OBRF method was implemented on the Trimble's eCognition Developer version 9 software to identify mangrove species.

OBRF classification method takes the image object as the fundamental unit of analysis, incorporating spectral features, texture, shape, and contextual information to distinguish landscape types. In this method, multi-scale segmentation was used to establish a hierarchical network structure of image objects, including the shape, compactness, and shape parameters [35]. The scale parameter is the determining parameter for the segmentation effect. The optimal scale parameter depends on the sensor type, resolution, segmentation objectives, and target objects. In this study, the optimal scale parameter for mangrove species classification was determined by visual experience based on the principle of "homogeneity" of the spectral and spatial. Finally, the shape value, compactness value and scale value were set to 0.2, 0.8, and 75, respectively.

We extracted the spectral reflectance, spectral indices, and texture information as the classification features based on UAV multi-spectral images. The specific classification features included mean and standard deviation of the four spectral bands (blue, green, red, and NIR), six spectral indices (NDVI, DVI, FDI, Cig, EVI, and SR), two geometry indices, and seven texture indices (Table 1). Then, the RF algorithm was applied to classify mangrove species based on the homogeneous image objects. RF algorithm can handle high and non-normal distribution data, and can integrate image data sources of different dimensions with auxiliary data into the workflow of image classification, which is a mature machine classification model at present. Ntree and mtry are two important parameters that affect the accuracy and operational efficiency of RF model. The variables ntree and mtry indicate the number of decision trees needed to construct the RF model and the maximum number of features randomly selected from the dataset for split decisions made during each decision tree creation, respectively. In this RF model, we set mtry and ntree to 5 and 500, respectively.

Table 1. The formulas of the 47 features for mangrove species height retrieval.

Features			Formula
Features based on GF-2	Spectral bands	B1, B2, B3, B4	-
	Vegetation indices	Cig [36]	$(B4/B2) - 1$
		DVI [37]	$B4 - B3$
		EVI [38]	$\frac{2.5\times(B4-B3)}{B4+6\times B3-7.5\times B1+1}$
		FDI [39]	$B4 - (B2 + B3)$
		NDVI [40]	$(B4 - B3)/(B4 + B3)$
		SR [41]	$B4/B3$
		TNDVI [42]	$\sqrt{0.5 + \frac{B4-B3}{B4+B3}}$
	Texture features	Mean [43]	$\frac{1}{MN}\sum_{i=0}^{M-1}\sum_{j=0}^{N-1} f(i,j,d,\theta)$
		Variance [43]	$\sum_{i=0}^{M-1}\sum_{j=0}^{N-1} (i-\mu)^2 f(i,j,d,\theta)$
		Homogeneity [43]	$\sum_{i=0}^{M-1}\sum_{j=0}^{N-1} \frac{f(i,j,d,\theta)}{1+(I-J)^2}$
		Contrast [43]	$\sum_{i=0}^{M-1}\sum_{j=0}^{N-1} (i-j)^2 f(i,j,d,\theta)$
		Dissimilarity [43]	$\sum_{i=0}^{M-1}\sum_{j=0}^{N-1} \lvert i-j\rvert f(i,j,d,\theta)$
		Entropy [43]	$-\sum_{i=0}^{M-1}\sum_{j=0}^{N-1} f(i,j) lg f(i,j,d,\theta)$
		Second Moment [43]	$\sum_{i=0}^{M-1}\sum_{j=0}^{N-1} f(i,j,d,\theta)^2$
		Correlation [43]	$\sum_{i=0}^{M-1}\sum_{j=0}^{N-1} \frac{(i-\mu)(j-\mu)f(i,j,d,\theta)^2}{\delta^2}$
Features based on GF-3	Polarization parameters	HH	-
		HV	-
		VV	-
		VH	-

2.3.2. Mangrove Species Height Retrieval

To estimate the mangrove species height within the FZNNR, we extracted 47 features from GF-2 and GF-3 data for building the height model. Based on GF-2, we extracted four band reflectance features (blue, green, red, and NIR), seven vegetation index features (CIg, DVI, EVI, FDI, NDVI, SR, TNDVI), and 32 texture features (Mean, Variance, Homogeneity, Contrast, Dissimilarity, Entropy, Second Moment, and Correlation for each single band). Based on GF-3 SAR image, we obtained four polarization features (HH, HV, VV, and VH). The 47 features are shown in Table 1.

2.3.3. Random Forest Regression Model

The height of mangrove species in FZNNR was estimated using the RF regression algorithm. The RF regression algorithm is a nonparametric statistical estimation method that explores the complex nonlinear relationships between dependent variables and predictor variables [44]. The basic theory of RF regression model is the building of numerous decision trees [45], and subsequently calculating the final predicted value by averaging the predicted values of all trees [46]. Usually, the algorithm randomly selects 2/3 of the original sample to build the decision tree, and the remaining 1/3 of the data is used as out-of-bag data (OOB) to validate the decision tree.

Large amounts of features can affect the accuracy of RF models. Therefore, it is one of the most important steps to filter out the effective features for building mangrove species model [47]. In this study, we use SHapley Additive exPlanations (SHAP) value to evaluate and filter feature variables. SHAP is a methodology employed to elucidate the contribution of features in model predictions. It employs the concept of Shapley Value from game theory to determine the contribution value of each feature in model prediction. It incorporates existing machine learning algorithms, like neural networks and RF, and calculates all possible feature combinations in the model to accurately obtain relative importance information for each feature [48]. The computation method of SHAP value surpasses other indicators like Gini importance and information entropy by considering all combinations of features, which provided a more precise assessment of highly correlated features and nonlinear combination features [49].

2.3.4. Accuracy Assessment

For accuracy assessment of the mangrove species classification result, we calculated the overall accuracy (OA), the producer's accuracy (PA), the user's accuracy (UA), and Kappa coefficient based on the confusion matrix (the rows of the matrix represent ground truth classes and the columns of the matrix represent predicted classes). The equations are as follows:

$$OA = \sum_{i=1}^{n} P_{ii}/P \tag{1}$$

$$PA_i = P_{ii}/P_{i+} \tag{2}$$

$$UA_i = P_{ii}/P_{+i} \tag{3}$$

$$Kappa = \frac{P\sum_{i=1}^{n} P_{ii} - \sum_{i=1}^{n}(P_{i+}P_{+i})}{P^2 - \sum_{i=1}^{n}(P_{i+}P_{+i})} \tag{4}$$

In the above categories, p represents the sum of samples, P_{ii} represents the number of samples in row i and column i of the error matrix, P_{+i} represents the sum of class i in the classification results, P_{i+} represents the sum of class i in the ground reality data, and n represents the number of categories.

For accuracy assessment of the mangrove species height result and the ability of the model, we used the 10-fold cross-validation method. The 10-fold cross-validation divides the dataset equally (or approximately) into 10 subsets, uses nine of them to build the model, and leaves one as the validation set. The coefficient of determination (R^2) and root-mean-square-error (RMSE) between the predicted mangrove specie height and the CHM data were calculated to evaluate the accuracy of the model. The equations are as follows:

$$R^2 = 1 - \frac{(y_i - \hat{y}_i)^2}{(y_i - \overline{y})^2} \tag{5}$$

$$RMSE = \sqrt{\frac{\sum_{i=1}^{n}(y_i - \hat{y}_i)^2}{n}} \tag{6}$$

where n is the number of samples, y_i and $\hat{y}_i$ represent the true and predicted values of mangrove specie height at sample point i, respectively, and $\overline{y}$ represents the average of the true values of tree height.

3. Results

3.1. Accuracy Assessment

The result of accuracy assessment of the mangrove species classification is displayed in Table 2. The OA of the mangrove species result was 91.43% (kappa = 0.89). KO reached the highest producer's accuracy (94.74%), followed by AC with 90.32% and AM with 88.89%. The error was mainly due to the misclassification between KO, AC, and AM at the center of the research area and the misclassification between AM and SA.

Table 2. The confusion matrix of the mangrove community classification results.

Community Type	Validation Samples							UA (%)
	KO	AC	AM	SA	Mudflat	Water	Total	
KO	36	2	1	0	0	0	39	92.31
AC	0	28	1	0	0	0	29	96.55
AM	2	1	24	1	0	0	28	85.71
SA	0	0	1	20	0	0	21	95.24
Mudflat	0	0	0	1	10	1	12	83.33
water	0	0	0	0	1	10	11	90.91
Total	38	31	27	22	11	11	140	
PA (%)	94.74	90.32	88.89	90.91	90.91	90.91	94.74	
Kappa: 0.89		OA: 91.43%						

We randomly selected 600 sample points based on the CHM derived from the LiDAR data to evaluate the mangrove species height model constructed by RF regression method. The RMSE of the result was 0.91 m, and R^2 was 0.71. Figure 3 shows the accuracy assessment of the mangrove species height result.

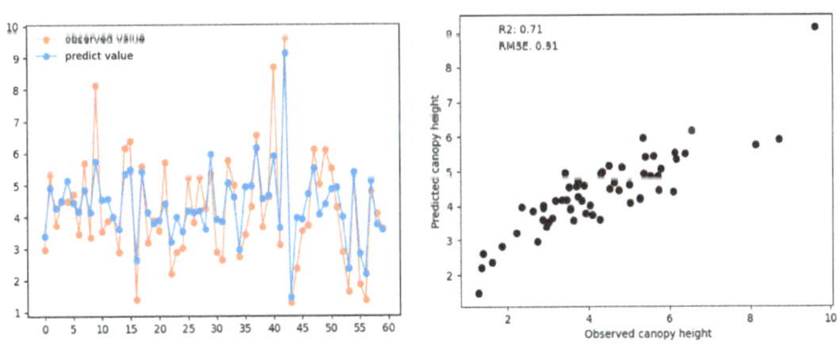

(a) (b)

Figure 3. (**a**) Correlation plot of the mangrove species height between observed and predicted by RF regression model; (**b**) the accuracy assessment of the mangrove species height model.

3.2. The Distribution of Mangrove Species of FZNNR

The distribution of mangrove species in FZNNR is shown in Figure 4, which was obtained based on the UAV multi-spectral image and OBRF method. The total area of mangrove forest in FZNNR was 50.92 ha. KO was the dominant mangrove species in FZNNR, with a total area of 32.73 ha. The total area of AC was 11.04 ha, mainly distributed on both sides of the water; and the total area of AM was 7.15 ha, mainly distributed in the central and southern part of the study area.

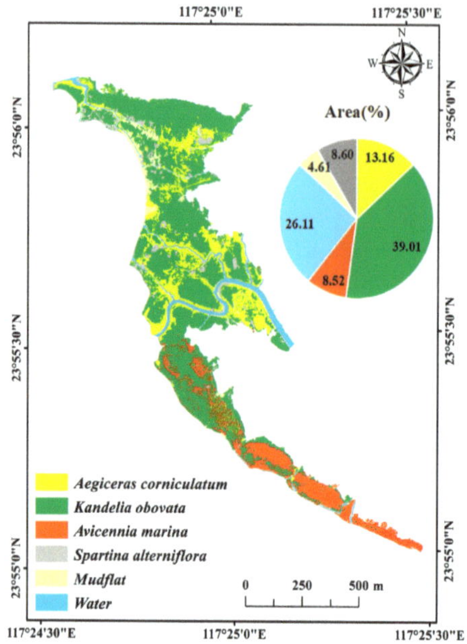

Figure 4. The spatial distribution of mangrove species in FZNNR.

3.3. Feature Importance for Mangrove Species Height Retrieval

In this study, the 47 features extracted from GF-2 PMS and GF-3 SAR images were filtered using the SHAP value in machine learning, and the results are presented in Figure 5. The RF model picked ten features for the purpose of estimating the height of mangrove species. Of these ten characteristics, the most important feature was the backward scattering coefficients of HH polarization from GF-3 SAR images, followed by VV polarization feature and the texture features of mean_1 (Calculated based on the blue band) from GF-2 images. The results indicated that the backscattering coefficient features of GF-3 images and the texture features of GF-2 images are the effective features for the mangrove species height estimation.

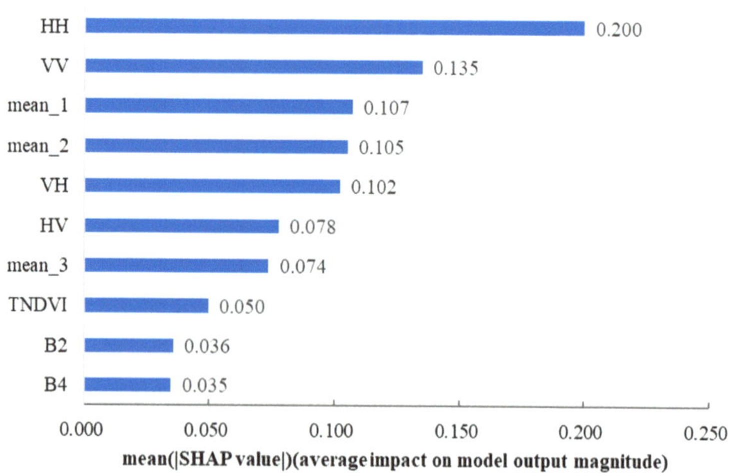

Figure 5. The mean SHAP absolute value of the top ten features.

3.4. The Spatial Distribution of Mangrove Species Height in FZNNR

To retrieve the mangrove species height in FZNNR (Figure 6), we used the RF regression model based on the UAV-LiDAR, GF-2, and GF-3 images. Our findings suggested that mangrove forests varied in height from 0.4 to 9.6 m (Figure 7). Heights exceeding 5.5 m were primarily found in the central zone of the study area, while heights less than 2 m were observed at the edges of the study area. The distribution map indicated that the height ranges for AC, KO, and AM were, respectively, between 4 and 7 m. The height map of the mangrove species, presented at a resolution of 0.8 m, is shown in Figure 6. For the species AC, the largest area (2.876 ha) is observed in the height interval of 4 m to 4.5 m. The second and third largest areas are found in the intervals of 3.5 m to 4 m (2.427 ha) and 4.5 m to 5 m (1.716 ha), respectively. In the case of KO, the largest area is within the 4 m to 4.5 m range, covering approximately 8.386 ha. The next largest areas are within the ranges of 4.5 m to 5 m (7.838 ha) and 3.5 m to 4 m (4.932 ha). For AM, the largest area (2.885 ha) is within the 4 m-to-4.5 m height interval, followed by the intervals of 4.5 m to 5 m (1.490 ha) and 3.5 m to 4 m (1.294 ha).

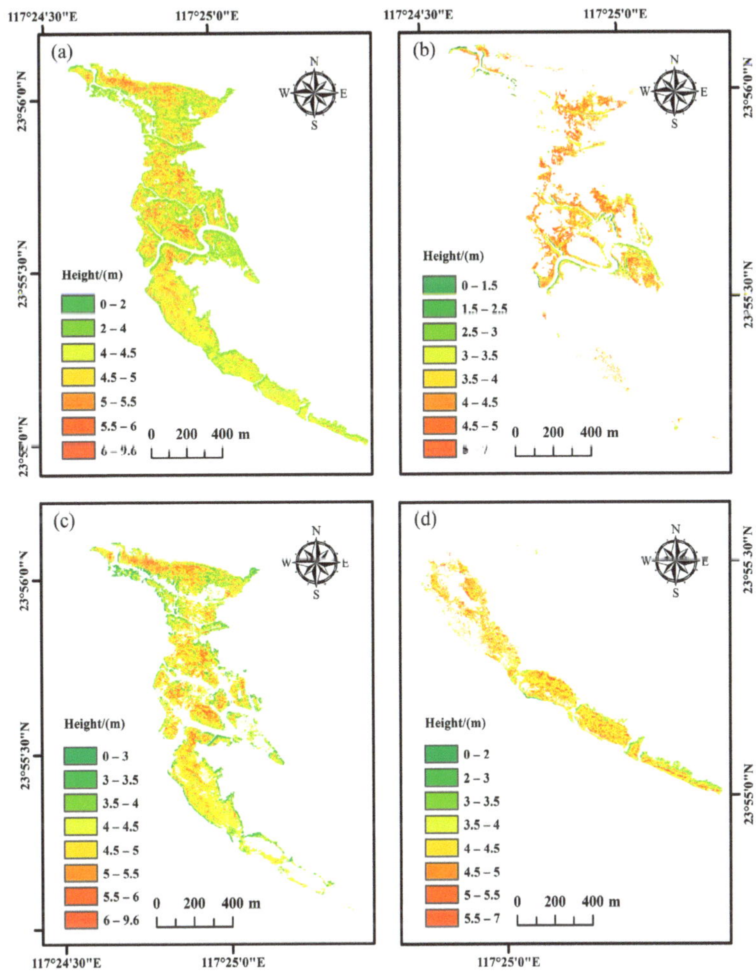

Figure 6. The mangrove species height map using the RF model in FZNNR. (**a**) The height map of all mangrove species; (**b**) the height map of AC; (**c**) the height map of KO; (**d**) the height map of AM.

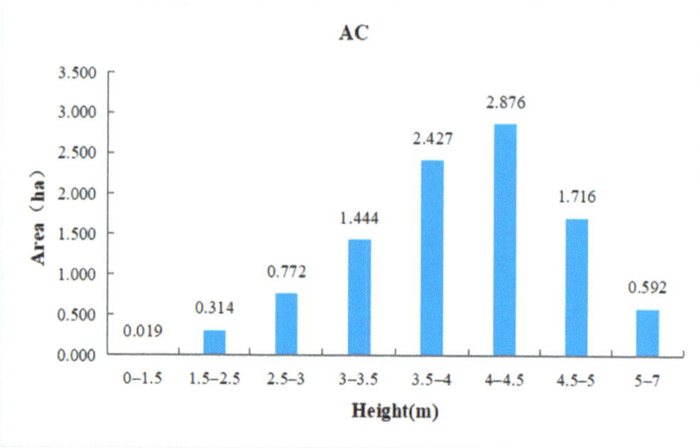

(**a**)

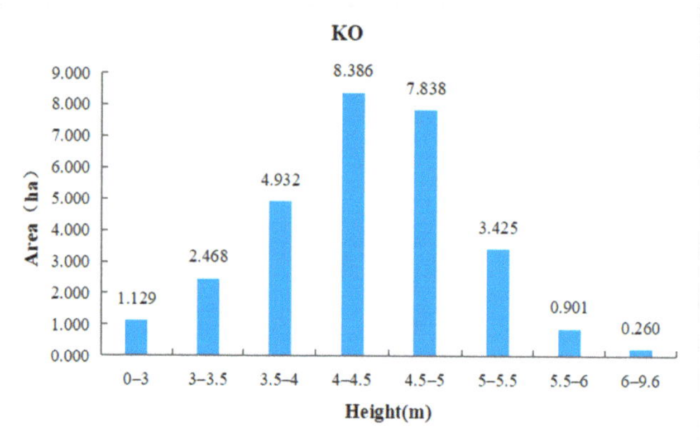

(**b**)

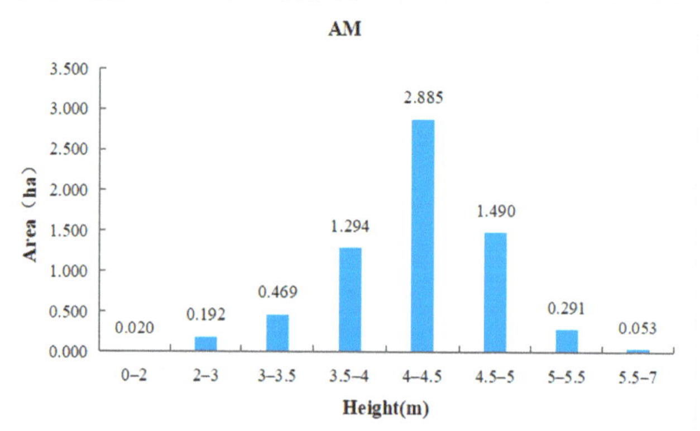

(**c**)

Figure 7. Statistics of mangrove species area at different height intervals. (**a**) The height statistic of AC; (**b**) the height statistic of KO; (**c**) the height statistic of AM.

4. Discussion

In our research, we combined UAV-LiDAR data, GF-2 Optical, and GF-3 SAR images to map the height of mangrove species at a resolution of 0.8 m in FZNNR, utilizing an RF regression model. We extracted a total of 47 feature variables and selected the 10 most relevant features which have a strong correlation with the height of mangrove species. These 10 selected features were used to construct the prediction model for mangrove forest height retrieval. The results demonstrated that these selected features effectively predict the height of the mangrove forest. Furthermore, the results validated the exceptional effectiveness of GF-3 SAR and GF-2 PMS in estimating the height of the mangrove species.

4.1. Feature Variable Importance

In this study, the machine learning SHAP value was utilized to filter the 47 features that were extracted from GF-2 PMS and GF-3 SAR data. The SHAP value calculates the contribution of each feature to the model output, considering the interactions between features [50]. Consequently, it provides a more precise depiction of feature impact on predicted outcomes, particularly for nonlinear models, surpassing other methods (such as feature importance) in terms of accuracy. Therefore, we choose the SHAP value to evaluate the importance of 47 features in RF model. Finally, ten features were selected by the SHAP value in order to estimate mangrove species height. The analysis of feature importance revealed that the HH of GF-3 SAR data was pivotal in predicting mangrove species height, followed by the VV of GF-3 SAR data. SAR imagery can emit radiation to identify and measure branches and trunks in order to determine the structure of the canopy [51]. This finding aligns with Liu et al., who affirmed the efficacy of backscattering variables (VV, VH) from Sentinel-1B images in estimating the mean height of forest stands at a 10 m resolution [20]. Zhu et al. [52] estimated the AGB of mangrove plantations by integrating fixed-wing UAV-based DSM, GF-2 optical, and GF-3 SAR data. They found that HH and HV played significant roles in AGB prediction. The texture features of mean_1 and mean_2 was ranked third and fourth in importance, respectively. These features represent the mean of the texture indices calculated by the blue and red bands of GF-2. This observation mirrors Kayitakire et al., who also discovered that texture indices extracted from IKONOS-2 were effective in inverting forest structure parameters [13]. Similarly, Sarker and Nichol [53] confirmed that texture indices of ALOS AVNIR-2 can enhance their estimation accuracy of forest biomass. Texture indices from high-resolution remote sensing images augment the discrimination of spatial information based on the original brightness of images, thereby playing a crucial role in inverting forest structure parameters.

4.2. Accuracy of Random Forest Regression Model

This study set out to construct a robust and precise model for estimating the height of mangrove species. In order to build non-linear connections between the height and features, we utilized an RF regression model, owing to its resilience to outliers in the training data, capability to handle non-parametric data, ability to reveal complex non-linear relationships between variables, and ease of model parameter tuning.

Previous studies have utilized machine learning algorithms such as stepwise multiple regression and multiple regression for forest height estimation. For instance, Zhu et al. [19] developed forest height models using both RF and stepwise multiple regression algorithms. Their findings revealed that estimates of forest height produced by the stepwise multiple regression model were less accurate than those generated by the RF model. Similarly, for the purpose of estimating forest properties, Luther et al. [54] used multiple regression and RF models, with the RF model outperforming the multiple regression model. These studies collectively demonstrated that the RF algorithm provides more accurate regression prediction results than other machine learning algorithms.

In our study, we constructed an RF model to estimate mangrove species height, achieving a good performance (R^2 = 0.71, RMSE = 0.91 m). Compared with previous studies on tree height prediction, our study demonstrated some improvement in prediction accuracy. For example, Simard et al. [55] successfully estimated mean tree height and biomass in mangrove forests in the Everglades National Park using Shuttle Radar Topography Mission (SRTM) elevation data, with an RMSE of 2.0 m for mean mangrove height estimation. Our study improved upon this precision, potentially due to the high spatial resolution of the data we used. Wang et al. [18], who retrieved the height of Hainan Island's mangrove forest using Sentinel-2 imagery, achieved an RMSE of 1.90 m and an R^2 of 0.67—a lower R^2 than that in our study and a higher RMSE. This could be because Sentinel-2 images have a lower spatial resolution than GF-2 images. The higher RMSE in Wang's study could be caused by the reality that the average height of mangroves in Hainan Island is greater than that in Zhangjiang Estuary. Overall, our model achieved a good performance (R^2 = 0.71, RMSE = 0.91 m), demonstrating that the RF algorithm holds significant potential for estimating mangrove forest height.

4.3. Limitations and Potentials

In our study, the RF regression model for estimating mangrove species height was influenced by raster matching issues between multiple data sources (GF-2, GF-3). The collection time of various data sources also had an impact on model accuracy, but this was within the error range, given the slow growth and relatively stable height of mangrove forests over several years.

The inherent limitations of GF-2 optical data may have contributed to the discrepancy between predicted height and the CHM. Despite its high spatial resolution, GF-2 image only has four bands and lacks the red-edge band and short-wave infrared band, which are sensitive to mangrove growth. Active sensors such as SAR and UAV-LiDAR could have mitigated this problem. The GF-3 SAR data used in this study address this limitation to mangrove species height prediction.

The RF algorithm can effectively handle high-dimensional and nonlinear regression problems, demonstrating high generalization ability and stability. However, it has some limitations. The RF model assumes a normal or approximate normal distribution of the data due to its reliance on the Mean Squared Error (MSE) as the loss function. This assumption could lead to large bias and variance in the model if not met [56]. Moreover, the RF model is sensitive to noise data and outliers, which may affect its accuracy and stability [57,58]. In addition to the limitations of the RF model, only spectral features, texture features, and SAR features were considered in this study, and other factors such as species and climate were not considered.

5. Conclusions

Combining GF-2, GF-3, and UAV-LiDAR data, our research investigated the possibility of estimating mangrove species height in FZZNR. We constructed a height retrieval model by integrating both GF-2 and GF-3 imagery with the CHM derived from UAV-LiDAR, based on the RF regression algorithm. The suggested the height retrieval model had an R^2 of 0.71 and an RMSE of 0.91 m, indicating good performance. Of the all features, the texture features and SAR features were crucial in mangrove height estimations. This method can overcome the accuracy limitations of a single remote sensing data source and improve the accuracy and reliability of mangrove forest height estimation. This study also demonstrated the advantage of using the CHM derived from UAV-LiDAR as the training data, which can reduce the cost and time of field measurements. Our study provided an effective technical means for further monitoring and managing mangrove ecosystems using multi-source remote sensing data. In future studies, we will explore the potential of using other remote sensing data, such as hyperspectral data, for mapping mangrove forest height and explore the ability of GF-2 and GF-3 data for mapping other structural information.

Author Contributions: Conceptualization, R.C., R.Z. and M.J.; Data curation, R.Z.; Formal analysis, R.C.; Funding acquisition, M.J.; Investigation, R.C.; Methodology, R.C.; Resources, M.J.; Software, R.C.; Supervision, R.Z., C.Z., Z.W. and M.J.; Visualization, C.Z. and Z.W.; Writing—original draft, R.C.; Writing—review and editing, R.Z. and M.J. All authors have read and agreed to the published version of the manuscript.

Funding: This work is jointly supported by the Youth Scientist Project National Key R&D Program of China (2023YFF1305600), the National Natural Science Foundation of China (42171379 and 42171372), the Youth Innovation Promotion Association of Chinese Academy of Sciences (2021227), the Open Research Program of the International Research Center of Big Data for Sustainable Development Goals (Grant No. CBAS2022ORP06), and the National Earth System Science Data Center (www.geodata.cn, accessed on 20 July 2023).

Data Availability Statement: The data presented in this study are available on request from the corresponding author. The data are not publicly available due to the funding project is not finished yet.

Acknowledgments: We express our gratitude to everyone who helped us to successfully complete this research.

Conflicts of Interest: The authors declare no conflict of interest.

References

1. Zhang, Z.; Pan, J.; Pan, Y.; Li, M. Biogeography, assembly patterns, driving factors, and interactions of archaeal community in mangrove sediments. *Msystems* **2021**, *6*, e01381-20. [CrossRef] [PubMed]
2. Wang, L.; Jia, M.; Yin, D.; Tian, J. A review of remote sensing for mangrove forests: 1956–2018. *Remote Sens. Environ.* **2019**, *231*, 111223. [CrossRef]
3. Wu, L.; Wang, L.; Shi, C.; Yin, D. Detecting mangrove photosynthesis with solar-induced chlorophyll fluorescence. *Int. J. Remote Sens.* **2022**, *43*, 1037–1053. [CrossRef]
4. Zeng, H.; Jia, M.; Zhang, R.; Wang, Z.; Mao, D.; Ren, C.; Zhao, C. Monitoring the light pollution changes of China's mangrove forests from 1992-2020 using nighttime light data. *Front. Mar. Sci.* **2023**, *10*, 1187702. [CrossRef]
5. Donato, D.C.; Kauffman, J.B.; Murdiyarso, D.; Kurnianto, S.; Stidham, M.; Kanninen, M. Mangroves among the most carbon-rich forests in the tropics. *Nat. Geosci.* **2011**, *4*, 293–297. [CrossRef]
6. Friesen, S.D.; Dunn, C.; Freeman, C. Decomposition as a regulator of carbon accretion in mangroves: A review. *Ecol. Eng.* **2018**, *114*, 173–178. [CrossRef]
7. Giri, C.; Ochieng, E.; Tieszen, L.L.; Zhu, Z.; Singh, A.; Loveland, T.; Masek, J.; Duke, N. Status and distribution of mangrove forests of the world using earth observation satellite data. *Glob. Ecol. Biogeogr.* **2011**, *20*, 154–159. [CrossRef]
8. Jia, M.; Wang, Z.; Mao, D.; Ren, C.; Song, K.; Zhao, C.; Wang, C.; Xiao, X.; Wang, Y. Mapping global distribution of mangrove forests at 10-m resolution. *Sci. Bull.* **2023**, *68*, 1306–1316. [CrossRef]
9. Wu, W.; Zhi, C.; Gao, Y.; Chen, C.; Chen, Z.; Su, H.; Lu, W.; Tian, B. Increasing fragmentation and squeezing of coastal wetlands: Status, drivers, and sustainable protection from the perspective of remote sensing. *Sci. Total Environ.* **2022**, *811*, 152339. [CrossRef]
10. Bathmann, J.; Peters, R.; Reef, R.; Berger, U.; Walther, M.; Lovelock, C.E. Modelling mangrove forest structure and species composition over tidal inundation gradients: The feedback between plant water use and porewater salinity in an arid mangrove ecosystem. *Agric. For. Meteorol.* **2021**, *308*, 108547. [CrossRef]
11. Hadadi, F.; Moazenzadeh, R.; Mohammadi, B. Estimation of actual evapotranspiration: A novel hybrid method based on remote sensing and artificial intelligence. *J. Hydrol.* **2022**, *609*, 127774. [CrossRef]
12. Hyyppä, J.; Hyyppä, H.; Inkinen, M.; Engdahl, M.; Linko, S.; Zhu, Y. Accuracy comparison of various remote sensing data sources in the retrieval of forest stand attributes. *For. Ecol. Manag.* **2000**, *128*, 109–120. [CrossRef]
13. Kayitakire, F.; Hamel, C.; Defourny, P. Retrieving forest structure variables based on image texture analysis and IKONOS-2 imagery. *Remote Sens. Environ.* **2006**, *102*, 390–401. [CrossRef]
14. Chopping, M.; Moisen, G.G.; Su, L.; Laliberte, A.; Rango, A.; Martonchik, J.V.; Peters, D.P. Large area mapping of southwestern forest crown cover, canopy height, and biomass using the NASA Multiangle Imaging Spectro-Radiometer. *Remote Sens. Environ.* **2008**, *112*, 2051–2063. [CrossRef]
15. Coops, N.C.; Tompalski, P.; Goodbody, T.R.; Queinnec, M.; Luther, J.E.; Bolton, D.K.; White, J.C.; Wulder, M.A.; van Lier, O.R.; Hermosilla, T. Modelling lidar-derived estimates of forest attributes over space and time: A review of approaches and future trends. *Remote Sens. Environ.* **2021**, *260*, 112477. [CrossRef]
16. Simard, M.; Pinto, N.; Fisher, J.B.; Baccini, A. Mapping forest canopy height globally with spaceborne lidar. *J. Geophys. Res. Biogeosciences* **2011**, *116*, 1–12. [CrossRef]
17. Lefsky, M.A.; Hudak, A.T.; Cohen, W.B.; Acker, S. Geographic variability in lidar predictions of forest stand structure in the Pacific Northwest. *Remote Sens. Environ.* **2005**, *95*, 532–548. [CrossRef]
18. Wang, D.; Wan, B.; Qiu, P.; Zuo, Z.; Wang, R.; Wu, X. Mapping height and aboveground biomass of mangrove forests on Hainan Island using UAV-LiDAR sampling. *Remote Sens.* **2019**, *11*, 2156. [CrossRef]
19. Zhu, X.; Wang, C.; Nie, S.; Pan, F.; Xi, X.; Hu, Z. Mapping forest height using photon-counting LiDAR data and Landsat 8 OLI data: A case study in Virginia and North Carolina, USA. *Ecol. Indic.* **2020**, *114*, 106287. [CrossRef]
20. Liu, Y.; Gong, W.; Xing, Y.; Hu, X.; Gong, J. Estimation of the forest stand mean height and aboveground biomass in Northeast China using SAR Sentinel-1B, multispectral Sentinel-2A, and DEM imagery. *ISPRS-J. Photogramm. Remote Sens.* **2019**, *151*, 277–289. [CrossRef]

21. Huang, K.; Yang, G.; Yuan, Y.; Sun, W.; Meng, X.; Ge, Y. Optical and SAR images Combined Mangrove Index based on multi-feature fusion. *Sci. Remote Sens.* **2022**, *5*, 100040. [CrossRef]
22. Tsokas, A.; Rysz, M.; Pardalos, P.M.; Dipple, K. SAR data applications in earth observation: An overview. *Expert Syst. Appl.* **2022**, *205*, 117342. [CrossRef]
23. Abdullahi, S.; Kugler, F.; Pretzsch, H. Prediction of stem volume in complex temperate forest stands using TanDEM-X SAR data. *Remote Sens. Environ.* **2016**, *174*, 197–211. [CrossRef]
24. Cloude, S.R.; Pottier, E. A review of target decomposition theorems in radar polarimetry. *IEEE Trans. Geosci. Remote Sens.* **1996**, *34*, 498–518. [CrossRef]
25. Naidoo, L.; Mathieu, R.; Main, R.; Wessels, K.; Asner, G.P. L-band Synthetic Aperture Radar imagery performs better than optical datasets at retrieving woody fractional cover in deciduous, dry savannahs. *Int. J. Appl. Earth Obs. Geoinf.* **2016**, *52*, 54–64. [CrossRef]
26. Zhen, J.; Liao, J.; Shen, G. Mapping mangrove forests of Dongzhaigang nature reserve in China using Landsat 8 and Radarsat-2 polarimetric SAR data. *Sensors* **2018**, *18*, 4012. [CrossRef] [PubMed]
27. Ghosh, S.M.; Behera, M.D.; Paramanik, S. Canopy height estimation using sentinel series images through machine learning models in a mangrove forest. *Remote Sens.* **2020**, *12*, 1519. [CrossRef]
28. Zhao, C.; Jia, M.; Wang, Z.; Mao, D.; Wang, Y. Toward a better understanding of coastal salt marsh mapping: A case from China using dual-temporal images. *Remote Sens. Environ.* **2023**, *295*, 113664. [CrossRef]
29. Stojanova, D.; Panov, P.; Gjorgjioski, V.; Kobler, A.; Džeroski, S. Estimating vegetation height and canopy cover from remotely sensed data with machine learning. *Ecol. Inform.* **2010**, *5*, 256–266. [CrossRef]
30. Smola, A.J.; Schölkopf, B. A tutorial on support vector regression. *Stat. Comput.* **2004**, *14*, 199–222. [CrossRef]
31. Wilkes, P.; Jones, S.D.; Suarez, L.; Mellor, A.; Woodgate, W.; Soto-Berelov, M.; Haywood, A.; Skidmore, A.K. Mapping forest canopy height across large areas by upscaling ALS estimates with freely available satellite data. *Remote Sens.* **2015**, *7*, 12563–12587. [CrossRef]
32. Yang, G.; Huang, K.; Sun, W.; Meng, X.; Mao, D.; Ge, Y. Enhanced mangrove vegetation index based on hyperspectral images for mapping mangrove. *ISPRS-J. Photogramm. Remote Sens.* **2022**, *189*, 236–254. [CrossRef]
33. Zhao, X.; Guo, Q.; Su, Y.; Xue, B. Improved progressive TIN densification filtering algorithm for airborne LiDAR data in forested areas. *ISPRS-J. Photogramm. Remote Sens.* **2016**, *117*, 79–91. [CrossRef]
34. Yin, D.; Wang, L. Individual mangrove tree measurement using UAV-based LiDAR data: Possibilities and challenges. *Remote Sens. Environ.* **2019**, *223*, 34–49. [CrossRef]
35. Kavzoglu, T.; Yildiz, M. Parameter-based performance analysis of object-based image analysis using aerial and Quikbird-2 images. *ISPRS Ann. Photogramm. Remote Sens. Spat. Inf. Sci.* **2014**, *2*, 31–37. [CrossRef]
36. Gitelson, A.A.; Gritz, Y.; Merzlyak, M.N. Relationships between leaf chlorophyll content and spectral reflectance and algorithms for non-destructive chlorophyll assessment in higher plant leaves. *J. Plant Physiol.* **2003**, *160*, 271–282. [CrossRef]
37. Tucker, C.J. Red and photographic infrared linear combinations for monitoring vegetation. *Remote Sens. Environ.* **1979**, *8*, 127–150. [CrossRef]
38. Korhonen, L.; Packalen, P.; Rautiainen, M. Comparison of Sentinel-2 and Landsat 8 in the estimation of boreal forest canopy cover and leaf area index. *Remote Sens. Environ.* **2017**, *195*, 259–274. [CrossRef]
39. Blonigen, B.A. A review of the empirical literature on FDI determinants. *Atl. Econ. J.* **2005**, *33*, 383–403. [CrossRef]
40. Valderrama-Landeros, L.; Flores-de-Santiago, F.; Kovacs, J.M.; Flores-Verdugo, F. An assessment of commonly employed satellite-based remote sensors for mapping mangrove species in Mexico using an NDVI-based classification scheme. *Environ. Monit. Assess.* **2017**, *190*, 23. [CrossRef]
41. Wicaksono, P.; Danoedoro, P.; Hartono; Nehren, U. Mangrove biomass carbon stock mapping of the Karimunjawa Islands using multispectral remote sensing. *Int. J. Remote Sens.* **2016**, *37*, 26–52. [CrossRef]
42. Castillo, J.A.A.; Apan, A.A.; Maraseni, T.N.; Salmo, S.G. Estimation and mapping of above-ground biomass of mangrove forests and their replacement land uses in the Philippines using Sentinel imagery. *ISPRS-J. Photogramm. Remote Sens.* **2017**, *134*, 70–85. [CrossRef]
43. Haralick, R.M.; Shanmugam, K.; Dinstein, I.H. Textural features for image classification. *IEEE Trans. Syst. Man Cybern.* **1973**, *6*, 610–621. [CrossRef]
44. Belgiu, M.; Drăguţ, L. Random forest in remote sensing: A review of applications and future directions. *ISPRS-J. Photogramm. Remote Sens.* **2016**, *114*, 24–31. [CrossRef]
45. Zhao, C.; Qin, C.-Z.; Wang, Z.; Mao, D.; Wang, Y.; Jia, M. Decision surface optimization in mapping exotic mangrove species (*Sonneratia apetala*) across latitudinal coastal areas of China. *ISPRS-J. Photogramm. Remote Sens.* **2022**, *193*, 269–283. [CrossRef]
46. Breiman, L. Random forests. *Mach. Learn.* **2001**, *45*, 5–32. [CrossRef]
47. Wang, D.; Wan, B.; Qiu, P.; Tan, X.; Zhang, Q. Mapping mangrove species using combined UAV-LiDAR and Sentinel-2 data: Feature selection and point density effects. *Adv. Space Res.* **2022**, *69*, 1494–1512. [CrossRef]
48. Ekanayake, I.; Meddage, D.; Rathnayake, U. A novel approach to explain the black-box nature of machine learning in compressive strength predictions of concrete using Shapley additive explanations (SHAP). *Case Stud. Constr. Mater.* **2022**, *16*, e01059. [CrossRef]
49. Lundberg, S.M.; Lee, S.-I. A unified approach to interpreting model predictions. In Proceedings of the Advances in Neural Information Processing Systems, Long Beach, CA, USA, 4–9 December 2017; Volume 30.

50. Lundberg, S.M.; Erion, G.; Chen, H.; DeGrave, A.; Prutkin, J.M.; Nair, B.; Katz, R.; Himmelfarb, J.; Bansal, N.; Lee, S.-I. From local explanations to global understanding with explainable AI for trees. *Nat. Mach. Intell.* **2020**, *2*, 56–67. [CrossRef]
51. Omar, H.; Misman, M.A.; Kassim, A.R. Synergetic of PALSAR-2 and Sentinel-1A SAR polarimetry for retrieving aboveground biomass in dipterocarp forest of Malaysia. *Appl. Sci.* **2017**, *7*, 675. [CrossRef]
52. Zhu, Y.; Liu, K.; Myint, S.W.; Du, Z.; Li, Y.; Cao, J.; Liu, L.; Wu, Z. Integration of GF2 optical, GF3 SAR, and UAV data for estimating aboveground biomass of China's largest artificially planted mangroves. *Remote Sens.* **2020**, *12*, 2039. [CrossRef]
53. Sarker, L.R.; Nichol, J.E. Improved forest biomass estimates using ALOS AVNIR-2 texture indices. *Remote Sens. Environ.* **2011**, *115*, 968–977. [CrossRef]
54. Luther, J.E.; Fournier, R.A.; van Lier, O.R.; Bujold, M. Extending ALS-based mapping of forest attributes with medium resolution satellite and environmental data. *Remote Sens.* **2019**, *11*, 1092. [CrossRef]
55. Simard, M.; Zhang, K.; Rivera-Monroy, V.H.; Ross, M.S.; Ruiz, P.L.; Castañeda-Moya, E.; Twilley, R.R.; Rodriguez, E. Mapping height and biomass of mangrove forests in Everglades National Park with SRTM elevation data. *Photogramm. Eng. Remote Sens.* **2006**, *72*, 299–311. [CrossRef]
56. Strobl, C.; Boulesteix, A.-L.; Kneib, T.; Augustin, T.; Zeileis, A. Conditional variable importance for random forests. *BMC Bioinform.* **2008**, *9*, 1–11. [CrossRef]
57. Aria, M.; Cuccurullo, C.; Gnasso, A. A comparison among interpretative proposals for Random Forests. *Mach. Learn. Appl.* **2021**, *6*, 100094. [CrossRef]
58. Segal, M.; Xiao, Y. Multivariate random forests. *Wiley Interdiscip. Rev. Data Min. Knowl. Discov.* **2011**, *1*, 80–87. [CrossRef]

remote sensing

Article

Multi-Scale Remote Sensing Assessment of Ecological Environment Quality and Its Driving Factors in Watersheds: A Case Study of Huashan Creek Watershed in China

Yajing Liao [1,*], Guirong Wu [2] and Zhenyu Zhang [1]

1 College of the Environment and Ecology, Xiamen University, Xiamen 361005, China; zhangzy@xmu.edu.cn
2 Fujian Geologic Surveying and Mapping Institute, Fuzhou 351005, China
* Correspondence: liaoyajing@stu.xmu.edu.cn

Abstract: The Huashan Creek watershed is the largest water source and the main production area of honeydew in Pinghe County, whose extensive cultivation of honeydew has exacerbated soil and water pollution. However, the spatial application of remote sensing ecological index (RSEI) in this watershed and key driving factors are not clear considering the applicability of data quality and the diversity of methodological scales. To explore the RSEI and driving factors at distinct scales in Huashan Creek watershed, this study constructed the RSEI based on the environmental balance matrix at seven scales in 2020, revealed its spatial response characteristics at different scales, and analyzed the key drivers. The results show that the 240 m grid as well as rural and watershed scale convergence analyses satisfy the assessment of RSEI, whose Moran indexes are 0.558, 0.595, and 0.146, respectively. The RSEIs at different scales have significant spatial aggregation characteristics, but the overall status is moderate. The central town–riparian area with poor RSEI contrasts with the western mountainous area, which has comparatively better quality. Population has a major influence on RSEI at multiple scales (0.8), with elevation and patch index acting significantly at the village and grid scales, respectively. These findings help to identify the spatial distribution of quality and control mechanisms of RSEI in the Huashan Creek watershed and provide new insights into key scales and drivers of ecological restoration practices in the watershed.

Keywords: remote sensing ecological index (RSEI); Huashan Creek watershed; spatiotemporalchange; geographically weighted regression

Citation: Liao, Y.; Wu, G.; Zhang, Z. Multi-Scale Remote Sensing Assessment of Ecological Environment Quality and Its Driving Factors in Watersheds: A Case Study of Huashan Creek Watershed in China. *Remote Sens.* **2023**, *15*, 5633. https://doi.org/10.3390/rs15245633

Academic Editor: Pradeep Wagle

Received: 28 September 2023
Revised: 7 November 2023
Accepted: 24 November 2023
Published: 5 December 2023

1. Introduction

Assessing the ecological quality of watersheds is paramount in assessing the health of natural ecosystems and understanding the effects of human activities on vital components of ecosystems, such as vegetation, water, soil, and climate [1]. This understanding is essential for scientific research and effective watershed management, providing valuable insights into ecological processes, ecosystem services, environmental risks, and conservation strategies [2]. However, the rapid growth of socioeconomic activities and population expansion has led to increased human intervention, resource degradation, and unsustainable exploitation [3,4]. Approximately 40% of land on Earth suffers from degradation or desertification, 70% of freshwater resources are overused or contaminated, and 60% of wetlands are degraded or have disappeared in the past century [2]. These challenges not only harm human well-being but also jeopardize the Earth's life-support systems. Therefore, studying the ecological quality of rapidly developing watersheds and the factors affecting them are urgent requirements.

The Huashan Creek watershed is highly typical, hence its proposed rational management measurement. The honeydew industry and town expansion in the watershed will generate a series of ecological and environmental problems, exerting an impact on the ecological quality of the region.

The Huashan Creek watershed in southeastern coastal China presents unique challenges and opportunities due to its agricultural activity, coastal proximity, and diverse land use patterns. This region is emblematic of the challenges faced by rapidly developing coastal areas, where ecological status directly impacts local communities and essential ecosystem services [5]. Cai [6] studied the measures used for water environment protection in Huashan Creek in Pinghe County. Moreover, Huang et al. [7] reported that in recent decades, anthropogenic production activities, the loss of native vegetation in the Huashan Creek watershed, and the massive loss of the soil surface layer have severely constrained the development of local agricultural production. Furthermore, Han [8] reported that the only drinking water source for 80,000 people in Pinghe County is Huashan Creek, which is under honeydew cultivation that endangers the water quality.

In recent years, models and statistical techniques based on remote sensing data have gained prominence for assessing ecological environmental quality across diverse landscapes [9–13]. Remote Sensing Earth Satellite Unity, with its high-resolution, long-term data series, and extensive monitoring capabilities, can observe different ecosystems in their radiometric bands as a means of constructing a remotely sensed ecological index (RSEI) and revealing spatial distribution and drivers of different landscapes [14,15]. The indicator system and method selection are key to assessing RSEI.

Currently, commonly used indicator systems include land use/cover type; vegetation cover; leaf area index; primary productivity; soil organic carbon content, which reflects the watershed subsurface elements and ecosystem structural characteristics [16]; precipitation; evapotranspiration; surface runoff; groundwater level; and hydrological connectivity, which reflects the input, output, and storage of moisture in the watershed [11,12,15]. Among these systems, the use of principal component analysis (PCA), the construction of a comprehensive remote sensing index based on greenness, humidity, dryness, and warmth; the determination of weighting coefficients of ecological indicators; as well as the quantitative characterization of RSEI have become popular methods for assessing ecological environmental quality [11,13,17].

Measurement and statistical techniques usually include the single or composite index method, model-based simulation method, and statistical analysis method based on remote sensing data. Among these methods, the data analysis strategy of using spatial autocorrelation analysis to determine the spatial distribution of RSEI based on remotely sensed data of ecological indicators combined with geodetectors and geographically weighted regression to determine driving factors has been widely used [18,19]. Comprehensive remote sensing indices amalgamate multiple indicators to effectively assess ecological conditions [16]. For instance, Xu et al. [20] employed PCA to derive RSEI based on factors such as greenness, humidity, dryness, and warmth. Moreover, RSEI has found wide-ranging applications in regional ecological monitoring [21–25], reflecting environmental changes driven by human activities, shifts in vegetation cover, and climate fluctuations. RSEI offers scalability, comparability across scales, and high reliability [20].

While existing research primarily addresses the distribution of RSEI at a single spatial scale, long-term trends, and driver evaluation, it often overlooks spatial variations at different scales. Assessments at large scales can obscure differences between ecological subsystems, while small scales may not be suitable for broad ecological regulation applications. Hence, the selection of an appropriate assessment scale unit is critical [26]. This selection enhances the objectivity, credibility, and practicality of results, thereby facilitating decision-making in ecological quality regulation and security [27]. Most RSEI studies focus on single scales and static analyses but lack a systematic exploration of suitable scales for regional assessments and their underlying mechanisms.

Geographically weighted regression (GWR) is a spatial analysis technique that is widely used in geography and related disciplines involving the analysis of spatial patterns. GWR explores the spatial changes of a study object at a certain scale and related drivers by creating local regression equations at each point in the spatial scale and can be used for the prediction of future results [19].

Since GWR considers the local effects of spatial objects, it has the advantage of high accuracy. In the Huashan Creek watershed in Pinghe County, GWR can be used to study soil and water environmental pollution. For example, GWR can be utilized to explore how honeydew cultivation, fertilizer use, soil erosion, and other factors affect changes in water quality in the watershed by establishing local regression equations at each point. This approach can be borrowed and applied to the assessment of ecological quality drivers to clearly understand the impact of these factors in different geographic locations, thus providing a basis for developing targeted management strategies [28].

A key feature of GWR compared to other data analysis techniques is that it considers the effect of spatial location. While traditional global regression models usually assume that the effect of an explanatory variable on the dependent variable is the same in all locations, GWR allows for this effect to differ across locations [29]. In addition, GWR allows the computation of local parameter estimates for each explanatory variable at each location, thus revealing the characteristics of the spatial distribution of the effect of the explanatory variable on the dependent variable [30]. Therefore, the application of GWR in assessing factors and mechanisms of ecological quality at different scales is crucial for conservation and sustainable development.

This research concentrated on the agriculturally disturbed Huashan Creek watershed in southeastern China. Using the RSEI model, we assessed ecological quality at different grid resolutions, including the administrative village level and sub-watershed level. We employed spatial correlation (Moran index) and GWR to analyze ecological quality across these scales. The objectives of this study were to (1) characterize ecological quality at an appropriate watershed scale, (2) explore spatial responses across scales, and (3) quantitatively assess the impact of various driving forces. By achieving these goals, this study aimed to provide insights into ecological changes that support conservation and sustainable development.

2. Materials and Methods

2.1. Study Area

A comprehensive multiscale remote sensing approach was used to evaluate the quality of the ecological environment and to identify its associated driving factors in the Huashan Creek watershed, which is in Pinghe County, Zhangzhou City, Fujian Province, China (24°11′–24°31′N; 117°05′–117°25′E; shown in Figure 1). The watershed, which covers an area of approximately 864 km^2 and includes 158 administrative villages has a subtropical monsoon climate, with an average annual temperature of approximately 20.9 °C. Rainfall in the region is highly variable, ranging from 980 mm to 2100 mm per year. The region experiences sunshine and solar radiation for a moderate amount of time, the temperature difference between spring, fall, and winter is small, while the climate is warm and is suitable for growing various agricultural crops and making industrial investment. The region is abundant in water resources and is mainly characterized by subtropical broadleaf evergreen forests.

The landscape of the Huashan Creek watershed is characterized by mountains, valleys, and urban areas. The soil types in the creek are representative of subtropical soils such as humid-thermo ferralitic, lateritic red earth, red earth, and yellow earth, which are gradually acidified and sloughed under the influence of human activities such as honeydew production. The total length of the catchment is approximately 88 km, and the catchment boasts an average gradient of 2.8‰. This watershed, which is geographically located in the hilly areas of southeastern Fujian Province, plays a critical role in soil and water conservation and the preservation of biodiversity for the southwestern portion of Fujian Province and the Taiwan Strait.

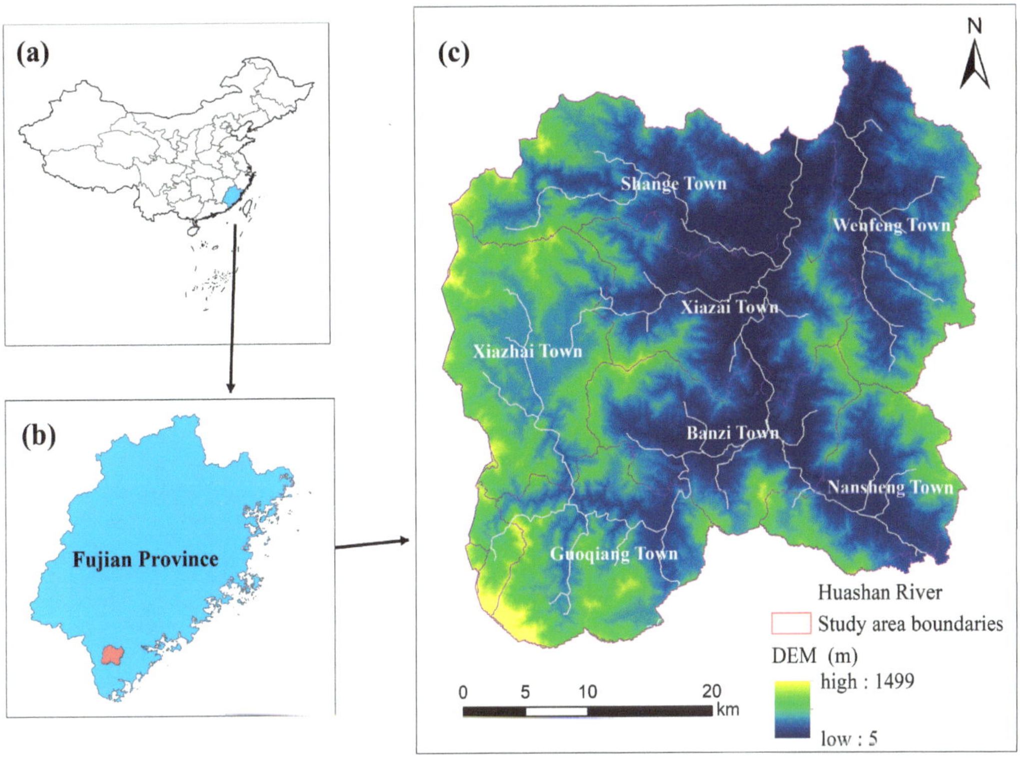

Figure 1. Digital elevation model (DEM)-based map of Huashan Creek Watershed in China. (**a**) The location of Fujian Province in China. (**b**) The location of Huashan Creek Watershed in Fujian Province. (**c**) The boundaries of Huashan Creek Watershed.

Notably, Pinghe County, which is within this watershed, is the primary producer of China's Guanxi honey pomelo and contributes significantly to the region's economy, which has an annual GDP of CNY 17.6 billion, of which 67% is attributed to agricultural output. Pomelo orchards cover approximately 60% of the total land area in the Huashan Creek watershed. In 2020, Pinghe honey pomelo production reached 2.1058 million tons, accounting for 21.1% of the national total for citrus fruits. The Huashan Creek watershed is the only drinking water source for nearly 80,000 people in Pinghe County, supplying approximately 42,500 t/d of drinking water. In 2019, 10.44 million m^3 of domestic water and 4.99 million m^3 of industrial water were supplied, and most of the remaining water was used to supply the 700,000-acre honeydew growing area. Over the past three decades, the expansion of agricultural land and the growth of the Guanxi honey pomelo industry in Pinghe have increased the vulnerability of the Huashan Creek watershed ecological environment to human activities [31].

Given this context, our study focused on the ecological environment quality of the primary Guanxi honey pomelo region, namely the Huashan Creek watershed. The objectives of our research were to investigate spatial variations in ecological quality within this watershed and to identify the key driving forces responsible for these variations.

2.2. Data Collection and Processing

In this investigation, we harnessed the capabilities of the Google Earth Engine (GEE) platform to access Landsat 8 OLI series satellite imagery from the year 2020. This selection of imagery was predicated on the minimal cloud cover of the Huashan Creek watershed, which does not exceed 10%, rendering GEE our primary data source. Subsequently, this acquired imagery was leveraged to facilitate the computation of RSEI. Notably, the satellite remote sensing images presented an average temporal resolution of 24 days.

The potential driving factors under consideration encompassed a comprehensive suite of land use indices, population density, road network density, landscape pattern indices, elevation, and proximity to water bodies. To ensure uniformity and facilitate comparative analysis, we standardized these diverse factors to a spatial resolution of 500 m, employing ArcGIS 10.8 as our tool of choice for this purpose. A comprehensive summary of our principal data sources is furnished in Table 1 for reference.

Table 1. List of products in the catalog of remote sensing data used in this study.

Data Name	Data Source	Website	Date	Resolution
Landsat8 OLI	United States Geological Survey	https://earthexplorer.usgs.gov/	Accessed on 1 May 2020.	30 m
Landcover data	GlobeLand30 Dataset	https://www.globallandcover.com/	Accessed on 1 May 2020.	30 m
Population density data	Socio-economic Data and Applications Center	https://sedac.ciesin.columbia.edu/	Accessed on 1 May 2020.	1 km
DEM	Geospatial Data Cloud	https://www.gscloud.cn/sources/	Accessed on 1 May 2020.	30 m
Scope of Huashan Creek Watchment	Resource and Environment Science and Data Center	https://www.resdc.cn	Accessed on 1 May 2020.	-

The software instruments enlisted for processing these data included SPSS 25 [32], ArcGIS 10.8 [33], and Fragstats 4.2 [34]. The deployment of this comprehensive suite of tools was undertaken with a deliberate objective: to augment the precision and resilience of our subsequent analytical endeavors.

2.2.1. Options for Different Scales

This study addressed the prevailing limitations of singularity and staticity inherent in the selection of evaluation units within RSEI investigations. Leveraging ArcGIS 10.3 grid sampling and magnitude transformation techniques, we systematically delineated a comprehensive framework comprising three distinctive categories of evaluation units, spanning a total of seven hierarchical levels. These units encompassed a range of isotropic scales (480 m × 480 m, 240 m × 240 m, 120 m × 120 m, 60 m × 60 m, and 30 m × 30 m). The selection of sample unit magnitudes within this range, spanning from 30 m to 480 m, is primarily influenced by data resolution parameters and the typical area dimensions of township administrative units within the Huashan Creek watershed.

Leveraging this diversified array of assessment units, we conducted a multifaceted assessment of RSEI within the Huashan Creek watershed. This holistic approach enabled us to conduct a comparative analysis of the spatial distribution of RSEI, discerning variations in heterogeneity across different scales within the study region. Additionally, this analysis unveiled the distinctive attributes characterizing the responsiveness of RSEI to alterations in spatial scales, shedding light on the underlying mechanisms influencing RSEI within the Huashan Creek watershed. By adopting this multifaceted approach, we aimed to mitigate against potential errors arising from a singular scale-centric perspective and ameliorate the issue of inadequately addressing spatial heterogeneity in the subjective delineation of evaluation units. See Figure 2 below for details.

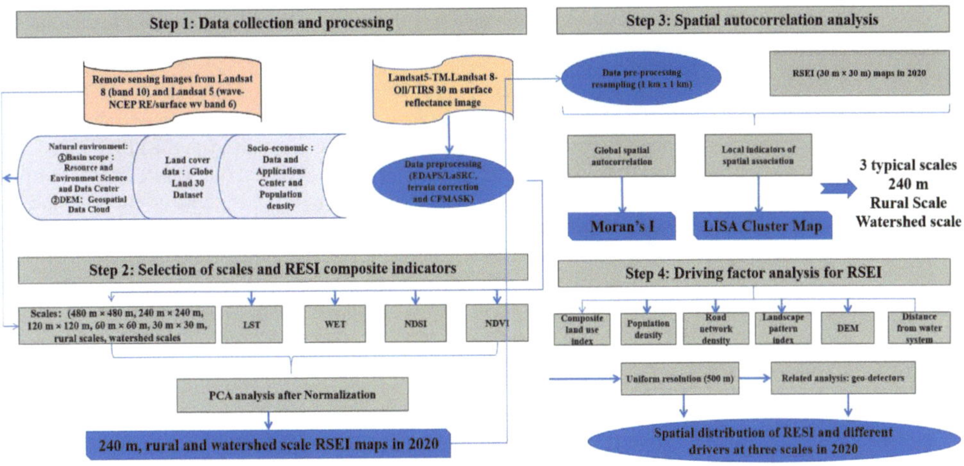

Figure 2. The framework of the research idea and methodology.

2.2.2. RSEI

RSEI is a recently introduced comprehensive ecological assessment tool tailored to examine ecological conditions exclusively through remote sensing techniques [20]. RSEI comprises four distinct metrics, namely greenness, humidity, dryness, and heat (see Table 2), each of which plays a critical role in ecological quality and can be directly perceived by human observers [35–38]. These metrics are derived from remote-sensing data, thereby ensuring a unified data source and minimizing the potential for errors stemming from disparate data sources. Prior research has substantiated the consistency, reliability, interpretability, and scalability of RSEI across diverse spatial and temporal scales [20]. Consequently, the computation of these metrics serves as a valuable tool for assessing ecological status based on readily available remote sensing data, often with minimal human intervention [39]. The formulas and significance of these indicators are presented in Figure S1.

Table 2. Formulas and explanations for the four eco-quality indicators WET, NDVI, NDSI, and LST.

Index	Formula	Explanation
WET	$\beta_1 B_{blue} + \beta_2 B_{green} + \beta_3 B_{red} + \beta_4 B_{nir} + \beta_5 B_{swir1} + \beta_6 B_{swir2}$	B_{blue}, B_{green}, B_{red}, B_{nir}, B_{swir1}, B_{swir2} represent the blue (0.45–0.51 μm), green (0.53–0.59 μm), red (0.64–0.67 μm), near-infrared (0.85–0.88 μm), short-wave infrared 1 (1.57–1.65 μm) and short-wave infrared 2 (2.11–2.29 μm) bands of Landsat 5 TM and Landsat 8 OLI/TIRS, respectively. β_i is the corresponding band parameter [35,36];
NDVI	$(B_{nir} - B_{red})/(B_{nir} + B_{red})$	
NDSI	$(SI + IBI)/2$	SI and IBI represent soil index and building index respectively [37];
LST	$K1/\ln(K1/D(t) + 1)$	K1 and D(t) represent Planck's constant and the radiative luminosity of a blackbody [38];

In the initial step of the analysis, Landsat 8 satellite imagery underwent preprocessing involving removing clouds and identifying water bodies, accomplished through the GEE Platform. Concurrently, the four ecological indicators were subjected to wavelet transformation to derive normalized proxies. Ultimately, RSEI was obtained through PCA, which synthesized the contributions of each component. This comprehensive evaluation function ascertains the weighting of individual principal components, reflecting the proportion of information contributed by each component relative to the total information contained in the original dataset. This approach effectively mitigates the pitfalls associated with arbitrary weighting assignments and circumvents the impact of correlations among indicators, leading to a rational and objective determination of weights.

Moreover, the load weight of Principal Component 1 (PC1), as extracted from each indicator in the database through PCA, indicates significant changes. This method dispels the subjectivity inherent in conventional weighting approaches, yielding results that objectively, reliably, and expeditiously monitor and appraise the quality of the regional ecological environment. The functional expression of RSEI can be formulated as follows [40]:

$$\text{RSEI} = \text{PC}_1[f(NDVI, \text{Wet}, \text{LST}, NDSI)] \tag{1}$$

The applicable range of RSEI is [0,1] [33]. The proximity of the RSEI value to 1 signifies a high level of environmental quality. Within the context of this research, a comprehensive classification of the RSEI was conducted to gain deep insights into the spatial distribution patterns associated with remote sensing of ecological and environmental quality within the Huashan Creek watershed. This classification framework enables a quantitative and visually informative analysis [11]. To clarify, the RSEI of the Huashan Creek watershed was classified into 5 distinct grades at 0.2 isometric intervals, namely, "poor" (0–0.2), "fair" (0.2–0.4), "moderate" (0.4–0.6), "good" (0.6–0.8), and "excellent" (0.8–1) [23,33].

2.2.3. Characterization of RSEI

In the context of our investigation, PCA showed that the cumulative variance explained by PCA1 and PC2 accounted for 65.84% and 91.21% of the total variance, respectively. Deconstructing the contributions of the four constituent indicators within PC1, it was observed that the Green Index (NDVI) and the Wetness Index (WET) made positive contributions, while the Dryness Index (NDBSI) and the High-Temperature Index exhibited negative influences [41]. These findings affirm that enhanced greenness and wetness favorably impact ecosystem integrity, whereas increased dryness and elevated temperatures exert detrimental effects, aligning with real-world observations [37]. Conversely, the interpretation of positive and negative signals within the other principal components (PC2, PC3, and PC4) proved to be volatile and challenging to ascribe to ecological phenomena [33]. Given that PC1 encapsulates the majority of pertinent features and that the loadings of each indicator on the first principal component were uniformly distributed, the utilization of information extracted from PC1 for the characterization of RSEI in our study is a logical and sound approach.

2.2.4. Spatial Autocorrelation Analyses

The primary objective of Exploratory Spatial Data Analysis (ESDA) is to unveil the spatial configurations and associations among geographical entities, wherein geospatial location and arrangement play pivotal roles in determining the correlation among attribute values of these entities. ESDA elucidates geographical phenomena spatial distribution patterns and traits by employing diverse statistical metrics and analytical tools, including spatial autocorrelation, aggregation, and discrete indicators. Consequently, exploratory spatial statistics serve as a means to identify and investigate the clustering distribution pattern, spatial heterogeneity, and the presence or absence of spatial spillover effects within the realm of RSEI.

(1) Global spatial autocorrelation (Global Moran's I: MI)

The MI index measures the overall extent of global autocorrelation, encompassing the degree of spatial correlation heterogeneity and the differentiation in the distribution of RSEI. The vector formula for calculating Moran's Index is presented below [3]:

$$\text{MI} = \frac{n}{\sum_i \sum_j G_{ij}} * \frac{\sum_i \sum_j G_{ij}(x_i - \overline{x})(x_j - \overline{x})}{\sum_i (x_i - \overline{x})^2}, \tag{2}$$

In this equation, MI denotes the global index, n is the total number of grid cells; *Gij* signifies the (*i, j*) element of the spatial weight matrix G (based on the variable K nearest neighbors, K = 4), following the principles of the first law of geography and inverse distance weighting. The values x_i and x_j correspond to the RSEI values of samples i and j, respectively, with $\overline{x}$ denoting the average RSEI values across the study area. MI is normalized by the spatial weight matrix G, yielding values within the range of [−1,1].

Regarding confidence levels, positive MI values indicate spatial clustering of RSEI in the province, with varying degrees of similarity and typicality. A close proximity to +1 signifies a positively correlated, significant clustering pattern, while a distant proximity from +1 implies a substantial divergence in atypical clustering, denoting a significant negative correlation with dispersion. Notably, MI values approximating or equal to zero signify a spatial distribution characterized by randomness, with no discernible correlation [42]. Additionally, the significance of the MI index was assessed using the Z-Score standardized statistic [42]:

$$Z = \frac{MI - E(MI)}{SD(MI)}, \tag{3}$$

where *E(I)* and *SD(I)* represent the theoretical mean and standardized variance, respectively, and MI signifies the global index. Specifically, when $MI > 0$ and Z-Score > 1.96, a considerable value indicates a vital and significant spatial positive correlation; conversely, a small value indicates a strong negative correlation and significant spatial variability [43].

(2) Local spatial autocorrelation (Local Moran's I: LI)

The RSEI MI solely captures the collective degree of divergence apparent within the broad spatial distribution of its density: it fails to elucidate localized disparities or pinpoint areas of concentration, as previously noted [44]. While LI is the decomposition of MI, the combination of LISA (local spatial association index to measure the level of significance) and the Moran scatterplot can reveal the scale or intensity of RSEI at the local, regional scale and its spatial autocorrelation with its neighboring places and its degree of significance. The formula for the LI is as follows [45]:

$$LI_i = Z_i * \sum_{j=1}^{n} G_{ij} * Z_j, \tag{4}$$

$$Z_i = \frac{x_i - \overline{x}}{S^2}, \tag{5}$$

$$Z_j = \left(x_j - \overline{x}\right), \tag{6}$$

where S^2 is the discrete variance of province *i*; Z_i and Z_j are the standardized mean values of RSEI for grids *i* and *j*, respectively, and the other variables have the same meanings as in Equation (2).

The Moran scatterplot form, which describes the correlation between variable A_i and its spatial lag vector $B_{(Ai)}$, is dissected into distinct quadrants and categorized into five typologies [46], namely: random distribution of the origin (non-significant), clustering of high values (H–H type), low values surrounded by high values (L–H type), clustering of low values (L–L type), and high values surrounded by low values (H–L type). Among these values, under the given confidence level, if LI_i significantly > 0 and $Z_i > 0$, grid i is in the H–H quadrant; if LI_i significantly > 0 and $Z_i < 0$, grid *i* is located in the L–H quadrant; if LI_i significantly < 0 and $Z_i > 0$, grid *i* is located in the H–L quadrant; if LI_i significantly < 0 and $Z_i < 0$, grid *i* is located in the L–L quadrant [46].

2.2.5. Driver Analysis

(1) Driver selection

The dynamics of RSEI are subject to a multifaceted interplay of factors encompassing natural attributes, land utilization, and socio-economic influences [47,48]. In this complex framework, natural determinants are paramount over extended time scales. Therefore, our investigation primarily highlights the pivotal role of specific natural elements affecting ecological traits within the focal area, which serves as the primary area for honeydew production. Among these salient natural determinants are elevation, which embodies considerations related to terrain, and proximity to water bodies, reflecting the importance of water sources.

Altitude significantly modulates the illumination, thermal conditions, hydrologic parameters, and nutrient status within the study domain, thus exerting a dominant influence on the distribution patterns of the landscape [49]. Simultaneously, rivers function as vital hydrologic conduits, attracting and reshaping nearby distributions of biodiversity, human activities, and ecosystem services, thus exerting a notable influence on the ecological milieu [50,51].

On the other hand, anthropogenic forces exert immediate and pronounced impacts at short temporal scales, ultimately directing the trajectory of local ecosystems in the long term. This study relies on indicators such as patterns of land use, socioeconomic parameters, and the spatial configuration of the landscape to gauge anthropogenic disturbance [14,52–55]. Among these indicators, changes in land use—manifested through shifts in structure, composition, and management practices—exert a pivotal influence on the functional and spatial balance of RSEI [56,57].

Socioeconomic factors such as the density of the population (an underlying determinant of human activities) and gross domestic product (GDP; acting as a visual gauge of ecosystem service value influenced by anthropogenic actions), tend to precipitate alterations in patterns of RSEI depletion as well as the quantity and quality of ecological amenities. This displacement results from social policies and developmental processes, further exacerbating the disruption of RSEI [58–60]. On the other hand, alterations in the spatial configuration of landscapes give rise to specific and discernable consequences by influencing the composition, structure, and ecological processes of ecosystems [61], thus molding RSEI [62,63].

In this study, we select indicators of landscape patterns designed to characterize landscape heterogeneity and levels of fragmentation, with the detailed meaning of these indicators elucidated in Figure S1. Therefore, a composite index consisting of land-use extent (La), population density (Pop), road distance (Dis_{road}), aggregation index (AI), maximum patch index (LPI), patch density (PD), Shannon's diversity index (SHDI), distance from water (Dis_{water}), and elevation (DEM) is adopted, while its components serve as independent variables driving the factors in the RSEI model.

All data are normalized to a uniform spatial resolution of 500 m prior to analysis. In order to mitigate issues arising from collinearity between factors, we perform a correlation analysis and principal components analysis and select relevant scales (at the basin level and village level, covering an area of 240 km × 240 km) in order to ensure robust findings that are not affected by covariance between factors.

(2) GWR

GWR is a valuable extension to the technique of local linear regression, designed to characterize spatially varying relationships within a given geographic region. This approach results in a regression model that elucidates local associations in discrete locations within a study area, effectively capturing localized spatial relationships and the inherent heterogeneity of variables. Finally, it should be noted that GWR has the distinct advantage of accounting for spatial dependence as well as variability within data, which is a departure from conventional global parameter estimation [3].

Another distinguishing feature of GWR is the inclusion of a weighting function, as it minimizes sample redundancy compared to traditional least squares methods, thereby increasing the accuracy of regression results and allowing spatially non-stationary relationships to be examined simultaneously [64,65]. The calculation model of GWR is

$$y_i = \beta_0(u_i, v_i) + \sum_k^m \beta_k(u_i, v_i) * x_{ik} + \epsilon_i \tag{7}$$

where y_i represents the dependent variable's value at position i, k stands for the total number of grid cells, x_{ik} (k = 1, 2,..., m) signifies the value of the kth independent variable at point i, while (u_i, v_i) denotes the spatial coordinate of sampling point i. Furthermore, β_0 (u_i, v_i) corresponds to the intercept term of sample i, and β_k (u_i, v_i) embodies the regression coefficient for the k-th independent variable, x_{ik}, at sampling point i. This coefficient is a function of geographic location, ascertained through a weighted function method in the estimation process, and ϵ_i is the error term.

The accuracy of the GWR model is highly dependent on the choice of kernel function and bandwidth. We adopt the AIC method to determine the optimal bandwidth in this study given the ability of AIC to address accuracy quickly and efficiently. It is imperative to point out that a low AIC value in the model corresponds to a superior fit to the observed data, which indicates superior performance [66]. The degree to which various factors influence RSEI diverges across different spatial scales. Therefore, we use RSEI as a dependent variable and each influencing factor as an independent variable by building GWR models on three representative scales, which were selected based on the similarities and differences observed through spatial clustering, the availability and precision of the driving factors, the specificity of the administrative villages, the natural characteristics of the watershed, and the economy of the overall effort. The regression coefficients from these GWR models offer insights into the magnitude of influence exerted by the influencing factors on RSEI.

3. Result

3.1. Patterns of Spatial Differentiation in Ecosystem Quality at Multiple Scales

In order to elucidate the characteristics of ecological quality responses within the Huashan Creek catchment in relation to variations in spatial scale, we meticulously examine RSEI values across three separate scale units encompassing seven gradations in total. Correspondingly, we undertake a comprehensive analysis of the magnitude and inherent trends within these values. Figure 3 reveals distinctive spatial patterns in the distribution of RSEI across a variety of scales while retaining overall similarity. It should be noted that across scales, the predominant proportion of the surface exhibits a state of moderate RSEI, with favorable conditions seen in the eastern and western regions, juxtaposed with poorer quality conditions in the central region. As the scale decreases from 480 km to 60 km, the spatial granularity of the RSEI information within the grid cell types becomes finer, with relatively marginal differences exhibiting some degree of similarity. Regions characterized by excellent and good RSEI are scattered in the extreme east and west of the study area, respectively, creating a spot-like pattern.

The moderate RSEI region assumes a pentagram-like shape that is roughly positioned in the center of the region although areas showing poor RSEI concentrations are notably located in the north–central part of the study area, mainly at sub rural scales, with sporadic occurrences in the north-western and south-central areas. At the rural and catchment scale, the distribution of RSEI occurs predominantly within the central core and periphery, showcasing moderate to high-quality performance. The extreme low zones (0–0.2) lie in the central watershed, which is in the core area of Pinghe County. Furthermore, taking advantage of the increased resolution of the 240 m grid, this study conducts an assessment of ecological environment quality scores at this scale, classifying them into 5 separate grades based on the size of the area: excellent (41.1%), good (43.7%), moderate (8.4%), fair (6.7%), and poor (0.1%). The results in Figure 4 reveal that the quality of the ecological environment within the Huashan Creek watershed predominantly occupies the moderate rating category at 8.4%, which is indicative of an overall moderate ecological status. To summarize, the overall distribution pattern of RSEI, whether at the administrative village level (Figure 4f) or at the grid scale (Figure 4a–e), highlights a prevalence of moderate and good conditions in contrast to a concentration of poor conditions in the urban center.

The magnitude of moderate RSEI coverage at the catchment scale (Figure 4g) exceeds that at the administrative village scale, with a noted concentration in the west–central part of the study area. Furthermore, the grid-scale delineates fine-scale areas with inferior RSEI distribution.

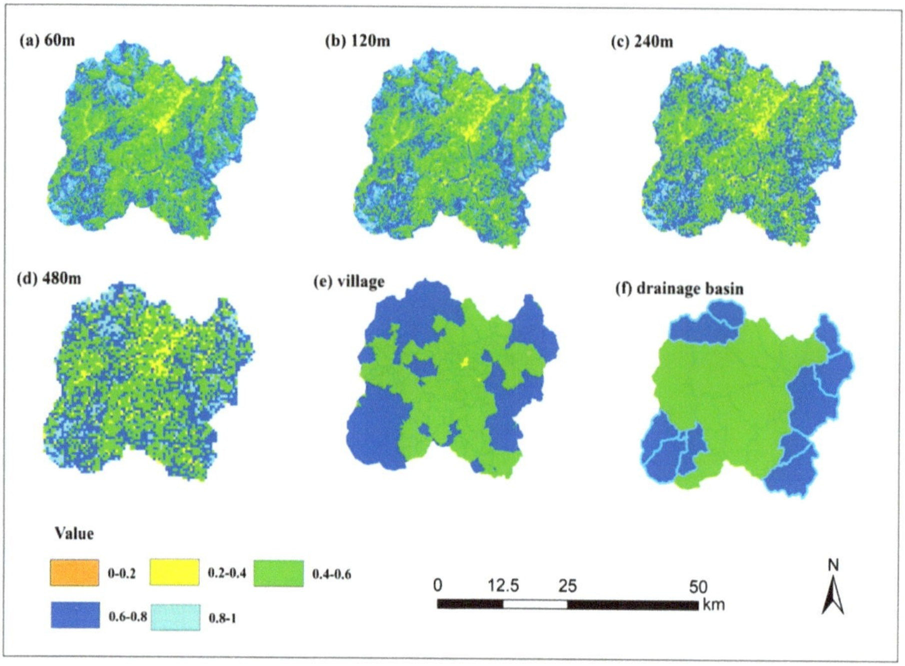

Figure 3. Characteristics of the spatial distribution of RSEI values in the Huashan Creek watershed in response to changes in scale.

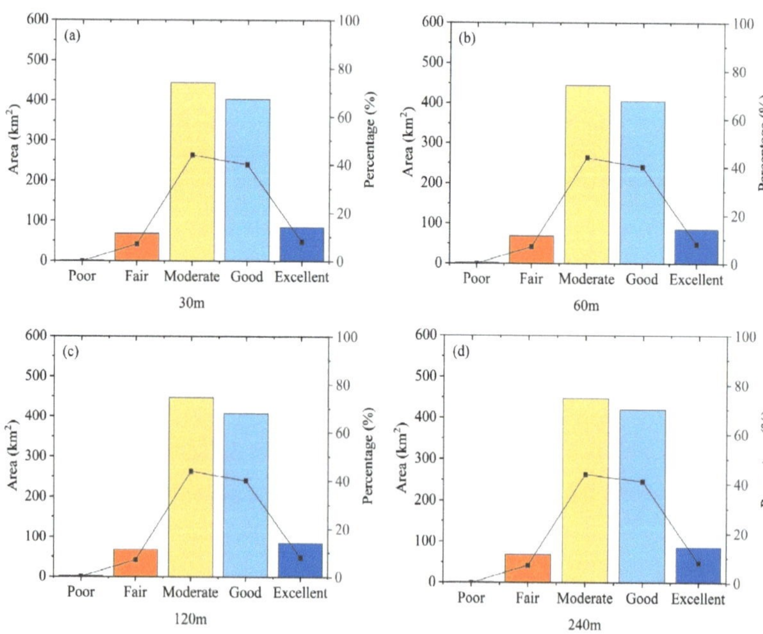

Figure 4. *Cont.*

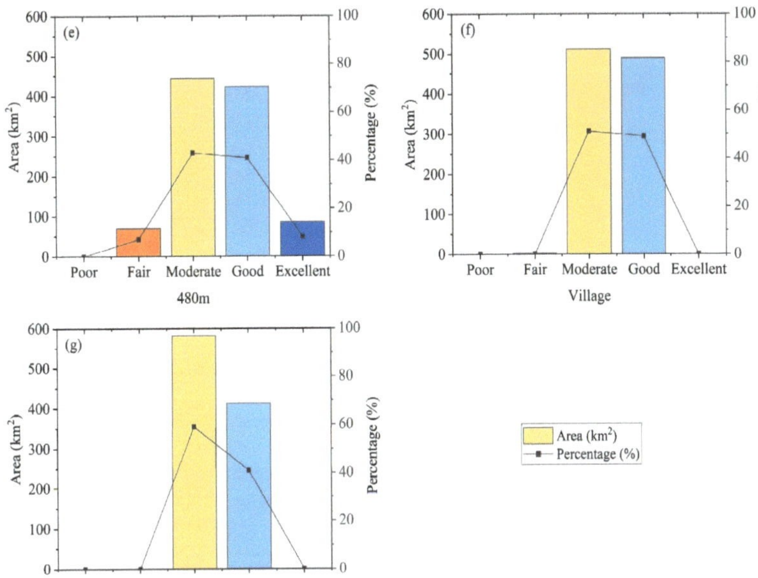

Figure 4. Area statistics for the five classes of RSEI at seven spatial scales.(**a**) 30 m scale (**b**) 60 m scale (**c**) 120 m scale (**d**) 240 m scale (**e**) 480 m scale (**f**) Village scale (**g**) Watershed scale.

3.2. Spatially Dependent Characterization of Ecosystem Quality at Multiple Scales

3.2.1. Global Spatial Correlation Multi-Scale Spatial Response Characteristics of Remotely Sensed Ecological Quality

To analyze the spatial distribution differences of RSEI at three types of seven scales, this study describes the equilibrium pattern of RSEI using statistical parameters. Therefore, global indices of spatial autocorrelation derived from the Spatial Data Exploration and Analysis Methodology (SDEAM) are used as described in Equation (3). As illustrated in Figure 5, Moran's I values for RSEI, computed at seven different scales over the course of 2020, showed a fluctuating pattern described as "increasing–decreasing–increasing" as the scale was decreased, revealing the presence of two scales characterized by indices of dissimilar spatial autocorrelation.

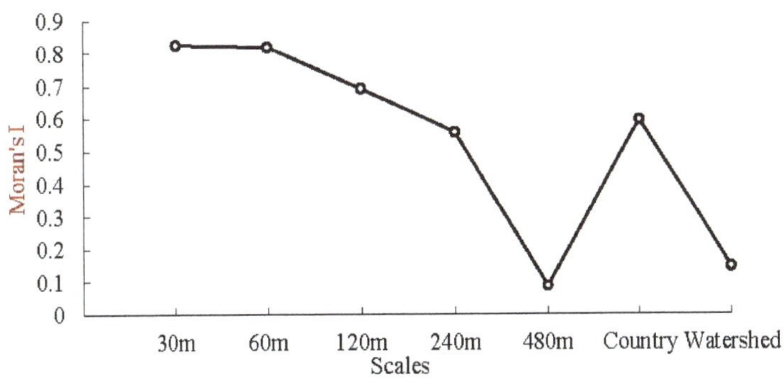

Figure 5. Moran's I values for RSEI at different scales.

The overall pattern of fluctuations with respect to scale reduction follows a "bottom-up" pattern, indicating the existence of two scale cutoffs at the 480 km × 480 km scale (the lower limit) and the village scale (the steepest). Note that the transition from a grid scale of

30 m to 240 m results in a gradual decrease in the Moran I value, accompanied by minimal variations in its amplitude, which remained persistently greater than 0.6, indicative of a conspicuous spatial clustering. Then when the scale reached 140 m, the Moran I value showed a steep decline to 0.1 followed by a rapid rise to 0.6 at the village scale and finally decreasing to approximately 0.15 at the basin scale.

Overall, global Moran's I values for RSEIs at the various scales consistently exceeded zero, with p values mostly below 0.001, with the Z-Score clearly exceeding 1.96, which clearly means the prevalence of highly significant positive spatial clustering in the spatial distribution of RSEIs across the seven separate scales of varying magnitudes. However, it is pertinent to note that the positive correlation features of agglomeration within the same grid scale type and across different scale types (grid, village, and watershed) manifest discernible differences.

In particular, when comparing the eigenvalues of the 5-level RSEI scale within grid scale types, there is a gradual increase in the overall Moran's I value from 0.086 for the 480 m × 480 m case to 0.825 for the 30 km × 30 km case, with the former approaching minimum values near 0 and the latter approaching unity. Correspondingly, the Z-Score displays a rising trajectory as the range of scales decreases (ranging from 8 to 3379). This progression accentuates the significance and strength of the clustering of positive correlations within the RSEI at the grid scale, which becomes pronounced as the scale decreases. In addition, there is a remarkable jump in the Moran's I value during the transition from 480 m × 480 m to 240 m × 240 m, while subsequent shifts between the remaining scales experience a change of roughly an order of magnitude, albeit with Moran's I value dropping below 0.69 beyond 120 m × 120 m, indicative of decreasing significance in the clustering of positive spatial correlations.

Additionally, when contrasting village and watershed scale-type 2 scales, the global Moran's I value for the former RSEI (0.595) substantially exceeds that of the latter (0.146), while the Z-Score exhibits a 13-fold increase (14.319 > 1.426). This divergence highlights the increased and statistically significant clustering of positive correlations evident at the village level. In addition, compared to the grid scale, the RSEI at the village scale displays similar Moran's I values and Z-Scores to those observed at 240 m × 240 m, while the largest catchment scale stands out as unique, characterized by a Z-Score of less than 1.96, signifying clustering by weakly significant spatial correlation of RSEI.

With a Moran's I value of 0.146, the watershed scale demonstrates a lack of statistically significant spatial clustering across cells, diverging from the pattern observed in the other assessments. As a result, it is discernible that within assessment cells of the same grid type, at microscopic scales of assessment, high Moran's I values correspond to a great degree of spatial clustering in RSEI. In contrast, spatial clustering at the macroscopic catchment scale, similar to the case of 480 m × 480 m, lacks statistical significance. On the other hand, the mesoscopic scale of the village, analogous to the case of 240 m × 240 m, exerts a noteworthy influence on the spatial clustering of RSEI. Thus, the next section will focus on characterizing and analytically exploring the spatial distribution of RSEI at the microscale of 240 m × 240 m, at the meso-village scale, and at the macroscale of the watershed.

3.2.2. Local Spatial Correlation Multi-Scale Spatial Response Characteristics of Remotely Sensed Ecological Quality

To further elucidate the local spatial autocorrelation and multiscale response characteristics of RSEI, this study employed a cross-scale assessment approach. In particular, the 240 m × 240 m grid scales, village administrative districts, and catchment scales exhibiting similar Moran's I values are carefully chosen. The indices of local spatial autocorrelation derived from Equation (4) were then used to assess and analyze heterogeneity in the spatial correlation of RSEI across the study region as it experienced changes in the scales of the geographical unit types. Finally, it should be noted that there were both shared patterns as well as significant disparities in the local spatial differences in RSEI observed across varying scales.

As illustrated in Figure 6, except at the meso-village and macro-catchment scales, the spatial clustering pattern of RSEI at the grid scales was characterized by a non-random spatial distribution, while the non-significant regions increased in extent as the scales expanded. In addition, areas characterized as L–L and H–H types showed variation between scales. It is interesting to note that L–L and H–H regions consistently covered larger areas than those of the H–L type, and no regions exhibited the L–H type. Specifically, at the macro watershed scale, a multi-region spread displayed a combination of southwestern H–H and northern L–L types.

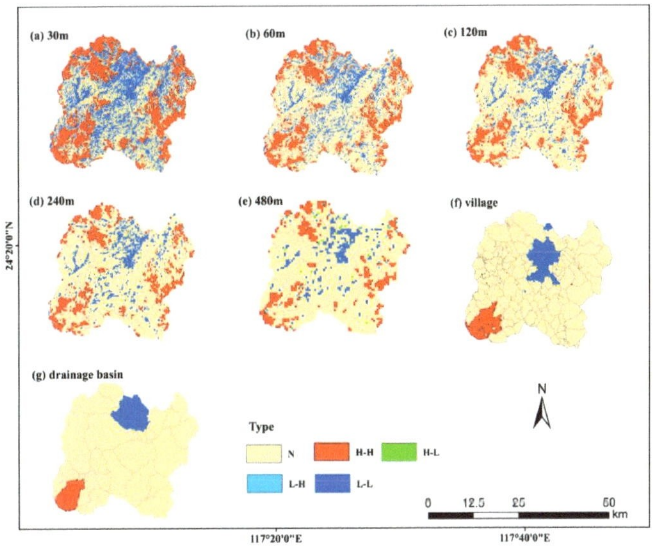

Figure 6. Spatial response characteristics of the local Moran's index to different scales: (**a**) 30 m, (**b**) 60 m, (**c**) 120 m, (**d**) 240 m, (**e**) 480 m, (**f**) Village, and (**g**) Watershed.

Similar distribution patterns were observed at a range of scales. On a meso-village scale, H–H continuous villages were prominent in the north–central region. At the 480 m × 480 m scale, there were few H–L types in the north–central region, with scattered H–H types in the SE and NW regions. At the 240 m × 240 m scale, there was a scattered H–L distribution, with a gradual concentration of H–H types in both the SE and NW. At the 120 km × 120 km scale, the H–L pattern was absent, with the central L–L pattern becoming continuous and the H–H pattern notably present in the NW, SW, and eastern parts of the country. At the 60 m × 60 m scale, an alternating pattern of H–H and L–L was observed in the east, center, and west. The alternating pattern between H–H and L–L was most distinct at the fine scale of 30 m × 30 m, especially in the upper, middle, and lower regions of the west. The H–H type was mainly located in the mountainous and vegetation-rich areas of the east and west, while the L–L type was prevalent in the central plains.

3.3. Response and Difference Analysis of Multiscale Drivers of Remotely Sensed Ecological Quality

A 240 m × 240 m grid was chosen for the identification of extraction points, informed by the ability of the grid to effectively mitigate outcome bias arising from covariance effects between factors as determined by correlation analysis and principal component analysis. A total of 11,141 sample points were then identified, each associated with specific driver values. Through an extensive filtering process employing correlation analysis, five independent variables were identified as suitable drivers, namely, distance from road (Dis_{Road}), population density (POP), density of patches (PD), distance from a water body

(Dis_{Water}), and elevation (DEM). We then meticulously evaluated the impact of these variables on the RSEI dependent variable through GWR, which was conducted at multiple spatial scales. In order to provide deep insights into the factors that influence the local spatial distribution of RSEI, 3 separate geospatial cell scales: micro (240 m × 240 m), meso (village administrative district), and macro (watershed)-based were selected on the basis of their similarity in Moran's I values. We then used the local GWR spatial coefficients derived from Equation (5) in order to dissect the variations in the effects of these five drivers across these diverse scales of geospatial cells.

3.3.1. Local Spatial Correlation: Multi-Scale Spatial Response Characteristics of Remotely Sensed Ecological Quality

Overall, the GWR results revealed a corrected R^2 of 0.42 and a condition number that was less than 0. These findings reflect the robustness of the model, which successfully withstood rigorous evaluations for multiple covariance considerations. The corrected R^2 surpasses the threshold of 0.6, indicating a relatively strong model fit. The regression coefficients within the GWR model provided insights into the individual indicators' respective impacts on RSEI. We conducted a comparative assessment of these regression coefficients, focusing on the primary driving variables for RSEI within the Huashan Creek catchment at various spatial scales (i.e., watershed, village, and 240 m × 240 m) (see Figure 7). The driving variables for RSEI were observed to have positive influences at all three scales, with values greater than 0 indicating substantial positive effects. Among these, the socioeconomic variable POP displayed the most significant influence, consistently exceeding 0.7.

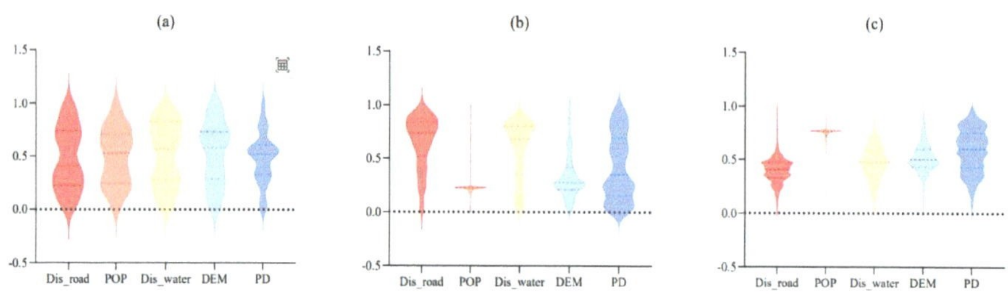

Figure 7. Regression coefficients for the drivers of the five RSEI (**a**) watershed scale, (**b**) village scale (**c**) 240 m raster scale.

In addition to POP, the variables DEM and PD demonstrated noteworthy influences on RSEI. However, disparities emerged when comparing the magnitudes of these influences across different spatial scales. Notably, Dis_{Road} and POP maintained consistent coefficients across watershed scales, with approximate values of 0.4 and 0.8, respectively. Conversely, Dis_{Water} had the least impact at the village scale, registering a coefficient of 0.2 while displaying intermediate effects at other scales (0.5). In contrast, DEM exhibited its highest influence at the village scale (0.7) and intermediate impacts at other scales (0.5).

3.3.2. Analysis of Spatial Differences in Multi-Scale Remote Sensing Ecosystem Quality Drivers Based on GWR

Figure S1 depicts the spatial distribution of the five drivers, highlighting significant spatial variability in the relationship between RSEI and these factors, contingent on the analysis scale. Given this spatial layout, we categorized the regression coefficients into five distinct levels of intensity: 0 to 0.2 (low), 0.2 to 0.4 (low), 0.4 to 0.6 (medium), 0.6 to 0.8 (high), and 0.8 to 1 (high). In the same scale, specifically at the micro-grid level of 240 m × 240 m, DEM exhibited a systematic distribution of coefficient classes, with a concentration of coefficients ranging from moderate to high intensity extending from the central area of low-intensity coefficients to the periphery.

Approximately two-thirds of the area was dominated by moderate and high-intensity classes, with river valleys, characterized by lower topography, making a smaller contribution to RSEI at the grid scale. POP displayed four classes of coefficients above the low-intensity range, with the high and medium intensities being widespread and the high intensity contiguous. Moderate, weak, and strong intensities were mostly clustered in the southwestern corner of the study area. Both Dis_{Road} and Dis_{Water} methods covered all coefficient classes. The distribution of Dis_{Road} appeared patchy, primarily in the north–central to south–central areas, with the moderate intensity area predominating. On the other hand, Dis_{Water} exhibited an alternating cluster-band pattern, with the moderate intensity area being the most extensive and the lower intensity low appearing as scattered spots.

The high-intensity area was primarily located to the northwest and south. PD extended across all gradient levels, with coefficients increasing toward the southeast in the northwest corner. In the rest of the region, the high-intensity center retreated outward, with the moderate area mostly in the southwest-to-eastern margins. At the rural mesocosmic scale, DEM exhibited three high-intensity coefficient classes, covering most of the region in an east-west and south-north direction, while the moderate-intensity class dotted the edges from east to west. On this scale, POP had four coefficients lower than the high intensity, with the low-intensity class spanning the region, and the moderate to high-intensity classes primarily found in the southwest corner.

Dis_{Road}, Dis_{Water}, and PD showed the full range of classes. Dis_{Road} featured a substantial area of high intensity that was contiguous to the west, with intensity gradually increasing from south to the northeast. Dis_{Water} displayed a gradual increase in intensity from the southwest to the northeast, with an area of high intensity to the northeast. PD showed a distinct increase in intensity from north to south, with a notable presence of average intensity in the northeast corner. At the macro watershed scale, each factor consisted of five classes and displayed a more even distribution. DEM, Dis_{Water}, and PD showed a gradual increase from the west to the east, with Dis_{Water} showing the most pronounced increase. DEM showed a high-intensity region in the southeast northeast corner, while PD displayed a high-intensity area in the southeast-northeast corner. On the other hand, POP demonstrated an increase in intensity from the south to the north, whereas Dis_{Road} showed an increasing distribution from east to west.

With respect to the impact of the same driver at different spatial scales, the influence of DEM on RSEI increased as the scale transitioned from the micro-grid scale to the meso-grid scale, subsequently exhibiting a slight attenuation at the macroscale of the watershed. The impact of POP on RSEI was most pronounced at the micro-scale and minimized at the meso scale, with relatively uniform contributions observed at the macro-catchment scale and within sub-catchments. The strength of the effects of Dis_{Road} and Dis_{Water} also intensified as the scale broadened, whereas the spatial distribution shifted from a pattern of dispersed clusters to a hierarchical east-west distribution. Of note, the range of PD's high-intensity impacts steadily decreased with the increase in scale.

4. Discussion

4.1. RSEI Response to Spatial Scales

For the three scale categories, as the scale resolution increases from 60 m to the catchment scale, the response of the ecological environmental suitability scale at seven distinct levels primarily shifts in attributes related to RSEI, ranging from complex and detailed spatial distributions to more straightforward and clearer models. However, there is a risk that the perception of a shift might obscure critical information [28]. Notably, RSEI spatial correlation reaches its lowest point as the scale expands to 480 m. Moreover, at the village scale, the spatial distribution results diverge markedly from those at the earlier scale, with the disparity growing significantly at the catchment scale (see Figure 3).

The study area was examined using the same research methods and data. The spatial distribution characteristics exhibit distinct patterns due to variations in the scale of the assessment units. This behavior reveals significant variability between scale categories but limited variability within each category. This variability occurs because evaluations of differences in environmental quality are sensitive to changes in the size of the assessment units involved. Large units reflect the local composition of the landscape or specific terrestrial ecological subsystems. Furthermore, as the unit size increases, RSEI evaluations tend to minimize differences in landscape elements within the unit [28].

In summary, this study highlights the importance of conducting RSEI response analyses at various spatial scales. Reliance on single-scale assessments alone often fails to accurately assess regional ecological quality since it does not comprehensively represent the true state of an area and desired outcomes. Thus, employing RSEI to analyze responses at different spatial scales is advantageous. Such an approach can mitigate errors arising from single-scale divisions and address issues related to the incomplete consideration of spatial heterogeneity in the division of assessment units. The choice of scale, however, should reflect analytic goals, available information, and decision-making requirements [67].

In this study, the alteration of spatial scope yields distinct characteristics in the global autocorrelation spatial analysis of ecological quality and functional control across three scale levels: sample grid, administrative boundary, and watershed scales, all based on natural ecosystems. The Moran's I value of the RSEI exhibits a fluctuating "decreasing-increasing-decreasing" pattern as the scale expands, revealing notable variations. In particular, as the scale of the grid increases, the information becomes less detailed, resulting in a gradual decrease in Moran's I.

The most significant difference in Moran's I value of RSEI was observed between the 240 m scale and the 480 m scale. This is because, at the 240 m scale, RSEI's spatial performance is similar to that calculated in rural areas, resulting in a higher spatial autocorrelation of 0.6. This result is much higher than the spatial autocorrelation of 0.15 at the watershed cell scale and 0.1 at the 480 m grid sample scale. Therefore, the 240 m and rural scales are useful for identifying spatial differences in ecological quality within the Huashan Creek watershed as well as for further analysis. The central valley examined in this study, characterized by poor drainage, has a more significant influence than elevation alone, primarily due to population growth and socio-economic development. Original forested areas and water bodies were cleared for agriculture and construction, increasing the amount of land used for these purposes. The change led to the fragmentation of once-contiguous forests and the formation of only a few isolated forested patches and small lakes. The actions exacerbated fragmentation, affecting the RSEI of the watershed and contributing to the land's unsuitability for honeydew cultivation. The western and eastern areas are closer to transportation routes that fragment the original forest cover and water bodies, resulting in areas that are at a greater distance from water bodies. However, the highly impacted area in the northwest corner may be used as a nature reserve, creating an ecological barrier that fosters biodiversity, raising the RSEI.

At the 480 m scale, spatial correlation is lowest because of its masking effect on finer ecological features, such as small lakes and vegetation. In the Huashan Creek watershed, where a significant portion of the land is used for honeydew production, the rural and the watershed scales are both appropriate for RSEI assessments. The approach is justified by the practicalities of collecting data and establishing ecological control models within administrative units, making it more feasible than utilizing larger watershed units spanning multiple regions. Administrative data at the village level offers greater accuracy for higher-level units and, when combined with grid-scale data, provides detailed support for village and catchment planning and protection efforts [68].

We utilized 240 m, rural, and watershed scales to assess RSEI within the district because we recognized the importance of natural catchment boundaries for the assessment of land use impacts and water quality change. Local autocorrelation spatial analysis was used to examine RSEI clustering. At larger scales, the finer grid scales fail to establish spatial correlation because they average high RSEI values across the eastern and western portions of the district.

Analysis using the 240 m, the village, and catchment scales revealed areas of excellent ecological quality (H–H), which are clustered in mountainous areas with dense vegetation cover, notably in the south-west. In contrast, poor ecological quality (L–L) areas are in the central–northern region, marked by low relief, degraded vegetation, urbanization, and high population densities. Mixed ecological quality areas are found in flat areas where urban and natural ecosystems co-exist. The central area, marked by low elevation and high population density, exhibits moderately poor ecological quality [23].

4.2. Response of RSEI Drivers to Spatial Scales

4.2.1. Key Drivers of RSEI

RSEI within the Huashan Creek watershed is influenced by various factors, including DEM, POP, Dis_{Road}, Dis_{Water}, and PD. However, the extent of these influences and their spatial distribution varies across scales and locations. Collectively, POP, DEM, and PD determine the magnitude and spatial pattern of RSEI at all scales, with POP having the most significant impact (approaching 0.8). Natural population growth and migration can result in urban expansion, reduced vegetative cover, and frequent changes in land use. Anthropogenic activity contributes to increased water use and elevated greenhouse gas emissions, affecting urban climate factors such as temperature, humidity, and aridity [69–71]. Elevation has a significant influence on factors such as rainfall distribution, temperature, radiation intensity, soil quality, plant ecology, growth, and overall ecological quality. It shapes human activities, including agriculture, industry, and transportation, with lower elevations generally experiencing a greater human impact, which increases the RSEI. Specifically, areas at lower elevations usually have higher population densities, road densities, patch densities, and distances to water bodies. Areas with these types of integrated landscapes are more susceptible to soil erosion and pollutant transport from upstream areas, which can reduce ecological quality [69]. The areas where DEM had the highest influence primarily consisted of those at lower elevations that were suitable for honeydew cultivation, where greenness, temperature, humidity, and desiccation had a significant influence. As Xiong et al. [33] indicate, ecological restoration or agro-industrial planning at the larger scale of the Minjiang River should first be done on steep north-facing slopes at low elevations. Shao et al. [72] also demonstrated that the combination of topographic superimposition of conditions such as humidity and greenness and anthropogenic activities is the key factor influencing the ecological quality of Qingdao city.

Regarding the other variables, there was significant spatial heterogeneity in the influencing factors in different regions and at different scales [73]. Li et al. [34] suggest that human activities contribute to the improvement of eco-environmental quality by reducing the degree of landscape fragmentation and the intensity of land use. Xiong et al. [33] indicate that the relationship between NDVI and topography is scale-dependent. Wu et al. [74] reveal that precipitation and land use are the main factors affecting the Sichuan-Yunnan Ecological Barrier Area, but they have different interpretations at different unit scales. Ling et al. [75] pointed out that vegetation, anthropogenic disturbances, and aridity index were the main factors influencing ecosystem services in the Manas River basin.

Furthermore, changes in landscape patterns, such as fragmentation, have implications for functions such as productivity, connectivity, climate regulation, and ecosystem health. Comparing key drivers at different scales, it is clear that proximity to roads and water bodies has a relatively similar impact on catchment scales, highlighting the rivers' vital role as water conduits. At the village level, proximity to water bodies has a more substantial effect than road distance because of the importance of irrigation, which affects water quantity and quality [20]. The construction of government road networks in rural areas highlights the importance of road distances to the characterization of an ecological environment. The performance of the 240 m grid scale is similar to that of watershed scales. Overall, the quality of an ecological environment is shaped by natural factors and changes in land use. Therefore, at lower elevations, anthropogenic activities will outweigh the effects of elevation. The increase of cultivated land, transportation, and construction land exacerbates the fragmentation of existing landscape patches such as forests and water bodies, whereas the establishment of nature reserves restores the environmental characteristics. These impacts will inevitably alter the RSEI.

4.2.2. Driven Response of RSEI at Different Scales

At the microgrid scale, regions with smaller contributions to RSEI are clustered in the central lowlands. Areas with higher populations and diverse land-use patches extend outward from the central region, forming a "pentagram" pattern. This pattern arises because of the significant influence of human activity, especially in poorly drained areas. Although drainage is limited, the central valleys provide an environment conducive to development. The increase in population and economic development leads to the expansion of agriculture and construction in the core areas, fragmenting the landscape and affecting the measured ecological quality. The transformation of forests and aquatic ecosystems into cultivation land leads to the fragmentation of forests and small lakes, thereby negatively impacting the ecological quality of watersheds.

Furthermore, the proximity of transportation routes to the western and eastern regions, facilitated by external links, intensifies the reduction of forest cover and the fragmentation of water bodies [76]. Factors associated with dispersed clusters of water bodies at varying distances are observed, with many areas showing negligible impact due to water resource depletion caused by human activity. In contrast, specific high-impact regions in the northwest corner play a crucial role because water bodies function as ecological obstacles and biodiversity reserves. Villages at the meso-rural level were influenced by the composition of rural populations and the fragmentation of land, reducing local human disturbances by minimizing the impact of rural out-migration on RSEI at larger scales. Additionally, villages in environmentally sensitive areas, especially at higher elevations in the southwest, may have a substantial influence on RSEI. At this scale, natural factors such as elevation and water bodies dominate. The "Y"-shaped area in the northern and central regions, which is close to water bodies and roads, provides ample resources for honeydew production and attracts residents.

Unsustainable practices and chemical inputs can contaminate and harm biodiversity and ecosystems [77]. At the macro watershed scale, the five influencing factors show a more even spatial distribution because watershed analysis smooths the impact of subtle features such as slope, water masses, and patch density. As we move eastward into the Huashan Creek drainage basin, decreasing elevation and increasing water masses result in the fragmentation of patches, intensifying the combined effect of these factors. Significant differences can be observed in the northern region, which houses a central urbanized catchment area where anthropogenic disturbance is more pronounced. Conversely, the high-elevation, sparsely populated western region features minor human activity in vulnerable mountainous areas, particularly in less developed regions such as Xiazhai Town and Guoqiang Town. In more isolated regions, the expansion of roadways can lead to the depletion of resources, permanently damaging the natural environment.

Notably, the smaller and larger scales of POP influence have a more pronounced effect on the spatial distribution of RSEI. However, the spatial pattern of the influence of DEM is apparent at larger rural scales [33], indicating that the higher the elevation, the more pronounced the spatial spread of RSEI. When the grid size employed in the analysis is smaller than typical micro geomorphic unit sizes, the finding highlights the importance of regional landscape, population density, and patch fragmentation. In contrast, when the grid size employed in the analysis is larger than the mean grid size, this approach is more likely to identify the impact of broader topographic patterns, water masses, and the distribution of transport routes while potentially neglecting finer-scale subtleties [78,79]. Overall, conducting a multi-scale driver analysis is more beneficial to our understanding of the mechanisms driving RSEI, offering valuable information on selecting environmental motivators at various levels.

4.3. Selection of Scale

This study examines the spatial response patterns of RSEI at different scales, such as 30 m, 60 m, 120 m, 240 m, 480 m, rural, and watershed scales. Previous research has demonstrated that the Huashan Creek watershed, where human activities and natural elements affect each other, is subject to considerable ecological strain [80,81]. Significantly, the spatial patterns of RSEI exhibit incongruous regional disparities in scale and location. Although detailed micro-level research can provide exact information, a more thorough examination may be needed [82]. Although administrative village scales effectively provide information on overall distribution and they may help decision-makers in precise management based on administrative boundaries, they may need to identify micro-spatial intricacies to make better decisions. Huang et al. and other researchers note the need for deeper understanding when evaluating the Chengdu-Chongqing Urban Agglomeration on a larger scale, indicating it is common for assessments to fail to acknowledge the variations in small-scale landscape components and structures [67].

Consequently, depending solely on a single-scale evaluation provides an inadequate approach for identifying varied ecological characteristics, suggesting there is no universally applicable scale for comprehensive assessments. Instead, it is essential to employ global and local spatial autocorrelation to assess attribute values and spatial discrepancies among geographic units that are created while paying attention to the unique qualities of various sizes [83,84]. Figure 6 highlights the substantial influence of spatial clustering at the scale of the mesocosmic village, such as the 240 m $\times$ 240 m scale, exhibiting a relatively large Moran's I of 0.6. This observation is important in both grid and administrative scale evaluations. The watershed-scale also demonstrates distinctive representativeness, with natural ecosystems delineating watershed boundaries, providing a more intuitive and scientifically sound basis for assessing land use and water quality change. It is worth noting that the Huashan River, a major tributary of the West Creek of the Jiulong River that is located in an area that engages primarily in honeydew production, has been the focus of significant local government efforts relating to economic development and environmental management [31].

By examining the similarities and differences in the observed spatial clustering, the availability and precision of the driving factors at different scales, the social specificity of the administrative areas, the overall typicality of the Huashan Creek Watershed, the variability of Moran's I, and the economy and sustainability of the overall workload, the 240-metre grid, the villages, and the watershed scales were used as the criteria for assessment. Moran's I at 240 m was found to be relatively high and close to 0.6 at the village scale (Figure 5), reflecting more information spatially than at the village scale. This comprehensive assessment methodology facilitates the examination of spatial discrepancies in geographical characteristics, topography, and administrative area size within the district. Additionally, the assessment approach supports the development of ecological management and compensation strategies, the extraction of accurate village-level data, and the production of individual watershed maps. Ultimately, this method can help researchers create an all-encompassing database containing both natural and socio-economic data. In summary, it is essential to ensure that the scale chosen for RSEI management policies is in line with analytical objectives, available information, and scale-related data, meeting stakeholders' decision-making needs to achieve maximum relevance.

4.4. Policy Implications

Gaining insight into the elements contributing to disparities in ecological quality in the Huashan Creek watershed is critical for formulating plans to manage ecosystem operations and coordinate efforts to ensure ecological preservation and enhance economic activities suitable for local circumstances. Examining the factors influencing RSEI spatial differentiation using geographically weighted regression across different scales demonstrates that considering the collective impacts of human activities, natural elements, and landscape patterns significantly enhances our ability to identify RSEI spatial disparities.

Landscape fragmentation can have a multitude of impacts on the ecological quality of the Huashan Creek watershed, influencing population dynamics, elevation, watershed water management, and honeydew production zones. These impacts can have cascading effects on the water, air, land, soils, and biological diversity of the environment.

The idea of 'patches' illustrates how land-use changes can drastically alter the ecological balance of a given area, which is frequently linked to the numerous honeydew gardens that constitute almost 60% of the Huashan Creek watershed. To ensure the successful management of ecological services, it is essential to implement specific strategies to minimize other human-caused disruptions and enhance the regional landscape structure and patterns, particularly considering the simultaneous impact of diverse landscape arrangements and activities associated with the thriving growth of the honeydew sector.

For instance, in high-altitude western mountainous regions characterized by intensive land development, efforts should reduce socio-economic activities while promoting environmentally friendly sectors such as sightseeing, eco-agriculture, and recreational demonstration facilities. Additionally, when expanding the honeydew industry, it is imperative to bolster ecological control and protection initiatives to prevent the exacerbation of problems such as soil erosion, forest cover reduction, water scarcity, and desertification. The optimization of regional landscape structures is particularly important because it enhances ecological security and diminishes the risk of compromising ecosystem functions due to the combined influences of elevation and human interference.

Enhancing ecological environmental quality requires optimizing activities based on landscape patterns in areas marked by intense human economic activities, such as the central plain zone within the area under study. This approach helps mitigate the potentially adverse effects of the interaction between anthropogenic interference factors and landscape pattern index factors on regional environmental quality, thereby preventing a decline in ecological environmental quality. It is essential to strictly adhere to arable land protection policies, which should be enforced in grain-producing regions to prevent arable land degradation and promote high-quality green development. During periods of population growth, urban expansion should adhere to well-defined boundaries between rural and urban areas. This approach can lead to efficient land use practices, restraining haphazard expansion and minimizing land resource wastage. Furthermore, the establishment of ecological corridors, green buffer zones, and ecological parks in and around central cities can enhance livability and support sustainable urban development.

4.5. Limitations and Prospects

In this study, we use human–land interaction theory and landscape ecology to develop a comprehensive framework consisting of two distinct approaches and three modes of operation. Our findings can be used to regulate the quality of the ecological environment within the Huashan Creek watershed. First, this analysis is conducted at the scale of township administrative units, where the Huashan Creek watershed could be effectively managed through two strategies: "change in land-use pattern and improve land-based ecosystem service function" and "adjust land-use structure and improve land ecosystem service function". These strategies are implemented within the overarching framework of "synergizing regional economic development, protecting the ecological environment, and precisely alleviating poverty".

In the context of "coordinated regional economic development, ecological, environmental protection, and precise poverty alleviation", optimizing ecological environment quality is achieved through "changing the land use mode to improve the function of land ecosystem services" and "modifying the land use structure to optimize the allocation of ecological landscape resources". This work was facilitated by a regulatory approach characterized by "precision, differentiation, and diversification". Conducting comparative assessments across multiple administrative unit scales, specifically focusing on the Huashan Creek watershed, is crucial for future research. This will help determine the most suitable administrative unit scale that aligns with the distinctive attributes of the study area, facilitating further investigations. Exploring the feasibility of gathering information on specific drivers of change, especially for assessing long-term ecological quality developments in the research region, could be a promising approach for forecasting future changes.

It is essential that a more comprehensive range of independent variables is considered when making decisions, including economic regulation, ecological compensation, zoning adjustments, and regional monitoring. The impact of employing these elements needs to be assessed. Therefore, a thorough comprehension of their intricate connections with natural, socio-economic, land-use, and policy elements and their effects on ecological quality is needed. Because current ecological quality models may not fully capture the complexity of the regional ecosystem quality, it is crucial that more suitable and representative indicators are identified to aid the integration of a broader array of data sources to enhance the assessment of regional ecological quality [41].

This research provides a useful examination of how, across different areas and scales, specific drivers of change impact RSEI in the Huashan Creek watershed. The primary objective of this study is to establish a reliable and effective system for continuously assessing the ecological quality of a region. By exploring the elements that may influence changes in the ecological environment and delving deeply into how various elements shape ecological environmental quality, this research provides a foundation for implementing well-informed ecological protection and urban development strategies within the study area [85]. Empirical evidence has confirmed the effectiveness of remote sensing ecological indices in effectively depicting the ecological environment. Ongoing efforts to enhance correlation indices and understand the connections between these index factors are expanding the horizons of research [81,86].

5. Conclusions

Based on the environmental equilibrium matrix of the Huashan Creek watershed at three types and seven scales in 2020, this study constructed a regional environmental index (RSEI), revealed spatial response characteristics of the RSEI to different scales, and analyzed the key drivers of the RSEI. The results are as follows:

(1) Global autocorrelation revealed that the size and area information associated with the RSEI gradually transitioned from a complex-detailed to a more intuitive and clearer pattern as the scale resolution decreased. The fusion analysis of 240 m unit, rural, and watershed scales can not only encompass the rich information on RSEI spatial differences but can also provide data support and a more intuitive scientific assessment of the impacts of land use and water quality changes;

(2) Local autocorrelation showed that RSEI exhibits relatively apparent spatial aggregation characteristics at different scales. The RSEI of the Huashan Creek watershed in 2020 was generally moderate. Areas with poor RSEI were concentrated in the built-up areas and riparian zones of the towns and cities in the central plain area. In contrast, areas with higher RSEI were concentrated in the western mountains with greater vegetation cover;

(3) Using PCA, we identified five key factors affecting RSEI: DEM, POP, Dis_{Road}, Dis_{Water}, and PD. Moreover, using GWR and controlling for confounding geographic factors, we found that human activities had a significant effect on environmental quality. The population was found to have a significant effect at all three scales. Elevation was significant at the administrative village level, while the patch landscape index significantly affected the grid and catchment scales. Complex interactions between natural features, land use, and socio-economic factors exacerbate the fragmentation of landscape patches, amplifying their effects on ecosystem quality in a non-linear manner. These complex interactions highlight the critical role of elevation, population, proximity to roads, and proximity to water bodies in shaping the ecological integrity of the region.

Supplementary Materials: The following supporting information can be downloaded at: https://www.mdpi.com/article/10.3390/rs15245633/s1, Figure S1: Characteristics of the spatial response to RESI drivers at typical scales.

Author Contributions: Conceptualization, Z.Z., Y.L. and G.W.; methodology, Z.Z.; software, Z.Z. and G.W.; validation, Y.L., Z.Z. and G.W.; formal analysis, Y.L.; investigation, Y.L.; resources, G.W.; data curation, Y.L. and G.W.; writing—original draft preparation, Y.L. and Z.Z.; writing—review and editing, Y.L. and Z.Z.; visualization, Y.L. and G.W.; supervision, Y.L.; project administration, Y.L.; funding acquisition, G.W. All authors have read and agreed to the published version of the manuscript.

Funding: This research was funded by Pinghe County Land Space Planning (2021–2035) was funded by Pinghe County Government.

Data Availability Statement: Data is available in a publicly accessible repository that does not issue DOIs. Publicly available datasets were analyzed in this study. This data can found here: https://earthexplorer.usgs.gov/. Data Valiability time: 12 July 2023.

Conflicts of Interest: The authors declare no conflict of interest.

References

1. Venter, O.; Sanderson, E.W.; Magrach, A.; Allan, J.R.; Beher, J.; Jones, K.R.; Possingham, H.P.; Laurance, W.F.; Wood, P.; Fekete, B.M. Sixteen years of change in the global terrestrial human footprint and implications for biodiversity conservation. *Nat. Commun.* **2016**, *7*, 12558. [CrossRef] [PubMed]
2. Costanza, R.; d'Arge, R.; De Groot, R.; Farber, S.; Grasso, M.; Hannon, B.; Limburg, K.; Naeem, S.; O'neill, R.V.; Paruelo, J. The value of the world's ecosystem services and natural capital. *Nature* **1997**, *387*, 253–260. [CrossRef]
3. Liu, Z.; Wu, R.; Chen, Y.; Fang, C.; Wang, S. Factors of ecosystem service values in a fast-developing region in China: Insights from the joint impacts of human activities and natural conditions. *J. Clean. Prod.* **2021**, *297*, 126588. [CrossRef]
4. McDonnell, M.J.; MacGregor-Fors, I. The ecological future of cities. *Science* **2016**, *352*, 936–938. [CrossRef] [PubMed]
5. Lin, J.; Huang, J.; Hadjikakou, M.; Huang, Y.; Li, K.; Bryan, B.A. Reframing water-related ecosystem services flows. *Ecosyst. Serv.* **2021**, *50*, 101306. [CrossRef]
6. Cai, J. Study on water environmental protection countermeasures of Huashan Stream in Pinghe County. *Leather Manuf. Environ. Technol.* **2021**, *2*, 18–20.
7. Huang, W.; Chen, G.; Hu, C. Real time regional soil erosion evaluation based on RS technique—Case study in Huashanxi Watershed, Fujian Province. *Chin. J. Geol. Hazard Contr.* **2003**, *14*, 118–120, 126.
8. Han, S. Optimal allocation of water resources in Pinghe County and its surrounding towns. *Hydr. Sci. Technol.* **2023**, *1*, 14–17.
9. Wen, X.; Lin, Z.; Tang, F. Remote sensing analysis of ecological change caused by construction of the new island city: Pingtan Comprehensive Experimental Zone, Fujian Province. *J. Appl. Ecol.* **2015**, *26*, 541–547.
10. Wang, G.Y.; Mang, S.; Cai, H.S.; Liu, S.R.; Zhang, Z.Q.; Wang, L.G.; Innes, J.L. Integrated watershed management: Evolution, development and emerging trends. *J. For. Res.* **2016**, *27*, 967–994. [CrossRef]
11. Song, H.M.; Xue, L. Dynamic monitoring and analysis of ecological environment in Weinan City, Northwest China based on RSEI model. *J. Appl. Ecol.* **2016**, *27*, 3913–3919.
12. Wu, Z.; Wang, M.; Chen, S.; Zou, D. Monitoring and evaluation of ecological environments spatio-temporal variation in mine based on RSEI a case of Yongding mine. *Ecol. Sci.* **2016**, *35*, 200.
13. Zhang, H.; Du, P.; Luo, J.; Li, E. Ecological change analysis of Nanjing city based on remote sensing ecological index. *Geosp. Inf.* **2017**, *15*, 58–62.
14. Xu, M.; Chen, C.; Deng, X. Systematic analysis of the coordination degree of China's economy-ecological environment system and its influencing factor. *Environ. Sci. Pollut. Res.* **2019**, *26*, 29722–29735. [CrossRef] [PubMed]
15. Xiao, W.; Guo, J.; He, T.; Lei, K.; Deng, X. Assessing the ecological impacts of opencast coal mining in Qinghai-Tibet Plateau-a case study in Muli coal field, China. *Ecol. Indic.* **2023**, *153*, 110454. [CrossRef]
16. Sishodia, R.P.; Ray, R.L.; Singh, S.K. Applications of remote sensing in precision agriculture: A review. *Remote Sens.* **2020**, *12*, 3136. [CrossRef]
17. Wang, S.; Zhang, X.-X.; Zhu, T.; Yang, W.; Zhao, J. Assessment of ecological environment quality in the Changbai Mountain Nature Reserve based on remote sensing technology. *Prog. Geogr.* **2016**, *35*, 1269–1278.
18. Liu, C.; Wu, X.; Wang, L. Analysis on land ecological security change and affect factors using RS and GWR in the Danjiangkou Reservoir area, China. *Appl. Geogr.* **2019**, *105*, 1–14. [CrossRef]
19. Zhang, R.; Li, C.K.; Yao, S.; Li, W. Study on the change factors of construction land in Taiyuan by integrating geographic detector and geographically weighted regression. *Bull. Surv. Mapp.* **2022**, *5*, 106–109, 119.
20. Xu, H.; Wang, M.; Shi, T.; Guan, H.; Fang, C.; Lin, Z. Prediction of ecological effects of potential population and impervious surface increases using a remote sensing based ecological index (RSEI). *Ecol. Indic.* **2018**, *93*, 730–740. [CrossRef]

21. Gao, P.; Kasimu, A.; Zhao, Y.; Lin, B.; Chai, J.; Ruzi, T.; Zhao, H. Evaluation of the temporal and spatial changes of ecological quality in the Hami oasis based on RSEI. *Sustainability* **2020**, *12*, 7716. [CrossRef]
22. Hang, X.; Li, Y.; Luo, X.; Xu, M.; Han, X. Assessing the ecological quality of Nanjing during its urbanization process by using satellite, meteorological, and socioeconomic data. *J. Meteorolog. Res.* **2020**, *34*, 280–293. [CrossRef]
23. Xiong, Y.; Xu, W.; Lu, N.; Huang, S.; Wu, C.; Wang, L.; Dai, F.; Kou, W. Assessment of spatial–temporal changes of ecological environment quality based on RSEI and GEE: A case study in Erhai Lake Basin, Yunnan province, China. *Ecol. Indic.* **2021**, *125*, 107518. [CrossRef]
24. Yang, C.; Zhang, C.; Li, Q.; Liu, H.; Gao, W.; Shi, T.; Liu, X.; Wu, G. Rapid urbanization and policy variation greatly drive ecological quality evolution in Guangdong-Hong Kong-Macau Greater Bay Area of China: A remote sensing perspective. *Ecol. Indic.* **2020**, *115*, 106373. [CrossRef]
25. Zhang, T.; Yang, R.; Yang, Y.; Li, L.; Chen, L. Assessing the urban eco-environmental quality by the remote-sensing ecological index: Application to Tianjin, North China. *ISPRS Int. J. Geo-Inf.* **2021**, *10*, 475. [CrossRef]
26. Zhang, H.; Ouyang, Z.; Zheng, H. Spatial scale characteristics of ecosystem services. *Chin. J. Ecol.* **2007**, *26*, 1432–1437.
27. Fu, B.; Zhou, G.; Bai, Y.; Song, C.; Liu, J.; Zhang, H.; Lv, Y.; Zheng, H.; Xie, G. The Main Terrestrial Ecosystem Services and Ecological Security in China. *Adv. Earth Sci.* **2009**, *24*, 571–576.
28. Jiang, F.; Zhang, Y.; Li, J.; Sun, Z. Research on remote sensing ecological environmental assessment method optimized by regional scale. *Environ. Sci. Pollut. Res. Int.* **2021**, *28*, 68174–68187. [CrossRef]
29. Geng, J.; Yu, K.; Xie, Z.; Zhao, G.; Ai, J.; Yang, L.; Yang, H.; Liu, J. Analysis of spatiotemporal variation and drivers of ecological quality in Fuzhou based on RSEI. *Remote Sens.* **2022**, *14*, 4900. [CrossRef]
30. Jin, Y.; Ge, Y.; Wang, J.; Heuvelink, G.B.M.; Wang, L. Geographically weighted area to-point regression kriging for spatial downscaling in remote sensing. *Remote Sens.* **2018**, *10*, 579. [CrossRef]
31. Huang, Y.; Huang, J.; Ervinia, A.; Duan, S.; Kaushal, S.S. Land use and climate variability amplifies watershed nitrogen exports in coastal China. *Ocean Coastal Manag.* **2021**, *207*, 104428. [CrossRef]
32. Leece, J.; Parker, F. Use and Misuse of SPSS. *Software Pract. Exper.* **1978**, *8*, 301–311. [CrossRef]
33. Xiong, Y.; Li, Y.; Xiong, S.; Wu, G.; Deng, O. Multi-scale spatial correlation between vegetation index and terrain attributes in a small watershed of the upper Minjiang River. *Ecol. Indic.* **2021**, *126*, 107610. [CrossRef]
34. Li, Y.; Li, Z.; Wang, J.; Zeng, H. Analyses of driving factors on the spatial variations in regional eco-environmental quality using two types of species distribution models: A case study of Minjiang River Basin, China. *Ecol. Indic.* **2022**, *139*, 108980. [CrossRef]
35. Baig, M.H.A.; Zhang, L.; Shuai, T.; Tong, Q. Derivation of a tasselled cap transformation based on Landsat 8 at-satellite reflectance. *Remote Sens. Lett.* **2014**, *5*, 423–431. [CrossRef]
36. Crist, E.P. A TM tasseled cap equivalent transformation for reflectance factor data. *Remote Sens. Environ.* **1985**, *17*, 301–306. [CrossRef]
37. Hu, X.; Xu, H. A new remote sensing index for assessing the spatial heterogeneity in urban ecological quality: A case from Fuzhou City, China. *Ecol. Indic.* **2018**, *89*, 11–21. [CrossRef]
38. Kumar, B.P.; Babu, K.R.; Anusha, B.; Rajasekhar, M. Geo environmental monitoring and assessment of land degradation and desertification in the semi-arid regions using Landsat 8 OLI/TIRS, LST, and NDVI approach. *Environ. Chall.* **2022**, *8*, 100578. [CrossRef]
39. Xu, H. A remote sensing index for assessment of regional ecological changes. *China Environ. Sci.* **2013**, *33*, 889–897.
40. Jing, Y.; Zhang, F.; He, Y.; Johnson, V.C.; Arikena, M. Assessment of spatial and temporal variation of ecological environment quality in Ebinur Lake Wetland National Nature Reserve, Xinjiang, China. *Ecol. Indic.* **2020**, *110*, 105874. [CrossRef]
41. Pabortsava, K.; Lampitt, R.S.; Benson, J.; Crowe, C.; McLachlan, R.; Le Moigne, F.A.; Mark Moore, C.; Pebody, C.; Provost, P.; Rees, A.P. Carbon sequestration in the deep Atlantic enhanced by Saharan dust. *Nat. Geosci.* **2017**, *10*, 189–194. [CrossRef]
42. Tepanosyan, G.; Sahakyan, L.; Zhang, C.; Saghatelyan, A. The application of Local Moran's I to identify spatial clusters and hot spots of Pb, Mo and Ti in urban soils of Yerevan. *Appl. Geochem.* **2019**, *104*, 116–123. [CrossRef]
43. Yuan, Y.; Cave, M.; Zhang, C. Using Local Moran's I to identify contamination hotspots of rare earth elements in urban soils of London. *Appl. Geochem.* **2018**, *88*, 167–178. [CrossRef]
44. Zhao, Z.; Gao, J.; Wang, Y.; Liu, J.; Li, S. Exploring spatially variable relationships between NDVI and climatic factors in a transition zone using geographically weighted regression. *Theor. Appl. Climatol.* **2015**, *120*, 507–519. [CrossRef]
45. Anselin, L. Local indicators of spatial association—LISA. *Geog. Anal.* **1995**, *27*, 93–115. [CrossRef]
46. Chen, Y.-G. Reconstructing the mathematical process of spatial autocorrelation based on Moran's statistics. *Geogr. Res.* **2009**, *28*, 1449–1463.
47. Chen, W.; Chi, G.; Li, J. The spatial aspect of ecosystem services balance and its determinants. *Land Use Policy* **2020**, *90*, 104263. [CrossRef]
48. Liu, Z.; Wang, S.; Fang, C. Spatiotemporal evolution and influencing mechanism of ecosystem service value in the Guangdong-Hong Kong-Macao Greater Bay Area. *J. Geog. Sci.* **2023**, *33*, 1226–1244. [CrossRef]
49. Wu, X.; Liu, S.; Zhao, S.; Hou, X.; Xu, J.; Dong, S.; Liu, G. Quantification and driving force analysis of ecosystem services supply, demand and balance in China. *Sci. Total Environ.* **2019**, *652*, 1375–1386. [CrossRef]
50. Arthington, A.H.; Bunn, S.E.; Poff, N.L.; Naiman, R.J. The challenge of providing environmental flow rules to sustain river ecosystems. *Ecol. Appl.* **2006**, *16*, 1311–1318. [CrossRef]

51. Liu, T.; Wang, H.-Z.; Wang, H.-Z.; Xu, H. The spatiotemporal evolution of ecological security in China based on the ecological footprint model with localization of parameters. *Ecol. Indic.* **2021**, *126*, 107636. [CrossRef]
52. Wang, C.; Li, X.; Yu, H.; Wang, Y. Tracing the spatial variation and value change of ecosystem services in Yellow River Delta, China. *Ecol. Indic.* **2019**, *96*, 270–277. [CrossRef]
53. Wang, S.-X.; Yao, Y.; Zhou, Y. Analysis of ecological quality of the environment and influencing factors in China during 2005–2010. *Int. J. Environ. Res. Public Health* **2014**, *11*, 1673–1693. [CrossRef] [PubMed]
54. Peng, J.; Xie, P.; Liu, Y.; Ma, J. Urban thermal environment dynamics and associated landscape pattern factors: A case study in the Beijing metropolitan region. *Remote Sens. Environ.* **2016**, *173*, 145–155. [CrossRef]
55. Di, X.; Hou, X.; Wang, Y.; Wu, L. Spatial-temporal characteristics of land use intensity of coastal zone in China during 2000–2010. *Chin. Geogr. Sci.* **2015**, *25*, 51–61. [CrossRef]
56. Chen, W.; Chi, G. Urbanization and ecosystem services: The multi-scale spatial spillover effects and spatial variations. *Land Use Policy* **2022**, *114*, 105964. [CrossRef]
57. Pan, Z.; Wang, J. Spatially heterogeneity response of ecosystem services supply and demand to urbanization in China. *Ecol. Eng.* **2021**, *169*, 106303. [CrossRef]
58. Song, Y. Relationships Between On-road $FFCO_2$ Emission and Socio-economics/Urban Form Factors. Ph.D. Dissertation, Arizona State University, Phoenix, AZ, USA, 2018.
59. Deng, C.; Liu, J.; Liu, Y.; Li, Z.; Nie, X.; Hu, X.; Wang, L.; Zhang, Y.; Zhang, G.; Zhu, D. Spatiotemporal dislocation of urbanization and ecological construction increased the ecosystem service supply and demand imbalance. *J. Environ. Manag.* **2021**, *288*, 112478. [CrossRef]
60. Wilkerson, M.L.; Mitchell, M.G.; Shanahan, D.; Wilson, K.A.; Ives, C.D.; Lovelock, C.E.; Rhodes, J.R. The role of socio-economic factors in planning and managing urban ecosystem services. *Ecosyst. Serv.* **2018**, *31*, 102–110. [CrossRef]
61. Olsson, E.G.A.; Austrheim, G.; Grenne, S.N. Landscape change patterns in mountains, land use and environmental diversity, Mid-Norway 1960–1993. *Landscape Ecol.* **2000**, *15*, 155–170. [CrossRef]
62. Haddad, N.M.; Brudvig, L.A.; Clobert, J.; Davies, K.F.; Gonzalez, A.; Holt, R.D.; Lovejoy, T.E.; Sexton, J.O.; Austin, M.P.; Collins, C.D. Habitat fragmentation and its lasting impact on Earth's ecosystems. *Sci. Adv.* **2015**, *1*, e1500052. [CrossRef]
63. O'Neill, R.V.; Krummel, J.R.; Gardner, R.H.E.A.; Sugihara, G.; Jackson, B.; DeAngelis, D.L.; Milne, B.T.; Turner, M.G.; Zygmunt, B.; Christensen, S.W. Indices of landscape pattern. *Landscape Ecol.* **1988**, *1*, 153–162. [CrossRef]
64. Wang, H.; Liu, Y.; Zhang, B.; Xu, S.; Ja, K.; Hong, S. Analysis of driving forces of urban land expansion in Wuhan metropolitan area based on Logistic-GTWR model. *Trans. Chin. Soc. Agr. Eng.* **2018**, *34*, 256–265.
65. Zhang, J.; Yang, Y.; Jiang, T.; Xu, S. The evolution analysis of Shanghai housing price based on spatiotemporal heterogeneity detection. *Sci. Survey. Mapp.* **2020**, *45*, 175–180, 196.
66. Gao, Y.; Huang, J.; Li, S.; Li, S. Spatial pattern of non-stationarity and scale-dependent relationships between NDVI and climatic factors—A case study in Qinghai-Tibet Plateau, China. *Ecol. Indic.* **2012**, *20*, 170–176. [CrossRef]
67. Huang, M.; Yue, W.; Fang, B.; Feng, S. Scale response characteristics and geographic exploration mechanism of spatial differentiation of ecosystem service values in Dabie Mountain area, central China from 1970 to 2015. *Res. Marx. Aesthet.* **2019**, *74*, 1904–1920.
68. Liu, Y.; Bi, J.; Lv, J.S.; Ma, Z.W.; Wang, C. Spatial multi-scale relationships of ecosystem services: A case study using a geostatistical methodology. *Sci. Rep.* **2017**, *7*, 9486. [CrossRef] [PubMed]
69. Huang, W.; Zhang, L.; Furumi, S.; Muramatsu, K.; Daigo, M.; Li, P. Topographic effects on estimating net primary productivity of green coniferous forest in complex terrain using Landsat data: A case study of Yoshino Mountain, Japan. *Int. J. Remote Sens.* **2010**, *31*, 2941–2957. [CrossRef]
70. Muñoz, J.D.; Kravchenko, A. Deriving the optimal scale for relating topographic attributes and cover crop plant biomass. *Geomorphology* **2012**, *179*, 197–207. [CrossRef]
71. Wang, Y.; Hou, X.; Wang, M.; Wu, L.; Ying, L.; Feng, Y. Topographic controls on vegetation index in a hilly landscape: A case study in the Jiaodong Peninsula, eastern China. *Environ. Earth Sci.* **2013**, *70*, 625–634. [CrossRef]
72. Shao, Z.; Tun, X.; Zhang, J.; Zhang, Z.; Gao, P.; Xu, J.; Zhang, Y. Temporal and spatial changes of ecological quality in Qingdao city based on remote sensing. *Sci. Soil Water Conserv.* **2022**, *20*, 34–40.
73. Zhang, Z.; Cai, Z.; Yang, J.; Guo, X. Ecological environmental quality assessment of Chinese estuarine wetlands during 2000–2020 based on a remote sensing ecological index. *Front. Mar. Sci.* **2022**, *9*, 981139. [CrossRef]
74. Wu, Y.; Wu, Y.; Li, C.; Gao, B.; Zheng, K.; Li, C. Multi-Scale Spatial Differentiation and Geographic Detection Response of Ecosystem Service Value in Sichuan-Yunnan Ecological Barrier Based on the Modifiable Areal Unit Problem. *Res. Soil Water Conserv.* **2023**, *30*, 333–342.
75. Ling, H.; Yan, J.; Xu, H.; Guo, B.; Zhang, Q. Estimates of shifts in ecosystem service values due to changes in key factors in the Manas River basin, northwest China. *Sci. Total Environ.* **2019**, *659*, 177–187. [CrossRef] [PubMed]
76. Yuan, B.; Fu, L.; Zou, Y.; Zhang, S.; Chen, X.; Li, F.; Deng, Z.; Xie, Y. Spatiotemporal change detection of ecological quality and the associated affecting factors in Dongting Lake Basin, based on RSEI. *J. Clean. Prod.* **2021**, *302*, 126995. [CrossRef]
77. Huang, J.L.; Lin, J.; Zhang, Y.Z.; Li, Q.S.; Hong, H.S. Analysis of phosphorus concentration in a subtropical river basin in southeast China: Implications for management. *Ocean Coastal Manag.* **2013**, *81*, 29–37. [CrossRef]

78. Xie, J.; Liu, T. Characterization of spatial scaling relationships between vegetation pattern and topography at different directions in Gurbantunggut desert, China. *Ecol. Complex.* **2010**, *7*, 234–242. [CrossRef]
79. Zhou, Q.; Wei, X.; Zhou, X.; Cai, M.; Xu, Y. Vegetation coverage change and its response to topography in a typical karst region: The Lianjiang River Basin in Southwest China. *Environ. Earth Sci.* **2019**, *78*, 191. [CrossRef]
80. Wang, J.; Tian, J.; Li, X.; Ma, Y.; Yi, W. Evaluation of concordance between environment and economy in Qinghai Lake Watershed, Qinghai-Tibet Plateau. *J. Geogr. Sci.* **2011**, *21*, 949–960. [CrossRef]
81. Qiu, L.; He, Y.; Zhang, L.; Wang, W.; Tang, Y. Spatiotemporal variation characteristics and influence factors of MODIS LST in Qilian Mountains. *Arid. Land Geogr.* **2020**, *43*, 726–737.
82. Chen, Y.; Yang, J.; Yu, W.; Ren, J.; Xiao, X.; Xia, J.C. Relationship between urban spatial form and seasonal land surface temperature under different grid scales. *Sustain. Cities Soc.* **2023**, *89*, 104374. [CrossRef]
83. Goodchild, M.F. The validity and usefulness of laws in geographic information science and geography. *Ann. Am. Assoc. Geogr.* **2004**, *94*, 300–303. [CrossRef]
84. Wang, B.; Zhang, Q.; Cui, F. Scientific research on ecosystem services and human well-being: A bibliometric analysis. *Ecol. Indic.* **2021**, *125*, 107449. [CrossRef]
85. Miao, X.; Liang, Q. Analysis of ecological environment changes in Yongjiang River basin based on remote sensing ecological index. *Resour. Environ. Yangtze Basin* **2021**, *30*, 427–438.
86. Wang, F.; Li, W.-H.; Lin, Y.-M.; Nan, X.-X.; Hu, Z.-R. Spatiotemporal Pattern and Driving Force Analysis of Ecological Environmental Quality in Typical Ecological Areas of the Yellow River Basin from 1990 to 2020. *Environ. Sci.* **2023**, *44*, 2518–2527.

remote sensing

Article

A Comparative Study of Landslide Susceptibility Mapping Using Bagging PU Learning in Class-Prior Probability Shift Datasets

Lingran Zhao [1,2], Hangling Ma [3], Jiahui Dong [3], Xueling Wu [3], Hang Xu [4,5] and Ruiqing Niu [3,*]

1 School of Automation, China University of Geosciences, Wuhan 430074, China; lrzhao@cug.edu.cn
2 Hubei Key Laboratory of Advanced Control and Intelligent Automation for Complex Systems, China University of Geosciences, Wuhan 430074, China
3 School of Geophysics and Geomatics, China University of Geosciences, Wuhan 430074, China; mhl@cug.edu.cn (H.M.); ivy.dong@cug.edu.cn (J.D.); wuxl@cug.edu.cn (X.W.)
4 Hubei Key Laboratory of Resources and Eco-Environment Geology, Hubei Geological Bureau, Wuhan 430000, China; xuhangkaren@163.com
5 Hubei Geological Environment Station, Wuhan 430000, China
* Correspondence: niuruiqing@cug.edu.cn

Abstract: Landslide susceptibility mapping is typically based on binary prediction probabilities. However, non-landslide samples in modeling datasets are often unlabeled data, and the phenomenon of class-priori shift, that is, the proportion of landslide samples frequently deviates from real-world scenarios and is spatially heterogeneous. By comparing the classification performance and predicted probability distributions across multiple unbalanced datasets with known and unknown sample proportions, this study assesses the landslide susceptibility model's generalization ability in the context of class-prior shifts. The study investigates the potential of Bagging PU Learning, a semi-supervised learning approach, in improving the generalization performance of landslide susceptibility models and proposes the Bagging PU-GDBT algorithm. Our findings highlight the effectiveness of Bagging PU Learning in enhancing the recall of landslides and the generalization capabilities of models on unbalanced datasets. This method reduces prediction uncertainties, especially in high and very high susceptibility zones. Furthermore, results emphasize the superiority of models trained on balanced datasets with 1:1 sample ratio for landslide susceptibility mapping over those trained on unbalanced datasets.

Keywords: landslide susceptibility mapping; class-prior probability shift; generalization performance evaluation; Bagging PU learning

Citation: Zhao, L.; Ma, H.; Dong, J.; Wu, X.; Xu, H.; Niu, R. A Comparative Study of Landslide Susceptibility Mapping Using Bagging PU Learning in Class-Prior Probability Shift Datasets. *Remote Sens.* **2023**, *15*, 5547. https://doi.org/10.3390/rs15235547

Academic Editor: Domenico Calcaterra

Received: 20 October 2023
Revised: 20 November 2023
Accepted: 27 November 2023
Published: 28 November 2023

1. Introduction

Landslides, characterized by their powerful geological disruptive forces, high frequency, and significant hazards, have led to substantial economic losses and casualties worldwide. Therefore, the monitoring and prevention of landslide disasters is particularly crucial. The improvement of monitoring devices such as GPS, Time Domain Reflectometry (TDR) [1], flexible inclinometers [2], and the development of sensor technologies like Fiber Bragg Grating (FBG) [3] and Red Green Blue+Depth (RGB-D) [4] have increased the difficulty of landslide monitoring due to the abundance of monitoring data and complex data structures. Moreover, constraints such as budgetary limitations, economic considerations, and existing engineering technological conditions make it impractical to conduct comprehensive surveys and real-time monitoring for all regions and landslide points. At this point in time, conducting susceptibility mapping for landslides within a specific regional scope, narrowing down the monitoring and prevention areas, proves to be a proactive, effective, and economical strategy.

Landslide susceptibility mapping modeling integrate the spatial characteristics of both regional remote sensing and non-remote sensing data, selecting a specific number of landslide and non-landslide calculation units. It employs either probability statistics or machine learning methods for modeling. The model is then used to predict the probability values of landslides occurring across the entire region, which are subsequently categorized into numerical intervals. Through this quantitative-to-qualitative analytical approach, regions at risk of landslides are delineated from high to low susceptibility. It involves classifying the probability of each computational unit in the study area as a landslide category within specific susceptibility intervals [5]. One of the challenges in this approach is the assumption that all samples are explicitly labeled as positive or negative. Computational units in the study area are typically considered negative samples unless explicitly labeled as landslides [6–9]. However, units not labeled as landslides might still represent potential landslide areas or undocumented landslide occurrences. In addition to oversampling the positive sample, many susceptibility studies have adjusted their sampling strategies to minimize errors arising from unreliable negative samples [10]. Previous studies employed techniques such as collecting non-landslide samples outside the buffer zone of landslide edges [11,12], selecting negative samples based on specific characteristics [13,14], constructing collaborative model of supervised and unsupervised learning to select reliable negative samples [15]. Alternatively, pre-classification methods were used to select negative samples from locations with very low pre-classification probability [16,17]. Despite these efforts, these methods often introduce subjectivity or additional complexity. Addressing these challenges, researchers have turned to the positive-unlabeled (PU) learning method for landslide susceptibility classification [18,19].

PU classification is designed to make two-class predictions, akin to the traditional binary classification problem, but with only labeled positive samples in the training set and unlabeled samples [20,21]. This method's adaptability to various domains highlights its potential to mitigate data labeling challenges and enhance the reliability of predictive models, including those related to landslide susceptibility mapping (LSM).

When addressing positive and negative samples in PU Learning, these methods assume either that the class-prior probability (the ratio of positive samples) is known or estimate it through iterative processes [22,23]. However, the challenge in landslide susceptibility mapping lies in the class-prior probability shift of the modeling dataset to reality. Class-prior probability shift refers to the shift of the distribution of the dataset in the source domain and the target domain [21]. The data distribution of the test set in susceptibility modeling, which is often assumed to be consistent with the training set, is assumption that may be violated in practice. The true class-prior for landslides is much smaller than the modeling dataset [24]. The distribution of landslides and non-landslides in nature is challenging to accurately obtain statistics from a preliminary survey. This discrepancy means that the data distribution in the test set does not accurately represent the wide range of labeled unknown areas in nature. In addition, when the established LSM model is extended to other regions, the regional heterogeneity of landslide distribution density will also lead to the class-priori shift problem.

The distribution of landslides exhibits heterogeneity and scale effects [25,26]. When studying and making decisions in different geographical regions and scales, the varying proportions of landslides to non-landslides can impact the reusability of LSM models. Considering that bagging PU learning adapts to different positive and negative sample distributions in the learning process, we aim to explore whether bagging PU learning can enhance the LSM model's generalization ability. Although previous studies have utilized PU-BaggingDT for landslide susceptibility assessment [18,19], there has been limited discussion on generalization performance. Additionally, Wright [27] demonstrated that bagging PU learning with SVM classifiers is more robust than PU-BaggingDT for unbalanced datasets. We apply Bagging PU-SVM to LSM modeling, and, considering that the GDBT algorithm, integrating decision trees, is more rapid, efficient, and robust, we propose the Bagging PU-GDBT algorithm. Several classical supervised learning models, including SVM, GDBT, LR, and RF, are employed for comparison. To explore reasonable LSM modeling dataset sample positive-to-negative ratios under the scenario of class-prior shift, we conducted tests on datasets with different sample ratios. The recall of landslide samples quantitatively assesses the model's classification ability, and the standard error qualitatively contrasts the model's uncertainty in predicting susceptibility values. Additionally, an analysis of the model's generalization ability across different geographical regions is performed. We hope that this exploration can contribute valuable insights to landslide early warning and subsequent mitigation efforts.

2. Materials and Methods

2.1. Study Area

The Three Gorges Reservoir area, located in central China, is characterized by a subtropical monsoon climate with mild and humid conditions, abundant rainfall, high mountains, deep canyons, complete stratigraphic outcrops, complex geomorphic structures, and extensive landslide hazard development. For our research, we chose two specific regions within the Three Gorges Reservoir area to establish datasets, as illustrated in Figure 1.

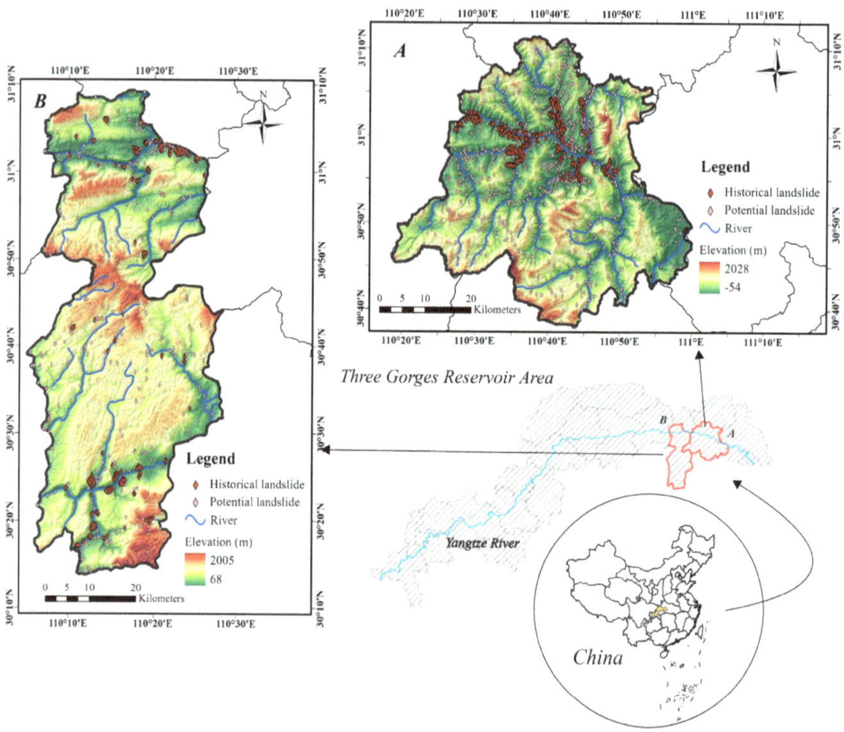

Figure 1. Location of the study areas.

Dataset A was collected from the high-incidence landslide areas in the Three Gorges Region, specifically from Zigui County, Hubei Province, covering an approximate area of 2274 km^2 (110°26′E~110°50′E, and 30°55′N~31°5′N). Zigui County is traversed by the Yangtze River from west to east, and it includes eight main tributaries organized into perennial stream, with a total flow length of 247.8 km.

Dataset B was collected from Region B, situated in the Enshi Tujia and Miao Autonomous Prefecture, Hubei Province, China, spanning from 110°18′E to 111°0′E longitude and 30°38′N to 31°11′N latitude, covering a total area of 2427 km^2. In the northern part of the study area, the main stem of the Yangtze River runs through Guandukou Town and Xinling Town from west to east, covering a total length of 24.89 km within the region. In the southern portion of the study area, one of the primary tributaries of the Yangtze River, the Qingjiang River, flows through Jingooping Township, Qingtaiping Town, and Shuibuya Town from west to east.

Both regions share similar climatic characteristics and altitude ranges. However, region A exhibits a greater diversity in geological formations compared to region B. Consequently, region A is selected as the primary research area, while region B is utilized to validate the model's generalization performance.

2.2. Landslide Inventories and Influence Factors

A landslide inventory is vital in susceptibility estimation, serving as labeled positive samples [28]. Our landslide inventory primarily comprised shallow landslides within the study area, identified through remote sensing image interpretation based on the landslide cataloging database provided by the Hubei Geological Environmental Center. Field surveys supplemented and validated this data. The inventory comprised two types of vector data: potential landslide points, indicating signs of deformation without a complete slide occurrence, and historical landslide polygons. The study area was partitioned using a grid unit with a size of 30 m, consistent with the resolution of most remote sensing images obtained for this study. Each grid unit was treated as a sample. In Region A, the landslide inventory comprises 201 historical landslide polygons and 876 potential landslide points, resulting in a total of 24,986 landslide units and 2,496,683 unlabeled units. Similarly, in Region B, there are 85 historical landslide polygons and 402 potential landslide points, yielding 17,221 landslide units and 2,873,629 unlabeled units. As landslide units were predominantly concentrated within historical landslide polygons, random sampling was conducted on the grid units in these polygons to reduce the likelihood of potential landslide points being mistaken as noise by the model, as illustrated in Figure 1. Ultimately, 3835 positive samples were obtained in Region A, and 3317 positive samples were obtained in Region B.

The evolution of landslides is a complex nonlinear dissipative system with an open structure, and the factors controlling and inducing landslides are still an area of active research [29,30]. The factors considered in this study encompass topographic, geological, environmental, and human-related aspects, as outlined in Table 1.

Table 1. Landslide susceptibility influence factors.

Category	Factors	Data Source
Topography	Elevation, aspect, slope, profile curvature, plan curvature, terrain surface texture, relative slope position, topographic wetness index (TWI), topographic roughness index (TRI), valley depth	ASTER GDEM data of the Geospatial Data Cloud (http://www.gscloud.cn/, accessed on 1 March 2023)
Geology	Lithology, strata, distance from faults, slope structure	1:200,000 scale geological map (http://dcc.ngac.org.cn/, accessed on 1 March 2023)
Environment	NDVI, land use, distance from rivers, magnitude, Annual average rainfall	Land use was extracted from GlobeLand30 (http://globeland30.org/, accessed on 1 March 2023). The NDVI was calculated in the Google Earth Engine platform. Rivers were derived from the 1:100,000 basic geographic database of China national catalogue service for geographic information. Magnitude and rainfall were provided by Hubei Geological Environment Station.
Human activity	Distance from roads, POI kernel density	1:100,000 basic geographic database of China national catalogue service for geographic information

2.3. Methodology

2.3.1. Landslide Susceptibility Mapping Models

Most PU Learning algorithms simplify the problem of learning classifiers from positive and unlabeled samples into a binary classification problem, relying on the approximate premise that the ratio of positive to unlabeled samples increases monotonically with the positive-negative sample ratio [20,31,32]. However, the occurrence of landslides in extensive geographic regions is a low-probability event. The dataset for susceptibility modeling, as a result of random sampling according to preset ratios, introduces uncertainty regarding the quantity of landslide samples in the unlabeled sample set. The data distribution of unlabeled samples may strongly influence the classifier, and the distribution of positive and negative samples in the test set often fails to accurately represent the sample distribution across the entire study area, making the problem more challenging in practice. Susceptibility evaluation ultimately necessitates probability estimates across the entire study area. Determining the actual proportion of positive and negative samples in the study area is unfeasible. In such circumstances, a bagging strategy might enhance the stability of the classifier in the face of diverse sample distributions [33,34]. Bagging PU learning, as proposed by Mordelet and Vert [31], employs an ensemble approach. This method employs bootstrapping to estimate a series of classifiers from a training set consisting of positive sample set P and unlabeled sample set U, and the results of these classifiers are aggregated through simple averaging. While this method introduces bias, it is not reliant on weights. It depends on the size of the subsample set and adopts a non-approximation setting to minimize the impact of positive samples in the unlabeled sample set. Given these characteristics of Bagging PU learning, it proves more suitable for landslide susceptibility research compared to other PU learning methods.

The specific operation of bagging PU learning is to randomly collect T subset U_t containing k unlabeled samples from U. A weak classifier, denoted as f_t, is trained using P and U_t to distinguish between P and U_t. For any $x \in \chi$ outside the training set (where $x \notin U$), f_t assigns a probability $f_t(x)$ of being a positive sample:

$$f(x) = \frac{1}{T}\sum_{t=1}^{T} f_t(x). \quad (1)$$

Ultimately, these T classifiers are aggregated to produce the result.

Bagging PU learning treats P as noise within U, where U represents empirical contamination. In other words, the higher the proportion of P within U, the more unstable the classifier may become. The motivation behind bagging is to obtain diverse set of classifiers by randomly sampling U_t with varying contamination rates, thus generating a series of classifiers with diversity. The size k of the subsample set generated from U can play a crucial role in balancing accuracy and the stability of individual classifiers. Since subsets with less random contamination in part train better-performing base classifiers, adjusting k allows for a performance gain from these superior base classifiers that can outweigh the losses incurred by training each classifier on a smaller dataset.

We use $\hat{\gamma}$ to denote the probability of hidden positive samples in U, and correspondingly, $\hat{\gamma}_t$ represents the probability of hidden positive samples in U_t. The mean and standard deviation of the positive sample rate can then be expressed as:

$$E(\hat{\gamma}_t) = \hat{\gamma}, \quad (2)$$

$$V(\hat{\gamma}_t) = \frac{1}{k}\hat{\gamma}(1 - \hat{\gamma}). \quad (3)$$

The intuition is that the stability of the classifier can be balanced by adjusting the size, k, of the subset of unlabeled samples to control the disturbance rate of positive samples in U_t.

On the one hand, a larger subset generally results in better classifiers, as any classification method typically improves with more training samples. On the other hand, empirical contamination has a greater impact on smaller subsets. Here, we follow the analysis of Mordelet and Vert [31]. When P is not too large, k may not need meticulous tuning since it does not consistently exert a substantial influence on performance. Setting $k = N_P$ appears to be a prudent choice. Additionally, the value of T does not significantly affect the overall accuracy of the model when k is greater than 200, relieving Bagging PU learning of the additional burden of parameter adjustment during training. In our experiments, the value of k aligns with the number of positive samples in the training set, and T is set to 30.

2.3.2. Measure the Uncertainty of the Predicted Value

The generalization ability of the model is primarily reflected by the prediction accuracy of the unseen data, serving as the most direct expression of the model's predictive capability.

Due to the limitations of the landslide inventory, it cannot encompass all historical landslide distributions across the entire study area. Consequently, the actual ratio of positive to negative samples in the study area remains unknown. Clearly, the predictive efficacy of a specific model fluctuates in datasets with varying distributions of positive and negative samples. Without measured values, determining the actual accuracy of the prediction probability for the entire region is unattainable. However, through numerous experiments, we can obtain the standard deviation of the mean estimate. By repeating experiments, we can establish a range and calculate the probability that the true overall mean falls within a certain range.

The uncertainty of the probability estimate is qualitatively summarized through the distribution of the standard error (SE) σ_n of the landslide susceptibility value across the susceptibility interval for n replicate experiments:

$$\mu = \frac{1}{n}\sum_{i=1}^{n} y_i, \tag{4}$$

$$\sigma = \sqrt{\frac{1}{n}\sum_{i=1}^{n} (y_i - \mu)^2}, \tag{5}$$

$$\sigma_n = \frac{\sigma}{\sqrt{n}}. \tag{6}$$

In Equations (4)–(6), y_i represents the estimated susceptibility value for the ith time, μ is the mean of the multiple repeated estimates results, and σ is the standard deviation of the multiple repeated estimates results. This distribution can be fitted using the polynomial obtained through the least square method.

All methods were implemented in Python 3.9. and executed on a Windows machine equipped with a 6-core Intel 2.9 GHz processor and 12 GB RAM.

3. Results

3.1. Classification Performance of Models

Out of the 3835 positive samples in region A, 70% were utilized as training samples, and the remaining 30% were designated as test samples for training and testing the Bagging PU models. These models were then compared with three supervised learning models: logistic regression (LR), support vector machines (SVM), and random forests (RF). An equivalent number of unlabeled sample subsets were collected as negative samples in addition to the positive samples. Due to the biased method used to obtain the landslide modeling dataset, the ratio of positive to negative samples does not align with the actual sample space in the study area. In the real world, there are significantly more negative samples than positive ones in the study area. One of our objectives is to assess whether imbalanced datasets with varying sample ratios significantly impact the classification accuracy of the models. Four positive and unlabeled sample ratios of 1:1, 1:5, 1:10, and

1:20 were adopted to establish the Test Set 1. For each ratio, non-landslide samples were randomly sampled ten times. The Receiver Operating Characteristic Curve (ROC) was utilized to compare the overall classification accuracy of the algorithmic models [35]. The average ROC curve of the model for ten tests is depicted in Figure 2.

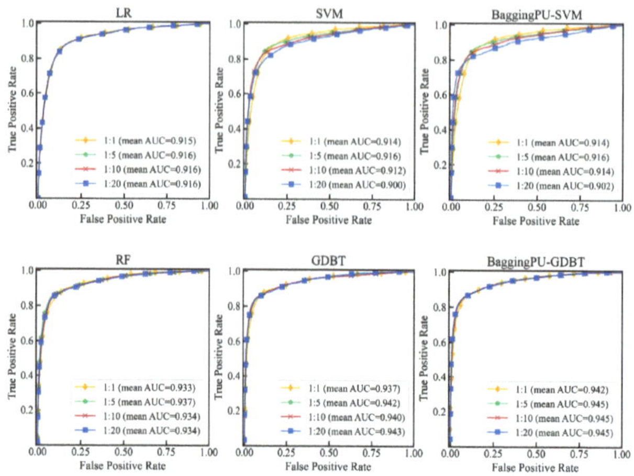

Figure 2. Average receiver operating characteristic curves and mean areas under the curve for Test Set 1 of models.

All models demonstrated satisfactory overall classification results. Notably, it seems that appropriately increasing the negative sample size can enhance the Area Under the Curve (AUC) of the LSM model. In this study, when the ratio of positive to negative samples is 1:5, all models except GDBT achieved their highest AUC. The AUC of 1:5 GDBT model is also second only to that of 1:20. However, the enhancement of bagging PU learning on the models' classification performance might not be readily apparent. ROC measures the overall performance of the model without considering the distribution of positive and negative samples. When the distribution of positive and negative samples changes, the shape of the ROC curve remains basically unchanged. However, in the real world, the recall and precision of the dataset may have changed. A higher recall indicates that more landslide samples are detected correctly. Precision, on the other hand, represents the proportion of correctly predicted landslide samples among all predicted landslide samples. When comparing the recall and precision of multiple predictions, bagging PU learning models generally exhibited improvements in both the recall and precision of landslide samples, as illustrated in Figure 3. This improvement becomes increasingly evident with the expansion of the negative sample size.

In the context of landslide susceptibility classification, the accuracy of the historical landslide list heavily relies on the validation range and accuracy of field surveys. The shift in prior probability complicates the ability of some trained models to adapt to changes in data distribution during actual predictions [36]. We cannot anticipate the extent of the class-prior shift in advance, which means that a classifier must adapt to an unknown test distribution.

To assess the generalization performance of the model predicted on the unknown test set, 10 test subsets were randomly sampled from computational units outside the training and initial test sets, forming Test Set 2. Each subset comprises 2514 samples, and the distribution of each subset is unknown. The models with 1:1 and 1:5 ratios, which demonstrated superior classification performance on Test Set 1, were employed to predict the samples in Test Set 2.

Figure 4 shows the recall and precision of models with training set positive and negative sample ratios of 1:1 and 1:5 on Test Set 2. The models generally exhibit high recall for landslide samples, and the recall of the 1:1 models of each algorithm being close to 1, and the recall of the 1:5 models exceeding 80%. However, precision is notably low. This discrepancy arises because the subsets contained very few landslide samples, resulting in a high recall and low precision scenario. Furthermore, it is evident that the 1:5 models utilizing Bagging PU learning exhibit enhancements in recall.

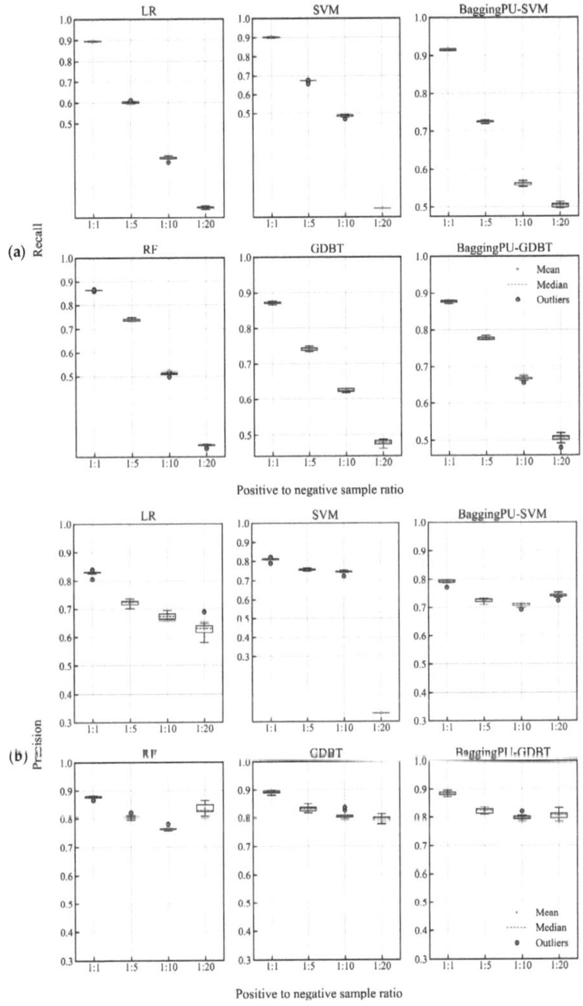

Figure 3. Classification performance for landslide samples in Test set 1. The proportion of positive and negative samples in Test Set 1 is the same as in the training set. (**a**) Recall of models with different positive to negative sample ratios; (**b**) Precision of models with different positive to negative sample ratios.

3.2. Uncertainty in the Predicted Susceptibility Values

The average values and SEs for the estimated probabilities were calculated. These probabilities were categorized into five susceptibility intervals: very low [0.01, 0.20], low (0.2, 0.45], moderate (0.45, 0.55], high (0.55, 0.80], and very high (0.80, 1.00], corresponding to five partitions with spatial susceptibility ranging from low to high [37]. The standard error of the mean (SEM) for the predicted probabilities in susceptibility zones of Test Set 1 are presented in Table 2.

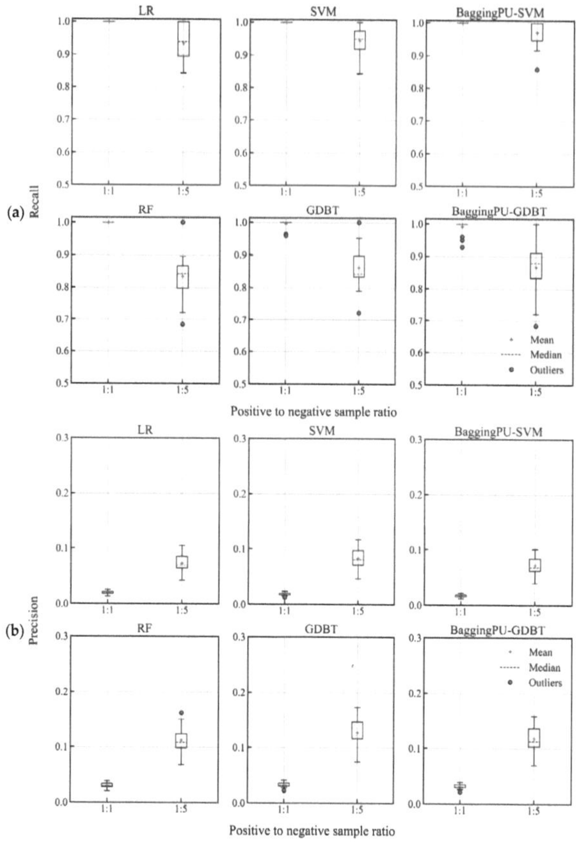

Figure 4. Classification performance for landslide samples in Test Set 2. The proportion of positive and negative samples in Test set 2 is unknown. (**a**) Recall of models with training set sample ratios of 1:1 and 1:5; (**b**) Precision of models with training set sample ratios of 1:1 and 1:5.

Table 2. The standard error of the mean of Test Set 1 in susceptibility zones.

Positive and Unlabeled Sample Ratio	Model	SEM				
		Very Low	Low	Moderate	High	Very High
1:1	LR	0.00211	0.00207	0.00463	0.00362	0.0016
	SVM	0.00208	0.0021	0.00465	0.00427	0.00158
	BaggingPU-SVM	0.00244	0.00209	0.00417	0.00366	0.00155
	RF	0.00216	0.00207	0.00443	0.00477	0.00142
	GDBT	0.00198	0.00207	0.0049	0.00529	0.0014
	BaggingPU-GDBT	0.00186	0.00205	0.00491	0.00503	0.00136
1:5	LR	0.0006	0.00395	0.00212	0.00303	0.00411
	SVM	0.00058	0.00423	0.0026	0.00356	0.00322
	BaggingPU-SVM	0.00058	0.00367	0.00297	0.00334	0.00287
	RF	0.00057	0.00425	0.00299	0.00335	0.00248
	GDBT	0.00055	0.00479	0.00303	0.00337	0.00227
	BaggingPU-GDBT	0.00055	0.0051	0.00309	0.00335	0.00212

For both 1:1 and 1:5 positive and unlabeled sample ratio, the SEMs of BaggingPU-SVM and BaggingPU-GDBT in the high and very high susceptibility zones are lower than those of SVM and GDBT, respectively. This suggests that Bagging PU learning effectively reduces uncertainty in the high susceptibility range.

Figure 5 illustrates the distribution of SEs for the predicted values of a model with training set positive and negative sample ratios of 1:1 and 1:5 on Test set 2. Due to the limited number of landslide samples in Test Set 2, some models lack predictive value in the high and very high susceptibility zones. A least squares curve can be used to fit the relationship between mean and SE. The coefficient of determination (r^2) indicates that Bagging PU learning reduces the uncertainty of the predicted values.

3.3. Generalization Prediction of Models

We employed the same set of influence factors used in Region A to construct a dataset for Region B and trained a model with a positive and unlabeled sample ratio of 1:1. To assess the generalization performance of the model beyond the study area, both the Region A based BaggingPU-GDBT model and the Region B based BaggingPU-GDBT model were utilized for LSM, as depicted in Figure 6.

There were noticeable disparities in the susceptibility zones between the two maps. To quantitatively compare these differences, we tallied the number of computing units and landslide units in each susceptibility zone, as presented in Table 3.

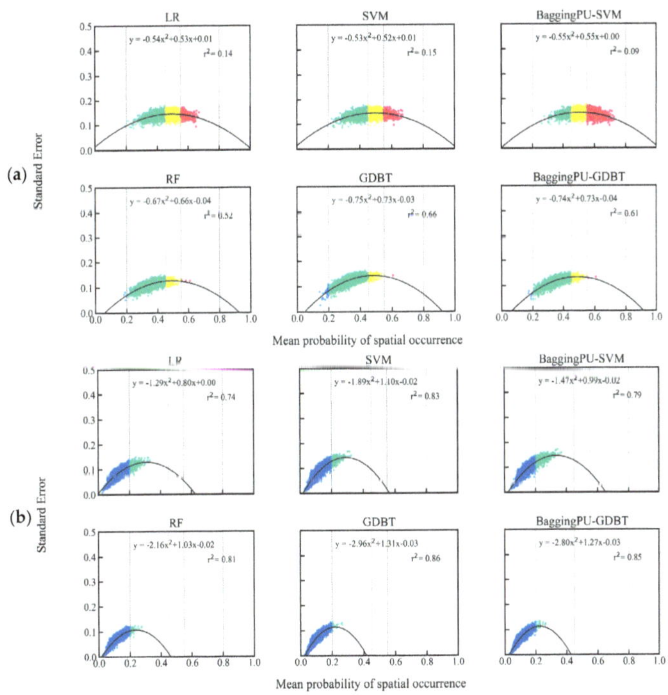

Figure 5. The spatial occurrence of standard errors (SEs) for the predicted susceptibility values of Test Set 2. (**a**) SEs of models with training set sample ratios of 1:1; (**b**) SEs of models with training set sample ratios of 1:5.

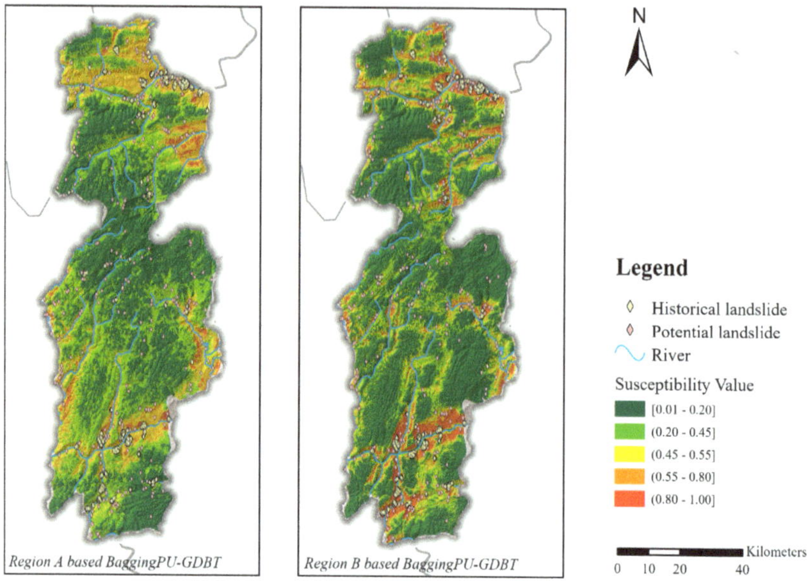

Figure 6. Landslide susceptibility mappings of region B with the 1:1 positive and unlabeled sample ratio.

Table 3. Zonal statistics of landslide frequency ratio.

Model	Susceptibility	Unite Proportion	Landslide Unit Proportion	Frequency Ratio
Region B based BaggingPU-GDBT	0.00–0.20	52.42%	0.37%	0.007
	0.20–0.45	23.44%	4.01%	0.171
	0.45–0.55	6.01%	3.46%	0.576
	0.55–0.80	10.64%	13.10%	1.231
	0.80–1.00	7.49%	79.06%	10.557
Region A based BaggingPU-GDBT	0.00–0.20	42.51%	6.87%	0.162
	0.20–0.45	34.32%	19.44%	0.566
	0.45–0.55	6.66%	7.81%	1.173
	0.55–0.80	14.86%	59.40%	3.997
	0.80–1.00	1.65%	6.48%	3.917

For the Region B based BaggingPU-GDBT model, the distribution of moderate and above susceptibility zones aligns with 95.62% of the known landslide samples, with most landslide samples concentrated in the very high susceptibility zone. For the Region A based BaggingPU-GDBT model, the distribution of moderate and above susceptibility zones corresponds to 73.69% of the known landslide samples.

While the recognition rate of landslide samples is not as high as that of the Region B based model, the landslide samples in the Region A based model are predominantly concentrated in the high susceptibility zone. The relatively small prediction area of the moderate and above susceptibility zones leading to a higher frequency ratio, and the susceptibility prediction results still provide reasonable reference values.

4. Discussion

4.1. Classification Ability of Models

In the context of landslide susceptibility assessment, we acknowledge a certain level of tolerance for misclassifying non-landslide units, provided that it aligns with overall accuracy and enhances landslide recognition rates. As depicted in Figures 3 and 4, it is evident that bagging PU learning enhances the recognition rate of landslide samples while maintaining satisfactory overall classification accuracy. Moreover, the increase in recall persists even when there are alterations in the distribution ratio of positive and negative samples within the dataset. This improvement is undeniably advantageous for landslide prevention and control efforts [38]. Furthermore, as dataset imbalances intensify, Bagging PU learning proves valuable in assisting SVM, a model sensitive to dataset disturbances, in maintaining a certain level of classification precision. While the GDBT model inherently exhibits robustness in datasets with varying sample ratios, bagging PU learning still enhances its recall slightly. By training classifiers using diverse subsets of unlabeled data and known landslide samples with different positive and negative sample distributions, varying disturbance rates of unlabeled landslide samples in the test set can be accommodated. This process renders the final ensemble classifier more flexible and stable [31].

From Figure 3, it can be observed that the recall of landslide samples in Test Set 1 de-creases monotonically as the positive-to-negative sample ratio decreases. When the posi-tive-to-negative sample ratio is 1:10, the recall is already below 70%. Therefore, we specu-late that, contrary to some previous studies [38–40], collecting more nega-tive samples than positive samples may not be a favorable choice in landslide suscepti bility modeling. Past studies examining the positive-to-negative sample ratio typically uti-lize ROC for direct comparison, rarely considering landslide recall at the sampling stage. Especially when employing complex algorithms, models built on imbalanced landslide datasets may tend to predict negative samples more, as these algorithms can better fit the training set. Consequently, the resulting susceptibility mapping may exhibit a disproportionately large area of very low susceptibility zones [41]. We utilize models trained with positive-to-negative sample ratios of 1:1 and 1:5 to generate susceptibility maps for Region A. As shown in Figure 7, this supports our viewpoint.

The 1:1 models and 1:5 models were used to predict all computing units in region A, generating susceptibility maps as displayed in Figure 8. The regions predicted as moderate and above susceptibility by the 1:5 models appear notably more concentrated, with fewer instances of misclassifying unlabeled samples as landslides. This corresponds to an improvement in precision. However, unlike typical binary classification problems, unlabeled samples are not definitively negative classes in landslide susceptibility estimation. Using evaluation metrics that treat unlabeled samples as negative classes cannot serve as the sole basis for assessing model quality. Our goal is not only to correctly classify historical landslides but also to estimate the spatial probability of landslides in unlabeled areas. Although the 1:1 models exhibit lower precision, fewer potential landslide points are classified in low and very low susceptibility areas. This suggests that these models have better captured the characteristics of such landslide samples, leading to more reasonable mapping results. In contrast, the 1:5 models increase precision at the cost of classifying more potential landslide points as noise. Moreover, in Figure 7b, logistic regression, serving as a lower-complexity algorithm in this study, predicts the smallest area of very low susceptibility zones. This observation indicates that higher-complexity models are more likely to excessively focus on non-landslide samples in imbalanced datasets. This issue has often been overlooked in the past and merits further exploration.

4.2. Generalization Ability of Models

The generalization performance of LSM models is typically verified through three approaches: 1. Examining the accuracy on the test set splinted from modeling dataset; 2. Applying the model to the region where the modeling dataset originated, excluding the modeling dataset; and 3. Selecting a different study area for validation [42]. In this study, we have tested all three approaches. Test Set 1 and Test Set 2, respectively, represent the test set with the same sample distribution as the modeling dataset and the test set with the same origin of the modeling dataset but different positive-to-negative sample ratios.

Region B serves as an additional study area to independently validate the generalization capability of the best model on other datasets.

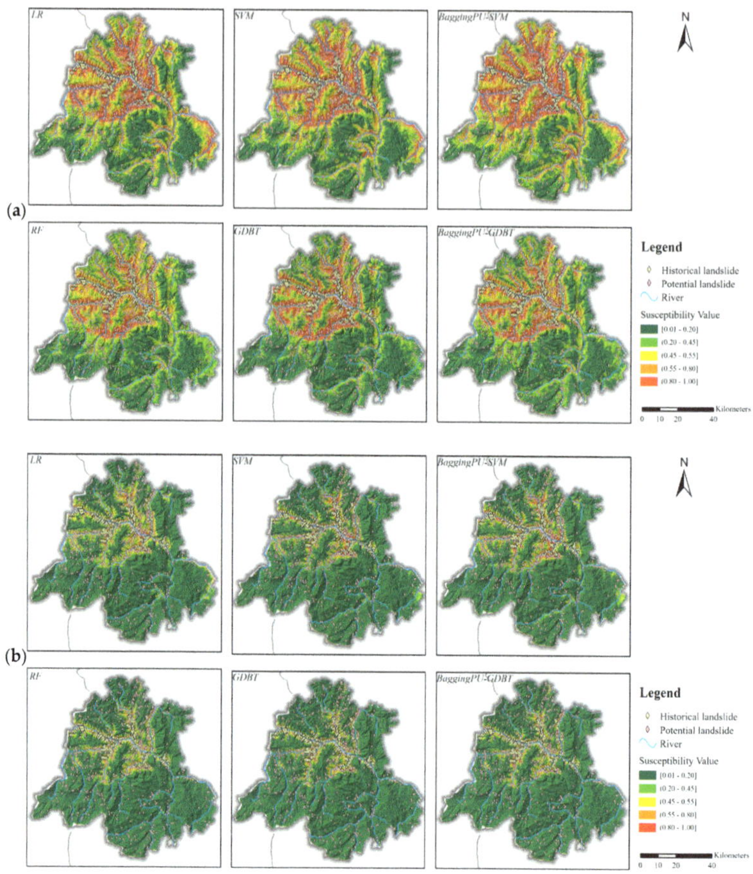

Figure 7. Landslide susceptibility Mappings (LSMs) of region A. (**a**) LSM of region A generated by model trained with 1:1 positive and unlabeled sample ratio; (**b**) LSM of region A generated by model trained with 1:5 positive and unlabeled sample ratio.

Figures 3a and 4a illustrate the stability of the bagging PU learning model. Across test sets with different sample ratio distributions, both Bagging PU-SVM and Bagging PU-GDBT exhibit higher recall compared to SVM and GDBT. Figure 8 provides a visualization of the prediction SEs for models trained with positive and unlabeled sample ratios of 1:1 and 1:5 in the training set, tested on Test Set 1. Comparing Figures 5 and 8, the SEs for predictions in the moderate and higher susceptibility zones of Test Set 1 are small, indicating a comprehensive learning of the model for known landslide samples. The overall SEs for predictions on Test Set 2 are larger, particularly in the moderate and higher susceptibility zones. This may be attributed to the uncertainty in sample distribution. However, the SE fitting curves for Bagging PU-SVM and Bagging PU-GDBT are slightly shifted to the right compared to those of SVM and GDBT, suggesting that Bagging PU learning assists in reducing the uncertainty in predicting values for very high susceptibility zones. Therefore, bagging PU methods are considered helpful for the model to overcome class-prior probability shift in the landslide sample distribution to some extent, thereby enhancing the model's generalization capability.

From Figure 6, it is evident that the Bagging PU-GDBT model trained on data from Region A exhibits limitations in predicting landslide susceptibility in Region B. We hypothesize that this discrepancy is due to variations in the influencing factors leading to landslides in the two regions. Figure 9 summarizes the contribution of each factor within the 1:1 Bagging PU-GDBT model for both regions.

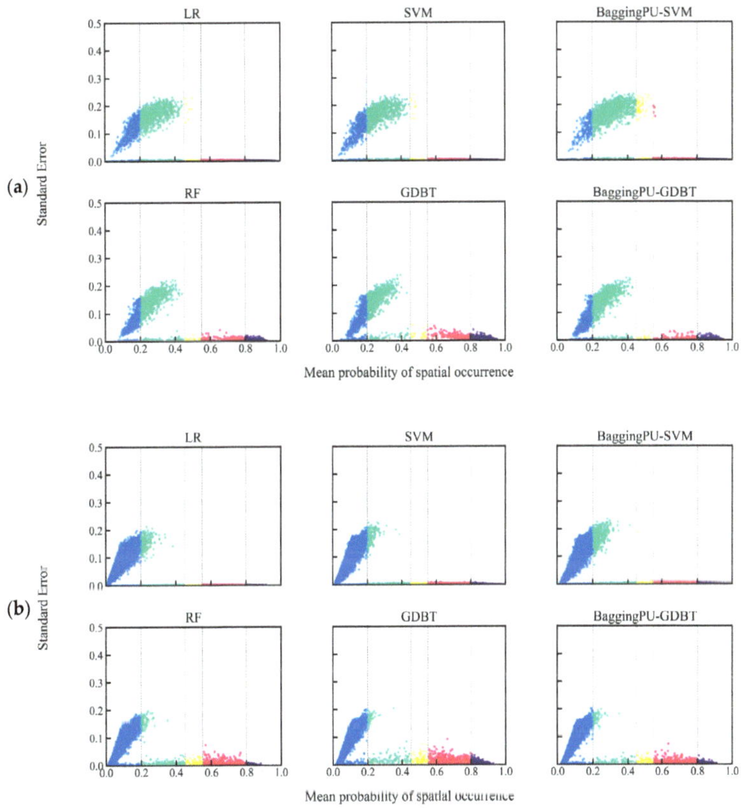

Figure 8. The spatial occurrence of standard errors (SEs) for the predicted susceptibility values of Test Set 1. (**a**) SEs of models with positive to negative sample ratios of 1:1; (**b**) SEs of models with positive to negative sample ratios of 1:5.

For the Region A model, elevation and rainfall contribute significantly more than other factors. For the Region B model, the factors with higher contributions are elevation, valley depth, and distance from road. The differences in the combinations of primary influencing factors between the two regions may be a crucial reason leading to the decreased generalization performance of the Region A-based BaggingPU-GDBT model when predicting in Region B.

In terms of factor contributions, it is evident that Region B is more susceptible to triggering factors when compared to Region A. In Region B, the impact of roads and seismic activity on landslides is more pronounced. Road construction disrupts slope stability through induced slope cutting, and the reservoir-induced earthquakes increased sensitivity of slope stability. There is relatively higher seismic activity in the eastern and southern parts of Region B, in contrast to the weaker seismic activity in the seismic zone where Region A is located.

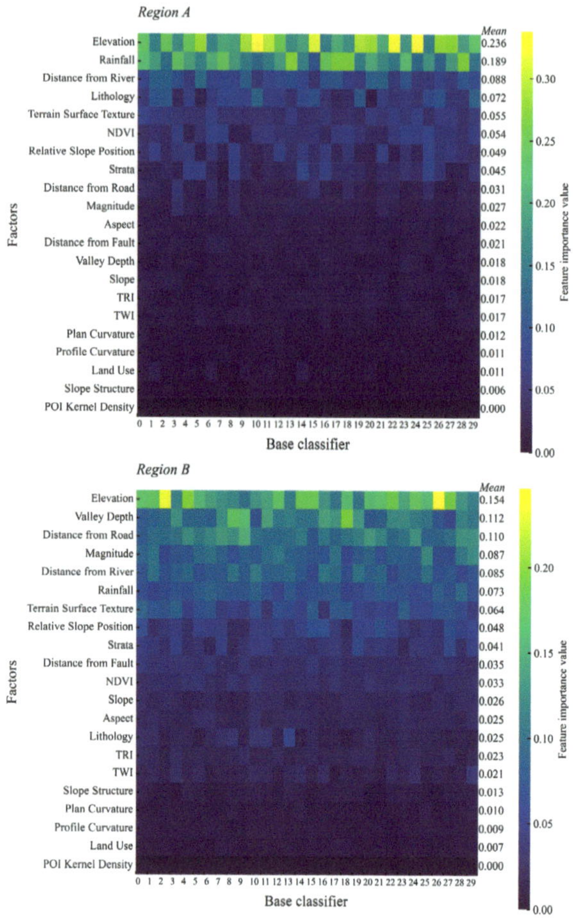

Figure 9. Factor Importance in the BaggingPU-GDBT model (Vertical axis organized from highest to lowest based on the average importance of factors across 30 base classifiers).

Despite this, the Region A-based BaggingPU-GDBT model manages to maintain a certain degree of generalization ability in Region B area due to the similarities in elevation and hydrological features. Elevation serves as the primary controlling factor for landslide occurrence in both regions, determining dominant topography, vegetation characteristics, soil weathering, and reflecting, to some extent, human economic and engineering activities [42]. The factors 'distance from river' and 'rainfall' hold relatively high importance in both regions. The mechanical properties of loose rock mass will change under the action of water [43,44]. The fragile surface materials of slopes exacerbate their instability during intense rainfall, seismic events, and drastic fluctuations in river water levels [45–47]. As depicted in Figures 6 and 7, in Region A, high susceptibility areas for landslides predominantly align along both sides of watercourses, while in Region B, these areas primarily cluster along the Yangtze River, Qing River, and their tributaries. The rock formations in both regions consist mainly of carbonate rocks, known for their robust resistance to weathering. Loose deposits are concentrated along the riverbanks, providing the material foundation for landslide occurrence. The water levels in these rivers experience seasonal fluctuations influenced by rainfall and groundwater undergoes changes. This pronounced variability in hydrological characteristics creates advantageous external dynamic conditions for the incubation and development of geological hazards in mountainous regions.

In summary, the spatial heterogeneity of influencing factors in different geographical regions significantly impacts the model's generalization performance [48]. Although we have preliminary discussed the importance of factors in predictive models in this study, further exploration of the contribution trends of features to landslide susceptibility should be considered in future research for model reuse in regions with high homogeneity in influencing factors.

Landslide Susceptibility Mapping (LSM), as a rapid technique for predicting the spatial distribution of disasters, provides guidance for rational spatial management. Landslides exert an influence on sediment supply in river basins. Landslides within reservoirs introduce a considerable amount of loose material, thereby contributing to subsequent debris flows within the reservoir [49]. The LSM in reservoir areas plays a pivotal role in anticipating potential debris flow occurrences and contributes to catchment planning. The LSM model proposed in this paper, adaptable to varying landslide distribution ratios in different regions, exhibits generalizability. It can be applied to other regions with geological, geographical, hydrological, and climatic similarities, generating landslide sensitivity mappings in situations where historical landslide inventories are incomplete. When applied to global landslide risk monitoring and disaster prevention, it can serve as a foundation for formulating appropriate natural disaster management policies. This is particularly crucial in low- to middle-income countries with high population density, where landslides in densely populated mountainous areas can lead to economic losses, casualties, and even direct and indirect costs for urban-scale constructions or infrastructure [50–53].

5. Conclusions

Landslide susceptibility modeling encounters challenges arising from imbalanced data distributions and class-prior probability shifts. This study introduces an approach utilizing Bagging PU Learning, a semi-supervised learning method, to enhance the performance of landslide susceptibility models. In addition, our study refines the estimation method of landslide susceptibility uncertainty, employing standard error, and conducts comprehensive evaluations encompassing classification accuracy and prediction probability uncertainty. Our findings suggest that:

1. The selection of positive-to-negative sample ratio profoundly affects the classification performance of LSM model. While models trained on unbalanced datasets exhibit superior overall binary classification performance, those trained on balanced datasets demonstrate higher landslide recall. This emphasizes the importance of the trade-off between precision and recall in LSM modeling, differing from typical binary classification problems.
2. The positive-to-negative sample ratio significantly impacts the mapping results. Models trained on unbalanced datasets tend to predict negative samples, lowering the overall landslide susceptibility probability in the region. Conversely, balanced datasets yield more reasonable for prevention and control planning.
3. Utilizing Bagging PU Learning in classifiers has the potential to boost recall in the context of class-prior probability shift, thereby enhancing the overall generalization performance of the model. This method can reduce the uncertainty of model predictions in high susceptibility areas. In this study, the BaggingPU-GDBT model shows the best performance.

Our research provides new ideas in landslide susceptibility modeling for reliable predictions in unbalanced data scenarios. However, given the spatial heterogeneity of influencing factors, further exploration on how to improve the generalization ability of the model in additional scenarios is needed.

Author Contributions: Conceptualization, L.Z. and R.N.; methodology, L.Z.; software, H.M.; validation, L.Z.; formal analysis, L.Z.; investigation, L.Z., J.D. and H.M.; resources, H.X.; data curation, J.D. and H.X.; writing—original draft preparation, L.Z.; writing—review and editing, L.Z. and X.W.;

visualization, L.Z.; supervision, R.N.; project administration, R.N.; funding acquisition, X.W. All authors have read and agreed to the published version of the manuscript.

Funding: This research was funded by the National Natural Science Foundation of China, grant number 42071429, and 111 project under Grant B17040.

Data Availability Statement: Data are contained within the article.

Conflicts of Interest: The authors declare no conflict of interest.

References

1. Su, M.-B.; Chen, I.-H.; Liao, C.-H. Using TDR Cables and GPS for Landslide Monitoring in High Mountain Area. *J. Geotech. Geoenviron. Eng.* **2009**, *135*, 1113–1121. [CrossRef]
2. Zhang, Y.; Tang, H.; Li, C.; Lu, G.; Cai, Y.; Zhang, J.; Tan, F. Design and Testing of a Flexible Inclinometer Probe for Model Tests of Landslide Deep Displacement Measurement. *Sensors* **2018**, *18*, 224. [CrossRef]
3. Zhu, H.-H.; Shi, B.; Zhang, C.-C. FBG-Based Monitoring of Geohazards: Current Status and Trends. *Sensors* **2017**, *17*, 452. [CrossRef] [PubMed]
4. Caviedes-Voullième, D.; Juez, C.; Murillo, J.; García-Navarro, P. 2D Dry Granular Free-Surface Flow over Complex Topography with Obstacles. Part I: Experimental Study Using a Consumer-Grade RGB-D Sensor. *Comput. Geosci.* **2014**, *73*, 177–197. [CrossRef]
5. Cao, Y.; Wei, X.; Fan, W.; Nan, Y.; Xiong, W.; Zhang, S. Landslide Susceptibility Assessment Using the Weight of Evidence Method: A Case Study in Xunyang Area, China. *PLoS ONE* **2021**, *16*, e0245668. [CrossRef] [PubMed]
6. Aditian, A.; Kubota, T.; Shinohara, Y. Comparison of GIS-Based Landslide Susceptibility Models Using Frequency Ratio, Logistic Regression, and Artificial Neural Network in a Tertiary Region of Ambon, Indonesia. *Geomorphology* **2018**, *318*, 101–111. [CrossRef]
7. Fang, Z.; Wang, Y.; Peng, L.; Hong, H. A Comparative Study of Heterogeneous Ensemble-Learning Techniques for Landslide Susceptibility Mapping. *Int. J. Geogr. Inf. Sci.* **2021**, *35*, 321–347. [CrossRef]
8. Felicísimo, Á.M.; Cuartero, A.; Remondo, J.; Quirós, E. Mapping Landslide Susceptibility with Logistic Regression, Multiple Adaptive Regression Splines, Classification and Regression Trees, and Maximum Entropy Methods: A Comparative Study. *Landslides* **2013**, *10*, 175–189. [CrossRef]
9. Yao, X.; Tham, L.G.; Dai, F.C. Landslide Susceptibility Mapping Based on Support Vector Machine: A Case Study on Natural Slopes of Hong Kong, China. *Geomorphology* **2008**, *101*, 572–582. [CrossRef]
10. Liu, M.; Liu, J.; Xu, S.; Zhou, T.; Ma, Y.; Zhang, F.; Konečný, M. Landslide Susceptibility Mapping with the Fusion of Multi-Feature SVM Model Based FCM Sampling Strategy: A Case Study from Shaanxi Province. *Int. J. Image Data Fusion* **2021**, *12*, 349–366. [CrossRef]
11. Nefeslioglu, H.A.; Gokceoglu, C.; Sonmez, H. An Assessment on the Use of Logistic Regression and Artificial Neural Networks with Different Sampling Strategies for the Preparation of Landslide Susceptibility Maps. *Eng. Geol.* **2008**, *97*, 171–191. [CrossRef]
12. Peng, L.; Niu, R.; Huang, B.; Wu, X.; Zhao, Y.; Ye, R. Landslide Susceptibility Mapping Based on Rough Set Theory and Support Vector Machines: A Case of the Three Gorges Area, China. *Geomorphology* **2014**, *204*, 287–301. [CrossRef]
13. Kavzoglu, T.; Sahin, E.K.; Colkesen, I. Landslide Susceptibility Mapping Using GIS-Based Multi-Criteria Decision Analysis, Support Vector Machines, and Logistic Regression. *Landslides* **2014**, *11*, 425–439. [CrossRef]
14. Rabby, Y.W.; Li, Y.; Hilafu, H. An Objective Absence Data Sampling Method for Landslide Susceptibility Mapping. *Sci. Rep.* **2023**, *13*, 1740. [CrossRef] [PubMed]
15. Su, C.; Wang, B.; Lv, Y.; Zhang, M.; Peng, D.; Bate, B.; Zhang, S. Improved Landslide Susceptibility Mapping Using Unsupervised and Supervised Collaborative Machine Learning Models. *Georisk Assess. Manag. Risk Eng. Syst. Geohazards* **2023**, *17*, 387–405. [CrossRef]
16. Huang, F.; Yin, K.; Jiang, S.; Huang, J.; Cao, Z. Landslide Susceptibility Assessment Based on Clustering Analysis and Support Vector Machine. *Chin. J. Rock Mech. Eng.* **2018**, *37*, 156–167. [CrossRef]
17. Sun, D.; Wu, X.; Wen, H.; Gu, Q. A LightGBM-Based Landslide Susceptibility Model Considering the Uncertainty of Non-Landslide Samples. *Geomat. Nat. Hazards Risk* **2023**, *14*, 2213807. [CrossRef]
18. Fang, Z.; Wang, Y.; Niu, R.; Peng, L. Landslide Susceptibility Prediction Based on Positive Unlabeled Learning Coupled With Adaptive Sampling. *IEEE J. Sel. Top. Appl. Earth Obs. Remote Sens.* **2021**, *14*, 11581–11592. [CrossRef]
19. Wu, B.; Qiu, W.; Jia, J.; Liu, N. Landslide Susceptibility Modeling Using Bagging-Based Positive-Unlabeled Learning. *IEEE Geosci. Remote Sens. Lett.* **2021**, *18*, 766–770. [CrossRef]
20. Elkan, C.; Noto, K. Learning Classifiers from Only Positive and Unlabeled Data. In Proceedings of the 14th ACM SIGKDD International Conference on Knowledge Discovery and Data Mining, Las Vegas, NV, USA, 24 August 2008; ACM: New York, NY, USA, 2008; pp. 213–220.
21. Nakajima, S.; Sugiyama, M. Positive-Unlabeled Classification under Class-Prior Shift: A Prior-Invariant Approach Based on Density Ratio Estimation. *Mach. Learn.* **2023**, *112*, 889–919. [CrossRef]
22. Li, X.; Liu, B. Learning to Classify Texts Using Positive and Unlabeled Data. In Proceedings of the 18th International Joint Conference on Artificial Intelligence, Acapulco, Mexico, 9–15 August 2003; Morgan Kaufmann Publishers Inc.: San Francisco, CA, USA, 2003; pp. 587–592.

23. Yu, H.; Han, J.; Chang, K.C.-C. PEBL: Positive Example Based Learning for Web Page Classification Using SVM. In Proceedings of the Eighth ACM SIGKDD International Conference on Knowledge Discovery and Data Mining, Edmonton, AB, Canada, 23 July 2002; ACM: New York, NY, USA, 2002; pp. 239–248.
24. Tang, L.; Yu, X.; Jiang, W.; Zhou, J. Comparative Study on Landslide Susceptibility Mapping Based on Unbalanced Sample Ratio. *Sci. Rep.* **2023**, *13*, 5823. [CrossRef]
25. Wang, Y.; Feng, L.; Li, S.; Ren, F.; Du, Q. A Hybrid Model Considering Spatial Heterogeneity for Landslide Susceptibility Mapping in Zhejiang Province, China. *Catena* **2020**, *188*, 104425. [CrossRef]
26. Wu, L. *The Multi-Fractal of the Spatial Distribution of Landslide*; Deng, W., Ed.; Chongqing Normal University: Chongqing, China, 2011; pp. 99–102.
27. Wright, R. Positive-Unlabeled Learning. 2017.
28. Ullah, K.; Wang, Y.; Fang, Z.; Wang, L.; Rahman, M. Multi-Hazard Susceptibility Mapping Based on Convolutional Neural Networks. *Geosci. Front.* **2022**, *13*, 101425. [CrossRef]
29. Liao, M.; Wen, H.; Yang, L. Identifying the Essential Conditioning Factors of Landslide Susceptibility Models under Different Grid Resolutions Using Hybrid Machine Learning: A Case of Wushan and Wuxi Counties, China. *Catena* **2022**, *217*, 106428. [CrossRef]
30. Wang, D.; Hao, M.; Chen, S.; Meng, Z.; Jiang, D.; Ding, F. Assessment of Landslide Susceptibility and Risk Factors in China. *Nat. Hazards* **2021**, *108*, 3045–3059. [CrossRef]
31. Mordelet, F.; Vert, J.-P. A Bagging SVM to Learn from Positive and Unlabeled Examples. *Pattern Recognit. Lett.* **2014**, *37*, 201–209. [CrossRef]
32. Scott, C.; Blanchard, G. Novelty Detection: Unlabeled Data Definitely Help. In *Artificial Intelligence and Statistics, Proceedings of the Twelfth International Conference on Artificial Intelligence and Statistics, Clearwater Beach, FL, USA, 15 April 2009*; Van Dyk, D., Welling, M., Eds.; PMLR: New York, NY, USA, 2009; Volume 5, pp. 464–471.
33. Breiman, L. Bagging Predictors. *Mach. Learn.* **1996**, *24*, 123–140. [CrossRef]
34. Breiman, L. Random Forests. *Mach. Learn.* **2001**, *45*, 5–32. [CrossRef]
35. Mandrekar, J.N. Receiver Operating Characteristic Curve in Diagnostic Test Assessment. *J. Thorac. Oncol.* **2010**, *5*, 1315–1316. [CrossRef]
36. Kouw, W.M.; Loog, M. An Introduction to Domain Adaptation and Transfer Learning. *arXiv* **2018**, arXiv:1812.11806. [CrossRef]
37. Guzzetti, F.; Reichenbach, P.; Ardizzone, F.; Cardinali, M.; Galli, M. Estimating the Quality of Landslide Susceptibility Models. *Geomorphology* **2006**, *81*, 166–184. [CrossRef]
38. Zhao, L.; Wu, X.; Niu, R.; Wang, Y.; Zhang, K. Using the Rotation and Random Forest Models of Ensemble Learning to Predict Landslide Susceptibility. *Geomat. Nat. Hazards Risk* **2020**, *11*, 1542–1561. [CrossRef]
39. Pourghasemi, H.R.; Kornejady, A.; Kerle, N.; Shabani, F. Investigating the Effects of Different Landslide Positioning Techniques, Landslide Partitioning Approaches, and Presence-Absence Balances on Landslide Susceptibility Mapping. *Catena* **2020**, *187*, 104364. [CrossRef]
40. Yang, C.; Liu, L.-L.; Huang, F.; Huang, L.; Wang, X.-M. Machine Learning-Based Landslide Susceptibility Assessment with Optimized Ratio of Landslide to Non-Landslide Samples. *Gondwana Res.* **2023**, *123*, 198–216. [CrossRef]
41. Gao, H.; Fam, P.S.; Tay, L.T.; Low, H.C. Comparative Landslide Spatial Research Based on Various Sample Sizes and Ratios in Penang Island, Malaysia. *Bull. Eng. Geol. Environ.* **2021**, *80*, 851–872. [CrossRef]
42. Sun, D.; Xu, J.; Wen, H.; Wang, Y. An Optimized Random Forest Model and Its Generalization Ability in Landslide Susceptibility Mapping: Application in Two Areas of Three Gorges Reservoir, China. *J. Earth Sci.* **2020**, *31*, 1068–1086. [CrossRef]
43. Chu, H.-J.; Chen, Y.-C.; Ali, M.; Höfle, B. Multi-Parameter Relief Map from High Resolution DEMs: A Case Study of Mudstone Badland. *Int. J. Environ. Res. Public Health* **2019**, *16*, 1109. [CrossRef]
44. Guo, Y.; Li, X.; Ju, S.; Lyu, Q.; Liu, T. Utilization of 3D Laser Scanning for Stability Evaluation and Deformation Monitoring of Landslides. *J. Environ. Public Health* **2022**, *2022*, 8225322. [CrossRef] [PubMed]
45. Mantovani, J.R.; Bueno, G.T.; Alcântara, E.; Park, E.; Cunha, A.P.; Londe, L.; Massi, K.; Marengo, J.A. Novel Landslide Susceptibility Mapping Based on Multi-Criteria Decision-Making in Ouro Preto, Brazil. *J. Geovisualization Spat. Anal.* **2023**, *7*, 7. [CrossRef]
46. Tesfa, C. GIS-Based AHP and FR Methods for Landslide Susceptibility Mapping in the Abay Gorge, Dejen–Renaissance Bridge, Central, Ethiopia. *Geotech. Geol. Eng.* **2022**, *40*, 5029–5043. [CrossRef]
47. Millán-Arancibia, C.; Lavado-Casimiro, W. Rainfall Thresholds Estimation for Shallow Landslides in Peru from Gridded Daily Data. *Nat. Hazards Earth Syst. Sci.* **2023**, *23*, 1191–1206. [CrossRef]
48. Zhang, J.; Ma, X.; Zhang, J.; Sun, D.; Zhou, X.; Mi, C.; Wen, H. Insights into Geospatial Heterogeneity of Landslide Susceptibility Based on the SHAP-XGBoost Model. *J. Environ. Manag.* **2023**, *332*, 117357. [CrossRef] [PubMed]
49. Jin, W.; Cui, P.; Zhang, G.; Wang, J.; Zhang, Y.; Zhang, P. Evaluating the Post-Earthquake Landslides Sediment Supply Capacity for Debris Flows. *Catena* **2023**, *220*, 106649. [CrossRef]
50. Carrión-Mero, P.; Montalván-Burbano, N.; Morante-Carballo, F.; Quesada-Román, A.; Apolo-Masache, B. Worldwide Research Trends in Landslide Science. *Int. J. Environ. Res. Public Health* **2021**, *18*, 9445. [CrossRef] [PubMed]

51. Alcántara-Ayala, I.; Garnica-Peña, R.J. Landslide Warning Systems in Low-And Lower-Middle-Income Countries: Future Challenges and Societal Impact. In *Progress in Landslide Research and Technology, Volume 1 Issue 1, 2022*; Progress in Landslide Research and Technology; Sassa, K., Konagai, K., Tiwari, B., Arbanas, Ž., Sassa, S., Eds.; Springer International Publishing: Cham, Germany, 2023; pp. 137–147. ISBN 978-3-031-16897-0.
52. Bucała-Hrabia, A.; Kijowska-Strugała, M.; Śleszyński, P.; Rączkowska, Z.; Izdebski, W.; Malinowski, Z. Evaluating the Use of the Landslide Database in Spatial Planning in Mountain Communes (the Polish Carpathians). *Land Use Policy* **2022**, *112*, 105842. [CrossRef]
53. Garcia-Delgado, H.; Petley, D.N.; Bermúdez, M.A.; Sepúlveda, S.A. Fatal Landslides in Colombia (from Historical Times to 2020) and Their Socio-Economic Impacts. *Landslides* **2022**, *19*, 1689–1716. [CrossRef]

remote sensing

MDPI

Article

Mapping of the Spatial Scope and Water Quality of Surface Water Based on the Google Earth Engine Cloud Platform and Landsat Time Series

Haohai Jin [1], Shiyu Fang [1] and Chao Chen [2,*]

1 Marine Science and Technology College, Zhejiang Ocean University, Zhoushan 316022, China; hulkkin@163.com (H.J.); fsy2486260170@163.com (S.F.)

2 School of Geography Science and Geomatics Engineering, Suzhou University of Science and Technology, Suzhou 215009, China

* Correspondence: chenchao2009@pku.edu.cn

Abstract: Surface water is an important parameter for water resource management and terrestrial water circulation research that is closely related to human production and livelihood. With the rapid development of remote sensing technology and cloud computing platforms, the use of remote sensing technology for large-scale and long-term surface water monitoring and investigation has become a research trend. Based on the Google Earth Engine (GEE) cloud platform and Landsat series satellite data, in this study, the Emergency Geomatics Service (EGS) operational surface water mapping algorithm and water index masking were utilized to extract the spatial scope of the water body. The validated models of the Secchi disk depth (SDD), chlorophyll-a (Chl-a) and suspended solids (SS) concentration were applied to water quality parameter inversion and water quality evaluation. Surface water extent extraction and water quality maps were created to analyze the spatial distribution of the water body and the spatial–temporal evolution characteristics of the water quality parameters. A verification experiment was carried out with the surface water in Zhejiang Province as the research object. The results show that the surface water in the study area from 1990 to 2022 could be accurately extracted. The kappa coefficients were all greater than 0.90, and the overall accuracies of the extractions were greater than 95.31%. From 1990 to 2022, the total surface water area in Zhejiang Province initially decreased and then increased. The minimum water area of 2027.49 km^2 occurred in 2005, and the maximum water area of 2614.96 km^2 occurred in 2020, with an annual average variation of 193.92 km^2. Since 2015, the proportion of high SS and Chl-a concentrations, and low SDD water bodies in Zhejiang Province have decreased, and the proportion with better water quality has increased significantly. The spatial distribution map of the surface water and the inversion results of the water quality parameters obtained in this study provide a valuable reference and guidance for regional water resource management, disaster monitoring and early warning, environmental protection, and aquaculture.

Keywords: GEE; Landsat; EGS operational surface water mapping algorithm; surface water; water quality parameter inversion; long sequence time series remote sensing monitoring

Citation: Jin, H.; Fang, S.; Chen, C. Mapping of the Spatial Scope and Water Quality of Surface Water Based on the Google Earth Engine Cloud Platform and Landsat Time Series. *Remote Sens.* **2023**, *15*, 4986. https://doi.org/10.3390/rs15204986

Academic Editor: Gabriel Senay

Received: 19 August 2023
Revised: 27 September 2023
Accepted: 13 October 2023
Published: 16 October 2023

1. Introduction

Surface water refers to the water that exists on the surface of the Earth's crust and is exposed to the atmosphere, including rivers, lakes, reservoirs, and ponds. It plays a crucial role in the balance between ecosystems and socio-economic development and provides essential resources for the survival of living organisms and the development of human society [1–3]. Surface water resources have a variety of ecosystem service values such as water storage, flood control, aquaculture, power generation, and transportation. With the gradual shortage of water resources, the social function of surface water is becoming increasingly significant, and the economic importance of surface water is continuously increasing [4,5]. The area of surface water resources has shrunk, and the pollution of the water environment has substantial negative impacts on the ecological environment and human development [6–8]. Therefore, understanding the spatial extent and time series features of surface water resources and investigating the variations in water quality parameters are significant for ecological protection, social production, and life development.

With the progress of remote sensing technology and the rapid development of cloud computing technology, the traditional hardware environment dependence problem has been solved, which makes large-scale and long-term water body monitoring more efficient [9]. Based on the characteristics of water spectral curves, scholars have proposed a variety of water information extraction methods, such as the single band threshold-based method, the spectrum photometric-based method, and the water index-based method. The single-band threshold-based method uses the characteristics of water's strong absorption and low spectral reflectance in the near-infrared and shortwave infrared bands to distinguish water from other ground objects by setting a threshold [10]. This method has the benefits of simplicity of operation and easy implementation. However, it does not make full use of multi-band information and is often affected by other ground objects with similar spectral characteristics. The spectrum photometric-based method identifies the spectral characteristics of water information by analyzing the spectral characteristic curves of ground objects in the original band or transformation space of remote sensing images. Then, it uses logical judgment rules to extract water information [11]. The spectrum photometric-based method has a high precision and wide application and has good adaptability to the complex distribution of ground objects. Nevertheless, the model construction process is cumbersome and the computational complexity is high [12]. The most commonly used water index-based method is based on the ratio operation between characteristic bands to enhance the contrast between water and other features, which is used to identify water information. Mcfeters (1996) first proposed the normalized difference water index (NDWI), using the ratio of band 2 (green band) to band 4 (near-infrared band) of Landsat thematic imager (TM) to recognize the water body. Xu (2006) used Landsat TM bands 2 (green band) and 5 (mid-infrared band) to evaluate the modified NDWI (MNDWI) [13,14]. Feyisa et al. developed a new index (the automated water extraction index, AWEI) using the five spectral bands of the Landsat 5 TM to increase the contrast between water and other dark surfaces and to maximize the separability of water and non-water pixels through band difference solution, addition, and the application of different coefficients [15]. With the development of computer technology, such as supervised classification, decision tree algorithms, the random forest method, and other machine learning algorithms have also been widely used in remote sensing water information extraction. Tulbure et al. used the random forest model to study the scope and dynamic changes of surface water and floods (1986–2011) in the Murray-Darling Basin (MDB). In addition to band reflectance, the explanatory variables of the model also include the brightness temperature, normalized difference vegetation index (NDVI), enhanced vegetation index (EVI), digital elevation model (DEM) derivative variables, and multiple water body indexes [16]. Wang et al. proposed a multi-dimensional and densely connected new deep convolutional neural network (DenseNet) for identifying water bodies from high-resolution multi-spectral images. Its extraction accuracy is higher than those of the water body index method and classical convolutional neural networks [17].

Water quality parameters refer to indicators that reflect the specific substances in or physical properties of the water, such as the dissolved oxygen content (DOC), pH value, turbidity, total nitrogen content, and total phosphorus content. Water quality parameters provide data support for the evaluation of water quality conditions, and shift the evaluation criteria from subjective judgment to quantitative description, which is more objective, scientific, and reliable. For water color remote sensing, any water component parameter with significant spectral characteristics or optical properties, namely, photoactive compounds, can be quantitatively inverted via remote sensing, such as the chlorophyll content, suspended solids, and DOC. The inversion methods mainly include empirical/semi-empirical methods and analytical/semi-analytical methods [18]. The empirical/semi-empirical model mainly realizes the inversion of water quality parameters based on the statistical relationships between the water quality parameters and remote sensing parameters. The semi-empirical method adds the theoretical analysis of the relationships between remote sensing parameters and water quality parameters to the statistical relationship. The core of the analytical/semi-analytical method is the bio-optical model. Thus far, numerous experts and scholars have constructed models using different water quality parameters based on Landsat data. Based on Landsat data and actual measured water quality data, Brivio et al. used regression analysis to establish an inversion model of the chlorophyll-a concentration in Lake Garda [19]. Sharaf El Din et al. extracted the principal components to develop algorithms for turbidity and total suspended solids (TSS) based on Landsat-8 surface reflectance data [20]. According to Landsat operational land imager (OLI) data, Song et al. established a remote sensing model for estimating transparency using the ratio of the red to blue bands and achieved the inversion of the global water transparency [21].

The main purpose of this study is to extract the surface water coverage based on the Google Earth Engine (GEE) cloud platform and long-term Landsat remote sensing images and to retrieve water quality parameters, including chlorophyll-a (Chl-a) and suspended solids (SS) concentration, and transparency, which are characterized by the Secchi disk depth (SDD). The spatial extent of the surface water and the spatial–temporal distribution of the water quality in Zhejiang Province from 1990 to 2022 were mapped, and the characteristics of the spatial–temporal variations in the water area and water quality parameters were analyzed. The results of this study provide data and technical support for the scientific management of regional water resources.

2. Study Area and Datasets

2.1. Study Area

Zhejiang Province, China, is located in the south wing of the Yangtze River Delta along the southeast coast of China (27°02′–31°11′N, 118°01′–123°10′E). It has a land area of 105,500 km^2. The terrain of Zhejiang Province is diverse, mainly including mountains and hills. The terrain elevation decreases from southwest to northeast in a stair-step manner. Zhejiang Province is located in the subtropical monsoon zone and has a mild and humid climate, abundant water resources, and a well-developed water system. The geographical location and topographic map of the study area is shown in Figure 1.

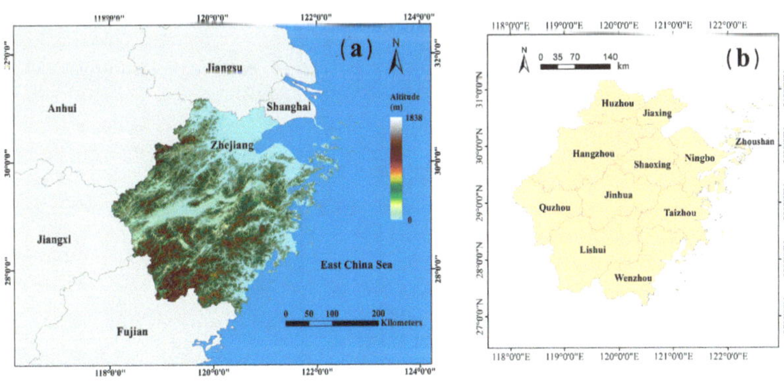

Figure 1. Study area: (**a**) geographical distribution and topographic map of Zhejiang Province, and (**b**) locations of the main districts and cities in Zhejiang Province.

2.2. Datasets

Landsat is an earth observation satellite mission jointly planned by National Aeronautics and Space Administration (NASA) and the United States Geological Survey (USGS). The project provides valuable data for a wide range of applications, including land use planning, resource management, and environmental monitoring. All Landsat images are publicly available through the USGS Earth Explorer. The 30 m resolution remote sensing images (including surface reflectance data and top of atmosphere data) of Zhejiang Province from 1990 to 2022 originated from the remote sensing data of the Landsat 5, 7, and 8 satellites. The European Commission Joint Research Centre (JRC) provides a rich, global long-term surface water data set, which provides data support for the investigation and research of water resources [22]. The spatial and temporal distribution information of global surface water from 1984 to 2021, as well as statistical data about the scope and variation, such as the occurrence (the frequency with which water was present), was obtained from the JRC global surface water dataset. The nearly global elevation data were obtained from the Shuttle Radar Topography Mission (SRTM) global digital elevation model released by NASA. Due to the stability of the elevation data and the span of the study period, in this study, we only used the digital elevation data for 2000 instead of data for all of the research years. Landsat Global Inland Water, a global inland water body dataset based on Landsat satellite data provided by GLCF (Global Land Cover Facility), includes the changes in inland water bodies around the world over the last 35 years. The datasets we used in this study are also accessible on the GEE platform [23]. The first 1 m resolution national-scale land-cover map of China, SinoLC-1, was established using a deep learning-based framework and open-access data including global land-cover (GLC) products, open street map (OSM), and Google Earth imagery [24]. The SinoLC-1 was created by the Low-to-High Network (L2HNet), which can be found at: https://www.sciencedirect.com/science/article/abs/pii/S0924271622002180?via%3Dihub (accessed on 15 March 2023). The datasets used in this study is shown in Table 1.

Table 1. Landsat time series and GEE datasets.

Landsat Time Series		Datasets		
Years	Satellite Sensor	Years	Data	Resolution
1990–1995	Landsat5 TM	1984–2021	JRC Global Surface Water Mapping Layers, v1.4	30 m
2000–2010	Landsat7 ETM+	2000	NASA SRTM DEM	
2015–2022	Landsat8 OLI	2000	GLCF: Landsat Global Inland Water	
		2021	SinoLC-1	1 m

3. Methods

Based on the GEE cloud platform and long-term Landsat remote sensing data, in this study, the Emergency Geomatics Service (EGS) operational surface water mapping algorithm was used to extract the surface water coverage area, and the verified model was utilized to retrieve the Chl-a, SS, and SDD. Water body distribution and water quality maps were created to analyze the temporal and spatial variation characteristics of the water body area and water quality parameters. The technical workflow of this study is shown in Figure 2.

3.1. Data Pre-Processing

In this study, the top of atmosphere (TOA) data and surface reflectance (SR) data from Landsat data, which had been geometrically corrected, were obtained from the (GEE) cloud platform. The existence of clouds has a great impact on remote sensing information extraction and target recognition. To ensure the quality of the remote sensing images, TOA data with less than 20% cloud coverage were selected when filtering data, and the quality assessment band was used for the cloud removal processing to generate long-term Landsat image data from 1990 to 2022 with an interval of 5 years using the median synthesis method. Based on the Zhejiang border shape file, the remote sensing images were cropped and the TOA data and SR data covering the study area were retained.

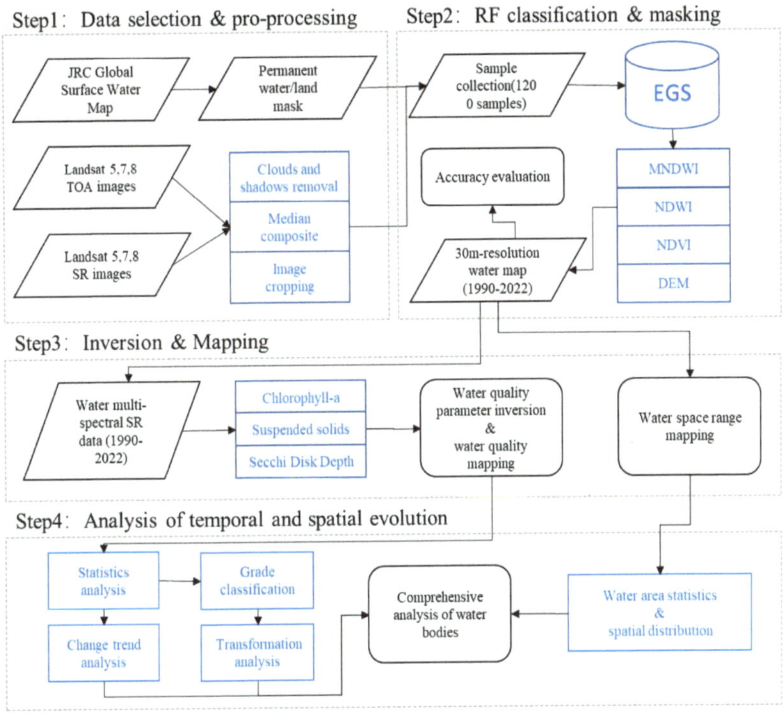

Figure 2. Surface water extent extraction and water quality parameter inversion process.

3.2. Water Spatial Range Mapping Based on EGS Operational Surface Water Mapping Algorithm

The EGS operational surface water mapping algorithm is a direct and fully automatic surface water extent extraction method. This algorithm uses historical inundation frequency products to mark land and water areas and sample the feature information; then, it implements land and water classification using the random forest algorithm to realize the extraction of the surface water [21]. Random forest is an ensemble algorithm, which belongs to the bagging type. It takes the decision tree as the basic unit and forms a random forest by integrating a large number of decision trees. Each time, it samples with replacements from the training set to form a new training set, and uses the new training set to train sub-models. The final results are obtained by voting or averaging, so that the results of the overall model have higher accuracy and generalization performance.

The JRC Global Surface Water Mapping Layers v1.4 dataset obtained from the GEE cloud platform contains the location and temporal distribution information about global surface water from 1984 to 2021 and includes statistical data on the scope of and changes in these water surfaces [25]. For the purpose of ensuring that all of the water sampling points are water samples, the algorithm generates permanent water data (water occurrence frequency of >80%) based on JRC global surface water data for masking. It extracts six bands of information (blue, green, red, near-infrared, and shortwave infrared 1 and 2) from the TOA data using the permanent water mask, combined with the permanent land (water occurrence frequency of 0%) as training samples. The number of samples in each category was greater than 600, with a ratio of 7:3 for the training and validation datasets. In the random forest algorithm, the number of decision trees was set to 100 and the separation of water and land was realized.

When only using band attributes for random forest training to extract the water bodies, the characteristics of the same spectrum of remote sensing images have a certain impact on the extraction results. For example, in this study, some mountain shadows were misjudged as water bodies. In order to further improve the classification accuracy and eliminate the influence of misclassification and singular values, the MNDWI and NDWI were used as masks to enhance the water body information. Slope calculation was conducted based on NASA's SRTM DEM (30 m), and the data with slopes of <20° were selected for masking. According to the high vegetation coverage characteristic in the shadowed part of the mountains, which indicates that the NDVI value was also high, we used the NDVI as a mask to remove the misclassification caused by the shadows of the mountains. The comparison of surface water extent extraction efficiency using RF method is shown in Figure 3.

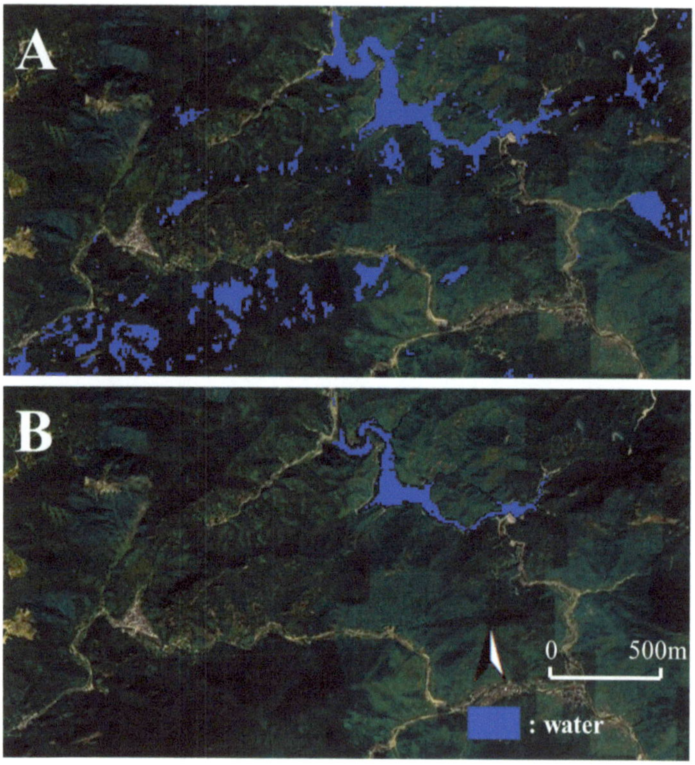

Figure 3. Comparison of surface water extent extraction efficiency: (**A**) RF direct extraction of water bodies, and (**B**) RF combined with NDVI, NDWI, MNDWI, and DEM mask to obtain water bodies.

3.3. Accuracy Evaluation of Surface Water Extent Extraction Results

In this study, the accuracy evaluation of surface water extent extraction results is divided into two parts. The first is the accuracy evaluation of the random forest training model, which is measured by two indexes of kappa and overall accuracy (OA) [26]. The Kappa coefficient is an indicator for consistency tests and can also be used to measure the effect of classification. Because of the classification problem, consistency is based on whether the model prediction results are consistent with the actual classification results. The calculation of the kappa coefficient is based on the confusion matrix, and the value is between −1 and 1.

The kappa coefficient calculation formula based on the confusion matrix is as follows:

$$Kappa = \frac{p_o - p_e}{1 - p_e}$$

where

$$p_0 = \frac{Sum\ of\ diagonal\ elements}{Sum\ of\ the\ whole\ matrix\ elements},$$

$$P_e = \frac{\sum_i Sum\ of\ the\ elements\ in\ row\ i \times Sum\ of the\ elements\ in\ column\ i}{(\sum All\ elements\ of\ the\ matrix)^2}$$

Overall accuracy: the sum of correctly classified pixels divided by the total number of pixels is used to describe the overall classification accuracy.

$$OA = \frac{N_a}{N_t}$$

where N_a is the number of correctly classified pixels, and N_t is the total number of pixels.

The greater the kappa and OA index values of the results obtained by statistical calculation, the better the model classification results and the higher the accuracy. In addition, the outcomes of this study are compared with the other two surface water data products and remote sensing source images in the same region. The accuracy and error of surface water extent extraction can be more intuitively and clearly shown through the display of the effect diagrams in small areas.

3.4. Inversion and Mapping of Water Quality Parameters

Water resources are an indispensable natural resource for humans and are a fundamental condition for human survival and development. However, with the development of society, the pollution of water resources and the environment has become increasingly serious. As a basic component of water pollution control and treatment, water quality detection has become vitally significant. Water conditions can be evaluated using various water quality parameters, including the water transparency, suspended solids concentration, chlorophyll concentration, and dissolved organic matter. Therefore, the inversion of water quality parameters is of great significance for water resource management, water environment protection, and water pollution prevention and control. Regarding water quality parameter inversion methods, the empirical/semi-empirical algorithm is generally more accurate and convenient to use, but it lacks a certain physical basis. The physical basis of the analysis model is powerful; nonetheless, the overly complex physical mechanism leads to the demand for a large amount of field data and cumbersome operation processing in the inversion of the model. In this study, a large number of previous studies were selected based on spatial–temporal conditions and some actual site data. Based on the water mask after the land–water separation, the Landsat SR data were masked and combined with the screened models to retrieve three water quality parameters in Zhejiang Province, including the Chl-a, SS, and SDD. The models of Chl-a, SS and SDD are shown in Table 2.

Table 2. Information about the water quality parameter inversion models.

Parameters	Models	References
Chl-a	$Y_{Chl-a} = 124.3 \times R_{red}{}^2 + 15.28 \times R_{red} + 0.914$	[27]
SDD	$Y_{SDD} = 6715.539 \times (R_{blue} + R_{green})^2 - 2794.743 \times (R_{blue} + R_{green}) + 336.465$	[28]
SS	$\ln_{SS} = 14.656 \times (R_{green} + R_{red})/(R_{green}/R_{red}) + 1.661$	[29]

3.5. Analysis of the Spatial Distribution of the Water Bodies and the Evolution of the Temporal-Spatial Patterns of the Water Quality Parameters

Based on the spatial mapping results of the surface water, statistical analysis of the area in the entire region and in different districts and cities in Zhejiang Province in different years was conducted, and the temporal variations in the spatial distribution of the water bodies were obtained. In order to characterize the degree of the dynamic changes in the surface water area in each region, the coefficient of variation (*CV*) of each district was calculated separately.

$$CV_{(i)} = \frac{\mu_i}{\overline{x_i}}, \quad (1)$$

where μ_i is the standard deviation, and $\overline{x_i}$ is the average value. If the *CV* value is high, there is a large difference between the individual data and the overall average value (or central tendency), and the data fluctuation is more intense; in other words, the water area in the region changes greatly.

Grade division was performed on the inversion results of the water quality parameters according to the different concentration ranges. The proportion of water bodies with different concentrations by year was compiled to obtain the spatial–temporal variation trends of the different concentration water bodies under different water quality parameters. Through calculations and drawing the Sankey diagram for the different concentrations in the same research phases, the transformation of the water bodies with different concentrations was vividly depicted.

4. Results and Analysis

4.1. Spatial Distribution Mapping of Surface Water

The spatial distribution characteristics of the surface water in each region and the extraction efficiency of the water body details are shown in Figure 4. The integrity and continuity of the surface water extent extraction were good, and the main river system characteristics in Zhejiang Province were well preserved. The overall shape and edge characteristics of Qiandao Lake (Figure 4D), the largest lake in Zhejiang Province, were effectively extracted. Lakes such as Hangzhou West Lake (Figure 4B) and Ningbo Dongqian Lake (Figure 4F) were accurately and completely extracted. Moreover, the small island in the lake was accurately distinguished and retained. The lakes, ponds, and crisscrossing rivers in the Hangzhou–Jiaxing–Huzhou Plain in northern Zhejiang Province (Figure 4A) and the scattered small water bodies in the Jinqu Basin (Figure 4E) were successfully distinguished and extracted. It was found that the algorithm utilized in this study has the ability to accurately identify the river network within the city, and it retained the continuity characteristics of the river while extracting the water body (Figure 5).

The Kappa coefficients for each year were greater than 0.90, and the OA values were greater than 0.95 (Table 3), indicating that the water information extraction results are credible. Furthermore, the surface water extent extraction results were compared with the effect pictures of GLCF, SinLC-1 products, and remote sensing source images in the same small area. In contrast with GLCF products, the products in this study have better effects on river continuity, integrity, and water bodies shape feature extraction. Despite SinLC-1 product with 1 m resolution, they also have high consistency.

Table 3. The performance of the surface water extent extraction accuracy for each year.

	1990	1995	2000	2005	2010	2015	2020	2022
OA (%)	0.9531	0.9589	0.9677	0.9795	0.9736	0.9736	0.9648	0.9735
Kappa	0.9061	0.9179	0.9354	0.9589	0.9471	0.9472	0.9295	0.9470

Based on the extraction of the water body information in each year, the surface water area of Zhejiang Province was counted with eleven districts as units (Table 4).

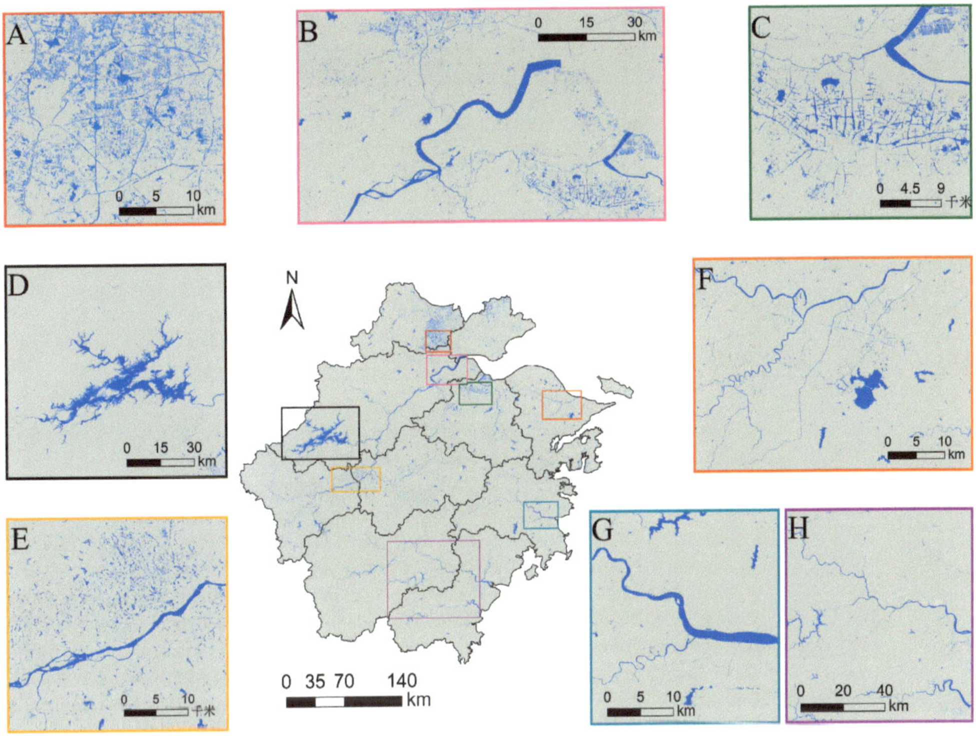

Figure 4. Surface water extent extraction results for 2020: (**A**) Hangzhou–Jiaxing–Huzhou Plain, (**B**) Qiantang River, (**C**) Shaoxing-Cao'e jiang, (**D**) Qiandao Lake, (**E**) Jinqu Basin, (**F**) Ningbo-Dongqian Lake, (**G**) Taizhou-Jiaojiang River, and (**H**) Wenzhou-Lishui.

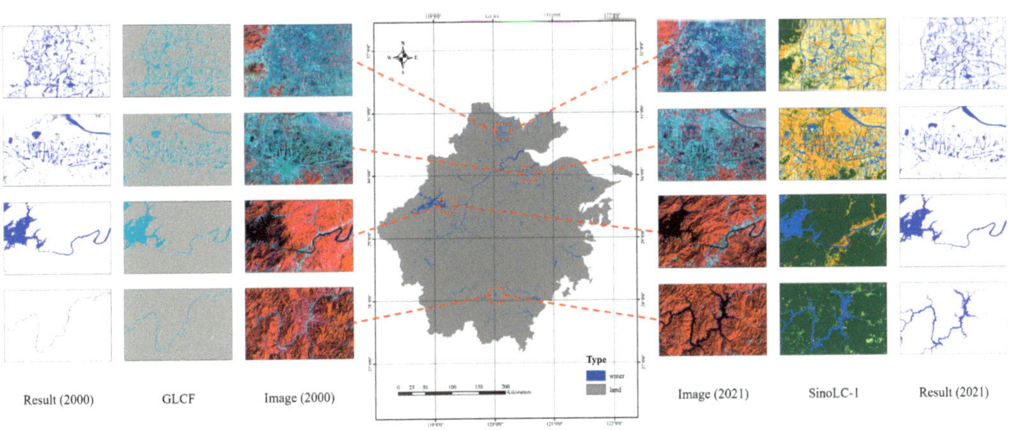

Figure 5. The extraction results of surface water range information are compared with the effect maps of GLCF, SinLC-1, and remote sensing source images in the same small area.

Table 4. Statistics of surface water area in 11 districts of Zhejiang Province.

Area (km^2)	1990	1995	2000	2005	2010	2015	2020	2022	$\overline{x_i}$
Hangzhou	897.96	872.13	828.72	754.06	825.89	873.74	854.20	834.56	842.65
Ningbo	168.02	150.65	162.21	126.08	159.87	172.30	184.95	187.71	163.97
Wenzhou	167.39	150.48	173.69	150.66	159.36	162.66	158.49	160.63	160.42
Jiaxing	167.42	145.16	148.27	113.37	127.42	143.11	150.33	153.58	143.58
Huzhou	268.88	228.04	254.23	203.76	243.47	298.63	317.94	304.89	264.98
Shaoxing	216.26	183.96	253.02	209.88	220.45	252.27	252.35	230.56	227.34
Jinhua	169.91	159.81	178.10	166.71	177.69	219.24	210.76	187.20	183.67
Quzhou	102.84	99.88	111.96	80.73	124.89	154.86	160.41	146.22	122.72
Zhoushan	7.94	5.82	7.09	5.19	4.81	6.78	7.01	7.35	6.49
Taizhou	156.68	139.55	146.02	134.50	149.47	155.42	162.86	163.83	151.04
Lishui	101.13	88.64	106.06	82.56	145.37	171.30	155.65	155.71	125.80
Total area	2424.41	2224.12	2369.3	2027.49	2338.70	2610.30	2614.96	2532.24	2392.69

As can be seen from Table 4 and Figure 6, the average surface water coverage area in Zhejiang Province from 1990 to 2022 was 2392.69 km^2. The total surface water area exhibited a decreasing trend from 1990 to 2005 and reached the minimum value of 2027.49 km^2 in 2005. Subsequently, the surface water coverage area continuously increased, and the maximum value of 2614.96 km^2 was reached in 2020. The average annual change in the area was 193.92 km^2. According to the statistics of the surface water data for eleven districts and cities in Zhejiang Province, the interannual variation in the surface water area was roughly the same as that of the total surface water area in Zhejiang, but there were some differences in individual cities. The average annual surface water area of Hangzhou is 842.65 km^2, far exceeding that of other districts, while Zhoushan only has 6.49 km^2, which is only 0.77% of Hangzhou. The surface water area in other districts is between 100–300 km^2.

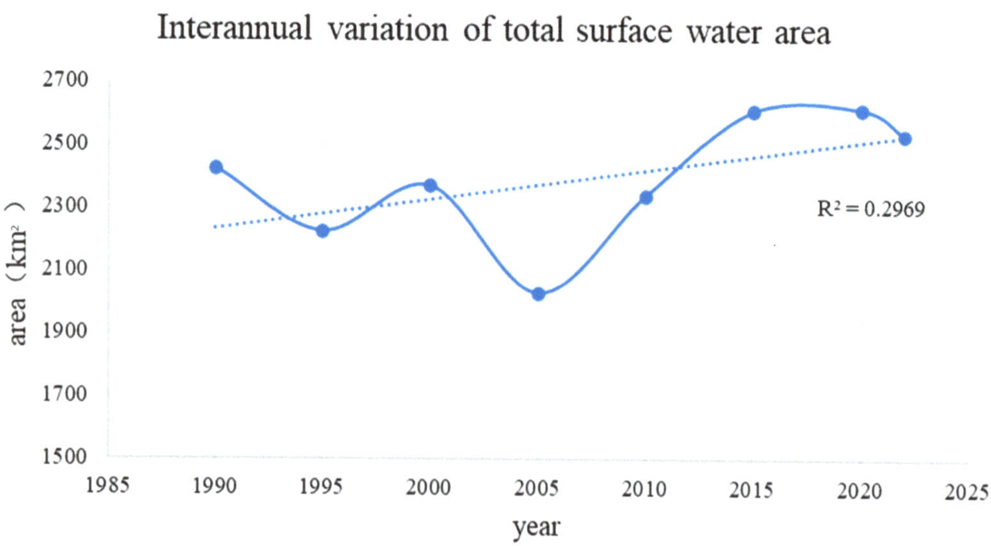

Figure 6. Interannual variation in total surface water area in Zhejiang during 1990–2022.

4.2. Inversion of Water Quality Parameters and Water Quality Map

4.2.1. Chlorophyll-a Concentration

A Landsat SR median synthetic image for the study year was acquired through the GEE cloud platform; then, the Chl-a concentration was inverted using the validated model.

Finally, the spatial and temporal distributions of the Chl-a concentration were mapped using the water mask data (Figure 7).

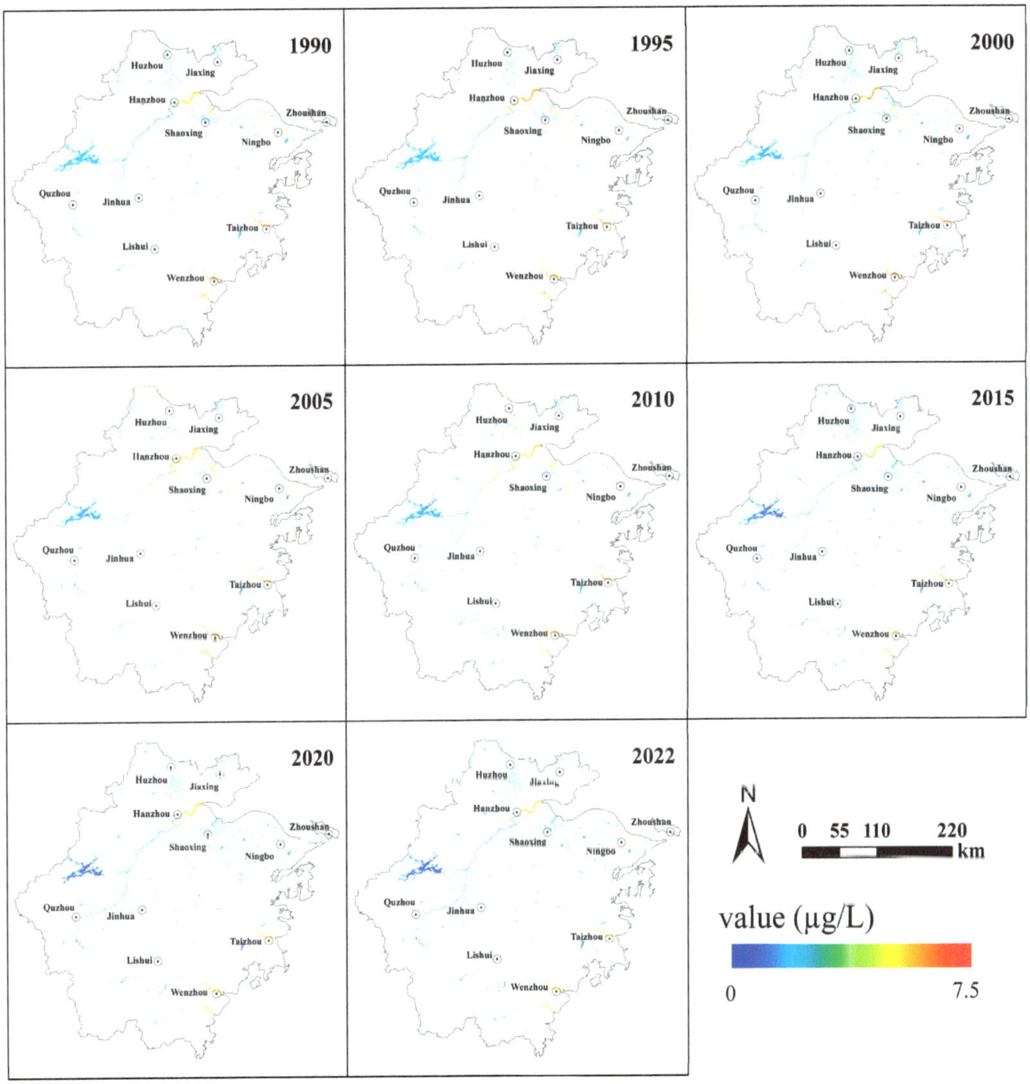

Figure 7. Distribution of Chl-a concentration in Zhejiang Province during 1990–2022.

Based on the data presented in Figure 7, the Chl-a concentrations of the water bodies in Zhejiang Province exhibited a decreasing trend from 1990 to 2022. The distribution pattern indicates that the concentration was lower in the mountainous regions, followed by the hilly and basin areas in the central region, and the highest concentration occurred in the coastal areas. The locations with rapidly increasing Chl-a concentrations strongly coincided with the distributions of several major cities.

To grade the results of the Chl-a concentration inversion, the outliers in the first and last 1% of the dataset were removed, and the water bodies were divided into five different gradations: low (0–1.5 μg/L), medium–low (1.5–3.0 μg/L), medium (3.0–4.5 μg/L), medium–high (4.5–6.0 μg/L), high (6.0–7.5 μg/L). The Sankey diagram was created to visualize the transition between the different grades of the Chl-a concentration (Figure 8).

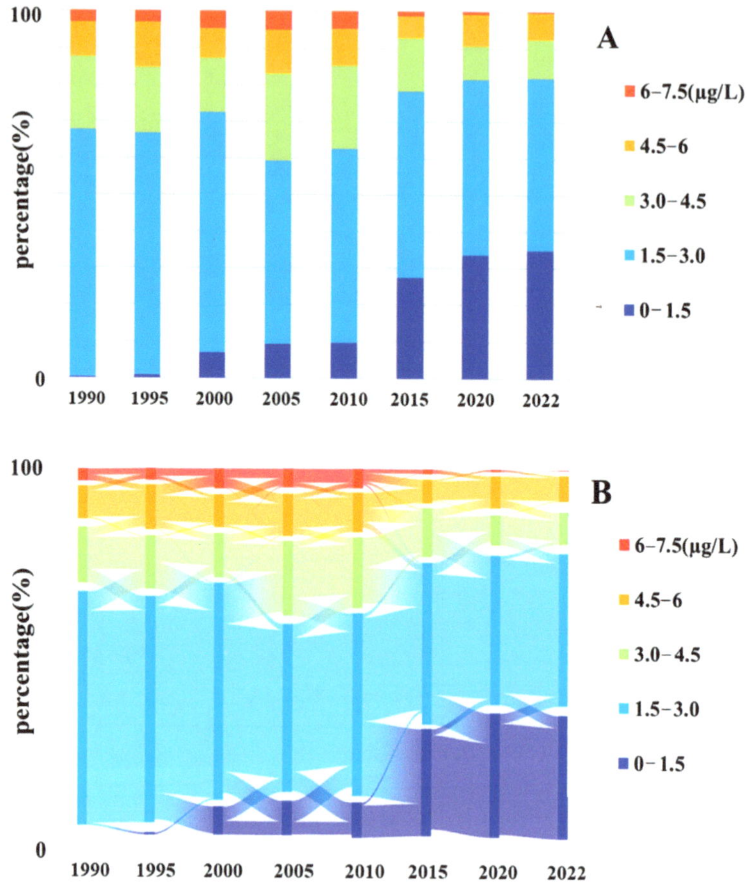

Figure 8. The percentage and the transition between the different grades of the Chl-a concentration, (**A**) hierarchical graph and (**B**) Sankey diagram of Chl-a concentrations of water bodies during 1990–2022.

As can be seen from Figure 8, the percentage of low Chl-a concentrations increased annually, and the proportion of high concentrations increased until 2010 and then rapidly decreased. However, the percentage of low Chl-a concentrations did not exceed 10% in all of the periods studied. The transformation between the low and medium–low Chl-a concentration categories was much stronger than in the high-value region. The concentration variations were mainly between the low, medium–low, and medium categories. A large number of medium–low Chl-a concentration water bodies shifted to medium Chl-a concentration water bodies in the years when the Chl-a concentrations increased significantly, while the medium–low Chl-a concentration water bodies shifted to low Chl-a concentration water bodies in years when they decreased dramatically. A large proportion of the medium–high concentrations changed, but throughout the time series, they remained in dynamic equilibrium.

4.2.2. Secchi Disk Depth

A Landsat SR median synthetic image for the study year was acquired through the GEE cloud platform; then, the SDD was inverted using the validated model. Finally, the spatial and temporal distributions of the SDD were mapped using the water mask data (Figure 9).

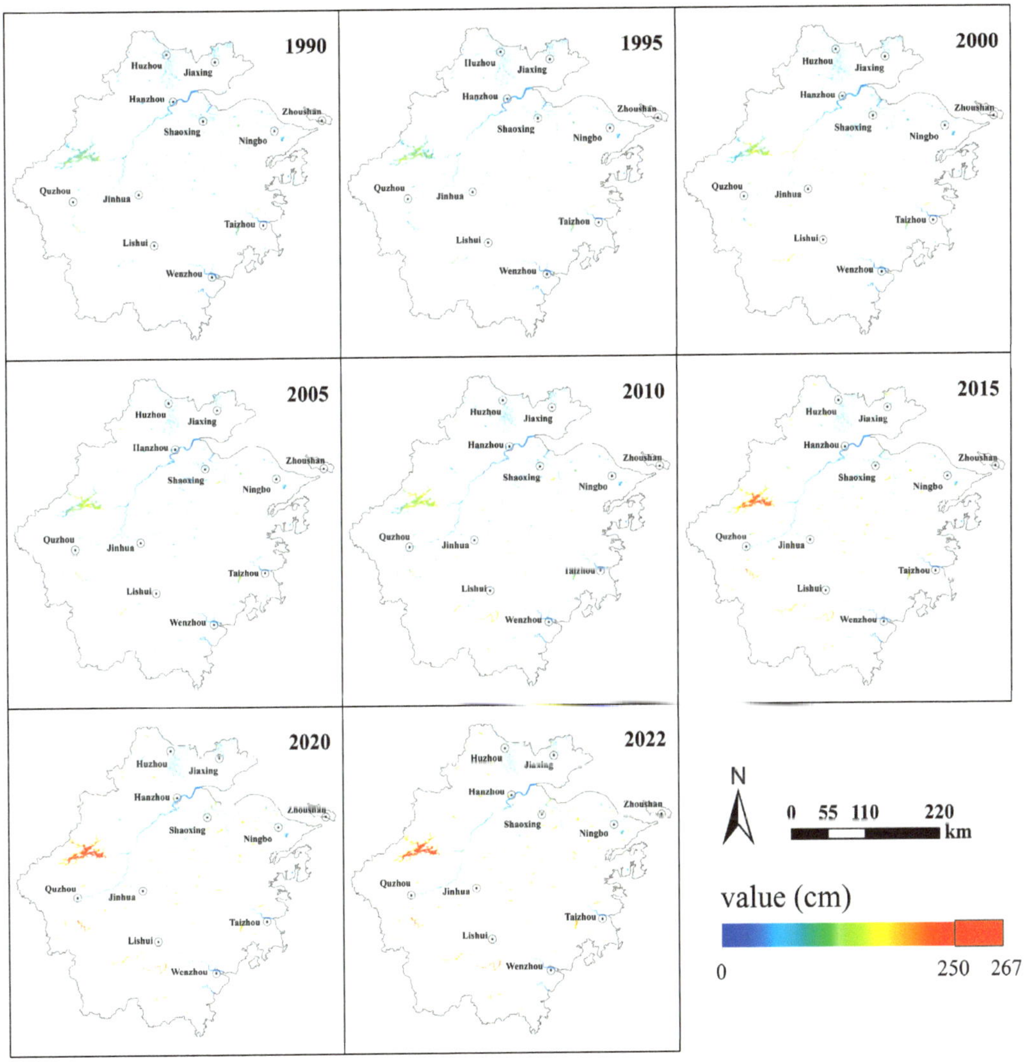

Figure 9. Distribution of SDD in Zhejiang Province during 1990–2022.

As can be seen from Figure 9, the SDD of the water in Zhejiang was highly dependent on the water depth. In most areas, except for the major rivers and reservoirs, the SDD was less than 70 cm. The downstream SDD of the large rivers was less than 50 cm, while the SDD in the upper reaches was around 100 cm due to the water quality and suspended sediment. On the province-wide scale, the areas with SDDs of 200 cm or greater were the large water bodies located in the mountainous terrain, such as the Xin'anjiang Reservoir in Hangzhou.

To eliminate abnormal SDD values, in this study, the first and last 1% of the data were removed, and then, the data were processed and rounded down to the nearest integer. Then, the water bodies were divided into five categories with equal intervals: low (0–50 cm), medium–low (50–100 cm), medium (100–150 cm), medium–high (150–200 cm), and high (200–267 cm). The data regarding transfers between levels were collected and plotted on a Sankey diagram (Figure 10). The percentage of low SDD (0–50 cm) water bodies was relatively stable; the high SDD water bodies were more obviously variable. Before 1995, there were virtually no areas of water with SDD values of 150 cm or above. During 2000–2010, only a few areas had medium–high SDD (150–200 cm). After 2010, SDD values of greater than 200 cm gradually began to appear.

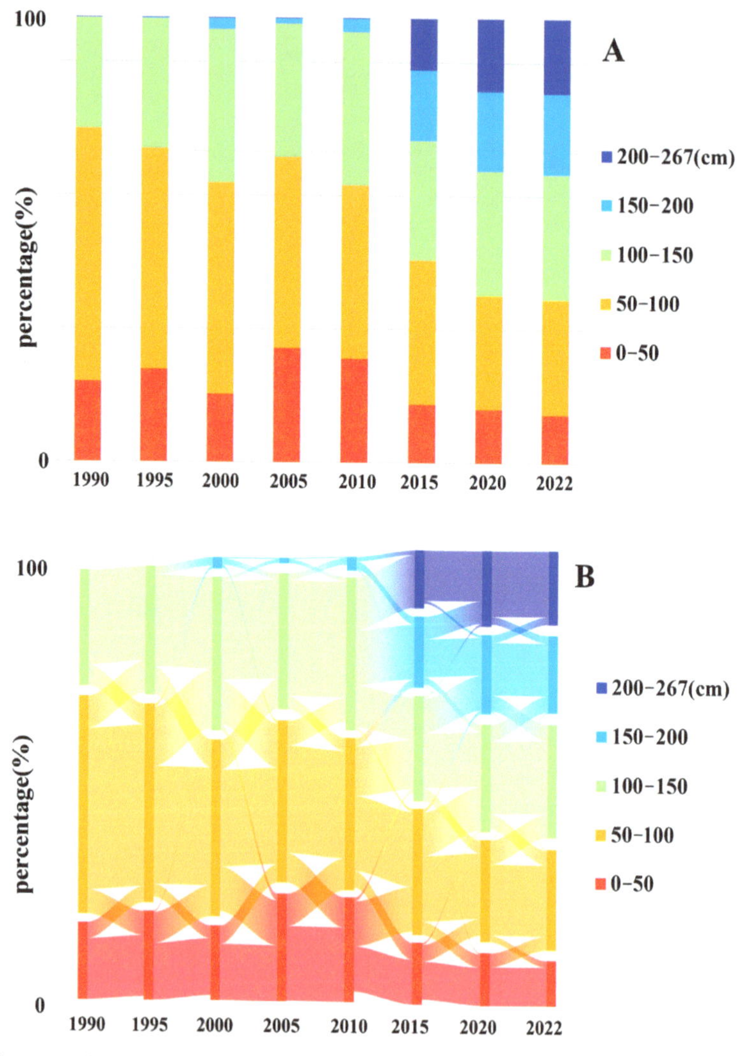

Figure 10. The percentage and the transition between the different grades of the Secchi disk depth, (**A**) hierarchical graph and (**B**) Sankey diagram of SDD values of water bodies during 1990–2022.

As can be seen from the Sankey diagram (Figure 7), numerous water bodies with medium–low SDD values transitioned to the medium–high SDD grade after 2010. Between

2010 and 2015, almost 60% of the medium SDD water bodies transitioned to the high and medium–high grades. Additionally, about 50% of the low and medium–low SDD water bodies transitioned to the medium and medium–high grades during this period. Among the different classes of water bodies, those with the largest transformed proportion were the medium–low and medium SDD water bodies, with nearly 50% of these water bodies changing each year, indicating that the stability of the SDD of this class of water bodies was worse.

4.2.3. Suspended Solids Concentration

Using the validated model and a Landsat SR median composite image of the study year, which were obtained based on the GEE cloud platform, the SS concentrations of the water bodies were inverted and a spatial–temporal distribution graph of the concentration was drawn using the water mask data (Figure 11).

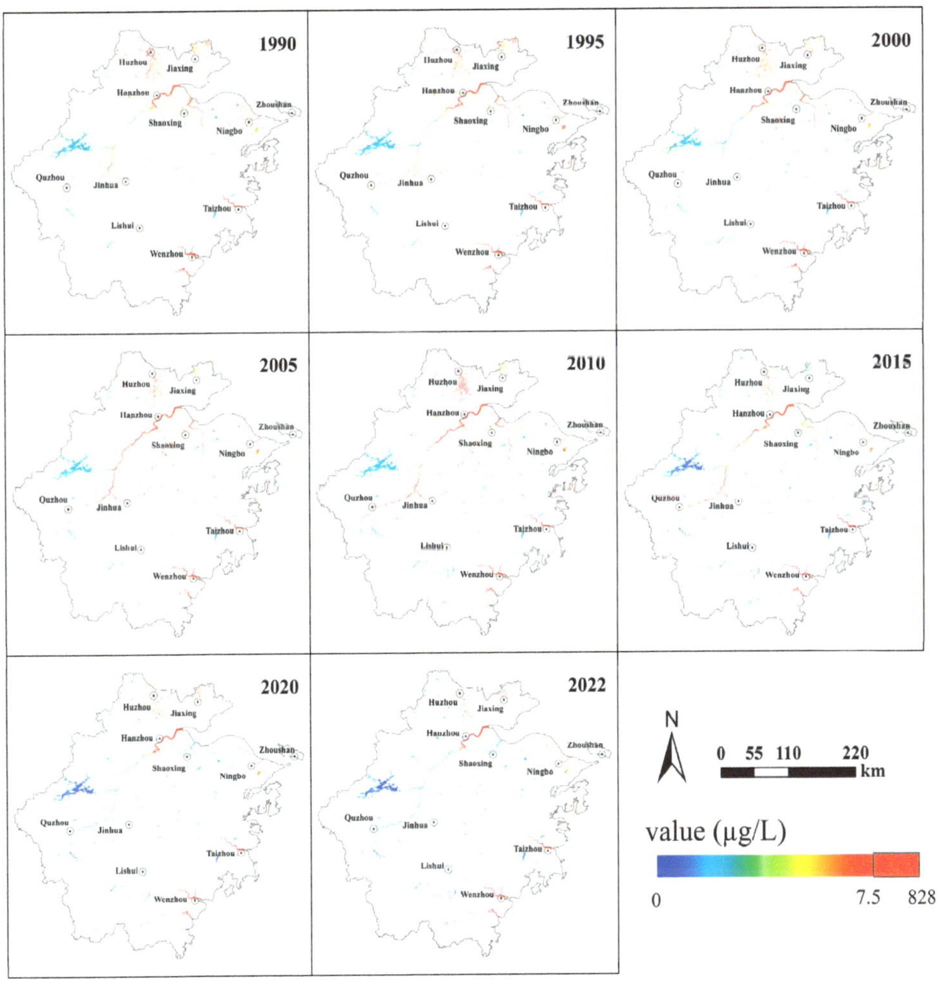

Figure 11. Distribution of SS concentrations of water bodies in Zhejiang Province during 1990–2022.

As can be seen from Figure 11, the SS concentrations of the water bodies in Zhejiang Province varied considerably between the two poles, and the regional differences were obvious. The SS concentrations in the upstream areas of the large rivers and the various types of reservoirs were less than 20 mg/L. The SS concentrations in the middle and upper reaches of the rivers remained within the range of 40–50 mg/L. The SS concentrations in the lower reaches of the rivers were greater than 200 mg/L, and a small portion of the water bodies within the high-value area had maximum concentrations of greater than 600 mg/L. In the remaining areas, the SS concentrations were mostly in the range of 20–50 mg/L.

Due to the great regional differences in the SS concentration, 40 mg/L was the middle of the largest span of the data, while more than 83% of the data were within 0–60 mg/L (the smallest span of data). Therefore, to distinguish the different concentrations of the water bodies better, in this study, we selected the data with the largest concentration span in the study period, removed the outlier values in the first and last 1%, and classified them according to an interval of 50 mg/L: low (0–10 mg/L), medium–low (10–20 mg/L), medium (20–30 mg/L), medium–high (30–40 mg/L), and high (40–828 mg/L). Then, we investigated the transfers between the different types (Figure 12).

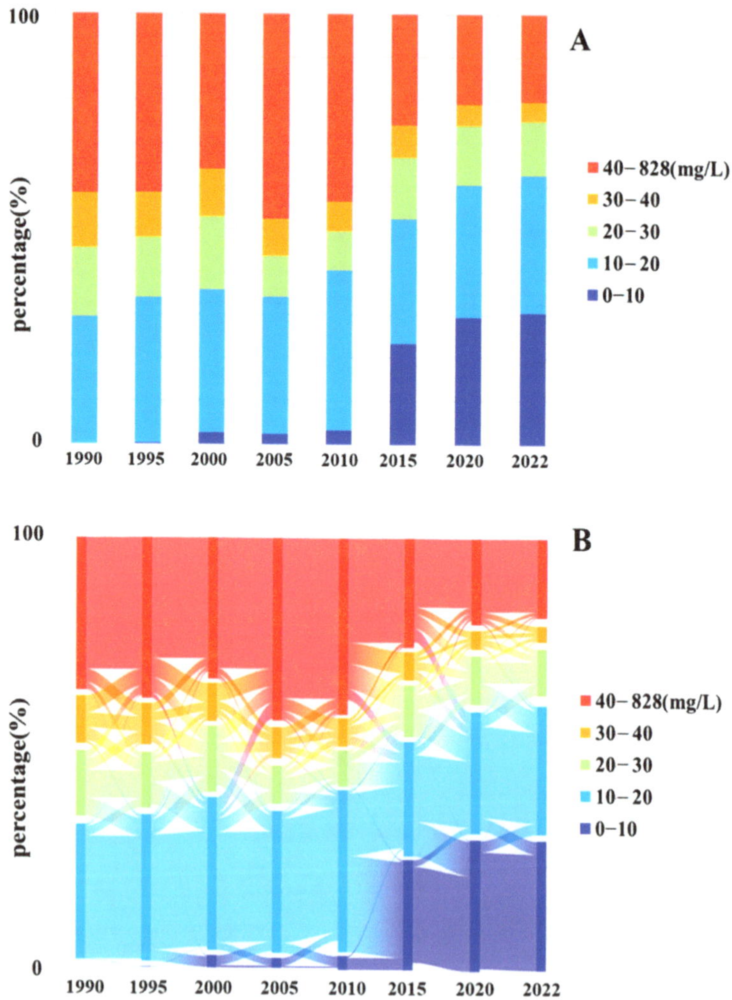

Figure 12. The percentage and the transition between the different grades of the SSC, (**A**) hierarchical graph and (**B**) Sankey diagram of SS concentrations of water bodies during 1990–2022.

As can be seen from Figure 12, the SS concentrations were concentrated below 20 mg/L and between 40 and 828 mg/L, and the sum of the two concentrations in the middle category accounted for no more than 30% of the total water bodies. Low SS concentration water bodies were virtually absent in 1990 and 1995, and they only had a small distribution from 2000 to 2005. They began to rapidly increase after 2010, accounting for more than 20% of the total water bodies. The percentage of the water bodies with SS concentrations of less than 20 mg/L exhibited a fluctuating yearly increase between 1990 and 2022, with only a slight decrease between 2005 and 2010.

The most significant changes in the SS concentration occurred between 2010 and 2015, during which over 50% of the medium–low SS water bodies transitioned to the low SS category, while nearly 40% of the high SS water bodies transitioned to the medium–low category. The combined proportion of the medium SS and medium–high SS water bodies was less variable and remained in a state of dynamic equilibrium. The reduction in the high SS water bodies was approximately the same as the increase in the combined area of the low and medium–low SS water bodies.

5. Discussion

5.1. Reliability of Research Results

Using permanent water points as a training set, the water areas were identified based on the random forest method and were post-processed using NASA's SRTM DEM data to obtain water–land classification images. The accuracy validation of the classification results revealed that the kappa coefficients were greater than 0.90, and the overall accuracy (OA) was greater than 0.95% for each study year. Compared with previous surface water extent extraction studies (Table 5), which employed a variety of different methods such as the NDWI, MNDWI, automatic water body extraction index (AWEIsh), and water identification (WI2015) [17,30–33], this study applied the EGS operational surface water mapping algorithm, which delivers a better performance in terms of overall accuracy and the kappa coefficient than the four index methods mentioned above [34,35].

Table 5. Comparison of the accuracies of the methods using different satellite data.

Satellite	Research Method	OA (%)	Kappa Coefficient
Landsat 8	RF (this study)	97.36	0.9472
	NDWI	90.81	0.8133
	MNDWI	93.24	0.8610
	$AWEI_{sh}$	94.52	0.8862
	WI_{2015}	94.48	0.8855

In this study of water quality parameter inversion, the inland water body band model utilized in previous studies, in which the Chl-a, SDD, and SS were modeled via regression using Landsat's Rred, (Rblue + Rgreen), and (Rblue + Rred)/(Rblue/Rred) as the factors, yielded R2 values of 0.846, 0.824, and 0.808, respectively. The inversion results can better reflect the spatial and temporal variations in the water quality parameters in Zhejiang Province and provide a better distinction between regional water bodies.

5.2. Distribution of Surface Water and Water Quality Condition

Hangzhou had a wide area and the largest river and lake in Zhejiang Province, and it possessed the largest surface water area, with an average annual coverage area of 842.65 km^2. Huzhou and Shaoxing belong to the typical Jiangnan Water Town with a dense network of crisscrossed river courses, and the coverage area was more than 200 km^2, second only to that in Hangzhou. Quzhou and Lishui in the western part of Zhejiang are located in mountainous and hilly areas. The water system in these areas was underdeveloped, and the area was smaller, only about 120 km^2. Zhoushan is a coastal archipelago with a narrow geographical area, and the surface water coverage area was less than 10 km^2. Table 6 reveals that the CV values of the cities were significantly different. The CV values of Hangzhou and Wenzhou were low (0.05), and the CV values of Lishui and Quzhou were higher (0.28 and 0.23, respectively).

Table 6. Statistical table of total area and water area change index of each region.

	Area of Districts	Standard Deviation	Area of Water	CV
Hangzhou	16,597	43.78	842.66	0.05
Ningbo	9816	19.72	163.97	0.12
Wenzhou	11,773	7.83	160.42	0.05
Jiaxing	4474	16.53	143.58	0.12
Huzhou	5596	40.06	264.98	0.15
Shaoxing	8778	24.71	227.35	0.11
Jinhua	11,830	21.13	183.68	0.12
Quzhou	8716	28.80	122.72	0.23
Zhoushan	1472	1.10	6.50	0.17
Taizhou	9185	10.61	151.04	0.07
Lishui	17,738	34.82	125.80	0.28

CV is a value with relative meaning. The standard deviation data of Hangzhou is very large, but it has the largest surface water area in Zhejiang Province (large cardinal), which makes its CV index very low. However, the average annual surface water coverage area of Wenzhou is only 160.42 km^2, and it is mainly concentrated between the two major rivers. Because of the exchange of water bodies between rivers and coastal waters, the water body area presents a relatively stable state. Lishui and Quzhou are located in mountainous and hilly regions. The water system is narrow and long, covering a small area, and there are many small ponds, which are easily affected by climatic factors and cause relatively large area changes. Two large reservoirs were built in Lishui and Quzhou in 2009 and 2001, respectively, with areas of the second and tenth largest in the province (most of the top ten were built before 1990). Therefore, the CV index of these two regions is higher.

The SS concentration was significantly higher in the downstream part of the river in the estuary region than in the other inland water bodies due to the convergence of several rivers and the accumulation of flowing water. Similarly, The Chla was higher and the SDD was lower. Before 2015, the magnitude of the changes in the various types of water in Zhejiang Province was relatively small, but there was a tendency to increase the number of water bodies with better quality conditions. Under the appeal of a policy, the water quality in Zhejiang Province has improved significantly since 2015. In other words, the proportion of the water bodies with high SS and Chla concentrations decreased yearly, the proportion with high Chla concentrations gradually converged to zero, and the proportion with high SSD values gradually increased and reached more than 10%.

5.3. Limitations of the Study

The remote sensing data acquired in this study were synthesized using the median of the Landsat annual satellite data. The data merging methods utilized can affect the original data, thus creating different errors in the study results [36–38]. For example, when using the maximum value composite, even if the merged image has been pre-processed by removing clouds, there will still be a large amount of noise in the composite image due to abnormal cloud boundary pixels. When using the minimum value composite, the water region extracted using the random forest method will yield misclassification of mountain shadows, and the most difficult issue to cope with during the post-processing is the misjudgment of the boundaries of the mountain reservoirs and holes in the centers of the water bodies. Therefore, the median value composite of the image was chosen for use in this study, but the median value composite also has the problem of inadequate synthesis of time series image pixels.

The limitation of the ground resolution of the image, the influence of river bridges, and the lack of comprehensive data collected at the sample points resulted in some of the narrow water bodies not being extracted completely, and intermittent patches of water were often formed [39–42]. In addition, the large scale of the study area, which included a wide variety of water body morphologies in various areas, made it impossible to perform a uniform mathematical morphology operation to fill the entire area.

6. Conclusions

This study explored the extraction of surface water extent based on the EGS operational surface water mapping algorithm and inversion of water quality variables based on validated models using Landsat time series, and created spatial distribution maps of surface water extent and water quality variables in Zhejiang Province at 5 year intervals from 1990 to 2020. The main conclusions of this study are as follows:

(1) The water body extraction accuracy of the EGS operational surface water mapping algorithm, combined with Landsat data, is superior, with kappa coefficients of >0.90 in each of the studied time phases, and the lowest overall accuracies exceed 95.31%. Fragmented mountain water bodies, intricate urban water networks, and meandering rivers can all be extracted well, and the comprehensive mask consisting of multiple indices greatly reduces the potential for the misclassification of land areas.

(2) The inter-annual variations in the surface water bodies in Zhejiang Province exhibited an initially decreasing and then increasing trend, with a minimum value of 2027.49 km^2 in 2005 and a maximum value of 2610.30 km^2 in 2015. Based on the statistics on the coefficient of variation (CV) of the water body area in each city in Zhejiang Province, Quzhou and Lishui experienced the most dramatic changes during 1990–2022, and Hangzhou and Wenzhou exhibited the smallest variations in water body area.

(3) According to the multi-spectral data for the water bodies extracted from the results, combined with the water quality band model utilized in previous studies, the water quality parameters in Zhejiang Province were inverted and mapped. In the watershed, the changes in the water quality conditions in Zhejiang Province during 1990–2022 exhibited increases after 2015. The water quality conditions of all of the types of water bodies changed slightly before 2015, while they improved significantly after 2015, and the proportion of the water bodies with better water quality conditions increased.

Author Contributions: Conceptualization, H.J. and C.C.; methodology, C.C.; software, H.J. and S.F.; writing—original draft preparation, H.J. and S.F.; writing—review and editing, C.C.; project administration, C.C.; funding acquisition, C.C. All authors have read and agreed to the published version of the manuscript.

Funding: This work was supported by the National Natural Science Foundation of China (Grant No. 42171311), the College Students' Science and Technology Innovation Project of Zhejiang Ocean University (No. 2022-A-005).

Data Availability Statement: The data used in this paper are available from Google Earth Engine upon reasonable request, and the model, products, or codes generated in this paper are available from the corresponding author upon reasonable request.

Acknowledgments: The authors would like to thank the editors and the anonymous reviewers for their outstanding comments and suggestions, which greatly helped to improve the technical quality and presentation of this manuscript.

Conflicts of Interest: The authors declare that they have no known competing interests that could have appeared to influence the work reported in this paper.

References

1. Jia, K.; Jiang, W.; Li, J.; Tang, Z. Spectral Matching Based on Discrete Particle Swarm Optimization: A New Method for Terrestrial Water Body Extraction Using Multi-Temporal Landsat 8 Images. *Remote Sens. Environ.* **2018**, *209*, 1–18. [CrossRef]
2. Sun, G.; Pan, Z.; Zhang, A.; Jia, X.; Ren, J.; Fu, H.; Yan, K. Large kernel spectral and spatial attention networks for hyperspectral image classification. *IEEE Trans. Geosci. Remote Sens.* **2023**, *61*, 5519915. [CrossRef]
3. Fu, H.; Sun, G.; Zhang, L.; Zhang, A.; Ren, J.; Jia, X.; Li, F. Three-dimensional singular spectrum analysis for precise land cover classification from UAV-borne hyperspectral benchmark datasets. *ISPRS J. Photogramm. Remote Sens.* **2023**, *203*, 115–134. [CrossRef]

4. Ten Brinke, W.B.M.; Knoop, J.; Muilwijk, H.; Ligtvoet, W. Social Disruption by Flooding, a European Perspective. *Int. J. Disaster Risk Reduct.* **2017**, *21*, 312–322. [CrossRef]
5. Zhang, Q.; Gu, X.; Shi, P.; Singh, V.P. Impact of Tropical Cyclones on Flood Risk in Southeastern China: Spatial Patterns, Causes and Implications. *Glob. Planet. Chang.* **2017**, *150*, 81–93. [CrossRef]
6. Marton, D.; Kapelan, Z. Risk and Reliability Analysis of Open Reservoir Water Shortages Using Optimization. *Procedia Eng.* **2014**, *89*, 1478–1485. [CrossRef]
7. Jia, M.; Wang, Z.; Mao, D.; Ren, C.; Song, K.; Zhao, C.; Wang, C.; Xiao, X.; Wang, Y. Mapping global distribution of mangrove forests at 10-m resolution. *Sci. Bull.* **2023**, *68*, 1306–1316. [CrossRef]
8. Chen, C.; Wang, L.; Yang, G.; Sun, W.; Song, Y. Mapping of ecological environment based on Google earth engine cloud computing platform and landsat long-term data: A case study of the zhoushan archipelago. *Remote Sens.* **2023**, *15*, 4072. [CrossRef]
9. Mutanga, O.; Kumar, L. Google Earth Engine Applications. *Remote Sens.* **2019**, *11*, 591. [CrossRef]
10. Chen, C.; Liang, J.; Yang, G.; Sun, W. Spatio-temporal distribution of harmful algal blooms and their correlations with marine hydrological elements in offshore areas, China. *Ocean. Coast. Manag.* **2023**, *238*, 106554. [CrossRef]
11. Ding, L.; Wu, H.; Wang, C.J.; Qin, Z.; Zhang, Q. Study of the water body extracting from MODIS images based on spectrum-photometric method. *Geomat. Spat. Inf. Technol.* **2006**, *29*, 25–27.
12. Chen, J.; Chen, S.; Fu, R.; Li, D.; Jiang, H.; Wang, C.; Peng, Y.; Jia, K.; Hicks, B.J. Remote Sensing Big Data for Water Environment Monitoring: Current Status, Challenges, and Future Prospects. *Earth's Future* **2022**, *10*, e2021EF002289. [CrossRef]
13. McFeeters, S.K. The Use of the Normalized Difference Water Index (NDWI) in the Delineation of Open Water Features. *Int. J. Remote Sens.* **1996**, *17*, 1425–1432. [CrossRef]
14. Xu, H. Modification of Normalised Difference Water Index (NDWI) to Enhance Open Water Features in Remotely Sensed Imagery. *Int. J. Remote Sens.* **2006**, *27*, 3025–3033. [CrossRef]
15. Feyisa, G.L.; Meilby, H.; Fensholt, R.; Proud, S.R. Automated Water Extraction Index: A New Technique for Surface Water Mapping Using Landsat Imagery. *Remote Sens. Environ.* **2014**, *140*, 23–35. [CrossRef]
16. Tulbure, M.G.; Broich, M.; Stehman, S.V.; Kommareddy, A. Surface Water Extent Dynamics from Three Decades of Seasonally Continuous Landsat Time Series at Subcontinental Scale in a Semi-Arid Region. *Remote Sens. Environ.* **2016**, *178*, 142–157. [CrossRef]
17. Wang, G.; Wu, M.; Wei, X.; Song, H. Water Identification from High-Resolution Remote Sensing Images Based on Multidimensional Densely Connected Convolutional Neural Networks. *Remote Sens.* **2020**, *12*, 795. [CrossRef]
18. Li, W.; Zhang, W.; Li, Z.; Wang, Y.; Chen, H.; Gao, H.; Zhou, Z.; Hao, J.; Li, C.; Wu, X. A New Method for Surface Water Extraction Using Multi-Temporal Landsat 8 Images Based on Maximum Entropy Model. *Eur. J. Remote Sens.* **2022**, *55*, 303–312. [CrossRef]
19. Brivio, P.A.; Giardino, C.; Zilioli, E. Determination of Chlorophyll Concentration Changes in Lake Garda Using an Image-Based Radiative Transfer Code for Landsat TM Images. *Int. J. Remote Sens.* **2001**, *22*, 487–502. [CrossRef]
20. Sharaf El Din, E. Enhancing the Accuracy of Retrieving Quantities of Turbidity and Total Suspended Solids Using Landsat-8-Based-Principal Component Analysis Technique. *J. Spat. Sci.* **2021**, *66*, 493–512. [CrossRef]
21. Song, K.; Liu, G.; Wang, Q.; Wen, Z.; Lyu, L.; Du, Y.; Sha, L.; Fang, C. Quantification of Lake Clarity in China Using Landsat OLI Imagery Data. *Remote Sens. Environ.* **2020**, *243*, 111800. [CrossRef]
22. Pekel, J.F.; Cottam, A.; Gorelick, N.; Belward, A.S. High-resolution mapping of global surface water and its long-term changes. *Nature* **2016**, *540*, 418–422. [CrossRef] [PubMed]
23. Peng, J.; Huang, Y.; Sun, W.; Chen, N.; Ning, Y.; Du, Q. Domainadaptation inremote sensing imageclassification: A survey. *IEEE J. Sel. Top. Appl. Earth Obs. Remote Sens.* **2022**, *15*, 9842–9859. [CrossRef]
24. Feng, M.; Sexton, J.O.; Channan, S.; Townshend, J.R. A Global, High-Resolution (30-m) Inland Water Body Dataset for 2000: First Results of a Topographic–Spectral Classification Algorithm. *Int. J. Digit. Earth* **2016**, *9*, 113–133. [CrossRef]
25. Olthof, I.; Rainville, T. Dynamic Surface Water Maps of Canada from 1984 to 2019 Landsat Satellite Imagery. *Remote Sens. Environ.* **2022**, *279*, 113121. [CrossRef]
26. Chen, C.; Chen, Y.; Jin, H.; Chen, L.; Liu, Z.; Sun, H.; Hong, J.; Wang, H.; Fang, S.; Zhang, X. 3D model construction and ecological environment investigation on a regional scale using UAV remote sensing. *Intell. Autom. Soft Comput.* **2023**, *37*, 1655–1672. [CrossRef]
27. Huang, L.; Yu, F.; Zhang, D.; Lin, L. Quantitative Retrieval of Chlorophyll a Concentration Based on Landsat-8 OLI in the Lakes. *J. Jiangxi Sci.* **2016**, *34*, 441–456.
28. Liu, Y.; Zhang, B.; Yao, X.; Zhang, H.; Feng, R. Remote sensing inversion of water transparency in Dongping Lake. *J. Surv. Mapp. Sci.* **2018**, *43*, 72–78.
29. Jie, G.; Yuchun, W.; Jiazhu, H.; Yunmei, L.; Jianguang, W.; Jiangping, G. Suspended Sediment Estimating Models in Lake Taihu Using Remote Sensing Data. *J. Lake Sci.* **2007**, *19*, 241–249. [CrossRef]
30. Yang, H.; Wang, Z.; Zhao, H.; Yu, G. Water Body Extraction Methods Study Based on RS and GIS. *Procedia Environ. Sci.* **2011**, *10*, 2619–2624.
31. Jiang, W.; He, G.; Long, T.; Ni, Y.; Liu, H.; Peng, Y.; Lv, K.; Wang, G. Multilayer Perceptron Neural Network for Surface Water Extraction in Landsat 8 OLI Satellite Images. *Remote Sens.* **2018**, *10*, 755. [CrossRef]
32. Olthof, I. Mapping Seasonal Inundation Frequency (1985–2016) along the St-John River, New Brunswick, Canada Using the Landsat Archive. *Remote Sens.* **2017**, *9*, 143. [CrossRef]
33. Rishikeshan, C.A.; Ramesh, H. An Automated Mathematical Morphology Driven Algorithm for Water Body Extraction from Remotely Sensed Images. *ISPRS J. Photogramm. Remote Sens.* **2018**, *146*, 11–21. [CrossRef]

34. Sheng, Y.; Song, C.; Wang, J.; Lyons, E.A.; Knox, B.R.; Cox, J.S.; Gao, F. Representative Lake Water Extent Mapping at Continental Scales Using Multi-Temporal Landsat-8 Imagery. *Remote Sens. Environ.* **2016**, *185*, 129–141. [CrossRef]
35. Taloor, A.K.; Thakur, P.K.; Jakariya, M. Remote Sensing and GIS Applications in Water Science. *Groundw. Sustain. Dev.* **2022**, *19*, 100817. [CrossRef]
36. Tang, H.; Lu, S.; Ali Baig, M.H.; Li, M.; Fang, C.; Wang, Y. Large-Scale Surface Water Mapping Based on Landsat and Sentinel-1 Images. *Water* **2022**, *14*, 1454. [CrossRef]
37. Wu, G.; De Leeuw, J.; Skidmore, A.K.; Prins, H.H.T.; Liu, Y. Comparison of MODIS and Landsat TM5 Images for Mapping Tempo–Spatial Dynamics of Secchi Disk Depths in Poyang Lake National Nature Reserve, China. *Int. J. Remote Sens.* **2008**, *29*, 2183–2198. [CrossRef]
38. Xing, W.; Guo, B.; Sheng, Y.; Yang, X.; Ji, M.; Xu, Y. Tracing Surface Water Change from 1990 to 2020 in China's Shandong Province Using Landsat Series Images. *Ecol. Indic.* **2022**, *140*, 108993. [CrossRef]
39. Yang, X.; Qiu, S.; Zhu, Z.; Rittenhouse, C.; Riordan, D.; Cullerton, M. Mapping understory plant communities in deciduous forests from Sentinel-2 time series. *Remote Sens. Environ.* **2023**, *293*, 113601. [CrossRef]
40. Yue, L.; Li, B.; Zhu, S.; Yuan, Q.; Shen, H. A Fully Automatic and High-Accuracy Surface Water Mapping Framework on Google Earth Engine Using Landsat Time-Series. *Int. J. Digit. Earth* **2023**, *16*, 210–233. [CrossRef]
41. Huang, Y.; Peng, J.; Sun, W.; Chen, N.; Du, Q.; Ning, Y.; Su, H. Two-branch attention adversarial domain adaptation network for hyperspectral image classification. *IEEE Trans. Geosci. Remote Sens.* **2022**, *60*, 5540813. [CrossRef]
42. Li, Z.; He, W.; Cheng, M.; Hu, J.; Yang, G.; Zhang, H. SinoLC-1: The First 1-Meter Resolution National-Scale Land-Cover Map of China Created with the Deep Learning Framework and Open-Access Data. *Earth Syst. Sci. Data Discuss.* **2023**, 1–38. [CrossRef]

remote sensing

MDPI

Article

A Quick-Look Software for In Situ Magnetic Field Modeling from Onboard Unmanned Aircraft Vehicles (UAVs) Measurements

Erwan Thebault *,†,‡ and Lydie-Sarah Gailler †,‡

Laboratoire Magmas et Volcans, OPGC, IRD, CNRS, Université Clermont Auvergne, 63000 Clermont-Ferrand, France; lydie.gailler@uca.fr

* Correspondence: erwan.thebault@uca.fr

† Current address: Campus Universitaire des Cézeaux, 6 Av. Blaise Pascal, 63178 Aubière, France.

‡ These authors contributed equally to this work.

Abstract: UAVs represent a tremendous opportunity to perform geophysical and repeated experiments, particularly in volcanic contexts. Their ability to be deployed rapidly and fly at various altitudes and the fact that they are easy to operate despite complex field conditions make them attractive for magnetic surveys. Detailed maps of the magnetic field in turn bring key constraints on the rocks' composition, thermal anomalies, intrusive systems, and crustal contrast evolution. Yet, raw magnetic field measurements require careful processing to minimize directional, positional, and crossover errors. Moreover, stitching together adjacent or overlapping surveys acquired at different times and altitudes is not a trivial task. Therefore, it is challenging in remote areas to directly evaluate the consistency of a survey and to ascertain the success of the field mission. In this paper, we present a fast algorithm allowing for a quick-look modeling of scalar magnetic intensity measurements. The approach relies on rectangular harmonic analysis (RHA). The field measurements are automatically corrected for a global main field. Then, they are projected along this main field and modeled in terms of RHA functions. The software can exploit the quality indices provided with data and a procedure is applied to mitigate the effect of outliers. Maps for the scalar and the vector anomaly fields are readily built on an interpolated regular grid leveled at a constant altitude. In order to assess the modeling and the inversion procedures, analyses are carried out with synthetic measurements derived from a high-resolution global lithospheric magnetic field model estimated on the French aeromagnetic grid and at UAV locations with some added nonrandom noise. These analyses indicate that RHA is efficient for first-order and direct mapping of the crustal magnetic field structures measured by UAVs but that it could be applied on airborne and marine magnetic intensity data covering dense and large geographical extensions.

Keywords: UAV; rectangular harmonic analysis; geomagnetism

Citation: Thebault, E.; Gailler, L.-S. A Quick-Look Software for In Situ Magnetic Field Modeling from Onboard Unmanned Aircraft Vehicles (UAVs) Measurements. *Remote Sens.* **2023**, *15*, 4549. https://doi.org/10.3390/rs15184549

Academic Editors: Yanni Dong, Tao Chen and Chao Chen

Received: 12 July 2023

Revised: 9 September 2023

Accepted: 11 September 2023

Published: 15 September 2023

1. Introduction

The use of unmanned aerial vehicles (UAVs) has become ubiquitous in geophysical prospecting. Rapid to deploy at various altitudes above the topography, they are especially useful in volcanic and extreme contexts for high-resolution magnetic field mapping. Such maps are then used for inferring information on the structure and composition heterogeneities, thermal anomalies, and intrusive systems, as well as their evolution through time [1].

Magnetometry onboard UAVs for crustal field investigations is challenging since the UAV moving platform is prone to magnetic field measurement errors. These errors arise from the electronics such as antenna, battery, or communication systems, and from varying topographic or wind conditions that induce magnetic field time variations from the engine's rotation. These UAV contaminations are mostly characterized by transient magnetic fields over short time duration or appearing as flight line offsets, crossover errors, drag effects, and biases that complicate their direct reading. These effects then superimpose to natural global magnetic fields, masking the local crustal magnetic fields. These latter geophysical fields such as the Earth's core field and the diurnal and/or rapid magnetospheric field variations occurring at the time of the survey [2] are themselves difficult to separate [3]. In a favorable situation, the total contributions of the geophysical internal and external magnetic fields are conveniently corrected for with measurements made at a nearby fixed station (or at a magnetic observatory). The natural Earth field corrections and the proper characterization and compensations for the UAV magnetic noises that depend on the specific environment, its type, and the instrumental setup [4] are usually applied in the laboratory.

In remote field areas, it is fundamental to monitor and to evaluate the success of a survey, to ensure its consistency with adjacent and/or previous surveys, and to quickly build onsite the first-order maps showing the crustal magnetic field structures. This information is the condition ensuring reactivity, onsite reiteration, or complementary UAV surveys. The autonomy of a light UAV is another limit that hampers the possibility to prospect large areas. Compiling different surveys acquired at different times, altitudes, conditions, or even epochs is another challenge that must sometimes be addressed in the field for data quality assessment.

Our primary incentive is to develop software allowing us to directly acquire onsite a spatially consistent crustal magnetic field model from a series of incomplete, unequally spaced, and noisy magnetic field scalar raw measurements. This software should also help identifying and downweighting discrepant measurements. In order to address this problem, we follow a regional modeling strategy coupled with an iterative data reweighting scheme. Among the wide class of usable approaches (see [5] for a review), the rectangular harmonic double Fourier series appears to be the most convenient one to tackle regional problems in a semiautomatic way with a limited number of numerical difficulties. The introduction of the so-called rectangular harmonic analysis (RHA) in geomagnetism is generally attributed to [6], even though, to the author's knowledge, the general expressions can already be found in [7] and, years later, in [8] for applications on magnetic field scalar anomalies. The RHA proposed by [6] underwent a series of adjustments [9] because the theoretical framework suffered from fundamental drawbacks [10]. However, even these empirical adjustments remain flawed, as the solution given for the potential field does not solve the Laplace equation.

In the first section, we rewrite the RHA theoretical framework, stressing that these functions solve a particular boundary value problem that is suited only to a restricted category of natural magnetic fields. This section aims at better defining the range of practical applications of RHA. In the second section, we set up the algebra for solving the least-squares inverse problem in an automatic manner. We consider additional *a priori* statistical constraints provided by the magnetometer, such as the individual data quality. The inverse problem then is solved iteratively using the Huber statistics [11] for the identification and the leverage reduction of discrepant measurements. In order to assess the efficiency of the RHA modeling procedure, we perform a series of end-to-end studies. For this, we simulate real airborne and UAV measurements, apply the software to synthetic magnetic anomalies without and with noise, and compare the results with the initial data. Finally, we discuss the efficiency of this approach and interest for developing further applications.

2. Method

We recall some fundamental properties of the magnetic field measured outside the magnetic sources. The magnetic potential is the solution of the Laplace equation. Let us denote Ω as the domain under study within which the magnetic field is measured. A boundary value problem consists of setting boundary conditions on all surfaces n of the closed surface $\partial\Omega$ such that $\partial\Omega = \partial_1\Omega \cup \partial_2\Omega .. \cup \partial_i\Omega ... \cup \partial_n\Omega$ is completely defined on each of the domain boundary $\partial_i\Omega$. This is the well-known Sturm–Liouville framework that can be set up as

$$\nabla^2 V(\mathbf{r}_o) = 0 \text{ within } \Omega, \tag{1a}$$

$$\alpha_i V(\mathbf{r}_o) + \beta_i \frac{\partial V(\mathbf{r}_o)}{\partial n_i} = G_i \text{ on each } \partial\Omega_i, \tag{1b}$$

with ∇^2 as the Laplace operator, n_i as the normal vector at the boundary $\partial_i\Omega$, α_i, β_i as two real functions or constants, and $\mathbf{r}_o = (r_o\mathbf{e_r}, \theta_o\mathbf{e}_\theta, \varphi_o\mathbf{e}_\varphi)$ as the vector where magnetic observations are available, for instance, here in the geographic reference frame (Figure 1). This very general formulation illustrates that the magnetic field can be represented in any complex geometry and reference frame, either analytically or numerically, by solving this boundary value problem. It is important to stress that the magnetic divergence free equation (Maxwell's equation) implies that the divergence of the magnetic vector field **B** integrated over the volume Ω is equal to the null flux of the magnetic field integrated over the surface of the volume, which is written as

$$\iiint \operatorname{div} \mathbf{B} \cdot \mathbf{d}\Omega = \iint \mathbf{B} \cdot \mathbf{d}\partial\Omega = 0. \tag{2}$$

This null flux equation can be restrictive in practice depending on the chosen boundary conditions. As a result, boundary conditions must be chosen with care, as the mathematical basis functions may in some circumstances not allow the representation of all components of the magnetic field and/or may have a slow convergence rate.

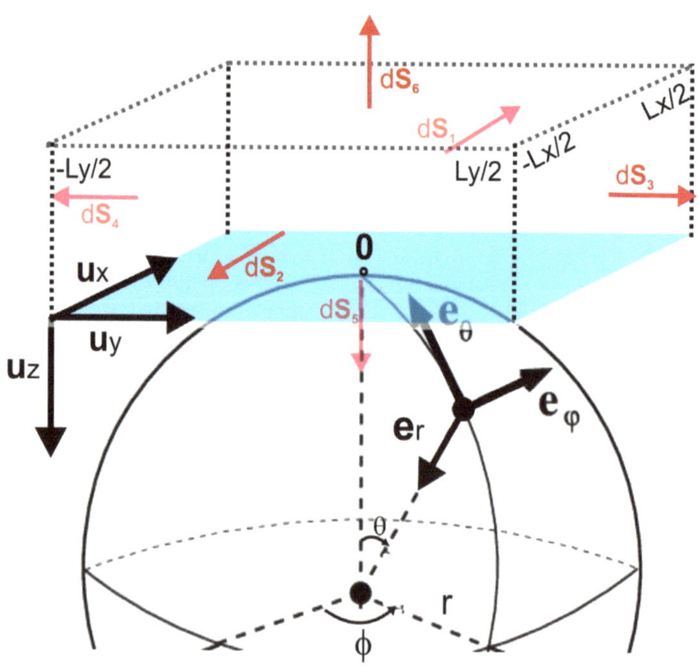

Figure 1. Cartesian and geocentric reference frames and coordinate systems discussed in this paper. The rectangular prism is tangent to the Earth's surface. Its center in the geocentric coordinate system is defined as the center of the available measurements. L_x and L_y are its dimensions in the north and east directions, respectively, and are defined according to the minimum and maximum geocentric data locations.

2.1. Solution in the Cartesian Reference Frame

We now consider representing the magnetic field in a Cartesian orthonormal reference frame where $\mathbf{u_x}$ is oriented northward, $\mathbf{u_y}$ is eastward, and $\mathbf{u_z}$ is oriented downward (Figure 1). The Laplace equation is now as follows:

$$\nabla^2 V(x,y,z) = \frac{\partial^2 V}{\partial^2 x} + \frac{\partial^2 V}{\partial^2 y} + \frac{\partial^2 V}{\partial^2 z} = 0.$$

Following the classical method of variable separation and writing

$$V(x,y,z) = f(x)g(x)h(z), \tag{3}$$

the Laplace equation will reduce to ordinary differential equations. Assuming further that the potential is defined within a bounded horizontal plane and possesses finite real valued boundary conditions, the formal solution for each variable is

$$f(x) = \alpha \sin(x\sqrt{\mu}) + \beta \cos(x\sqrt{\mu}), \tag{4a}$$

$$g(y) = \gamma \sin(y\sqrt{\lambda - \mu}) + \delta \cos\left(y\sqrt{\lambda - \mu}\right), \tag{4b}$$

$$h(z) = \psi \sinh(z\sqrt{\lambda}) + \phi \cosh(z\sqrt{\lambda}), \tag{4c}$$

where the eigenvalues μ and λ are such that $\lambda \geq 0$, $\lambda - \mu \geq 0$, and $(\alpha, \beta, \gamma, \delta, \zeta, \eta)$ are, at this stage, arbitrary real valued constants. The eigenvalues μ and λ of the ordinary differential equations are to be explicited by solving a particular boundary value problem. There are various ways of setting conditions in Equation (1b) on each of the boundaries, among which are the homogeneous Dirichlet ($\beta_i = G_i = 0$), Neumann ($\alpha_i = G_i = 0$), mixed, periodic, etc., depending on the properties of the field under study.

We now assume that the domain under study is a rectangular prism defined by $[-L_x/2, L_x/2] \times [-L_y/2, L_y/2] \times]-\infty, z_p]$, with L_x and L_y as the typical horizontal dimensions of the rectangle along x and y, respectively, and z_p as an arbitrary finite altitude above the Earth's surface (considering that z is oriented downward). In this representation, the domain is semi-infinite in the vertical dimension. Since $\mathbf{B} = -\nabla V$, it can be verified easily that none of the homogeneous boundary conditions (Neumann or Dirichlet) on V are suited for representing the three magnetic field components everywhere within the rectangular domain. For example, the homogeneous Dirichlet conditions on the lateral boundaries would imply within the domain that the magnetic field components $\partial_x V(x,0,z) = \partial_y V(0,y,z) = \partial_z V(0,y,z) = \partial_z V(x,0,z) = 0$. This solution is not able to represent any magnetic field everywhere in the rectangular prism. The magnetic potential in general verifies neither Neumann nor Dirichlet nor mixed homogeneous nor periodic boundary conditions. In this case, the problem is no longer self-adjoint (i.e., the basis is not orthonormal), and its analytical resolution could become, by far, more difficult ([12], chapter 12). However, we aim at deriving a semiautomatic software so that the orthogonality of the basis functions is an important property for numerically solving the inverse problem. With this constraint, the most practical and least restrictive boundary conditions among the ones discussed above are the periodic boundary conditions.

The periodic lateral boundary conditions give rise to a double Fourier series in the (x, y) dimensions. If we further assume that the magnetic field cancels out at $-\infty$, then Equation (4c) reduces to $e^{\sqrt{\lambda}z}$ and the product in Equation (3) now becomes

$$V(x,y,z) = \sum_{n=0}^{\infty} \sum_{m=0}^{\infty} [\alpha_n^m A_n^m(x,y) + \beta_n^m B_n^m(x,y) + \gamma_n^m C_n^m(x,y) + \delta_n^m D_n^m(x,y)] e^{\sqrt{\lambda}z}, \tag{5}$$

with

$$A_n^m(x,y) = \cos(\frac{2\pi nx}{L_x})\cos(\frac{2\pi my}{L_y}), \quad (6a)$$

$$B_n^m(x,y) = \cos(\frac{2\pi nx}{L_x})\sin(\frac{2\pi my}{L_y}), \quad (6b)$$

$$C_n^m(x,y) = \sin(\frac{2\pi nx}{L_x})\cos(\frac{2\pi my}{L_y}), \quad (6c)$$

$$D_n^m(x,y) = \sin(\frac{2\pi nx}{L_x})\sin(\frac{2\pi my}{L_y}), \quad (6d)$$

and

$$\sqrt{\lambda} = k(n,m) = \sqrt{\left(\frac{2\pi n}{L_x}\right)^2 + \left(\frac{2\pi m}{L_y}\right)^2}. \quad (7)$$

When the infinite series is truncated to maximum indices N and M (not necessarily equal), the minimum wavelength represented by the basis functions is

$$l = \sqrt{\left(\frac{L_x}{N}\right)^2 + \left(\frac{L_y}{M}\right)^2}. \quad (8)$$

The solutions for the three magnetic field components are obtained by taking the gradient of the potential:

$$\partial_x V(x,y,z) = \sum_{n=0}^{\infty}\sum_{m=0}^{\infty}[-\alpha_n^m C_n^m(x,y) - \beta_n^m D_n^m(x,y) + \gamma_n^m A_n^m(x,y) + \delta_n^m B_n^m(x,y)]e^{k(n,m)z}, \quad (9a)$$

$$\partial_y V(x,y,z) = \sum_{n=0}^{\infty}\sum_{m=0}^{\infty}[-\alpha_n^m B_n^m(x,y) + \beta_n^m A_n^m(x,y) - \gamma_n^m D_n^m(x,y) + \delta_n^m C_n^m(x,y)]e^{k(n,m)z}, \quad (9b)$$

$$\partial_z V(x,y,z) = \sum_{n=0}^{\infty}\sum_{m=0}^{\infty}k(n,m)[\alpha_n^m A_n^m(x,y) + \beta_n^m B_n^m(x,y) + \gamma_n^m C_n^m(x,y) + \delta_n^m D_n^m(x,y)]e^{k(n,m)z}. \quad (9c)$$

Since the choice for the boundary conditions may be restrictive for magnetic field representation, we explore the limitations induced by the periodic ones.

Let us consider B_x, B_y, and B_z, the three magnetic field components in the Cartesian reference frame of any true vector magnetic field that we expand in terms of the rectangular harmonic basis functions. Defining the six surfaces of the rectangular prism (Figure 1) such that $d\mathbf{S_1} = dS_1(L_x/2,y,z)\mathbf{u}_x$, $d\mathbf{S_2} = -dS_2(-L_x/2,y,z)\mathbf{u_x}$, $d\mathbf{S_3} = dS_3(x,L_y/2,z)\mathbf{u_y}$, $d\mathbf{S_4} = -dS_4(x,-L_y/2,z)\mathbf{u_y}$, $d\mathbf{S_5} = dS_5(x,y,z_0)\mathbf{u_z}$, $d\mathbf{S_6} = -dS_6(x,y,-\infty)\mathbf{u_z}$, where z_0 is the arbitrary altitude reference of the bottom surface, the null flux field condition across all surfaces is

$$\begin{aligned}&\iint (B_x(L_x/2,y,z) - B_x(-L_x/2,y,z))dydz\\ &+\iint (B_y(x,L_y/2,z) - B_y(x,-L_y/2,z))dxdz\\ &+\iint (B_z(x,y,z_d) - B_z(x,y,-\infty))dxdy = 0.\end{aligned}$$

From Equations (6a)–(6d) and (9a)–(9c) it is straightforward to obtain that the null flux equation reduces to

$$\sum_{n=1}^{\infty}\left(\frac{\gamma_n^0 L_y}{k(n,0)}e^{k(n,0)z_d}-\frac{\gamma_n^0 L_y}{k(n,0)}e^{k(n,0)z_d}\right) \tag{10}$$

$$+\sum_{m=1}^{\infty}\left(\frac{\beta_0^m L_x}{k(0,m)}e^{k(0,m)z_d}-\frac{\beta_0^m L_x}{k(0,m)}e^{k(0,m)z_d}\right) \tag{11}$$

$$+\left(k(0,0)L_xL_y\gamma_0^0 e^{k(0,0)z_d}-0\right)=0. \tag{12}$$

The first two sums are the direct consequence of the 2π periodic trigonometric functions that impose that the magnetic field flux across surfaces $\mathbf{S_1}$ and $\mathbf{S_2}$ and across $\mathbf{S_3}$ and $\mathbf{S_4}$, respectively, are term-by-term equal. More restrictive is the final two terms implying that the magnetic field flux across the surface $\mathbf{S_5}$ should be null. This corresponds to a zero average vertical magnetic field component $\langle \partial_z V(x,y,z)\rangle = 0$ on all horizontal surfaces of arbitrary altitude z. The rectangular harmonic analysis is therefore suited only if B_z averages out on all horizontal surfaces. Assuming that the rectangular harmonic analysis (RHA) is a valid approach for the representation of an anomaly vector field is, therefore, not sufficient. The magnetic vertical field that can be represented by these functions must have a zero mean within the rectangular domain. The fundamental reason is that the boundary value problem is incomplete. The RHA is not a true 3D but, rather, a partial 2D representation of the Earth's magnetic field. As a result, it does not represent the behavior of Newtonian potential fields with altitude except under these restrictive conditions.

According to Equation (1), the full Sturm–Liouville boundary value problem involves setting boundary conditions on all surfaces, including the bottom and the top surfaces. However, as in 2D FFT analysis, boundary conditions were set up on the lateral surfaces only. The reason is that solving within a finite rectangular prism would involve an extra trigonometric basis function in the vertical direction. Numerically determining the parameters of this extra basis function would require sufficient magnetic field sampling in altitude, which is a situation almost never met in practice. For an open domain such as the semi-infinite rectangular prism (with the upper boundary at infinity), the solution is even less convenient because it involves no more discrete series but, rather, continuous parameter functions. The horizontal double Fourier solution is therefore not complete and we should ensure that the data agree with the RHA constraint that the vertical magnetic field averages out. In general, we do not measure the vector field, just its intensity, and we cannot directly verify whether the data agree with this condition. However, imposing this constraint to the measured intensity anomaly should be a sufficient remedy. This inconvenience is a small price to pay when the purpose is precisely to unveil small scale crustal magnetic fields. Similar limitations are discussed in the spherical geometry and the reader will find in [13] practical examples for core and crustal Earth's magnetic field modeling in the 3D space using the incomplete spherical cap analysis [14], the semi-infinite cone analysis [15], and the revised spherical cap harmonic analysis [16].

2.2. Expression for the Scalar Anomaly Field

It is well established that the anomaly field intensity dF can be considered as a perturbation in the direction of the ambient field, provided that $\|d\mathbf{B}_c\|^2 \ll \|\mathbf{B}\|^2$, where $d\mathbf{B}_c$ and $\mathbf{B}$ are the anomaly and the Earth's ambient vector fields, respectively. Defining F as the modulus of vector $\mathbf{B}$ expressed in the local Cartesian reference frame, this is [17]

$$dF = \frac{\partial_x V.B_x + \partial_y V.B_y + \partial_z V.B_z}{F}. \tag{13}$$

This expression holds true even when both the anomaly and the main field vary in space. In the special approximation that the ambient field $\mathbf{B}$ is constant throughout the region of interest with $B_x/F \simeq C_x$, $B_y/F \simeq C_y$ and $B_z/F \simeq C_z$, we show, after replacing $\partial_x V$,$\partial_y V$ and $\partial_z V$ by their rectangular harmonic expansion in Equations (9a)–(9c), that

$$dF \simeq \sum_{n=0}^{\infty}\sum_{m=0}^{\infty}\left[\widetilde{\alpha}_n^m A_n^m(x,y) + \widetilde{\beta}_n^m B_n^m(x,y) + \widetilde{\gamma}_n^m C_n^m(x,y) + \widetilde{\delta}_n^m D_n^m(x,y)\right] e^{k(n,m)z}, \tag{14}$$

with the linear relationships between parameters:

$$\widetilde{\alpha}_n^m \simeq C_z k(n,m)\alpha_n^m + C_y \frac{2\pi m}{L_y}\beta_n^m + C_x \frac{2\pi n}{L_x}\gamma_n^m, \tag{15a}$$

$$\widetilde{\beta}_n^m \simeq -C_y \frac{2\pi m}{L_y}\alpha_n^m + C_z k(n,m)\beta_n^m + C_x \frac{2\pi n}{L_x}\delta_n^m, \tag{15b}$$

$$\widetilde{\gamma}_n^m \simeq -C_x \frac{2\pi n}{L_x}\alpha_n^m + C_z k(n,m)\gamma_n^m + C_y \frac{2\pi m}{L_y}\delta_n^m, \tag{15c}$$

$$\widetilde{\delta}_n^m \simeq -C_x \frac{2\pi n}{L_x}\beta_n^m - C_y \frac{2\pi m}{L_y}\gamma_n^m + C_z k(n,m)\delta_n^m. \tag{15d}$$

In the constant ambient main field approximation, the anomaly field intensity therefore also satisfies the Laplace equation and can be expanded by Equation (14). One could therefore directly model the intensity anomaly with a 2D double Fourier expansion and, if needed, relate the estimated coefficients to the coefficients describing the vector magnetic field using Equations (15a)–(15d). The direct modeling of intensity was proposed without demonstrating these limitations in the seminal paper of [8]. This approximation requires that the size of the region is small enough and/or that the spatial gradient of the Earth's vector main field can be neglected. If the ratios B_x/F, B_y/F, and B_z/F cannot be assumed constant, the solution obtained from Equation (14) is not a potential field. For this reason, we argue that systematically modeling the anomaly field dF by linearization along the constant or varying ambient magnetic field using Equation (13) is physically the most relevant procedure. It can be performed at almost no additional numerical cost and directly provides the coefficients for the vector field. Note that, as is the case for any geomagnetic linearization of scalar anomaly along the Earth's ambient field, a special treatment is necessary in magnetic equatorial regions in order to account for the Backus effect [18].

The RHA approach is different from the 2D FFT commonly used in geomagnetic prospecting. An RHA vector field model can be obtained by least-squares inversion of an unequally spaced discrete set of anomaly field measurements without interpolation and, thus, incorporation of fictitious field values. Contrary to 2D FFT, which requires this interpolation step for data reduction to a common altitude, the RHA modeling can also deal with measurements acquired at different altitudes. This represents a valuable advantage over the 2D FFT because there is no need to remove the average field value for each survey, but only to remove the average field for all surveys. This better preserves the large wavelengths and opens a way to iteratively incorporate new UAV surveys to the full dataset. This possibility also shows that RHA can deal with measurements acquired over (reasonably) large geographical extensions for which constant altitude measurements in a geographic reference frame correspond to non constant altitudes in the Cartesian reference frame. Therefore, RHA as it is defined here is not a planar approximation because the geographical coordinate system can be converted into the Cartesian one while accounting for the curvature of the Earth in the variations of z. Similarly, the spatially varying ambient magnetic field can be projected exactly from the geographic to the Cartesian frames. To benefit from these advantages, the solution must be obtained by solving an inverse problem that can, in some situations, be numerically ill posed and that must be set up with some care.

3. Least-Squares Inverse and Forward Problems

The inverse problem for the intensity anomaly is linear and solved by iterative weighted least-squares with $\mathbf{W}$ the diagonal data quality matrix (assuming uncorrelated data errors) at the first iteration. The design matrix $\mathbf{A}$ is built from the RHA basis functions at the measurement location d projected along the Earth's core magnetic field values. Since the geographical data coverage is generally nonhomogeneous, the inverse problem can be numerically ill posed despite the orthonormality of the trigonometric basis functions. The numerical solution for the model parameters p is obtained by generalized singular value decomposition. We estimate

$$p = \left(\mathbf{A}^T\mathbf{W}\mathbf{A}\right)^{-1}\mathbf{A}^T\mathbf{W}d,$$

by approximating $\mathbf{G} = \left(\mathbf{A}^T\mathbf{W}\mathbf{A}\right)^{-1}$ by its Moore–Penrose pseudoinverse $\mathbf{G} \simeq \mathbf{V}\mathbf{\Sigma}^+\mathbf{V}^T$, where $\mathbf{V}$ and $\mathbf{\Sigma}^+$ are the numerical eigenvectors and nonzero eigenvalues, respectively. The weighting matrix $\mathbf{W}$ is then updated according to the misfit residuals following the implementation advocated by Huber ([11], chapter 7), and successive inversions are performed until the model parameters converge. Once the model parameters p are obtained, the forward problem is computed on a regular grid and at a constant altitude.

In this paper, magnetic field component maps are displayed at the minimum altitude of the full dataset, but this could be modified in the code. During this forward computation on the regular grid, we address the well-known drawbacks occurring with periodic functions. When the functions to be represented (here, the magnetic field components) are themselves not periodic, the solutions are convergent in the mean square sense, subject to ringing near the boundaries, and prone to the Gibbs effect (e.g., [10]). Numerical strategies could be followed in order to mitigate these issues, such as arbitrarily extending the domain [19], applying a window taper, padding the extended domain with zero values, or filling it with synthetic measurements. For the purpose of this work, these empirical solutions are not fully satisfactory as they can hardly be fully automated. For example, enlarging the dimension of the region implies that the orthonormality is artificially degraded. This may lead to additional numerical issues with respect to an inverse problem often already ill conditioned due to the scarcity of the measurements. Moreover, enlarging the dimension of the region implies that the minimum resolved wavelength in Equation (8) is artificially increased, so this tends to smooth out useful small-scale magnetic field structures. A chosen alternative for alleviating the issues related to the nonuniform convergence, while keeping the integrity of the true measurements and the dimensions of the rectangular prism, is to approximate the solution by a uniformly convergent solution.

This can be achieved by applying the σ-factor on each of the parameters in front of the double Fourier series ([20], chapter 4). If $\mathbf{G}$ is the design matrix on the nodes of a regular grid at a constant altitude, with p the estimated parameters, b the magnetic model component, and $\mathbf{\Lambda}$ the matrix of Lanczos factors, the forward problem is computed with $b = \mathbf{G}\mathbf{\Lambda}p$. Analytically, this is written, for the magnetic potential,

$$V(x,y,y) = \sum_{n=0}^{\infty}\sum_{m=0}^{\infty}\sigma(n,m)[\alpha_n^m A_n^m(x,y) + \beta_n^m B_n^m(x,y) + \gamma_n^m C_n^m(x,y) + \delta_n^m D_n^m(x,y)]e^{k(n,m)z}, \tag{16}$$

with

$$\sigma(n,m) = sinc\left(\frac{n}{N+1}\right)sinc\left(\frac{m}{M+1}\right). \tag{17}$$

Each magnetic field component of the model on the regular grid can therefore be computed and displayed after applying the σ-factor on the model parameters. This approach does not eliminate all cumbersome Gibbs oscillations but considerably smooths out their amplitude within the rectangular prism by pushing back the discrepancies near the edges of the domain. The counterpart is that the σ-factor can in turn also dampen genuine magnetic field signals. Properly setting the σ-factor for each case study and for in-depth analysis requires a scientist in the loop. We recall that our purpose is to quickly obtain first-order consistent maps at constant altitudes directly on the survey site. Such maps do not necessarily correspond to the optimum solutions that can be obtained in the laboratory after careful UAV compensation, data processing, and inversion.

4. Algorithm and Synthetic Analyses

4.1. Algorithm and Processes

The data processing chain is illustrated in Figure 2. The data conversion from direct magnetometers and GPS readings to the RHA software input format are left to the user. The input data format for the software is latitude, longitude in the geocentric reference frame, altitude in km, time in modified Julian day (MJD), intensity in the units of nT, and data quality index when available. However, the software can also deal with a simplified input format such as latitude, longitude, altitude, and intensity only when no other information is available. The measurements are first corrected for the ambient core intensity field with the auxiliary CHAOS-7 model [21] whose coefficients should be downloaded and updated, if necessary, depending on the data epochs.

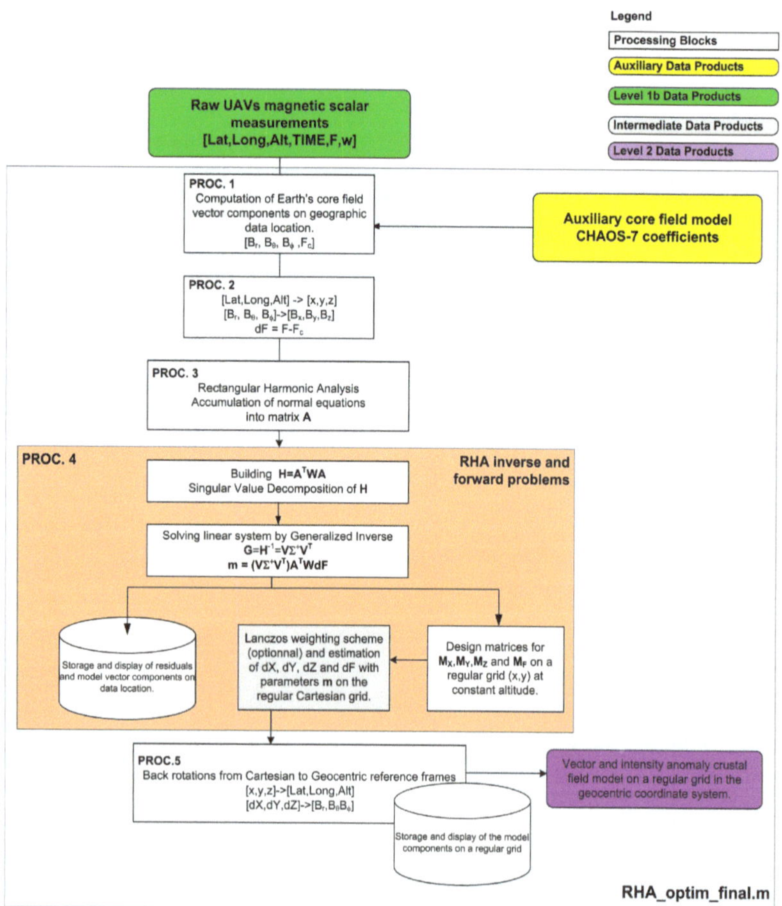

Figure 2. General flowchart of the five main processing blocks generating a regional magnetic vector crustal field from raw UAV scalar intensity measurements. See text for details.

The vector core field components computed at the data location are stored and appended to the original data (Proc. 1). Measurement locations are then rotated and projected in a local Cartesian coordinate system accounting for the natural curvature of the Earth, where x is north, y is east, and z is oriented downward (Proc. 1), and the vector core field components at each location are then converted from the geocentric to the Cartesian reference frame (Proc. 2). The RHA basis functions are next assembled at the Cartesian data coordinates with the core field components projected in the Cartesian reference frame (Proc. 3). In processing block 4, the inverse problem is estimated using generalized singular value decomposition (GSVD). Normalized eigenvalues smaller than 10^{-4} are arbitrarily rejected to ensure that the matrix is numerically invertible (this number can be modified). This precaution allows setting of the maximum degree expansion indices $N = M$ almost arbitrarily, provided they correspond to a minimum wavelength (in Equation (8)) that corresponds roughly to the maximum spacing between the flight lines. The minimum wavelength should not be too much smaller than this distance. At this stage, the original and the modeled intensity values and the residual maps are displayed on the data location. The estimated model parameters are then used to estimate the vector field and the intensity Cartesian components on a regular grid defined at the minimum data altitude (Proc. 5). Finally, the gridded vector components and their locations are converted back and displayed from the local Cartesian to the original geocentric reference frame (Proc. 5). Note that measurements are generally provided in the geodetic coordinates. However, for the purpose of a quick-look estimation, we do not distinguish between geocentric and geodetic systems. It should not be difficult to perform the transformations from geodetic to Cartesian and vice versa.

4.2. Computation of Synthetic Datasets

We test the efficiency of RHA and its ability to model the Earth's magnetic crustal field. Three closed-loop numerical experiments are carried out.

In the first test, we consider the original aeromagnetic grid of the French metropolitan territory [22] and compute a total field intensity for the main and the lithospheric fields. The original French anomaly intensity grid, not in the public domain, was included in the World Magnetic Anomaly Map (WDMAM-2, [23]). We compute the Earth's core vector field with the CHAOS-7 model [21] expanded to spherical harmonic (SH) degree 15 at epoch 2023.0 in decimal year. The vector crustal field contributions are computed from SH degree 16 to 1319 using the extended dedicated lithospheric field model derived with the WDMAM-2 map and the European Swarm satellite measurements [24]. The maximum spatial resolution of the combined core and crustal field models is about 30 km at the Earth's mean radius. The anomaly intensity is then computed by subtracting the total intensity of the core field from the combined core and crust intensity values. For this synthetic dataset, hereafter named SYN-AERO-PERFECT, we do not assign any weight and the measurements are perfect (no noise added). The locations of the synthetic data anomaly intensity projected in the Cartesian reference are shown in Figure 3a.

For the second analysis we use a series of UAV real surveys to generate a realistic data distribution. Part of these surveys were acquired over the Puy de Dôme volcano (Massif Central, France) by our group during the validation/calibration phase of UAVs as an efficient platform for magnetometry [25]. However, we purposefully stretch both the altitudes and the areas in order to test the efficiency of RHA to model significant altitude differences over dimensions larger than individual surveys. This is also a way to test the ability of RHA to patch together small-dimension surveys in a consistent way, and to secure wavelengths larger than the dimensions of the individual surveys. The full dataset, hereafter named SYN-UAVS-PERFECT, contains 24 surveys covering altitudes between 5.2 and 20.0 km. The total intensity measurements are computed in a similar way to before. The synthetic measurements have no weight and contain no noise. The data locations are displayed in the Cartesian reference frame in Figure 3b.

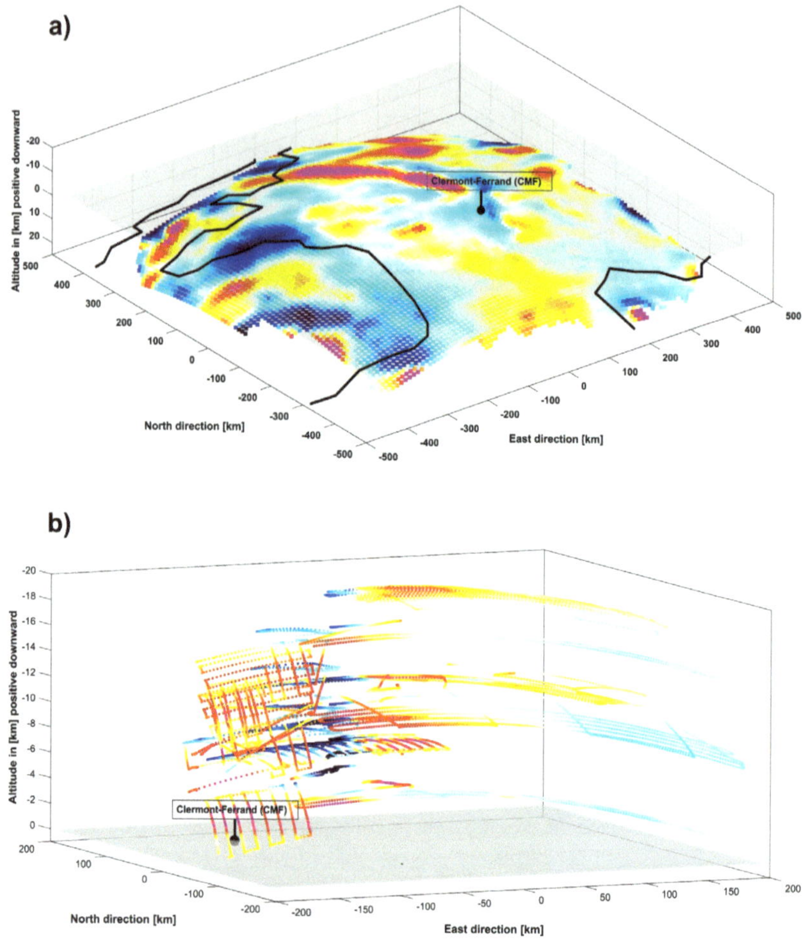

Figure 3. (**a**) Crustal intensity anomaly field (in nT) over France and data location used for building the SYN-AERO-PERFECT dataset. (**b**) Crustal intensity anomaly field (in nT) at UAV data locations used for building the SYN-UAVS-PERFECT and SYN-UAVS-NOISED datasets. See text for details.

Finally, we simulate a series of noisy scalar measurements by adding a geophysical noise to the SYN-UAVS-PERFECT dataset. The noise is computed by selecting the time series of the permanent magnetic ground station installed at Clermont-Ferrand in November 2021 (International Association of Geomagnetism and Aeronomy code "CMF" available at Bureau Central de Magnétisme Terrestre—BCMT: http://www.bcmt.fr, accessed on 10 September 2023, located in Figure 3). This station delivers minute vector and scalar measurements. We extract from the database the intensity channel between 23 April 2023 0:00 LT and 24 April 2023 23:59 LT. This corresponds to a period during which a magnetic storm classified as severe was measured worldwide (G4, following the National Oceanic and Atmospheric Administration classification). The intensity signal is high-pass filtered in order to remove the global magnetospheric field variations with periods larger than 20 min, which is typically the duration of a UAV survey. The remaining rapid geophysical variations considered as the natural noise (Figure 4) are then added to the crustal field signal computed at each data location and altitude. This noise that is correlated in time and not stationary is more realistic than pure random noise. When superimposed to the crustal field signals it introduces magnetic field variations along the UAV trajectory that mimic well noise contamination and flight-line offsets. The corresponding dataset is hereafter named SYN-UAVS-NOISED.

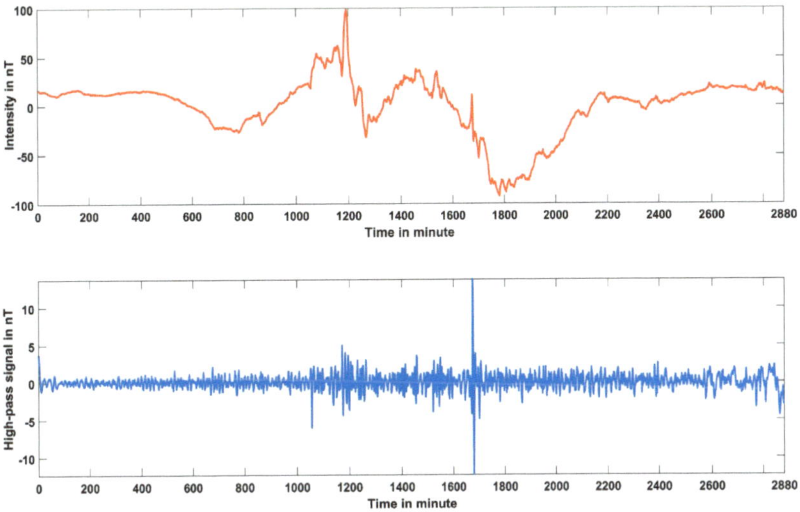

Figure 4. Top: Magnetic intensity of the 23–24 April 2023 magnetic storm (in nT) recorded at the CMF ground magnetic station. **Bottom**: Residual signal after high-pass filtering that is added as a natural noise to build the SYN-UAVS-NOISED dataset.

5. Results and Discussion

RHA models are derived from the three datasets described above. For the test case computed with the dataset SYN-AERO-PERFECT, we derive a model over France from the regular scalar anomaly grid to maximum expansion indices $N = M = 30$ in the infinite series (Equations (9a)–(9c)). For the inversion of datasets SYN-UAVS-PERFECT and SYN-UAVS-NOISED, the maximum truncation indices are set to $N = M = 15$. According to Equation (8), the maximum spatial resolution for all three cases is about 30 km in the east and north directions. We remind the reader that the synthetic measurements were also computed with a global model to the maximum spatial resolution of about 38 km. Therefore, these RHA truncation indices correspond to the spatial information content of the synthetic measurements.

The residuals between the input scalar anomaly field computed in SYN-AERO-PERFECT and the RHA scalar anomaly model are displayed in Figure 5a after correcting for the constant field value. The global standard deviation of the inversion is 0.1 nT and most residuals correspond to oscillations typical of truncated double Fourier series. There is no increase of residuals near the edges, footprint of numerical ringing, or geometric distortion. The maps for three magnetic field components are predicted in the Cartesian reference frame at a constant altitude of 4 km above the Earth's mean radius and converted back to the north (Figure 5b), east (Figure 5c), and vertical directions (Figure 5d) in the initial geocentric reference frame. The figures show that the model is stable and that it resolves the input measurements to the spatial resolution of the magnetic signal. One primary conclusion of this test is that RHA, which is a Cartesian representation, can be used to derive magnetic field models in the geocentric reference frame even when the curvature of the Earth cannot be neglected over extensions as large as 1000 km.

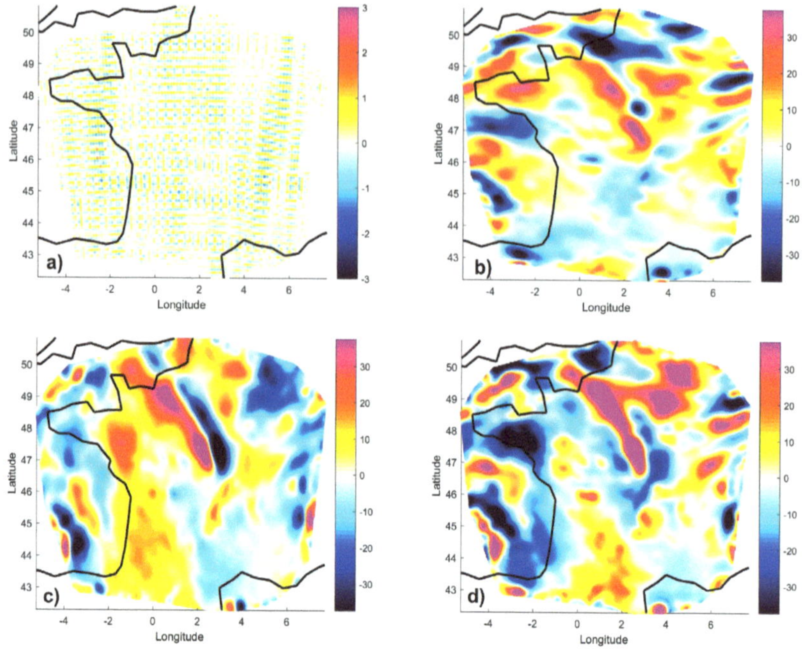

Figure 5. (**a**) Residual map (in nT) between the data and the RHA scalar anomalies predicted over France after modeling the SYN-AERO-PERFECT dataset. (**b**) Reconstruction of the magnetic north (**b**), the east (**c**), and the vertical component (**d**) of the crustal field. All maps are displayed in the geocentric reference frame at 4 km altitude above the Earth's radius.

Regarding the realistic UAV data location experiments, the standard deviation between the synthetic measurements at their actual location and the RHA model predictions derived from the SYN-UAVS-PERFECT and SYN-UAVS-NOISED datasets are 0.2 nT and 1.5 nT, respectively. The misfit value for the RHA model based on the noisy dataset corresponds to the level of noise introduced in the dataset (Figure 4). The scalar and vector field models are then predicted on a regular grid at a constant altitude of 5 km above the Earth's mean radius. Both residual maps between the intensity anomaly RHA predictions and the expected synthetic anomaly field values computed with the global spherical harmonic model [24] are displayed in Figures 6a and 7a, respectively. The residuals show almost identical misfit patterns, with a comparatively larger amplitude for the model obtained with the noisy measurements. The vector field model derived from the SYN-UAVS-NOISED dataset is slightly noisier than the vector field model obtained with the SYN-UAVS-PERFECT one.

Figure 8a shows the initial geophysical noise (displayed as a time series in Figure 4 bottom) in the geographical frame that was added to the intensity data for the numerical simulation with the SYN-UAVS-NOISED dataset. This noise manifests itself as rapid spatial variations, along-flight errors, and data incompatibility in overlapping areas, but the algorithm identifies and downweights most of the noisy data (Figure 8b) during the inversion process. It should be noted that the geophysical noise we added comes from a magnetic field time series during a magnetic storm. With this signal being a time varying potential field, some of its contributions leak into the crustal field model in the geographic reference frame.

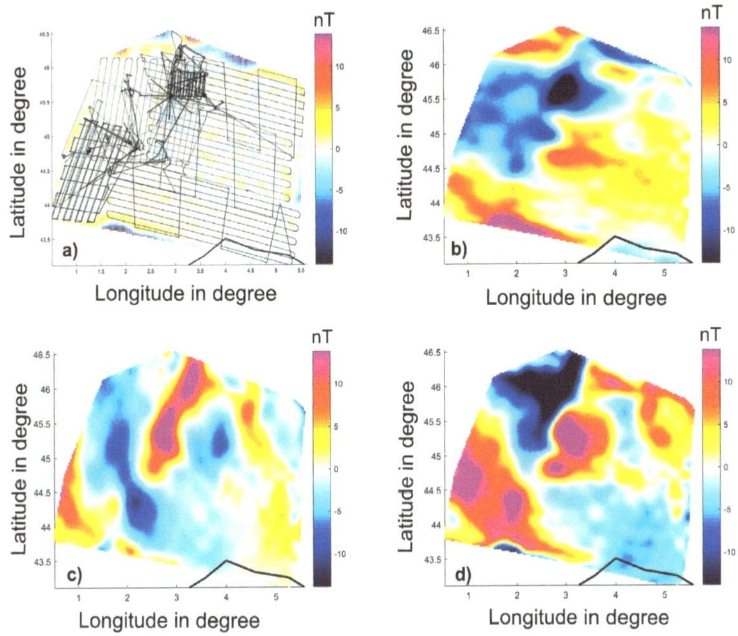

Figure 6. (**a**) Residual map between the true and the RHA scalar anomalies over Massif-Central, France, predicted after modeling the SYN-UAVS-PERFECT dataset. (**b**) Reconstruction of the magnetic north (**b**), the east (**c**), and the vertical component (**d**) at the constant altitude of 5 km.

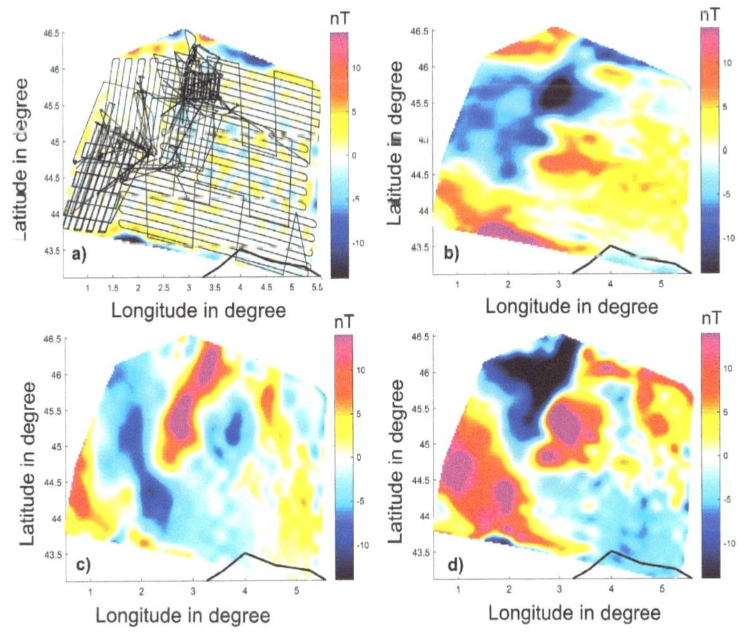

Figure 7. (**a**) Residual map between the true and the RHA scalar anomalies over Massif-Central, France, predicted after modeling the SYN-UAVS-NOISED dataset. (**b**) Reconstruction of the magnetic north (**b**), the east (**c**), and the vertical component (**d**) of the crustal field at the constant altitude of 5 km.

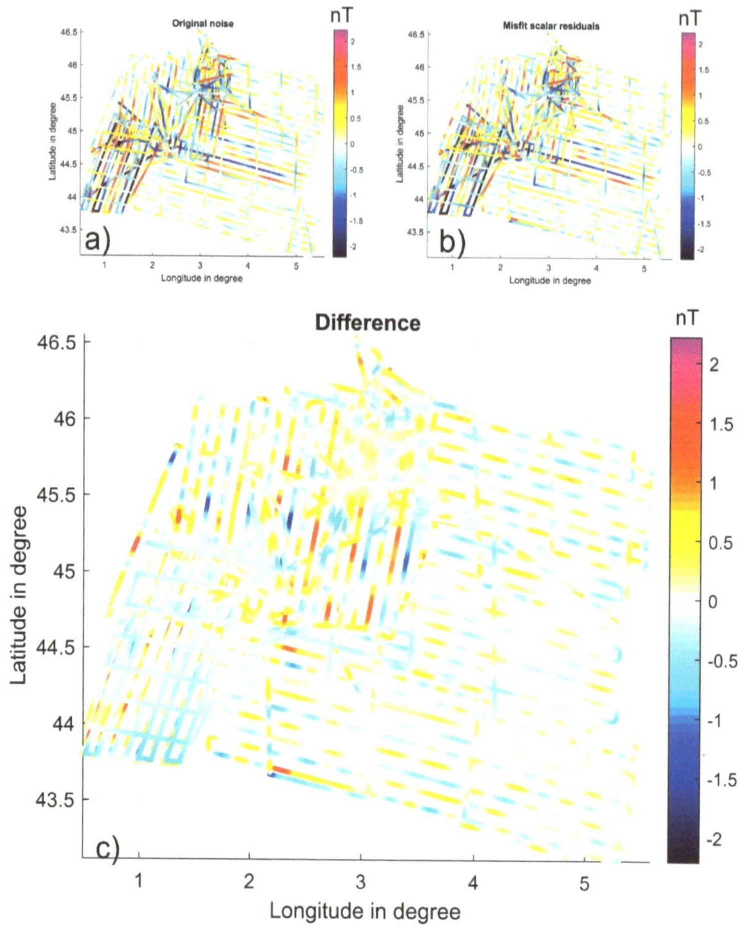

Figure 8. (**a**) Original noise introduced in the SYN-UAVS-NOISED dataset. (**b**) Misfit residuals between the model and the noisy data. (**c**) Difference between (**a**,**b**).

For both cases on UAV real trajectories there is no spatially coherent residuals between the model and the expected field, except near the edges where no measurement was available to build the RHA models. This indicates that the RHA models did not miss either large- or small-scale structures contained in the data. We remind the reader that intensity anomaly and vector field maps are leveled at the constant altitude of 5.2 km over a regular interpolation grid. The leveling altitude corresponds to the minimum altitude present in the dataset. Since the initial data altitudes comprise between 5.2 km and 20 km according to Figure 3b, the computation of the vector model corresponds to a downward continuation in all locations except for a small part of the southwest quadrant. Mapping the results at the minimum altitude was chosen for illustration purposes and does not correspond to the most advantageous scenario since downward continuation of potential fields is known to be prone to increased errors (e.g., [26]). The increased errors are due to the radial function that amplifies the smaller more than the larger length scales. We could have chosen another altitude for mapping the RHA model at a constant level (for example, at the median or the maximum data altitude); however, the maps obtained in these synthetic analyses from scalar measurements only are also stable for the three vector magnetic field components, and the magnetic field structures are spatially consistent.

6. Conclusions

The main incentive of this work was to process and to stitch together individual UAV surveys obtained in situ in order to build consistent magnetic field vector maps interpolated and leveled at a constant altitude. For this, an RHA mathematical representation was developed. Synthetic analyses based on closed-loop tests for three different case scenarios simulating real data distribution were carried out in order to demonstrate the ability of RHA to achieve this aim. It meets our initial needs to monitor and to quickly evaluate onsite the success of a magnetic survey, to filter out nonpotential field contamination signals, to verify the consistency between adjacent and/or surveys available at differing altitudes and epochs, to estimate the data noise level, and to identify in the geographical domain the series of problematic data acquisition. The synthetic analyses also demonstrate that the RHA approach can provide vector magnetic field models from scalar measurements acquired over large spatial extents such as aeromagnetic, airborne, and dense offshore marine magnetic surveys.

The RHA modeling method includes a processing block designed to minimize the effect of outliers and to reject measurements that do not obey the potential theory; however, the algorithm is not able to inform us about the origin of all possible contaminations. The reason is that UAV magnetic effects depend too strongly on each individual setup to allow the development of an automatic procedure (e.g., [27]). If it can readily be used for first-order in situ modeling, final robust magnetic field models must be obtained in the laboratory in a second step after further data analyses to correct for the systematic UAV magnetic bias and noise. In that regard, the algorithm provides useful guidance for their identification.

One limitation of RHA is its inability to separate the internal from the external magnetic fields and to deal with signals with wavelengths larger than the typical dimensions of the area under study. Such issues are, however, common to many regional potential field methods [5]. Beyond this standard limitation, one major outcome is that RHA is a mathematical potential field modeling approach. As such, it could be related analytically to physical parameters expanded in terms of double Fourier series such as pseudo-gravity anomalies, magnetic susceptibility distributions, or integrated magnetization models. This would directly, in situ, provide further and complementary information about the apparent physical properties of the magnetic sources.

Author Contributions: Conceptualization E.T. and L.-S.G.; mathematical formulation E.T.; methodology E.T. and L.-S.G.; data preparation E.T. and L.-S.G.; software E.T. and L.-S.G.; writing E.T. and L.-S.G. Both authors contributed equally to this research. All authors have read and agreed to the published version of the manuscript.

Funding: This work was funded partly funded by CNES in the framework of the project "Imagerie géophysique multi-échelles des édifices volcaniques" and the Agence National Research project Scan4Volc https://anr.fr/Project-ANR-21-CE49-0015, accessed on 10 September 2023. This is contribution no. 616 of the ClerVolc program of the International Research Center for Disaster Sciences and Sustainable Development of the University of Clermont Auvergne.

Data Availability Statement: The lithospheric magnetic field used for the synthetic analyses is available at https://swarm-diss.eo.esa.int/#swarm/Level2longterm/MLI/SW_OPER_MLI_SHA_2E_00000000T000000_99999999T999999_0801.ZIP, accessed on 10 September 2023. The auxiliary core field model CHAOS-7 is available at https://www.space.dtu.dk/english/research/scientific_data_and_models/magnetic_field_models, accessed on 10 September 2023. The synthetic datasets generated for this study and the code are available as an archive at https://www.opgc.fr/DOI/DataSoft/1qg6-qh39/RHA_optim_final.zip, accessed on 10 September 2023, under the Creative Commons licence with doi: https://doi.org/10.25519/1QG6-QH39, accessed on 10 September 2023.

Acknowledgments: We would like to thank the three anonymous referees for their constructive comments.

Conflicts of Interest: The authors declare no conflict of interest.

References

1. Antoine, R.; Lopez, T.; Tanguy, M.; Lissak, C.; Gailler, L.; Labazuy, P.; Fauchard, C. Geoscientists in the sky: Unmanned aerial vehicles responding to geohazards. *Surv. Geophys.* **2020**, *41*, 1285–1321. [CrossRef]
2. Hulot, G.; Sabaka, T.; Olsen, N. The present field. *Treatise Geophys.* **2007**, *5*, 33–75.
3. Finlay, C.; Lesur, V.; Thébault, E.; Vervelidou, F.; Morschhauser, A.; Shore, R. Challenges handling magnetospheric and ionospheric signals in internal geomagnetic field modelling. *Space Sci. Rev.* **2017**, *206*, 157–189. [CrossRef]
4. Zheng, Y.; Li, S.; Xing, K.; Zhang, X. Unmanned aerial vehicles for magnetic surveys: A review on platform selection and interference suppression. *Drones* **2021**, *5*, 93. [CrossRef]
5. Schott, J.J.; Thébault, E. Modelling the earth's magnetic field from global to regional scales. In *Geomagnetic Observations and Models*; Springer: Berlin/Heidelberg, Germany, 2010; pp. 229–264.
6. Alldredge, L. Rectangular harmonic analysis applied to the geomagnetic field. *J. Geophys. Res. Solid Earth* **1981**, *86*, 3021–3026. [CrossRef]
7. Vestine, E.H.; Davids, N. Analysis and interpretation of geomagnetic anomalies. *Terr. Magn. Atmos. Electr.* **1945**, *50*, 1–36. [CrossRef]
8. Bhattacharyya, B. Two-dimensional harmonic analysis as a tool for magnetic interpretation. *Geophysics* **1965**, *30*, 829–857. [CrossRef]
9. Alldredge, L.R. Geomagnetic local and regional harmonic analyses. *J. Geophys. Res. Solid Earth* **1982**, *87*, 1921–1926. [CrossRef]
10. Haines, G. Regional Magnetic Field Modelling A Review. *J. Geomagn. Geoelectr.* **1990**, *42*, 1001–1018. [CrossRef]
11. Huber, P.J. *Robust Statistics*; John Wiley & Sons: Hoboken, NJ, USA, 2004; Volume 523.
12. Coddington, E.A.; Levinson, N.; Teichmann, T. *Theory of Ordinary Differential Equations*; McGraw-Hill Inc.: New York, NY, USA, 1956.
13. Thébault, E.; Gaya-Piqué, L. Applied comparisons between SCHA and R-SCHA regional modeling techniques. *Geochem. Geophys. Geosyst.* **2008**, *9*, 7. [CrossRef]
14. Haines, G. Spherical cap harmonic analysis. *J. Geophys. Res. Solid Earth* **1985**, *90*, 2583–2591. [CrossRef]
15. Thébault, E. A proposal for regional modelling at the Earth's surface, R-SCHA2D. *Geophys. J. Int.* **2008**, *174*, 118–134. [CrossRef]
16. Thébault, E.; Schott, J.; Mandea, M. Revised spherical cap harmonic analysis (R-SCHA): Validation and properties. *J. Geophys. Res. Solid Earth* **2006**, *111*, B01102. [CrossRef]
17. Blakely, R.J. *Potential Theory in Gravity and Magnetic Applications*; Cambridge University Press: Cambridge, UK, 1996.
18. Backus, G.E. Non-uniqueness of the external geomagnetic field determined by surface intensity measurements. *J. Geophys. Res.* **1970**, *75*, 6339–6341. [CrossRef]
19. Jiang, T.; Li, J.; Dang, Y.; Zhang, C.; Wang, Z.; Ke, B. Regional gravity field modeling based on rectangular harmonic analysis. *Sci. China Earth Sci.* **2014**, *57*, 1637–1644. [CrossRef]
20. Lanczos, C. *Applied Analysis*; Dover Publication, Inc.: New York, NY, USA, 1988.
21. Finlay, C.C.; Kloss, C.; Olsen, N.; Hammer, M.D.; Tøffner-Clausen, L.; Grayver, A.; Kuvshinov, A. The CHAOS-7 geomagnetic field model and observed changes in the South Atlantic Anomaly. *Earth Planets Space* **2020**, *72*, 1–31. [CrossRef]
22. Le Mouel, J. Sur la Distribution des Elements Magnetiques en France. Ph.D. Thesis, University de Paris, Paris, France, 1969.
23. Lesur, V.; Hamoudi, M.; Choi, Y.; Dyment, J.; Thébault, E. Building the second version of the world digital magnetic anomaly map (WDMAM). *Earth Planets Space* **2016**, *68*, 1–13. [CrossRef]
24. Thébault, E.; Hulot, G.; Langlais, B.; Vigneron, P. A spherical harmonic model of Earth's lithospheric magnetic field up to degree 1050. *Geophys. Res. Lett.* **2021**, *48*, e2021GL095147. [CrossRef]
25. Gailler, L.; Labazuy, P.; Régis, E.; Bontemps, M.; Souriot, T.; Bacques, G.; Carton, B. Validation of a new UAV magnetic prospecting tool for volcano monitoring and geohazard assessment. *Remote Sens.* **2021**, *13*, 894. [CrossRef]
26. Nabighian, M.N. Additional comments on the analytic signal of two-dimensional magnetic bodies with polygonal cross-section. *Geophysics* **1974**, *39*, 85–92. [CrossRef]
27. Qiao, Z.K.; Yuan, P.; Wang, L.F.; Zhang, Z.H.; Huang, Y.S.; Zhang, J.J.; Li, L.L.; Zhang, Z.Y.; Wu, B.; Lin, Q. Research on aeromagnetic compensation of a multi-rotor UAV based on robust principal component analysis. *J. Appl. Geophys.* **2022**, *206*, 104791. [CrossRef]

remote sensing

MDPI

Article

Study on Road Network Vulnerability Considering the Risk of Landslide Geological Disasters in China's Tibet

Yunchang Yao [1,2,3], Liang Cheng [1,2,3,*], Song Chen [1,2,3], Hui Chen [1,2,3], Mingfei Chen [4], Ning Li [1,2,3], Zeming Li [1,2,3], Shengkun Dongye [1,2,3], Yifan Gu [1,2,3] and Junfan Yi [1,2,3]

1 School of Geography and Ocean Science, Nanjing University, 163 Xianlin Road, Nanjing 210023, China; dg20270045@smail.nju.edu.cn (Y.Y.); chensong@nju.edu.cn (S.C.); dg20270003@smail.nju.edu.cn (H.C.); lining@smail.nju.edu.cn (N.L.); dg21270026@smail.nju.edu.cn (Z.L.); 602022270005@smail.nju.edu.cn (S.D.); mg20270070@smail.nju.edu.cn (Y.G.); 502022270091@smail.nju.edu.cn (J.Y.)

2 Jiangsu Provincial Key Laboratory of Geographic Information Science and Technology, Nanjing University, Nanjing 210023, China

3 Collaborative Innovation Center of Novel Software Technology and Industrialization, Nanjing University, Nanjing 210023, China

4 School of Electrical and Information Engineering, Beijing University of Civil Engineering and Architecture, Beijing 100044, China; 202007010106@stu.bucea.edu.cn

* Correspondence: lcheng@nju.edu.cn

Abstract: Road traffic is occasionally blocked by landslide geological disasters in remote mountainous areas, causing obstruction to economic society and national defense construction. It is vital to conduct landslide geological disaster risk assessment and vulnerability research on the road network. Based on landslide geological disaster risk on the road network, this study analyzed the potential effects of the main environmental elements. Due to the lack of previous research works, this study proposed an effective, rational, and understandable multicriteria heuristic analytical hierarchy process model, fuzzy comprehensive evaluation, and frequency ratio-interactive fuzzy stack analysis for vulnerability assessment of road networks in large and complex networks. Based on the comprehensive use of geographic information technology, the road network vulnerability of Tibet in China was evaluated by introducing slope, topographic relief, normalized difference vegetation index (NDVI), annual mean precipitation, distance from river drainage, glaciers and snow, habitation, seismic center and geological fault zone, and soil erosion intensity. According to the findings of the study, the three-stage framework proposed in this study can provide correct inferences and explanations for the potential phenomena of landslide geological disasters; the geological disaster risk are unevenly distributed in the study area; the distribution of the road network vulnerability in China's Tibet significantly differs among different cities; the high-vulnerability section presents significant regional characteristics, which overlap with the area with a high risk of landslide geological disasters, and its distribution is mostly located in traffic arteries, link aggregations, and relatively frequent human activity.

Keywords: landslide geological disasters; risk assessment; road network vulnerability; Tibet

Citation: Yao, Y.; Cheng, L.; Chen, S.; Chen, H.; Chen, M.; Li, N.; Li, Z.; Dongye, S.; Gu, Y.; Yi, J. Study on Road Network Vulnerability Considering the Risk of Landslide Geological Disasters in China's Tibet. *Remote Sens.* **2023**, *15*, 4221. https://doi.org/10.3390/rs15174221

Academic Editor: Michele Saroli

Received: 19 May 2023
Revised: 7 August 2023
Accepted: 15 August 2023
Published: 28 August 2023

1. Introduction

Geological disasters occur frequently in China annually. As the main type of collapse and landslide (hereinafter referred to as "landslide") these geological disasters have caused huge losses to the country, economy, and society [1–4]. According to a report by the Highway Bureau of the Ministry of Transport in 2020, the total mileage of road disasters will exceed 6000 km and the direct loss of road production costs will reach CNY 22.8 billion [5]. The road failure caused by landslide blocking or even burying, the vulnerability of the road network is constantly exposed. Landslide geological disasters may greatly impact the transport capacity of nearby roads. Keeping major highways open, in particular, is crucial for supporting the functioning of the economy, society, and people. Therefore, the risk assessment of road landslide geological disasters is particularly important. The topography of Tibet in China is very complex, with various types of landform. The natural conditions of the surface are particularly obvious, and landslide geological disasters are widely experienced. The occurrence of landslide geological disasters is closely related to topography, geological structure, stratigraphic lithology, hydrology condition, human activity, and other factors. Owing to the complexity of the formation mechanism, no set of risk assessment systems is suitable for all areas, and factor selection is generally determined by the specific conditions of the study area [6,7]. Over the past few decades, various risk assessment methods for geological disasters have been developed worldwide, including analytical hierarchy process (AHP) [8], field surveys [8], satellite remote sensing [9], statistical models [10], deterministic coefficients, and machine learning [11–13]. It was found that previous studies were mostly from single models or combinations of several models, which obtained the risk susceptibility degree and partition but lacked research models with excellent hierarchical measures and high efficiency.

Vulnerability is a measure of the degree of disaster-bearing damage, which is an important link in disaster damage estimation and risk assessment and a bridge between disaster factors and situations. Through a literature review and analysis, it was found that the vulnerability evaluation factor mainly focuses on state and structural vulnerabilities [14–16], and vulnerability analysis methods include single analysis, multidimensional evaluation, optimization methods, and multiobjective algorithms [17–21]. Vulnerability has been analyzed and applied in power lines [22], communication networks, transportation [23], natural gas pipelines [24], urban flooding [25], emergency rescue [26], world heritage sites [27], and many more. The research on the vulnerability of road networks originated from the 1995 Kobe earthquake in Japan and the terrorist incidents on 11 September 2001. The 7.2-magnitude earthquake in Kobe, Japan damaged almost all transportation systems around the city, and the terrorist attacks of September 11th exposed transportation vulnerability to road network disruptions [28]. The concept of road network vulnerability was first proposed by Berdica, who believed that the vulnerability of a road transport system is a sensitive factor subject to a significant decrease in the service level of the road network [29]. Taylor et al. also defined the vulnerability of road networks from a similar point of view [30]; they believe that the vulnerability of the network depends on whether the accessibility of the network is greatly reduced after a few nodes (sections of road) are removed. Subsequently, other scholars insisted on an expression of vulnerability and risk, and vulnerability needs to be defined in the scope of risk degree. Some scholars such as Sohn tried to explain the vulnerability of road network more precisely. They connected the vulnerability of road networks with the probability and consequences of risks, and believed that the vulnerability, reliability, and risk of road networks were closely related concepts [31]. A statistical model was developed to reveal and quantify the main factors defining transport related consequences [32]; the integrated equilibrium model of traffic demand change [33], the attacker–defender model [34], which this research on road networks has promoted, has progressed road network research to a certain extent. It should be pointed out that these scenario-based evaluation methods, strategy-based evaluation methods, simulation-based evaluation methods and mathematical model evaluation methods have their own advantages and disadvantages. How to use them together or propose more effective methods to better evaluate the vulnerability of the road network is also worth studying [35]. Research on road network vulnerability has gained increasing attention in recent years. Significant changes have taken place in the study of road network vulnerability, including the study and application of multidimensional assessment, multiobjective algorithms, and optimization methods [19]. Road network vulnerability is an inherent attribute of the traffic network, which reflects the degree of overall service level decline when the road network is affected by natural disasters, traffic accidents or road maintenance. There is no unified definition of road network vulnerability, owing to a lack of clear measures. Although the global attack–defense model and map of landslide sensitivity, analysis of the road transportation network vulnerability, travel mode, and travel time affect the capacity and emissions [36,37], capacity-weighted spectrum analysis is used to identify potential key links, cloud model-based urban road network traffic system congestion status, and many road vulnerability and risk assessment studies [14,20,21,38,39]. However, vulnerability analysis application in road disasters research is still in the exploration and development stages, and its theory and methods require further improvement. The precise assessment of road network vulnerability is difficult because of the uncertainty and complexity of environmental differences and vulnerability-related events along roads [28]. As the road network system is confronted with multiple disturbances, it is necessary to propose more effective, rational, and understandable methods for vulnerability assessment of road networks in large and complex networks [30]. It is worth noting that road network vulnerability evaluation is particularly urgent and meaningful in remote mountainous areas [4,40].

In the above context, this study proposed a reasonably efficient and well-measured hierarchical research paradigm and model to predict vulnerability to geological disasters in high mountainous areas; exploring the correlation between landslide geological disaster risk and different environmental elements [41] and predicting the impact of regional landslide geological disaster risks on road networks to provide a scientific basis and technical support for taking measures to reduce vulnerability and loss. The remainder of this article is structured as follows. Section 2 describes the study area overview, datasets, and processing. In Section 3, the methods and models used in the case study are described. Section 4 introduces the risk assessment of geological hazards and the results of the road network vulnerability. Section 5 is the discussion followed by the conclusion.

2. Materials

2.1. Overview of the Study Area

The Tibet Autonomous Region is in the southwest of the Qinghai–Tibet Plateau, between north latitude 26°50′ to 36°53′ and east longitude 78°25′ to 99°06′. The region has jurisdiction over six prefecture-level cities and one region (Figure 1) with a land area of 1,202,800 km^2, accounting for approximately one-eighth of China's total area. The climate of Tibet is unique, complex, and diverse, showing characteristics of cold and dry conditions in the northwest and warm and humid conditions in the southeast. The vertical zonality is very significant. Mountains, glaciers, lakes, rivers, large terrain fluctuations, rock layer weathering and crushing, and geological conditions are complex. Tibet has a fragile ecological environment and poor economic and social development. As an area sensitive to global climate change, its strategic position is extremely important. In the context of the current century of change and the epidemic situation, road traffic not only represents infrastructure connectivity, but also represents the smooth flow of trade and people. By the end of 2020, the length of highway open to traffic in Tibet exceeded 117,000 km. In 2020, Tibet's annual cargo transport turnover reached 15.818 billion ton/km, of which 11.673 billion ton/km were transported by road. Simultaneously, Tibet's natural environment is complex, and mountain collapses and landslides occur frequently, with widespread distribution and serious harm, resulting in dangerous roads and necessitating higher requirements for high-quality transportation infrastructure construction in Tibet and economic and social development promotion.

2.2. Data Sources and Preprocessing

2.2.1. Data Collection

Raw data were used for road network vulnerability evaluation, including occurrence of landslide geological disasters in the study area, national to prefectural roads, ASTER GDEM(DEM), normalized difference vegetation index(NDVI), annual mean precipitation, river systems, China's second glacier cataloging dataset [42], geological fault zone, seismic center, soil erosion, and habitation (Table 1).

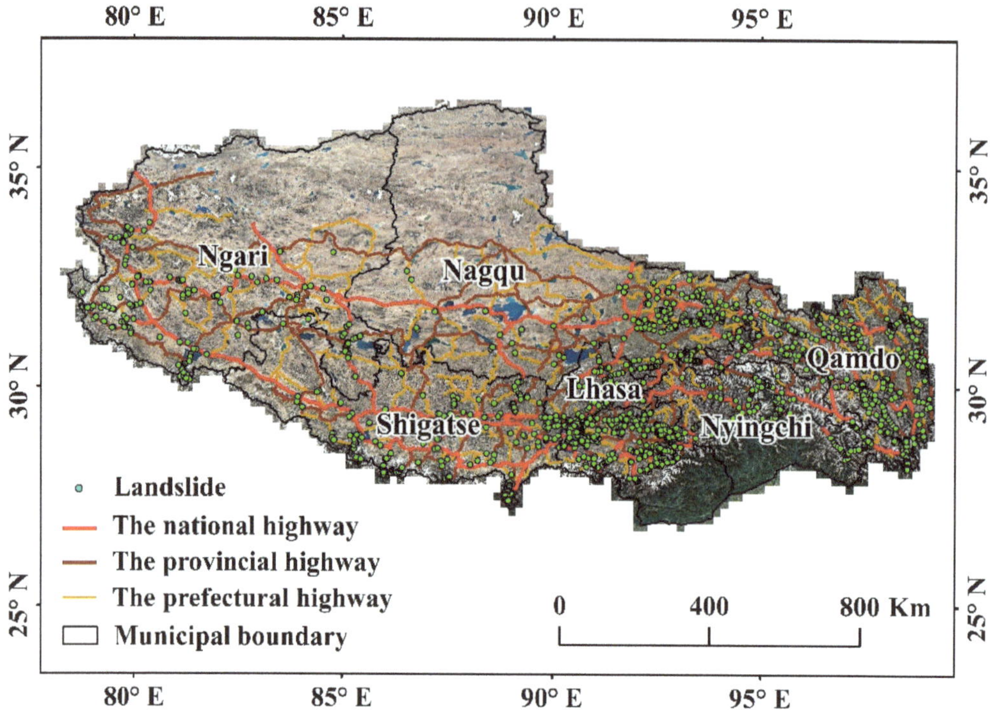

Figure 1. Geographical location map of study area.

Table 1. Data used in this study.

Item	Data Format	Time (Year)	Scale or Resolution	Sources
Road network	Vector	2021	1:1,000,000	National geographic information resource directory service system (https://www.webmap.cn/main.do?method=inde (accessed on 18 November 2022))
Landslide	Vector	-	-	Environmental Sciences and Data, Chinese Academy of Sciences (https://www.resdc.cn (accessed on 12 November 2022))
DEM	Raster	2019	30 m	Computer Network Information Center of Chinese Academy of Sciences, Geography (https://www.cnic.cn/front/pc.html?_1608531541135#/cnicSite/arpKnowledge/324/168/324/TRUE (accessed on 15 November 2022))
NDVI	Raster	2001–2020	30 m	The National Data Center for Ecological Sciences Science Data Bank (http://www.nesdc.org.cn/sdo/detail?id=60f68d757e28174f0e7d8d49 (accessed on 19 November 2022))
Precipitation	Raster	2001–2020	1000 m	Science Data Bank (https://www.scidb.cn/en/cstr/31253.11.sciencedb.01607 (accessed on 18 November 2022))
River system	Vector	2021	1:1,000,000	National geographic information resource directory service system (https://www.webmap.cn/main.do?method=inde (accessed on 4 December 2022))
Glacier snow	Vector	2014	-	Cold and arid region scientific data center (http://westdc.westgis.ac.cn (accessed on 18 November 2022))
Geological fault zone	Vector	-	1:1,000,000	Data Sharing Infrastructure of National Earthquake Data Center (http://data.earthquake.cn (accessed on 21 November 2022))
Seismic center	Vector	1908–2019	-	Environmental Sciences and Data, Chinese Academy of Sciences (https://www.resdc.cn (accessed on 4 December 2022))
Soil erosion	Raster	-	1000 m	Environmental Sciences and Data, Chinese Academy of Sciences (https://www.resdc.cn (accessed on 14 November 2022))
Habitation	Vector	2021	-	National geographic information resource directory service system (https://www.webmap.cn/main.do?method=inde (accessed on 7 December 2022))

2.2.2. Data Preprocessing

To achieve consistency in the multisource data spatial benchmark, we uniformly adopted or converted the data layer into the WGS_1984_World_Mercator for processing and analysis [43]. As shown in Table 1, the resolution of the basic data used in the research is not completely consistent. This research used the resampling tool in GIS to unify the resolution of different data to 30 m. Among them, the cubic convolution map algebra technique was used for continuous data, such as slope, and the nearest neighbor method was used for categorical data, such as soil erosion intensity [44]. The continuous variable type environmental factors were graded, while the discrete variable types were classified and quantified. The classification quantification (Table 2) and thematic map (Figure 2) of the environmental evaluation factors of each landslide geological hazard were generated.

Table 2. Quantitative classification of landslide environmental evaluation factors.

Primary Variable	Environmental Evaluation Factors	Sources/Quantization Level
Topography	Slope (°)	(0–8), (8–18), (18–29), (29–41), (41–87)
	Topographic relief (m)	(0–14), (14–29), (29–48), (48–76), (76–1101)
Soil-vegetation conditions	Soil erosion intensity	Slight, mild, moderate, intense, violent
	NDVI	(0–0.07), (0.07–0.19), (0.19–0.36), (0.36–0.59), (0.59–1)
Hydrology	Annual mean precipitation (mm)	(103–307), (307–414), (414–558), (558–702), (702–1292)
	Distance from river system (m)	(0–2683), (2683–5161), (5161–7892), (7892–12369), (12369–32160)
	Distance from glacier snow (m)	(0–22,847), (22,847–50,636), (50,636–84,646), (84,646–133,843), (133,843–254,081)
Crustal motion	Distance from geological fault zone (m)	(0–11,650), (11,650–22,926), (22,926–37,011), (37,011–55,852), (55,852–126,975)
	Distance from seismic center (m)	(0–18,107), (18,107–33,688), (33,688–52,430), (52,430–76,820), (76,820–139,894)
Human activity	Distance from habitation (m)	(0–27,044), (27,044–79,403), (79,403–150,741), (150,741–233,019), (233,019–357,650)

Linear vector data, such as rivers and ditches of road networks and drainage elements, were classified, extracted, stitched, cut, stored, and grid into single values. The European distance from each cell grid to the nearest source was calculated by using the spatial distance analysis module of ArcGIS 10.8. The same method was also applied to the single value raster processing of the spatial distribution vector data of glaciers and snow, geological fault zones, seismic center, and habitation in the study area to calculate the Euclidean distance from each cell to the nearest source. In the process of DEM preprocessing, we used the grid surface analysis and a map algebra module to calculate and derive slope and topographic relief. After smoothing and normalizing the data, we obtained the average value of each pixel relative to NDVI over the years. For China's monthly precipitation data set from 2001 to 2020 in NetCDF format, the annual mean precipitation data of the study area in GeoTiff format were obtained through Python programming processing. Based on the general requirements of soil Erosion Classification Standard SL190-96 of the People's Republic of China, the degree of soil erosion in the study area was divided into five grades: slight, mild, moderate, intense, and violent. According to experimental requirements, of the 2521 geological disaster points in Tibet in recent years, 70% were randomly extracted as training sample points [37], and the remaining 30% were used to test and verify model accuracy.

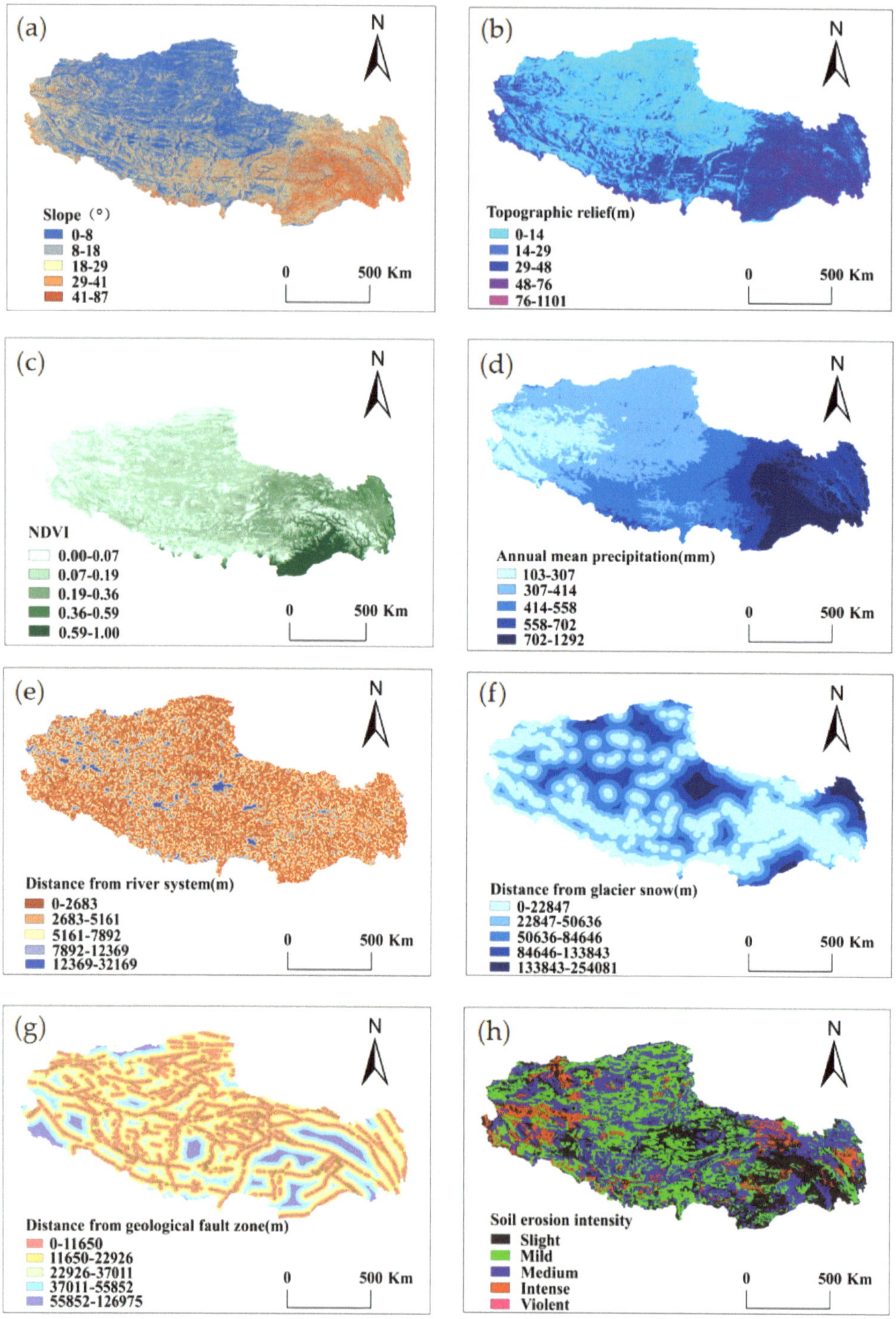

Figure 2. *Cont.*

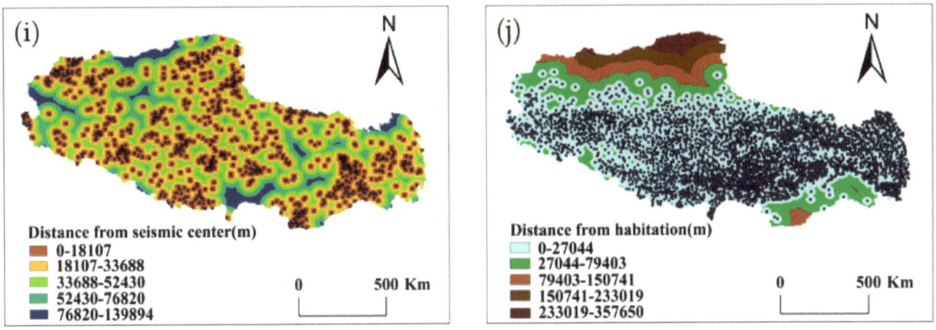

Figure 2. Landslide environmental factors: (**a**) slope; (**b**) topographic relief; (**c**) NDVI; (**d**) annual mean precipitation; (**e**) distance from river system; (**f**) distance from glacier snow; (**g**) distance from geological fault zone; (**h**) soil erosion intensity; (**i**) distance from seismic center; and (**j**) distance from habitation.

3. Methods

3.1. Multicriteria Heuristic Analytical Hierarchy Process Model

By comprehensively collecting relevant data, the environmental factor mechanism of road landslide geological hazards in the research area was analyzed, and a heuristic hierarchical analysis model with five criteria and ten indicators was established. The multicriteria heuristic analytical hierarchy process model integrates qualitative and quantitative system analyses, simulates human thinking and decision-making processes [45], excavates and extracts the data of multifactor complex systems, and has many advantages, such as clear thinking and strong systems. Based on prior knowledge of professionals or cross fields, the method model can be applied in hierarchical clustering and combinations according to its influence and affiliation when solving specific problems to form an orderly hierarchical structure analysis and evaluation model.

In this study, the target layer was the road network vulnerability evaluation of the main environmental factors in Tibet: terrain changes, vegetation–soil conditions, regional hydrological conditions, geological activity structure, and habitation intensity. The index layer comprises slope, topographic fluctuation, annual average vegetation coverage, annual average precipitation, distance from river system, distance from glacier and snow, distance from geological fault zone, distance from seismic center, distance from habitation and soil erosion intensity. Constructing a judgment matrix is a key step in the heuristic analytical hierarchy process. The judgment matrix represents the relative importance of each element at the previous level. Based on this criterion, the options were compared pairwise and graded according to their importance. The nth-order judgment matrix is as follows:

$$A = \begin{pmatrix} \mathrm{b1/b1} & \mathrm{b1/b2} & \ldots & \mathrm{b1/bn} \\ \mathrm{b2/b1} & \mathrm{b1/b1} & \ldots & \mathrm{b1/bn} \\ \vdots & \vdots & & \vdots \\ \mathrm{bn/b1} & \mathrm{bn/b2} & \ldots & \mathrm{bn/bn} \end{pmatrix} \quad (1)$$

where A denotes the judgment matrix. The scales of the judgment matrix elements were defined in Table 3. Based on multicriteria, prior knowledge, and expert-assisted decisions, the following judgment matrix was formed for the target layer (Table 4).

Table 3. Quantitative classification of geological disaster.

Factor i over Factor j	Scaling
Equal	1
Moderate	3
Strong	5
Very strong	7
Extreme	9
Median value of two adjacent judgments	2, 4, 6, 8
Opposites	Reciprocals

Table 4. Judgment matrix and weight.

	B1	B2	B3	B4	B5	W
B1	1	2	3	1/3	1/2	0.17
B2	1/2	1	3	1/3	1	0.14
B3	1/3	1/3	1	1/5	1/3	0.06
B4	3	3	5	1	3	0.44
B5	2	1	3	1/3	1	0.19

B1-terrain change status; B2-soil–vegetation conditions; B3-regional hydrological conditions; B4-crustal motion; B5-human activity intensity; W is the weight value.

To investigate whether the results of the analytical hierarchy process are reasonable, a consistency test for the determination matrix is required. To test the consistency index *CI* of the judgment matrix, it is necessary to divide the consistency index *CI* by the same order as the average random consistency index *RI* table (Table 5), and the consistency ratio of the judgment matrix should be recorded as *CR*.

$$CI = \frac{\lambda - \mathrm{n}}{n-1} \tag{2}$$

$$CR = \frac{CI}{RI} \tag{3}$$

Table 5. Random consistency index (RI).

n	1	2	3	4	5	6	7	8	9	10	11	12	13	14
RI	0	0	0.52	0.89	1.12	1.24	1.32	1.41	1.45	1.49	1.52	1.54	1.56	1.58

The judgment matrix is considered to have satisfactory consistency when CR < 0.1; otherwise, when CR $\geq$ 0.1, the judgment matrix is adjusted until it is satisfactory. After MATLAB programming calculation, the maximum feature root of the above judgment matrix is λ = 5.19, and the feature vector is the weight vector W in the table, CI = 0.05, RI = 1.12, and CR = 0.04 < 0.10. This judgment matrix exhibited satisfactory consistency. Regarding the importance of each index layer element to the corresponding criterion layer, the evaluation results are prone to overfitting or underfitting phenomena when analyzing and mining the landslide geological disasters distribution law. Therefore, in this study, the fuzzy logic model of quantitative statistics was used for improvement.

3.2. Fuzzy Comprehensive Evaluation Model

The fuzzy comprehensive evaluation method transforms qualitative evaluations into quantitative evaluations according to the membership degree theory of fuzzy mathematics.

It has the advantages of being simple, practical, systematic, and can comprehensively evaluate fuzzy problems [46]. The basic idea of the fuzzy comprehensive evaluation method is based on the principles of fuzzy linear transformation and maximum membership, analysis of the factors affecting the objects to be evaluated, and comprehensive evaluation through fuzzy operation. It is suitable for evaluating objects that are restricted by a variety of factor attributes; however, a reasonable overall evaluation result can be comprehensively judged through the information characteristics of the attributes of each index.

In a complex risk evaluation system for landslide geological disasters, there are many factors, and there are different levels between them [47]. It was necessary to divide the evaluation factors into several categories according to certain attributes, conduct a comprehensive evaluation of each category, and then conduct a high-level comprehensive evaluation of the various types of evaluation results.

Step 1. Determine the factor set for the evaluation object. The object to be evaluated is considered as a finite domain U = {u1,u2, ..., up}, where p is the number of factors involved in the evaluation, and set U is the factor set. The factor set in this study is the entire element of the index layer.

Step 2. Determine the evaluation level domain; that is, the set of V levels. Each level can correspond to a fuzzy subset V = {v1, v2, vi, ..., vm}, and the evaluation set in this study is the entire element of the above criterion layer. The grades were divided into five classes: terrain changes were minor, small, medium, large, and greater; vegetation–soil conditions were worse, poor, medium, good, and superior; regional hydrological conditions were worse, poor, medium, good, and superior; distance from crustal motion were near, nearer, medium, far, and farther; and human activity intensity were slight, mild, moderate, intense, and violent.

Step 3. Construct the fuzzy relationship matrix, R. The grade fuzzy subset is established and needs to consider objects individually from each factor ui (i = 1, 2, ..., p) on quantification, namely from the perspective of single factors, evaluated in the grade fuzzy subset of membership (R | ui), where m represents the level number of the fuzzy set. To reveal and excavate the nonlinear relationship between environmental factors and induced landslide geological disasters, the membership function was determined using the fuzzy statistics method, and the membership frequency method was used to obtain the fuzzy relationship matrix. The membership function of the corresponding hierarchical fuzzy subset is.

$$R \mid u_{slope} = \sum_{i \geq 0}^{n} Ui_{slope} / \sum_{x \geq 1}^{m} V\,x_{slope} \tag{4}$$

$$R \mid u_{topographic\ relief} = \sum_{i \geq 0}^{n} Ui_{topographic\ relief} / \sum_{x \geq 1}^{m} V\,x_{topographic\ relief} \tag{5}$$

$$R \mid u_{NDVI} = \sum_{i \geq 0}^{n} Ui_{NDVI} / \sum_{x \geq 1}^{m} V\,x_{NDVI} \tag{6}$$

$$R \mid u_{annual\ mean\ precipitation} = \sum_{i \geq 0}^{n} Ui_{annual\ mean\ precipitation} / \sum_{x \geq 1}^{m} V\,x_{annual\ mean\ precipitation} \tag{7}$$

$$R \mid u_{distance\ from\ river\ system} = \sum_{i \geq 0}^{n} Ui_{distance\ from\ river\ system} / \sum_{x \geq 1}^{m} V\,x_{distance\ from\ river\ system} \tag{8}$$

$$R \mid u_{distance\ from\ glacier\ snow} = \sum_{i \geq 0}^{n} Ui_{distance\ from\ glacier\ snow} / \sum_{x \geq 1}^{m} V\,x_{distance\ from\ glacier\ snow} \tag{9}$$

$$R \mid u_{distance\ from\ geological\ fault\ zone} = \sum_{i \geq 0}^{n} Ui_{distance\ from\ geological\ fault\ zone} / \sum_{x \geq 1}^{m} V\,x_{distance\ from\ geological\ fault\ zone} \tag{10}$$

$$R \mid u_{\text{soil erosion}} = \sum_{i \geq 0}^{n} Ui_{\text{soil erosion}} / \sum_{x \geq 1}^{m} V\, x_{\text{soil erosion}} \quad (11)$$

$$R \mid u_{\text{distance from seismic center}} = \sum_{i \geq 0}^{n} Ui_{\text{distance from seismic center}} / \sum_{x \geq 1}^{m} V\, x_{\text{distance from seismic center}} \quad (12)$$

$$R \mid u_{\text{distance from habitation}} = \sum_{i \geq 0}^{n} Ui_{\text{distance from habitation}} / \sum_{x \geq 1}^{m} V\, x_{\text{distance from habitation}} \quad (13)$$

Step 4. Fuzzy comprehensive evaluation results. Based on the above weight allocation and fuzzy transformation operation of each evaluation factor, the fuzzy comprehensive evaluation results were obtained.

$$B = A \cdot R = [0.17\ 0.14\ 0.06\ 0.44\ 0.19] \cdot \begin{bmatrix} 0.14 & 0.23 & 0.18 & 0.08 & 0.50 & 0.42 & 0.35 & 0.30 & 0.32 & 0.99 \\ 0.25 & 0.32 & 0.31 & 0.08 & 0.32 & 0.33 & 0.25 & 0.23 & 0.26 & 0.01 \\ 0.25 & 0.27 & 0.33 & 0.36 & 0.13 & 0.17 & 0.17 & 0.33 & 0.24 & 0.00 \\ 0.23 & 0.16 & 0.12 & 0.35 & 0.05 & 0.04 & 0.17 & 0.14 & 0.15 & 0.00 \\ 0.12 & 0.02 & 0.06 & 0.13 & 0.01 & 0.04 & 0.06 & 0.00 & 0.04 & 0.00 \end{bmatrix} \quad (14)$$

where A is the vector of the weight coefficient of each fuzzy subset level, and R is the decision matrix of the fuzzy relationship. To consider various factors, ordinary matrix multiplication was used for the synthesis operation because the model is suitable for sorting multiple factors. B is the fuzzy comprehensive evaluation result of each evaluation index, which was normalized to

$$B_{\text{Normalized weights}} = [0.11\ 0.10\ 0.09\ 0.13\ 0.09\ 0.08\ 0.11\ 0.09\ 0.10\ 0.10] \quad (15)$$

3.3. Frequency Ratio-Interactive Fuzzy Overlay Analysis

Because existing related studies have the problem of training the model with the training set, the test set was used to test the model generalization ability and adjust the model according to the accuracy of the test set. To improve the operational efficiency and generalization ability of the presentation model, this study proposed a ubiquitous machine-learning model that attempts to supervise the model training stage based on the feature vector and regional fuzzy measure operator. The ubiquitous machine learning model is based on set theory [3,48]. In the "fuzzy superposition" analysis, the set corresponds to the subordinate class, and the data value is reclassified or transformed to the same informational feature level of each element. The results of the frequency ratio interactive fuzzy superposition analysis quantified the likelihood that each position in the input grid belonged to the fuzzy set of each evaluation index grade. Figure 3 shows the landslide frequency ratios [49] of the grid values for each environmental assessment factor. For example, the slope of a region indicates the possibility of transformation and fuzzy superposition.

The grid layer of multiple landslide geological hazard risk indicators was comprehensively analyzed, and a regional vulnerability analysis was conducted by grading the combination of nonlinear superposition operators. Because the membership frequency ratio of each environmental risk factor in this study is between [0, 1] and is stored in grade, a nonlinear superposition operator should be used for fuzzy stack analysis [50]. Through reference to related studies and experimental analysis, it is found that the combination of Fuzzy Gamma and Fuzzy Product superposition operators, especially the value of $0.95 < \gamma < 1$, has a good measure for the dispersion of local risk values in the study area.

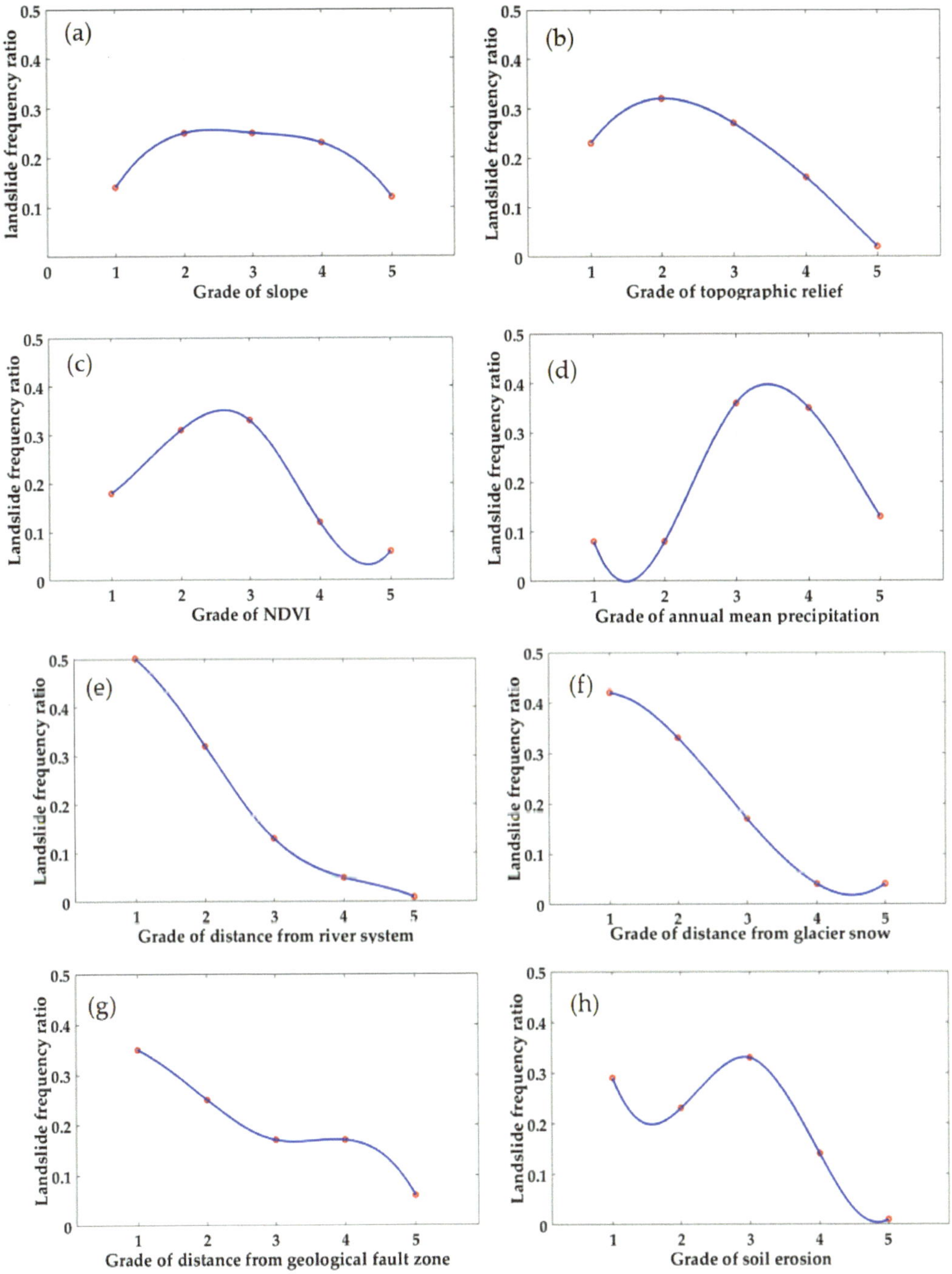

Figure 3. *Cont.*

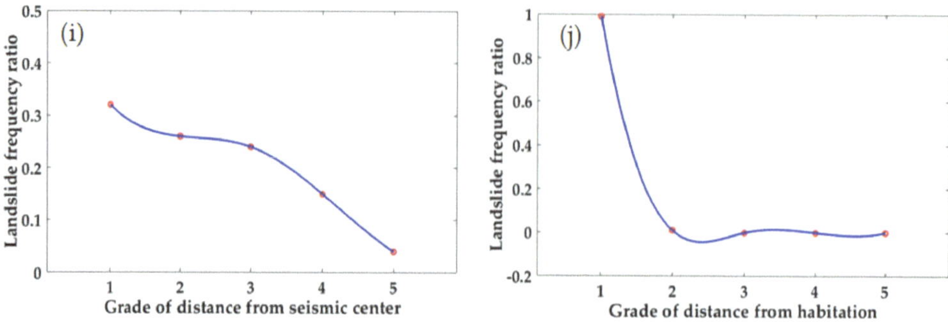

Figure 3. Landslide frequency ratio curve of each evaluation index: (**a**) grade of slope; (**b**) grade of topographic relief; (**c**) grade of NDVI; (**d**) grade of mean annual precipitation; (**e**) grade of distance from river system; (**f**) grade of distance from glacial snow; (**g**) grade of distance from geological fault zone; (**h**) grade of soil erosion; (**i**) grade of distance from seismic center; and (**j**) grade of distance from habitation.

4. Results

4.1. Frequency Ratio-Interactive Fuzzy Overlay Analysis

In this study, we adopted the natural breakpoint method to classify the risk index into five categories [51]: very low (0.00–0.68), low (0.68–0.73), medium (0.73–0.77), high (0.77–0.82), and very high (0.82–1.00). As can be seen from Figure 4, the high hazard risk is in the eastern and southern regions of the study area, relatively dispersed in the middle, western, and northern regions of the area, the median risk is relatively evenly distributed, and the low hazard risk is distributed in the midland, northern, and western regions. Areas with very low, low, medium, high, and very high risk indices accounted for 6.29%, 19.07%, 25.80%, 26.21%, and 22.63% of the study area, respectively.

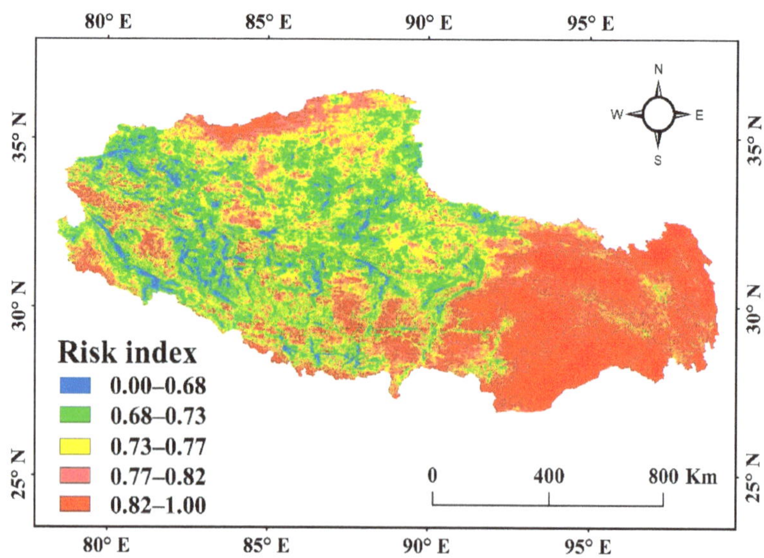

Figure 4. Landslide geological disaster risk distribution in the study area.

From the distributed regional environment perspective, landslide development is not only affected by regional hydrological conditions, geological environment, topography, vegetation coverage, soil erosion, and other natural conditions, but also by the frequency of human engineering activity near towns and along highways. The Himalayan–Hengduan Mountain valley with great topography and a developed water system, located in the alpine and canyon areas of eastern Tibet, with its strong precipitation, broken structure, and crushed fault zone, is a high-risk area for landslide geological disasters. However, the vast areas in midland, western, and northern Tibet, with less human engineering activity, and valleys and basins are mild- and medium-risk areas for landslide geological disasters.

4.2. Test of Evaluation Results

In this study, the rationality of the evaluation results was tested by selecting the distribution of the sample points at risk levels (Figure 5) to ensure the objectivity and robustness of the evaluation model [51]. From the test results, the risk division results, and the distribution of the test sample points were analyzed by spatial statistics. The results showed that 291 geological disaster test sites are in very high-risk areas, 270 in high-risk areas, 123 in medium-risk areas, and only 41 and 31 in low-risk and very low-risk areas, respectively. That is 90.48% of the geological disaster test sites were located above the medium-risk zone and 74.21% were located above the high-risk areas. With an increase in risk level of landslide geological disasters, the geological disaster occurrence rate increases significantly. The diagnosis indicated the excellent accuracy of the evaluation model.

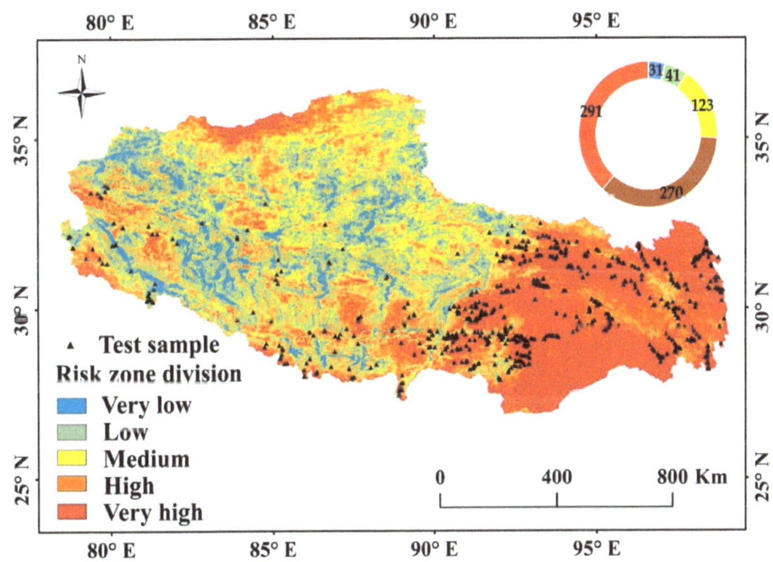

Figure 5. Test sample distribution of landslide geological disasters at each level of risk.

4.3. Analysis of Road Network Vulnerability

The linear data of the main road network from national to prefectural roads in the study area were associated with the landslide geological hazard-prone risk results obtained from the comprehensive evaluation. The vulnerability distribution of the regional road network in Tibet, China, is shown in Figure 6. The vulnerability index was divided into five grades: very low (0.52–0.66), low (0.66–0.71), medium (0.71–0.76), high (0.76–0.80), and extremely high (0.80–0.90). In addition, we proposed a method to calculate the vulnerability distribution percentage of the interurban regional road network.

$$\mathrm{Pvul} = \mathrm{M}(v_h + v_{eh}) / \mathrm{M}_{total} \quad (16)$$

where Pvul is the vulnerability distribution percentage of the regional road network, $M(v_h + v_{eh})$ is the mileage value of vulnerability grade of high and extremely high level of the road network in a city, and M_{total} is the total mileage value of all studied roads in the city. Table 6 presents the distribution of road network vulnerabilities among the cities in the study area. Overall, more than two-thirds of the roads in Qamdo and Nyingchi had high and extremely high vulnerability levels, but less than a third of the roads in Ngari and Xigazehad had the same vulnerability. More than half of the roads in Lhasa and Shannan cities had high and extremely vulnerability levels while Naqu city is the opposite.

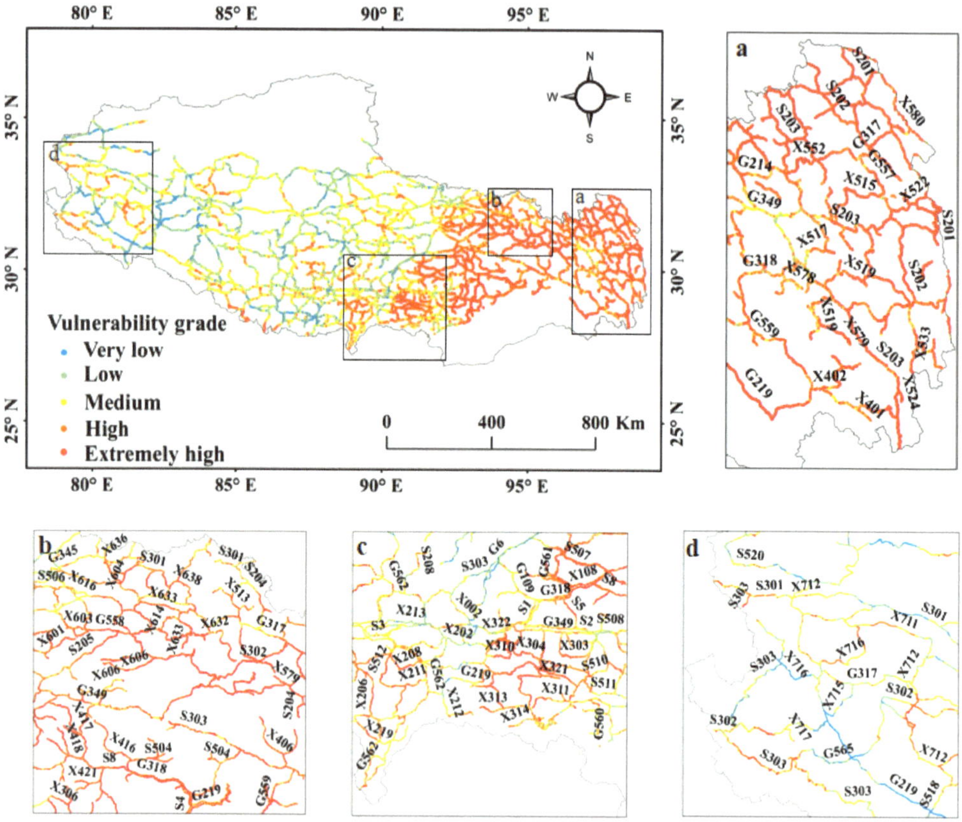

Figure 6. Vulnerability level distribution of the main road network in the study area ((**a**–**d**) are enlarged maps of typical high-risk road clusters, corresponding to the east, north, south and west of the study area respectively).

Table 6. Road network vulnerability distribution of seven Tibetan cities.

Risk Level (km)	Ngari Prefecture	Qamdo City	Lhasa City	Nyingchi City	Nagqu City	Xigaze City	Shannan City
Very low	1590	1	153	0	660	945	28
Low	2734	91	319	11	2833	2470	322
Medium	3001	579	504	268	3763	3206	976
High	1394	2059	714	1597	2785	1972	1467
Extremely high	144	3861	525	1981	1316	421	1043
Total mileage	8862	6591	2215	3858	11357	9015	3836

We selected four high risk road aggregation areas (Figure 6a–d). From the perspective of the distribution of high (including extremely high) vulnerability sections, the eastern and western sections of national highway G219 (Xin–Zang highway), G317 (north line of Sichuan–Tibet highway), the eastern and middle sections of G318 (southern line of Sichuan–Tibet highway), G214 (Yunnan–Tibet highway), G559, the provincial highways S201, S303, S8, and the prefectural highways X108, X321, X402, X508, and X606 all belong to this vulnerability level. The road network had denser branch lines than the road network structure. From a geographical perspective, these highly vulnerability sections exhibit distinct regional characteristics. In addition to overlapping with the high risk of landslide geological disasters, they occur in traffic arteries, link aggregations, and areas of relatively frequent human activity. For example, the G219 line and the Prochacha highway extending to Yunnan Province have been incorporated into the new G219, which serves the construction of the new Yunnan–Tibet highway and connects with the Sichuan–Tibet highway, namely the G318. In addition, G317, G318, G214, G559, S303, and other highways into Tibet have become important links for transportation, social production, border trade, cultural tourism, national defense construction, and ethnic unity. The S201 and S302, X108 and X401, and other branch roads in the area are the only sections for intercounty and rural trade, medical treatment, schools, and commuting. The vulnerability of these typical sections was prominent because of the landslide geological disaster risk.

5. Discussion

There is increasing concern about lifeline systems (e.g., water and energy supply, sewage, communication, and transport systems) and their vulnerability to damage and cause disruption during disasters such as geological hazards [1,17,52]. Vulnerability can be characterized across road networks or between risk scenarios for different networks [53]. Currently, most vulnerability studies focus on one aspect of vulnerability of the road network and lack an integrated consideration of indicators at different levels. In addition, there were more studies on the vulnerability of road networks in urban transport systems, while few studies have been conducted on transport networks in border areas [54]. Therefore, in this study, in response to the characteristic that road network vulnerability is mainly affected by natural environmental factors. We achieved qualitative and quantitative analysis and evaluation by combining a multicriteria heuristic analytical hierarchical process model [55] and fuzzy set theory [56,57], which is used for geological disaster risk assessment and mapping [58], on the basis of the existing studies and expert knowledge. In addition, we proposed a frequency ratio interactive fuzzy superposition algorithm. The landslide geological disaster risk map was generated by integrating the selected thematic layers through the comprehensive analysis of multiple landslide geological disaster risk indicators. Then, the credibility and accuracy analyses were carried out and the results were verified. Finally, the road network data in the study area were correlated with the landslide geological hazard susceptibility risk results obtained from the comprehensive evaluation to obtain the susceptibility distribution. The results show that the proposed three-stage framework based on the multicriteria heuristic analysis hierarchical process model, fuzzy comprehensive evaluation, and frequency ratio-interactive fuzzy superposition analysis is a reliable risk classification method, which is effective for vulnerability evaluation and zoning of road networks [59].

Whether single knowledge driven or data-driven [2,60,61], the evaluation results are prone to over or underfitting. Although this study analyzes and excavates the distribution law of landslide geological disasters and adopts a fuzzy logic model of quantitative statistics with hierarchical thought, some deficiencies remain. For instance, only the evaluation factors for road network vulnerability were selected, and traffic accidents were not included. In the frequency ratio-interactive fuzzy superposition of the model training stage, the γ values of the nonlinear superposition operators need to be adjusted within a reasonable range to obtain evaluation results with good measurement significance and astringency. Although the relevant evaluation model has moved toward complexity, regional environmental differences combined with road network vulnerability analysis is weak, and prediction accuracy and versatility require further improvement. Over the past years, observational, environmental, and instrumental data and the method of index verification was analyzed and verified by concerned study. These methods include support vector machine models [62], logistic regression models [63], random forest models [64], artificial neural networks [65], and deep learning algorithms [48]. In addition, uncertainty quantification (UQ) is an important benchmark to assess the performance of artificial intelligence (AI) and particularly deep learning ensembled-based models [66]. In the fairness of decision making, we should also pay attention to the characteristics of the road network itself and the local disaster relief capacity, and then combine the local economy, environment, and population to quantify the evaluation indicators. Meanwhile, because of the complex internal mechanism of the landslide geological disasters factors, there is no unified selection standard. Challenges still exist regarding how AI can be better applied to the vulnerability assessment of landslide geological disasters. Therefore, determining a set of general factor selection criteria and evaluation models should be the subject of future research and development. The three-stage framework proposed in this study can provide correct inferences and explanations for the potential phenomena. Firstly, on the basis of constructing a target-based vulnerability assessment system of the mountain road network, the preliminary vulnerability assessment of the mountain road network was completed. Then, using the preliminary results as a sample set, a road vulnerability prediction model can be constructed using the multiple machine learning algorithms. Finally, the model with the highest prediction accuracy was selected to complete the vulnerability assessment of road network for landslide geological disasters in high and remote mountainous areas.

The strategic position of the study area is important, and human activity is mainly concentrated in towns and plateau valleys with relatively dense road networks. The road network vulnerability research results were based on the risk of landslide geological disasters, and for different vulnerability sources, different measures or techniques can be taken to reduce the risks and consequences of vulnerability. For example, increasing the connection strength of low-grade and high-grade roads; exploring scientifically feasible three-dimensional traffic in plateau mountainous areas; improving the operational rate of road networks; construction and regular inspection of slope protection engineering facilities; scientific and reasonable allocation of emergency rescue facilities, personnel, and materials; difficult highway construction and maintenance in the study area; and strengthening traffic control and diversion based on the carrying capacity of roads and big data of traffic flow in key areas with frequent human activity [67,68].

6. Conclusions

The three-stage framework proposed in this study can provide correct inferences and explanations for the potential phenomena of landslide geological disasters. Landslide geological disasters are among the most destructive disasters and can cause traffic blockages related to road facilities, as well as direct and indirect economic losses. This study, guided by objectivity and operability, started with the main environmental elements of geological disasters in the breeding area, adopted a multicriteria heuristic analytical hierarchy process model, fuzzy comprehensive evaluation, and frequency ratio-interactive fuzzy stack analysis of the hierarchical thinking mode, and comprehensively evaluated the road vulnerability network in Tibet, China. In this study, based on relevant research results and expert consultation, regional characteristics, evaluation factors, and an index system were selected, and multisource data were investigated and collected. The environmental factors of the continuous variable type were classified and discretized, while those of the discrete variable type were classified and quantified. First, a heuristic hierarchical analysis model with five criteria and ten indexes was established. Subsequently, according to the evaluation object being restricted by various factors, a fuzzy comprehensive evaluation logic model of quantitative statistics was used, and the fuzzy comprehensive evaluation results of each evaluation index were obtained. Moreover, through frequency ratio interactive fuzzy superposition, the risk evaluation of the hierarchical superposition combination of the nonlinear superposition operator was adopted, and the spatial distribution characteristics were obtained. Finally, the road network vulnerability in the study area was evaluated. This research proposed a reasonably efficient and well-measured hierarchical research paradigm and model to predict road network vulnerability to geological disasters in high mountainous areas. The limitation is that the values of the nonlinear superposition operators need to be adjusted within a reasonable range to obtain evaluation results with good measurement significance and astringency. The main conclusions were summarized as follows.

(1) The high-value areas of geological hazard risk are in the eastern and southern parts of the study area and are relatively dispersed in the western part of the study area; the medium geological hazard risk is relatively evenly distributed in the north; and the low-value area of geological hazard risk is distributed in the midland, northern, and western regions. The Himalayan–Hengduan Mountains in southern Tibet, located in the alpine and canyon areas of Tibet, with great topography and a developed water system with strong precipitation, broken structures, and crushed fault zones, are high-risk areas for geological landslide disasters. However, the vast areas in midland, western, and northern Tibet, with less human engineering activity, and valleys and basins, are mild- and medium-risk areas for landslide geological disasters.

(2) The distribution of road network vulnerability in China's Tibet has the following characteristics. In general, more than two-thirds of the roads in Qamdo and Nyingchi have high and extremely high vulnerability levels, but less than a third of the roads in AliNgari, Xigaze, and Nagqu have more than half the same vulnerability. More than half of the roads in Lhasa and Shannan Cities have high and extremely high vulnerability levels while Naqu city is the opposite.

(3) In the study area, the high-vulnerability section presents significant regional characteristics. In addition to overlapping with the area with a high risk of landslide geological disasters, its distribution is mostly located in traffic arteries, link aggregations, and relatively frequent human activity.

Author Contributions: Conceptualization, Y.Y.; methodology, Y.Y. and S.C.; software, Y.Y., M.C. and S.D.; validation, M.C. and J.Y.; formal analysis, H.C. and Z.L.; investigation, Y.Y. and S.C.; resources, L.C., data curation, Y.Y. and L.C.; writing—original draft, Y.Y.; writing—review and editing, L.C., S.C., H.C. and N.L.; visualization, Y.Y. and Y.G.; supervision, L.C. and S.C.; project administration, L.C. and S.C. All authors have read and agreed to the published version of the manuscript.

Funding: This research was supported by the National Natural Science Foundation of China (grant numbers 42201450) and the National Key Research and Development Program of China (grant numbers 2022YFB3903600).

Data Availability Statement: The data presented in this study are available on request from the corresponding author.

Acknowledgments: We would like to extend special thanks to the editor and the anonymous reviewers for their valuable comments on the previous version of the manuscript.

Conflicts of Interest: The authors declare no conflict of interest.

References

1. Qiang, Y.; Xu, J. Empirical Assessment of Road Network Resilience in Natural Hazards Using Crowdsourced Traffic Data. *Int. J. Geogr. Inf. Sci.* **2020**, *34*, 2434–2450. [CrossRef]
2. Zhang, Q.; Yu, H.; Li, Z.; Zhang, G.; Ma, D.T. Assessing Potential Likelihood and Impacts of Landslides on Transportation Network Vulnerability. *Transp. Res. Part D Transp. Environ.* **2020**, *82*, 102304. [CrossRef]
3. Ha, H.; Bui, Q.D.; Khuc, T.D.; Tran, D.T.; Pham, B.T.; Mai, S.H.; Nguyen, L.P.; Luu, C. A Machine Learning Approach in Spatial Predicting of Landslides and Flash Flood Susceptible Zones for a Road Network. *Model. Earth Syst. Environ.* **2022**, *8*, 4341–4357. [CrossRef]
4. Amlan, H.A.; Hassan, S.A.; Alias, N.E. Discovering the Global Landscape of Vulnerability Assessment Method of Transportation Network Studies: A Bibliometric Review. *Phys. Chem. Earth* **2023**, *129*, 103336. [CrossRef]

5. Fan, H.; Lu, Y.; Hu, Y.; Fang, J.; Lv, C.; Xu, C.; Feng, X.; Liu, Y. A Landslide Susceptibility Evaluation of Highway Disasters Based on the Frequency Ratio Coupling Model. *Sustainability* **2022**, *14*, 7740. [CrossRef]
6. Ayalew, L.; Yamagishi, H.; Marui, H.; Kanno, T. Landslides in Sado Island of Japan: Part II. GIS-Based Susceptibility Mapping with Comparisons of Results from Two Methods and Verifications. *Eng. Geol.* **2005**, *81*, 432–445. [CrossRef]
7. Tang, R.X.; Kulatilake, P.H.S.W.; Yan, E.C.; Cai, J. Sen. Evaluating Landslide Susceptibility Based on Cluster Analysis, Probabilistic Methods, and Artificial Neural Networks. *Bull. Eng. Geol. Environ.* **2020**, *79*, 2235–2254. [CrossRef]
8. Maghsoudi Moud, F.; Abbaszadeh Shahri, A.; van Ruitenbeek, F.; Hewson, R.; van der Meijde, M. Evaluation of the Modified AHP-VIKOR for Mapping and Ranking Copper Mineralized Areas, A Case Study from the Kerman Metallogenic Belt, SE Iran. *Arab. J. Geosci.* **2022**, *15*, 1756. [CrossRef]
9. El-Maissi, A.M.; Argyroudis, S.A.; Kassem, M.M.; Leong, L.V.; Mohamed Nazri, F. An Integrated Framework for the Quantification of Road Network Seismic Vulnerability and Accessibility to Critical Services. *Sustainability* **2022**, *14*, 12474. [CrossRef]
10. Guzzetti, F.; Carrara, A.; Cardinali, M.; Reichenbach, P. Landslide Hazard Evaluation: A Review of Current Techniques and Their Application in A Multi-Scale Study, Central Italy. *Geomorphology* **1999**, *31*, 181–216. [CrossRef]
11. Moayedi, H.; Mehrabi, M.; Mosallanezhad, M.; Rashid, A.S.A.; Pradhan, B. Modification of Landslide Susceptibility Mapping Using Optimized PSO-ANN Technique. *Eng. Comput.* **2019**, *35*, 967–984. [CrossRef]
12. Xie, W.; Nie, W.; Saffari, P.; Robledo, L.F.; Descote, P.Y.; Jian, W. Landslide Hazard Assessment Based on Bayesian Optimization–Support Vector Machine in Nanping City, China. *Nat. Hazards* **2021**, *109*, 931–948. [CrossRef]
13. Huang, S.; Dou, H.; Jian, W.; Guo, C.; Sun, Y. Spatial Prediction of the Geological Hazard Vulnerability of Mountain Road Network Using Machine Learning Algorithms. *Geomat. Nat. Hazards Risk* **2023**, *14*, 2170832. [CrossRef]
14. Liu, Y.Q.; Wu, X. Analysis and Research on the Influencing Factor of the Road Transportation Network Vulnerability Based on the Interpretative Structural Model. In Proceedings of the 2015 IEEE International Conference on Grey Systems and Intelligent Services (GSIS), Leicester, UK, 18–20 August 2015; pp. 296–300. [CrossRef]
15. Zhang, H.; Yao, Y. An Integrative Vulnerability Evaluation Model to Urban Road Complex Network. *Wirel. Pers. Commun.* **2019**, *107*, 193–204. [CrossRef]
16. Cignetti, M.; Godone, D.; Bertolo, D.; Paganone, M.; Thuegaz, P.; Giordan, D. Rockfall Susceptibility along the Regional Road Network of Aosta Valley Region (Northwestern Italy). *J. Maps* **2021**, *17*, 54–64. [CrossRef]
17. Wang, Z.; Chan, A.P.C.; Yuan, J.; Xia, B.; Skitmore, M.; Li, Q. Recent Advances in Modeling the Vulnerability of Transportation Networks. *J. Infrastruct. Syst.* **2015**, *21*, 1–9. [CrossRef]
18. Pellicani, R.; Argentiero, I.; Spilotro, G. GIS-Based Predictive Models for Regional-Scale Landslide Susceptibility Assessment and Risk Mapping along Road Corridors. *Geomat. Nat. Hazards Risk* **2017**, *8*, 1012–1033. [CrossRef]
19. Hassan, S.A.; Amlan, H.A.; Alias, N.E.; Ab-Kadir, M.A.; Sukor, N.S.A. Vulnerability of Road Transportation Networks under Natural Hazards: A Bibliometric Analysis and Review. *Int. J. Disaster Risk Reduct.* **2022**, *83*, 103393. [CrossRef]
20. Alsabhan, A.H.; Singh, K.; Sharma, A.; Alam, S.; Pandey, D.D.; Rahman, S.A.S.; Khursheed, A.; Munshi, F.M. Landslide Susceptibility Assessment in the Himalayan Range Based along Kasauli–Parwanoo Road Corridor Using Weight of Evidence, Information Value, and Frequency Ratio. *J. King Saud Univ.-Sci.* **2022**, *34*, 101759. [CrossRef]
21. El-Maissi, A.M.; Argyroudis, S.A.; Kassem, M.M.; Mohamed Nazri, F. Integrated Seismic Vulnerability Assessment of Road Network in Complex Built Environment toward More Resilient Cities. *Sustain. Cities Soc.* **2023**, *89*, 104363. [CrossRef]
22. Ouyang, M.; Hong, L.; Mao, Z.J.; Yu, M.H.; Qi, F. A Methodological Approach to Analyze Vulnerability of Interdependent Infrastructures. *Simul. Model. Pract. Theory* **2009**, *17*, 817–828. [CrossRef]
23. Redzuan, A.A.H.; Zakaria, R.; Anuar, A.N.; Aminudin, E.; Yusof, N.M. Road Network Vulnerability Based on Diversion Routes to Reconnect Disrupted Road Segments. *Sustainability* **2022**, *14*, 2244. [CrossRef]
24. Wang, S.; Hong, L.; Ouyang, M.; Zhang, J.; Chen, X. Vulnerability Analysis of Interdependent Infrastructure Systems under Edge Attack Strategies. *Saf. Sci.* **2013**, *51*, 328–337. [CrossRef]
25. Borowska-Stefańska, M.; Kowalski, M.; Wiśniewski, S.; Dulebenets, M.A. The Impact of Self-Evacuation from Flood Hazard Areas on the Equilibrium of the Road Transport. *Saf. Sci.* **2023**, *157*, 105934. [CrossRef]
26. Aslan, E.; Çelik, M. Pre-Positioning of Relief Items under Road/Facility Vulnerability with Concurrent Restoration and Relief Transportation. *IISE Trans.* **2019**, *51*, 847–868. [CrossRef]
27. Guerriero, L.; Napoli, M.D.; Novellino, A.; Martire, D.D.; Rispoli, C.; Lee, K.; Bee, E.; Harrison, A.; Calcaterra, D. Multi-Hazard Susceptibility Assessment Using Analytic Hierarchy Process: The Derwent Valley Mills UNESCO World Heritage Site Case Study (United Kingdom). *J. Cult. Herit.* **2022**, *55*, 339–345. [CrossRef]
28. Chen, A.; Yang, C.; Kongsomsaksakul, S.; Lee, M. Network-Based Acessibility Measures for Vulnerability Analysis of Degradable Transportation Networks. *Netw. Spat. Econ.* **2007**, *7*, 241–256. [CrossRef]
29. Berdica, K. An Introduction to Road Vulnerability: What Has Been Done, Is Done and Should Be Done. *Transp. Policy* **2002**, *9*, 117–127. [CrossRef]
30. Taylor, M.A.P.; Sekhar, S.V.C.; D'Este, G.M. Application of Accessibility Based Methods for Vulnerability Analysis of Strategic Road Networks. *Netw. Spat. Econ.* **2006**, *6*, 267–291. [CrossRef]
31. Sohn, J. Evaluating the Significance of Highway Network Links under the Flood Damage: An Accessibility Approach. *Transp. Res. Part A Policy Pract.* **2006**, *40*, 491–506. [CrossRef]

32. Erath, A.; Birdsall, J.; Axhausen, K.W.; Hajdin, R. Vulnerability Assessment Methodology for Swiss Road Network. *Transp. Res. Rec.* **2009**, *2137*, 118–126. [CrossRef]
33. Du, Z.P.; Nicholson, A. Degradable Transportation Systems: Sensitivity and Reliability Analysis. *Transp. Res. Part B Methodol.* **1997**, *31*, 225–237. [CrossRef]
34. Cheung, K.F.; Bell, M.G.H. Attacker–Defender Model against Quantal Response Adversaries for Cyber Security in Logistics Management: An Introductory Study. *Eur. J. Oper. Res.* **2021**, *291*, 471–481. [CrossRef]
35. Murray, A.T.; Matisziw, T.C.; Grubesic, T.H. A Methodological Overview of Network Vulnerability Analysis. *Growth Chang.* **2008**, *39*, 573–592. [CrossRef]
36. Wu, Q.; Han, Z.; Cui, C.; Liu, F.; Zhao, Y.; Xie, Z. Vulnerability Identification and Cascading Failure Spatiotemporal Patterns on Road Network under the Rainstorm Disaster. *ISPRS Int. J. Geo-Inf.* **2022**, *11*, 564. [CrossRef]
37. Zhang, H.; Ouyang, M.; Sun, W.; Hong, L. An Approach for Accessibility Assessment and Vulnerability Analysis of National Multimodal Transport Systems. *Risk Anal.* **2023**, 1–18. [CrossRef] [PubMed]
38. Drejza, S.; Bernatchez, P.; Marie, G.; Friesinger, S. Quantifying Road Vulnerability to Coastal Hazards: Development of a Synthetic Index. *Ocean Coast. Manag.* **2019**, *181*, 104894. [CrossRef]
39. Ahmad, M.N.; Shao, Z.; Aslam, R.W.; Ahmad, I.; Liao, M.; Li, X.; Song, Y. Landslide Hazard, Susceptibility and Risk Assessment (HSRA) Based on Remote Sensing and GIS Data Models: A Case Study of Muzaffarabad Pakistan. *Stoch. Environ. Res. Risk Assess.* **2022**, *36*, 4041–4056. [CrossRef]
40. Rupi, F.; Bernardi, S.; Rossi, G.; Danesi, A. The Evaluation of Road Network Vulnerability in Mountainous Areas: A Case Study. *Netw. Spat. Econ.* **2015**, *15*, 397–411. [CrossRef]
41. Kasmalkar, I.G.; Serafin, K.A.; Miao, Y.; Bick, I.A.; Ortolano, L.; Ouyang, D.; Suckale, J. When Floods Hit the Road: Resilience to Flood-Related Traffic Disruption in the San Francisco Bay Area and Beyond. *Sci. Adv.* **2020**, *6*, 1–8. [CrossRef]
42. Guo, W.; Liu, S.; Xu, J.; Wu, L.; Shangguan, D.; Yao, X.; Wei, J.; Bao, W.; Yu, P.; Liu, Q.; et al. The Second Chinese Glacier Inventory: Data, Methods and Results. *J. Glaciol.* **2015**, *61*, 357–372. [CrossRef]
43. Mallick, J.; Singh, R.K.; AlAwadh, M.A.; Islam, S.; Khan, R.A.; Qureshi, M.N. GIS-Based Landslide Susceptibility Evaluation Using Fuzzy-AHP Multi-Criteria Decision-Making Techniques in the Abha Watershed, Saudi Arabia. *Environ. Earth Sci.* **2018**, *77*, 276. [CrossRef]
44. Shahri, A.A.; Spross, J.; Johansson, F.; Larsson, S. Landslide Susceptibility Hazard Map in Southwest Sweden Using Artificial Neural Network. *Catena* **2019**, *183*, 104225. [CrossRef]
45. Jena, R.; Pradhan, B.; Beydoun, G.; Alamri, A.; Shanableh, A. Spatial Earthquake Vulnerability Assessment by Using Multi-Criteria Decision Making and Probabilistic Neural Network Techniques in Odisha, India. *Geocarto Int.* **2021**, *37*, 8080–8099. [CrossRef]
46. Zhang, P.; Feng, G. Application of Fuzzy Comprehensive Evaluation to Evaluate the Effect of Water Flooding Development. *J. Pet. Explor. Prod. Technol.* **2018**, *8*, 1455–1463. [CrossRef]
47. Liu, Q.; Tang, A.; Huang, D.; Huang, Z.; Zhang, B.; Xu, X. Total Probabilistic Measure for the Potential Risk of Regional Roads Exposed to Landslides. *Reliab. Eng. Syst. Saf.* **2022**, *228*, 108822. [CrossRef]
48. Lv, L.; Chen, T.; Dou, J.; Plaza, A. A Hybrid Ensemble-Based Deep-Learning Framework for Landslide Susceptibility Mapping. *Int. J. Appl. Earth Obs. Geoinf.* **2022**, *108*, 102713. [CrossRef]
49. Saravanan, S.; Istijono, B.; Jennifer, J.J.; Abijith, D.; Sivaranjani, S. Landslide Susceptibility Assessment Using Frequency Ratio Technique—A Case Study of NH67 Road Corridor in the Nilgiris District, Tamilnadu, India. *IOP Conf. Ser. Earth Environ. Sci.* **2021**, *708*, 012017. [CrossRef]
50. Fang, Z.; Wang, Y.; Peng, L.; Hong, H. A Comparative Study of Heterogeneous Ensemble-Learning Techniques for Landslide Susceptibility Mapping. *Int. J. Geogr. Inf. Sci.* **2021**, *35*, 321–347. [CrossRef]
51. He, Y.; Xiong, J.; Cheng, W.; Yang, J.; He, W.; Yong, Z.; Duan, Y.; Liu, J.; Yang, G.; Wang, N. Assessing the Risk Posed by Flash Floods to the Transportation Network in Southwestern China. *Geocarto Int.* **2022**, *37*, 13210–13228. [CrossRef]
52. Lin, J.; Lin, W. Transportation System Vulnerability Assessment Considering Environmental Impact. *J. Adv. Transp.* **2022**, *2022*, 8711894. [CrossRef]
53. Liu, M.; Agarwal, J.; Blockley, D. Vulnerability of Road Networks. *Civ. Eng. Environ. Syst.* **2016**, *33*, 147–175. [CrossRef]
54. Alexakis, D.D.; Agapiou, A.; Tzouvaras, M.; Themistocleous, K.; Neocleous, K.; Michaelides, S.; Hadjimitsis, D.G. Integrated Use of GIS and Remote Sensing for Monitoring Landslides in Transportation Pavements: The Case Study of Paphos Area in Cyprus. *Nat. Hazards* **2014**, *72*, 119–141. [CrossRef]
55. Bianchini, S.; Solari, L.; Casagli, N. A GIS-Based Procedure for Landslide Intensity Evaluation and Specific Risk Analysis Supported by Persistent Scatterers Interferometry (PSI). *Remote Sens.* **2017**, *9*, 1093. [CrossRef]
56. Leonardi, G.; Palamara, R.; Suraci, F. A Fuzzy Methodology to Evaluate the Landslide Risk in Road Lifelines. *Transp. Res. Procedia* **2020**, *45*, 732–739. [CrossRef]
57. Yin, Y.; Sheng, Y.; He, Y.; Qin, J. Modeling Vague Spatiotemporal Objects Based on Interval Type-2 Fuzzy Sets. *Int. J. Geogr. Inf. Sci.* **2022**, *36*, 1258–1273. [CrossRef]
58. Crawford, M.M.; Dortch, J.M.; Koch, H.J.; Zhu, Y.; Haneberg, W.C.; Wang, Z.; Bryson, L.S. Landslide Risk Assessment in Eastern Kentucky, USA: Developing a Regional Scale, Limited Resource Approach. *Remote Sens.* **2022**, *14*, 6246. [CrossRef]
59. Donnini, M.; Esposito, G.; Pisano, L.; Petrucci, O.; Lollino, P.; Reichenbach, P. National and Regional-Scale Landslide Indicators and Indexes: Applications in Italy. *Open Geosci.* **2022**, *14*, 763–775. [CrossRef]

60. Zeng, H.; Zhu, Q.; Ding, Y.; Hu, H.; Chen, L.; Xie, X.; Chen, M.; Yao, Y. Graph Neural Networks with Constraints of Environmental Consistency for Landslide Susceptibility Evaluation. *Int. J. Geogr. Inf. Sci.* **2022**, *36*, 2270–2295. [CrossRef]
61. Cengiz, L.D.; Ercanoglu, M. A Novel Data-Driven Approach to Pairwise Comparisons in AHP Using Fuzzy Relations and Matrices for Landslide Susceptibility Assessments. *Environ. Earth Sci.* **2022**, *81*, 1–23. [CrossRef]
62. Zhou, C.; Yin, K.; Cao, Y.; Ahmed, B.; Li, Y.; Catani, F.; Pourghasemi, H.R. Landslide Susceptibility Modeling Applying Machine Learning Methods: A Case Study from Longju in the Three Gorges Reservoir Area, China. *Comput. Geosci.* **2018**, *112*, 23–37. [CrossRef]
63. Umar, Z.; Pradhan, B.; Ahmad, A.; Jebur, M.N.; Tehrany, M.S. Earthquake Induced Landslide Susceptibility Mapping Using an Integrated Ensemble Frequency Ratio and Logistic Regression Models in West Sumatera Province, Indonesia. *Catena* **2014**, *118*, 124–135. [CrossRef]
64. Kim, J.C.; Lee, S.; Jung, H.S.; Lee, S. Landslide Susceptibility Mapping Using Random Forest and Boosted Tree Models in Pyeong-Chang, Korea. *Geocarto Int.* **2018**, *33*, 1000–1015. [CrossRef]
65. Pradhan, B.; Lee, S.; Buchroithner, M.F. A GIS-Based Back-Propagation Neural Network Model and Its Cross-Application and Validation for Landslide Susceptibility Analyses. *Comput. Environ. Urban Syst.* **2010**, *34*, 216–235. [CrossRef]
66. Shahri, A.A.; Shan, C.; Larsson, S. A Novel Approach to Uncertainty Quantification in Groundwater Table Modeling by Automated Predictive Deep Learning. *Nat. Resour. Res.* **2022**, *31*, 1351–1373. [CrossRef]
67. Li, X.; Liu, X. Research on Identification Method of Key Road Sections in the Road Network under Disaster Situation. In Proceedings of the 2020 5th International Conference on Electromechanical Control Technology and Transportation (ICECTT), Nanchang, China, 15–17 May 2020; pp. 575–578. [CrossRef]
68. Chamorro, A.; Echaveguren, T.; Allen, E.; Contreras, M.; Cartes, P.; Contreras, M.; Jimenez, G.; Pattillo, C.; Solminihac, H.D.; Vargas, J.; et al. Risk Management System for Road Networks Exposed to Natural Hazards. *Lifelines* **2022**, 166–177. [CrossRef]

remote sensing

MDPI

Article

A Spatial Data-Driven Approach for Mineral Prospectivity Mapping

Indishe P. Senanayake [1,2,3,*], Anthony S. Kiem [1,3], Gregory R. Hancock [1,3], Václav Metelka [3,4,†], Chris B. Folkes [5], Phillip L. Blevin [5] and Anthony R. Budd [3,6]

1 School of Environmental and Life Sciences, College of Engineering, Science and Environment, The University of Newcastle, Callaghan, NSW 2308, Australia; anthony.kiem@newcastle.edu.au (A.S.K.); greg.hancock@newcastle.edu.au (G.R.H.)
2 School of Engineering, College of Engineering, Science and Environment, The University of Newcastle, Callaghan, NSW 2308, Australia
3 Mineral Exploration Cooperative Research Centre, Kensington, WA 6151, Australia; vaclav.metelka@csiro.au (V.M.); anthony.budd@ga.gov.au (A.R.B.)
4 The Commonwealth Scientific and Industrial Research Organisation, Kensington, WA 6151, Australia
5 Geological Survey of New South Wales, Department of Regional NSW, Maitland, NSW 2310, Australia; chris.folkes@regional.nsw.gov.au (C.B.F.); phil.blevin@planning.nsw.gov.au (P.L.B.)
6 Geoscience Australia, Canberra, ACT 2609, Australia
* Correspondence: indishe.senanayake@newcastle.edu.au
† Current affiliation: AIC Mines Ltd., Subiaco, WA 6008, Australia.

Abstract: Mineral prospectivity mapping is a crucial technique for discovering new economic mineral deposits. However, detailed knowledge-based geological exploration and interpretations generally involve significant costs, time, and human resources. In this study, an ensemble machine learning approach was tested using geoscience datasets to map Cu-Au and Pb-Zn mineral prospectivity in the Cobar Basin, NSW, Australia. The input datasets (magnetic, gravity, faults, electromagnetic, and magnetotelluric data layers) were chosen by considering their association with Cu-Au and Pb-Zn mineralization patterns. Three machine learning algorithms, namely random forest (RF), support vector machine (SVM), and maximum-likelihood (MaxL) classification, were applied to the input data. The results of the three algorithms were ensembled to produce Cu-Au and Pb-Zn prospectivity maps over the Cobar Basin with improved classification accuracy. The findings demonstrate good agreement with known mineral occurrence points and existing mineral prospectivity maps developed using the weights-of-evidence (WofE) method. The ability to capture training points accurately and the simplicity of the proposed approach make it advantageous over complex mineral prospectivity mapping methods, to serve as a preliminary evaluation technique. The methodology can be modified with different datasets and algorithms, facilitating the investigations of mineral prospectivity in other regions and providing guidance for more detailed, high-resolution geological investigations.

Keywords: machine learning; maximum-likelihood classification; mineral prospectivity; random forest; support vector machine

Citation: Senanayake, I.P.; Kiem, A.S.; Hancock, G.R.; Metelka, V.; Folkes, C.B.; Blevin, P.L.; Budd, A.R. A Spatial Data-Driven Approach for Mineral Prospectivity Mapping. *Remote Sens.* **2023**, *15*, 4074. https://doi.org/10.3390/rs15164074

Academic Editors: Yanni Dong, Chao Chen and Tao Chen

Received: 25 July 2023
Revised: 14 August 2023
Accepted: 15 August 2023
Published: 18 August 2023

1. Introduction

Mineral prospectivity mapping aids the discovery of new ore deposits and plays a vital role in mineral exploration and the mining industry. Geological, geochemical, geophysical, and remotely sensed datasets are commonly employed in mineral prospectivity mapping approaches, by analysing them with various geo-computational techniques to identify key mineralisation patterns (e.g., Refs. [1–4]). Geophysical methods, such as seismic and electromagnetic (EM) surveys, help detect subsurface variations in the physical properties that may indicate mineralization [5,6]. Geochemical analyses of rock and soil samples unveil elemental compositions, helping in identifying anomaly patterns associated with mineral deposits [7,8]. Remote sensing methods, like satellite imagery and airborne surveys, enable the detection of surface alterations, mineral signatures, and topographic features, enhancing the identification of potential exploration targets [9–11]. The integration of these multidisciplinary datasets and methods offers a robust approach to mineral exploration, aiding in the delineation and assessment of prospective areas for further investigation [12].

Mineral prospectivity mapping techniques can be classified mainly into two categories: (i) expert-knowledge-driven (EKD, e.g., Refs. [13–15]); and (ii) data-driven (e.g., Refs. [16–18]) approaches. EKD approaches employ geological expertise who possess a deep understanding of the geological formations and mineral indicators to identify and interpret new deposits, whereas data-driven approaches employ the spatial associations of different geoscience datasets to map mineral potential. The EKD methods incorporate domain-specific insights, leading to targeted and informed exploration efforts. However, it might be limited by subjectivity and biases associated with human decision-making, potentially overlooking unconventional mineralization patterns or rare deposits. On the contrary, data-driven approaches employ advanced technologies such as remote sensing, machine learning, and geospatial analysis to process large datasets and identify spatial patterns and anomalies. While data-driven methods offer objectivity, speed, and the potential to find unrecognized mineral occurrences, they can be limited by data quality issues, algorithmic uncertainties, and the risk of false positives. Probabilistic methods, such as weights-of-evidence (WofE) and logistic regression, have been previously used for many mineral prospectivity mapping studies across the globe [19–22] and are now more widely accepted in the mining industry. Still, the prevalent method for generating new targets and evaluating the potential of mineralisation, especially in mid-tier and junior mineral houses, remains EKD in some aspects. However, these EKD methods require more specialised human resources, often implying higher costs, as well as longer project timeframes.

Classification can be identified as the main process used in data-driven mineral prospectivity mapping. The outputs of classification methods classify locations as prospective or non-prospective, as binary maps or maps showing a prospectivity confidence range. Classification approaches are primarily categorised into (i) supervised and (ii) unsupervised methods. In supervised methods, the model is trained with given training samples, whereas in unsupervised methods, clusters are generated based on the similarities and dissimilarities of the spatial data.

Often the association between mineralisation patterns and geospatial datasets is complex and non-linear. In this context, machine learning algorithms can be a viable option to capture these complex relationships between input data and mineralisation patterns. Many researchers have tested machine learning methods to map mineral prospectivity, which often outperform the traditional statistical and empirical explorative models [23–28]. Brown et al. [23] have used a multilayer feed-forward neural network, trained with a gradient descent back-propagation algorithm to map Au prospectivity in Tenterfield, NSW, Australia. The neural network performed well, even with one-third less data than a WofE method. Zuo and Carranza [29] tested a support vector machine (SVM) algorithm to map Au prospectivity in the western Meguma Terrain of Canada. Their results showed that the SVM method outperformed the WofE approach. McKay and Harris [25] tested random forest (RF) methods to map Au mineralisation in Hearn Geologic Province in Nunavut, Canada and observed a better performance than knowledge-driven methods. A RF-based data-driven method employed by Carranza and Laborte [16] outperformed the evidential belief modelling in mapping the hydrothermal Au-Cu prospectivity in Catanduanes Island, Philippines. However, the results of different classification methods can never be entirely identical due to the nature of the algorithms. Therefore, outputs from a single algorithm can inherit particular errors or biases.

This study aims to assess a simple data-driven machine learning approach to map mineral prospectivity, which can produce quick results on a large scale and can be used as an initial guide for more detailed geological evaluations. This data-driven approach has minimal involvement of geological expertise (i.e., it remains more objective with less potential for human-induced bias). Therefore, less human resources and time are required for this method. This study aims to investigate the performance of three machine learning algorithms, (i) RF, (ii) SVM, and (iii) maximum-likelihood classification (MaxL) in south-eastern Australia to map Cu-Au and Pb-Zn mineralisation patterns over the Cobar Basin, NSW, Australia with a unique combination of input geoscience datasets. Given the possible errors inherited in the outputs of a single classification algorithm, a simple ensemble method is employed in this study to combine the results of the three classification algorithms.

2. Study Area

The Cobar Superbasin in the Central Lachlan Orogen (Figure 1) is a mineral-rich Late Silurian to Early Devonian basin found in central-western NSW, about 700 km northwest of Sydney. The basin is NSW's most heavily mineralised Siluro-Devonian basin, enriched with precious and base metal mineral systems [30,31]. An overview of the geology of the Cobar Basin is described by Folkes et al. [32], and a detailed description of its structural framework is given by Glen et al. [33]. The Cobar area has a long history of mining and exploration dating back to 1869, when copper was discovered in the area. The Cobar Basin is a major polymetallic mineral province, including Au-only, Au- and Ag-rich, and Cu-and Pb-Zn-rich systems, with a total estimated metal endowment (past production plus known resources) exceeding 224.54 t Au, 2.519 Mt Cu, 2.801 Mt Pb, 4.760 Mt Zn, and 6924 t Ag [34,35].

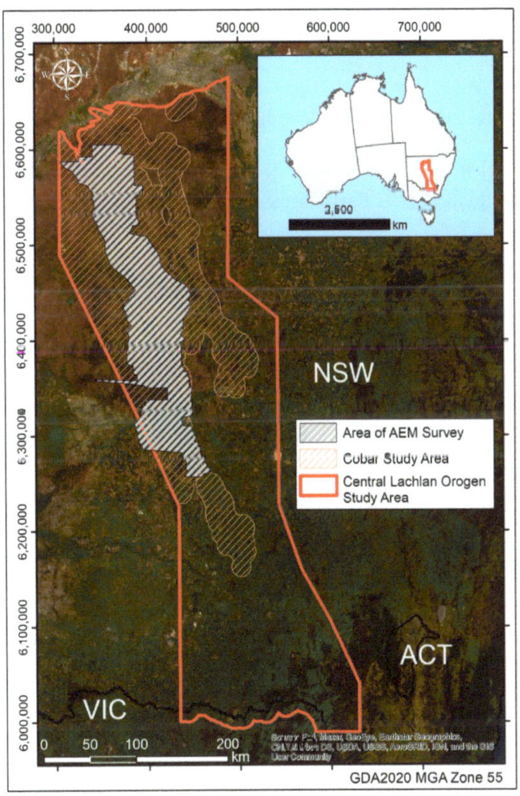

Figure 1. Study area, the Central Lachlan Orogen and the Cobar Superbasin, NSW.

Fitzherbert and Downes [35] and Ford et al. [31] provide good overviews of the tectonic and mineralisation history of the basin. The following is taken from these sources (and references therein). Geologically, the Cobar Basin rocks underwent three main tectonic phases,

I. Rift phase deposition and widespread volcanism in discrete volcanic troughs/belts ca. 420 Ma;
II. Transition to sag phase deposition at ca. 410 Ma, with stable sag phase deposition continuing until ca. 400 Ma. Bindian extension/contraction at ca. 410 Ma; and
III. Inversion phase during the Tabberabberan Orogeny ca. 390–380 Ma.

Fault-focused mineralisation is associated with these three tectonic phases, often resulting in the formation of complex polymetallic orebodies. The mineralisation mainly occurred after the transition to the sag phase basin deposition, whereas only a few orebodies have been formed at the rift phase volcanism/basin deposition.

Three major mineralisation systems can be found in and around the Cobar Basin: (i) Cu-Au-±Zn mineralisation; (ii) Pb-Zn mineralisation; and (iii) Sn-W mineralisation. The Cobar Cu-Au mineralisation extends mainly along the eastern boundary of the basin, with fewer occurrences in the western and southern parts of the basin. The style of these deposits varies from the massive sulphide-rich zones to the breccia, vein, or shear-hosted mineralisation. Some zones form multiple narrow ore lenses with short strike lengths and a more considerable vertical extent.

Cobar Pb-Zn-±Ag mineralisation occurs as individual orebodies predominantly adjacent to, or over parts of earlier formed Cobar Cu-Au(±Zn) systems. The fault/shear hosted Pb-Zn sulphide-rich zones often 'piggyback' on the previously formed Cu-Au(±Zn) orebodies (e.g., CSA Mine, Great Cobar Mines) along the major faults on the eastern boundary of the basin. The style of the deposits varies from low-temperature replacement to fault/shear-hosted massive sulphide orebodies with a similar deposit-scale morphology as the Cu-Au-(±Zn) orebodies.

Granite-related Sn-W deposits of the Central Lachlan Orogen form along a 250 km N-NW belt that extends from northern Victoria to the Tallebung region of central NSW. This belt is the second-largest tin field in NSW after New England. These deposits formed outside the Cobar Basin (sensu stricto) and are therefore not included in this study.

2.1. Mineral Prospectivity Mapping in Central Lachlan Orogen Using WofE Method

The WofE approach has been previously used to map mineral prospectivity over the Central Lachlan Orogen area by Kenex Pty Ltd., in colloboration with the Geological Survey of NSW (GSNSW) [31]. The expertise of the GSNSW staff was employed in developing mineral system models for three mineral systems in the central Lachlan Orogen, i.e., (i) Cobar Cu-Au, (ii) Cobar Pb-Zn, and (iii) granite-related Sn-W. The key predictive variables representing different critical ore-forming processes of these mineral systems (i.e., source, transport, trap, and deposition) were determined using these models. Between 138 to 196 predictive maps were developed using the pre-competitive geoscience data from the GSNSW, and analysed for the three mineral system models using the EKD method. Then, spatial data tables were developed by using the information about all the predictive maps, their relevance to the mineral systems and spatial correlation with known mineralisation. Based on multiple criteria (e.g., regional coverage of data, spatial correlation with training points, minimal duplication of predictive map patterns), between 8 to 10 predictive maps were chosen to develop mineral prospectivity maps for these three major mineral systems in the central Lachlan Orogen.

3. Methodology

3.1. Overview of the Methodology

A spatial data-driven ensemble machine learning (SDDEML) approach using three supervised learning algorithms, i.e., (i) RF, (ii) SVM and (iii) MaxL, was employed in this study to assess Cobar Cu-Au and Pb-Zn mineral prospectivity using geo-spatial data inputs. The method employed in this study is herein called the SDDEML approach. The produced prospectivity maps were compared against the results of the WofE method (see Section 2.1) and the existing mineral occurrence datasets for validation. ArcGIS 10.8 was used as the main data processing platform in this study.

3.2. Datasets

Data layers that are associated with the mineralisation patterns were chosen as the input datasets. Mineral occurrence points from the NSW MetIndEx database (https://datasets.seed.nsw.gov.au/dataset/nsw-mineral-occurrences, accessed on 1 October 2021) were used as training data in supervised classification. The datasets used in this work, with their sources, are given in Table 1.

Table 1. Datasets used in this study.

Dataset	Source
Total Magnetic Intensity with variable reduction to the pole (VRTP)	Geoscience Australia https://ecat.ga.gov.au/geonetwork/srv/eng/catalog.search#/metadata/131519, accessed on 5 August 2021
Total Magnetic Intensity–First Vertical Derivative (1VD)	Geoscience Australia https://ecat.ga.gov.au/geonetwork/srv/eng/catalog.search#/metadata/132275, accessed on 5 August 2021
Gravity anomaly	Geoscience Australia https://ecat.ga.gov.au/geonetwork/srv/eng/catalog.search#/metadata/133023, accessed on 5 August 2021
The Australian Lithospheric Architecture Magnetotelluric Project (AusLAMP) data	Geoscience Australia https://ecat.ga.gov.au/geonetwork/srv/eng/catalog.search#/metadata/131889, accessed on 20 August 2021
Cobar AEM data	GSNSW and Geoscience Australia https://minview.geoscience.nsw.gov.au/#/(dlmodal:geophys-survey-air/AIR0759)?lon=148.5&lat=-32.5&z=7&bm=bm1&l=gp118:y:100, accessed on 30 July 2021
Fault attribution of Zone 55W	GSNSW
AusMOHO	Australian Passive Seismic Server (AusPass) and the Australian National University Data Commons http://rses.anu.edu.au/seismology/AuSREM/AusMoho, accessed on 10 October 2021
Depth to basement	GSNSW https://geonetwork.geoscience.nsw.gov.au/geonetwork/srv/eng/catalog.search#/metadata/97772139-31b0-412b-ab0a-86a6b2166c2d, accessed on 20 August 2021
Mineral Occurrence (NSW MetIndEx) data	GSNSW http://portal.auscope.org.au/geonetwork/srv/eng/catalog.search#/metadata/fd59b712f10394e68f07261981ac6d771d72aacd, accessed on 1 October 2021

The association of gravity and magnetic layers with the mineral occurrences is well established; therefore, these layers were used in testing the many mineral exploration models (e.g., Ref. [36]). The variable reduction to the pole (VRTP) [37] and the first vertical derivative (1VD) magnetic grids [38] with an 80 m cell size obtained from Geoscience Australia were used for this work. Magnetic data provides vital information about the distribution of magnetic minerals in rocks. For example, most sedimentary rocks consist of negligible amounts of magnetic minerals, while igneous and metamorphic rocks consist of appreciable amounts of magnetic minerals. Magnetic data provides valuable information on the boundaries of extensive rock units due to the contrast of magnetic properties between different rocks. These magnetic gradients are enhanced in the derivatives of the total field and can not only map the lithological contacts, but also the fault zones that control fluid flow, and hence the location of many mineral systems.

Gravity anomaly data is commonly used in mineral exploration, considering its variability over different rocks. Gravity anomaly values are higher over the dense rocks compared to the lighter rocks. The Bouguer Anomaly gravity grid of Australia has been developed with airborne gravity/gravity gradiometry data from the National Australian Geophysical Database (NAGD), data from the Australian National Gravity Database (ANGD), and the Global Gravity Grid from National Oceanic and Atmospheric Administration (NOAA) over a 400 m grid [39]. Gravity data can again attribute to, for example, the hidden intrusions with particular shapes that are either less or more dense than the surrounding rocks. Gravity gradients can also show the deep-seated lithospheric structures that many studies have recognised to be the important controlling features of mineral systems [40,41].

The lithospheric-scale structural features can also be mapped by measuring the resistivity of the rocks through the deep looking magnetotelluric method. The Australian Lithospheric Architecture Magnetotelluric (MT) Project (AusLAMP) dataset for NSW [42–45], which was developed through a collaboration between Geoscience Australia and the GSNSW, was used to assess the deep conductivity/resistivity features that were previously recognised to delineate the corridors and anomalies associated with major mineral systems [46]. This MT dataset is available in an ascii grid format with 285 depth slices (from 2190 m to 337,290 m) over a half degree (~55 km) grid.

Data from an Airborne EM (AEM) survey conducted by Geoscience Australia and the GSNSW, to measure the natural electrical properties of rocks and soil in the Cobar–Lake Cargelligo area, was also used in this work as an input dataset. Electromagnetic surveying can, apart from directly detecting conductive ore minerals, enhance the mapping the lithostructural features in the magnetically quiet regions such as sedimentary basins. Only the footprint of the AEM survey (Figure 1) was chosen as the area of interest (AOI) for this study. This inversion of the AEM dataset consists of 32 depth slices extending from 5 m to 500 m. Currently, the AEM datasets only cover selected parts of Australia and do not provide continent-wide coverage. Therefore, the methodology was tested both with and without the AEM datasets to evaluate their impact on the results when expanding the methodology over other areas in future work.

The mineralisation patterns of the Lachlan Orogen region have a strong association with faults and intrusions [47]. Therefore, the fault attribution geodatabase of the area [35] was also used in this work. The AusMOHO [48,49] and depth to basement data layers [50] were also used as input data, because both the basin morphology and the deep structures that reach the base of the lithosphere could have played a role in the local and regional fluid flow.

The training data for the supervised classification were acquired from the GSNSW MetIndEx mineral occurrence dataset. MetIndex is the mineral occurrence database for NSW, with records of known mineral occurrences across the state. These records have been compiled by the GSNSW, with resource and production data acquired from both historical and current mining operations.

3.3. Data Pre-Processing

As the first step, the input data layers were prepared for the classification process. The datasets were clipped to the extent of the study' (i.e., the footprint of the Cobar AEM survey) including the gravity anomaly layer, and the magnetic-VRTP and 1VD layers.

The depth to Moho across the Cobar Basin varies from ~26 km to 42 km as per the AusMOHO dataset. A relationship between the MT data and gold mineralisation in western Victoria and south-eastern South Australia was found by Heinson et al. [46] across the low resistivity zone from ~20 km depth to the depth of Moho. Using similar conditions over the Cobar Basin, MT depth slices from 25 km to 42 km across the AOI (11 in total) were extracted. In order to reduce the dataset by choosing the layers exhibiting the highest significance, a principal component analysis (PCA) was applied to the depth slices, and subsequently, covariance and correlation matrices were developed. The first PC demonstrated an accumulated percentage of eigenvalues of 95.2%, which means that a significant portion of the variability of the dataset is captured by this component. The two input MT layers having high loading on the first PC were chosen for the classification process. Similarly, significant AEM layers were chosen after applying a PCA on the AEM depth slices.

The association between the Cu-Au and Pb-Zn mineral occurrences and faults were also considered in selecting the model inputs. The faults over the study area were classified into five separate layers (Figure 2a–e), based on the fault order that is an attribute of the data layer [47]. This association between the mineral occurrences and the faults is seen visually when the MetIndEx mineral occurrence (training) points are overlain on the faults layer (Figure 2a–e). Euclidean distances up to 50 km from the faults were calculated for these layers on a grid with the same size as the magnetic layers. These values were then inverted to have higher values for the pixels closer to the faults and lower values for the pixels far from the faults (Figure 2f–j). Based on the density of the mineral occurrences within a 5 km buffer from the faults, weights varying from 55 to 80 were assigned to each of these five layers. Then, the five weighted layers were combined together by choosing the highest pixel value at each point (Figure 2k). All the input layers were standardised to avoid any bias in classification caused by varied value ranges.

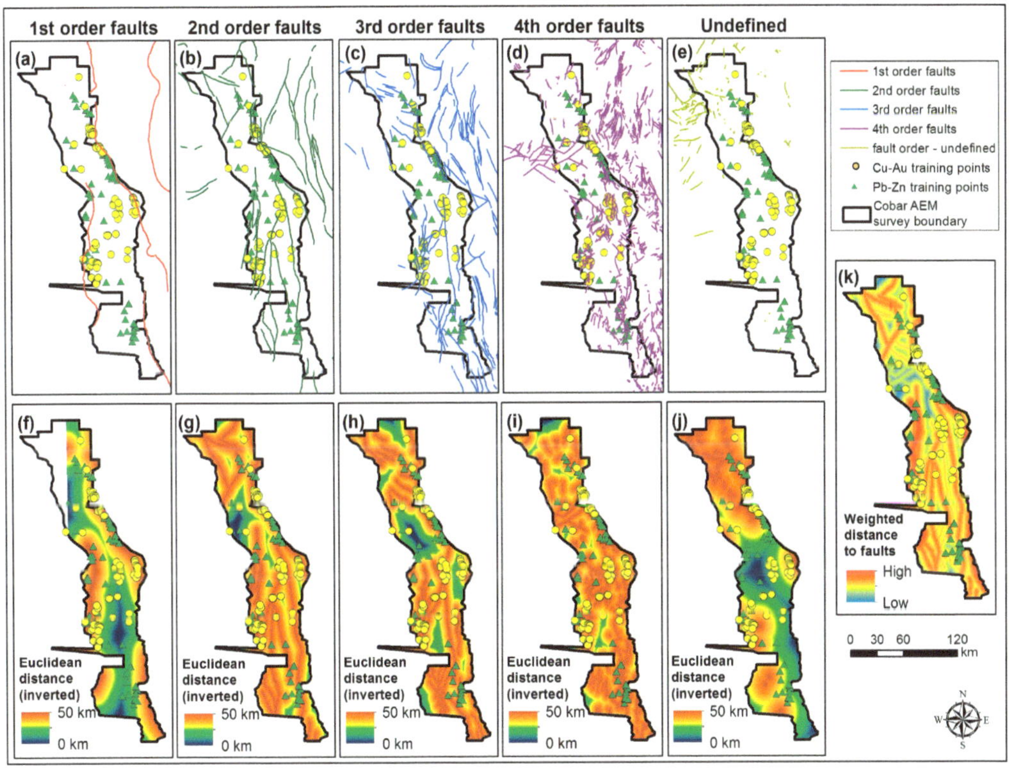

Figure 2. (**a–e**) The classification of faults over the Cobar AEM footprint based on the order of the faults [47]. (**f–j**) The inverted Euclidean distance from faults for each fault order group. (**k**) The merged, weighted distance from faults layer.

For training the model, ninety-seven Cu-Au and sixty-six Pb-Zn occurrence points over the study area were extracted from the MetIndEx dataset (see Figure 3). Given that the Cobar Basin is extensively explored for mineral occurrences, random negative points were created over the outcropping areas outside a 3 km buffer from the Cu-Au and Pb-Zn occurrence points. The outcropping regions were identified using the 'depth to basement layer' (i.e., data gaps or white regions within the depth to basement layer in Figure 3). Over 100 random negative points for Cu-Au and Pb-Zn were generated, separately.

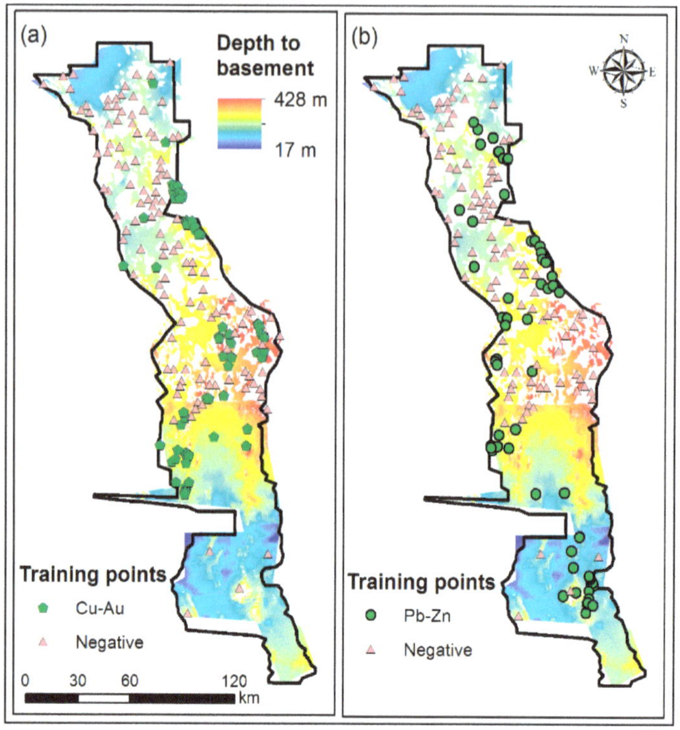

Figure 3. (**a**) Cu-Au, and (**b**) Pb-Zn training points used for classification overlaid on the depth to basement layer. The outcropping areas are shown as data gaps (white areas) within the depth to basement layer.

3.4. Classification

The supervised classification algorithms were applied to the magnetic-VRTP, magnetic-1VD, gravity anomaly, MT, distance to faults, and AEM data layers (Figure 4) based on the training points shown in Figure 3. The algorithms were tested both with and without the AEM layer to assess its impact as an input dataset (to take into account that the AEM datasets are not available for all parts of the country).

Three machine learning algorithms, (i) RF, (ii) SVM, and (iii) MaxL, were tested to produce binary maps showing the Cu-Au and Pb-Zn mineral prospectivity. The selection of these algorithms is made by considering their specific strengths and adaptability to the complex challenges posed by mineral mapping. RF's robustness in handling complex and noisy data, and small sample sizes suits the complex geological and geophysical variations associated with mineral mapping [51]. SVM's capability to handle both linear and nonlinear relationships [52] aids in classifying the mineral systems based on various properties. MaxL's probabilistic approach is valuable for estimating the class probabilities and uncertainty [53], which is crucial in providing confidence levels in mineral occurrence predictions. Here, 70% of the data points (Figure 3) were used for training, and 30% for validation. Subsequently, the three binary maps showing the mineral prospectivity (1 for prospective pixels and 0 for non-prospective pixels) for each mineralisation group (i.e., Cu-Au or Pb-Zn) were summed together to identify the regions classified as prospective areas by all three algorithms. i.e.,

$$\text{Map}_{\text{Final}} = \text{Map}_{\text{Random\_forest}} + \text{Map}_{\text{SVM}} + \text{Map}_{\text{MaxL}} \quad (1)$$

where Map_{Final} is the map showing the combined value of the mineral prospectivity from all three algorithms, and Map_{Random_forest}, Map_{SVM}, and Map_{MaxL} are the binary values of the mineral prospectivity maps produced by RF, SVM and MaxL algorithms, respectively. The collective integration of these carefully chosen algorithms enables the capture of intricate interrelationships between the input layers and the training data. The summary of this process is illustrated in Figure 5. The possible final pixel values and their mineral prospectivity classification based on the combinations of the resultant binary maps of RF, SVM and MaxL classification are shown in Table 2.

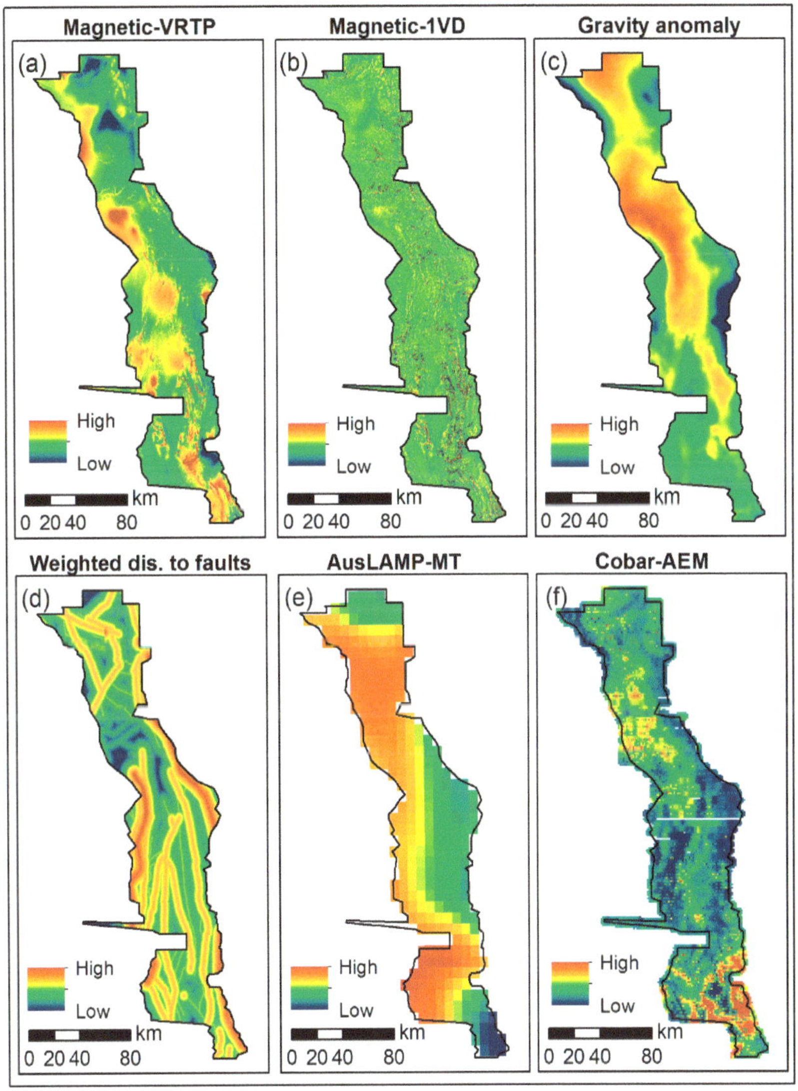

Figure 4. The input layers used for the classification (**a**) magnetic-VRTP, (**b**) magnetic-1VD, (**c**) gravity anomaly, (**d**) distance to faults (weighted), (**e**) AusLAMP magnetotelluric, and (**f**) Cobar AEM data over the Cobar AEM survey footprint.

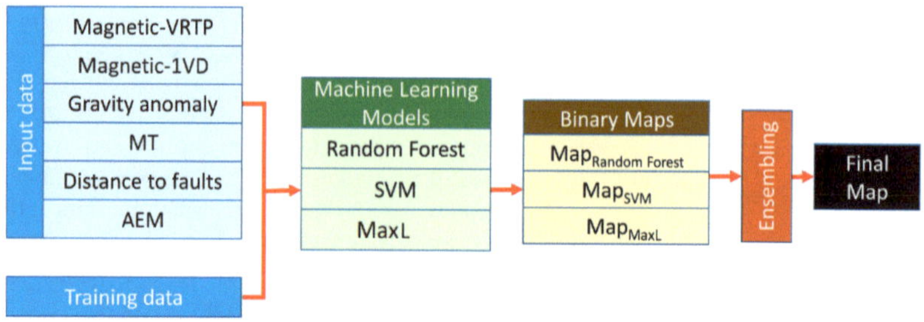

Figure 5. The ensemble approach of combining the mineral prospectivity maps produced using RF, SVM, and MaxL algorithms.

Table 2. The possible pixel values of the map showing the summed value of the RF-, SVM-, and MaxL-based binary mineral prospectivity maps.

Pixel Value				Mineral Prospectivity
$\text{Map}_{\text{Random\_Forest}}$	Map_{SVM}	MAP_{MaxL}	Final Map	
0	0	0	0	None
1	0	0	1	Low
0	1	0	1	
0	0	1	1	
1	1	0	2	Moderate
0	1	1	2	
1	0	1	2	
1	1	1	3	High

The binary map produced by the MaxL algorithm (Map_{MaxL}) was merged with its confidence ($\text{Map}_{\text{MaxL-confidence}}$) of classification using,

$$\text{Map}_{\text{MaxL-classified}} = \text{Map}_{\text{MaxL}} \times \text{Map}_{\text{MaxL-confidence}} \tag{2}$$

Here, $\text{Map}_{\text{MaxL-classified}}$ gives the classification confidence of the pixels only in prospective areas.

4. Results

The results of the MaxL classification algorithm-based Cu-Au and Pb-Zn mineral prospectivity mapping are shown in Figures 6 and 7, respectively. Here, panels a and b show the results of the MaxL classification performed by excluding and including the AEM data, respectively. The outputs of the MaxL algorithm-based binary maps (panels a(i) and b(i) of Figures 6 and 7) combined with the classification confidence layers (panels a(ii) and b(ii) of Figures 6 and 7) are shown in panels a(iii) and b(iii) of Figures 6 and 7 (see Equation (2)). Values of 0 in the binary maps show pixels with no mineral prospectivity, whereas values of 1 show prospective pixels. Equation (2) eliminates the areas having 0 values (i.e., no prospectivity) in the final maps. The classification confidence shown in these maps should not be mistaken for the confidence of the mineral prospectivity. However, the classification confidence can indirectly indicate the mineral prospectivity constrained by the input datasets. The areas with high classification confidence (>70%) show a good agreement with the mineral occurrence points (panels a(iii) and b(iii) of Figures 6 and 7). These figures depict that the prospective areas are confined to much smaller regions when the classification algorithm was run, including the AEM data. However, the two output maps produced with and without the AEM data showed similar spatial patterns. Three Cu-Au occurrence points over the northwest region of the AOI have not been well-captured by the MaxL algorithm (Figure 6a(iii),b(iii)). This should be further investigated in detailed explorations.

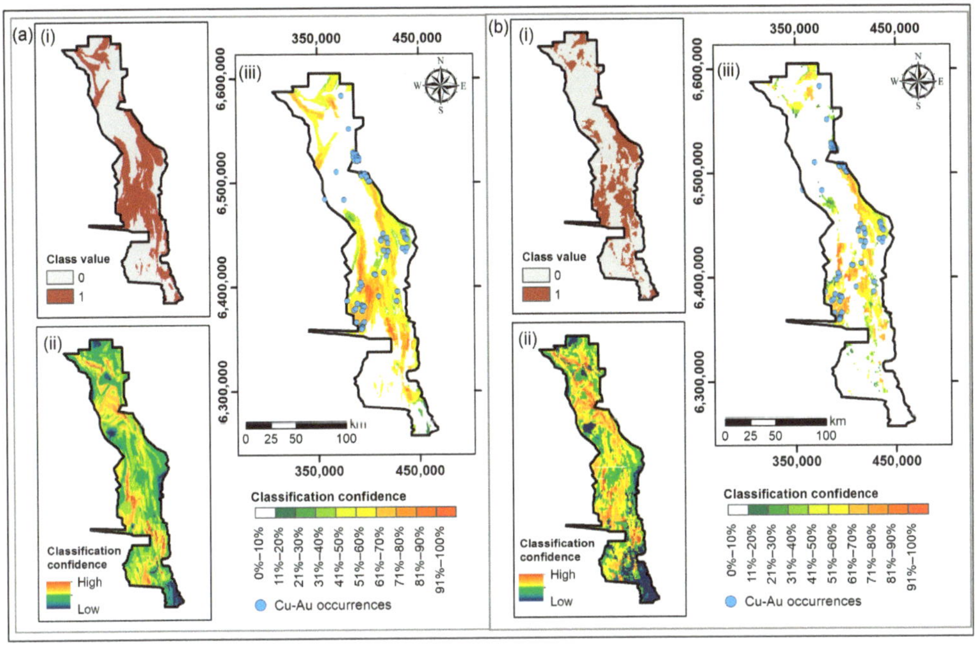

Figure 6. Panels (**a**) and (**b**), respectively, show the results of the Cu-Au prospectivity mapping tested with MaxL classification by excluding and including the AEM data over the Cobar AEM survey footprint. (**a**(**i**)) and (**b**(**i**)) show the binary maps showing the Cu-Au prospectivity (i.e., pixel value 1 = prospective and pixel value 0 = no prospectivity), (**a**(**ii**)) and (**b**(**ii**)) show the classification confidence over the entire area, and (**a**(**iii**)) and (**b**(**iii**)) shows the Cu-Au prospective areas combined with the classification confidence with the overlain Cu-Au mineral occurrence points.

The prospectivity of Cu-Au and Pb-Zn mineralisation as captured by the three machine learning algorithms, i.e., RF, SVM and MaxL, is shown in Figures 8–11. In allcases, the models demonstrated over 80% overall validation accuracy. Figures 8 and 10 show the result of the algorithms tested without the Cobar AEM data as an input, whereas Figures 9 and 11 show the results of the algorithms tested with the Cobar AEM data. In the binary output maps of the RF, SVM and MaxL (panels a–c of Figures 8–11), pixels assigned with a class value of 1 (i.e., class-1) show the prospective areas. Pixels assigned with a value of 0 (i.e., class-0) show areas with no prospectivity as per each machine learning algorithm. Panel d shows the combined result of the RF, SVM and MaxL after summing the three binary maps (i.e., panels a–c). The resultant map consists of four pixel values, i.e., 0 to 3 (see Table 1). Pixels captured by all three algorithms as prospective areas are assigned with a value of 3 (i.e., class-3, high prospectivity), pixels captured by only two algorithms as prospective areas are assigned with a value of 2 (i.e., class-2, moderate prospectivity) and pixels captured by prospective areas by only one algorithm are assigned with a value of 1 (i.e., class-1, low prospectivity). Pixels not captured by any of the three algorithms as prospective areas were assigned with a value of 0 (i.e., class 0, no prospectivity).

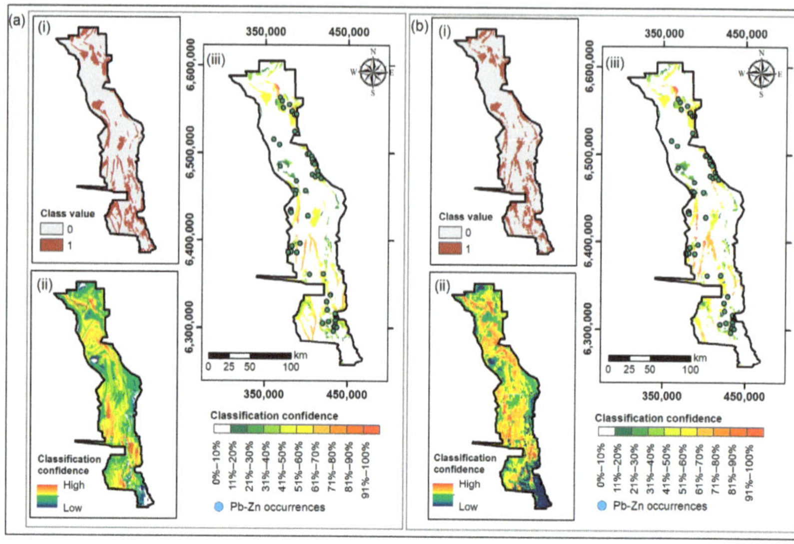

Figure 7. Panels (**a**) and (**b**), respectively, show the results of the Pb-Zn prospectivity mapping tested with MaxL classification by including and excluding the AEM data over the Cobar AEM survey footprint. (**a**(**i**)) and (**b**(**i**)) show the binary maps showing the Pb-Zn prospectivity (i.e., pixel value 1 = prospective and pixel value 0 = no prospectivity), (**a**(**ii**)) and (**b**(**ii**)) shows the classification confidence over the entire area, and (**a**(**iii**)) and (**b**(**iii**)) shows the Pb-Zn prospective areas combined with the classification confidence with the overlain Pb-Zn mineral occurrence points.

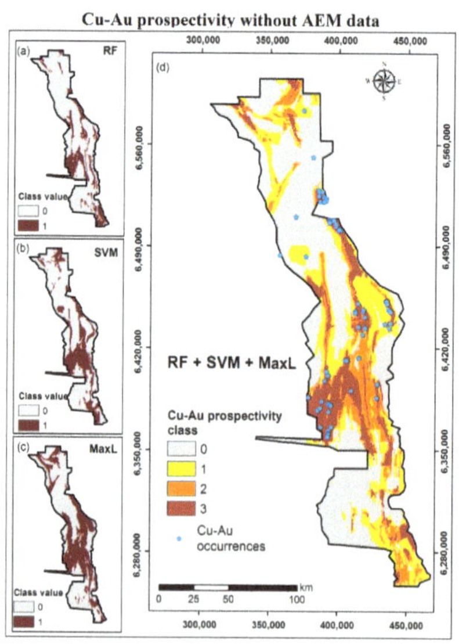

Figure 8. Cu-Au prospectivity maps developed with (**a**) RF, (**b**) SVM, and (**c**) MaxL without using Cobar AEM data as an input layer. Class values 1 and 0 show prospective and non-prospective areas, respectively. Map (**d**) shows the ensembled result after summing (**a**–**c**). Class value 3, 2, and 1 depict areas which were captured as prospective areas by (i) all three machine learning algorithms (high confidence), (ii) only two algorithms (moderate confidence), and (iii) only one algorithm (low confidence), respectively. Class value 0 depicts areas that are classified as non-prospective areas by all three algorithms.

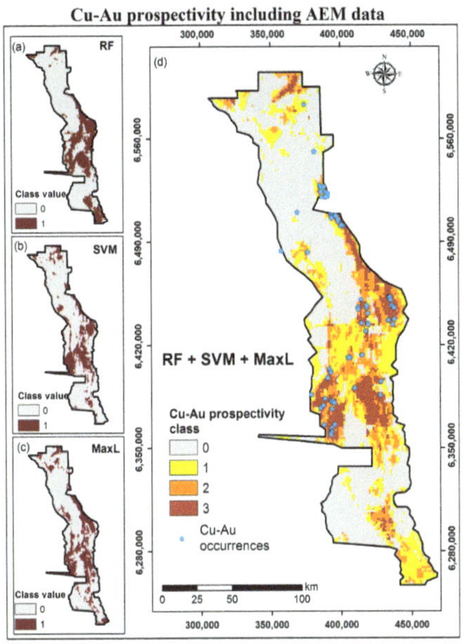

Figure 9. Cu-Au prospectivity maps developed with (**a**) RF, (**b**) SVM, and (**c**) MaxL, including Cobar AEM data as an input layer. Class values 1 and 0 show prospective and non-prospective areas, respectively. Map (**d**) shows the ensembled result after summing (**a**–**c**). Class value 3, 2, and 1 depict areas which were captured as prospective areas by (i) all three machine learning algorithms (high confidence), (ii) only two algorithms (moderate confidence), and (iii) only one algorithm (low confidence), respectively. Class value 0 depicts areas that are classified as non-prospective areas by all three algorithms.

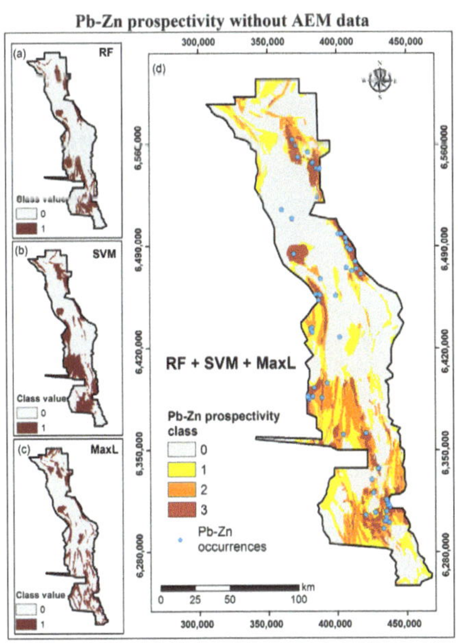

Figure 10. Pb-Zn prospectivity maps developed with (**a**) RF, (**b**) SVM, and (**c**) MaxL without using Cobar AEM data as an input layer. Class values 1 and 0 show prospective and non-prospective areas,

respectively. Map (**d**) shows the ensembled result after summing (**a**–**c**). Class value 3, 2, and 1 depict areas which were captured as prospective areas by (i) all three machine learning algorithms (high confidence), (ii) only two algorithms (moderate confidence), and (iii) only one algorithm (low confidence), respectively. Class value 0 depicts areas that are classified as non-prospective areas by all three algorithms.

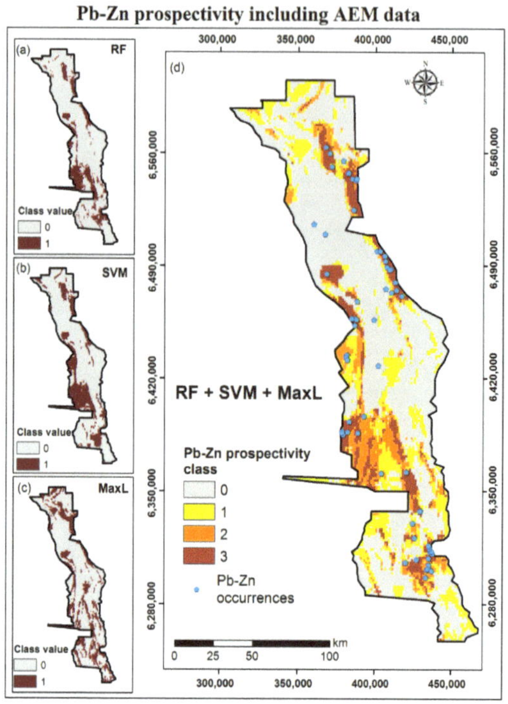

Figure 11. Cu-Au prospectivity maps developed with (**a**) RF, (**b**) SVM, and (**c**) MaxL, including Cobar AEM data as an input layer. Class values 1 and 0 show prospective and non-prospective areas, respectively. Map (**d**) shows the ensembled result of summing (**a**–**c**). Class value 3, 2, and 1 depict areas which were captured as prospective areas by (i) all three machine learning algorithms (high confidence), (ii) only two algorithms (moderate confidence), and (iii) only one algorithm (low confidence), respectively. Class value 0 depicts areas that are classified as non-prospective areas by all three algorithms.

The prospective regions became confined to slightly smaller regions when the AEM dataset was introduced as an input layer to the classification. However, the mineral prospectivity maps of Cu-Au and Pb-Zn prepared with and without the AEM data show similar spatial patterns.

5. Discussion

5.1. Interpretation and Comparison of Results with Existing Studies

Large areas of the northern- (>~6,560,000 mS) and southern-most (<~6,320,000 mS) parts of the AOI, as well as the eastern and western margins of the central area, have been identified as prospective regions for the Cobar Cu-Au mineral systems. The Cu-Au prospective areas show a general North-South orientation that correlates with the general strike of the major geological units in the Cobar Basin. The results of the Cu-Au prospectivity mapping using the SVM algorithm (Figures 9 and 10) showed a better agreement with the MetIndEx Cu-Au mineral occurrence/training points than the MaxL and RF methods. For example, the Cu-Au occurrences in the northwest of the study area have largely been captured by prospective areas using the SVM-based classification but were not well-captured by the RF and MaxL methods (Figure 8).

Including the AEM data as an input layer for the classification models confined the prospective areas to a slightly smaller extent. However, the classification results without the AEM data showed similar spatial patterns of mineral potential, suggesting that the SDDEML approach tested in this study can be effectively applied over regions where AEM datasets are unavailable.

Class 3 (i.e., highly prospective class where all three algorithms detected a pixel as a prospective area) areas of our Cu-Au mineral prospectivity maps (Figure 12) show a moderate to a strong agreement with the MetIndEx Cu-Au mineral occurrence points. A moderate agreement is observed between class 2 (i.e., pixels where the results of only two algorithms identified them as prospective areas) and the Cu-Au mineral occurrence points. Class 1 (i.e., pixels where the results of only one algorithm identified them as prospective areas) showed a low correlation compared to the MetIndEx data. This implies the robustness of ensembling the three machine learning models to get better results that are less affected by the errors and biases in the individual algorithms.

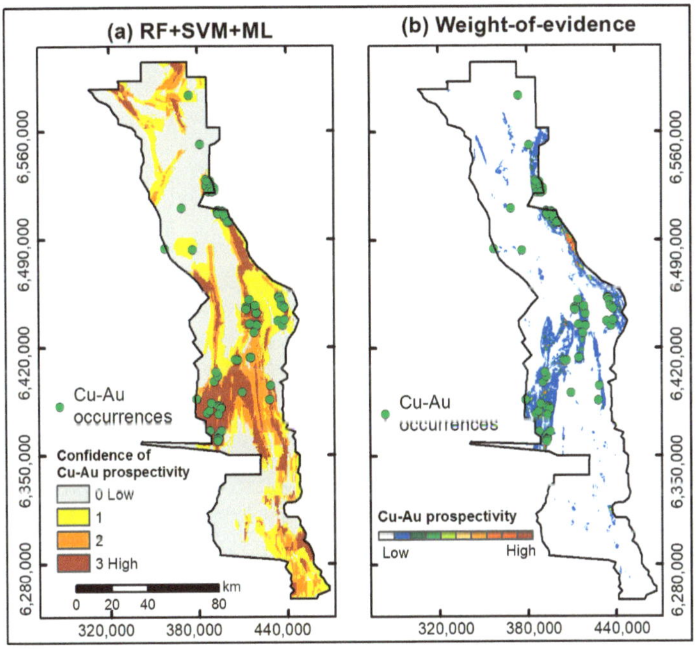

Figure 12. Comparison between the Cu-Au prospectivity mapping results from the (**a**) ensemble data-driven approach (this study), and (**b**) the WofE approach used by GSNSW [31].

A moderate to strong agreement was observed between the class 3 (highly prospective) areas of the SDDEML method-based Cu-Au prospectivity maps developed in this study and the prospective areas of the equivalent map produced with the existing WofE approach (Figure 12; Ford et al. [31]—see Section 2.1). A moderate agreement was observed between class 2 of the mineral prospectivity map using the SDDEML approach and the WofE-based results. The agreement between class 1 and the EKD method-based result is less significant.

Importantly, the results of the data-driven-ensemble method presented in this paper showed the Cu-Au prospectivity in the northern- and southern-most parts of the AOI, which were not captured by the WofE approach (Figure 12). Differences in the input data and the algorithms might have caused this disparity. However, it is worth performing a more detailed investigation of these regions to assess any further mineral potential.

Using the data-driven SDDEML methods in this study, prospective areas for the Cobar Pb-Zn mineral systems were observed in approximately North-South oriented bands in the AOI's north-eastern, south-eastern, and southern areas. Notably, many of the predicted prospective areas for the Pb-Zn mineral systems overlap with those for the Cu-Au mineral systems. A possible explanation for this is the documented 'piggyback' style of the mineral systems of many mineral occurrences for the Cu-Au and Pb-Zn mineral systems in the Cobar Basin (see Section 2 of this paper and Ford et al. [31]).

The existing EKD method-based mineral prospectivity maps and the data-driven SDDEML method-based mineral prospectivity maps produced by this study showed similar prospectivity patterns across the AOI. There is a strong agreement between the class 3 (highly prospective) areas of the SDDEML method-based Pb-Zn prospectivity maps and the prospective areas of the equivalent map produced by the existing EKD approach (Figure 13). A moderate agreement was observed between class 2 of the mineral prospectivity map using the SDDEML approach and EKD based results. The agreement between class 1 (least prospective) and the EKD method-based result is less significant.

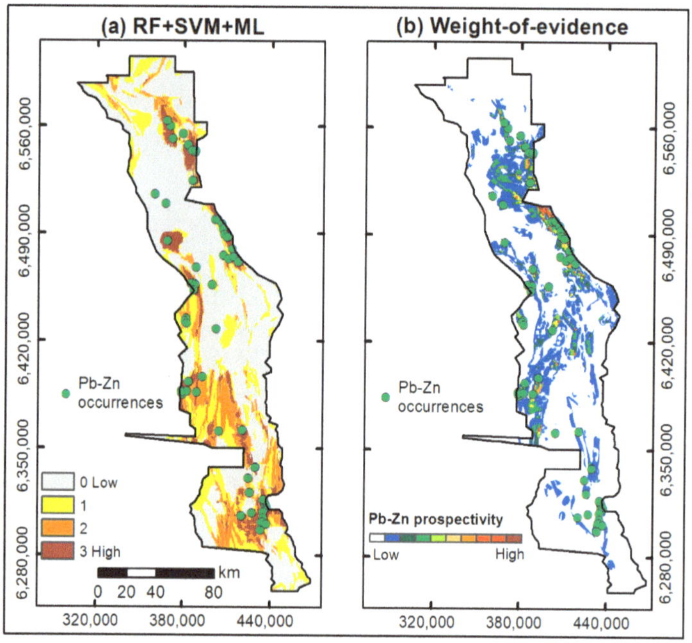

Figure 13. Comparison between the Pb-Zn prospectivity mapping results from the (**a**) ensemble data-driven approach (this study), and (**b**) the WofE approach used by GSNSW [31].

The mineral prospectivity maps produced in this study with the SDDEML methods showed a good agreement with the MetIndEx mineral occurrence (training) points (Figures 8–12). This further demonstrates that the SDDEML approach is an effective method to assess the mineral prospectivity of the major mineral systems in the Cobar Basin. The results of ensembling three robust machine learning models (specifically class 3 and 2) showed a better agreement to the results of the WofE method, compared to binary maps produced by using only one model. Although this is a simple approach, ensembling the results makes the SDDEML approach robust in identifying the prospective mineral zones. The results also indicate that the AEM data is not an essential requirement for this analysis; therefore, the SDDEML method could be used to assess mineral prospectivity over other areas where AEM datasets are unavailable.

5.2. Potential Future Work

The results of the spatial data-driven approach in this study are highly encouraging, and the method could be tested in other areas, over different spatial scales, to map the mineral prospectivity using the available geoscience datasets as a preliminary method of investigation. However, it is necessary to conduct a field validation of the results prior to a detailed exploration. When applying this method over other regions, the random negative data points used for training the algorithms could also be collected from drillholes showing no indication of mineralisation. The processing speed of the SDDEML method is much more effective than the more complex and time-consuming knowledge-driven mineral prospectivity mapping methods.

The SDDEML method can be further improved and modified by including different data combinations, classification algorithms, and their combinations. However, it would be worthwhile comparing the results of this method with more complex algorithms such as spatial random forest (e.g., Ref. [36]) and logistic regression methods over a similar AOI to get a better insight into the results. The algorithms for the SDDEML method could also be tested at different depths based on the effective depth resolution/confidence or DOI (depth of investigation) of the different input layers. Although this method should not replace detailed geological investigations, given the speed and simplicity of this data-driven method, the results could be used as an initial assessment and a validation method to compare the results with other mineral prospectivity mapping methods of more poorly investigated areas. Such results can be used as an initial, less biased, guiding step at regional or district scales before performing more detailed geological assessments, such as drilling, and exploration programs at the camp and prospect scales.

6. Conclusions

A spatial data-driven ensemble machine learning (SDDEML) approach was tested in this study to map Cu-Au and Pb-Zn mineral prospectivity in the Cobar Basin of New South Wales, Australia. Three machine-learning approaches, RF, SVM and MaxL, were used in identifying potential deposits and the results were ensembled to achieve maps with higher classification accuracy. Here, magnetic-VRTP, magnetic-1VD, gravity anomaly, distance to faults (weighted), AusLAMP magnetotelluric and, Cobar AEM datasets were used as input data considering their influence on Cu-Au and Pb-Zn mineralisation. Ensembling multiple models provided a better accuracy compared to the individual classification algorithms and provides a more robust approach in mapping mineral prospectivity over different spatial extents. The method was tested using AEM data and without the use of AEM data to examine its potential over areas where AEM data are unavailable. The results indicated that AEM data can somewhat narrow down the extent of mineral prospectivity into slightly smaller regions. The method tested here can be modified and further tested with different combinations of input layers and classification algorithms. The findings of this study are promising for expanding this method over other regions to identify mineral prospectivity with different input datasets. The results of the proposed method can provide an initial assessment and valuable insights for more detailed geological assessments and exploration in prospective mineralised terranes.

Author Contributions: Conceptualization, I.P.S., V.M., A.S.K. and G.R.H.; methodology, I.P.S. and V.M.; validation, I.P.S., C.B.F. and P.L.B.; formal analysis, I.P.S.; resources, A.S.K., G.R.H., A.R.B., C.B.F. and P.L.B.; data curation, I.P.S.; writing—original draft preparation, I.P.S.; writing—review and editing, I.P.S., A.S.K., G.R.H., V.M., C.B.F., P.L.B. and A.R.B.; visualization, I.P.S.; supervision, A.S.K., G.R.H. and V.M.; project administration, V.M. and A.R.B.; funding acquisition, A.S.K. and G.R.H. All authors have read and agreed to the published version of the manuscript.

Funding: The work has been supported by the Mineral Exploration Cooperative Research Centre, whose activities are funded by the Australian Government's Cooperative Research Centre Program. This is MinEx CRC Document 2023/27. The Article Processing Charges were supported by the Geological Survey of New South Wales.

Data Availability Statement: Sources of the datasets used in this study are provided in the 'datasets' section. The final maps are available on request from the corresponding author.

Acknowledgments: We would like to express our gratitude to Giovanni Spampinato from the Geological Survey of New South Wales, Adrian Hitchman from Geoscience Australia, Dave Cole and Hassan Talebi from CSIRO, and other personnel involved with the MinEx CRC project for their assistance and suggestions. Chris Folkes and Phil Blevin publish with the permission of the Executive Director, Geological Survey of New South Wales.

Conflicts of Interest: The authors declare no conflict of interest.

References

1. Marjoribanks, R. *Geological Methods in Mineral Exploration and Mining*; Springer: Berlin/Heidelberg, Germany, 2010.
2. Zou, H.; Han, R.; Liu, M.; Cromie, P.; Zaw, K.; Fang, W.; Huang, J.; Ren, T.; Wu, J. Geological, geophysical, and geochemical characteristics of the Ban Kiouchep Cu–Pb–Ag deposit and its exploration significance in Northern Laos. *Ore Geol. Rev.* **2020**, *124*, 103603. [CrossRef]
3. Ali, M.A.H.; Mewafy, F.M.; Qian, W.; Alshehri, F.; Ahmed, M.S.; Saleem, H.A. Integration of Electrical Resistivity Tomography and Induced Polarization for Characterization and Mapping of (Pb-Zn-Ag) Sulfide Deposits. *Minerals* **2023**, *13*, 986. [CrossRef]
4. Porwal, A.K.; Kreuzer, O.P. Introduction to the special issue: Mineral prospectivity analysis and quantitative resource estimation. *Ore Geol. Rev.* **2010**, *38*, 121–127. [CrossRef]
5. Hoover, D.B.; Klein, D.P.; Campbell, D.C.; du Bray, E. Geophysical methods in exploration and mineral environmental investigations. *Prelim. Compil. Descr. Geoenviron. Miner. Depos. Models USGS Open-File Rep.* **1995**, *95*, 19–27.
6. Kearey, P.; Brooks, M.; Hill, I. *An Introduction to Geophysical Exploration*; Wiley-Blackwell: Maldan, MA, USA, 2002; Volume 4.
7. Rose, A.W. Geochemical exploration. In *Geochemistry. Encyclopedia of Earth Science*; Springer: Dordrecht, The Netherlands, 1998.
8. Grunsky, E.C.; Caritat, P.D. State-of-the-art analysis of geochemical data for mineral exploration. Geochemistry: Exploration, Environment. *Analysis* **2020**, *20*, 217–232.
9. Agar, B.; Coulter, D. Remote sensing for mineral exploration—A decade perspective 1997–2007. In Proceedings of the Exploration 07: Fifth Decennial International Conference on Mineral Exploration, Toronto, ON, Canada, 9–12 September 2007; pp. 109–136.
10. Sabins, F.F. Remote sensing for mineral exploration. *Ore Geol. Rev.* **1999**, *14*, 157–183. [CrossRef]
11. Shirmard, H.; Farahbakhsh, E.; Müller, R.D.; Chandra, R. A review of machine learning in processing remote sensing data for mineral exploration. *Remote Sens. Environ.* **2022**, *268*, 112750. [CrossRef]
12. Lawley, C.J.; Tschirhart, V.; Smith, J.W.; Pehrsson, S.J.; Schetselaar, E.M.; Schaeffer, A.J.; Houlé, M.G.; Eglington, B.M. Prospectivity modelling of Canadian magmatic Ni (±Cu±Co±PGE) sulphide mineral systems. *Ore Geol. Rev.* **2021**, *132*, 103985. [CrossRef]
13. Harris, J.; Grunsky, E.; Behnia, P.; Corrigan, D. Data-and knowledge-driven mineral prospectivity maps for Canada's North. *Ore Geol. Rev.* **2015**, *71*, 788–803. [CrossRef]
14. Hosseini, S.A.; Abedi, M. Data envelopment analysis: A knowledge-driven method for mineral prospectivity mapping. *Comput. Geosci.* **2015**, *82*, 111–119. [CrossRef]
15. Ma, Y.; Zhao, J.; Sui, Y.; Liao, S.; Zhang, Z. Application of knowledge-driven methods for mineral prospectivity mapping of polymetallic sulfide deposits in the southwest Indian ridge between 46° and 52° E. *Minerals* **2020**, *10*, 970. [CrossRef]
16. Carranza, E.J.M.; Laborte, A.G. Data-driven predictive modeling of mineral prospectivity using random forests: A case study in Catanduanes Island (Philippines). *Nat. Resour. Res.* **2016**, *25*, 35–50. [CrossRef]
17. Sun, T.; Li, H.; Wu, K.; Chen, F.; Zhu, Z.; Hu, Z. Data-driven predictive modelling of mineral prospectivity using machine learning and deep learning methods: A case study from southern Jiangxi Province, China. *Minerals* **2020**, *10*, 102. [CrossRef]
18. Yousefi, M.; Nykänen, V. Data-driven logistic-based weighting of geochemical and geological evidence layers in mineral prospectivity mapping. *J. Geochem. Explor.* **2016**, *164*, 94–106. [CrossRef]
19. Fu, C.; Chen, K.; Yang, Q.; Chen, J.; Wang, J.; Liu, J.; Xiang, Y.; Li, Y.; Rajesh, H. Mapping gold mineral prospectivity based on weights of evidence method in southeast Asmara, Eritrea. *J. Afr. Earth Sci.* **2021**, *176*, 104143. [CrossRef]
20. Xiong, Y.; Zuo, R. GIS-based rare events logistic regression for mineral prospectivity mapping. *Comput. Geosci.* **2018**, *111*, 18–25. [CrossRef]
21. Yuan, F.; Li, X.; Zhang, M.; Jowitt, S.M.; Jia, C.; Zheng, T.; Zhou, T. Three-dimensional weights of evidence-based prospectivity modeling: A case study of the Baixiangshan mining area, Ningwu Basin, Middle and Lower Yangtze Metallogenic Belt, China. *J. Geochem. Explor.* **2014**, *145*, 82–97. [CrossRef]
22. Zeghouane, H.; Allek, K.; Kesraoui, M. GIS-based weights of evidence modeling applied to mineral prospectivity mapping of Sn-W and rare metals in Laouni area, Central Hoggar, Algeria. *Arab. J. Geosci.* **2016**, *9*, 373. [CrossRef]
23. Brown, W.M.; Gedeon, T.; Groves, D.; Barnes, R. Artificial neural networks: A new method for mineral prospectivity mapping. *Aust. J. Earth Sci.* **2000**, *47*, 757–770. [CrossRef]
24. Carranza, E.J.M.; Laborte, A.G. Random forest predictive modeling of mineral prospectivity with small number of prospects and data with missing values in Abra (Philippines). *Comput. Geosci.* **2015**, *74*, 60–70. [CrossRef]

25. McKay, G.; Harris, J. Comparison of the data-driven random forests model and a knowledge-driven method for mineral prospectivity mapping: A case study for gold deposits around the Huritz Group and Nueltin Suite, Nunavut, Canada. *Nat. Resour. Res.* **2016**, *25*, 125–143. [CrossRef]
26. Rodriguez-Galiano, V.; Sanchez-Castillo, M.; Chica-Olmo, M.; Chica-Rivas, M. Machine learning predictive models for mineral prospectivity: An evaluation of neural networks, random forest, regression trees and support vector machines. *Ore Geol. Rev.* **2015**, *71*, 804–818. [CrossRef]
27. Zhang, N.; Zhou, K.; Li, D. Back-propagation neural network and support vector machines for gold mineral prospectivity mapping in the Hatu region, Xinjiang, China. *Earth Sci. Inform.* **2018**, *11*, 553–566. [CrossRef]
28. Zuo, R. Geodata science-based mineral prospectivity mapping: A review. *Nat. Resour. Res.* **2020**, *29*, 3415–3424. [CrossRef]
29. Zuo, R.; Carranza, E.J.M. Support vector machine: A tool for mapping mineral prospectivity. *Comput. Geosci.* **2011**, *37*, 1967–1975. [CrossRef]
30. David, V. Cobar Deposits–Structural control. *ASEG Ext. Abstr.* **2018**, *2018*, 1–9. [CrossRef]
31. Ford, A.; Peters, K.; Downes, P.; Blevin, P.; Greenfield, J.; Fitzherbert, J. *Central Lachlan Orogen Mineral Systems Mineral Systems Report*; Report GS2020/0741; Kenex Pty Ltd.: Dongara, WA, Australia, 2020.
32. Folkes, C.B.; Carlton, A.; Eastlake, M.; Deyssing, L.; Trigg, S.; Montgomery, K.; Matthews, S.; Spampinato, G.; Roach, I.; Gilmore, P.; et al. *The Cobar AEM Survey Interpretation Report*; Report GS2021/1592; Mining, Exploration and Geoscience: Maitland, NSW, Australia, 2022.
33. Glen, R.; Drummond, B.; Goleby, B.; Palmer, D.; Wake-Dyster, K. Structure of the Cobar Basin, New South Wales, based on seismic reflection profiling. *Aust. J. Earth Sci.* **1994**, *41*, 341–352. [CrossRef]
34. Seccombe, P.K.; Jiang, Z.; Downes, P.M. Sulfur isotope and fluid inclusion geochemistry of metamorphic Cu–Au deposits, central Cobar area, NSW, Australia. *Aust. J. Earth Sci.* **2017**, *64*, 537–556. [CrossRef]
35. Fitzherbert, J.A.; Downes, P.M. *A Mineral System Model for Cu-Au-Pb-Zn-Ag Systems of the Cobar Basin, Central Lachlan Orogen, New South Wales 2020*; Report GS2021/0042; Regional NSW: Maitland, NSW, Australia, 2020.
36. Talebi, H.; Peeters, L.J.; Otto, A.; Tolosana-Delgado, R. A truly spatial Random Forests algorithm for geoscience data analysis and modelling. *Math. Geosci.* **2022**, *54*, 1–22. [CrossRef]
37. Poudjom Djomani, Y.; Minty, B.R.S. *Total Magnetic Intensity (TMI) Grid of Australia with Variable Reduction to Pole (VRTP) 2019*, 7th ed.; Geoscience Australia: Canberra, Australia, 2019. [CrossRef]
38. Poudjom Djomani, Y. *Total Magnetic Intensity Grid of Australia 2019—First Vertical Derivative (1VD)*; Geoscience Australia: Canberra, Australia, 2019. [CrossRef]
39. Lane, R.J.L.; Wynne, P.E.; Poudjom Djomani, Y.; Stratford, W.R.; Barretto, J.A.; Caratori Tontini, F. *2019 Australian National Gravity Grids: Free Air Anomaly, Complete Bouguer Anomaly, De-Trended Global Isostatic Residual, 400 m Cell Size (Includes Point Located Data)*; Geoscience Australia: Canberra, Australia, 2020.
40. McCuaig, T.C.; Beresford, S.; Hronsky, J.M.A. Translating the mineral systems approach into an effective exploration targeting system. *Ore Geol. Rev.* **2010**, *38*, 128–138. [CrossRef]
41. Occhipinti, S.; Metelka, V.; Lindsay, M.; Aitken, A.; Pirajno, F.; Tyler, I. The evolution from plate margin to intraplate mineral systems in the Capricorn Orogen, links to prospectivity. *Ore Geol. Rev.* **2020**, *127*, 103811. [CrossRef]
42. Kirkby, A.; Musgrave, R.J.; Czarnota, K.; Doublier, M.P.; Duan, J.; Cayley, R.A.; Kyi, D. Lithospheric architecture of a Phanerozoic orogen from magnetotellurics: AusLAMP in the Tasmanides, southeast Australia. *Tectonophysics* **2020**, *793*, 228560. [CrossRef]
43. Kirkby, A.L. *Resistivity Model of the Southeast Australian Mainland from AusLAMP Magnetotelluric Data*; Geoscience Australia: Canberra, Australia, 2020. [CrossRef]
44. Kyi, D.; Duan, J.; Kirkby, A.L.; Stolz, N. *Australian Lithospheric Architecture Magnetotelluric Project (AusLAMP): New South Wales: Data Release (Phase One). Record 2020/011*; Geoscience Australia: Canberra, Australia, 2020. [CrossRef]
45. Robertson, K.; Heinson, G.; Thiel, S. Lithospheric reworking at the Proterozoic–Phanerozoic transition of Australia imaged using AusLAMP Magnetotelluric data. *Earth Planet. Sci. Lett.* **2016**, *452*, 27–35. [CrossRef]
46. Heinson, G.; Duan, J.; Kirkby, A.; Robertson, K.; Thiel, S.; Aivazpourporgou, S.; Soyer, W. Lower crustal resistivity signature of an orogenic gold system. *Sci. Rep.* **2021**, *11*, 15807. [CrossRef] [PubMed]
47. Folkes, C.B.; Stuart, C. *Fault Attribution for the Western Lachlan Orogen of NSW*; Report GS2020/0955; Regional NSW: Maitland, NSW, Australia, 2020.
48. Kennett, B.L.N.; Salmon, M.; Saygin, E.; Group, A.W. AusMoho: The variation of Moho depth in Australia. *Geophys. J. Int.* **2011**, *187*, 946–958. [CrossRef]
49. Salmon, M.; Kennett, B.L.N.; Stern, T.; Aitken, A.R.A. The Moho in Australia and New Zealand. *Tectonophysics* **2013**, *609*, 288–298. [CrossRef]
50. Stolz, E.; Spampinato, G.; Davidson, J. A statewide 3D geological model for New South Wales. *ASEG Ext. Abstr.* **2019**, *2019*, 1–4. [CrossRef]
51. Biau, G.; Scornet, E. A random forest guided tour. *Test* **2016**, *25*, 197–227. [CrossRef]

52. Ghosh, S.; Dasgupta, A.; Swetapadma, A. A Study on Support Vector Machine based Linear and Non-Linear Pattern Classification. In Proceedings of the 2019 International Conference on Intelligent Sustainable Systems (ICISS), Palladam, India, 21–22 February 2019; pp. 24–28.
53. Singh, G.B.; Singh, G.B. Probabilistic Methods: Maximum Likelihood. In *Fundamentals of Bioinformatics and Computational Biology: Methods and Exercises in MATLAB*; Springer: Cham, Switzerland, 2015; pp. 273–286.

remote sensing

MDPI

Article

An Approach for Monitoring Shallow Surface Outcrop Mining Activities Based on Multisource Satellite Remote Sensing Data

Shiyao Li [1,2], Run Wang [3,*], Lei Wang [2], Shaoyu Liu [4], Jiang Ye [3], Hang Xu [3] and Ruiqing Niu [1]

1 School of Geophysics and Geomatics, China University of Geosciences, Wuhan 430074, China; lishiyao@cug.edu.cn (S.L.); niuruiqing@cug.edu.cn (R.N.)
2 Wuhan Center, China Geological Survey (Central South China Innovation Center for Geosciences), Wuhan 430205, China; wlei_b@mail.cgs.gov.cn
3 Geological Environmental Center of Hubei Province, Wuhan 430034, China; yejiang@cug.edu.cn (J.Y.); hangxu@cug.edu.cn (H.X.)
4 Ningxia Survey and Monitor Institute of Land and Resources, Yinchuan 750002, China; shaoyuliu@cug.edu.cn
* Correspondence: runwang@cug.edu.cn

Abstract: Monitoring mine activities can help management track the status of mineral resource exploration and mine rehabilitation. It is crucial to the sustainable development of the mining industry and the protection of the geological environment in mining areas. To monitor the mining activities of shallow surface outcrops in the arid and semi-arid regions of northwest China, this paper proposes a remote sensing monitoring approach of mining activities based on deep learning and integrated interferometric synthetic aperture radar technique. This approach uses the DeepLabV3-ResNet model to identify and extract the spatial location of the mine patches and then uses object-oriented analysis and spatial analysis methods to optimize the mine patch boundaries. SBAS-InSAR technique is used to obtain the time-series deformation information of the mine patches and is combined with the multi-temporal optical imagery to analyze the mining activities in the study area. The proposed approach has a recognition accuracy of 95.80% for the identification and extraction of mine patches, with an F1-score of 0.727 at the pixel level, and the average area similarity for all patches is 0.78 at the object-oriented level. The proposed approach possesses the capability to analyze mining activities, indicating promising prospects for engineering applications. It provides a reference for monitoring mining activities using multisource satellite remote sensing.

Keywords: mining activities monitoring; multisource remote sensing; InSAR; shallow surface outcrop; DeepLabV3-ResNet; object-oriented analysis

Citation: Li, S.; Wang, R.; Wang, L.; Liu, S.; Ye, J.; Xu, H.; Niu, R. An Approach for Monitoring Shallow Surface Outcrop Mining Activities Based on Multisource Satellite Remote Sensing Data. *Remote Sens.* **2023**, *15*, 4062. https://doi.org/10.3390/rs15164062

Academic Editor: Amin Beiranvand Pour

Received: 18 July 2023
Revised: 14 August 2023
Accepted: 15 August 2023
Published: 17 August 2023

1. Introduction

Northwest China is abundant in mineral resources, providing a large amount of raw materials for the manufacturing and energy industries. The mining industry has become the backbone of the local economy [1]. However, the geological environment of the region is relatively fragile due to the strong effects of both the endogenic force (e.g., volcanism and crustal movement) and the exogenic force (e.g., weathering, denudation, transportation, and accumulation). Mining activities, especially open-pit mining, further exacerbate the disturbance of the geological environment, destroying surface soils and natural vegetation and disrupting ecosystem service flows [2–4]. Monitoring mining activities and the progress of mine rehabilitation is essential to promote sustainability in China's manufacturing industry's upstream supply chain and preserve the mining area's geological environment.

The conventional approach to monitoring the geo-environment of mining involves field investigations, local inspections, and public supervision. However, limited by the topographic conditions of mines and the level of cooperation of the mining companies, these approaches suffer from low efficiency and are problematic in maintaining credibility and impartiality. The obtained mining activities information is often delayed and is susceptible to being omitted. It is not easy to monitor the geological environment of mines promptly and precisely. Remote sensing technology captures large-scale ground surface information and provides long-term dynamic observation, making it an indispensable data source for monitoring the geological environment of mines in recent years.

Optical remote sensing is a well-established technique for monitoring mining areas. Multi-temporal imageries can be utilized to extract mining-related patches, such as open pits, ore heaps, waste rock dumps, and tailings reservoirs, and allow for tracking ongoing mining activities, monitoring mining geo-environment problems, assessing environmental concerns, and evaluating the progress of mine rehabilitation efforts [5–9]. Optical remote sensing monitoring methods can be categorized into two main approaches: knowledge-based and data-driven.

The knowledge-based approach relies on a priori knowledge of the mine patch features. Spectral data of different mine features (e.g., coal, gangue, tailings, etc.) are first collected using the spectrometer. The spectral curves are then analyzed to develop inversion models identifying specific mine patches, such as coal heaps, gangue dumps, and tailing reservoirs [10–15]. However, spectral curve collection, analysis, and inversion model construction are time-consuming and labor-intensive when dealing with large spatial extents. Additionally, high-resolution satellite images often have limited spectral bands; for instance, the commonly used SPOT, GeoFen, and ZiYuan series [16,17] offer three or four bands that include only visible and near-infrared wavelengths. The sparse spectral information restricts the implementation of some inversion models.

The data-driven approach does not require prior knowledge but extracts the required information directly from the data. Researchers have already utilized SVM, RF, CNN, FCN, U-Net, etc. [17–23] to extract open pits tailings reservoirs and classify land use and land cover in mining areas. Over the past few years, deep learning techniques have become increasingly popular in mine environmental monitoring [24–26]. Unlike traditional machine learning algorithms that need to construct and select relevant features, deep learning methods can autonomously learn the most representative and separable features without manual feature construction or rules design [27]. The hierarchical end-to-end learning model of deep learning substantially enhances the level of automation and intelligence in remote sensing monitoring of mine environments.

However, the models often cannot produce satisfactory results directly in practical production. Creating a deep learning model for monitoring mining areas requires large and representative training samples. The mining area occupies only a small fraction of the overall surface, and the morphology of mine patches can vary greatly due to factors such as deposit structure, the spatial distribution of underground veins, geographic and geomorphic conditions, mineral type, mining processes, and the duration of mining. In such conditions, collecting sufficient samples and training a robust deep-learning model is difficult. The model will inevitably produce inaccurate mine patch boundaries. The object-based image analysis (OBIA) method can compensate for this shortcoming [28,29]. The OBIA divides the image scene into sub-regions based on the objects' spectral, geometric, and textural homogeneity or heterogeneity characteristics [30]. The divided boundaries can help to optimize the results generated by the deep learning model.

In this paper, an approach that combines deep learning and OBIA is presented to extract mine patches. Additionally, adverse weather conditions can disrupt the acquisition of optical remote sensing images, making it difficult to obtain uninterrupted data over consecutive months. This paper employs time-series synthetic aperture radar (SAR) images to generate continuous surface deformation. The surface dynamics information combined with optical imagery is used to monitor the activities of mine patches.

2. Method

2.1. Materials

2.1.1. Study Area

The study area is located at the border between Ningxia Hui Autonomous Region and Inner Mongolia Autonomous Region in northwest China, as illustrated in Figure 1. The geographic coordinates of the area span from 105.39°E to 105.64°E in longitude and from 37.59°N to 37.70°N in latitude. The study area covers an area of 272.57 km^2 and has an elevation ranging from 1253 to 1638 m.

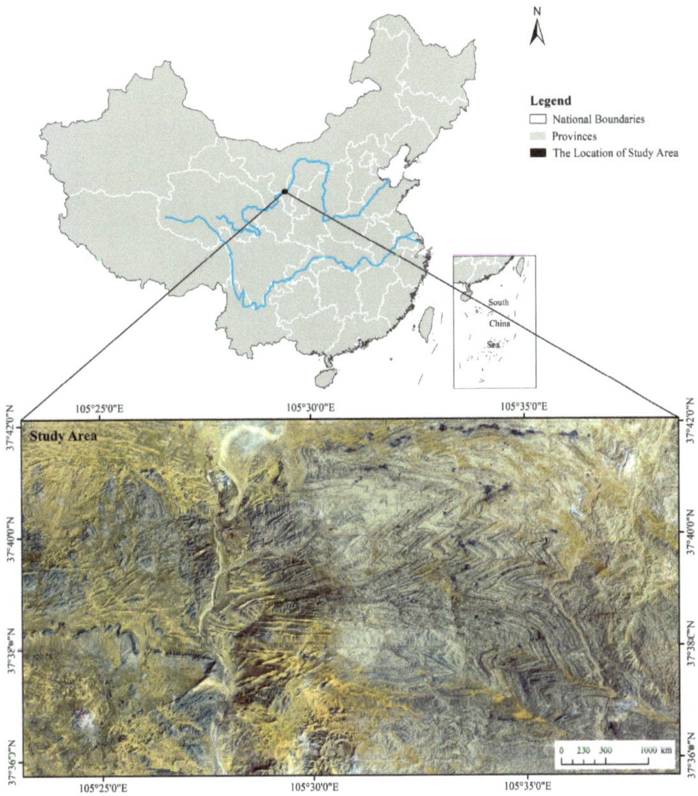

Figure 1. The location of the study area.

The area belongs to the Weiningbeishan region, situated in the transitional zone between the Tengger Desert and the Ningxia Plain. The topography of this region primarily comprises medium-low mountains and hills. It experiences a temperate continental arid climate with an average annual precipitation of less than 180 mm. The dominant vegetation type is the steppe desert. The outcropping strata in this region are primarily the Devonian and Carboniferous systems. The stratigraphy of this region represents a suite of clastic-carbonatite construction gypsum-bearing, indicative of a marine-continental interactional sedimentary environment. Magmatic activity in the region is relatively limited, with rare occurrences of magmatic rocks exposed. The quartz diorite porphyrite veins intruded into Upper Devonian and Lower Carboniferous strata during the Indosinian-Yanshanian Orogeny exhibit significant mineralization potential. The Upper Devonian Laojunshan Formation and the Lower Carboniferous Qianheishan Formation are the main ore-bearing strata in this region. The east–west trending faults formed during the Indosinian Orogeny and the secondary faults generated by subsequent tectonic movements are the main ore-controlling structures.

The extended period of uninterrupted mining has led to the emergence of numerous small-scale and dispersed mine patches within the study area. Illicit mining activities also exist in remote mountainous regions and shallow outcrop sites. These activities have resulted in the devastation of land resources, the degradation of delicate natural ecosystems, and the creation of concealed hazards to mine safety.

The mineral species are mainly iron, clay, sandstone, and coal. Except for coal, mined underground, all other minerals are mined in open-pit mining. The main focus of mining activity monitoring in this paper is on open pits and gangue dumps. The detailed description of the interpretation signs in true color remote sensing imagery is as follows (Figure 2):

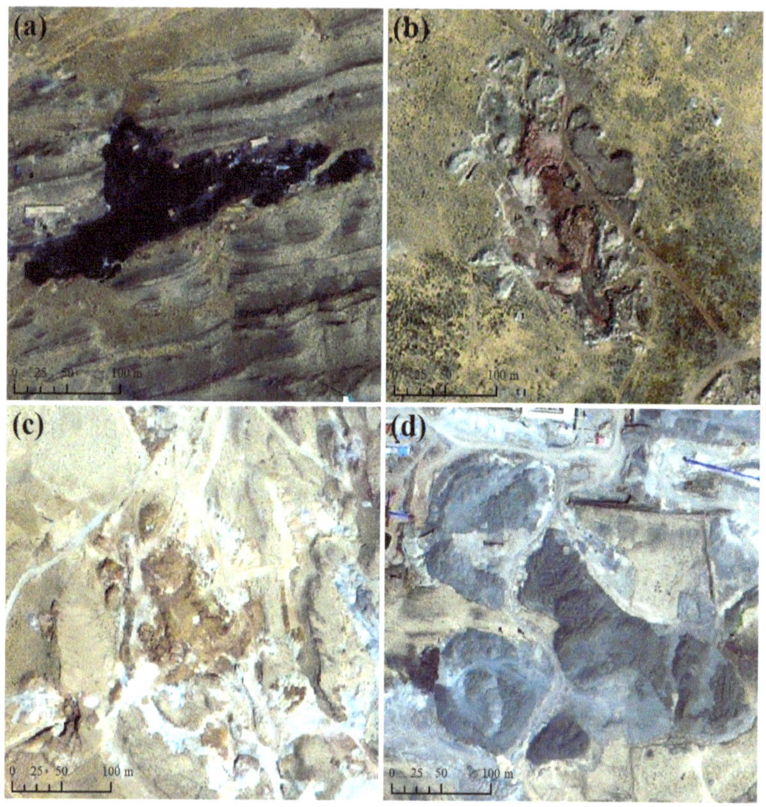

Figure 2. The remote sensing interpretation signs of mine patches in true color. (**a**) Coal gangue dumps; (**b**) Iron ore open pits; (**c**) Clay open pits; (**d**) Sandstone open pits.

(a) Coal gangue dumps: Coal gangue dumps are often distributed in the coal mine area. The color is gray to gray-black. The patches often have an irregular polygon shape with clear demarcation from the surrounding land. The artificial extraction and accumulation traces are obvious in the patches.

(b) Iron ore open pits: The color is grayish yellow, light red, reddish brown to brown. The texture of the patches is characterized by clumps or strips. The mining and excavation traces left by mechanical mining equipment are obvious.

(c) Clay open pits: The color is whitish gray to grayish yellow. The texture of the patches is blurred and shows a more homogeneous pattern. There are obvious artificial excavation and extraction traces in the spot, which is different from the surrounding natural mountains.

(d) Sandstone open pits: The color is gray to light gray, with high brightness. The texture of the patches is obvious and messy. The hillside pit exhibits positive topography. The depressed pit exhibits negative topography with near-circular or stratified terraces.

2.1.2. Data Sources

The datasets used in this study include high-resolution optical satellite imagery, time-series SAR data, and Land Use/Land Cover data products.

The Gaofen-2 (GF-2) satellite was the primary high-resolution optical satellite imagery source. It possesses a spatial resolution of 1 m for visible light bands. In this paper, digital orthophoto maps (DOMs) of the study area were generated for two distinct periods: 2019 and 2021.

The time-series Sentinel-1A Single Look Complex (SLC) SAR data were utilized for detecting mining activities. These data can be downloaded from the ASF Search Vertex platform at https://search.asf.alaska.edu/ (accessed on 17 December 2022). A total of 90 acquisitions in ascending track from 1 January 2019 to 30 December 2021 were collected, and detailed data information is presented in Table 1. Additionally, the POD Precise Orbit Ephemerides (AUX_POEORB) and the 30 m Shuttle Radar Topography Mission (SRTM) digital elevation model (DEM) were utilized to improve the orbit parameters and remove topographic phases. The GACOS (Generic Atmospheric Correction Online Service for InSAR) atmospheric data is also introduced on the same day as the SAR data to assist in atmospheric correction [31–33].

Table 1. Detailed information on Sentinel-1A SLC SAR data.

Beam Mode	Radar Wavelength	Spatial Resolution	Polarization Mode	Path/Frame Number	Incidence Angle
Interferometric Wide Swath mode (IW)	5.6 cm	5 m × 20 m	VV	157/117	37.28°

Land cover categories were obtained from ESA WorldCover Maps (https://viewer.esa-worldcover.org/worldcover/ (accessed on 7 August 2022)), which contain categories such as shrubland, built-up areas, cropland, bare/sparse vegetation, and a few water bodies in this study area. Those land cover categories were used to assist in identifying the categories of mine patches.

2.2. Methodology

The processing procedure of the mining activities monitoring approach proposed in this paper is summarized in Figure 3. Four key steps are included: (1) interpreting mine patches and sampling the training and testing samples; (2) identifying mine patches using DeepLabV3-ResNet and obtaining the initial spatial information of mine patches from high-resolution images taken in 2019; (3) extracting mine patch boundaries using multiresolution segmentation and spatial analysis; (4) monitoring mining activities from 2019 to 2021 using SBAS-InSAR.

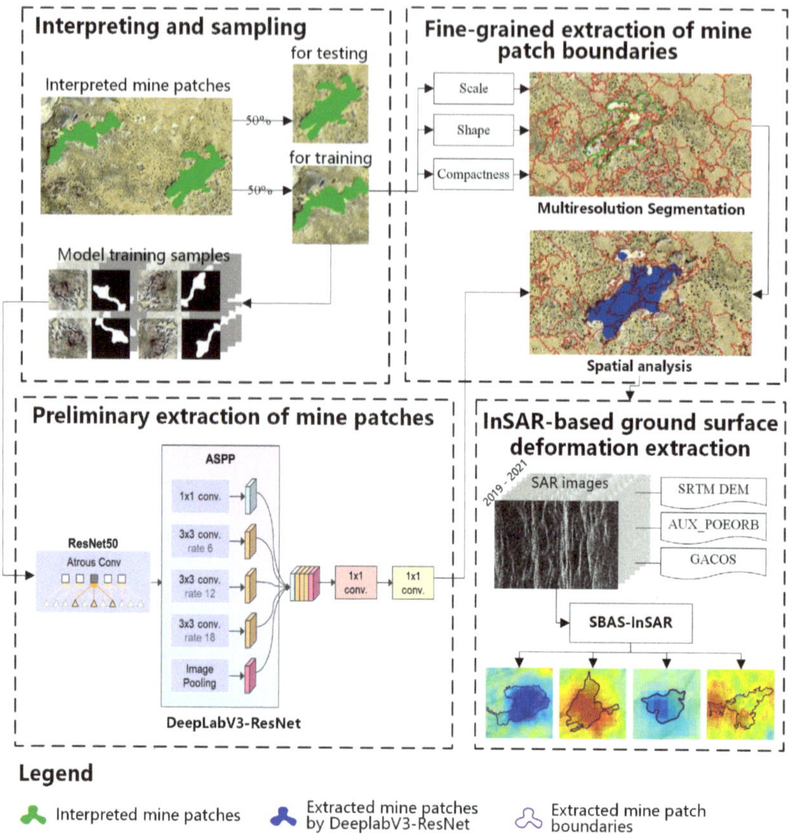

Figure 3. The proposed mining activities monitoring approach.

2.2.1. Interpreting and Sampling

A total of 238 mine patches were interpreted manually based on the established signs in Section 2.1, including 114 coal gangue dumps, 80 iron ore open pits, 30 clay ore open pits, and 14 sandstone open pits.

A total of 50% of the mine patches, evenly distributed throughout the study area, are randomly selected to generate training samples. The original image with interpreted mine boundaries is further split into small image patches of 256 × 256 pixels with a 128-pixel step for model training. Additionally, the angular rotation method is employed by rotating the split image patches by 90°, 180°, and 270° to supply more training samples. In addition, these mine patches are used as a reference for multiresolution segmentation and spatial analysis in extracting mine patch boundaries.

The remaining mine patches are used for accuracy assessment. The composition of mining and non-mining areas in this study area is imbalanced, with the ratio of mining to non-mining areas being approximately 0.015. In order to obtain realistic accuracy assessment results, the testing samples used for pixel-based accuracy assessment also adopt a corresponding imbalanced distribution.

2.2.2. Preliminary Extraction of Mine Patches

The deep learning semantic segmentation model is first used to extract the spatial information of mine patches. It can make pixel-wise predictions for a given image without requiring manual feature engineering and provide the spatial location of the mine patches while extracting the spatial extent. In this paper, we adopt Deeplabv3-ResNet [34], a robust and versatile semantic segmentation model that can be easily implemented and fine-tuned for various applications.

The architecture of the DeepLabV3-ResNet model in this paper is shown in Figure 3. ResNet50 is designed as the backbone of DeepLabV3 due to its powerful performance with reasonable computational power and processing time demands. The feature map obtained from block 4 is input into a modified version of the Atrous Spatial Pyramid Pooling (ASPP) module, which comprises five sub-modules. The outputs from each sub-module are concatenated and passed through a 1×1 convolutional block. The final prediction is generated by passing the resulting feature map through a 1×1 convolutional block and then upsampling it bilinearly. The advantage of this model is the use of the Atrous Convolution module. It can increase the receptive field of the model without reducing the spatial resolution of feature maps and obtain a larger range of contextual information.

2.2.3. Fine-Grained Extraction of Mine Patch Boundaries

1. Multiresolution segmentation

The multiresolution segmentation algorithm consecutively merges pixels with their neighbors based on relative homogeneity criteria, a combination of the color and shape properties [35]. Two parameters, i.e., the shape and the compactness, need to be determined to customize the criterion. The shape determines to what degree shape influences the segmentation compared to color. The compactness gives it a relative weighting against the smoothness of the segmentation. The strategy for optimization of multiresolution segmentation parameters in this paper is as follows:

(1) The scale parameter is pre-set as 50 [20]. The values of the shape and compactness parameters are optimized using a grid search method. Let shape = {0.1, 0.2, …, 0.9}, compactness = {0.1, 0.2, . . . , 0.9}. Each combination of shape and compactness parameters generates a segmentation result. The manually interpreted mine patches used for model training are employed as reference data to select the optimal parameters. The similarity between these reference patches and the segmentation results is used for determining the optimal parameters; a smaller area difference indicates better segmentation results. Multiple segmentation objects are merged when they correspond to the same reference mine patch. The optimal parameters for segmentation are determined by the average area difference between the reference mine patches and the merged objects. Specifically, the shape and compactness parameters corresponding to the smallest average area difference are considered the best.

(2) After determining the shape and compactness parameters, the optimal value of the scale parameter is estimated using the Estimation of Scale Parameters 2 tool [36,37]. This tool is based on the local variance of object heterogeneity within a scene. It generates the segmentation results corresponding to multiple scale parameters from the bottom up and calculates the local variance at each scale. When the local variance reaches an inflection point and turns to a decreasing trend, the corresponding scale parameter produces results with the maximum heterogeneity between objects. The optimal scale parameter is determined by the number of segmented objects covered within the reference mine patches. The aim is complete segments; thus, it is preferable to have fewer segmented objects. Based on the above strategy, the scale, shape, and compactness parameters determined in this paper are 79, 0.2, and 0.5, respectively.

2. Spatial Analysis

The preliminary extraction of mine patches is prone to spatial discontinuities, such as the generation of holes, fractional patches, etc. This paper adopts the segmented objects generated by the multiresolution segmentation algorithm to obtain fine-grained mine patch boundaries through spatial analysis. First, the preliminary extraction and segmentation results are superimposed to retain the intersecting objects. Then, a total area control strategy is employed to filter out unimportant objects, and the process is as follows:

(1) Calculate the area of each intersection region between the preliminary extraction and the segmentation, denoted as A_{intsct}, and the area proportion p_{intsct};

(2) Calculate the total area of the reference mine patches, denoted as $\sum A_{ref}$;

(3) Select the segmented objects intersected with the reference mine patches, arranging them in descending order of p_{intsct}, and calculating the corresponding cumulative area $\sum A_{\text{intsct}}$;

(4) Record the p_{intsct}, when the difference between $\sum A_{\text{intsct}}$ and $\sum A_{ref}$ is the smallest;

(5) Remove the object with an area proportion less than p_{intsct};

(6) Retain the remaining objects and merge the neighboring objects;

(7) Remove segmented objects that intersect with the known buildings, cultivated land, and water bodies using auxiliary data, and remove the objects with an area of less than 50 m^2.

Figure 4 demonstrates an example of filtering out unimportant objects. There are six segmented objects that intersect with the preliminarily extracted patches. The areas of each intersecting region and their proportions within their respective objects—R1 through R6—are initially calculated. Based on these proportions, the objects R1–R6 are subsequently ranked in descending order. For instance, the order established is as follows: R1 (99%), R3 (15%), R6 (12%), R4 (7%), R5 (5%), and R2 (2%). Next, we calculate the cumulative areas of intersecting regions, starting with R1 alone, then adding areas from R1 to R3, and so on, until the full area from R1 to R2 is accounted for. These cumulative areas are then compared with the interpreted area. We record the area ratio when the difference between the cumulative and interpreted areas is smallest. Finally, the area ratios of all intersecting regions within their corresponding segmented objects are calculated, and only the segmented objects with an area ratio greater than the optimal ratio are retained.

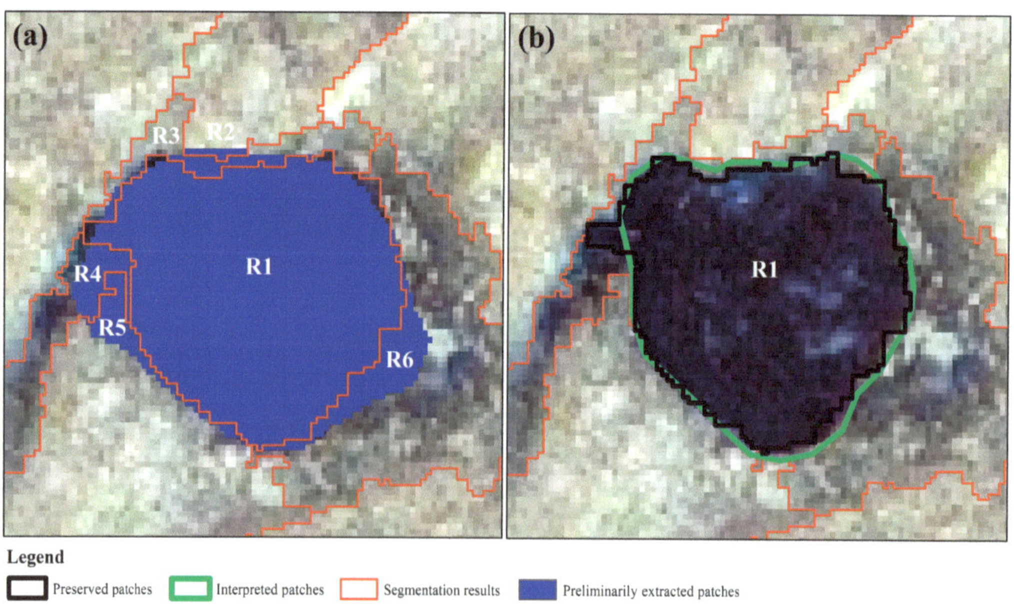

Figure 4. An example of filtering out unimportant objects. (**a**) The segmented objects R1–R6 that intersected with the preliminarily extracted patches; (**b**) The retained object R1.

Based on the above strategy, the value of P_{intsct} is 21.5% in this paper.

2.2.4. InSAR-Based Surface Deformation Extraction

The SBAS-InSAR technique divides the SAR dataset into several subsets based on the thresholds of temporal and spatial baselines. The individual SAR datasets constructed have smaller temporal and spatial baselines. The interferometric decorrelation problem caused by long time and spatial baselines is effectively suppressed. Differential interference is processed for each constructed interferometric image pair to obtain the unwrapped phase. To increase the data time sampling of the surface deformation variables and to avoid the possible rank loss problem of solving the set of equations during the least squares solution, the singular value decomposition (SVD) method is used to combine different small baseline sets to obtain the results of the surface time-series deformation for the whole period [38,39].

Let there be $N + 1$ scene of SAR images covering the study area. Generate a combination of interferometric image pairs according to the spatial and temporal baseline thresholds. The interferometric phase of any pixel in the interferogram obtained by differential interference of the master image and the slave images can be expressed as follows:

$$\partial\varphi_j = \varphi_{T_a} - \varphi_{T_b} \approx \frac{4\pi}{\lambda}\left[d_{T_b} - d_{T_a}\right] + \Delta\varphi_j'$$

The $d_{T_a}(d_{T_b})$ indicate cumulative deformation in the line-of-sight direction between $T_b(T_a)$ and T_0; $\Delta\varphi_j'$ indicates the sum of the atmospheric delay phase, the residual terrain phase, and other factors.

It is assumed that the surface deformation follows a segmentally linear pattern over time. Then, the phase of the jth differential interferogram can be expressed as follows:

$$\partial\varphi_{\mathrm{j}} = \sum_{K=\mathrm{T}_{a,j}+1}^{\mathrm{T}_{b,j}} (\mathrm{T}_K - \mathrm{T}_{K-1})V_K$$

$$\partial\varphi = \begin{pmatrix} X_{11} & \cdots & X_{1n} \\ \vdots & \ddots & \vdots \\ X_{m1} & \cdots & X_{mn} \end{pmatrix} V$$

The variable V_K represents the deformation rate at time K. The unwrapped phase can be combined into an $\mathrm{M} \times \mathrm{N}$ matrix through the time integral of the master and slave images. The final temporal deformation rate is calculated by the Moore–Penrose pseudoinverse of this matrix.

3. Result

3.1. *Accuracy Assessment for Mine Patches Extraction Results*

The extracted mine patches are shown in Figure 5. The proposed approach has identified 228 out of 238 manually interpreted patches with an identification accuracy of 95.80%. These patches comprise 112 coal gangue dumps, 74 iron ore open pits, 28 clay ore open pits, and 14 sandstone open pits. Ten small patches, averaging 4136 square meters each, have been omitted.

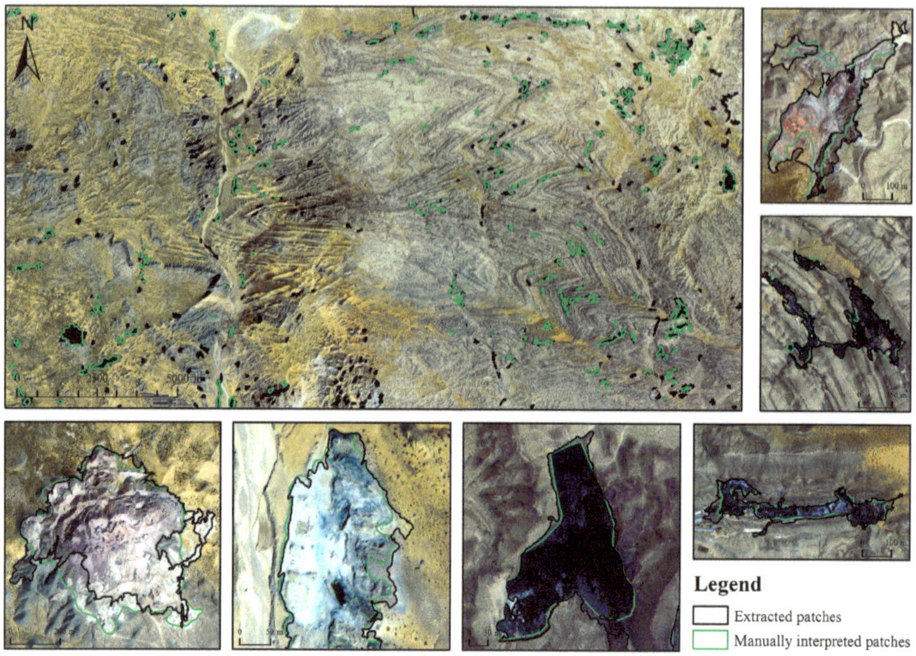

Figure 5. The mine patches extracted by the proposed scheme.

To further assess accuracy, 340,868 random samples have been extracted, including 5000 mining samples and 334,912 non-mining samples (the sampling method can be found in Section 3.1). However, using only overall accuracy (OA) to evaluate the samples is biased and misleading. It is because the non-mining samples make up a significantly larger portion, approximately 98.25%, of the total area compared to the mining samples. It is essential to contemplate alternative metrics to guarantee a just and precise assessment. A confusion matrix has been established, and the producer's accuracy, user's accuracy, kappa coefficient, and F1-score have been employed (as shown in Table 2).

Table 2. Error matrices and accuracy assessment for mining area and non-mining area.

Extracted Data	Reference Data			User's Acc.	
	Mining	Non-Mining	Total		
Mining	3985	1971	5956	0.669	
Non-mining	1015	332,941	333,964	0.997	0.727 (F_1-score)
Total	5000	334,912	340,868		0.723 (Kappa)
Producer's Acc.	0.797	0.994			99.12% (OA)

The F1-score for mine patches extraction is 0.727, and the kappa coefficient is 0.723. The producer's accuracy for mine patches (0.797) is higher than the user's (0.669), indicating that the proposed approach effectively extracts mine patches from the study area. However, there are some instances of misclassification, such as mountain shadows, wind turbines, and their surroundings being misidentified as mine patches.

To assess the accuracy of the extracted open-pit mine boundaries, this study measured the area similarity between accurately identified patches and manually interpreted boundaries. The area similarity coefficient is utilized. This coefficient measures the similarity between manually interpreted patches and the accurately extracted patches by calculating their area ratio. A ratio closer to 1 indicates a significantly higher degree of similarity.

Figure 6 shows the statistical plot of the similarity results. Over half of the patches (59.22%) have an area similarity greater than 0.80, while the average area similarity for all patches is 0.78. The similarity of different types of patches varies slightly. The extracted patches of coal gangue dumps and iron ore open pits are more similar to the manually interpreted boundaries, respectively, with a similarity of 0.81 and 0.80, respectively; the area similarity of clay ore open pits and sandstone open pits is closed at 0.78 and 0.76, respectively.

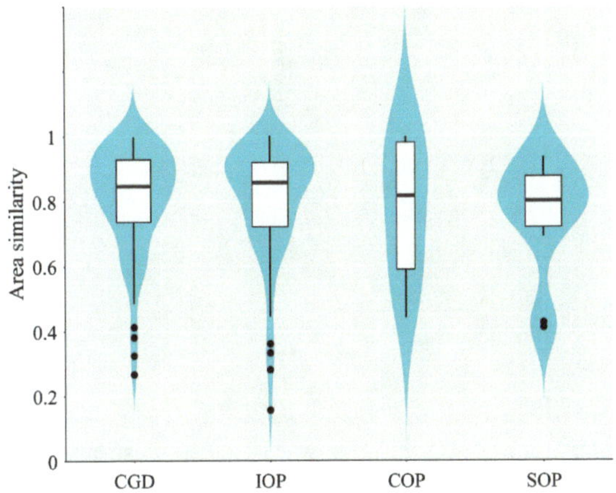

Figure 6. The statistical plot of area similarity for coal gangue dumps (CGD), iron ore open pits (IOP), clay ore open pits (COP), sandstone open pits (SOP).

3.2. Surface Deformation Information Extraction Results

This research utilized SBAS-InSAR to analyze the ground deformation between June 2019 and December 2021. Figure 7 shows the baseline distribution of Sentinel-1A acquisitions. The super master image was 2020/07/04, with a maximum absolute baseline of 111.46 m, a minimum absolute baseline of 1.97 m, and an average absolute baseline of 48.30 m. The maximum temporal baseline was 60 days, the minimum was 12 days, and the average was 35.96 days. SBAS-InSAR processing was performed on 334 interferometry pairs with an average connection per image of 7.51.

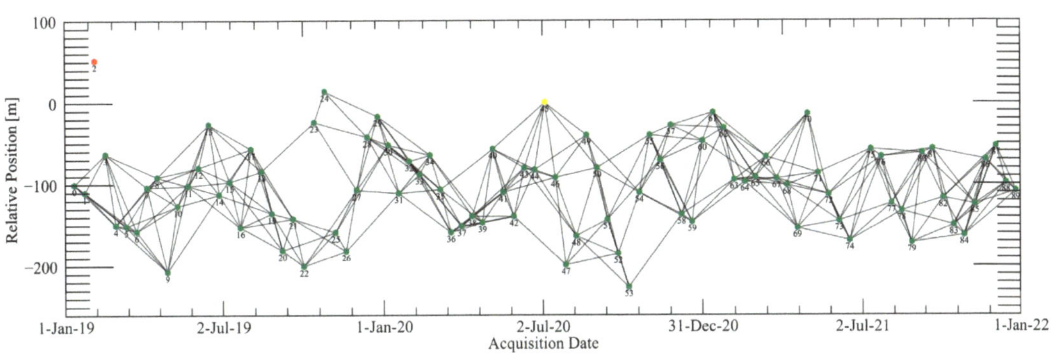

Figure 7. Baseline distribution of Sentinel-1A acquisitions. Yellow dot: Super-Master acquisition; Green dot: valid acquisitions; Red dot: discarded acquisitions; Numbers: order of acquisitions.

The vertical ground deformation of the study area is shown in Figure 8, with a spatial resolution of 15 m. The deformation rate in the study area ranges from −18.04 mm/year to 14.89 mm/year. Due to the absence of a historical monitoring record or concurrent precise leveling measurements, the surface deformation obtained from InSAR is only used to qualitatively analyze pixel deformation trends. Therefore, InSAR deformation measurements are mainly utilized for qualitative analysis of mine activities in combination with multi-temporal optical images.

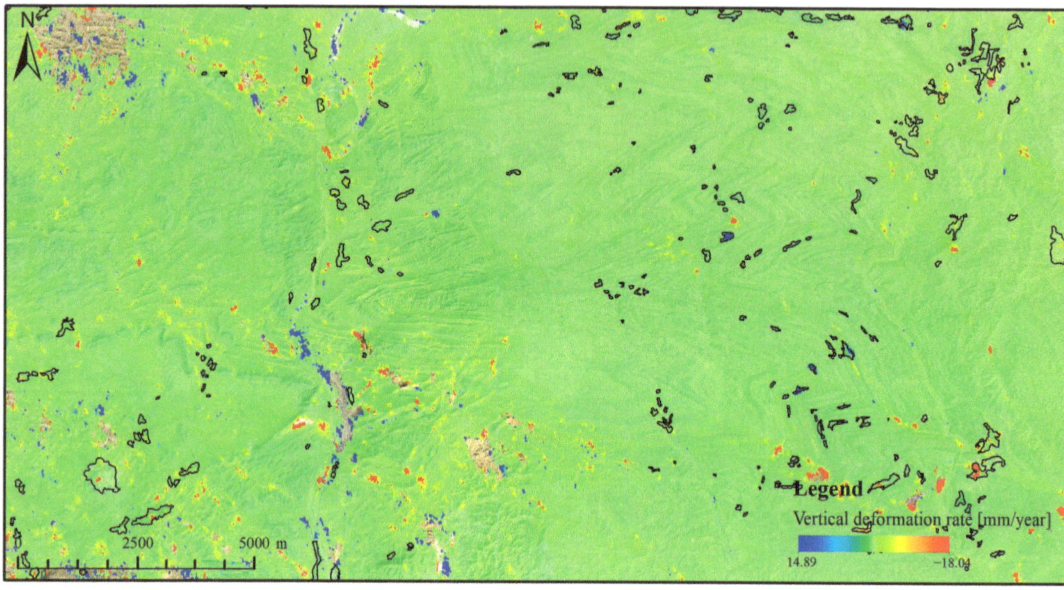

Figure 8. Vertical ground deformation rate (mm/year) from June 2019 to December 2021.

4. Discussion

4.1. Result Analysis for Mine Patches Extraction

4.1.1. The Advantages of the Proposed Approach

To further understand the performance of the proposed approach, we compared the results of DeepLabV3 with those in this paper. As shown in Figure 9, the proposed approach has better performance in handling holes (Figure 9a) and fragmented patches (Figure 9b) and can eliminate some small and isolated errors (Figure 9c).

The deep learning approach enables rapid identification and extraction of mine patches with only some manual labeling work. However, the mining area covers only a small part of the surface, but the scene is very complex. It is difficult to extract a complete patch using deep learning methods alone. The object-oriented segmentation algorithm leverages images' homogeneity and heterogeneity to extract spatially continuous geographic patches. Incorporating object-oriented image segmentation as a post-processing technique addresses the issue of insufficient spatial continuity in deep learning extraction outcomes. Furthermore, in practical applications, the utilization of historical mining patches can help minimize the reliance on manual intervention. The proposed approach has good prospects for engineering applications.

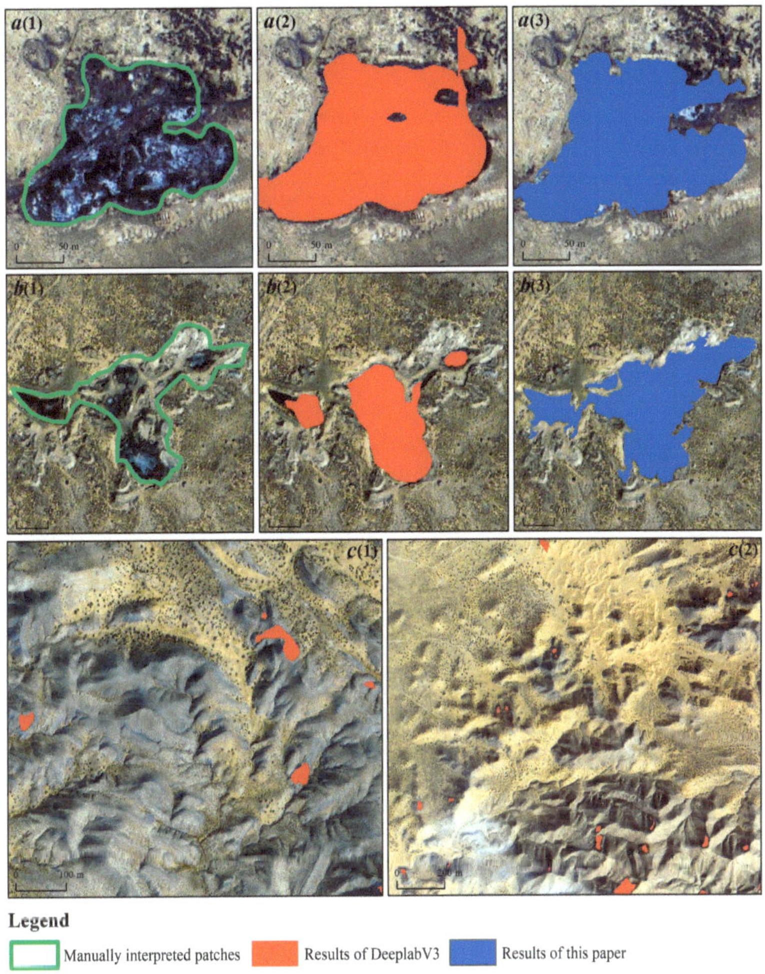

Figure 9. Mine patches extracted by manual interpretation, DeepLabV3, and the proposed approach in this paper. (**a1–a3**) Patch with holes; (**b1–b3**) Fragmented patches; (**c1**,**c2**) Some small and isolated errors.

4.1.2. The Omission and Commission Errors

The omission errors of patches are rare, and the omissions mainly occur with very small patches that closely resemble the surrounding features. The commission errors of mine patches are the main source of errors in results. In particular, the wind farm and the mountain shadows are misidentified as mine patches.

The construction process of a wind farm involves mountain excavation, land leveling, installation of wind turbines, and the construction of infrastructure such as roads. It exhibits spectral, geometric, and textural features similar to those of an open-pit mining area. In the preliminary extraction step based on the DeepLabV3-ResNet model, it was identified incorrectly.

The commission errors of mountain shadows are mainly due to spectral similarity. Due to the obstruction of solar radiation reflection pathways, mountain shadows have low values in optical images. Similarly, coal or coal gangue also exhibit relatively small numerical values due to the high absorption rate of visible light. In addition, the spatial location's consistency is also a cause of errors. Mountain shadows are mainly distributed in the folded areas in the northeastern study area, while the patches of the coal mining area are distributed along the strike of the coal-bearing strata located in the same folded area.

4.1.3. Area Variance Analysis

The difference in patch area mainly stems from the mine patches with larger scales and longer duration of mining. Large-scale open-pit mines contain multi-phase mining surfaces. The early-stage mining exposes rocks to weathering, erosion, accumulation, and wind–sand modification. The spectral and textural information of the mining patches has changed, making them difficult to extract and reducing the size of the extracted patches. Furthermore, several roads and mine patches are segmented into the same object during the multiresolution segmentation step. The subsequent steps recognize them as a whole, slightly increasing the size of the extracted mine patches.

4.2. Mining Activities Analysis Combined InSAR with Multi-Temporal Optical Remote Sensing

Monitoring open-pit mine activity solely with optical remote sensing imagery is challenging. The image quality can be affected by various factors, such as imaging conditions and data processing methodologies. For instance, high-resolution optical remote sensing satellites often incorporate the capability of lens tilt in their onboard cameras to reduce revisit intervals. Different lens angles during image capture can lead to variations in the shadows of objects in the imagery. In the absence of elevation change information, with the use of multi-temporal optical imagery to carry out interpretation processing, the shadow will affect the interpreter to determine the positive and negative terrain. In particular, mining activities that do not significantly change the spatial extent of mining patches, and small-scale illicit mining activities with shallow surface outcrop, are difficult to interpret directly from optical imagery. The time series of surface deformation measurements obtained by InSAR can visually reflect the mining activities around the mine patches.

In this paper, a combination of SBAS-InSAR and multi-temporal high-resolution optical imagery is employed to gather information on mining activities within the mine patches. As open-pit mining leads to the expansion of mining surfaces, any ground surface deformation occurring within a 0.5 km radius of the patches is attributed to the ongoing mining activity at the respective patch. A pixel in the measured image is classified as having no deformation if the annual deformation rate falls within the range of ± 5 mm/year. A positive deformation is assigned to an annual deformation rate exceeding 5 mm/year, while a negative deformation is assigned to an annual deformation rate below -5 mm/year.

Statistically, among the mine scenes associated with the 228 extracted mine patches, there are 22 which exhibited positive deformation, 38 showed negative deformation, 4 with interferometric decorrelation, and 164 with no significant change within the scenes. Through a comparison of the 2019 and 2021 optical imagery, the 64 deformations anomalies all reflect related mining activities, including continued mining, land leveling, and construction of other infrastructure.

The mine patches associated with positive deformation are all coal gangue dumps, with no noticeable difference in optical imagery between 2019 and 2021, which shows the surface accumulation process of coal gangue in underground coal mining areas. The mine patches associated with negative deformation are open pits and coal gangue dumps. Continuous mining in open-pit mines is the primary cause of negative deformation, which has also been confirmed in optical imagery. These mining scenes consist of 14 iron ore open pits, 7 clay open pits, and 5 sandstone open pits. In addition to the continuous mining, negative deformations were found in five open-pit iron ore mines that have been ecologically rehabilitated, including four sites of land leveling and one site where a photovoltaic solar power station has been constructed. Negative deformation is also observed at seven coal gangue dumps. Among these, one site undergoes land leveling, while the remaining six sites are associated with the transportation of coal gangue. In addition, the interferometric decorrelation caused by excessive deformation is an important characteristic of mine activities [40–45]. It is necessary to utilize optical imagery for the identification of such patches. In this study area, there are four patches with interferometric decorrelation, including one ongoing iron ore open pit, one ongoing sandstone open pit, and two iron ore open pits where land leveling has been conducted.

The boundaries of mine patches and time-series surface deformation can help regulatory authorities identify illicit mining activities. By comparing the gathered data with management records, such as mining rights and permits, such mining activities as mining operations that continue beyond the permitted time frame, encroachments beyond the approved spatial boundaries, and failures to fulfill the obligations for mine site reclamation and restoration can be easily found. For example, the Central Government's Fourth Ecological Environmental Protection Inspection Group inspected the Ningxia Hui Autonomous Region in December 2021 and found that illicit mining in the northern mountainous areas of Zhongning County, Zhongwei City, was prominent, seriously damaging the ecological environment. We use the proposed approach to retrospectively analyze the mining activities in the reported illicit mining cases from 2019 to 2021, as depicted in Figure 10 (the time phase of the optical image is 2021, while the time phase of the mine patches is 2019). Figure 10a represents Tongtiegou Ceramic Clay Mine, showcasing conspicuous mining activities in its central-eastern and southern regions. The time-series deformation at the ROI #1 point location exhibits an escalated mining intensity since August 2020. Figure 10b represents Pingtanghugou Baitugangzi Sandstone quarry, where substantial mining operations are prevalent in the southwestern part of the mine scene. The time-series deformation at the ROI #2 point location reveals a consistent pattern of illicit mining activities occurring beyond the authorized mining boundaries throughout the monitoring period.

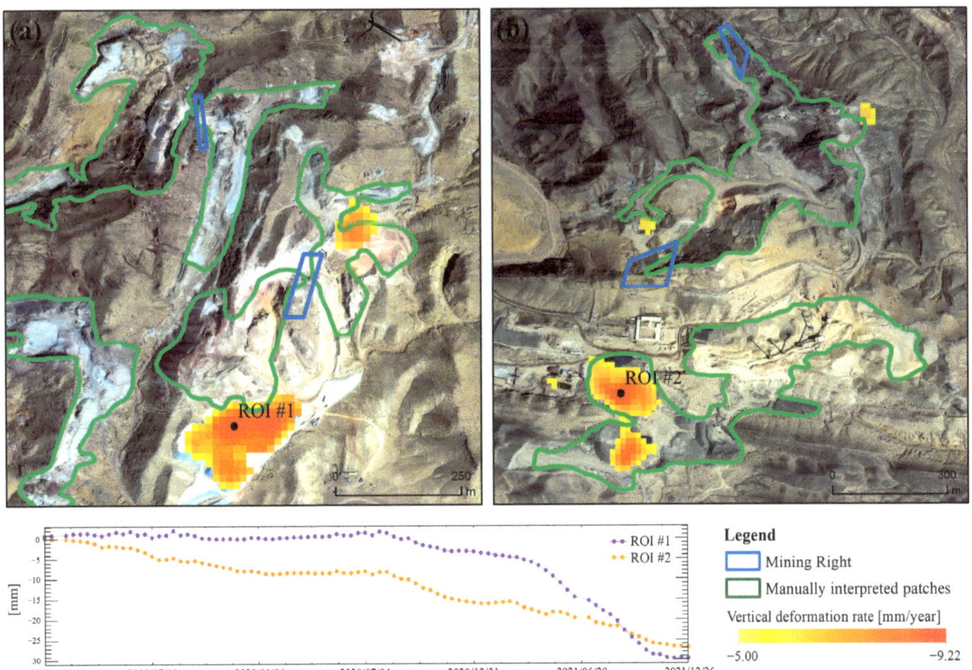

Figure 10. Examples of mine patches of illicit mining activities. (**a**) Tongtiegou Ceramic Clay Mine; (**b**) Pingtanghugou Baitugangzi Sandstone quarry.

5. Conclusions

In this paper, we propose a deep-learning-based approach for mine monitoring by combining InSAR and multi-temporal high-resolution optical imagery, which is used to monitor mining activities in the Weiningbeishan area of northwestern China. This approach first uses the DeepLabV3-ResNet model to preliminary localize and identify the spatial location of mine patches. On this basis, the multiresolution segmentation algorithm and spatial analysis method are utilized for post-processing to obtain the boundaries of the mine patches. Finally, the time-series deformation information acquired by SBAS-InSAR of mine patches and the multi-temporal optical imagery are combined to analyze mine activities. The proposed approach takes advantage of the fact that deep learning does not require the construction of complex feature engineering and thus reduces the manual involvement in the pre-survey of remote sensing of mines and utilizes the post-processing method of object-oriented analysis to make up for the lack of geospatial continuity of the extraction results of deep learning, and the use of time-series InSAR avoids the limitations of multi-temporal optical remote sensing which is difficult to be observed continuously due to the imaging conditions. The proposed approach provides a new way to comprehensively utilize satellite remote sensing techniques to monitor mining activities in arid and semi-arid areas. The proposed approach capitalizes on the inherent advantage of deep learning, eliminating the need for intricate feature engineering and reducing the manual effort involved in pre-surveying remote sensing of mines.

Additionally, it combines object-oriented analysis as a post-processing method to compensate for the lack of spatial consistency in the extraction results obtained by deep learning. Furthermore, the utilization of time-series InSAR overcomes the limitations of optical remote sensing, which cannot observe continuously due to weather conditions. The proposed approach presents a novel perspective on effectively leveraging satellite remote sensing techniques for monitoring mining activities in arid and semi-arid regions. Future work will explore two aspects. The first aspect involves optimizing the post-processing process to obtain more accurate boundaries of mine patches. The second aspect focuses on studying the surface processes from mining to abandonment to rehabilitation of open-pit mines in arid and semi-arid vegetation-exposed areas and constructing sample libraries of the different processes to improve the recognition accuracy of mine patches.

Author Contributions: Conceptualization, S.L. (Shiyao Li), R.W. and L.W.; methodology, S.L. (Shiyao Li) and R.W.; validation, S.L. (Shaoyu Liu), J.Y. and H.X.; formal analysis, S.L. (Shiyao Li) and R.W.; writing—original draft preparation, S.L. (Shiyao Li); writing—review and editing, S.L. (Shiyao Li) and R.W.; visualization, S.L. (Shiyao Li) and R.W.; supervision, R.N. and L.W.; funding acquisition, S.L. (Shiyao Li), L.W. and S.L. (Shaoyu Liu). All authors have read and agreed to the published version of the manuscript.

Funding: This research was funded by Geological Survey Projects of the China Geological Survey (Grant No. DD20211391), Natural Science Foundation of Ningxia Hui Autonomous Region (Grant No. 2021AAC03431).

Data Availability Statement: Readers who need data for an in-depth study of this paper can contact the corresponding author for access.

Conflicts of Interest: The authors declare no conflict of interest.

References

1. Yu, L.; Xu, Y.; Xue, Y.; Li, X.; Cheng, Y.; Liu, X.; Porwal, A.; Holden, E.; Yang, J.; Gong, P. Monitoring surface mining belts using multiple remote sensing datasets: A global perspective. *Ore Geol. Rev.* **2018**, *101*, 675–687. [CrossRef]
2. Li, C.; Wang, A.; Chen, X.; Chen, Q.; Zhang, Y.; Li, Y. Regional distribution and sustainable development strategy of mineral resources in China. *Chin. Geogr. Sci.* **2013**, *23*, 470–481. [CrossRef]
3. Sonter, L.J.; Moran, C.J.; Barrett, D.J.; Soares-Filho, B.S. Processes of land use change in mining regions. *J. Clean. Prod.* **2014**, *84*, 494–501. [CrossRef]
4. Gastauer, M.; Silva, J.R.; Junior, C.F.C.; Ramos, S.J.; Filho, P.W.M.S.; Neto, A.E.F.; Siqueira, J.O. Mine land rehabilitation: Modern ecological approaches for more sustainable mining. *J. Clean. Prod.* **2018**, *172*, 1409–1422. [CrossRef]

5. Van der Meer, F.D.; Van der Werff, H.M.; Van Ruitenbeek, F.J.; Hecker, C.A.; Bakker, W.H.; Noomen, M.F.; Van der Meijde, M.; Carranza, E.J.M.; De Smeth, J.B.; Woldai, T. Multi-and hyperspectral geologic remote sensing: A review. *Int. J. Appl. Earth Obs. Geoinf.* **2012**, *14*, 112–128. [CrossRef]
6. Li, J.; Pei, Y.; Zhao, S.; Xiao, R.; Sang, X.; Zhang, C. A Review of Remote Sensing for Environmental Monitoring in China. *Remote Sens.* **2020**, *12*, 1130. [CrossRef]
7. Song, W.; Song, W.; Gu, H.; Li, F. Progress in the Remote Sensing Monitoring of the Ecological Environment in Mining Areas. *Int. J. Environ. Res. Public Health* **2020**, *17*, 1846. [CrossRef]
8. Yang, Z.; Li, Z.; Zhu, J.; Wang, Y.; Wu, L. Use of SAR/InSAR in mining deformation monitoring, parameter inversion, and forward predictions: A review. *IEEE Geosci. Remote Sens. Mag.* **2020**, *8*, 71–90. [CrossRef]
9. McKenna, P.B.; Lechner, A.M.; Phinn, S.; Erskine, P.D. Remote Sensing of Mine Site Rehabilitation for Ecological Outcomes: A Global Systematic Review. *Remote Sens.* **2020**, *12*, 3535. [CrossRef]
10. Notesco, G.; Kopačková, V.; Rojík, P.; Schwartz, G.; Livne, I.; Dor, E.B. Mineral Classification of Land Surface Using Multispectral LWIR and Hyperspectral SWIR Remote-Sensing Data. A Case Study over the Sokolov Lignite Open-Pit Mines, the Czech Republic. *Remote Sens.* **2014**, *6*, 7005–7025. [CrossRef]
11. Kirsch, M.; Lorenz, S.; Zimmermann, R.; Tusa, L.; Möckel, R.; Hödl, P.; Booysen, R.; Khodadadzadeh, M.; Gloaguen, R. Integration of Terrestrial and Drone-Borne Hyperspectral and Photogrammetric Sensing Methods for Exploration Mapping and Mining Monitoring. *Remote Sens.* **2018**, *10*, 1366. [CrossRef]
12. He, D.; Le, B.T.; Xiao, D.; Mao, Y.; Shan, F.; Ha, T.T.L. Coal mine area monitoring method by machine learning and multispectral remote sensing images. *Infrared Phys. Technol.* **2019**, *103*, 103070. [CrossRef]
13. Padró, J.C.; Carabassa, V.; Balagué, J.; Brotons, L.; Alcañiz, J.M.; Pons, X. Monitoring opencast mine restorations using Unmanned Aerial System (UAS) imagery. *Sci. Total Environ.* **2019**, *657*, 1602–1614. [CrossRef]
14. Purwadi, I.; van der Werff, H.M.; Lievens, C. Targeting rare earth element bearing mine tailings on Bangka Island, Indonesia, with Sentinel-2 MSI. *Int. J. Appl. Earth Obs. Geoinf.* **2020**, *88*, 102055. [CrossRef]
15. Hu, F.; Hu, Y.; Cui, E.; Guan, Y.; Gao, B.; Wang, X.; Wang, K.; Liu, Y.; Yao, X. Recognition method of coal and gangue combined with structural similarity index measure and principal component analysis network under multispectral imaging. *Microchem. J.* **2023**, *186*, 108330. [CrossRef]
16. Zhang, Y.; Guindon, B.; Lantz, N.; Shipman, T.; Chao, D.; Raymond, D. Quantification of anthropogenic and natural changes in oil sands mining infrastructure land based on RapidEye and SPOT5. *Int. J. Appl. Earth Obs. Geoinf.* **2014**, *29*, 31–43. [CrossRef]
17. Chen, W.; Li, X.; He, H.; Wang, L. A Review of Fine-Scale Land Use and Land Cover Classification in Open-Pit Mining Areas by Remote Sensing Techniques. *Remote Sens.* **2018**, *10*, 15. [CrossRef]
18. Chen, W.; Li, X.; Wang, L. Fine Land Cover Classification in an Open Pit Mining Area Using Optimized Support Vector Machine and WorldView-3 Imagery. *Remote Sens.* **2020**, *12*, 82. [CrossRef]
19. Chen, W.; Li, X.; He, H.; Wang, L. Assessing Different Feature Sets' Effects on Land Cover Classification in Complex Surface-Mined Landscapes by ZiYuan-3 Satellite Imagery. *Remote Sens.* **2018**, *10*, 23. [CrossRef]
20. Chen, T.; Hu, N.; Niu, R.; Zhen, N.; Plaza, A. Object-Oriented Open-Pit Mine Mapping Using Gaofen-2 Satellite Image and Convolutional Neural Network, for the Yuzhou City, China. *Remote Sens.* **2020**, *12*, 3895. [CrossRef]
21. Chen, T.; Zheng, X.; Niu, R.; Plaza, A. Open-pit mine area mapping with Gaofen-2 satellite images using U-Net+. *IEEE J. Sel. Top. Appl. Earth Obs. Remote Sens.* **2022**, *15*, 3589–3599. [CrossRef]
22. Li, X.; Chen, W.; Cheng, X.; Wang, L. A Comparison of Machine Learning Algorithms for Mapping of Complex Surface-Mined and Agricultural Landscapes Using ZiYuan-3 Stereo Satellite Imagery. *Remote Sens.* **2016**, *8*, 514. [CrossRef]
23. Wang, C.; Chen, T.; Plaza, A. MFE-ResNet: A new extraction framework for land cover characterization in mining areas. *Future Gener. Comput. Syst.* **2023**, *145*, 550–562. [CrossRef]
24. Du, S.; Li, W.; Li, J.; Du, S.; Zhang, C.; Sun, Y. Open-pit mine change detection from high resolution remote sensing images using DA-UNet++ and object-based approach. *Int. J. Min. Reclam. Environ.* **2022**, *36*, 512–535. [CrossRef]
25. Du, S.; Xing, J.; Li, J.; Du, S.; Zhang, C.; Sun, Y. Open-Pit Mine Extraction from Very High-Resolution Remote Sensing Images Using OM-DeepLab. *Nat. Resour. Res.* **2022**, *31*, 3173–3194. [CrossRef]
26. Zhou, G.; Xu, J.; Chen, W.; Li, X.; Li, J.; Wang, L. Deep feature enhancement method for land cover with irregular and sparse spatial distribution features: A case study on open-pit mining. *IEEE Trans. Geosci. Remote Sens.* **2023**, *61*, 4401220. [CrossRef]
27. Chen, Y.; Jiang, H.; Li, C.; Jia, X.; Ghamisi, P. Deep feature extraction and classification of hyperspectral images based on convolutional neural networks. *IEEE Trans. Geosci. Remote Sens.* **2016**, *54*, 6232–6251. [CrossRef]
28. Blaschke, T. Object based image analysis for remote sensing. *ISPRS J. Photogramm. Remote Sens.* **2010**, *65*, 2–16. [CrossRef]
29. Hussain, M.; Chen, D.; Cheng, A.; Wei, H.; Stanley, D. Change detection from remotely sensed images: From pixel-based to object-based approaches. *ISPRS J. Photogramm. Remote Sens.* **2013**, *80*, 91–106. [CrossRef]
30. Bruzzone, L.; Carlin, L. A multilevel context-based system for classification of very high spatial resolution images. *IEEE Trans. Geosci. Remote Sens.* **2006**, *44*, 2587–2600. [CrossRef]
31. Yu, C.; Li, Z.; Penna, N.T.; Crippa, P. Generic atmospheric correction model for interferometric synthetic aperture radar observations. *J. Geophys. Res. Solid Earth.* **2018**, *123*, 9202–9222. [CrossRef]
32. Yu, C.; Li, Z.; Penna, N.T. Interferometric synthetic aperture radar atmospheric correction using a GPS-based iterative tropospheric decomposition model. *Remote Sens. Environ.* **2018**, *204*, 109–121. [CrossRef]

33. Yu, C.; Penna, N.T.; Li, Z. Generation of real-time mode high-resolution water vapor fields from GPS observations. *J. Geophys. Res. Atmos.* **2017**, *122*, 2008–2025. [CrossRef]
34. Chen, L.C.; Papandreou, G.; Schroff, F.; Adam, H. Rethinking atrous convolution for semantic image segmentation. *arXiv* **2017**, arXiv:1706.05587.
35. Benz, U.C.; Hofmann, P.; Willhauck, G.; Lingenfelder, I.; Heynen, M. Multi-resolution, object-oriented fuzzy analysis of remote sensing data for GIS-ready information. *ISPRS J. Photogramm. Remote Sens.* **2004**, *58*, 239–258. [CrossRef]
36. Drăguţ, L.; Tiede, D.; Levick, S.R. ESP: A tool to estimate scale parameter for multiresolution image segmentation of remotely sensed data. *Int. J. Geogr. Inf. Sci.* **2010**, *24*, 859–871. [CrossRef]
37. Drăguţ, L.; Csillik, O.; Eisank, C.; Tiede, D. Automated parameterisation for multi-scale image segmentation on multiple layers. *ISPRS J. Photogramm. Remote Sens.* **2014**, *88*, 119–127. [CrossRef]
38. Berardino, P.; Fornaro, G.; Lanari, R.; Sansosti, E. A new algorithm for surface deformation monitoring based on small baseline differential SAR interferograms. *IEEE Trans. Geosci. Remote Sens.* **2002**, *40*, 2375–2383. [CrossRef]
39. Li, S.; Xu, W.; Li, Z. Review of the SBAS InSAR Time-series algorithms, applications, and challenges. *Geod. Geodyn.* **2022**, *13*, 114–126. [CrossRef]
40. Forkuor, G.; Ullmann, T.; Griesbeck, M. Mapping and Monitoring Small-Scale Mining Activities in Ghana using Sentinel-1 Time Series (2015–2019). *Remote Sens.* **2020**, *12*, 911. [CrossRef]
41. Moon, J.; Lee, H. Analysis of Activity in an Open-Pit Mine by Using InSAR Coherence-Based Normalized Difference Activity Index. *Remote Sens.* **2021**, *13*, 1861. [CrossRef]
42. Zhao, G.; Wang, L.; Deng, K.; Wang, M.; Xu, Y.; Zheng, M.; Luo, Q. An Adaptive Offset-Tracking Method Based on Deformation Gradients and Image Noises for Mining Deformation Monitoring. *Remote Sens.* **2021**, *13*, 2958. [CrossRef]
43. Zhang, Y.; Lian, X.; Ge, L.; Liu, X.; Du, Z.; Yang, W.; Wu, Y.; Hu, H.; Cai, Y. Surface Subsidence Monitoring Induced by Underground Coal Mining by Combining DInSAR and UAV Photogrammetry. *Remote Sens.* **2022**, *14*, 4711. [CrossRef]
44. Moomen, A.-W.; Lacroix, P.; Benvenuti, A.; Planque, M.; Piller, T.; Davis, K.; Miranda, M.; Ibrahim, E.; Giuliani, G. Assessing the Applications of Earth Observation Data for Monitoring Artisanal and Small-Scale Gold Mining (ASGM) in Developing Countries. *Remote Sens.* **2022**, *14*, 2971. [CrossRef]
45. Alessi, M.A.; Chirico, P.G.; Sunder, S.; O'Pry, K.L. Detection and Monitoring of Small-Scale Diamond and Gold Mining Dredges Using Synthetic Aperture Radar on the Kadéï (Sangha) River, Central African Republic. *Remote Sens.* **2023**, *15*, 913. [CrossRef]

remote sensing

MDPI

Article

Mapping Alteration Minerals Using ZY-1 02D Hyperspectral Remote Sensing Data in Coalbed Methane Enrichment Areas

Li Chen, Xinxin Sui *, Rongyuan Liu, Hong Chen, Yu Li, Xian Zhang and Haomin Chen

China Areo Geophysical Survey and Remote Sensing Center for Natural Resources, Beijing 100083, China; cli_a@mail.cgs.gov.cn (L.C.); liurongyuan@mail.cgs.gov.cn (R.L.); chong@mail.cgs.gov.cn (H.C.); liyu@mail.cgs.gov.cn (Y.L.); zhangxian@mail.cgs.gov.cn (X.Z.); chenhaomin@mail.cgs.gov.cn (H.C.)
* Correspondence: suixinxin@mail.cgs.gov.cn

Abstract: As a clean energy resource, coalbed methane (CBM) is an important industry in China's dual-carbon strategic planning. Despite the immense potential of CBM resources in China, the current exploration level remains low due to outdated survey technology, impeding large-scale exploration and development. This study investigates the application of hyperspectral data in CBM enrichment areas, specifically focusing on the extraction of alteration minerals in the Hudi coal mine area of the Qinshui Basin using ZY-1 02D and Hyperion hyperspectral data. The hyperspectral alteration mineral identification methods are summarized and analyzed. A method that combines spectral feature matching and diagnostic characteristic parameters is proposed for mineral extraction based on the spectral characteristics of different minerals. The extraction results are verified through field samples using X-ray diffraction analysis. Results show that (1) both ZY-1 02D and Hyperion hyperspectral data yield favorable extraction results for clay and carbonate minerals; (2) the overall accuracy of clay and carbonate minerals extraction is higher using ZY-1 02D data compared with Hyperion data, with accuracies of 81.67% and 79.03%, respectively; (3) the proposed method effectively extracts alteration minerals in CBM enrichment areas using hyperspectral data, thereby providing valuable technical support for the application of hyperspectral data.

Keywords: hyperspectral remote sensing; ZY-1-02D; Hyperion; alteration mineral; coalbed methane enrichment areas

Citation: Chen, L.; Sui, X.; Liu, R.; Chen, H.; Li, Y.; Zhang, X.; Chen, H. Mapping Alteration Minerals Using ZY-1 02D Hyperspectral Remote Sensing Data in Coalbed Methane Enrichment Areas. *Remote Sens.* **2023**, *15*, 3590. https://doi.org/10.3390/rs15143590

Academic Editor: Geert Verhoeven

Received: 27 June 2023
Revised: 7 July 2023
Accepted: 11 July 2023
Published: 18 July 2023

1. Introduction

In China's dual-carbon strategic planning, coalbed methane (CBM) has emerged as an important clean energy resource. More than 90% of CBM comprises methane (CH_4), along with other heavier hydrocarbons, such as CO_2, N_2, C_3H_8, C_4H_{10}, and others [1]. Nevertheless, methane remarkably contributes to the greenhouse effect, ozone layer depletion, global warming, and environmental damage. As the world's largest energy producer, China has strongly supported and developed the exploration and development of CBM enrichment areas [2]. However, traditional exploration methods for CBM enrichment areas mainly rely on ground drilling, seismic exploration, geophysical techniques, and geochemical surveys [3,4]. These methods suffer from exploration efficiency, limited scope, and the need for substantial human, financial, and material resources. In particular, exploration risks increase dramatically in mountainous areas. Efficient, economical, and accurate identification of CBM enrichment areas has become an urgent problem.

Satellite remote sensing technology possesses the capability to detect and analyze surface rock and mineral composition with precision [5–10]. Specifically, hyperspectral data provide high-resolution images for ground object classification and enable detailed analysis of the chemical composition of certain minerals [11–15]. In recent years, hyperspectral remote sensing technology has played an important role in mineral identification [16–18], geological mapping [19], alteration anomaly zoning [20,21], and prospecting prediction [22,23]. For example, mineral mapping has been conducted in the Eastern Tien Shan, Xinjiang, and the Maherabad area in eastern Iran using HyMap data [24,25]. Fine mineral identification has been applied in Liuyuan, Gansu, and Cuprite, USA, using GF-5 hyperspectral imagery [26]. However, research on the hyperspectral characteristics of different alteration minerals in CBM enrichment areas is scarce. Studies have revealed that hydrocarbon microseepage phenomena exist in more than 85% of discovered oil and gas fields worldwide [4]. These phenomena subject the surface soil and rock to prolongred exposure to hydrocarbon substances, leading to alterations through various physical and chemical processes, including the enrichment of clay minerals, carbonate minerals, and iron minerals [27]. The application of hyperspectral remote sensing technology holds the potential to prospect CBM enrichment areas, effectively compensating for the limitations of traditional exploration methods and providing an important technical method for the efficient and rapid identification of CBM enrichment areas. While information about the abnormal alteration of minerals is rare and easily affected by external interference, the difficulty of identification of CBM enrichment areas has greatly increased.

The ZY-1 02D was successfully launched in China on 12 September 2019. It operates in a solar synchronous orbit with a ground coverage width of 60 km, an average orbital altitude of 778 km, and an inclination of 98.5° [28,29]. The satellite is equipped with a visible near-infrared camera and a hyperspectral camera (AHSI). The AHSI data provides a total of 166 visible near-infrared short-wave bands. This includes 76 visible near-infrared (VNIR) bands with a wavelength range of 395–1040 nm and a spectral resolution of 10 nm, and 90 short-wave infrared (SWIR) bands with a wavelength range of 1005–2501 nm and a spectral resolution of 20 nm [30,31]. At present, ZY1-02D hyperspectral data have been applied in land use classification and geological and mineral exploration [32,33].

ZY-1 02D hyperspectral data provide a basis for investigating the diagnostic spectral characteristics of alteration minerals in CBM enrichment areas. To solve the problems of tiny abnormal information extraction, the purpose of this study is to analyze the diagnostic spectral characteristics of different alteration minerals in CBM using ZY-1 02D hyperspectral data. This analysis aims to extract typical alteration minerals within the study area through a method that combines spectral feature matching and diagnostic characteristic parameters. Hyperion, the world's first successfully launched civilian satellite hyperspectral imaging spectrometer, is onboard the National Aeronautics and Space Administration Earth Observer-1 (EO-1) satellite [34,35]. The study demonstrates the capability of domestic data to extract alteration minerals through comparing the extraction results with Hyperion data and conducting field verification. This research provides valuable support for the identification of alteration minerals and the prediction of CBM enrichment areas.

2. Study Area and Data Sources

2.1. Study Area

Qinshui Basin, an important coal-bearing basin in China, is predominantly composed of anthracite and medium-high metamorphic bituminous coal. It boasts abundant coal resources, with a total reserves of approximately 2700 × 10^8 t [36]. The geological characteristics of this basin provide a favorable foundation for the formation of CBM reservoirs. The study area specifically encompasses the southeast margin of Qinshui Basin, located in Hudi Town, Qinshui County, Jincheng City, Shanxi Province [37]. The geographical extent of the study region spans from 112°30′E to 112°40′E and 35°35′N to 35°50′N, covering an total area of approximately 500 km^2 (Figure 1).

The study area is characterized by complex topography, featuring low mountains, hills, and distinct loess landform. It is situated within a sedimentary basin formed through the transformation and deformation of the Paleozoic Craton basin in North China. The stratigraphic sequence includes Ordovician (O), Carboniferous (C), Permian (P), Triassic (T), Paleogene (E), Neogene (N), and Quaternary (Q) formations, with a general trend of NNE and a dip angle of less than 10° in the study area [26]. The lithology consists of limestone, sandstone, siltstone, mudstone, sandy mudstone, and argillaceous sandstone. The structural complexity in the study area is relatively low, primarily characterized by folds, with a limited occurrence of faults and no magmatic rock activity. The main coal-bearing strata consist of the Carboniferous Taiyuan Formation and the Permian Shanxi Formation, with coal seams found at depths of less than 800 m. The No. 3 coal seam of the Shanxi Formation and the No. 15 coal seam of the Taiyuan Formation are the primary coal seams being mined in the study area, renowned for their stable development and high gas content [37].

In CBM enrichment areas, the presence of hydrocarbon microseepage leads to widespread mineral alterations, such as clayization, carbonatization, and mineralization. Therefore, the diagnostic hyperspectral response characteristics of clay and carbonate minerals are considered indicative features for distinguishing and extracting alteration information in CBM enrichment areas using hyperspectral remote sensing techniques.

Figure 1. Location of the study area.

2.2. Data

2.2.1. Hyperspectral Remote Sensing Data

The ZY-1 02D is China's first independently developed civilian hyperspectral satellite. AHSI is the one of the important payloads onboard the ZY-1 02D satellite. The SWIR spectrum of AHSI data, in particular, exhibits excellent spectral features, making it well-suited for distinguishing the diagnostic spectral features of ground objects. The ZY-1 02D AHSI data used in this study were captured on 21 October 2020 and have no cloud coverage.

Hyperion data are highly suitable for extracting alteration minerals in CBM enrichment areas due to their most useful spectral features in the NIR and SWIR bands. The Hyperion image used for verification in this research was acquired on 6 October 2014, under cloud-free conditions. A comparison of specific parameters between the ZY-1 02D AHSI and EO-1 Hyperion is presented in Table 1.

Table 1. Comparison of specific parameters for the ZY-1 02D AHSI and EO-1 Hyperion.

Parameters	ZY-1 02D AHSI	EO-1 Hyperion
Average orbital altitude/km	778	705
Inclination/(°)	98.5	98.7
Width/km	60	7.5
Spatial resolution/m	30	30
Band number	166	242
Band range/nm	395–1040 (VNIR); 1005–2501 (SWIR)	356–1058 (VNIR); 852–2577 (SWIR)
Spectral resolution/nm	10 (VNIR); 20 (SWIR)	10

2.2.2. Field Sample Data

Abundant field data, including various types of rock and soil samples, were collected in the study area in October 2014 and 2017, corresponding to the image acquisition time to verify the extraction results from the hyperspectral remote sensing images. The field measured spectra of these samples were obtained using an Analytical Spectral Devices FieldSpec3 spectrometer on cloudless days between 10:00 and 14:00. During measurement, the probe was held perpendicular to the detection target to eliminate the effects of various interference factors. A standard white panel was used for calibration before each measurement, and the average of five measurements was used as the final spectral curve after applying Savitzky–Golay smooth filtering to remove noise. A total of 300 spectral curves were obtained from 60 field samples, covering a wavelength range of 350 nm to 2500 nm in the study area. An example of the spectral curves is shown in Figure 2. X-ray diffraction (XRD) analysis was performed on the field samples for further quantitative analysis.

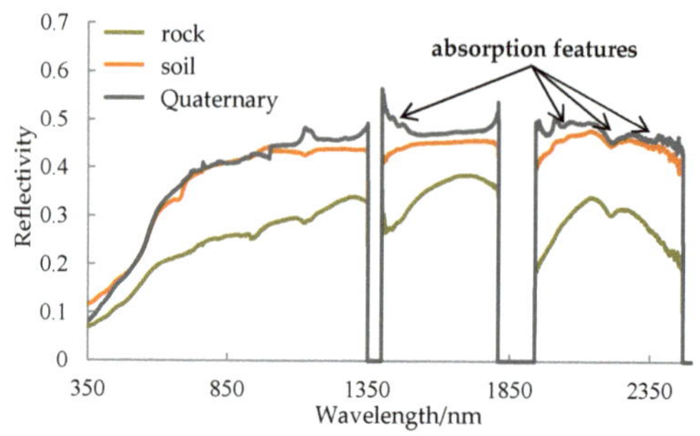

Figure 2. Field spectra curves of rock and soil.

3. Methods

This methodology includes the following four steps: theoretical basis, data preprocessing, alteration information extraction, and accuracy assessment. Firstly, the hydrocarbon microseepage theory was introduced, and diagnostic hyperspectral characteristics of alteration minerals, such as the typical clay and carbonate minerals, was analyzed. Secondly, data preprocessing was performed. Next, integrated extraction method of spectral feature matching and diagnostic characteristic parameters for alteration minerals was proposed. Finally, accuracy assessment was carried out. The technical flowchart is shown in Figure 3.

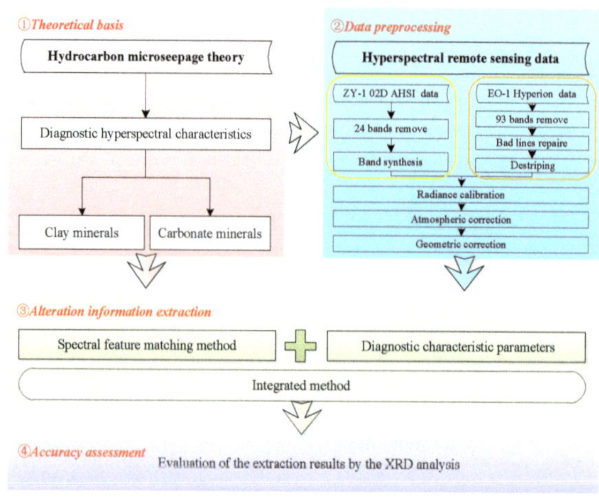

Figure 3. Technical flowchart of the methodology in this research.

3.1. Hydrocarbon Microseepage Theory and Diagnostic Hyperspectral Characteristics of Alteration Minerals

Hydrocarbon microseepage is a complex process that occurs in CBM enrichment areas. Under a specific temperature, pressure, and hydrodynamic conditions, CBM remains relatively stable. However, CBM undergoes diffusion and migration to the roof and floor when external conditions change. The double-pore structure of matrix pores and cracks in coal seams, along with the existence of fault structures, provides a channel for the migration of CBM, allowing it to penetrate the cap rock and reach the surface [38]. Prolonged hydrocarbon microseepage alters the surface redox environment and promotes the enrichment of altered minerals, such as ferrous iron mineralization, clayization, and carbonatization [39]. Hence, CBM enrichment areas can be identified using the diagnostic hyperspectral characteristics of alteration minerals based on hydrocarbon microseepage theory.

Different minerals exhibit distinctive diagnostic spectral characteristics. For instance, kaolinite, a clay mineral, displays a strong absorption feature at 2.2 μm in the SWIR region. Siderite, which is both a ferrous iron and carbonate mineral, exhibits diagnostic spectral characteristics at approximately 1.0–1.1 μm and 2.3 μm, respectively. The asymmetric diagnostic absorption feature at 2.3 μm is particularly important for the identification of carbonate minerals [40]. The spectral characteristics of typical clay and carbonate minerals provided by the USGS_MIN spectral library are shown in Figure 4.

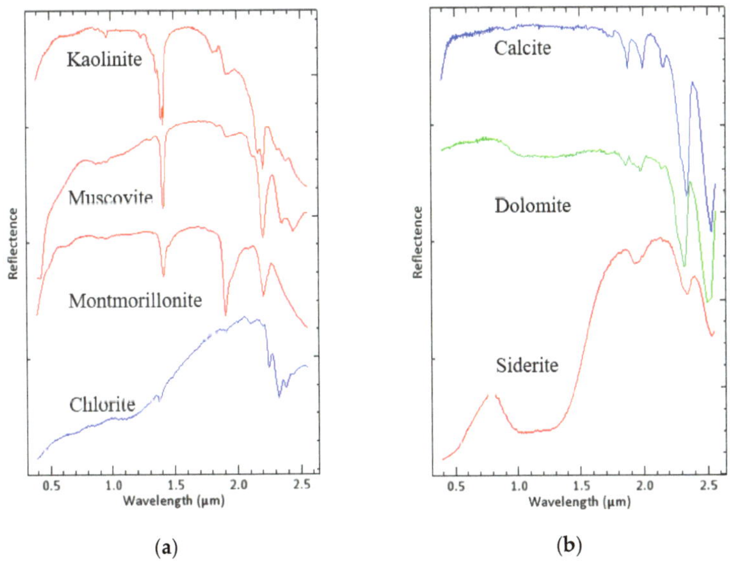

Figure 4. Spectral characteristics of typical clay and carbonate minerals in the USGS_MIN spectral library. (**a**) Clay minerals; (**b**) Carbonate minerals.

3.1.1. Clay Minerals

Clay minerals are mainly silicate minerals containing aluminum, iron, and magnesium, along with water. The long-term microseepage of hydrocarbons and hydrogen sulfide leads to the alteration of feldspar minerals into clay minerals, enriching them on the surface redox environment [41]. Typical clay minerals include kaolinite, montmorillonite, chlorite, and muscovite. The main absorption characteristics of clay minerals are observed around 1.4, 1.9, and 2.2–2.3 μm. However, each mineral possesses its unique absorption characteristics, as shown in Table 2.

Table 2. Spectral characteristics of typical clay minerals (μm).

Name	Wavelength of Absorption	Wavelength of Reflection
Kaolinite	1.403, 1.915, 2.205	0.720–1.263, 1.513–1.780
Muscovite	1.403, 1.925, 2.205	0.933–1.343, 1.478–2.075
Montmorillonite	1.418, 1.905, 2.225	0.785–1.308, 1.570–1.825
Chlorite	1.388, 1.985, 2.315	1.825, 2.135

The spectral characteristics of kaolinite and muscovite are similar, showing strong absorption spectra at 1.4 and 2.2 μm, along with weak absorption spectra at 1.9 μm. Montmorillonite exhibits adsorption depths comparable to kaolinite at 1.4 and 2.2 μm, while demonstrating strong absorption at 1.9 μm. However, the spectral characteristics of chlorite are relatively different from the aforementioned three types. Chlorite exhibits absorption at 2.3 μm, but does not show obvious adsorption at 1.4 and 1.9 μm.

3.1.2. Carbonate Minerals

The microseepage of hydrocarbons reaching the surface, combined with action of oxygen, leads to the production of carbon dioxide. In a long-term acidic environment, carbonate minerals, such as calcite, dolomite, and siderite, become enriched on the surface due to the precipitation of various metal ions, forming carbonate mineralization halos [42]. The spectral characteristics of carbonate are mainly between 1.7–2.5 μm, with at least 5 absorption features observed, particularly at 2.3 and 2.5 μm, which serve as typical bands for identifying carbonate minerals, as shown in Table 3.

Table 3. Spectral characteristics of typical carbonate minerals (μm).

Name	Wavelength of Absorption		Wavelength of Reflection	
	Weak Absorption	Strong Absorption	Weak Reflection	Strong Reflection
Calcite	1.875, 1.995, 2.155	2.335, 2.528	1.915, 2.065, 2.185	2.386
Dolomite	1.855, 1.985	2.315, 2.528	1.885, 2.025	2.375
Siderite	1.945	2.335, 2.528	1.855, 2.145	2.400

The absorption characteristics of calcite exhibit strong absorption near 2.335 and 2.528 μm, and weak absorption near 1.875, 1.995, and 2.155 μm. Both dolomite and siderite display consistent strong adsorption after 2.3 and 2.5 μm. However, siderite exhibits weak absorption characteristics specifically at 1.945 μm. In summary, the absorption characteristics of carbonate minerals are mainly observed in the range of 2.31–2.35 μm and 2.52 μm. The content of carbonate minerals demonstrates a strong correlation with the absorption depth, absorption area, and symmetry of the abovementioned spectral characteristics. Therefore, the identification and extraction of carbonate minerals can be achieved through considering the three absorption feature parameters of the characteristic spectrum.

3.2. Data Preprocessing

The main data preprocessing steps for ZY-1 02D AHSI data are as follows: (1) band synthesis, (2) radiance calibration, (3) atmospheric correction, and (4) geometric correction [43]. The VNIR and SWIR bands were combined to create a spectral coverage ranging from 395 nm to 2 501 nm. Three overlapping bands (band 77–79) in the near–infrared range and 21 noise bands (band 97–103, 125–133, 137, 163–166) were eliminated, resulting in a total of 142 bands. Radiance calibration was performed using the corresponding radiometric coefficients for each band. FLAASH, a module in ENVI software, was used for atmospheric correction. Geometric correction was applied to eliminate geometric distortions. For Hyperion data, in addition to the above processes, the following steps were performed due to water vapor and noise influence, bad lines, and striping: (1) Ninety-three bands were identified as invalid or useless and were removed, resulting in 149 retained bands. (2) Bad lines were repaired using the average value of neighbor lines. (3) A global destriping method was employed to reduce striping artifacts [44–47]. The preprocessed results of ZY-1 02D AHSI data and Hyperion data are shown in Figure 5.

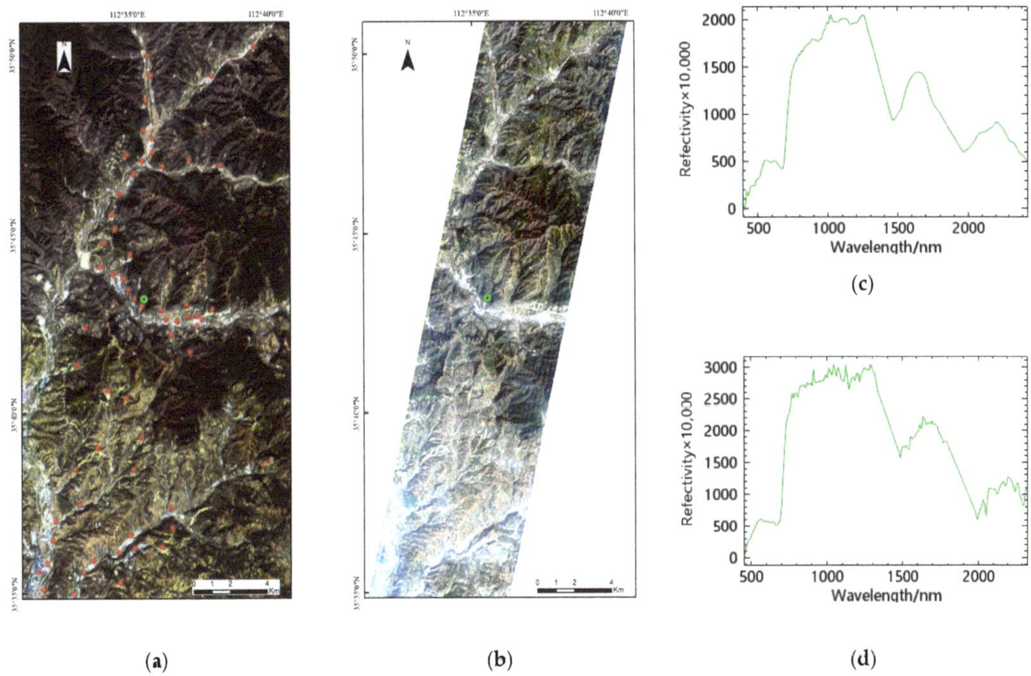

Figure 5. Preprocessed results. (**a**) Results of ZY-1 02D AHSI data, with a B29 (R), B19 (G), B10 (B) true color composite image. The red points represent field samples, and the green point indicates the position of the vegetation spectrum. (**b**) Results of Hyperion data, with a B29 (R), B20 (G), B12 (B) true color composite image. The green point indicates the position of the vegetation spectrum. (**c**) Vegetation spectrum of ZY-1 02D image at the green point. (**d**) Vegetation spectrum of Hyperion image at the green point.

3.3. Extraction Method of Alteration Minerals

Currently, three main methods are used for extracting hyperspectral mineral information: spectral matching, spectral characteristic parameters, and spectral unmixing [48]. Spectral matching methods include spectral angle mapping (SAM), matching filtering (MF), and mixture-tuned matched filtering (MTMF). Representative methods of spectral characteristic parameters include spectral feature fitting (SFF), spectral absorption index (SAI), and absorption band positioning analysis [49,50]. Spectral unmixing involves using linear or nonlinear unmixing methods to determine mineral content [51,52]. However, these methods often fall short of achieving precise mineral identification and suffer from low accuracy. The spectral matching method is not sensitive to tiny differences in mineral spectra, and is easily affected by external interference such as terrain and background, so it cannot highlight the importance of spectral absorption characteristics. Compared with the spectral matching method, the spectral characteristic parameter method enhances the ability of distinguishing ground objects, but in practical application, it is easily affected by spectral signal-to-noise ratio and mineral mixing, and the spectral shape of mineral characteristic absorption is not fully considered. Hence, in order to improve the recognition accuracy of alteration minerals and reduce the influence of other interferences, an integrated extraction method of hyperspectral mineral information that combines spectral feature matching and diagnostic characteristic parameters is proposed in this study, considering the spectral characteristics of different minerals.

3.3.1. Spectral Feature Matching Method

The spectral matching method plays an important role in the identification and extraction of alteration minerals through comparing the full spectral curve with the reference spectrum. In recent years, spectral feature matching has become more targeted, focusing on subsections of the characteristic spectral curve of different minerals. Different spectral matching methods involve selecting a similarity measure function, which is essential for calculating the matching degree. Examples of similarity measure functions include Euclidean distance, coefficient, and SAM. SAM calculates the spectral angle between the target spectrum and the reference spectrum in an N-dimensional vector space [8]. The formula is as follows:

$$\theta = \arccos \frac{\sum_{i=1}^{N} xy}{\sqrt{\sum_{i=1}^{N} (x)^2 \sum_{i=1}^{N} (y)^2}} \tag{1}$$

where θ is the spectral angle, x is the target spectrum, y is the reference spectrum, and N is the number of bands. A smaller angle indicates higher similarity. The MTMF method combines hybrid modulation and MT on the basis of linear mixing and signal processing [23]. The results of MTMF include the MF score image, which compares every endmember spectrum with every pixel spectrum, and an infeasibility image. One constraint of this method is that the content of each endmember must be positive and sum to 1. The advantages of MTMF are that it does not require background endmember spectra and provides high accuracy of in the extraction of alteration minerals.

Reference spectra typically come from three sources: laboratory-measured spectra in standard spectral libraries, field-measured spectra, and pure pixel spectra extracted from images. Endmember extraction involves identifying pure pixels in the image and comparing them with the spectral library to determine the mineral type. In this study, the following steps are performed using ENVI software: (1) applying minimum noise fraction to reduce dimensionality and remove noise, (2) using pixel purity index to identify pure pixels, (3) employing N-dimensional visualization to display the feature spectra of endmembers, and (4) utilizing spectral analysis to determine the mineral type.

3.3.2. Diagnostic Characteristic Parameters

The full spectral curve can be regarded as the superposition of spectral features caused by different components within an object. Therefore, various local spectral absorption characteristics can be utilized as diagnostic characteristic parameters to identify different minerals. A local absorption feature diagram is shown in Figure 6 [24]. The absorption characteristics of the spectral curve are mainly composed of two reflection points (S1 and S2) and an absorption valley point (M).

In the figure, λ_{S1}, R_{S1}, λ_{S2}, R_{S2}, λ_M, and R_M represent the wavelength and reflectivity of the points (S1, S2, and M). The main diagnostic characteristic parameters include (1) absorption position (λ_M), which is the wavelength of the maximum absorption value of the spectral curve; (2) absorption reflectivity (R_M), which is the reflectivity of the maximum absorption value of the spectral curve; (3) absorption width (W), which is the wavelength width between the two reflection points, S1 and S2; (4) absorption depth (H), which is the distance between the absorption valley point (M) and the connecting line between the two reflection points (S1 and S2), known as the nonabsorbing baseline; (5) absorption area (A), which is the area enclosed by the nonabsorbing baseline and the spectral curve; (6) absorption symmetry (S), which is the ratio of the area on the left side of the adsorption position to the area on the right side, divided by the vertical line of the absorption position.

The continuum removal algorithm is adopted for information extraction to further highlight the diagnostic spectral characteristics of alteration minerals.

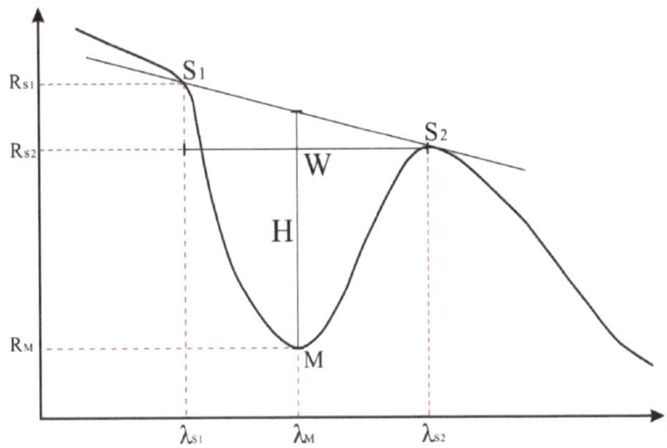

Figure 6. Local absorption feature. S1 and S2 are two reflection points and M is an absorption valley point. λ_{S1}, R_{S1}, λ_{S2}, R_{S2}, λ_M, and R_M are the wavelength and reflectivity of the points (S1, S2, and M), respectively. *W* is absorption width and *H* is absorption depth.

3.3.3. Integrated Method of Spectral Feature Matching and Diagnostic Characteristic Parameters

A method of comprehensive spectral features matching and diagnostic characteristic parameters based on mineral hierarchical identification theory was proposed to extract alternation mineral information in the study area. The main steps are as follows. (1) Analyze the position, width, depth, area, and symmetry of the main characteristic absorption based on the spectral characteristics of minerals. (2) Calculate the spectral feature matching degree between the spectral characteristics of different minerals and their corresponding standard spectra. (3) Obtain the main diagnostic characteristic parameters of different minerals. (4) Set threshold values for the above parameters to obtain the distribution of mineral information. The technical flowchart of this study is shown in Figure 7.

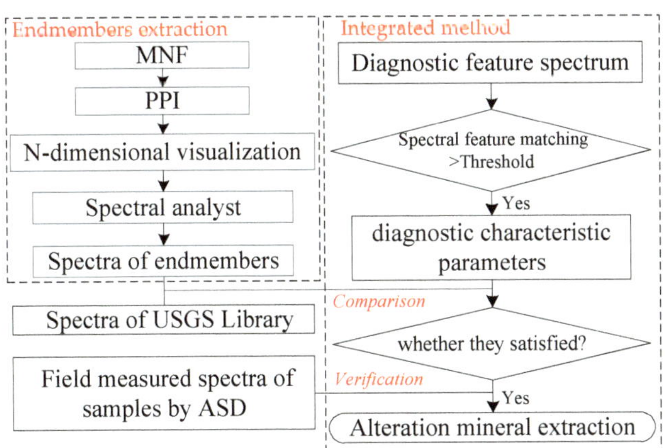

Figure 7. Technical flowchart of integrated method in this research.

3.4. Accuracy Assessment

Sixty field samples collected in the study area (Figure 5a) were used to evaluate the extraction results of alteration minerals based on the XRD analysis and quantitatively analyze the identification effects. Using the confusion matrix, the extracted results of alternation minerals were compared with the results of XRD analysis to calculate the producer's accuracy (PA), user's accuracy (UA), and overall accuracy (OA) [53–57].

4. Results and Analysis

4.1. Extraction Results of the Diagnostic Characteristic Parameters of Typical Minerals

4.1.1. Clay Mineral

Using the mentioned method of endmember extraction, the endmembers of the clay mineral were obtained from hyperspectral remote sensing imagery, as shown in Figure 8. The continuum removal algorithm was utilized to enhance the diagnostic spectral characteristics of clay minerals [58]. The spectral curve exhibits absorption characteristics of clay minerals at 2200 nm, which are similar to the characteristics of kaolinite and montmorillonite in the spectral library. The diagnostic absorption characteristics were further analyzed through calculating the diagnostic characteristic parameters using the IDL language as shown in Table 4.

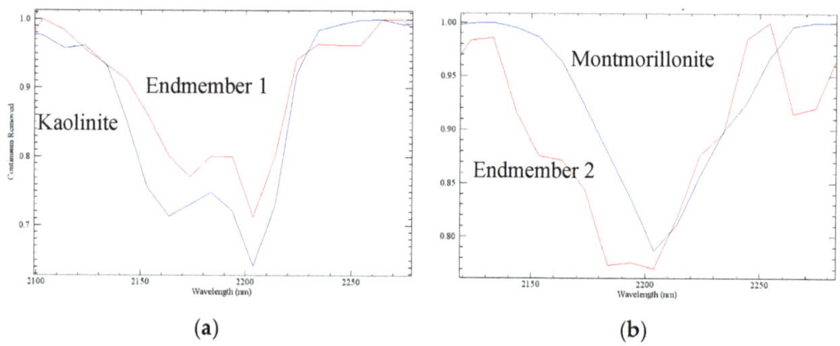

Figure 8. Comparison of the endmember spectra of clay minerals extracted from the hyperspectral imagery and the standard spectra in USGS_MIN spectral library. (**a**) Endmember 1 and kaolinite; (**b**) Endmember 2 and montmorillonite.

Table 4. Comparison of diagnostic characteristic parameters of the endmember spectra of clay minerals extracted from the hyperspectral imagery and standard spectra in USGS_MIN spectral library.

Name	λ_M/nm	R_M [1]	W/nm	H	A	S
Endmember 1	2203	0.1392	162	0.2882	141.62	1.58
Endmember 2	2203	0.1446	121	0.2297	103.97	1.33
Kaolinite	2203	0.3824	202	0.3589	175.70	2.23
Montmorillonite	2203	0.5230	141	0.2132	128.23	1.02

[1] Except for R_M, which represents the actual reflectance value at the absorption position, the other parameters are calculated after continuum removal.

The quantitative calculation results of the diagnostic characteristic parameters clearly indicate that the spectral absorption position is completely consistent. However, the spectral reflectance of the endmember spectra is slightly low because they are obtained from hyperspectral satellite remote sensing images and affected by atmospheric interference, affecting the absorption depth and area to some extent. However, the difference in absorption width and depth between endmember 2 and montmorillonite, obtained after continuum removal, is only 14.18% and 7.74%, respectively. Therefore, the diagnostic characteristic parameters of the extracted mineral endmember spectra are consistent with the spectra in the spectral library. Specifically, the absorption position, width, and depth of endmember 1 and endmember 2 are consistent with kaolinite and montmorillonite in the spectral library, respectively, allowing for a preliminary identification as clay minerals.

4.1.2. Carbonate Mineral

For carbonate minerals, the classical diagnostic characteristic parameters of the spectral curves mainly occur at 2300 nm, considering the remarkable influence of atmospheric water vapor. Therefore, noise bands at 1400 and 1900 nm in the hyperspectral images were eliminated. After continuum removal, the diagnostic absorption spectral characteristics of the endmembers extracted from the hyperspectral satellite remote sensing image were compared with typical carbonate minerals in the spectral library, as shown in Figure 9. The quantitative analysis results of the diagnostic characteristic parameters of carbonate minerals are shown in Table 5.

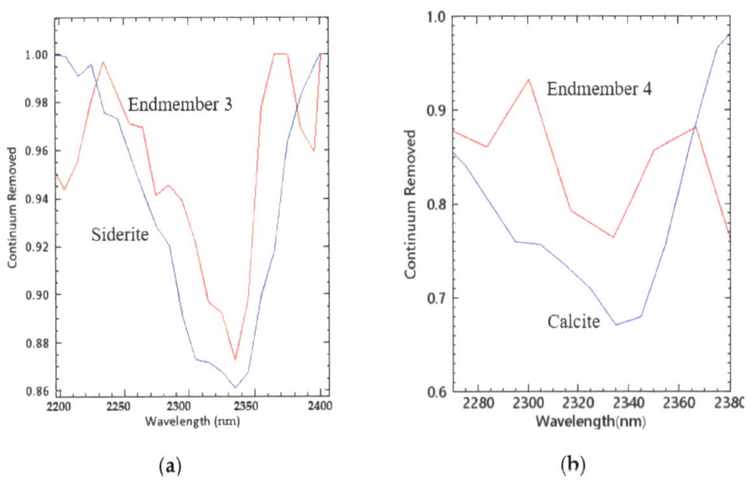

Figure 9. Comparison of the endmember spectra of carbonate minerals extracted from the hyperspectral imagery and the standard spectra in USGS_MIN spectral library. (**a**) Endmember 3 and siderite; (**b**) Endmember 4 and calcite.

Table 5. Comparison of diagnostic characteristic parameters of the endmember spectra of carbonate minerals extracted from the hyperspectral imagery and standard spectra in USGS_MIN spectral library.

Name	λ_M/nm	R_M [1]	W/nm	H	A	S
Endmember 3	2335	0.2013	131	0.1274	112.19	2.98
Endmember 4	2335	0.5301	101	0.2358	93.68	2.64
Siderite	2335	0.4119	202	0.1390	186.79	2.36
Calcite	2335	0.5718	192	0.3286	161.09	2.92

[1] Except for R_M, which represents the actual reflectance value of the absorption position, the other parameters are calculated after continuum removal.

The results show that the spectral absorption position of the endmember is exactly the same as that of calcite and siderite. However, the reflectivity of the endmember obtained from the hyperspectral imagery is low due to atmospheric interference during image acquisition, resulting in slightly lower absorption depth and absorption area. In particular, the absorption reflectivity of endmember 3 is only 51.13% of that of siderite in the spectral library. The absorption width of the endmember spectrum is narrow, and the absorption area is small overall. However, the absorption depth of endmember 3 is consistent with siderite, with a difference of only 8.34%. The absorption reflectivity and symmetry of endmember 3 are similar to that of calcite, with differences of only 7.29% and 9.59%, respectively. Therefore, the absorption position, depth, and symmetry of the extracted endmember 3 and endmember 4 are consistent with calcite and siderite in the spectral library, respectively, allowing for a preliminary determination that these endmembers represent carbonate minerals.

4.2. Extraction Results Using the Integrated Method of Spectral Feature Matching and Diagnostic Characteristic Parameters

4.2.1. Clay Minerals

Clay minerals were extracted using the proposed integrated method based on the indications of clay minerals, such as the distinct spectral feature from 2100 nm to 2300 nm and the diagnostic characteristic parameters, including the absorption position, width, and depth at 2203 nm. The spectral feature matching degrees were obtained using the SAM within the 2100–2300 range. The absorption position, width, and depth of pixels with a high matching degree were calculated at 2203 nm. If these thresholds of diagnostic characteristic parameters were met, then the pixels were identified as clay minerals. The distribution of clay minerals in the study area is illustrated in Figure 10, which compares the extraction results from ZY-1 02D data and Hyperion data.

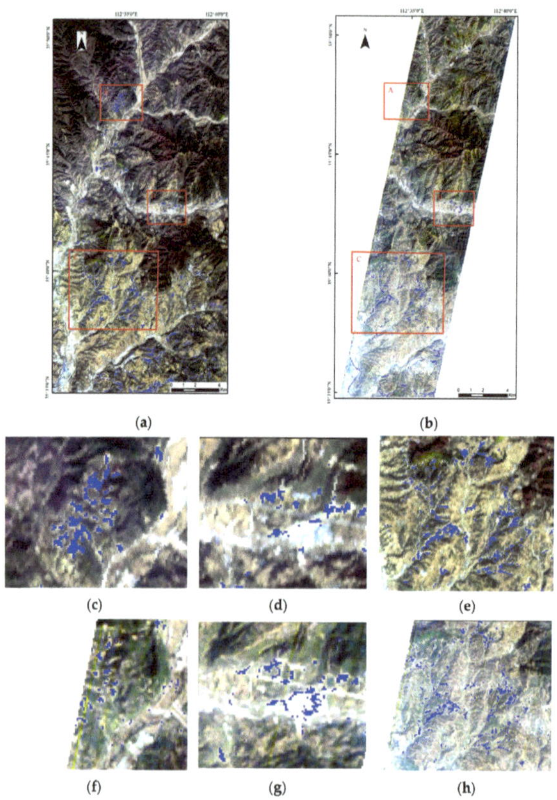

Figure 10. Comparison of the extraction results of clay minerals extracted from ZY-1 02D data and Hyperion data. The distributions of clay minerals are denoted in blue. (**a**) The extraction result of ZY-1 02D data; (**b**) The extraction result of Hyperion data; (**c**–**e**) The zoom-in detailed subfigures of region A, B, and C of ZY-1 02D data, respectively; (**f**–**h**) The zoom-in detailed subfigures of region A, B, and C of Hyperion data, respectively.

As shown in Figure 9, clay minerals are mainly concentrated in the central and southern areas of the study area, with sporadic distribution in the northwest part. Several obvious anomalies can be observed. In accordance with the actual geological data, the central region of the study area corresponds to the Hudi coal mine area of Qinshui Basin, Jincheng, Shanxi, China, which is consistent with the extraction results of clay minerals. In addition, the distribution of clay minerals is relatively consistent when comparing the extraction results from Hyperion data, with both datasets showing concentrations in the central and southern parts of the study area.

4.2.2. Carbonate Minerals

Using the proposed method, carbonate minerals were identified on the basis of the spectral feature matching degree within the 2200 nm to 2400 nm range and the diagnostic characteristic parameters, including the absorption position, depth, and symmetry at 2335 nm. First, the spectral feature matching degrees were acquired using the MTMF within 2200–2400 nm. Second, the absorption position, depth, and symmetry of pixels with a high matching degree were calculated at 2335 nm. Finally, carbonate minerals were extracted in the study area when the parameters reached the defined thresholds, as shown in Figure 11.

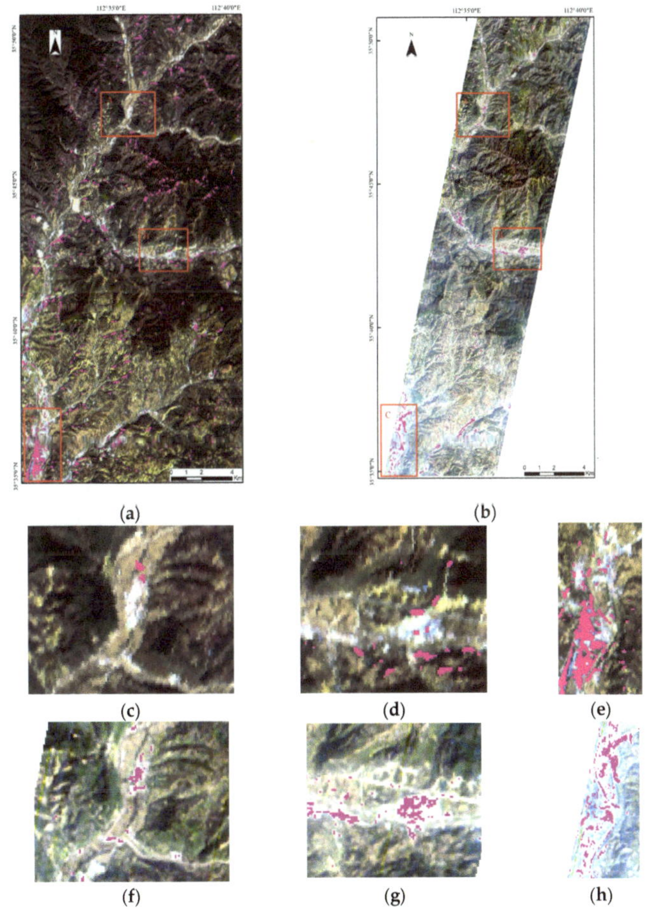

Figure 11. Comparison of the extraction results of carbonate minerals extracted from ZY-1 02D and Hyperion data. The distributions of carbonate minerals are denoted in magenta (**a**) The extraction result of ZY-1 02D data; (**b**) The extraction result of Hyperion data; (**c**–**e**) The zoom-in detailed subfigures of region A, B, and C of ZY-1 02D data, respectively; (**f**–**h**) The zoom-in detailed subfigures of region A, B, and C of Hyperion data, respectively.

Carbonate minerals are widespread in the central and southwest of the study area, with a thin distribution in the northern part. The central region of the study area corresponds to the Hudi coal mine area. However, a large number of anomalies in the southwest have not been effectively verified. Field surveys indicate that no CBM mining has been conducted although coal seams distributed underground exist. Overall, the extraction results of carbonate minerals using ZY-1 02D data are relatively consistent with the results obtained from Hyperion data in the middle and southern parts of the study area.

4.3. Accuracy Assessment

Four evaluation factors, such as PA, UA, and OA, were used to quantitatively analyze the extraction results of alteration minerals from ZY-1 02D and Hyperion data. These factors were compared with the results of XRD analysis, as shown in Table 6.

Table 6. Accuracy assessment of ZY-1 02D and Hyperion data.

Data	Minerals	*PA* (%)	*UA* (%)	OA (%)
ZY-1 02D	Clay	81.25	83.87	81.67
	Carbonate	79.41	81.82	79.03
Hyperion	Clay	78.12	80.64	78.33
	Carbonate	73.53	83.33	76.67

The overall accuracy of clay and carbonate mineral extraction results from ZY-1 02D data is 81.67% and 79.03%, respectively, which is better than the Hyperion data (78.33% and 76.67%). However, all accuracy values are greater than 75%, confirming the reliability of alteration mineral extraction using hyperspectral remote sensing technology. Furthermore, the user's accuracy of clay minerals from ZY-1 02D data and carbonate minerals from Hyperion data reached 83.87% and 83.33%, respectively. This further demonstrates that different data have their unique advantages for the extraction of different minerals.

5. Discussion

5.1. Analysis of the Influence of Different Hyperspectral Satellite Remote Sensing Data

In this research, both ZY-1 02D and Hyperion images were used to extract alteration minerals. Although the two datasets have the same spatial resolution and a large number of bands as hyperspectral satellite remote sensing data, several differences are still observed. Hyperion data contain 76 more bands than ZY-1 02D data due to the higher spectral resolution of SWIR bands in Hyperion data (10 nm) compared with ZY-1 02D data (20 nm) [59]. This results in Hyperion data having more spectral details but also containing a larger amount of redundant information. For example, although the vegetation spectral curves at the same location exhibit a similar waveform, the SWIR bands of Hyperion data may exhibit slight fluctuations, which can be either a spectral characteristic or noise, as shown in Figure 5c,d. In addition, the reflectivity of Hyperion data is universally higher than ZY-1 02D data, making the Hyperion images appear brighter (Figure 12). However, the advantage of ZY-1 02D data is their 60 km width, allowing for large-scale observation. Furthermore, the Hyperion data may contain bad lines and striping due to detector element malfunctions and systematic noise, which can affect the extraction results.

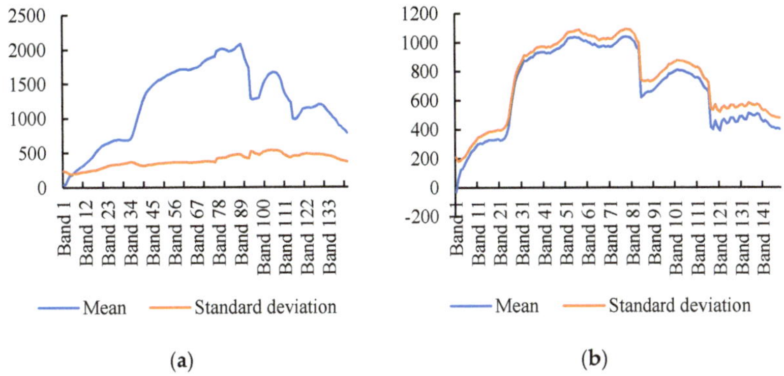

Figure 12. Data quality comparison of the ZY-1 02D (**a**) and Hyperion (**b**) images.

5.2. Analysis of Extraction Efficiency of Different Alteration Minerals

The normalized difference vegetation index was utilized to eliminate vegetation and reduce its influence on alteration mineral extraction. The extraction results of clay and carbonate minerals from the ZY-1 02D and Hyperion images were obtained in the bare areas of the study area using an integrated method of spectral feature matching and diagnostic characteristic parameters. The alteration minerals were predominantly distributed in the central and southern regions. On the basis of the regional geology, seismic data, and well logging information in the study area, the central region was identified as the Jincheng Hudi mine area of the Qinshui Basin [60]. The distribution of clay and carbonate minerals was consistent between the two different hyperspectral imageries in the middle of the study area. However, clay minerals still exhibited distribution in the northwest part of the study area, with several obvious anomalies in the ZY-1 02D image. They cannot be extracted in the Hyperion data due to the narrower width. Apparent false anomalies were observed along the river in the southwest of the Hyperion image. For the carbonate minerals, a large number of anomalies existed in the southwestern part of the study area, but they had not been effectively verified. Although coal seams were indeed distributed based on field surveys, CBM was not being extracted.

Quantitative assessment of the extraction results showed that the PA, UA, and OA of clay minerals extracted using ZY-1 02D data were higher than those obtained from Hyperion data. Specifically, the OA was 3.34% higher. Except for the UA of carbonate minerals extracted using ZY-1 02D data, which was 1.51% lower, the other factors were higher than Hyperion data. Overall, the extraction efficiency of ZY-1 02D data was better than that of Hyperion data for the two different minerals. In terms of different minerals, the extraction efficiency of clay minerals was better than that of carbonate minerals. This difference can be attributed to the spectral characteristics of carbonate minerals falling between 2200 and 2400 nm, where the data quality of these bands is poorer compared with the range from 2100 nm to 2300 nm, which severely limits the quality of the extraction results.

However, the extraction results of alteration minerals are highly affected by other interfering factors, such as fertilizer, acid rain, and chemical pollution, due to the feeble and imperceptible nature of the alteration minerals caused by CBM microseepages. Moreover, distributions of alteration minerals many not necessarily correlate the regions of CBM enrichment because microseepage can be controlled using drainage and structure [61]. Thus, the mineral alterations in this study area were only caused by CBM microseepages to the ground surface. Further experiments are required to investigate the spectral characteristics of alteration minerals in more CBM enrichment areas [62].

Consequently, some works should be studied in the future. First, the analysis and comparison of spectra from a larger number of field samples should be conducted. Second, further fieldwork is necessary to verify the universality and generalizability of this method in other regions. Specifically, geochemical analysis is required in the future to investigate the alteration minerals caused by CBM hydrocarbon microseepage. Finally, the study did not involve the extraction of altered vegetation due to the immaturity of extraction methods. The diagnostic spectral features of altered vegetation should be explored in future research. Therefore, additional investigation into the potential geological application of hyperspectral data is crucial and promising.

6. Conclusions

In this study, alteration minerals were successfully extracted from civilian ZY-1 02D hyperspectral data and Hyperion data in the Hudi coal mine area of the Qinshui Basin. A mineral identification method that combines spectral feature matching and diagnostic characteristic parameters was adopted on the basis of the spectral characteristics of clay minerals and carbonate minerals. The extraction results of ZY-1 02D and Hyperion data were compared with geological data to effectively extract alteration minerals in CBM enrichment areas. This research demonstrates the application potential of domestic hyperspectral data in CBM identification and provides valuable technical support for large-scale and rapid CBM exploration.

The following conclusions were obtained:

(1) The extraction of alteration minerals, including clay and carbonate minerals, was successfully achieved using ZY-1 02D and Hyperion data. The distribution of clay and carbonate minerals exhibited good accuracy (81.67% and 79.03%, respectively) when analyzed using XRD.
(2) Comparing the extraction results of Hyperion data and ZY-1 02D data, the PA, UA, and OA for clay mineral extraction were higher with ZY-1 02D data than with Hyperion data. However, the UA for carbonate mineral extraction using ZY-1 02D data was 1.51% lower than with Hyperion data.
(3) The study illustrated the potential geological application of hyperspectral satellite remote sensing data in identifying CBM enrichment regions. This method offers a large-scale, convenient, and highly efficient approach compared with traditional seismic exploration and drilling methods.

Author Contributions: Conceptualization, L.C. and X.S.; methodology, L.C.; software, L.C.; validation, L.C., H.C. (Hong Chen) and Y.L.; formal analysis, L.C.; investigation, X.S.; resources, H.C. (Hong Chen); data curation, X.Z.; writing—original draft preparation, L.C.; writing—review and editing, L.C.; visualization, H.C. (Haomin Chen); supervision, X.Z.; project administration, R.L.; funding acquisition, R.L. All authors have read and agreed to the published version of the manuscript.

Funding: This research was funded by the General Program of the National Natural Science Foundation of China, grant number 42271408.

Acknowledgments: ZY-102D AHSI imagery was obtained from China Areo Geophysical Survey and Remote Sensing Center for Natural Resources, and XRD data were measured by Beijing Research Institute of Uranium Geology. We express our sincere thanks. We are also grateful to the USGS Earth Explorer (https://earthexplorer.usgs.gov/ accessed on 25 October 2019) for providing Hyperion data. Thanks also go to editors and the anonymous reviewers for helping us to improve the quality of the manuscript.

Conflicts of Interest: The authors declare no conflict of interest.

References

1. Moore, T.A. Coalbed methane: A review. *Int. J. Coal Geol.* **2012**, *101*, 36–81. [CrossRef]
2. Fu, X.; Qin, Y.; Wang, G.G.X.; Rudolph, V. Evaluation of gas content of coalbed methane reservoirs with the aid of geophysical logging technology. *Fuel* **2009**, *88*, 2269–2277. [CrossRef]
3. Zhang, C.Y.; Qin, Q.M.; Chen, L.; Nan, W.; Zhao, S.S.; Hui, J. Rapid determination of coalbed methane exploration target region utilizing hyperspectral remote sensing. *Int. J. Coal Geol.* **2015**, *150–151*, 19–34. [CrossRef]
4. Song, X.; Ma, H.; Saalidong, B.M.; Li, K. Petrography, Mineralogy, and Geochemistry of Thermally Altered Coal in the Tashan Coal Mine, Datong Coalfield, China. *Minerals* **2021**, *11*, 1024. [CrossRef]

5. Gan, F.P.; Dong, X.F.; Yan, B.K.; Liang, S.N. Research progress of spectrometry geological remote sensing. *J. Nanjing Univ. Inf. Sci. Technol. Nat. Sci. Ed.* **2018**, *10*, 44–62. [CrossRef]
6. Zheng, X.W.; Peng, B.; Shang, K. Assessment of the interpretation ability of domestic satellites in geological remote sensing. *Remote Sens. Nat. Resour.* **2021**, *33*, 1–10. [CrossRef]
7. Yao, F.; Xu, X.; Yang, J.; Geng, X. A Remote-Sensing-Based alteration zonation model of the Duolong Porphyry Copper Ore District, Tibet. *Remote Sens.* **2021**, *13*, 5073. [CrossRef]
8. Maghsoudi Moud, F.; van Ruitenbeek, F.; Hewson, R.; van der Meijde, M. An approach to accuracy assessment of ASTER derived mineral maps. *Remote Sens.* **2021**, *13*, 2499. [CrossRef]
9. Chen, L.; Qin, Q.M.; Chen, C.; Jiang, H.B. Remote sensing information of mineralizing alteration extraction methods. In Proceedings of the 2012 IEEE International Geoscience and Remote Sensing Symposium, Munich, Germany, 22–27 July 2012; pp. 5947–5950. [CrossRef]
10. Chen, C.; Liang, J.; Yang, G.; Sun, W. Spatio-temporal distribution of harmful algal blooms and their correlations with marine hydrological elements in offshore areas, China. *Ocean. Coast. Manag.* **2023**, *238*, 106554. [CrossRef]
11. Chen, C.; Chen, Y.; Jin, H.; Chen, L.; Liu, Z.; Sun, H.; Hong, J.; Wang, H.; Fang, S.; Zhang, X. 3D model construction and ecological environment investigation on a regional scale using UAV remote sensing. *Intell. Autom. Soft Comput.* **2023**, *37*, 1655–1672. [CrossRef]
12. van der Meer, F.D.; van der Werff, H.M.A.; van Ruitenbeek, F.J.A.; Hecker, C.A.; Bakker, W.H.; Noomen, M.F.; Meijde, M.V.D.; Carranza, E.J.M.; de Smeth, J.B.; Woldai, T. Multi- and hyperspectral geologic remote sensing: A review. *Int. J. Appl. Earth Obs. Geoinf.* **2012**, *14*, 112–128. [CrossRef]
13. Mohamed, A.; Emam, A.; Zoheir, B. SAM-HIT: A simulated annealing multispectral to hyperspectral imagery data transformation. *Remote Sens.* **2023**, *15*, 1154. [CrossRef]
14. Islam, M.R.; Siddiqa, A.; Ibn Afjal, M.; Uddin, M.P.; Ulhaq, A. Hyperspectral image classification via information theoretic dimension reduction. *Remote Sens.* **2023**, *15*, 1147. [CrossRef]
15. Zhou, Z.L.; Wang, Y.L.; Wu, Q.M.J.; Yang, C.N.; Sun, X.M. Effective and efficient global context verification for image copy detection. *IEEE Trans. Inf. Forensics Secur.* **2017**, *12*, 48–63. [CrossRef]
16. Boubanga-Tombet, S.; Huot, A.; Vitins, I.; Heuberger, S.; Veuve, C.; Eisele, A.; Hewson, R.; Guyot, E.; Marcotte, F.; Chamberland, M. Thermal Infrared hyperspectral imaging for mineralogy mapping of a mine face. *Remote Sens.* **2018**, *10*, 1518. [CrossRef]
17. Yan, B.K.; Dong, X.F.; Wang, Z.; Yang, S.M.; Yu, J.C.; Li, N.; Gan, F.P. Mineral information extraction technology by airborne hyperspectral remote sensing and its application progress: An example of mineralization belts of western China. *Geol. Surv. China* **2016**, *3*, 55–62.
18. Ren, Z.; Zhai, Q.; Sun, L. A novel method for hyperspectral mineral mapping based on clustering-matching and nonnegative matrix factorization. *Remote Sens.* **2022**, *14*, 1042. [CrossRef]
19. Ye, B.; Tian, S.F.; Cheng, Q.M.; Ge, Y.Z. Application of lithological mapping based on advanced hyperspectral imager (AHSI) imagery onboard Gaofen-5 (GF-5) satellite. *Remote Sens.* **2020**, *12*, 3990. [CrossRef]
20. Xu, D.X.; Yang, B.; Shao, Z.G.; Liu, C.F.; Wang, D.; Wang, J.P. A hyperspectral analysis of alteration zoning in the Hadamengou gold deposit and its significance for ore prospecting. *Remote Sens. Nat. Resour.* **2023**, *35*, 123–131. [CrossRef]
21. Yang, X.; Chen, J.G.; Chen, Z.J. Classification of alteration zones based on drill core hyperspectral data using semi-supervised adversarial autoencoder: A case study in Pulang Porphyry Copper Deposit, China. *Remote Sens.* **2023**, *15*, 1059. [CrossRef]
22. Wang, Q.; Ren, G.L. Application of hyperspectral remote sensing data-based anomaly extraction in copper-gold prospecting in the Solake area in the Altyn metallogenic belt, Xinjiang. *Remote Sens. Nat. Resour.* **2022**, *34*, 277–285.
23. Chen, Q.; Zhao, Z.; Zhou, J.; Zhu, R.; Xia, J.; Sun, T.; Zhao, X.; Chao, J. ASTER and GF-5 satellite data for mapping hydrothermal alteration minerals in the Longtoushan Pb-Zn Deposit, SW China. *Remote Sens.* **2022**, *14*, 1253. [CrossRef]
24. Huo, H.Y.; Ni, Z.Y.; Jiang, X.G.; Zhou, P.; Liu, L. Mineral mapping and ore prospecting with HyMap data over Eastern Tien Shan, Xinjiang Uyghur Autonomous Region. *Remote Sens.* **2014**, *6*, 11829–11851. [CrossRef]
25. Molan, Y.E.; Refahi, D.; Tarashti, A.H. Mineral mapping in the Maherabad area, eastern Iran, using the HyMap remote sensing data. *Int. J. Appl. Earth Obs. Geoinf.* **2014**, *27*, 117–127. [CrossRef]
26. Dong, X.F.; Gan, F.P.; Li, N.; Yan, B.K.; Zhang, L.; Zhao, J.Q.; Yu, J.C.; Liu, R.Y.; Ma, Y.N. Fine mineral identification of GF-5 hyperspectral image. *J. Remote Sens.* **2020**, *24*, 454–464. [CrossRef]
27. Pour, A.B.; Zoheir, B.; Pradhan, B.; Hashim, M. Editorial for the special issue: Multispectral and hyperspectral remote sensing data for mineral exploration and environmental monitoring of mined areas. *Remote Sens.* **2021**, *13*, 519. [CrossRef]
28. Tan, W.; Wang, S.; He, H.; Qi, W. Reconstructing coastal blue with blue spectrum based on ZY-1 (02D) satellite. *Optik* **2021**, *242*, 166901.
29. Tian, J.; Su, S.; Tian, Q.; Zhan, W.; Xi, Y.; Wang, N. A novel spectral index for estimating fractional cover of non-photosynthetic vegetation using near-infrared bands of Sentinel satellite. *Int. J. Appl. Earth Obs. Geoinf.* **2021**, *101*, 102361. [CrossRef]
30. Zhong, Y.; Wang, X.; Wang, S.; Zhang, L. Advances in spaceborne hyperspectral remote sensing in China. *Geo Spat. Inform. Sci.* **2021**, *24*, 95–120.
31. Sun, W.; Liu, K.; Ren, G.; Liu, W.; Yang, G.; Meng, X.; Peng, J. A simple and effective spectral-spatial method for mapping large-scale coastal wetlands using China ZY1-02D satellite hyperspectral images. *Int. J. Appl. Earth Obs. Geoinf.* **2021**, *104*, 102572. [CrossRef]

32. Li, G.J.; Yang, X.S.; Zhang, X.; Li, X.M.; Li, D.L.; Du, C. Application and analysis of ZY1-02D hyperspectral data in geological and mineral survey. *Remote Sens. Land Resour.* **2021**, *33*, 134–140. [CrossRef]
33. Sun, Y.; Liu, J.J.; Zhai, D.G.; Liu, Z.J.; Zhang, F.F.; Zhao, Y.J.; Liu, P.F.; Wang, Z.T. Alteration minerals mapping and identification of gold mineralization belt based on ZY1-02D hyperspectral data in Fangshankounanshan area, Gansu Province. *Geol. Explor.* **2022**, *58*, 399–409. [CrossRef]
34. Farooq, S.; Govil, H. Mapping Regolith and Gossan for mineral exploration in the Eastern Kumaon Himalaya, India using Hyperion data and object oriented image classification. *Adv. Space Res.* **2014**, *53*, 1676–1685. [CrossRef]
35. Liu, L.; Zhou, J.; Han, L.; Xu, X.L. Mineral mapping and ore prospecting using Landsat TM and Hyperion data, Wushitala, Xinjiang, northwestern China. *Ore Geol. Rev.* **2017**, *81*, 280–295. [CrossRef]
36. Chen, L.; Yang, X.C.; Zhen, G.W. Potential of Sentinel-2 data for alteration extraction in coal-bed methane reservoirs. *Ore Geol. Rev.* **2019**, *108*, 134–146. [CrossRef]
37. Wang, N.; Qin, Q.M.; Chen, L.; Bai, Y.B.; Zhao, S.S. Dynamic monitoring of coalbed methane reservoirs using Super-Low Frequency electromagnetic prospecting. *Int. J. Coal Geol.* **2014**, *127*, 24–41. [CrossRef]
38. Pashin, J.C.; Mcintyre-Redden, M.R.; Mann, S.D.; Kopaska-Merkel, D.C.; Varonka, M.; Orem, W. Relationships between water and gas chemistry in mature coalbed methane reservoirs of the Black Warrior Basin. *Int. J. Coal Geol.* **2014**, *126*, 92–105. [CrossRef]
39. Noomen, M.F.; van der Werff, H.M.A.; van der Meer, F.D. Spectral and spatial indicators of botanical changes caused by long-term hydrocarbon seepage. *Ecol. Inf.* **2012**, *8*, 55–64. [CrossRef]
40. van der Meer, F.D.; van Dijk, P.V.; van der Werff, H.M.A.; Yang, H. Remote sensing and petroleum seepage: A review and case study. *Terra Nova* **2002**, *14*, 1–17. [CrossRef]
41. Hunt, G.R.; Ashley, R.P. Altered rock spectra in the visible and near infrared. *Econ. Geol.* **1979**, *74*, 1613–1629. [CrossRef]
42. Kühn, F.; Oppermann, K.; Horig, B. Hydrocarbon index—An algorithm for hyperspectral detection of hydrocarbons. *Int. J. Remote Sens.* **2004**, *25*, 2467–2473. [CrossRef]
43. Cun, C.L.; Zhang, H.W.; Fan, S.T.; Li, Y.F.; Yun, X.D. Atmospheric correction and application of ZY-1-02D hyperspectral data. *Remote Sens. Inf.* **2022**, *37*, 45–50.
44. Chen, S.B.; Chen, L.; Liu, Y.L.; Li, X.L. Experimental simulation on mixed spectra of leaves and calcite for inversion of carbonate minerals from EO-1 Hyperion data. GISci. *Remote Sens.* **2013**, *50*, 690–703. [CrossRef]
45. Bhattacharya, S.; Majumdar, T.J.; Rajawat, A.S.; Panigrahy, M.K.; Das, P.R. Utilization of Hyperion data over Dongargarh, India, for mapping altered/weathered and clay minerals along with field spectral measurements. *Int. J. Remote Sens.* **2012**, *33*, 5438–5450. [CrossRef]
46. Gersman, R.; Ben Dor, E.; Beyth, M.; Avigad, D.; Abraha, M.; Kibreab, A. Mapping of hydrothermally altered rocks by the EO-1 Hyperion sensor, Northern Danakil Depression, Eritrea. *Int. J. Remote Sens.* **2008**, *29*, 3911–3936. [CrossRef]
47. Kruse, F.A.; Boardman, J.W.; Huntington, J.F. Comparison of airborne hyperspectral data and EO-1 Hyperion for mineral mapping. *IEEE Trans. Geosci. Remote Sens.* **2003**, *41*, 1388–1400. [CrossRef]
48. Li, N.; Dong, X.F.; Gan, F.P.; Yan, B.K. Application evaluation of ZY-1-02D satellite hyperspectral data in geological survey. *Spacecr. Eng.* **2020**, *29*, 186–191. [CrossRef]
49. Zaini, N.; van der Meer, F.; van der Werff, H. Effect of grain size and mineral mixing on carbonate absorption features in the SWIR and TIR wavelength regions. *Remote Sens.* **2012**, *4*, 987–1003. [CrossRef]
50. Li, H.S.; Zhou, P.; Han, M.X.; Huo, H.Y.; Cui, Y.M.; Geng, L.P. Research and development of mineral recognition software module based on spectral absorption index. *Comput. Mod.* **2012**, *198*, 69–72. [CrossRef]
51. Chen, X.L. Research on Emission Spectral Feature Extraction of Typical Mineral and Rock Spectral Unmixing. Ph.D. Thesis, Northeastern University, Shenyang, China, 2013.
52. Sun, X.; Xu, L.L.; Wang, X.Y.; Tian, Y.; Wang, W.; Zhang, Z.Y. Mineral identification from hyperspectral images based on the optimized K-P-Means unmixing method. *Remote Sens. Nat. Resour.* **2022**, *34*, 43–49. [CrossRef]
53. Yang, X.C.; Chen, L. Evaluation of automated urban surface water extraction from Sentinel-2A imagery using different water indices. *J. Appl. Remote Sens.* **2017**, *11*, 026016. [CrossRef]
54. Chen, C.; Liang, J.; Xie, F.; Hu, Z.; Sun, W.; Yang, G.; Yu, J.; Chen, L.; Wang, L.H.; Wang, L.Y.; et al. Temporal and spatial variation of coastline using remote sensing images for Zhoushan archipelago, China. *Int. J. Appl. Earth Obs. Geoinf.* **2022**, *107*, 102711. [CrossRef]
55. Chen, C.; Fu, J.; Zhang, S.; Zhao, X. Coastline information extraction based on the tasseled cap transformation of Landsat-8 OLI images. *Estuar. Coast. Shelf Sci.* **2019**, *217*, 281–291. [CrossRef]
56. Liang, J.; Chen, C.; Song, Y.; Sun, W.; Yang, G. Long-term mapping of land use and cover changes using Landsat images on the Google Earth Engine Cloud Platform in bay area—A case study of Hangzhou Bay, China. *Sustain. Horiz.* **2023**, *7*, 100061. [CrossRef]
57. Chen, H.; Chen, C.; Zhang, Z.; Lu, C.; Wang, L.; He, X.; Chu, Y.; Chen, J. Changes of the spatial and temporal characteristics of land-use landscape patterns using multi-temporal Landsat satellite data: A case study of Zhoushan Island, China. *Ocean Coast. Manag.* **2021**, *213*, 1005842. [CrossRef]
58. Malenovský, Z.; Homolová, L.; Zurita-Milla, R.; Lukeš, P.; Kaplan, V.; Hanuš, J.; Gastellu-Etchegorry, J.P.; Schaepman, M.E. Retrieval of spruce leaf chlorophyll content from airborne image data using continuum removal and radiative transfer. *Remote Sens. Environ.* **2013**, *131*, 85–102. [CrossRef]

59. Sun, P.Y.; Ke, Y.H.; Zhong, R.F.; Zhao, S.H.; Liu, Y. Radiance quality assessment of ZY-1-02D VNIC/AHSI image data. *Remote Sens. Technol. Appl.* **2022**, *37*, 938–952.
60. Chen, L.; Li, W.; Chen, L.; Wang, M.F.; Jia, W.J. Potential prediction of coalbed methane anomaly region utilizing Landsat8 OLI and GF-2 images. In Proceedings of the 2019 IEEE International Geoscience and Remote Sensing Symposium, Yokohama, Japan, 28 July 2019; pp. 6819–6822. [CrossRef]
61. Qin, Q.M.; Zhang, Z.; Chen, L.; Wang, N.; Zhang, C. Oil and gas reservoir exploration based on hyperspectral remote sensing and super-low-frequency electromagnetic detection. *J. Appl. Remote Sens.* **2016**, *10*, 016017. [CrossRef]
62. Zhen, G.W.; Chen, L.; Chen, C.; Guo, B. Alteration anomaly information extraction using hyperspectral remote sensing in coalbed methane enrichment. In Proceedings of the 2016 IEEE International Geoscience and Remote Sensing Symposium, Beijing, China, 10 July 2016; pp. 5394–5397. [CrossRef]

remote sensing

Article

Research on Long-Term Tidal-Height-Prediction-Based Decomposition Algorithms and Machine Learning Models

Wenchao Ban [1], Liangduo Shen [1,*], Fan Lu [2], Xuanru Liu [3] and Yun Pan [2]

[1] School of Marine Engineering Equipment, Zhejiang Ocean University, Zhoushan 316022, China; banwenchao@zjou.edu.cn
[2] School of Naval Architecture and Maritime, Zhejiang Ocean University, Zhoushan 316022, China; panyunhk@zjou.edu.cn (Y.P.)
[3] Zhoushan City Land Reserve Center, Zhoushan 316022, China; s20081500012@zjou.edu.cn
* Correspondence: slduo@zjou.edu.cn

Abstract: Tidal-level prediction is crucial for ensuring the safety and efficiency of offshore marine activities, port and channel management, water transportation resource development, and life-saving operations. Although tidal harmonic analysis is among the most prevalent methods for predicting tidal water level fluctuations, it relies on extensive data, and its long-term prediction accuracy can be limited. To enhance prediction performance, this paper proposes a model that combines the variational mode decomposition (VMD) algorithm with the long short-term memory (LSTM) neural network. The initial step involves decomposing the original data using the VMD algorithm, followed by applying the LSTM to each decomposition component. Finally, all prediction results are superimposed and summed. The model is tested using the 2018 tidal time series data from the Lvsi station in Zhoushan City and the 2020 tidal time series data from the Ganpu station. The results are compared with those from the classical harmonic analysis model, the traditional machine learning model, and the decomposition-based machine learning method. The experimental outcomes demonstrate the superior predictive capabilities of the proposed model.

Keywords: tide level prediction; harmonic analysis; VMD algorithm; LSTM neural network; combined prediction model

Citation: Ban, W.; Shen, L.; Lu, F.; Liu, X.; Pan, Y. Research on Long-Term Tidal-Height-Prediction-Based Decomposition Algorithms and Machine Learning Models. *Remote Sens.* **2023**, *15*, 3045. https://doi.org/10.3390/rs15123045

Academic Editor: Chung-yen Kuo

Received: 21 April 2023
Revised: 8 June 2023
Accepted: 8 June 2023
Published: 10 June 2023

1. Introduction

Tides [1,2] are among the most crucial physical elements of the ocean. This periodic rising and falling of seawater is due to the gravitational forces exerted on the ocean's surface by the movement of celestial bodies. Analyzing and understanding ocean tides allows us to comprehend the laws of change and internal dynamic mechanisms, which in turn enable us to project future tidal variations through tidal predictions. Tidal prediction is an essential technical means for identifying, researching, and effectively utilizing marine resources.

Given China's vast oceanic area and complex tidal variations, it is imperative to establish a detailed, accurate, convenient, and flexible marine tidal protection system. Analyzing and mastering tidal movements and improving the accuracy of tidal predictions [3,4] are crucial for ensuring the safety of marine activities, preventing and mitigating disasters, and harnessing tidal power generation. Therefore, making accurate ocean tide predictions is of utmost importance [5]. By thoroughly researching tides, we can utilize them effectively and mitigate potential hazards. Tidal harmonic analysis [6,7] involves calculating future tidal changes for a specific period by using actual tidal data from the predicted sea area and then superimposing this data to obtain future tidal variations, ultimately enabling tidal predictions. The classical method of tidal harmonic analysis was initiated by Kelvin in England, who proposed the technique in 1868 and invented the tidal prediction machine based on this principle. Subsequently, Darwin further investigated tidal frequency division [8,9], followed by Dodson in the 1820s, who incorporated Brown's theory to increase the number of divisions and improve the accuracy of tidal summation analysis.

As generations of scholars have built upon their predecessors' work, the tidal harmonic analysis method has been continually refined [10,11], making significant contributions to tidal research in China. While the method can generally predict tidal height changes, it cannot guarantee the accuracy of tidal predictions. Unavoidable errors arise from factors such as the representation of astronomical image angles and the selection of tidal constituents, which often hinder our ability to achieve the desired level of accuracy in tidal forecast errors. In recent years, with the flourishing of deep learning technologies [12–17], novel methods have emerged for tide level prediction research. The primary distinction from traditional approaches is that machine learning time series forecasting methods do not necessitate large amounts of prior data or theoretical conjectures to determine model parameters. By learning from the input set's sample data, these methods can directly identify patterns among model arguments and data, enabling predictions. Their powerful data learning and information extraction capabilities facilitate high-precision prediction results. Consequently, machine learning techniques have effectively surpassed traditional forecasting methods in many areas involving modeling nonlinearity. For instance, Jung S used the LSTM [18] neural network to predict water level in a tidal river, Okwuashi O used SVM [19] to develop a tide level prediction model, and J. -C. Yin combined a discrete wavelet transform (DWT) with variable neural networks [20] to achieve real-time predictions for some tidal stations in the United States. The focus of this research is on time series machine learning models combined with decomposition algorithms [21–25]. It is acknowledged that improving the computational power of machine learning is challenging, but if the volatility and complexity in the original data can be reduced through decomposition algorithms, prediction difficulty in machine learning models can be diminished, ultimately enhancing the final prediction results. Currently, the more reliable decomposition algorithms include the complete ensemble empirical mode decomposition with adaptive noise (CEEMDAN) algorithm [26,27] and the variational mode decomposition (VMD) [28,29] algorithm, which are seldom applied to the field of tidal prediction. The CEEMDAN decomposition algorithm is a step-by-step improvement based on the ensemble empirical mode decomposition (EEMD) [30–32] algorithm, which more effectively addresses the modal mixing phenomenon generated by the EEMD, reduces computational speed limitations, and can completely decompose the time series data from high to low frequencies into several intrinsic mode function (IMF) components [33], thereby decreasing data complexity. The VMD decomposition algorithm employs a non-recursive method for signal processing, decomposing the original signal by solving the constrained variable problem. This approach effectively avoids issues such as boundary effects and modal confusion, resulting in high-precision decomposition outcomes for complex data.

We recognize that the predictive capacity of a single model is ultimately limited [34–37], so we shift our focus to developing a combined model. In this paper, we choose to combine the VMD algorithm with the LSTM neural network to form a VMD-LSTM model. Based on two different sets of tide level data, we establish the classical tidal harmonic analysis, LSTM, SVM, CEEMDANN-LSTM, and VMD-LSTM models for comparative analysis in five groups, with the expectation [38–41] of providing new ideas and methods for high-accuracy tide level data prediction.

2. Materials and Methods

The data employed in this study were sourced from the National Marine Information Center (https://www.cnss.com.cn/tide/ accessed on 12 December 2022) and include annual tidal information from the Lvsi station in 2018 and the Ganpu station in 2020. The sample sizes for these stations are 8760 and 8783, respectively, featuring an hourly sampling interval and tidal height measurements in meters. The observed tidal patterns are characterized as conventional semi-diurnal shallow tides with reciprocating flow properties.

To effectively compare the predictive performance of various methods, the first two months of the original data were designated as the training set, while the remaining ten months constituted the prediction set. The two-month data were employed to project the tidal trends throughout the subsequent ten months. Given the significant amount of data, it is impractical to visualize all of it, as this would result in a cluttered and unclear graph. Therefore, we have chosen to visualize tidal height data from periods with lesser tidal fluctuations (8100–8600 h) and greater tidal fluctuations (5500–6000 h), each spanning 500 h, for a more meaningful comparison. For the complete prediction set data, the error results will be directly analyzed in a tabular format. The unusual tidal height variations at both stations transpired in mid-August when the sun, moon, and earth were almost aligned, leading to the peak gravitational force on the seawater. This occurrence, in conjunction with the dominant southeast winds and tidal direction along the coast during summer, resulted in the abnormal tidal height shifts and an overall elevation in tidal height. The tidal time series data for the Lvsi and Ganpu stations in Zhoushan City are depicted in Figures 1 and 2.

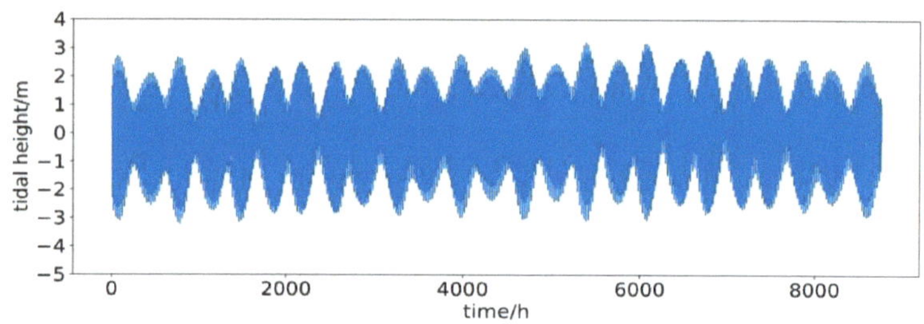

Figure 1. Timing diagram of tidal height data at Lvsi Station.

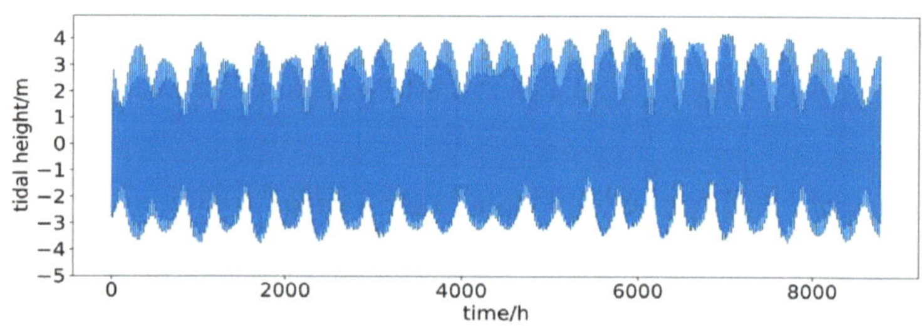

Figure 2. Timing diagram of tidal height data at Ganpu Station.

2.1. Classical Tidal Harmonic Analysis Model

The classical tidal harmonic analysis model employed in this research utilized the T_TIDE program, a widely recognized and effective tool for tidal analysis and prediction. In this study, a total of 36 subtidal tides were chosen for the analysis, with an emphasis on 11 major subtidal tides: K1, K2, M2, M4, M6, MS4, N2, O1, P1, Q1, and S2. These subtidal tides were deemed to be of critical importance due to their significant impact on tidal dynamics.

The harmonic analysis was conducted using observed real-time tidal height data, which allowed for the extraction of harmonic constants for the main sub-tides. These constants were then utilized for tidal prediction purposes. In the classical tidal harmonic analysis model, as described by Equation (1), each tidal constituent can be expressed as a trigonometric function. A linear superposition of these trigonometric functions is then used to represent the water level at any given time.

$$Z_{(t)} = S_0 + \sum_{j=1}^{J} [H_j cos(\sigma_j t - g_j)] \tag{1}$$

$Z_{(t)}$ is the observed water level at time t. σ_j, H_j, and g_j are the angular velocity, amplitude, and delay angle (°), respectively, corresponding to the jth subtide. j is the number of resolved subtides. J means the number of sub-tides, S_0 is the mean sea level, and H_j in Equation (2) can be linearized using the variables a_j and b_j. In the equation, $H_j = \sqrt{a_j{}^2 + b_j{}^2}$, $g_j = Arctan(b_j/a_j)$.

When doing linear regression, Equation (2) is required:

$$Z = AX \tag{2}$$

where Z is the observed water level matrix, A is the known coefficient matrix, and X is the unknown parameter matrix to be solved, as shown in Equations (3) and (4):

$$\mathbf{Z} = \begin{bmatrix} Z(t_1) \\ Z(t_2) \\ \cdots \\ Z(t_N) \end{bmatrix}, \mathbf{X} = \begin{bmatrix} S_0 \\ a_1 \\ \cdots \\ a_j \\ b_1 \\ \cdots \\ b_j \end{bmatrix} \tag{3}$$

$$A = \begin{bmatrix} 1 & cos\sigma_1 t_1 & \cdots & cos\sigma_j t_1 \; sin\sigma_1 t_1 \ldots sin\sigma_j t_1 \\ 1 & cos\sigma_1 t_2 & \cdots & cos\sigma_j t_2 \; sin\sigma_1 t_2 \ldots sin\sigma_j t_2 \\ \cdots & \cdots & \cdots & \cdots \quad \cdots \quad \cdots \quad \cdots \\ 1 & cos\sigma_1 t_N & \cdots & cos\sigma_j t_N sin\sigma_1 t_N \ldots sin\sigma_j t_N \end{bmatrix} \tag{4}$$

According to the least squares method, the solution of $\mathbf{X}$ is:

$$\mathbf{X} = \left(A^T A\right)^{-1} A^T \mathbf{Z} \tag{5}$$

By employing the classical tidal harmonic analysis model, the research aimed to achieve a comprehensive understanding of tidal dynamics at the study site. This approach effectively accounted for the influences of various tidal constituents on water level changes over time. Consequently, the methodology provided a robust foundation for reliable tidal predictions, which are crucial for efficient coastal zone management, navigation, and infrastructure development.

2.2. CEEMDAN Model

The CEEMDAN decomposition algorithm can effectively decompose the highly fluctuating raw time series data from high to low frequencies into several intrinsic mode function (IMF) components. This helps improve the prediction accuracy by reducing the data fluctuations. The most significant advantage of the CEEMDAN decomposition algorithm is its completeness; the decomposed data can be aggregated back into the raw data without loss, and the efficiency of the CEEMDAN decomposition algorithm is very high, providing rapid computation. When decomposing the tidal data with CEEMDAN, the white noise of the standard normal distribution is added to the observation data sequence $x(t) = (x_1, x_2 \ldots, x_t)$ in $\lambda(\lambda = 1, 2, \ldots, I)$ times $v^\lambda(t)$, where T is the length of the tidal data, x_t is the value of tidal at time t, I is the number of noise additions, and the reconstructed observation data sequence $x^\lambda(t)$ is expressed in Equation (6).

$$x^\lambda(t) = x(t) + v^\lambda(t) \tag{6}$$

The first EMD solution of $x^\lambda(t)$ yields the components IMF_1 and its residuals $m_1(t)$, which are expressed in Equations (7) and (8).

$$IMF_1 = \frac{1}{I} \times \sum_{\lambda=1}^{I} IMF_1^\lambda \tag{7}$$

$$m_1(t) = x(t) - IMF_1 \tag{8}$$

Add $v^\lambda(t)$ to $m_1(t)$ and perform EMD decomposition to obtain IMF_2 and $m_2(t)$; add $v^\lambda(t)$ to $m_{n-1}(t)$ after decomposition n − 1 time and perform EMD decomposition for $m_{n-1}^\lambda(t)$ for the nth time to obtain IMF_n and $m_n(t)$ as Equations (9) and (10), respectively.

$$IMF_n = \frac{1}{I} \times \sum_{\lambda=n-1}^{I} IMF_n^\lambda \tag{9}$$

$$m_n(t) = m_{n-1}(t) - IMF_n \tag{10}$$

After J decomposition, $m_n(t)$ cannot be decomposed by EMD continuously, andthe tidal observation data series $x(t)$ is expressed in Equation (11).

$$x(t) = \sum_{n=1}^{J} IMF_n + m_J(t) \tag{11}$$

Compared to EMD, CEEMDAN combines positive and negative white noise added to the original data. The white noise property is utilized to amplify the uncorrelated degree of the difficult-to-separate modes, enabling the extraction of these modes and reducing the reconstruction error through the cancellation of positive and negative white noise.

Figures 3 and 4 display the decomposition results of the CEEMDAN algorithm, which was implemented using the PyEMD package in Python 3.8. The original tidal height time series data from the records of the Lvsi station and the Ganpu station in Zhoushan were fully decomposed according to high and low frequencies. The results of each component were predicted using the LSTM model and then summed and reconstructed. The flow chart is shown in Figure 5.

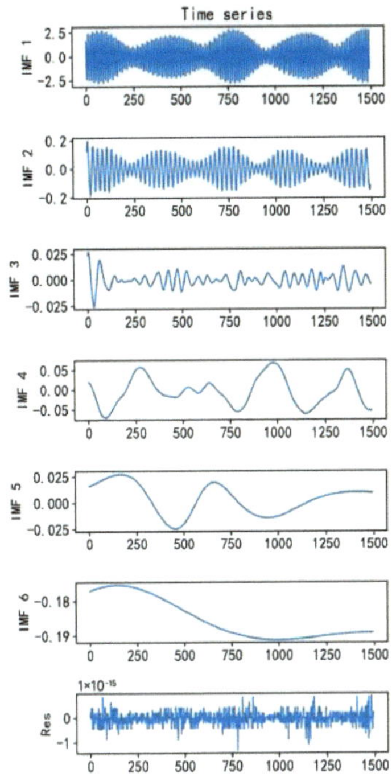

Figure 3. Decomposition results (Lvsi station).

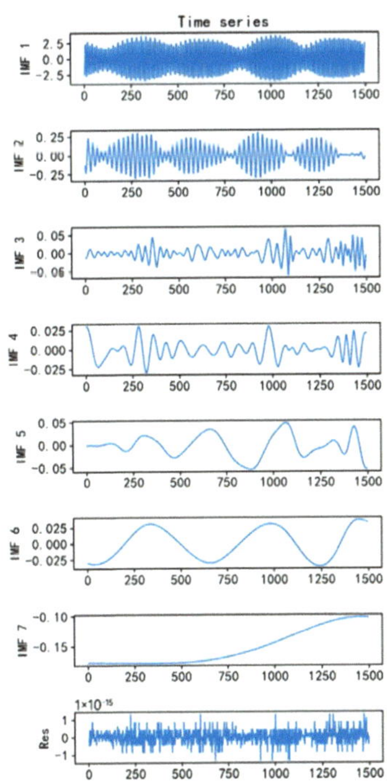

Figure 4. Decomposition results (Ganpu station).

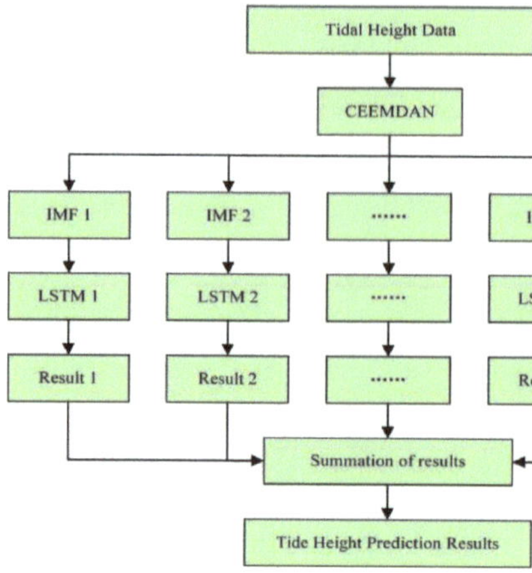

Figure 5. CEEMDAN-LSTM flow chart.

2.3. LSTM Model

LSTM is better equipped to handle time series data, as it processes a long span of time to capture the internal relationships within the data, thereby achieving long-term memory. Each LSTM unit consists of three gates: the forget gate, the input gate, and the output gate. The structure of a single neuron is shown in Figure 6.

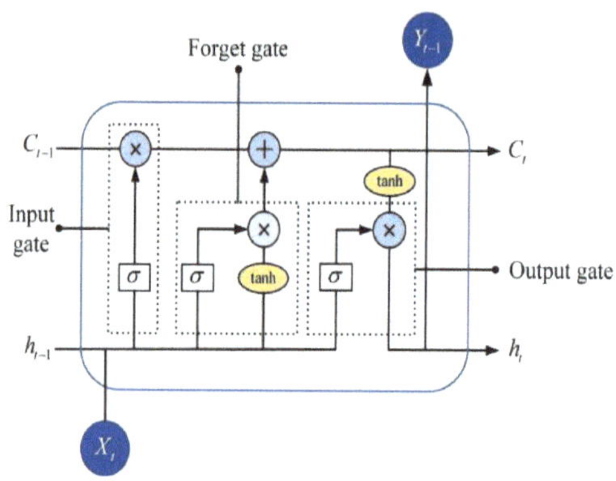

Figure 6. Neuronal structure.

$$i_t = \sigma_i(W_i \times [h_{t-1}, x_t] + b_i) \tag{12}$$

$$f_t = \sigma\left(W_f \times [h_{t-1}, x_t] + b_f\right) \tag{13}$$

$$C_t = f \times C_{t-1} + i_t \times \widetilde{C}_t \tag{14}$$

$$\widetilde{C}_t = \tanh(W_c \times [h_{t-1}, x_t] + b_c) \tag{15}$$

$$O_t = \sigma(w_0 \times [h_{t-1}, x_t] + b_0) \tag{16}$$

$$h_t = O_t * tanh(C_t) \tag{17}$$

In the above equation, i_t, f_t, c_t, and O_t represent the oblivion gate, the input gate, the cell state and the output gate, respectively; w_i, w_f, w_c, and w_o all imply different weight matrices; and b_f, b_i, b_c, and b_o represent the corresponding bias vectors, The activation function is tanh, h_t is the output at moment t, x_t is the input at moment t, and $\widetilde{C}_t$ is the new value of the candidate vector update.

The LSTM models are implemented using Python 3.8. Our LSTM network consists of three hidden layers, with each hidden layer followed by a dropout layer to prevent overfitting. The network has a total of 300 dimensions and uses "relu" as the activation function and "Adam" as the optimizer and takes six steps to test the value of one step, for a total of 100 epochs. These crucial parameters are determined through continuous experimentation, testing numerous combinations of these parameters with different values until a stable network and optimal training options are obtained. To ensure better comparative experiments, we maintain these optimal parameter values without change.

2.4. SVM Model

Support vector machines (SVM) is a widely used machine learning technique for regression analysis and statistical classification. The primary goal of SVM is to find the optimal hyperplane that maximizes the margin between different classes or minimizes the regression error. The SVM algorithm maps input vectors to a high-dimensional space using kernel functions, where it seeks to identify the optimal partitioning hyperplane and two parallel hyperplanes that best separate the data.

The total error of the SVM is minimized when the margin between the parallel hyperplanes is maximized. In this experimental comparison, stochastic gradient optimization is employed to optimize the model parameters, ensuring a more efficient convergence towards the optimal solution. MATLAB is used for the modeling process, with the following settings: KernelFunction is set to "linear"," KernelScale is set to 1, and the BoxConstraint range is defined as $[1 \times 10^{-3}, 1 \times 10^{3}]$.

The chosen settings for the experiment (linear KernelFunction, KernelScale of 1, and BoxConstraint range of $[1 \times 10^{-3}, 1 \times 10^{3}]$) ensure a balance between model complexity and generalization performance. By using stochastic gradient optimization and these specific settings, the experimental comparison aims to showcase the effectiveness of SVM in solving regression and classification problems, highlighting its potential applications in various domains.

2.5. VMD Algorithm

VMD is a novel time-frequency decomposition algorithm that aims to decompose the multi-component signal in the data into multiple single-component amplitude-modulated frequency signals. This approach helps to avoid issues such as pseudo-component problems and endpoint effects that may be encountered, resulting in more effective processing of nonlinear signals. The specific decomposition process is as follows: Initialize u_k^1, w_k^1, and λ_k^1 and $n = 0$.

(1) $n = n + 1$; enter the loop.
(2) Update according to the update formula of u_k and w_k until the number of decompositions is k.
(3) Update λ according to the update formula of λ.
(4) Given the accuracy ε, if the stopping condition is satisfied,

$$\sum_k \frac{\|u_k^{n+1} - u_k^n\|_2^2}{\|u_k^n\|_2^2} < \varepsilon \tag{18}$$

Terminate the cycle; otherwise, proceed to step (2) to continue the cycle.

In the above description, u_k represents the decomposed single-component AMF signal, w_k denotes the center frequency of each single-component AM/FM signal, λ is the Lagrangian multiplier, and n stands for the number of iterations.

When utilizing VMD to decompose the data, it is essential to set the parameters of the VMD algorithm (penalty factor α and modal component K) in advance. We reconstruct the corresponding time series by summing the sub-time series for each value of K and α. When the reconstructed time series exhibits a perfect correlation with the original time series, the values of K and α that yield the best efficiency for the VMD-based computational model are chosen as the optimal values. In this experiment, the penalty factor of the VMD decomposition algorithm is set to 2000, the specified number of decomposition modes is set to 4, and the articulation process of the LSTM neural network is identical to that of the CEEMDAN decomposition algorithm. The specific decomposition diagrams for the Lvsi and Ganpu stations are shown in Figures 7 and 8.

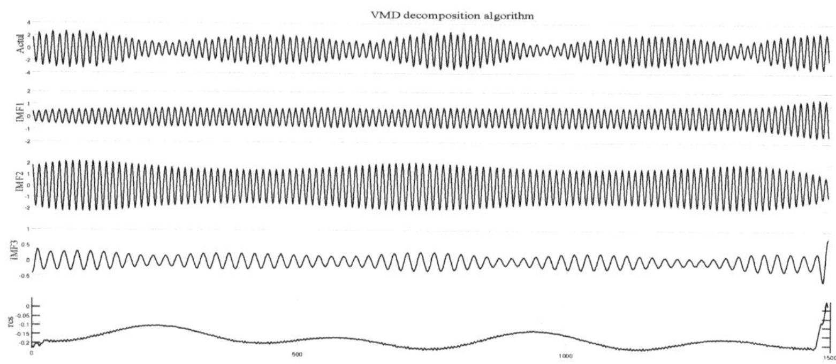

Figure 7. Splitting result based on variational modal decomposition (VMD) algorithm (Lvsi station).

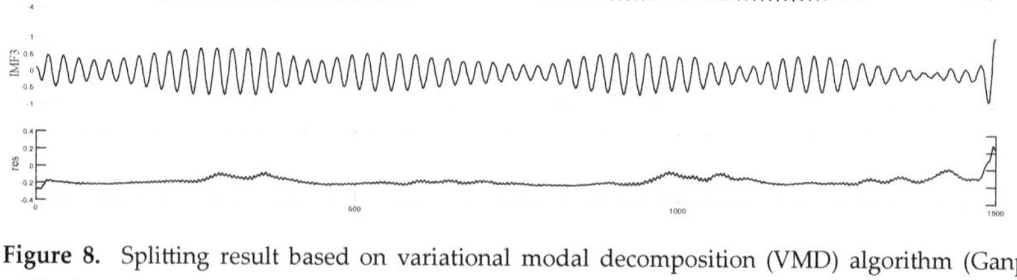

Figure 8. Splitting result based on variational modal decomposition (VMD) algorithm (Ganpu station).

2.6. Model Evaluation Indexes

To emphasize the performance differences among the compared models, we opted to use three error metrics, namely root mean square error (RMSE), mean absolute error (MAE), and goodness of fit (R^2), to prevent chance errors caused by individual metrics. The units of RMSE and MAE are meters, and smaller values indicate higher prediction accuracy. Moreover, the closer the value of R^2 is to 1, the better the fit. The expressions are shown in Equations (19)–(21):

$$RMSE = \sqrt{\frac{1}{m}\sum_{i=1}^{m}(y_i - z_i)^2} \quad (19)$$

$$MAE = \frac{1}{m}\sum_{i=1}^{m} / y_i - z_i / \quad (20)$$

$$R^2 = 1 - \frac{\sum_{i=1}^{m}(y_i - z_i)^2}{\sum_{i=1}^{m}(y_i - z_i)^2} \quad (21)$$

where m is the number of test samples, and y_i and z_i are the original and predicted values of the data, respectively.

3. Results

We initially chose to compare the predicted results of conventional tidal wave height. The conventional tidal wave height prediction results of the classical tidal harmonic analysis, SVM, LSTM, CEEMDAN-LSTM, and VMD-LSTM models for the Lvsi and Ganpu stations are displayed in Figures 9 and 10, respectively, while the error analysis results are shown in Tables 1 and 2. The raw data in the figures represent the 500 h of conventional tidal wave height data in the sample data range 8100 h–8600 h. It is evident that all five models perform well in conventional tidal wave height prediction, with high graphical fit and no significant errors, which means that all models can achieve the desired results in conventional tidal wave height prediction.

After comparing the error analysis metrics in Tables 1 and 2, it is concluded that the VMD-LSTM model exhibits smaller errors and better predictions. At both sites, the RMSE and MAE metrics of VMD-LSTM are improved by more than 50% compared to the comparison models, and the R^2 is closer to 1. We believe that the conventional tidal wave height prediction is satisfactory for each model, and even if we use the more advanced VMD decomposition algorithm to enhance the prediction accuracy, the significant numerical improvement in the accuracy index is also a valuable addition.

Table 1. Error analysis results of conventional tidal height prediction at Lvsi station.

	Harmonic Analysis	SVM	LSTM	CEEMDAN-LSTM	VMD-LSTM
RMSE (m)	0.284	0.143	0.144	0.134	0.068
MAE (m)	0.228	0.119	0.119	0.114	0.058
R^2	0.956	0.989	0.979	0.980	0.992

Table 2. Error analysis results of conventional tidal height prediction at Ganpu station.

	Harmonic Analysis	SVM	LSTM	CEEMDAN-LSTM	VMD-LSTM
RMSE (m)	0.304	0.144	0.143	0.147	0.058
MAE (m)	0.245	0.119	0.112	0.124	0.048
R^2	0.977	0.995	0.995	0.995	0.998

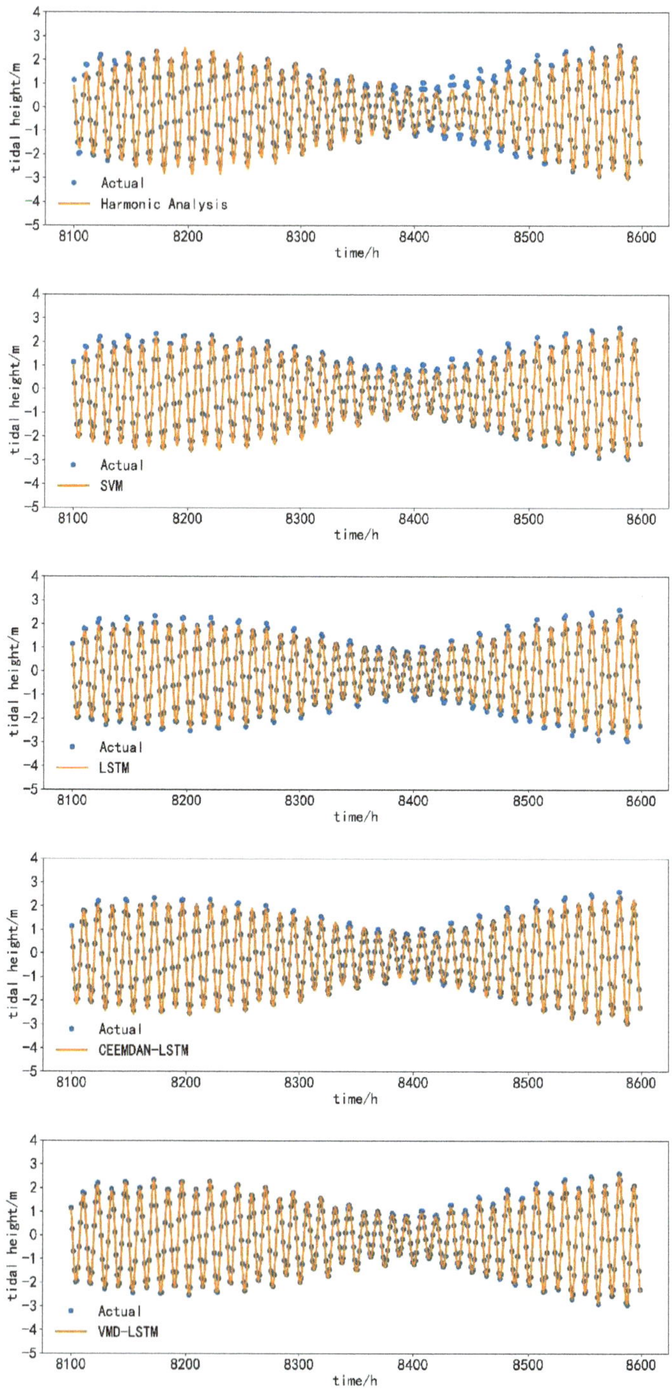

Figure 9. Comparison of prediction results of conventional tide level height (Lvsi station).

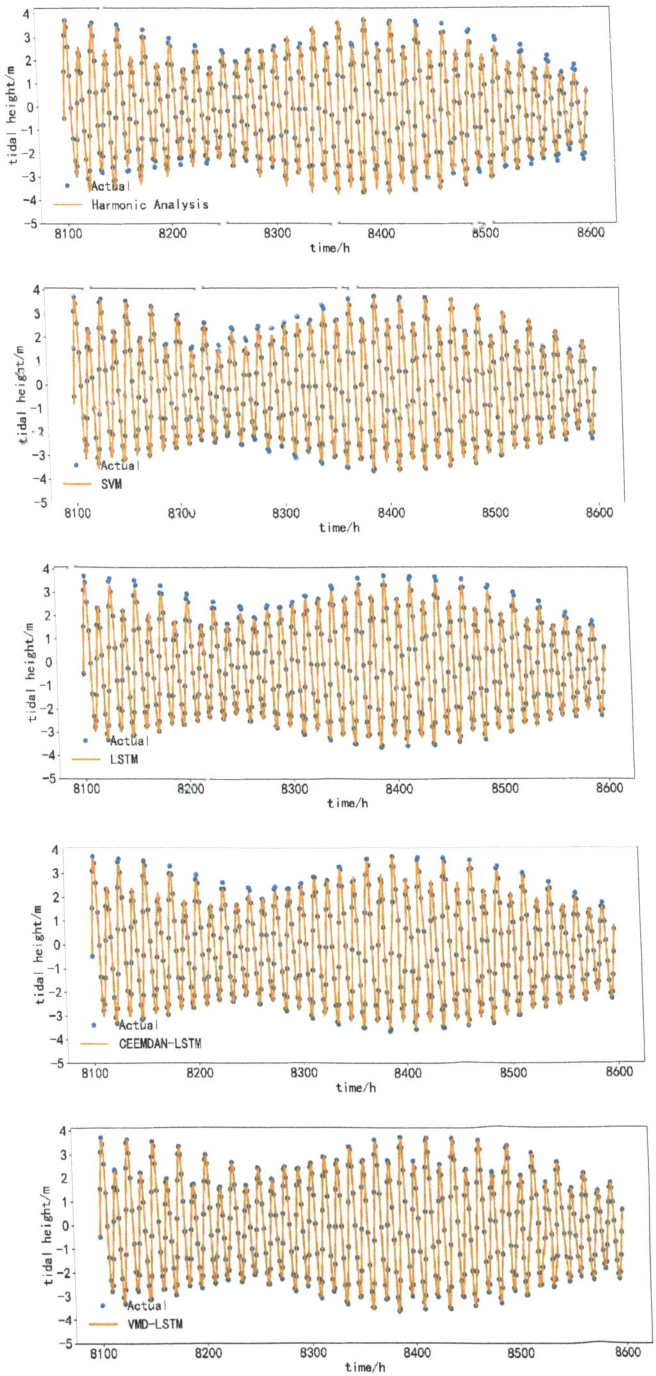

Figure 10. Comparison of prediction results of conventional tide level height (Ganpu station).

To better compare the performance of each model, we chose to continue analyzi the prediction results of the five models in abnormal tidal wave height conditions. We selected the abnormal tidal wave height data of 500 h from 5500 h to 6000 h in the original data for comparison. The prediction results of the five models in the abnormal tidal wave height are presented in Figures 11 and 12, and the error analysis results are shown in Tables 3 and 4, while the total tidal height prediction error results are presented in Tables 5 and 6.

In the comparison of anomalous tidal height prediction results, we can observe that the VMD-LSTM model's performance is superior. The four compared models have significant errors at the crests and troughs, with varying degrees of errors, but the VMD-LSTM model overcomes this problem. Its graphical fit is extremely high, and there are no obvious errors. After analyzing the errors of each model, the VMD-LSTM model shows a 36.5% improvement in both the lowest RMSE error metric and the lowest MAE error metric, compared to the four comparison models. The R^2 is above 0.99 for both stations.

When we compare the error analysis indicators of the entire ten-month forecast set, we can also conclude that the comprehensive prediction performance of the VMD-LSTM model remains superior.

Table 3. Error analysis results of abnormal tidal height prediction at Lvsi station.

	Harmonic Analysis	SVM	LSTM	CEEMDAN-LSTM	VMD-LSTM
RMSE (m)	0.428	0.509	0.361	0.189	0.120
MAE (m)	0.380	0.495	0.335	0.156	0.102
R^2	0.892	0.854	0.924	0.980	0.992

Table 4. Error analysis results of abnormal tidal height prediction at Ganpu station.

	Harmonic Analysis	SVM	LSTM	CEEMDAN-LSTM	VMD-LSTM
RMSE (m)	0.594	0.465	0.288	0.178	0.072
MAE (m)	0.5	0.428	0.224	0.141	0.060
R^2	0.932	0.959	0.983	0.994	0.998

Table 5. Error analysis results of total tidal height prediction at Lvsi station.

	Harmonic Analysis	SVM	LSTM	CEEMDAN-LSTM	VMD-LSTM
RMSE (m)	0.372	0.339	0.260	0.171	0.091
MAE (m)	0.299	0.288	0.215	0.138	0.075
R^2	0.925	0.945	0.967	0.987	0.996

Table 6. Error analysis results of total tidal height prediction at Ganpu station.

	Harmonic Analysis	SVM	LSTM	CEEMDAN-LSTM	VMD-LSTM
RMSE (m)	0.504	0.323	0.224	0.147	0.067
MAE (m)	0.402	0.266	0.164	0.117	0.055
R^2	0.935	0.976	0.989	0.995	0.998

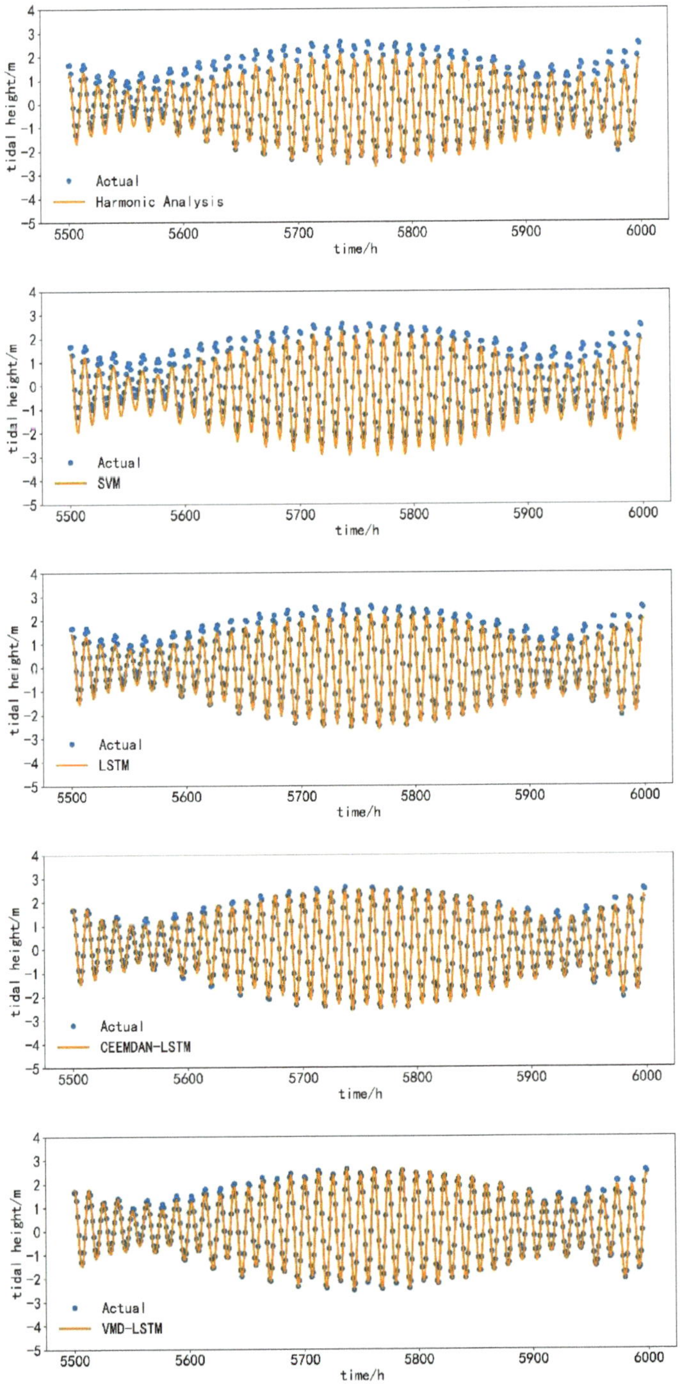

Figure 11. Comparison of abnormal tidal height prediction results (Lvsi station).

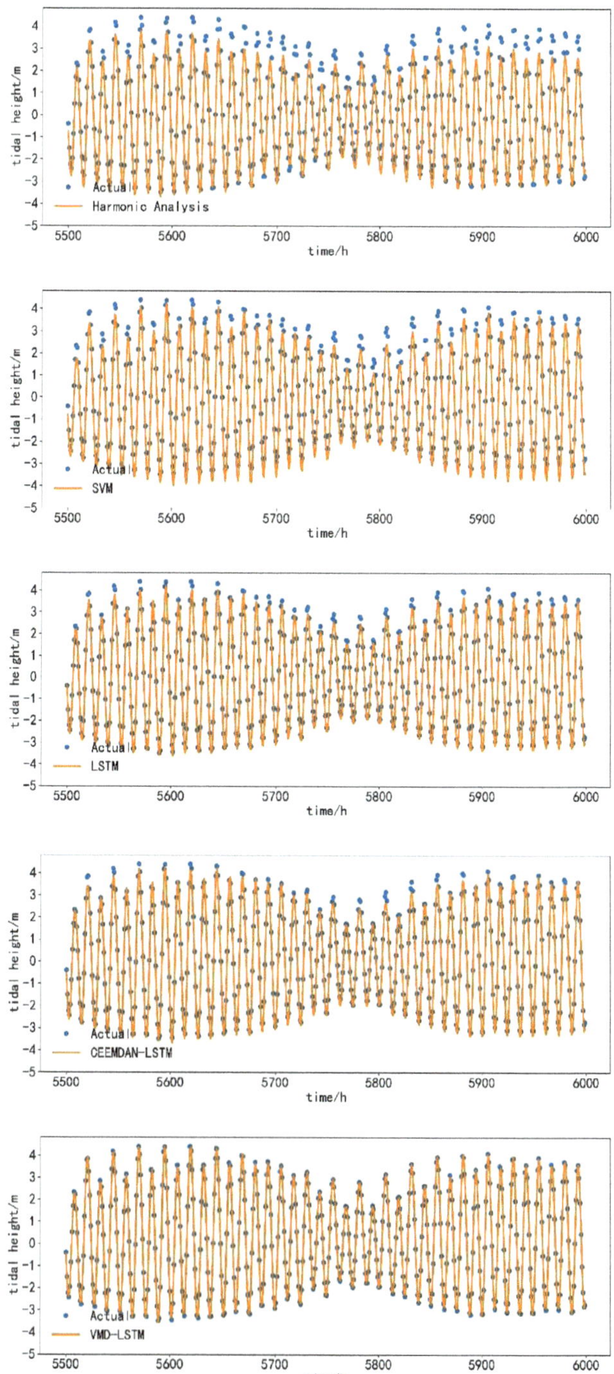

Figure 12. Comparison of prediction results of abnormal tide level height (Ganpu station).

4. Discussion

Taking into account all the experimental results, we believe that the VMD decomposition algorithm is indeed a significant improvement and an ideal method for tidal height prediction accuracy. Comparing various methods, the classical tidal harmonic analysis method has high requirements for each data parameter. As Pan [1] and others have pointed out, the prediction accuracy of this method can be affected by numerous factors, such as the choice of subtidal zone, the number of subtidal zones, topographic effects, and more.

The CEEMDAN algorithm has been used successfully by Zhang [26] and others; however, although the CEEMDAN algorithm has high decomposition integrity, the cumulative error increases relatively due to the high number of low-frequency components generated by this decomposition method. Despite the good decomposition effect, it is still lacking in processing anti-noise and outlier fluctuations, and the data value is reduced after decomposition, but the handling of fluctuation frequency is lacking, making it inferior to the VMD algorithm. The experimental results show that the CEEMDAN-LSTM model's prediction performance is better than the classical harmonic analysis model and simple machine learning model but inferior to the VMD-LSTM model.

The VMD decomposition algorithm can simplify the data's physical structure, decomposing the raw signal of complexity and non-smoothness into feature mode components and capturing multi-scale information at different resolution levels. As a result, the decomposed training data can be trained more efficiently, and prediction performance can be better when applied to individual models. Using training data with a simplified structure in model training can learn data patterns more efficiently and reduce the effect of noise contained in the original data [28].

In contrast to Yin [20], who used a combination of discrete wavelet transform (DWT) and variable neural network to predict tidal heights, and Wang [32], who combined a hybrid model of EMD/EEMD and ARIMA to predict long-term streamflow, the EEMD- and DWT-based models for high-frequency sub-time sequences are less efficient and accurate than the low-frequency model, while the VMD-based model yields good efficiency and accuracy for both high-frequency and low-frequency sub-time sequences.

DWT is a reliable method for non-stationary signal analysis, but it can only choose a fixed wavelet basis function when used, which is weakly adaptive. The VMD non-recursive decomposition approach effectively overcomes its shortcomings. Compared with the EMD, EEMD, and CEEMD methods, it allows for a more effective separation of tidal signals and noise. The highly adaptive decomposition characteristics of VMD, the rigorous theoretical basis, and the advantages of high-frequency noise suppression, combined with LSTM's long-term dependence, achieve high-accuracy predictions of tidal forecasts. Moreover, the CEEMDAN decomposition algorithm in this study decomposes the original time series into more sub-time series than VMD, so the error accumulation of the CEEMDAN-based model may be greater than that of the VMD-based model. Therefore, VMD is a more effective model than CEEMDAN for developing tidal forecasts.

All models can obtain better prediction results in conventional tidal wave height prediction. However, when the tidal height increases, the prediction ability of the classical tidal harmonic analysis method and a single machine learning model is limited. The CEEMDAN-LSTM model can maintain good prediction accuracy, while the VMD-LSTM model can still maintain high-accuracy long-term prediction results, overcoming the problems that these models cannot solve. This indicates that the VMD algorithm is an effective way to improve the accuracy of tidal height prediction.

5. Conclusions

(1) The combination of the VMD decomposition algorithm and the LSTM neural network effectively increases the precision for long-term conventional tidal forecasting and addresses the problem of inaccurate prediction results of anomalous tidal data in long-term tide prediction. Achieving high-precision long-term tide prediction is essential for realistic marine activities. However, prediction results based on reconcil-

iation analysis often cannot meet the current refinement requirements. The pairing of the VMD decomposition algorithm and the LSTM model may provide a new method for this purpose, thus meeting the practical application needs of the growing marine economic activities.

(2) The introduction of the VMD decomposition algorithm enables better processing of the original tide level data, effectively decomposing more stable modal components and improving the model's prediction accuracy. This approach provides new research ideas and methods for tide level prediction. This method can also be expanded, considering the combination of the VMD decomposition algorithm with more machine learning time prediction models to obtain a new combined model, or applying this model to more data types, such as waves.

Author Contributions: Conceptualization, F.L.; methodology, W.B.; software, Y.P.; validation, Y.P.; formal analysis, W.B.; investigation, W.B.; resources, L.S.; data curation, W.B.; writing—original draft preparation, W.B.; writing—review and editing, W.B.; visualization, X.L.; supervision, X.L.; project administration, L.S.; funding acquisition, L.S. All authors have read and agreed to the published version of the manuscript.

Funding: This study is financially supported by General Projects of Zhoushan Science and Technology Bureau and the National Natural Science Foundation of China (No. 52101330).

Data Availability Statement: The data used to support the findings of this study are included within the article.

Conflicts of Interest: The authors declare no conflict of interest.

References

1. Pan, H.; Lv, X.; Wang, Y.; Matte, P.; Chen, H.; Jin, G. Exploration of tidal-fluvial interaction in the Columbia river estuary using S_TIDE. *J. Geophys. Res. Oceans* **2018**, *123*, 6598–6619. [CrossRef]
2. Zhu, X.H.; Ma, Y.L.; Guo, X.; Fan, X.; Long, Y. Tidal currents in the Qiongzhou Strait estimated from shipboard ADCP data during spring 2013. In Proceedings of the EGU General Assembly Conference Abstracts, Vienna, Austria, 27 April–2 May 2014; p. 2504.
3. Lee, T.L. Back-propagation neural network for long-term tidal predictions. *Ocean. Eng.* **2004**, *31*, 225–238. [CrossRef]
4. Chang, H.K.; Lin, L.C. Multi-point tidal prediction using artificial neural network with tide-generating forces. *Coast. Eng.* **2006**, *53*, 857–864. [CrossRef]
5. Li, M.; Liu, M.; Liu, X.; Peng, T.; Wang, S. Decomposition of long time-series fraction of absorbed photosynthetically active radiation signal for distinguishing heavy metal stress in rice. *Comput. Electron. Agric.* **2022**, *198*, 107111. [CrossRef]
6. Pawlowicz, R.; Beardsley, B.; Lentz, S. Classical tidal harmonic analysis including error estimates in MATLAB using T_TIDE. *Comput. Geosci.* **2002**, *28*, 929–937. [CrossRef]
7. Foreman, M.; Henry, R. The harmonic analysis of tidal model time series. *Adv. Water Resour.* **1989**, *12*, 109–120. [CrossRef]
8. Sassi, H.G.; Hoitink, A.J.F.; De Brye, B.; Vermeulen, B.; Deleersnijder, E. Tidal impact on the division of river discharge over distributary channels in the Mahakam Delta. *Ocean. Dyn.* **2011**, *61*, 2211–2228. [CrossRef]
9. Jay, D.A.; Flinchem, E.P. Interaction of fluctuating river flow with a barotropic tide: A demonstration of wavelet tidal analysis methods. *J. Geophys. Res. Ocean.* **1997**, *102*, 5705–5720. [CrossRef]
10. Cacabelos, E.; Olabarria, C.; Incera, M.; Troncoso, J.S. Effects of habitat structure and tidal height on epifaunal assemblages associated with macroalgae. *Estuar. Coast. Shelf Sci.* **2010**, *89*, 43–52. [CrossRef]
11. Penna, N.T.; Stewart, M.P. Aliased tidal signatures in continuous GPS height time series. *Geophys. Res. Lett.* **2003**, *30*. [CrossRef]
12. Ji, L.; Xiao, Z.; Lei, Z. Research on the algorithm of education data mining based on big data. In Proceedings of the IEEE 2nd International Conference on Computer Science and Educational Informatization (CSEI), Xinxiang, China, 12–14 June 2020.
13. Guo, S.; Wang, Q. Application of knowledge distillation based on transfer learning of ERNIE model in intelligent dialogue intention recognition. *Sensors* **2022**, *22*, 1270. [CrossRef]
14. Ma, P.; Gao, Q. EEG signal and feature interaction modeling-based eye behavior prediction research. *Comput. Math. Methods Med.* **2020**, *2020*, 2801015. [CrossRef]
15. Xie, Y.; Zhao, J.; Qiang, B.; Mi, L.; Tang, C.; Li, L. Attention mechanism-based CNN-LSTM model for wind turbine fault prediction using SSN ontology annotation. *Wirel. Commun. Mob. Comput.* **2021**, *2021*, 6627588. [CrossRef]
16. Zhou, Y.; Li, T.; Shi, J.; Qian, Z. A CEEMDAN and XGBOOST-based approach to forecast crude oil prices. *Complexity* **2019**, *2019*, 4392785. [CrossRef]

17. Sakinah, N.; Tahir, M.; Badriyah, T.; Syarif, I. LSTM with adam optimization-powered high accuracy preeclampsia classification. In Proceedings of the International Electronics Symposium (IES), Surabaya, Indonesia, 27–28 September 2019; pp. 314–319. [CrossRef]
18. Jung, S.; Cho, H.; Kim, J.; Lee, G. Prediction of water level in a tidal river using a deep-learning based LSTM model. *J. Korea Water Resour. Assoc.* **2018**, *51*, 1207–1216.
19. Okwuashi, O.; Ndehedehe, C. Tide modelling using support vector machine regression. *J. Spat. Sci.* **2017**, *62*, 29–46. [CrossRef]
20. Yin, J.-C.; Perakis, A.N.; Wang, N. Ensemble Real-Time Tidal Level Prediction Mechanism Using Multiresolution Wavelet Decomposition Method. *IEEE Trans. Geosci. Remote. Sens.* **2018**, *56*, 4856–4865. [CrossRef]
21. Zhang, C.-Y.; Wang, Z.; Fei, C.-W.; Yuan, Z.-S.; Wei, J.-S.; Tang, W.-Z. Fuzzy multi-SVR learning model for reliability-based design optimization of turbine blades. *Materials* **2019**, *12*, 2341. [CrossRef]
22. Qu, B.; Xing, Z.; Liu, Y.; Chen, L. Research on short-term output power forecast model of wind farm based on neural network combination algorithm. *Wind. Energy* **2022**, *25*, 1710–1734. [CrossRef]
23. Cao, J.; Li, Z.; Li, J. Financial time series forecasting model based on CEEMDAN and LSTM. *Phys. A Stat. Mech. Its Appl.* **2019**, *519*, 127–139. [CrossRef]
24. Wang, W.; Chau, K.; Xu, D.; Chen, X.Y. Improving forecasting accuracy of annual runoff time series using ARIMA based on EEMD decomposition. *Water Resour. Manag.* **2015**, *29*, 2655–2675. [CrossRef]
25. Zhang, Y.; Zhao, Y.; Kong, C.; Chen, B. A new prediction method based on VMD-PRBF-ARMA-E model considering wind speed characteristic. *Energy Convers. Manag.* **2020**, *203*, 112254. [CrossRef]
26. Zhang, X.; Li, S.; Zhang, C. Short-Term Prediction of SDN Network Traffic Based on CEEMDAN and Mixed Kernel Least Squares Support Vector Machine. In Proceedings of the 5th International Conference on Pattern Recognition and Artificial Intelligence (PRAI), IEEE, Chengdu, China, 19–21 August 2022; p. 1201.
27. Rongbin, C.; Sanming, L. Research on wind power prediction method based on CEEMDAN SSA-GRU. In Proceedings of the 2021 IEEE International Conference on Electrical Engineering and Mechatronics Technology (ICEEMT), Qingdao, China, 2–4 July 2021; pp. 597–601. [CrossRef]
28. Wang, W.; Tong, M.; Yu, M. Blood glucose prediction with VMD and LSTM optimized by improved particle swarm optimization. *IEEE Access* **2020**, *8*, 217908–217916. [CrossRef]
29. Li, H.; Liu, T.; Wu, X.; Chen, Q. An optimized VMD method and its applications in bearing fault diagnosis. *Measurement* **2020**, *166*, 108185. [CrossRef]
30. Jin, X.-B.; Yang, N.-X.; Wang, X.-Y.; Bai, Y.-T.; Su, T.-L.; Kong, J.-L. Deep hybrid model based on EMD with classification by frequency characteristics for long-term air quality prediction. *Mathematics* **2020**, *8*, 214. [CrossRef]
31. Wang, Z.-Y.; Qiu, J.; Li, F.-F. Hybrid models combining EMD/EEMD and ARIMA for long-term streamflow forecasting. *Water* **2018**, *10*, 853. [CrossRef]
32. Lee, H.S. Estimation of extreme sea levels along the Bangladesh coast due to storm surge and sea level rise using EEMD and EVA. *J. Geophys. Res. Ocean.* **2013**, *118*, 4273–4285. [CrossRef]
33. Yan, K.; Hua, J. Deep learning technology for chiller faults diagnosis. In Proceedings of the IEEE International Conference on Dependable, Autonomic and Secure Computing, (DASC/PiCom/CBDCom/CyberSciTech), Fukuoka, Japan, 5–8 August 2019; pp. 72–79. [CrossRef]
34. Rosati, M.; Kelly, T.; Ringwood, J.V. Nonlinear data-based hydrodynamic modeling of a fixed oscillating water column wave energy device. *IEEE Access* **2021**, *9*, 149756–149765. [CrossRef]
35. Kumar, N.K.; Savitha, R.; Al Mamun, A. Regional ocean wave height prediction using sequential learning neural networks. *Ocean Eng.* **2017**, *129*, 605–612. [CrossRef]
36. Kavousi-Fard, A.; Su, W. A combined prognostic model based on machine learning for tidal current prediction. *IEEE Trans. Geosci. Remote Sens.* **2017**, *55*, 3108–3114. [CrossRef]
37. Liu, J.; Shi, G.; Zhu, K. High-precision combined tidal forecasting model. *Algorithms* **2019**, *12*, 65. [CrossRef]
38. Wei, J.; Jiang, G.-Q.; Liu, X. Parameterization of typhoon-induced ocean cooling using temperature equation and machine learning algorithms: An example of typhoon Soulik (2013). *Ocean Dyn.* **2017**, *67*, 1179–1193. [CrossRef]
39. Yen, P.H.; Jan, C.D.; Lee, Y.P.; Lee, H.F. Application of Kalman filter to short-term tide level prediction. *J. Waterw. Port Coast. Ocean. Eng.* **1996**, *122*, 226–231. [CrossRef]
40. Granata, F.; Di Nunno, F. Artificial Intelligence models for prediction of the tide level in Venice. *Stoch. Environ. Res. Risk Assess.* **2021**, *35*, 2537–2548. [CrossRef]
41. Wu, W.; Li, L.; Yin, J.; Zhang, W. A modular tide level prediction method based on a NARX neural network. *IEEE Access* **2021**, *9*, 147416–147429. [CrossRef]

MDPI AG
Grosspeteranlage 5
4052 Basel
Switzerland
Tel.: +41 61 683 77 34

Remote Sensing Editorial Office
E-mail: remotesensing@mdpi.com
www.mdpi.com/journal/remotesensing

www.ingramcontent.com/pod-product-compliance
Lightning Source LLC
Chambersburg PA
CBHW041432210926
51850CB00013B/3412
9783725835591